# Book of Abstracts of the 65th Annual Meeting of the European Federation of Animal Science

EAAP - European Federation of Animal Science

The European Federation of Animal Science wishes to express its appreciation to the
Ministero delle Politiche Agricole Alimentari e Forestali (Italy) and the
Associazione Italiana Allevatori (Italy)
for their valuable support of its activities.

# Book of Abstracts of the 65th Annual Meeting of the European Federation of Animal Science

Copenhagen, Denmark, 25-29 August, 2014

**EAAP Scientific Committee:**

H. Simianer
G. Van Duinkerken
H. Spoolder
M. Vestergaard
A. Bernues
M. Klopcic
L. Bodin
C. Lauridsen
N. Miraglia
G. Pollott

Proceedings publication and Abstract Submission System (OASES) by

EAN: 9789086862481
e-EAN: 9789086867998
ISBN: 978-90-8686-248-1
e-ISBN: 978-90-8686-799-8
DOI: 10.3920/978-90-8686-799-8

ISSN 1382-6077

First published, 2014


Wageningen Academic Publishers
P.O. Box 220
6700 AE Wageningen
The Netherlands
www.WageningenAcademic.com
copyright@WageningenAcademic.com

# Welcome to Copenhagen

The Danish Organising Committee welcomes you to the 65$^{th}$ Annual Meeting of the EAAP. The conference is a unique opportunity for scientists and others interested to meet and acquire new knowledge as well as to exchange experiences and ideas within animal science. The meeting takes place in Copenhagen from 25$^{th}$ to 29$^{th}$ August, 2014.

The main theme of the meeting is "Quality and sustainability in animal production" dealing with topics such as resource efficiency, animal welfare, diversification and agroecology aspects as well as product quality. The programme covers various disciplines such as animal genetics, animal health and welfare, animal nutrition, animal physiology, and livestock farming systems and also cattle, horse, pig, sheep and goat, and fur animal production.

During the meeting, the participants have the opportunity to attend theme sessions with oral presentations and posters presented by scientists from all over the world. Posters can be discussed with the presenters at the poster session on Tuesday 26$^{th}$ and Wednesday 27$^{th}$ and are accessible for viewing throughout the entire meeting. Results will be presented from successful international research projects including projects with stakeholder participation where focus is on knowledge exchange towards innovation.

Participants will also have the opportunity to take part in Challenge Programmes which provide fora to formulate proposals to meet future challenges, and Discovery Sessions with focus on new disciplines and technologies essential to the continued development of animal science as well as the Plenary Session entitled "Integrated human-animals relationships" and Industry Sessions.

We are sure that the EAAP 2014 Annual Meeting will be a productive meeting with excellent science and inspiring discussions and that you will also enjoy Copenhagen with its many cultural attractions and social events.

*Brian Bech Nielsen*

President of the Danish Organising Committee

# National Organisers of the 65th EAAP Annual Meeting

## Danish National Organising Committee

President
- **Brian Bech Nielsen**
  Aarhus University, Rektor@au.dk

Vice President
- **Birgit Nørrung**
  University of Copenhagen, Birgit.Noerrung@sund.ku.dk

Executive Secretaries
- **John Hermansen**
  Aarhus University, John.Hermansen@agrsci.dk
- **Vivi Hunnicke Nielsen**
  Aarhus University, ViviH.Nielsen@agrsci.dk

Members
- **Niels Christian Nielsen**
  Aarhus University, dean.scitech@au.dk
- **Klaus Lønne Ingvartsen**
  Aarhus University, KlausL.Ingvartsen@agrsci.dk
- **Kristen Sejrsen**
  Aarhus University, Kr.Sejrsen@agrsci.dk
- **Anders Klöcker**
  Ministry of Food, Agriculture and Fisheries of Denmark, andklo@naturerhverv.dk
- **Peder Philipp**
  Danish Agriculture and Food Council, PederPhilipp@mail.dk
- **Erik Larsen**
  Danish Agriculture and Food Council, ela@lf.dk
- **Jan Mousing**
  Knowledge Centre for Agriculture, jam@vfl.dk
- **Asbjørn Børsting**
  DLG Group, asb@dlg.dk
- **Per Falholt**
  NovoZymes, pf@novozymes.com
- **Erik Bisgaard Madsen**
  University of Copenhagen, proem@science.ku.dk
- **Jørgen Madsen**
  University of Copenhagen, jom@sund.ku.dk
- **Henrik C. Wegener**
  Technical University of Denmark, DTU-prorektor@adm.dtu.dk
- **Bjarne Nielsen**
  Pig Research Centre, bni@lf.dk
- **Claus Fertin**
  VikingGenetics, clfer@vikinggenetics.com
- **Peter Foged Larsen**
  Kopenhagen Fur, pfl@kopenhagenfur.com

## Danish Scientific Committee

President
- **Kristen Sejrsen**
  Aarhus University, Kr.Sejrsen@agrsci.dk

Executive Secretary
- **John Hermansen**, Aarhus University, John.Hermansen@agrsci.dk
- **Vivi H. Nielsen**, Aarhus University, ViviH.Nielsen@agrsci.dk

Animal Genetics
- **Peter Løvendahl**, Aarhus University, Peter.Lovendahl@agrsci.dk
- **Thomas Mark**, University of Copenhagen, thm@sund.ku.dk
- **Bjarne Nielsen**, Pig Research Centre, bni@lf.dk
- **Søren Borchersen**, VikingGenetics, sobor@vikinggenetics.com

Animal Health and Welfare
- **Lene Munksgaard**, Aarhus University, Lene.Munksgaard@agrsci.dk
- **Anders Ringgaard** Kristensen, University of Copenhagen, ark@sund.ku.dk
- **Finn Strudsholm**, Agrotech, fns@agrotech.dk

Animal Nutrition
- **Jacob Sehested**, Aarhus University, Jacob.Sehested@agrsci.dk
- **Peder Nørgaard**, University of Copenhagen, pen@sund.ku.dk
- **Niels Kjeldsen**, Pig Research Centre, njk@lf.dk

Animal Physiology
- **Mogens Vestergaard**, Aarhus University, Mogens.Vestergaard@agrsci.dk
- **Mette Olaf Nielsen**, University of Copenhagen, Mette.Olaf.Nielsen@sund.ku.dk

Livestock Farming Systems
- **Troels Kristensen**, Aarhus University, Troels.Kristensen@agrsci.dk
- **Hanne Hansen**, University of Copenhagen, hhh@sund.ku.dk
- **Peter Enemark**, Knowledge Centre for Agriculture, pse@vfl.dk

Cattle Production
- **Martin R. Weisbjerg**, Aarhus University, Martin.Weisbjerg@agrsci.dk
- **Jørgen Madsen**, University of Copenhagen, jom@sund.ku.dk
- **Ole Aaes**, Knowledge Centre for Agriculture, oea@vfl.dk

Horse Production
- **Jens Malmkvist**, Aarhus University, Jens.Malmkvist@agrsci.dk
- **Jørgen Kold**, Knowledge Centre for Agriculture, jrk@vfl.dk

Pig Production
- **Charlotte Lauridsen**, Aarhus University, Charlotte.Lauridsen@agrsci.dk
- **Christian Fink Hansen**, University of Copenhagen, cfh@sund.ku.dk
- **Martin Andersson**, Pig Research Centre, mea@lf.dk

Fur Animal Production
- **Steen Møller**, Aarhus University, SteenH.Moller@agrsci.dk
- **Vivi Hunnicke Nielsen**, Aarhus University, ViviH.Nielsen@agrsci.dk

# EAAP Program Foundation

**Aims**

EAAP aims to bring to our annual meetings, speakers who can present the latest findings and views on developments in the various fields of science relevant to animal production and its allied industries. In order to sustain the quality of the scientific program that will continue to entice the broad interest in EAAP meetings we have created the 'EAAP Program Foundation'. This Foundation aims to support:

- Invited speakers with a high international profile by funding part or all of registration and travel costs.
- Delegates from less favoured areas by offering scholarships to attend EAAP meetings.
- Young scientists by providing prizes for best presentations.

The '**EAAP Program Foundation**' is an initiative of the Scientific Committee (SC) of EAAP. The Foundation aims to stimulate the quality of the scientific program of the EAAP meetings and to ensure that the science meets societal needs. The Foundation Board of Trustees oversees these aims and seeks to recruit sponsors to support its activities.

**Sponsorships**

1. **Meeting sponsor – from 5000 €**
   - acknowledgements in the final booklet with contact address and logo
   - one page allowance in the final booklet
   - advertising/information material inserted in the bags of delegates
   - advertising/information material on a stand display (at additional cost to be negotiated)
   - acknowledgement in the EAAP Newsletter with possibility of a page of publicity
   - possibility to add session and speaker support (at additional cost to be negotiated)
2. **Session sponsor – from 3000 to 5000 €**
   - acknowledgements in the final booklet with contact address and logo
   - one page allowance in the final booklet
   - advertising/ information material in the delegate bag
   - slides at beginning of session to acknowledge support and recognition by session chair
   - acknowledgement in the EAAP Newsletter.
3. **Speaker sponsor – from 2000 € (cost will be defined according to speakers country of origin)**
   - half page allowance in the final booklet
   - recognition by speaker of the support at session (at additional cost to be negotiated)
   - acknowledgement in the EAAP Newsletter
4. **Registration Sponsor – from 600 to 1.200 €**
   - acknowledgements in the booklet with contact address and logo
   - advertising/information material in the delegate bag

**The Association**

EAAP (The European Federation of Animal Science) organises every year an international meeting which attracts between 900 and 1500 people. The main aims of EAAP are to promote, by means of active co-operation between its members and other relevant international and national organisations, the advancement of scientific research, sustainable development and systems of production; experimentation, application and extension; to improve the technical and economic conditions of the livestock sector; to promote the welfare of farm animals and the conservation of the rural environment; to control and optimise the use of natural resources in general and animal genetic resources in particular; to encourage the involvement of young scientists and technicians. More information on the organisation and its activities can be found at www.eaap.org.

**Contact and further information**

If you are interested to become a sponsor of the 'EAAP Program Foundation' or want to have further information, please contact the EAAP Secretariat (eaap@eaap.org, Phone +39 06 44202639).

# Acknowledgements

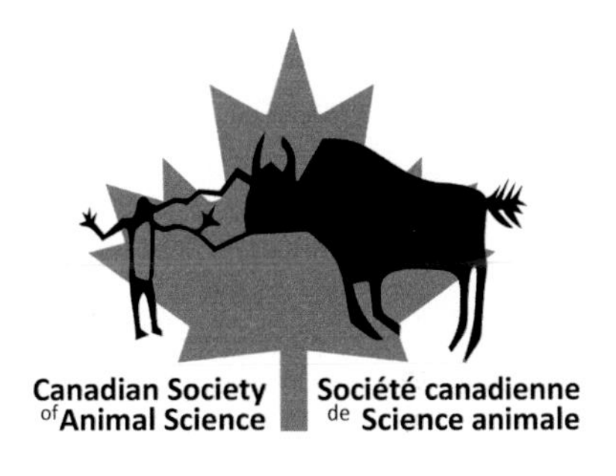

## Thank you
## to the 65th EAAP Annual Congress Sponsors and Friends

**Platinum**

**Gold**

KOPENHAGEN
FUR

**Silver**

**Bronze**

**App sponsor**

**Friends of EAAP**

# European Federation of Animal Science (EAAP)

*President:* P. Chemineau
*Secretary General:* A. Rosati
*Address:* Via G. Tomassetti 3, A/I
I-00161 Rome, Italy
*Phone:* +39 06 4420 2639
*Fax:* +39 06 4426 6798
*E-mail:* eaap@eaap.org
*Web:* www.eaap.org

---

## 66th EAAP Annual meeting of the European Federation of Animal Science

**Warsaw, Poland**

## Organising Committee

**President**:
- Prof. Dr Hab. Roman Niżnikowski, Warsaw University of Life Sciences – SGGW – roman_niznikowski@sggw.pl

**Secretary:**
- Dr Dorota Krencik, National Animal Breeding Centre (KCHZ) – d.krencik@kchz.agro.pl

**Members:**
- Dr Hab. Dorota Lewczuk, Institute of Genetic and Animal Breeding PAS
- Dr Hab. Robert Głogowski, Warsaw University of Life Sciences – SGGW
- Dr Marek Balcerak, Warsaw University of Life Sciences – SGGW
- Dr Marcin Gołębiewski, Warsaw University of Life Sciences – SGGW
- Dr Hab. Joanna Makulska, Agricultural University in Cracow
- Dr Wojciech Neja, University of Technology and Life Sciences in Bydgoszcz
- Dr Hab. Tomasz Strabel, Poznań University of Life Sciences
- Dr Artur Oprządek, Agricultural Property Agency (ANR)
- Dr Witold Rant, Warsaw University of Life Sciences – SGGW

**Conference website:** www.eaap2015.org

# Scientific Programme EAAP 2014

| Monday 25 August<br>8.30 – 12.30 | Monday 26 August<br>14.00 – 18.00 | Tuesday 26 August<br>8.30 – 12.30 | Tuesday 26 August<br>14.00 – 18.00 |
|---|---|---|---|
| **Session 1**<br>Horse keeping and society<br>Chair: M. Saastamoinen/ R. Evans | **Session 9**<br>Grass based farming: all aspects<br>Chair: A. de Veer | 8.30-9.30<br>**Session 18**<br>Challenge programme: Ethics teaching in animal science<br>Chair: B. Gremmen | **Session 20**<br>Sustainable cows, herd practices and product quality 1<br>Chair: G. Thaller |
| **Session 2**<br>Cellular physiology of the growth process<br>Chair: C.H. Knight/ M. Vestergaard | **Session 10**<br>Precision Livestock Farming (PLF) 2. Making sense of sensors to support farm management<br>Chair: I. Halachmi/ B. Early/M. Klopcic | 8.30-9.30<br>**Poster Session** | **Session 21**<br>ASAS session: Embryonic and foetal programming<br>Chair: K. Vonnahme |
| **Session 3**<br>Challenge programme: Overall assessment of perinatal lamb mortality and strategies to improve lamb survival<br>Chair: C. Dwyer | **Session 11**<br>Precision Livestock Farming (PLF) 3. Making sense of sensors to support farm management<br>Chair: I. Halachmi/M. Guarino/H. Spoolder | 9.30-10.15<br>**Session 19**<br>Plenary session: Part 1. Integrated human-animal relationships<br>Chair: P. Chemineau | **Session 22**<br>Challenge programme: Careers in the livestock industry<br>Chair: M. Gauly |
| **Session 4**<br>Resource use efficiency and improvement options from farm to global level<br>Chair: M. Zehetmeier | **Session 12**<br>Precision Livestock Farming (PLF) 4. Making sense of sensors to support farm management<br>Chair: I. Halachmi/ C. Lokhorst | 10.45-11.30<br>**Welcome Ceremony** | **Session 23**<br>Non-standard traits in genomic selection<br>Chair: J. Lassen |
| **Session 5**<br>Precision Livestock Farming (PLF) 1. Making sense of sensors to support farm management<br>Chair: I. Halachmi/T. Banhazi | **Session 13**<br>Appetite control – mechanisms and comparative aspects<br>Chair: J.L. Sartin | 11.30-12.30<br>**Poster Session 1** | **Session 24**<br>Observations at slaughter to improve welfare and health on pig farms<br>Chair: G. Bee |
| **Session 6**<br>Micronutrients and their impact on production, health and the environment<br>Chair: C. Lauridsen | **Session 14**<br>Sequence based analysis<br>Chair: J. Jensen | | **Session 25**<br>Sheep and Goat feeding and health<br>Chair: E. Ugarte |
| **Session 7**<br>Feeding (and management) to improve gut barrier function and immunity in livestock<br>Chair: E. Tsiplakou | **Session 15**<br>Industry session: New advances in exogenous enzymes in livestock production<br>Chair: P. Falholt; Novozymes | | **Session 26**<br>Organic livestock farming – challenges and future perspectives<br>Chair: T. Kristensen |
| **Session 8**<br>Young train session: Dairy innovative research and extension by young scientists<br>Chair: A. Kuipers/M. Gaully | **Session 16**<br>Discovery session: Proteomics in farm animals<br>Chair: A. Almeida | | **Session 27**<br>Discovery session and workshop: Insects for feed<br>Chair: T. Veldkamp |
| | **Session 17**<br>Welfare, behaviour and health in horse management<br>Chair: R. Evans | | |

| Wednesday 27 August<br>8.30 – 12.30 | Wednesday 27 August<br>14.00 – 18.00 | Thursday 28 August<br>8.30 – 12.30 | Thursday 28 August<br>14.00 – 18.00 |
|---|---|---|---|
| 8.30 – 9.30<br><br>**Session 28**<br>Discovery session: Ethical aspects of animal breeding<br>Chair: G. Gandini<br><br>8.30-9.30<br><br>**Poster Session 2**<br><br>9.30-10.30<br><br>**Session 29**<br>Plenary session:<br>Part 2.<br>Integrated human-animal relationships<br>Chair: P. Chemineau<br><br>10.50-11.45<br><br>**Awards Ceremony**<br><br>11.30 – 12.30<br><br>**Poster Session 2** | **Session 30**<br>Challenge programme: Environmental optimisation of the pig production system<br>Chair: H.D. Poulsen<br><br>**Session 31**<br>Competitiveness of European beef production<br>Chair: B. Fürst-Waltl/ P. Sarzeaud<br><br>**Session 32**<br>Discovery Session: The horse as key player of local development: looking to the future in the $3^{rd}$ millennium<br>Chair: N. Miraglia/ A.L. Holgerssons<br><br>**Session 33**<br>Interaction between stress and immune reactions<br>Chair: A. Prunier/E. Merlot<br><br>**Session 34**<br>Discovery session: Sustainable intensification to feed the world: fact or fiction?<br>Chair: M. Tichit/K. Eilers<br><br>**Session 35**<br>Genetics Commission Young Scientists competition<br>Chair: E. Awusu<br><br>**Session 36**<br>Fibre Working Group: Fur, fibre and skin animal products<br>Chair: V.H. Nielsen<br><br>**Session 37**<br>Industry session: Probiotics<br>Chair: N.E. Auclair<br><br>**Session 38**<br>Challenge programme: Improving the responsible use of antibiotics<br>Chair: L. Boyle | 8.30 – 12.30<br><br>**Session 39**<br>Cattle debate: Family v mega farm<br>Chair: M. Coffey<br><br>**Session 40**<br>Strategies for improving productivity in small ruminants<br>Chair: J. Conington<br><br>**Session 41**<br>Animal Genetics<br>Chair: H. Esfandyari<br><br>**Session 42**<br>Livestock effects on the environment<br>Chair: S. Ingrand<br><br>**Session 43**<br>Pig nutrition in gestation, lactation and progeny development<br>Chair: G. Bee<br><br>**Session 44**<br>Physiology – metabolism and digestion<br>Chair: N. Silanikove<br><br>**Session 45**<br>Animal nutrition<br>Chair: G. van Duinkerken<br><br>**Session 46**<br>Horse genetics<br>Chair: I. Cervantes<br><br>**Session 47**<br>Health and production diseases<br>Chair: G. Das<br><br>11.30-12.30<br><br>**Commission business meetings** | **Session 48**<br>Sustainable cows, herd practices and product quality 2<br>Chair: G. Thaller<br><br>**Session 49**<br>The role of small ruminants in meeting global challenges for sustainable intensification<br>Chair: T. Adnoy<br><br>**Session 50**<br>Analysis of longitudinal data<br>Chair: H. Mulder<br><br>**Session 51**<br>Challenge programme: Animal task Force – Appropriation of innovation by farmers in animal agriculture<br>Chair: M. Scholten<br><br>**Session 52**<br>Feed efficiency and feed resources<br>Chair: G. Savoini<br><br>**Session 53**<br>Industry session: Quality pays –There is no substitution for skills in efficient pig production<br>Chair: M. Andersen; Danish Agric. & Food Council<br><br>**Session 54**<br>Behaviour and welfare in farm animals<br>Chair: H. Spoolder<br><br>**Session 55**<br>Market-orientated pig production – conventional and non-conventional<br>Chair: C. Lauridsen<br><br>**Session 56**<br>Udder health, reproduction and longevity in cattle<br>Chair: M. Klopcic |

## Commission on Animal Genetics

| | | |
|---|---|---|
| Dr Simianer | President<br>Germany | University of Goettingen<br>hsimian@gwdg.de |
| Dr Baumung | Vice-President<br>Italy | FAO Rome<br>roswitha.baumung@fao.org |
| Dr Meuwissen | Vice-President<br>Norway | Norwegian University of Life Sciences<br>theo.meuwissen@umb.no |
| Dr Szyda | Vice-President<br>Poland | Agricultural University of Wroclaw<br>szyda@karnet.ar.wroc.pl |
| Dr Ibañez | Secretary<br>Spain | IRTA<br>noelia.ibanez@irta.es |
| Dr De Vries | Industry rep.<br>Netherlands | CRV<br>alfred.de.vries@crv4all.com |

## Commission on Animal Nutrition

| | | |
|---|---|---|
| Dr Van Duinkerken | President<br>Netherlands | Wageningen University<br>gert.vanduinkerken@wur.nl |
| Dr Savoini | Vice-President<br>Italy | University of Milan<br>giovanni.savoini@unimi.it |
| Dr Tsiplakou | Secretary<br>Greece | Agricultural University of Athens<br>eltsiplakou@aua.gr |
| Dr Auclair | Industry rep.<br>France | LFA Lesaffre<br>ea@lesaffre.fr |

## Commission on Health and Welfare

| | | |
|---|---|---|
| Dr Spoolder | President<br>Netherlands | ASG-WUR<br>hans.spoolder@wur.nl |
| Dr Krieter | Vice-President<br>Germany | University Kiel<br>jkrieter@tierzucht.uni-kiel.de |
| Dr. Boyle | Secretary<br>Ireland | Teagasc<br>laura.boyle@teagasc.ie |
| Dr Das | Secretary<br>Germany | University of Goettingen<br>gdas@gwdg.de |
| Mr Pearce | Industry rep.<br>United Kingdom | Pfizer<br>michael.c.pearce@pfizer.com |

## Commission on Animal Physiology

| | | |
|---|---|---|
| Dr Vestergaard | President | Aarhus University |
| | Denmark | mogens.vestergaard@agrsci.dk |
| Dr Driancourt | Vice president/ | |
| | Industry rep. | Intervet |
| | France | marc-antoine.driancourt@sp.intervet.com |
| Dr Silanikove | Vice-President | Agricultural Research Organization (ARO) |
| | Israel | nsilaniks@volcani.agri.gov.il |
| Dr Quesnel | Secretary | INRA Saint Gilles |
| | France | helene.quesnel@rennes.inra.fr |
| Dr Scollan | Secretary | Institute of Biological, Environmental and rural sciences |
| | United Kingdom | ngs@aber.ac.uk |

## Commission on Livestock Farming Systems

| | | |
|---|---|---|
| Dr Bernués Jal | President | Norwegian University of Life Sciences |
| | Norway | alberto.bernues@nmbu.no |
| Dr Ingrand | Vice-President | INRA |
| | France | ingrand@clermond.inra.fr |
| Dr Tichit | Vice-President | INRA |
| | France | muriel.tichit@agroparistech.fr |
| Dr Eilers | Secretary | Schuttelaar & Partners |
| | Netherlands | keilers@schuttelaar.nl |
| Mrs Zehetmeier | Secretary | Institute Agricultural Economics and Farm Management |
| | Germany | monika.zehetmeier@tum.de |

## Commission on Cattle Production

| | | |
|---|---|---|
| Dr Klopcic | Secretary/acting as President/ | |
| | Industry rep. | University of Ljublijana |
| | Slovenia | marija.klopcic@bf.uni-lj.si |
| Dr Thaller | Vice-President | Animal Breeding and Husbandry |
| | Germany | georg.thaller@tierzucht.uni-kiel.de |
| Dr Coffey | Vice president/ | |
| | Industry rep. | SAC, Scotland |
| | United Kingdom | mike.coffey@sac.ac.uk |
| Dr. Halachmi | Vice-President | Agricultural Research Organization (ARO) |
| | Israel | halachmi@volcani.agri.gov.il |
| Dr Hocquette | Secretary | INRA |
| | France | jean-francois.hocquet@clermont.inra.fr |
| Dr. Fürst-Waltl | Secretary | University of Natural Resources and Life Sciences |
| | Austria | birgit.fuerst-waltl@boku.ac.at |

## Commission on Sheep and Goat Production

| Name | Position / Country | Institution / E-mail |
|---|---|---|
| Dr Bodin | President | INRA-SAGA |
| | France | loys.bodin@toulouse.inra.fr |
| Dr Ringdorfer | Vice-President | LFZ Raumberg-Gumpenstein |
| | Austria | ferdinand.ringdorfer@raumberg-gumpenstein.at |
| Dr Conington | Vice-President | SAC |
| | United Kingdom | joanne.conington@sac.ac.uk |
| Dr Papachristoforou | Vice President | Agricultural Research Institute |
| | Cyprus | c.papachristoforou@cut.ac.cy |
| Dr Milerski | Secretary/ | |
| | Industry rep. | Research Institute of Animal Science |
| | Czech Republic | m.milerski@seznam.cz |
| Dr Ugarte | Secretary/ | |
| | Industry rep. | NEIKER-Tecnalia |
| | Spain | eugarte@neiker.net |

## Commission on Pig Production

| Name | Position / Country | Institution / E-mail |
|---|---|---|
| Dr Lauridsen | President | Aarhus University |
| | Denmark | charlotte.lauridsen@agrsci.dk |
| Dr Knol | Vice President/ | |
| | Industry rep. | TOPIGS |
| | Netherlands | egbert.knol@topigs.com |
| Dr Bee | Secretary | Agroscope Liebefeld-Posieux ALP |
| | Switzerland | giuseppe.bee@alp.admin.ch |
| Dr Velarde | Secretary | IRTA |
| | Spain | antonio.velarde@irta.es |
| Dr. Millet | Secretary | ILVO |
| | Belgium | sam.millet@ilvo.vlaanderen.be |

## Commission on Horse Production

| Name | Position / Country | Institution / E-mail |
|---|---|---|
| Dr Miraglia | President | Molise University |
| | Italy | miraglia@unimol.it |
| Dr Burger | Vice president | Clinic Swiss National Stud |
| | Switzerland | dominique.burger@mbox.haras.admin.ch |
| Dr Janssen | Vice president | BIOSYST |
| | Belgium | steven.janssens@biw.kuleuven.be |
| Dr Lewczuk | Vice president | IGABPAS |
| | Poland | d.lewczuk@ighz.pl |
| Dr Saastamoinen | Vice president | MTT Agrifood Research Finland |
| | Finland | markku.saastamoinen@mtt.fi |
| Dr Holgersson | Secretary | Swedish University of Agriculture |
| | Sweden | anna-lena.holgersson@hipp.slu.se |
| Dr Hausberger | Secretary | CNRS University |
| | France | martine.hausberger@univ-rennes1.fr |

# Scientific programme

## Session 01. Horse keeping and society

Date: 25 August 2014; 08:30 – 12:30 hours
Chairperson: M. Saastamoinen/R. Evans

### Theatres Session 01



## Session 02. Cellular physiology of the growth process

Date: 25 August 2014; 08:30 – 12:30 hours
Chairperson: C.H. Knight/M. Vestergaard

### Theatres Session 02

**Posters Session 02**



## Session 03. Challenge programme: Overall assessment of perinatal lamb mortality and strategies to improve lamb survival

Date: 25 August 2014; 08:30 – 12:30 hours
Chairperson: C. Dwyer

**Theatres Session 03**

## Session 04. Resource use efficiency and improvement options from farm to global level

Date: 25 August 2014; 08:30 – 12:30 hours
Chairperson: M. Zehetmeier

### Theatres Session 04

### Posters Session 04

## Session 05. Industry session: Precision Livestock Farming; making sense of sensors to support farm management – part 1

Date: 25 August 2014; 08:30 – 12:30 hours
Chairperson: I. Halachmi/T. Banhazi

### Theatres Session 05



### Posters Session 05

## Session 06. Micronutrients and their impact on production, health and the environment

Date: 25 August 2014; 08:30 – 12:30 hours
Chairperson: C. Lauridsen

### Theatres Session 06

**Posters Session 06**

## Session 07. Feeding (and management) to improve gut barrier function and immunity in livestock

Date: 25 August 2014; 08:30 – 12:30 hours
Chairperson: E. Tsiplakou

### Theatres Session 07

## Session 08. Young train session: dairy innovative research and extension by young scientists

Date: 25 August 2014; 08:30 – 12:30 hours
Chairperson: A. Kuipers/M. Gaully

### Theatres Session 08

### Posters Session 08



## Session 09. Grass based farming: all aspects

Date: 25 August 2014; 14:00 – 18:00 hours
Chairperson: A. de Veer

### Theatres Session 09



### Posters Session 09

## Session 10. Industry session: Precision Livestock Farming; making sense of sensors to support farm management – part 2

Date: 25 August 2014; 14:00 – 18:00 hours
Chairperson: I. Halachmi/B. Early/M. Klopcic

### Theatres Session 10

## Session 11. Industry session: Precision Livestock Farming; making sense of sensors to support farm management – part 3

Date: 25 August 2014; 14:00 – 18:00 hours
Chairperson: I. Halachmi/M. Guarino/H. Spoolder

### Theatres Session 11

## Session 12. Industry session: Precision Livestock Farming; making sense of sensors to support farm management – part 4

Date: 25 August 2014; 14:00 – 18:00 hours
Chairperson: I. Halachmi/C. Lokhorst

### Theatres Session 12



## Session 13. Appetite control – mechanisms and comparative aspects

Date: 25 August 2014; 14:00 – 18:00 hours
Chairperson: J.L. Sartin

### Theatres Session 13

## Session 14. Sequence based analysis

Date: 25 August 2014; 14:00 – 18:00 hours
Chairperson: J. Jensen

### Theatres Session 14



### Posters Session 14

## Session 15. Industry session: new advances in exogenous enzymes in livestock production

Date: 25 August 2014; 14:00 – 18:00 hours
Chairperson: P. Falholt; Novozymes

### Theatres Session 15



### Posters Session 15

## Session 16. Discovery session: Proteomics in farm animals

Co-organized with COST action FA1002
Date: 25 August 2014; 14:00 – 18:00 hours
Chairperson: A. Almeida

### Theatres Session 16



## Session 17. Welfare, behaviour and health in horse management

Date: 25 August 2014; 14:00 – 18:00 hours
Chairperson: R. Evans

### Theatres Session 17

### Posters Session 17



## Session 18. Challenge programme: Ethics teaching in animal science

Date: 26 August 2014; 08:30 – 09:30 hours
Chairperson: B. Gremmen

### Theatres Session 18



## Session 19. Plenary session – part 1: Integrated human-animal relationships

Date: 26 August 2014; 09:30 – 11:30 hours
Chairperson: P. Chemineau

### Theatres Session 19

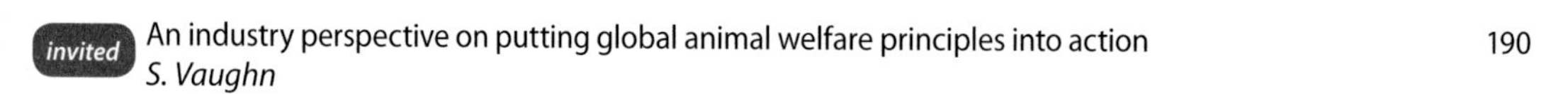

## Session 20. Sustainable cows, herd practices and product quality – part1

Date: 26 August 2014; 14:00 – 18:00 hours
Chairperson: G. Thaller

### Theatres Session 20



### Posters Session 20

## Session 21. Embryonic and foetal programming

Date: 26 August 2014; 14:00 – 18:00 hours
Chairperson: K. Vonnahme

### Theatres Session 21



### Posters Session 21



## Session 23. Non-standard traits in genomic selection

Date: 26 August 2014; 14:00 – 18:00 hours
Chairperson: J. Lassen

### Theatres Session 23

### Posters Session 23



## Session 24. Observations at slaughter to improve welfare and health on pig farms

Date: 26 August 2014; 14:00 – 18:00 hours
Chairperson: G. Bee

### Theatres Session 24

### Posters Session 24



## Session 25. Sheep and goat feeding and health

Date: 26 August 2014; 14:00 – 18:00 hours
Chairperson: E. Ugarte

### Theatres Session 25

## Posters Session 25

## Session 26. Organic livestock farming – challenges and future perspectives

Date: 26 August 2014; 14:00 – 18:00 hours
Chairperson: T. Kristensen

### Theatres Session 26



### Posters Session 26

## Session 27. Discovery session: Insects for feed

Date: 26 August 2014; 14:00 – 18:00 hours
Chairperson: T. Veldkamp

### Theatres Session 27



## Session 29. Plenary session – part 2: Integrated human-animal relationships

Date: 27 August 2014; 09:30 – 11:30 hours
Chairperson: P. Chemineau

### Theatres Session 29



## Session 31. Competitiveness of European beef production

Date: 27 August 2014; 14:00 – 18:00 hours
Chairperson: B. Fürst-Waltl/P Sarzeaud

### Theatres Session 31

**Posters Session 31**

## Session 32. Discovery session: The horse as key player of local development: looking to the future in the 3rd millennium

Date: 27 August 2014; 14:00 – 18:00 hours
Chairperson: N. Miraglia/A.L. Holgerssons

### Theatres Session 32



## Session 33. Interactions between stress, metabolism and immunity

Date: 27 August 2014; 14:00 – 18:00 hours
Chairperson: A. Prunier/E. Merlot

### Theatres Session 33

**Posters Session 33**



## Session 34. Discovery session: Sustainable intensification to feed the world: fact or fiction?

Date: 27 August 2014; 14:00 – 18:00 hours
Chairperson: M. Tichit/K. Eilers

**Theatres Session 34**



## Session 35. Genetics commission: young scientists' competition

Date: 27 August 2014; 14:00 – 18:00 hours
Chairperson: E. Awusu

**Theatres Session 35**

### Posters Session 35



## Session 36. Fur, fibre and skin animal production

Date: 27 August 2014; 14:00 – 18:00 hours
Chairperson: V.H. Nielsen

### Theatres Session 36

## Session 37. Industry session: probiotics

Date: 27 August 2014; 14:00 – 18:00 hours
Chairperson: N.E. Auclair

### Theatres Session 37

**Posters Session 37**



## Session 38. Challenge programme: Improving the responsible use of antibiotics

Date: 27 August 2014; 14:00 – 18:00 hours
Chairperson: L. Boyle

**Theatres Session 38**



## Session 39. Cattle debate session: family versus mega farm

Date: 28 August 2014; 08:30 – 11:30 hours
Chairperson: M. Coffey

**Theatres Session 39**

## Session 40. Strategies for improving productivity in small ruminants

Date: 28 August 2014; 08:30 – 11:30 hours
Chairperson: J. Conington

### Theatres Session 40



### Posters Session 40

## Session 41. Free communications on animal genetics

Date: 28 August 2014; 08:30 – 11:30 hours
Chairperson: H. Esfandyari

### Theatres Session 41

**Posters Session 41**

## Session 42. Livestock effects on the environment

Date: 28 August 2014; 08:30 – 11:30 hours
Chairperson: S. Ingrand

### Theatres Session 42

### Posters Session 42



## Session 43. Pig nutrition in gestation, lactation and progeny development

Date: 28 August 2014; 08:30 – 11:30 hours
Chairperson: G. Bee

### Theatres Session 43

**Posters Session 43**



## Session 44. Physiology metabolism and digestion

Date: 28 August 2014; 08:30 – 11:30 hours
Chairperson: N. Silanikove

**Theatres Session 44**

### Posters Session 44



## Session 45. Free communications animal nutrition

Date: 28 August 2014; 08:30 – 11:30 hours
Chairperson: G. van Duinkerken

### Theatres Session 45

## Posters Session 45

## Session 46. Horse genetics free communications

Date: 28 August 2014; 08:30 – 11:30 hours
Chairperson: I. Cervantes

### Theatres Session 46

### Posters Session 46



## Session 47. Health and production diseases

Date: 28 August 2014; 08:30 – 11:30 hours
Chairperson: G. Das

### Theatres Session 47

## Posters Session 47

## Session 48. Sustainable cows, herd practices and product quality – part2

Date: 28 August 2014; 14:00 – 18:00 hours
Chairperson: G. Thaller

### Theatres Session 48



### Posters Session 48

## Session 49. The role of small ruminants in meeting global challenges for sustainable intensification

Date: 28 August 2014; 14:00 – 18:00 hours
Chairperson: T. Adnoy

### Theatres Session 49



### Posters Session 49

## Session 50. Analysis of longitudinal data

Date: 28 August 2014; 14:00 – 18:00 hours
Chairperson: H. Mulder

### Theatres Session 50



### Posters Session 50

## Session 52. Feed efficiency and feed resources

Date: 28 August 2014; 14:00 – 18:00 hours
Chairperson: G. Savoini

### Theatres Session 52



### Posters Session 52

## Session 53. Industry session: quality pays

Date: 28 August 2014; 14:00 – 18:00 hours
Chairperson: M. Andersen; Danish Agric. & Food Council

### Theatres Session 53



### Posters Session 53



## Session 54. Behaviour and welfare in farm animals

Date: 28 August 2014; 14:00 – 18:00 hours
Chairperson: H. Spoolder

### Theatres Session 54

### Posters Session 54



## Session 55. Market-orientated pig production – conventional and non-conventional

Date: 28 August 2014; 14:00 – 18:00 hours
Chairperson: C. Lauridsen

### Theatres Session 55

**Posters Session 55**

## Session 56. Udder health, reproduction and longevity in cattle

Date: 28 August 2014; 14:00 – 18:00 hours
Chairperson: M. Klopcic

### Theatres Session 56

## Posters Session 56

**The equine industry in Canada: developing resiliency in a diverse and complex landscape**

*H. Sansom and B.L. Wilton*
*University of Guelph, School of Environmental Design and Rural Development, 50 Stone Rd. East, Guelph, N1G 2W1, Canada; hrsansom@hotmail.com*

The equine industry in Canada is diverse and complex, in part due to the geographical scale of the country of Canada and also due to the wide variety of equine disciplines and activities that are popular in Canada. While equine sports and activities remain popular in Canada, there are signs that the demographics are shifting to an older participant base, raising questions regarding the long-term sustainability of the industry. The distribution of the population in Canada also contributes to challenges in terms of developing and sustaining an active base for equine sports and horse-keeping in Canada. This leads to localized regions of equine activity making it difficult to share resources and to build national programs. This presentation will provide an overview of the equine industry in Canada, current challenges facing the industry, and the development of resiliency for the industry through the examination of localized case studies. Topics covered will include equine numbers, disciplines, equine welfare, current trends in the sector, and innovations developed through research. This session will draw upon data from Equine Canada, provincial equestrian organizations, the racing industry, as well as key informant interviews. The relationship between the equine sector and other agricultural sectors will also be included as this is an important factor in terms of government support for research and innovation in the equine industry in Canada.

---

Session 01 Theatre 2

**Horse breeding in Europe – quo vadis?**

*S. Wägeli, H. Jörg and C. Herholz*
*Bern University of Applied Sciences, School of Agricultural, Forest and Food Sciences HAFL, Länggasse 85, 3052 Zollikofen, Switzerland; salome.waegeli@bfh.ch*

Recently, the role of the horse in the European society changed. Sport and performance are increasingly less important for horse riders and keepers. Moreover, many equestrian and breeding associations have problems with decreasing numbers of members. This development is strongly influencing the future of horse breeding in Europe. How do the horse breeders in Europe have to face the new market trends? And importantly, who is going to be the future horse breeder? The author's and several other studies in European countries have dealt with this topic and similar trends could be identified. The studies showed that the future horse buyer is going to be female and around thirty years old. She buys ! a horse to have fun and due to her love for animals and nature. Purchase criteria will be the horses' character, its social behavior, its 'willingness' and its ease to ride. Summing up, the role of horses' character will strongly increase. Accordingly, breeders will have to adjust their marketing and to develop new customer-oriented and individualized approaches to compete in the global horse market. Thus, the selection criteria of breeders and breeding associations has also to change to meet customers demand. Horses are so called 'credence goods' – hence objective and standardized selection procedures to measure the character of a horse will be an important marketing instrument to sell horses in the future. A similar development regarding gender composition can be expected in the breeding business. Furthermore, studies showed that the average age of breeders is rather high and most of the future breeders will not stem from traditional breeder families. Hence, young breeder programs, with for example, sponsorship by older experienced breeders, are necessary to preserve the knowledge of generations. The breeding associations have to counter the current development with innovative ideas to ensure sustainable successful horse breeding in Europe.

## Session 01 Theatre 3

**Equine society or not – the cultural hegemony of horse keeping at home versus at livery**

*J. McKeown*
*Aberystwyth University, School of Management and Business, Llanbadarn Fawr, Aberystwyth, SY23 2AL, United Kingdom; juml@aber.ac.uk*

An autoethnographical piece investigating equine society from the aspect of keeping horses at home versus in a livery yard. The paper researches coping mechanisms when things go wrong; responsibility for decision making; the power structure in yards and the role of equestrian friends. Part of equestrian subculture is 'yard culture' where horse owners form groups, ride together and follow the latest equine trends. Yards have power structures and opinion leaders with some owners feeling subordinate. However there is always someone around to ask for advice and to ride with. Things are different when your horses are at home. Is there an equine subculture at this level or not? Whilst there is more responsibility is it easier to make decisions without the pressure of conforming to the latest trends? What happens when things go wrong? The author has kept horses in livery and at home and uses her experience to investigate the cultures of horse keeping; contrasting equestrian life in a livery yard with that at home. The research focuses on understanding the behaviour changes of the owner in the contrasting environments. Is a livery yard an example of cultural hegemony where one or more owners impose their will on the others, manipulating the yard culture? How does this contrast with keeping horses at home where the owner is the yard culture? From this author's experience of being on three very different yards and keeping horses at home for many years, there are many contrasts. As a novice horse owner, keeping a horse on an adult only, experienced breeding and training yard was extremely beneficial in the long run but made her lack confidence at the time as she felt inferior to the professionals. She worked to learn and grow in confidence, gaining a sense of belonging. Now with full responsibility for her horses she feels free to keep them as she sees fit. Friends also have their horses at home and form an extended yard culture that is much more relaxed than on a livery yard.

## Session 01 Theatre 4

**Why do you have so many horses? Horse culture in Iceland**

*G. Helgadóttir and I. Sigurðardóttir*
*Hólar University College, Rural Tourism, Hólar, 551 Sauðárkrókur, Iceland; gudr@holar.is*

This is a recurring question but the first person to pose it to me was a German hitchhiker travelling in Northwest Iceland. Gazing at the landscape he asked: 'Why are there so many horses in the hills above each farm?' I had no sensible answer so I replied: 'For sentimental reasons'. This remains a viable hypothesis. The number of horses in Europe declined at the end of the $19^{th}$ century but is rising again. The ratio of horses to people in Iceland is 240 horses per 1000 residents, while the corresponding ratio for Europe is 13/1000 and 15 horses for every 1000 inhabitants in France, renowned for its horse culture. In the northwest region of Iceland this ratio was even higher or 2,600/1000. In 2012 there were 77,380 horses registered in Iceland, compared to 77,158 in 2011. So the horse population was at that point growing. Questions have been raised about the economic and environmental sustainability of the high number of horses in Iceland. Given the economic yield this does not seem feasible, even in cases of keeping working horses such as in equestrian tourism enterprises. In the world literature on environmental impacts of horses, the presence of horses is often associated with trampling, erosion and weed infestation. So why do Icelanders keep so many horses? The reasons are not economic and not environmental which suggests that the reasons are cultural. Data was collected with in-depth interviews with horse owners, horse breeders and equestrian tourism operators in Iceland on their horse keeping and particularly how they decide on the number of horses they keep and what they consider an ideal number of horses and for what purpose.

**Horses in the Society 2012-2014-project in Finland**

*S. Mäki-Tuuri and A. Laitinen*
*Hippolis, Opistontie 10a2, 32100 Ypäjä, Finland; info@hippolis.fi*

The equine sector is linked to municipal activities in many diverse ways. In addition to land use, zoning and construction, it has an influence on environment, economics, farming, physical education and social services. The characteristics of the equine industry are variably familiar at the municipality level, and there is only little proactive dialogue among the parties involved. Creating fact-based decision-making requires facilitated cooperation and understanding between officials, entrepreneurs and societies. The project covers provinces of Häme and Uusimaa in the Southern Finland. These provinces are the most densely horse-populated areas in Finland. The aim of the project is to provide information for the municipal decision-makers, and to encourage co-operation between different actors within the equine industry. Municipal demand for the information will be covered by co-operation with the local actors. The project has arranged seminars and regional discussion meetings, has participated in fairs and discussion meetings and is coordinating networks. Counselling for municipalities has been experienced as necessary. One action is to publish 'A horse in society' -book and brochures regarding the impacts of the equine industry. Also a video 'Well-being with horse -models of equine-assisted therapy and social pedagogical procedures' was created. Parties involved pay attention to the problems only if there is concrete influence on their business or operations. Officials mostly need information about land use and zoning, manure handling and storage, horse trails and equine assisted activities. If the equine industry's opinions are ignored, the quality of the laws and regulations will deteriorate and restrict the overall development of the industry and its profitability. The Social, Health Care and Education sectors require more information about equine assisted activities. There is a need for research, information and best practice examples. Lack of business skills among micro-entrepreneurs and part-time employment restricts the development of the equine assisted activities.

---



**Sport horse breeding and agriculture: the end of a relationship?**

*S. Wägeli[1], T. Wülser[1], J. Grossniklaus[1], C. Herholz[1], H. Jörg[1], H.R. Bracher[2], H.R. Häfliger[2] and A. Lüth[2]*
*[1]Bern University of Applied Sciences, School of Agricultural, Forest and Food Sciences HAFL, Equine Sciences, Länggasse 85, 3052 Zollikofen, Switzerland, [2]Swiss Sport Horse Breeding Association, Les Longs Prés, Postfach 125, 1580 Avenches, Switzerland; salome.waegeli@bfh.ch*

Historically, horse breeding in Switzerland has been strongly characterized by agriculture. Therefore horse and cattle breeding were equivalently judged and financially supported by the Swiss government. However, in the last thirty years many Swiss horse breeders changed their objectives and tried to breed a sport horse which can compete on an international level. In this regard the Swiss government states that there is no direct relationship between sport horse breeding and agriculture anymore, and cut the funding for performance tests for studhorses (regulations on breeding TZV 916.310; entered into force on 1.1.2013). The aim of this study was to analyze the relation between sport horse breeding and agriculture in Switzerland. A written survey with 147 active members of the Swiss breeding federation of sport horses (ZVCH; 8.6% of the ZVCH active members) was conducted. The sample was representative with regard to geographical origin. The gender and age composition of Swiss horse breeders was unknown prior to the survey. The data was analyzed with SPSS 20. The results showed that 65% of sport studhorses remain owned by Swiss farmers. Moreover, a growing amount of studhorses in horse boarding facilities are held on Swiss farms (90.5%). Around 18.1 million CHF in cash flows directly to Swiss farmers due to Swiss horse breeding – generated through horse boarding and providing feed and litter. The profit of Swiss farmers is 59% (of total 18.1 million CHF) from boarding sport studhorses. As the study showed, there is still a strong relationship between sport horse breeding and agriculture in Switzerland. In conclusion, sport horse breeding does have an economic and social impact to Swiss agriculture.

### Equine triads: horse, owner and partner – who loves ya baby?

*J. McKeown[1], K. Dashper[2] and C. Wallace[3]*
*[1]Aberystwyth University, Aberystwyth, SY23 3AL, United Kingdom, [2]Leeds Metropolitan University, Leeds, LS1 3HE, United Kingdom, [3]Manchester Metropolitan University, Crewe, CW1 5DU, United Kingdom; jum1@aber.ac.uk*

A multi introspective paper researching the attitude and behaviour of the 'horse husband' towards their partner's horse(s) and equestrian activities. Each of the authors own one or more horses kept in different environments and this paper aims to understand the feelings of our partners towards our equine companions. Each author interviewed another's husband/partner with a set of semi-structured questions aimed at generating introspective data. This was then reviewed by the wife/partner and introspective comments produced. Finally, the three husbands' data were compared and analysed within the situational context. The insights generated adds an extra dimension to knowledge of the horse-human relationship. Horse owners at this level are pre-dominantly women and the time taken in caring for horses as well as riding and taking part in the equestrian social scene can have a large impact on family life. Findings suggest that conflict can arise between horse owners and 'horse husbands', regarding the temporal aspects of caring for and exercising horses versus the fulfilment of domestic, relational and child-care commitments. A further cause of tension relates to the financial implications of horse ownership which can constrain both domestic and horse related expenditure. In sum, partners' perceptions impact upon the ability of horse owners to fulfil their own horse related commitments in full, and finance and time are frequently used as bargaining tools in the fulfilment of both partners' leisure lifestyles. Time, commitment and the investment of resources (financial and emotional) are constantly negotiated between horse owner, their horse(s) and 'horse husbands'.

---



### Home off the range: managing wild horse herds and public land users' rights

*C.J. Stowe*
*University of Kentucky, Agricultural Economics, 307 CE Barnhart, Lexington, KY 40546, USA; jill.stowe@uky.edu*

In the United States, wild horses conjure images of the Wild West and freedom and signify the historical importance of the horse to the settlement of the country. With the safety of herds of wild horses threatened, an act of the United States Congress in 1971 delegated the management of wild horse and burro herds on public lands at a size consistent with the land's capacity to the Bureau of Land Management (BLM), at the same time requiring it to preserve and maintain a natural ecological balance and multiple-use relationship on those lands. Since herd size can double every four years, the BLM has undertaken a number of strategies to manage this population, including finding private homes for wild horses and burros. The BLM is under increasing pressure from the public lands' multiple users, including cattle farmers and recreationists; underscoring the importance of appropriately managing multi-user relationships is that recently, local governments in one state have sued the federal government for mismanagement leading to overpopulation at the expense of the range and the animals themselves. To this end, since finding private homes is a critical component of the BLM's management strategy, understanding consumer demand for wild horses will help the BLM develop more efficient herd management strategies and ultimately improve multi-user relationships. Data for the study came from BLM internet auctions in 2012 and 2013. Multivariate regression was used to estimate a hedonic pricing model, which identified individual characteristics that predicted demand; the market value of those characteristics was then estimated. Results suggested that color, height, amount of training, and number of months the wild horse has been in captivity were the most significant drivers of demand. Capitalizing on these results may allow the federal government to improve its overall management of public lands.

**Perspective of the use of cold-blooded horses in Poland – illusions or reality?**

*G. Polak and J. Krupinski*
*National Research Institute of Animal Production, National Focal Point, Wspolna 30, 00-930 Warsaw, Poland; grazyna.polak@izoo.krakow.pl*

The local cold-blooded horse breeds Sokolski and Sztumski have been bred in Poland from the late nineteenth century. In the second half of the twentieth century, the development of mechanization caused a change in the breeding goal. A genetic resources conservation program for the horses, carried out from 2008, aims to preserve all the typical characteristics of the population. The program indicates that both breeds Sokolski and Sztumski should be used primarily as draft animals in these niches of the economy where they can make important social, economic and environmental contributions: in agro-tourism, organic farming, forestry, and protection of the landscape. The European Draught Horse Federation (FECTU) brings together organizations, associations and individuals interested in the use of modern methods of working horses and intensively promotes the organic sector. The Federat ion brings together 5,000 members from 11 European countries, including Poland. The main objectives are popularization of the use of working horses and the protection of local breeds, as part of European culture and tradition. In 2012 FECTU representatives met with members of the Commission and the Parliament of the European Union. The aim of the project was to present current perspectives and the use of Polish cold-blooded horse local breeds, in the light of current European trends. Conducted in 2001 and in 2013 a survey among farmers participating in the conservation program indicates decrease from 33% to 24% of the number of breeders who are interested in working use of horses. On the other hand the results indicate an increase of the percentage of breeders for whom the main reason for breeding them is the sale of broodmares and possibility of receiving payments under agri-environmental measures (from 63% to 72%). Currently (2014) the conservation program covers 470 animals, incl uding 261 Sokolski and 209 Sztumski mares. It is expected that by 2020 populations will increase to over 3,000 mares in both populations.

---



**Natural values on semi-natural and permanent pastures grazed by horses**

*I. Herzon[1], M.T. Saastamoinen[2], S. Särkijärvi[2], M. Myllymäki[2] and C. Schreurs[1]*
*[1]University of Helsinki, Department of Agricultural Sciences, P.O. Box 27, 00014 University of Helsinki, Finland, [2]MTT Agrifood Research Finland, Animal Production, Opistontie 10, 32100 Ypäjä, Finland; markku.saastamoinen@mtt.fi*

Composition of ground vegetation on pastures determines the value of the forage for grazing animals as well as biodiversity values for species associated with the pastoral ecosystems. A study on three case pastures grazed for a long-term exclusively by horses was conducted to evaluate how grazing affected the pasture structure and vegetation in respect to biodiversity and feeding values and horses' welfare. Also, 50 horse and pasture owners were interviewed. Biological values were particularly high in a semi-natural pasture while on an intensively managed permanent pasture these were sustained by under-grazed and non-managed areas such as ditches and stone-heaps. Through selective grazing horses create highly heterogeneous vegetation and landscape structure that is beneficial for biodiversity. There were more meadow indicator species among plants not consumed by horses than among those that are preferred. Value for pollinators of plant species avoided or under-grazed by horses was as high as of those that are preferred. In all pasture types, specific management is needed to avoid accumulation of species that both reduce forage and biological values of pastures. The mean cover of plant species preferred by horses grazed on the semi-natural pasture was 76%, and comparable to that in the intensively managed pasture. The nutritional value of the vegetation meets the needs of most horse categories, but grazing pressure, under-grazed area and the growth stage of the vegetation have to be considered. Further, some mineral supplementation is needed to balance the mineral intake. No injuries or health problems were observed in the horses. The survey respondents appreciated welfare benefits for horses, e.g. increased exercise, reduction in behavior problems and possibility for social and species-specific behavior.

## Session 01 Theatre 11

**A web tool helping horse stables to improve manure management**

*I. Riipi[1], A. Reinikainen[1], S. Särkijärvi[2], M. Saastamoinen[2], M. Myllymäki[2], M. Järvinen[3] and L. Rantamäki-Lahtinen[3]*

*[1]MTT Agrifood Research, Biotechnology and Food Research, Latokartanonkaari 9, 00790 Helsinki, Finland, [2]MTT Agrifood Research, Animal Production Research, Opistontie 10 A 1, 32100 Ypäjä, Finland, [3]MTT Agrifood Research, Economic Research, Latokartanonkaari 9, 00790 Helsinki, Finland; markku.saastamoinen@mtt.fi*

Manure management is the biggest environmental challenge in horse economy and it also has a great impact on the image of the sector. Manure management is problematic for many horse entrepreneurs, particularly in or near urban areas. Storage and utilization of manure is especially challenging for those stables which do not have opportunities to use manure as a fertilizer in their own fields. For these stables, manure represents a sizeable expenditure, and in some cases, manure management is neglected. If manure is not managed properly it can create heavy negative loads on the environment, particularly locally. There is no unambiguous solution of how to organize a workable manure management. The operational environment, location and resources available affect the selection of manure management options for each stable. Thus, there is a need for knowledge on different management practices and existing solutions in order to help developing manure management. This information has been collected in InnoEquine –project, which was funded by EU Central Baltic Interreg IV A program. The information has been integrated into a web tool in four categories: Handling of manure, Storing of manure, Utilization of manure and Paddocks. Both general information and examples of practical solutions are presented. Data collection for web tool was done by interviewing horse entrepreneurs in Finland and in Sweden and also through literature research. Some case examples have been presented in more detail. The InnoEquine web tool is freely available for horse entrepreneurs and everyone interested in Finnish, Swedish and English in www.hippolis.fi/innohorse.

---

## Session 02 Theatre 1

**Mesenchymal progenitor cells in intramuscular connective tissue development**

*M. Du*

*Washington State University, Department of Animal Sciences, 100 Dairy Road, Pullman, WA 99164, USA; min.du@wsu.edu*

The abundance and cross-linking of intramuscular connective tissue contributes to the background toughness of meat, thus undesirable. Connective tissue is mainly synthesized by intramuscular fibroblasts. Myocytes, adipocytes and fibroblasts are derived from a common pool of progenitor cells during the early fetal development. It appears that multipotent mesenchymal progenitors first diverge into either myogenic or adipogenic/fibrogenic lineages; myogenic progenitor cells further develop into muscle fibers and satellite cells, while adipogenic/fibrogenic lineage cells develop into the stromal-vascular fraction of skeletal muscle where resides adipocytes, fibroblasts and resident fibro-adipogenic progenitor cells (FAP, the counterpart of satellite cells). Strengthening FAP proliferation enhances both intramuscular adipogenesis and fibrogenesis, leading to the elevation of both marbling and connective tissue content in the resulting meat. Due to the bipotential of FAP, enhancing FAP differentiation to adipogenesis reduces fibrogenesis correspondingly, which likely results in the overall improvement of marbling (more intramuscular adipocytes) and tenderness (less connective tissue) of meat. Recent studies show that Zfp423 has a critical role in committing FAPs to adipogenesis. Nutritional, environmental and genetic factors shape progenitor cell commitment and differentiation; however, up to now, our knowledge regarding mechanisms governing progenitor cell differentiation remains rudimentary. Such knowledge is critical for us to develop strategies to prime mesenchymal progenitor cell differentiation, in order to improve animal growth efficiency and meat quality.

## Session 02 Theatre 2

**Modulating Sox2 gene expression in cattle fibroblasts by using TALE-TFs**

*Y. Shamshirgaran[1], M. Tahmoorespour[2], H. Dehghani[3], H. Sumer[4] and P.J. Verma[4]*
*[1]University of Birjand, Animal Science Department, Birjand, 97191-13944, Iran, [2]Ferdowsi University of Mashhad, Animal Science Department, Mashhad, 91775-1163, Iran, [3]Ferdowsi University of Mashhad, Department of Basic Science, Mashhad, 91775-1163, Iran, [4]Monash Institute of Medical Research, Centre for Reproduction and Development, Clayton, VIC 3168, Australia; m.tahmoorespur@gmail.com*

Sox2 is a transcription factor that plays a key role in the stemness of the pluripotent cells. Although conventional methods have been used for generating livestock induced Pluripotent Stem Cells (iPSCs) by introducing an exogenous Sox2, however, artificial transcription factors (ATFs) have more advantages for affecting gene expression when compared to the conventional methods. The ability of ATFs for modulating the endogenous expression of genes is less explored and has not been reported in livestock. Transcription Activator Like Effectors (TALEs) presented biologists with a new class of DNA binding proteins which can be easily customized for targeting a specific sequence through the genomes. In this study we developed a Golden-Gate Cloning system for generating engineering TALEs fused with VP64, a transcription activation domain, to be expressed and function in mammalian cells. The functionality of the generated TALE-TFs were confirmed by using a fluorescent based reporter system in HEK293 cells. Further, four target sites around the transcription start site of cattle Sox2 gene were selected to be targeted by using designed TALE-TFs. Retroviruses were used for transducting the Sox2-TALE-TFs into the bovine fibroblasts. Sox2 gene expression was quantified in all of the cells which received either an individual TALE-TF or a combination of different TALE-TFs. Our results show that TALE-TFs are functional in cattle cells and can upregulate the Sox2 expression by more than 5 fold. Moreover, a combination of multiple TALE-TFs shows synergistic effects on the level of Sox2 transcript enabling researchers for fine tuning the gene expression. Thus, TALE-TFs can have numerous applications in biology and agriculture.

---

## Session 02 Theatre 3

**The muscle satellite cell: Importance for growth and regeneration**

*N. Oksbjerg*
*Aarhus University, Foulum, Blichers Allé 20, 8830 Tjele, Denmark; niels.oksbjerg@agrsci.dk*

The aim of the presentation is to review the results on satellite cells (SCs) in hyperplastic muscle. SCs, residing in the niche between the sarcolemma and the basal membrane, and together with their progeny, are the source of additional myonuclei during the rapid phase of growth. Thus, in various types of meat animals, the SCs are the main source of muscle DNA (50-90%) at slaughter. In this period of growth, irradiation of muscle decreases SC proliferation and muscle mass. Several studies have shown that increased muscle growth by either breeding or increased feed intake increase SC proliferation, and intrauterine growth restriction decreases SC proliferation postnatally. Following ablation of muscle, the increased loading on synergistic muscles, results in increased muscle hypertrophy and SC proliferation whereas unloading by tail suspension causes atrophy and reduced SC proliferation. Re-growth following unloading is attenuated when muscles are irradiated prior to re-growth. Although muscle tissue has a slow protein turnover, the regenerative capacity (i.e. SC activation, content, proliferation, and differentiation) is high in response to wear and tear. In adults, SCs are dormant and express Pax7, but become activated by muscle stretch. The molecular basis involves nitrogen oxide signalling, myostatin, and HGF. Notch signaling is also involved in activation of SCs. Following activation, SCs enter the G1 phase and Myf5, MyoD and Myogenin are temporally up-regulated. This leads to differentiation, which may result in donation of new nuclei to the injured muscle fibre or fusion with each other and form new muscle fibres. Proliferation and differentiation is regulated by a vast number of hormones and growth factors in endocrine and para/autocrine ways. A few SCs escape the cell cycle and re-enter their quiescent state. In conclusion, SCs may limit muscle fibre hypertrophy, and the molecular regulation of SC behaviour is emerging, which is important for developing new strategies to improve muscle growth.

**Cellular and molecular basis of adipose tissue development: from stem cells to adipocyte physiology**

*I. Louveau, M.H. Perruchot and F. Gondret*
*INRA, UMR1348 Pegase, Domaine de la Prise, 35590 Saint Gilles, France; isabelle.louveau@rennes.inra.fr*

White adipose tissue plays a key role in the regulation of energy balance in vertebrates. Its primary function is to store and release energy. It is also recognized to secrete a variety of factors called adipokines that are involved in a wide range of physiological and metabolic functions. Unlike other tissues, adipose tissue mass has large capacity to expand and can be seen as a dynamic tissue able to adapt to a variety of environmental and genetic factors. The aim of this review is to update the cellular and molecular mechanisms associated with adipose tissue growth and development in domestic animals, with a special focus on the pig. In contrast to other tissues, the embryonic origin of adipose cells remains the subject of debate. Among the variety of cell types contained in adipose tissue, we highlight recent data dealing with adipose-derived stromal stem cells that contribute to tissue homeostasis. The relationships between the proportions of the different stem cells and other molecular indicators of adipose tissue functions are investigated. The current knowledge on the dynamic expression of genes involved in cell cycle arrest, differentiation, and physiology during fetal development is summarized to determine how these changes may be linked to newborn survival. Both genetics and nutrients can affect these processes; they also influence the postnatal growth of adipose tissue. In brief, a better understanding of the developmental biology of adipose tissue is of great importance for targeting this tissue in the animal production field.

---



**Effects of dietary methionine deficiency on adipose tissue growth and oxidative status in piglets**

*R. Castellano[1], M.H. Perruchot[1], J.A. Conde-Aguilera[1], J. Van Milgen[1], Y. Mercier[2], S. Tesseraud[3] and F. Gondret[1]*
*[1]INRA, UMR1348 PEGASE, Domaine de la Prise, 35590 Saint-Gilles, France, [2]Adisseo, France SAS, 92160 Antony, France, [3]INRA, UR83 Recherches Avicoles, 37380 Nouzilly, France; rosa.castellano@rennes.inra.fr*

Adipose tissue plays a crucial role in energy balance, but excessive fat deposition may be linked to systemic oxidative stress. Methionine (Met) is the second limiting amino acid (AA) for growth in pigs and a precursor for cysteine, one of the three AA of the antioxidant glutathione, so that the Met supply could modify the antioxidant status. This study aimed to determine the effects of dietary Met deficiency on the development of adipose tissue and redox metabolism. Weaned piglets (9 kg BW) received a diet providing either an adequate (CTRL) or a deficient supply in Met (-33%, REST) during 10 days (n=6 per group). At a similar BW, the dietary Met restriction led to a 29% increase in perirenal fat proportion, an 11% increase in lipid content in subcutaneous adipose tissue, and greater activities of lipogenic enzymes in both sites. These changes were associated with greater expression levels of genes involved in glucose uptake (GLUT4), lipogenesis (FASN, ME1) and lipolysis (HSL, ATGL and PLIN2) in the two adipose tissues. The activities of the anti-oxidative enzymes superoxide dismutase (SOD), catalase and gluthatione reductase (GRX) were increased in both adipose tissues of REST piglets compared to CTRL piglets. The oxidized glutathione content in adipose tissues was also greater in REST piglets, but the reduced form did not vary between groups. The total antioxidant activities in plasma (as assessed by free radical scavenging and ferric reducing ability) were lower in REST piglets. In conclusion, a dietary Met deficiency changed the partitioning of energy towards a greater adiposity and modified the redox status of adipocytes with consequences on the antioxidant capacity of piglets.

**A dynamic mathematical model to study the flexibility of energy reserves in adipose and muscle cells**

*M. Taghipoor, J. Van Milgen and F. Gondret*
*INRA, UMR1348 Pegase, Domaine de la Prise, 35590 Saint Gilles, France; florence.gondret@rennes.inra.fr*

Animals need energy for their basal metabolism and productive functions. When dietary energy intake exceeds energy expenditure, the excess is stored as glycogen and triacylglycerols (TG) in cells. The dynamics of uptake, oxidation and storage of nutrients at the cellular level are very important to understand tissue composition and to predict the effects of different sources of energy supply to animals. A dynamic model was developed to investigate the plasticity of energy stores into a generic cell. This model was based on a system of coupled ordinary differential equations, with each equation representing a single biochemical reaction or successive transformations of intermediate metabolites. Reaction rates were regulated by the activities of enzymes involved in the reactions. The ATP/ADP ratio was assumed to be the main intracellular signal to switch between storage and oxidation pathways. Cellular uptake of glucose and free fatty acids was modeled in a continuous or discontinuous phenomenon during the day. Simulations included the time-dependent evolution of glycogen and TG in the cell. Because of its generic aspect, the model can be used to study muscle or adipocyte metabolism by changing the basic parameters. Further development would include hormone signals regulating the partitioning of nutrients between muscle and adipose cells, and the coupling between hypertrophy and hyperplasia in tissue development. In conclusion, the model provides a dynamic framework to investigate relevant hypotheses in farm animal production, including the postprandial metabolism (fast dynamics) and flexibility of energy reserves in response to changes in feed intake, physical activity or growth (slow dynamics).



**Adipose tissue hypoxia is related to increased mtDNA copies and decreased VEGF-A in fat dairy cows**

*L. Laubenthal[1], L. Locher[2], J. Winkler[3], U. Meyer[3], J. Rehage[2], S. Dänicke[3], H. Sauerwein[1] and S. Häussler[1]*
*[1]University of Bonn, Institute of Animal Science, Katzenburgweg 7-9, 53115 Bonn, Germany, [2]University for Veterinary Medicine, Clinic for Cattle, Bischofsholer Damm 15, 30173 Hannover, Germany, [3]Friedrich-Loeffler-Institute, Institute of Animal Nutrition, Bundesallee 50, 38116 Braunschweig, Germany; lilian.laubenthal@uni-bonn.de*

Due to the potentially inadequate vascularization, as previously indicated by the pro-angiogenic vascular endothelial growth factor A (VEGF-A) in adipose tissue (AT) of overconditioned cows, we hypothesized that cell oxygen supply is insufficient and will increase the hypoxia-inducible-factor-1α (HIF-1α) indicating local hypoxia. We aimed to examine whether mitochondrial DNA (mtDNA) copy numbers, reflecting the abundance of mitochondria and thus the energetic status of a cell, are related to HIF-1α. Non-pregnant, non-lactating cows (n=8) were fed diets with increasing concentrate feed proportions during the first 6 weeks of the trial until a proportion of 60% was reached, which was maintained for another 9 weeks. Body weight and condition score increased from 540±20 to 792±29 kg and from 2.31±0.12 to 4.53±0.14 throughout the trial, respectively. Subcutaneous AT biopsies from the beginning and after 15 weeks of trial were immunohistochemically evaluated for HIF-1α-positive cells or analyzed for mtDNA copy number and VEGF-A protein expression via multiplex qPCR and semi-quantitative Western blot, respectively. Data were analyzed using the pairwise Student´s t-test (HIF-1α) and the Wilcoxon test (mtDNA, VEGF-A; SPSS). HIF-1α-positive cells tended to be increased 3.3-fold throughout the trial (P=0.065) and were correlated to mtDNA copies (ρ=0.55, P=0.034), and to VEGF-A amounts (r=-0.654, P=0.008). Decreasing VEGF-A together with excessive fat accumulation indicates inadequate angiogenesis leading to local hypoxia in AT of overconditioned cows. The increase of mtDNA copies might be a compensatory reaction to prolong adipocyte life in face of hypoxic conditions.

**mRNA abundance of the lactate receptor HCA1 in two different adipose tissue depots of dairy cows**

*M. Weber[1], L. Locher[2], K. Huber[3], J. Rehage[2], S. Dänicke[4], H. Sauerwein[1] and M. Mielenz[1,5]*
*[1]University of Bonn, Institute of Animal Science, Katzenburgweg 7, 53115 Bonn, Germany, [2]University of Veterinary Medicine, Clinic for Cattle, Bischofsholer Damm 15, 30173 Hannover, Germany, [3]University of Veterinary Medicine, Institute of Physiology, Bischofsholer Damm 15, 30173 Hannover, Germany, [4]Friedrich Loeffler Institute, Institute of Animal Nutrition, Bundesallee 50, 38116 Braunschweig, Germany, [5]Present address: Leibniz Institute for Farm Animal Biology, Institute of Nutritional Physiology, Wilhelm-Stahl-Allee 2, 18196 Dummerstorf, Germany; sauerwein@uni-bonn.de*

The G-protein coupled receptor HCA1 mediates antilipolytic effects via binding lactate, an intermediate of energy metabolism. HCA1 is mainly located on adipocytes and enables them to respond to metabolic changes through fine-tuning of lipolysis. Thus we hypothesized that different amounts of concentrate in the diet, resulting in differences in the energy balance, will affect the HCA1 mRNA abundance in bovine adipose tissue (AT). Furthermore, we aimed to determine whether the mRNA abundance between subcutaneous (SC) and retroperitoneal (RP) AT differs among late pregnancy and lactation. Biopsy samples of SCAT and RPAT from 20 pluriparous Holstein Friesian cows were collected at day (d) -42, 1, 21 and 100 relative to parturition. Until d -42 all cows were fed the same silage-based diet. Between d -42 and d -1, 10 animals each were assigned to either a high-concentrate (HC, 60:40 concentrate:roughage) or a low concentrate group (LC, 30:70). From d 1 to 24 the concentrate proportion was increased from 30 to 50% for all cows. HCA1 mRNA abundance was not affected by treatment, but showed a time-dependent increase in SCAT from d -42 to d 100 ($P<0.05$). In RPAT, no time-dependent changes were observed. Comparing the fat depots, HCA1 mRNA abundance was higher in RPAT than in SCAT on days -42 and 21 ($P<0.001$). The presented results indicate that HCA1 mRNA is regulated independently of the feeding regimen. Differences between SCAT and RPAT suggest differences in the lipolytic regulation.



**Adult stem cells in adipose tissue and skeletal muscle of pigs differing in body composition**

*M.H. Perruchot, L. Lefaucheur, F. Gondret and I. Louveau*
*INRA, UMR 1348 Pegase, Domaine de la Prise, 35590 Saint-Gilles, France; marie-helene.perruchot@rennes.inra.fr*

The control of body composition is of the upmost importance for animal production. It is now recognized that adult stem cells are essential for tissue homeostasis. The current study was undertaken to determine whether changes in the relative proportion of muscle and adipose tissue mass in pigs are controlled by the plasticity of adult stem cells. Between 10 and 20 weeks of age, 24 Large White barrows from two lines divergently selected for residual feed intake (RFI), an indicator of feed efficiency, were fed *ad libitum* either a high-starch (LF) diet or a high-fat/high-fiber (HF) diet (n=6 per line and diet). At 20 weeks of age, stroma-vascular cells were isolated by collagenase treatment from subcutaneous adipose tissue and longissimus muscle. Cell immuno-phenotypes were then determined by flow cytometry using cell surface markers (CD34, CD56 and CD90). Irrespective of diet, low RFI pigs were leaner than high RFI pigs. In the two lines, pigs fed the HF diet exhibited a reduced body fatness compared with pigs fed the LF diet. Whatever the experimental group considered, adipose and muscle cells were predominantly positive for CD56 (muscle precursor marker) and CD90 (mesenchymal stem cell marker) markers. Cells expressing the CD34 hematopoietic stem cell marker were identified in muscle but not in adipose tissue. The proportion of CD90+ and CD56+ cells in adipose tissue did not differ between the experimental groups. In muscle, the proportion of CD34+ cells was higher ($P<0.005$), whereas that of CD90+ cells was lower ($P<0.005$) in the low than in the high RFI pigs, whatever the diet. The current study indicates that divergent selection for RFI in growing pigs had strongly affected the relative proportions of adult stem cell populations in skeletal muscle but not in adipose tissue. Altogether, there is no simple relationship between the proportion of adult stem cells and body adiposity.

**Effects of sulfur amino acids on porcine preadipocyte differentiation in culture**

*R. Castellano[1], M.H. Perruchot[1], F. Mayeur[1], S. Tesseraud[2], Y. Mercier[3] and F. Gondret[1]*
*[1]INRA, UMR1348 Pegase, Domaine de la Prise, 35590 Saint Gilles, France, [2]INRA, UR83 Recherches Avicoles, 37380 Nouzilly, France, [3]ADISSEO, France SAS, 92160 Antony, France; rosa.castellano@rennes.inra.fr*

Adipogenesis is characterized by two temporally distinct processes: (1) proliferation during which pre-adipocytes replicate and increase in number and (2) differentiation during which pre-adipocytes change in morphology and accumulate lipids. Deregulation of the intracellular glutathione, the most important non-enzymatic intracellular anti-oxidant, has recently been shown to affect murine preadipocyte differentiation. Methionine (Met) is the essential precursor of Cysteine (Cys), which is the rate-limiting factor in glutathione synthesis. Here, we investigated whether low levels of Met+Cys may alter porcine adipogenesis. Preadipocytes were isolated from a subcutaneous fat depot of growing piglets and cultured with different doses of Met (0, 40, 70 and 100 µM (adequate physiological dose) in a media deprived of Cys. Very low levels of Met (0 or 40 µM) in the culture media dramatically lowered preadipocyte proliferation and reduced lipid accumulation in differentiating cells. Preadipocyte proliferation was similar at doses of 70 µM and 100 µM of Met, but lipid droplets in differentiating cells after 13, 21 and 24 days (d) of culture were more numerous at a dose of 70 µM than at a dose of 100 µM. At d13 of differentiation, the expression levels of CEBPalfa PPARgamma, two genes regulating the transcription of adipocyte-specific genes, and of LPL and HSL, two enzymes of lipid metabolism, were higher at a dose of 70 µM than at a dose of 100 µM. At d24 of differentiation, expression levels of these genes were much higher at a dose of 100 µM. These results suggest a delay between time-related changes in gene expressions and lipid accumulation in cells. They show that very strong deficiency in Met decreases both proliferation and differentiation, whereas a mild deficiency of Met alters only differentiation of porcine preadipocytes.

---



**Expression of microRNA-196a and -885 in bovine skeletal muscles**

*S. Muroya, M. Oe, I. Nakajima and K. Ojima*
*NARO Inst. Livestock and Grassland Science, Animal Products Research Division, Ikenodai 2, Tsukuba, Ibaraki, 305-0901, Japan; muros@affrc.go.jp*

MicroRNAs (miRNA) are non-coding small RNAs that attenuate their target gene expression in various biological processes. Using 5 cattle, we analyzed the expression of miR-196a, myosin heavy chain (MyHC) isoforms, and the predicted target genes of miR-196a, collagen α 1 chain of the type I (Col1a1) and III (Col3a1) in the masseter (MS), diaphragm (DP), semispinalis (SP), semitendinosus (ST), and longissimus dorsi (LD) muscles to investigate effect of miRNA on muscle type specification. According to the quantitative RT-PCR results, miR-196a is exclusively expressed in MyHC-2x isoform-abundant ST and LD muscles compared to the others ($P<0.001$). On the other hand, expression of miR-885, a ST-enriched miRNA compared to MS in our previous study, was not significantly different among the muscles tested ($P=0.178$). Especially, the miR-196a expression was correlated with those of MyHC-slow and -2x isoforms ($r=-0.530$ and $r=0.976$, respectively). The Col1a1 and Col3a1 expression were correlated with the miR-196a expression ($r=-0.311$ and $r=0.354$, respectively), and showed a higher level in the slow type than in the fast type muscles with a high correlation to the MyHC-slow expression ($r=0.806$ and $r=0.801$, respectively). These results indicate that the miR-196a expression is associated with the muscle type with relation to MyHC isoform expression in the bovine skeletal muscles, suggesting a close relationship in transcriptional regulation between MyHC-2x isoform and miR-196a in those muscles. These results also indicate that Col1a1 and Col3a1 expression is highly dependent on MyHC isoform type rather than the post-transcriptional regulator miR-196a.

## Session 03 Theatre 1

**Lamb mortality: current knowledge and research perspectives**

*F. Corbiere[1,2] and J.M. Gautier[2,3]*
*[1]National Veterinary School of Toulouse, Ruminants Medicine, 23 Chemin des Capelles, 31076 Toulouse, France, [2]UMT Small Ruminants Heard Health Management, 23 Chemin des Capelles, 31076 Toulouse, France, [3]Institut de l'Elevage, BP 42118, 31321 Castanet Tolosan, France; f.corbiere@envt.fr*

The overall lamb mortality rate before weaning is reported to around 10 to 20%, depending on the production system. The age related distribution of lamb mortality indicates that early death (within the first week of life) contributes to the majority of these losses. Numerous infectious and non-infectious origins are well known, but their precise implication and interactions are often difficult to establish in the field. Indeed, the lamb vigor appears as a central component of its early viability and its ability to face infectious threats. This vigor may be influenced by numerous risk factors linked to flock management practices but also seems to be partially dependent of genetic determinants. Beyond the control of major infectious diseases, it appears that the ewe's management from mating to lambing is as important as the management of the neonate. This implies that research studies should not only focus on the pathophysiology of early infectious or metabolic disorders, but also get interested on animal welfare, farmers' work from a psycho-sociological point of view since all these factors may be involved in lamb vigor. The present paper will propose a rapid overview of current knowledge on lamb mortality and focus on research perspectives in the area.

## Session 03 Theatre 2

**The importance of nutrition during gestation for lamb vigour and survival**

*J.A. Rooke[1], G. Arnott[2], C.M. Dwyer[1] and K.M.D. Rutherford[1]*
*[1]SRUC, West Mains Road, EH9 3JG, Edinburgh, United Kingdom, [2]Queens University, Belfast, BT9 7BL, United Kingdom; john.rooke@sruc.ac.uk*

The prenatal period is of critical importance in defining how individuals respond to their environment throughout life. Stress experienced by pregnant females has been shown to have detrimental effects on offspring behaviour, health and productivity. Sub-optimal nutrition is one of the stressors ewes are most likely to be exposed to and indeed may indeed be part of the normal production cycle. Because of the well-known relationship between low lamb birth-weight and chances of survival, birth-weight is a useful proxy for lamb survival. A systematic review showed that maternal under-nutrition in the last third of pregnancy consistently impaired lamb birth-weight and subsequent vigour and performance. Earlier under-nutrition had a variable effect on performance. Feeding the ewe above requirements did not have positive effects on lamb performance and welfare. However, there is little evidence as to how any responses are dependant on sheep breed, maternal condition score and litter size. Effects of maternal micro-nutrient deficiency and excess on lamb mortality and vigour are also variable and dependant upon the status of the dam. There are considerable gaps in knowledge particularly in respect of how other stressors such as disease and environment interact with nutrition and the possibility of trans-generational impacts of nutrition.

**Using genetic selection to improve lamb survival in extensive sheep production systems**

*J. Conington, K. Moore and C. Dwyer*
*SRUC, W. Mains Rd., Edinburgh, EH93JG, United Kingdom; jo.conington@sruc.ac.uk*

Improving lamb survival through breeding is relevant for extensively-managed flocks or for flocks where minimal intervention by man is practised. Having lamb survival integrated into existing sheep breeding programmes facilitates the monitoring of it alongside that of other traits in the breeding objective. Industry data on Blackface sheep comprised 173,895 lamb records from 1976-2011 including 49,917 lamb birth weight records. Lamb survival (SURV01) was coded 0=dead if any lamb recorded at birth had no live weight recorded at either 8 weeks or subsequent weighing occasions, and 1=alive, if it survived at least until 8 weeks with a record of live weight. Lamb survival accounting for lambs known to be born dead (SURV12) from 27,111 lambs born 2007-2011, was coded 0=lambs known to be born dead, 1=born alive but no subsequent live weights (so assumed dead) and 2=lambs born with subsequent live weights. Generalised Linear Mixed Model (GLMM) for binomial trait distribution was used to fit a logit transformation function to predict the probability of the effect of each factor for SURV01. Main fixed effects were flock, year/season, age of dam, sex of lamb, birth type and birth weight (covariate). The heritability ($h^2$) of SURV01 and SURV12 was estimated with (+) and without (-) maternal heritability ($m^2$) accounted for, using ASREML software. Lamb survival was 87.8% for the SURV01 and SURV12 traits, with 5.5% lambs born dead in the SURV12 2007+ data set. Compared to male lambs, female lambs have a 1.3 times better survival odds. Lambs from 3 year old ewes are 1.4 times more likely to survive compared to lambs from 2 year old ewes and have the highest survival odds compared to all other ewe ages. The direct $h^2$(s.e) of SURV01(-) was 0.05(0.015) but was 0.01(0.006) when $m^2$ was included in the model (0.08)(0.03) giving a total $h^2$ of 0.05 and 0.09 for SURV01(-) and SURV01(+) respectively. New parameters for lamb survival in 3 other UK sheep breeds will be estimated before lamb survival is included in the UK Sheepbreeder recording scheme in 2015.

---



**The role of ewe and lamb behaviour in lamb survival**

*C.M. Dwyer[1], M. Nath[2] and S.M. Matheson[1]*
*[1]SRUC, West Mains Road, Edinburgh EH9 3JG, United Kingdom, [2]BioSS, Mayfield Road, Edinburgh EH9 3JZ, United Kingdom; cathy.dwyer@sruc.ac.uk*

The survival of the neonatal lamb is critically dependent on the coordinated expression of ewe and lamb behaviour at birth. The role of the mother is to nurture the lamb and to develop an exclusive attachment to her own lambs, whereas the lamb needs to begin independent breathing and thermoregulation, and locate the udder to drink colostrum. The lamb also learns to recognise its own mother within about 24 hours of birth. Whilst the ewe can facilitate the lambs udder-seeking activities the lamb is primarily responsible for processing sensory and other cues from the ewe and reaching the udder to suck. Studies using embryo-transfer have shown that these early behaviours of the lamb are not significantly affected by the maternal behaviour of the ewe, but are more related to the abilities of the lamb at birth. Furthermore, survival analysis has shown that significant contributors to lamb survival to weaning, in addition to birth weight and litter size, are the ease of the birth process and the ability of the lamb to suck from the ewe shortly after birth. Significant breed, line and sire within breed effects on lamb sucking behaviour have been shown to exist, suggesting that this behaviour may be partially under genetic control. Recent research has focussed on developing methods to record these important behaviours on commercial farms in sufficient numbers to allow analyses of genetic parameters. Scoring systems for lamb traits were developed and collected for more than 11,000 lamb records. Genetic analyses revealed that lamb behavioural traits had moderate heritability suggesting that progress in improving lamb survival may be achieved by genetic selection for improved lamb vigour. For outdoor lambing flocks, where these scores are difficult to measure, our recent studies suggest that lamb rectal temperature recorded in the first 24 hours of life, is also significantly associated with improved survivability, and may be correlated with lamb ability to reach the udder quickly after birth.

**Physiological factors contributing to lamb mortality**

*R. Nowak[1], M. Hammadi[2] and M. Chniter[2]*
*[1]INRA, Physiologie de la Reproduction et des Comportements, 37380 Nouzilly, France, [2]Institut des Régions Arides, Laboratoire Elevage et Faune Sauvage, 4119 Médenine, Tunisia; raymond.nowak@tours.inra.fr*

New-born lambs have limited energy reserves and need a rapid access to colostrum to maintain homeothermy and survive. In addition, colostrum provides immunoglobulins which ensure passive systemic immunity. Recent work has focused on increasing prolificacy with the aim to reach optimal lamb production per ewe. Yet, despite years of research, newborn lambs still suffer a high mortality rate which shows how difficult it is to solve this problem. Three aspects should be considered to face this matter. Firstly, the level of nutrition during pregnancy influences plasma concentrations of hormones and metabolites related to lactogenesis. Progesterone concentrations decrease before lambing but should ewes be under-fed the fall is too small to initiate the onset of colostrum production. β-hydroxybutyrate concentrations are higher in under-fed than in well-fed ewes, suggesting that the former mobilizes more adipose tissue but they still do not meet their metabolisable energy requirements for colostrum production, which may be critical in twin-bearing ewes. Secondly, neonatal metabolic factors and survival are intimately linked. Glucose, proteins, IgG, cholesterol and triglycerides plasma levels are low in newborn lambs, and increase over the first 3 days. Rectal temperature increases too, with significant effects of litter size and birth weight. Usually, rectal temperatures, glucose and protein levels are higher in lambs that survive beyond one month of age in comparison to those that die. Smaller lambs have lower rectal temperature and metabolites plasma level in the first 3 days which impairs their chance of survival. Finally, delayed lactation or insufficient colostrum yield may impair mother-young relationship since suckling has strong rewarding properties in the establishment of a bond with the mother. Lambs which are slow to get attached to their mother may see their chance of survival jeopardized.

---

Session 03 Theatre 6

**Experiences from commercial farms: limits and interests**

*J.M. Gautier[1,2], L. Sagot[3] and F. Corbière[1,4]*
*[1]UMT Maîtrise de la santé des petits ruminants, 23 chemin des Capelles, BP 87614, 31076 Toulouse Cedex, France, [2]Institut de l'Elevage, BP 42118, 31321 Castanet Tolosan Cedex, France, [3]Institut de l'Elevage/ Ciirpo, Ferme expérimentale du Mourier, 87800 Saint Priest Ligoure, France, [4]École Nationale Vétérinaire de Toulouse, 23 chemin des Capelles, BP 87614, 31076 Toulouse Cedex, France; jean-marc.gautier@idele.fr*

All experiences from commercial farms had to face to several problems or questions: (1) Prospective or retrospective approach? In prospective studies the quality of data collection can be monitored and finally be more adequate to answer some specific questions. However this approach need more time to yield results (the trial must be finished). Based on already available data, retrospective studies can give answers within a short time with generally less reliability due to recall of facts from memory. (2) The motivation of farmers and technicians to collect data. This part must not be minimized. (3) Exposure to risk factors: a lambing period or farm level collection? In some rearing system, there are different lambing periods with different level of exposure to risk factors. This makes really tough the field approach. (4) Statistical power? With data collected at the individual level (lamb or ewe such as litter size, sex ...) the statistical power is often strong enough. But for data collected at the lambing period or farm level (practices, raring conditions) it's sometimes difficult to identify any statistically significant risk factors. The number of farms and the number of lambing periods will determinate the statistical power. A prospective study was conducted in 60 French commercial farms during two years to describe the rate and age of lamb mortality and the exposure to some known or suspected risk factors. After two years of study only 54 farms data were fully collected and suitable for a statistical approach. Despite the quality of individually and collectively based data and the large amount of records, only a few risk factors were found to be significantly associated with lamb mortality.

## Session 03 Theatre 7

**Identifying the obstacles for achieving improvements in lamb mortality rates**

*I.H. Holmøy and K. Muri*
*Norwegian University of Life Sciences, Faculty of Veterinary Medicine and Biosciences, Department of Production Animal Clinical Sciences, P.O. Box 8146 Dep., 0033 Oslo, Norway; karianne.muri@nmbu.no*

Large differences in mortality rates between flocks and rather strong correlation between rates in two consecutive years within the flocks clearly suggest that flock level risk factors are important and to some extent stable from year to year within a flock. To address flock level risk factors in Norwegian sheep production, data from a questionnaire survey on sheep housing and management were combined with data on flock level neonatal lamb mortality. Results indicated that the farmers with most experience and those who provided close monitoring of lambing ewes and active support of newborn lambs failing to suck were most successful with respect to neonatal survival. None of the housing factors showed any relation to flock level neonatal mortality, clearly indicating that management practices should be the main focus when preventive flock level factors are addressed. Despite research efforts elucidating the causes of lamb mortality, the problem remains persistent. The reasons for this are likely to be multifactorial, ranging from issues such as knowledge transfer, economic incentives, practical constraints, the attitudes of the farmers, and their relationship with the people providing them with advice. Therefore, in order to identify the main obstacles for achieving the required changes in management at farm-level, there is a need for multi-disciplinary approaches involving methods from the social sciences. Some of the likely constraints and methods to investigate them will be discussed.

## Session 04 Theatre 1

**Approaches to defining sustainable intensification**

*D. Moran*
*SRUC, Land Economy, Environment & Society, Kings Buildings, EH9 3JG, United Kingdom; dominic.moran@sruc.ac.uk*

Sustainable intensification (SI) has been advanced as a global objective for agriculture, involving increased productivity, resource use efficiency and the associated objectives for minimising harmful externalities including greenhouse gas emissions, water use and biodiversity loss. This presentation will consider the competing definitions of SI and the extent to existing metrics and models may allow us to judge whether, how and where agriculture can actually sustainably intensify.

**Towards sustainable animal diets**

*H. Makkar and P. Ankers*
*FAO, AGA, 00153 Rome, Italy; harinder.makkar@fao.org*

Animal feed and feeding is the foundation of livestock production systems. Not only animal productivity, health and welfare, product quality and safety, producers' income, household security, but also land use and land use change, water pollution and greenhouse gas emission are impacted to a great extent by the feed and how a feed is fed to livestock. Through an iterative process a concept of sustainable animal diets (StAnD) has been developed. The concept is based on the Three-P dimensions of sustainability (Planet, People and Profit), complemented by another vital aspect, namely the ethics of using a particular feed. The analysis reported here derives from 1,195 respondents worldwide. Respondents ranged across academia, industry, farmers' associations, government organizations, non-governmental organizations and inter-governmental organizations. This study has identified directions for positive change that should be followed in the production and feeding of StAnD. That positive change is dictated by higher importance assigned to the Planet, People and Ethics dimensions, and lower to the Profit dimension. An important message from the study was that enhancing the quality of the environment and socio-economic conditions of people, and respecting the ethics of rearing livestock, would be acceptable to the stakeholders, even if it implied constrained profit. The study has also prioritized elements of the sustainability dimensions of StAnD, and identified sectors that should take the initiative, and has also presented modalities for incorporating the StAnD concept into practice. These could be the focus in follow-up studies and actions. Sustainable animal diets are expected to be beneficial for the animal, the environment and society, and are likely to generate socio-economic benefits, furthering resource use efficiency, poverty alleviation and food security efforts. This requires active participation of researchers, extension workers, science managers, policy-makers, industry and farmers.

---

Session 04 Theatre 3

**Productivity, efficiency and environmental load from cattle production since 1950**

*T. Kristensen[1], O. Aaes[2] and M.R. Weisbjerg[3]*
*[1]Aarhus University, Agroecology, P.O. Box 50, 8830 Tjele, Denmark, [2]Knowledge Centre Agriculture, Cattle, Agro Food Park 15, 8200 Århus N, Denmark, [3]Aarhus University, Animal Science, P.O. Box 50, 8830 Tjele, Denmark; troels.kristensen@agrsci.dk*

Cattle production during the last century has changed dramatically with steady raising production per animal and increasing herd and farm size. The effect of these changes on total production, herd efficiency, surplus of nitrogen (N) at herd and farm level and emission of greenhouse gasses (GHG) per kg product has been evaluated for the Danish cattle sector (dairy and beef). Typical farms were modelled in (A) 1950, representing the first part of the century before mechanization and introduction of new technologies and a more global marked; (B) 1980, representing a period with heavily use of external resources like fertilizer and protein; and (C) 2010, today with focus on balancing production and risk of environmental damage. In A the family farm was based on 8 dairy cows producing 3400 kg ECM fed primarily on pasture and hay, only limited supplemented with imported protein. In B the farm was still family based, but herd size had increased to 20 dairy cows producing 5,000 kg ECM and feeding was changed towards silage and larger proportion of imported feed, while in C herd has increased to 134 dairy cows producing 9,000 kg ECM fed indoor all year. Key figures for efficiency (A, B, C): N-efficiency herd (19, 16, 23%), NE (net energy) used to produce milk and meat in % of intake (49, 54, 57%) and for environment N surplus (64, 228, 147 kg N per ha farmland), GHG (1.48, 1,94, 1,17 kg $CO_2$ eq. per kg ECM, and land use (0.34, 0.23, 0,15 ha global land per 1000 kg ECM). During the period investigated dairy production changed based on improved genetic potential, biological knowledge and not least management, leading to higher efficiency, less land use and emission per kg product, while surplus of N in relation to the area needed for feed production at the farm increased from A to B, while decreasing toward C although stocking rate increased (2,960, 4,960, 6,960 kg ECM/ha).

**Monitoring sheep breed changes in Britain over 40 years: breed choice responds to economic pressures**

*G.E. Pollott*
*RVC, Royal College Street, London, NW1 0TU, United Kingdom; gpollott@rvc.ac.uk*

Between 1971 and 2012 there have been 5 ad hoc surveys carried out in Britain to monitor sheep breed numbers and use. These have been largely conducted by postal surveys to a random sample of British sheep breeders. A 30% response rate has been the norm. Over this 40 year period there have been dramatic changes in breed use with a rise in crossbred ewe numbers from 31% in 1971 to 55% in 2012. This reduction in purebreds has been paradoxically linked to a dramatic rise in the number of breeds found; from 60 in 1971 to 90 in 2012. This increase has been a combination of imported breeds (30) and newly derived composite breeds (12). The most dramatic example of a change in breed use is seen in the Texel. This breed was not found in 1971 and is now the major Terminal Sire breed in the UK, sires more crossbred ewes than any breed and contributes most to market lamb production. This rise is due to the lean content and size of carcase produced, which fits well with the UK market requirements. At the same time several traditional British breeds have declined over this 40 year period; the main reason being that their carcases are too small or too fat. Two British breeds have risen in numbers during this period. The Bluefaced Leicester is the sire of the major crossbred ewe type, which has increased in numbers from 311,000 in 1971 to over 2 million in 2012. The Lleyn is a local Welsh breed. Both breeds are known for their better efficiency and hence profitability. Few of the newly imported breeds or composites have had much impact, with the exception of the Texel and Charollais. These changes in breed numbers have to be seen against a background of changing government and EU policy, which encouraged sheep production between 1980 and 2000 and led to a dramatic increase in ewe numbers nationally (25 million sheep in 1971, 44 million in 1996 and 30 million in 2012). Sheep breeders collectively respond to market and political signals in both sheep numbers and breed composition of the national flock.

---



**A novel approach to assess efficiency of land use by livestock to produce human food**

*H. Mollenhorst[1], C.W. Klootwijk[1], C.E. Van Middelaar[1], H.H.E. Van Zanten[1,2] and I.J.M. De Boer[1]*
*[1]Animal Production Systems group, Wageningen University, P.O. Box 338, 6700 AH Wageningen, the Netherlands, [2]Wageningen UR Livestock Research, P.O. Box 135, 6700 AC Wageningen, the Netherlands; erwin.mollenhorst@wur.nl*

The increasing demand for animal-source food will intensify the claim on land. Land use efficiency, therefore, is increasingly important for livestock production. Life cycle assessment (LCA) is commonly used to assess land use of animal-source food along the chain. Current LCA studies show that land use per kg of protein is similar for eggs and milk. These studies, however, ignore that compared to diets of dairy cattle, diets of laying hens generally contain more plant products that humans could consume directly. Moreover, current LCA studies do not account for the suitability of land to grow human-edible plants. Aim of this study was to develop a method to determine the efficiency of land use by livestock to produce food. We illustrated our novel approach for the case of Dutch egg and milk production systems. First, we used an LCA to assess land occupation from feed production required for one kg of human-digestible (HD) protein from eggs or milk. Second, this land was assessed for its potential to directly produce HD protein. Finally, we defined land use efficiency as the ratio of the HD protein production from livestock over the potential to directly produce HD protein from food crops. Land occupation for egg production was about 26 $m^2 \times yr$/kg HD protein. On this land, about twice as much HD protein from food crops could be produced, resulting in a ratio of 0.5. For milk production, the ratio depended strongly on soil type. Dairy farms on peat soils, which are most suitable to grow grass, have a ratio higher than 1, implying efficient land use. Dairy farms on sandy soils, however, have a ratio of about 0.4. Results show that conclusions on land use efficiency of animal production change when the alternative production is taken into account. This novel approach helps to identify livestock systems that could contribute to efficient use of land.

## Optimization of energy and water consumption on dairy farms

*M.D. Murphy[1], J. Upton[2] and M.J. O'Mahony[1]*
*[1]Cork Institute of Technology, Process Energy & Transport, Bishopstown Cork, NA, Ireland, [2]Teagasc Moorepark, AGRIC, Fermoy, Cork, NA, Ireland; michaeld.murphy@cit.ie*

In this study the relationship between energy and water consumption for the purpose of milk cooling was investigated. A controller was developed and tested for a milk cooling system. An ice on coil bank with air agitation and a dual stage plate heat exchanger with variable flow pumping control were manufactured for testing. The controller was designed to facilitate the near instant cooling of milk using vary amounts of water and cold thermal storage. Two principle controller configurations were developed and eight individual controllers setting for water and cold thermal energy consumption were tested. The introduction of a disturbance rejection mechanism was analysed for the controller. The control system was able to achieve instant milk cooling over a wide range of controller settings. An energy cost model was employed to calculate the optimum controller setting for specific water and electricity costs. From the results of this work it can be concluded that potentially large energy and cost saving can be made by applying a control scheme to a conventional milk cooling system.

---



## Evolving metrics used in evaluating livestock and poultry systems: a North America focus

*R.O.Y. Black*
*Michigan State University, Agriculture, Food, and Resource Economics, Justin S. Morrill Hall of Agriculture, 446 West Circle Drive, East Lansing, MI 48824, USA; blackj@msu.edu*

The structure and institutional environment the value chain livestock and poultry farms are members of has evolved in nearly unimaginable ways in North America over the last half century and the rate of change remains unabated. Variants of local foods exist alongside an egg production firm that supplies a significant proportion of the eggs to a major fast food chain. As in Western Europe, animal welfare considerations and use of growth and milk production regulators are part of the research and regulatory environment, business strategies to define and capture markets based upon production system attributes are becoming common, strategies of various groups for provision of pertinent 'information' fill the media, and ballot initiatives as an alternative to standard political mechanism are widely used. This paper reviews the evolving metrics for system evaluation associated with this myriad of forces.

**Mixed crop-livestock farming systems: a sustainable way to produce beef?**

*P. Veysset, M. Lherm, D. Bébin and M. Roulenc*
*INRA, UMRH, Clermont-Theix, 63122 Saint Genès-Champanelle, France; veysset@clermont.inra.fr*

Mixed crop-livestock farming has gained broad consensus as an economically and environmentally sustainable farming system. Working on a Charolais-area suckler cattle farms network, we subdivided the 66 farms of a constant sample for 2 years (2010 and 2011) into four groups: (1) specialized conventional livestock farms (100% grassland farms, n=7); (2) integrated conventional crop–livestock farms (specialized farms that only market animal products but that grow cereal crops on-farm for animal feed, n=31); (3) mixed conventional crop-livestock farms (farms that sell beef and cereal crops to market, n=21); (4) organic farms (n=7). We analyze the differences in structure and in drivers of technical, economic and environmental performances. The figures for all the farms over two years (2010 and 2011) were pooled into a single sample for each group. The farms that sell crops alongside beef miss out on potential economies of scale. These farms are larger than specialized beef farms (with or without on-farm feed crops) and all types of farms show comparable economic performances. The large mixed crop-livestock farms make heavier and consequently less efficient use of inputs. This less efficient inputs use also weakens their environmental performances. This subpopulation of suckler cattle farms appears unable to translate a mixed crop–livestock strategy into economies of scope. Organic farms most efficiently exploit the diversity of herd feed resources, thus positioning organic agriculture as a prototype mixed crop–livestock system meeting the core principles of agroecology.



**Natural resource use efficiency gap analysis, insights from global modeling**

*P.J. Gerber, A. Mottet, A. Uwizeye, A. Falcucci and G. Tempio*
*Food and Agriculture Organization of the United Nations, Viale delle Terme di Caracalla, 00153 Rome, Italy; pierre.gerber@fao.org*

The livestock sector is a major user of natural resources, such as land and water, using about 35% of total land and representing about 8% of total water withdrawals, mostly for feed production. Livestock production also has a strong impact on nutrient cycles. Natural resource use efficiency (NRUE) is defined as the amount of natural resource engaged per unit of product. It is a concept of particular relevance to livestock supply chains, which are typically longer than crop supply chains; involve more biophysical processes and often include the recycling and re-use of by-products from other sectors. NRUE is also a concept of global relevance. It applies to the most affluent areas of the globe, where the sector is requested to minimise its environmental impact, and to emerging economies, where livestock production expands rapidly in a context of weak environmental regulations and often wastes natural resources. It is also relevant to the poorest regions of the world where there is a need for maximizing production out of limited resources. Drawing on the Global Livestock Environment Assessment Model (GLEAM), this communication explores gaps in NRUE. GLEAM uses geo-referenced data to computed NRUE for each production units in the model (cells of a raster map). NRUE gaps are defined by the distribution of values computed for production units within a given system or region. Results for land, nutrients, water and net biomass appropriation show important gaps in all production systems and regions. For example, the fraction of feed nitrogen found in milk ranges from 2.6 to 36.9% for lactating dairy cows in Asia. Bridging such gaps would allow for significant efficiency gains and reduced pressure on natural resources. Tentative results show that bridging 90% the NRUE gap within production systems could reduce natural resources mobilisation by about 30%. This calls for specific policy environments that foster innovation and technology transfer.

**A mechanistic model to explore potential beef production of cattle breeds in contrasting climates**

*A. Van Der Linden[1,2], S.J. Oosting[2], G.W.J. Van De Ven[1], M.K. Van Ittersum[1] and I.J.M. De Boer[2]*
*[1]Wageningen University, Plant Production Systems, Droevendaalsesteeg 1, 6700 AK Wageningen, the Netherlands, [2]Wageningen University, Animal Production Systems, De Elst 1, 6700 AH Wageningen, the Netherlands; aart.vanderlinden@wur.nl*

Sustainable intensification of livestock production systems is a way to realise the increasing global demand for meat. Current empirical studies reveal meat production levels obtained by best practices, but do not clarify the theoretically achievable (i.e. potential) beef production, which is defined by animal genotype and climate only. In crop production, the production ecological concepts of potential, limited and actual production are generally used to give insight in the scope to increase production from their current levels, and to quantify options for sustainable intensification. This research aims to apply these concepts to beef production to gain insight in potential beef production levels and yield gaps. A mechanistic, dynamic model is developed to simulate beef cattle growth based on genotype and climate. Thermoregulation, energy partitioning, feed digestion and herd dynamics are included. Potential beef production is assessed in France and Ethiopia, for two breeds: Charolais (native to France) and Boran cattle (native to Ethiopia). Potential beef production of Charolais cattle in France and Ethiopia in free grazing systems was 328 and 224 g beef per kg total body weight (TBW) per year, and 251 and 280 g beef per kg TBW per year for Boran cattle. Feed conversion ratio (FCR), defined as kg DM feed per kg beef, of Charolais cattle in France and Ethiopia was 25.3 and 33.5, and FCR for Boran cattle was 32.1 and 27.6. Potential production was greater in stables than in free grazing systems, because of more favourable climatic conditions. Simulations indicate that potential beef production is greater in a climate to which a breed is adapted than a sub-optimal climate. The results provide scope to benchmark actual beef production relative to potential beef production.



**Cost comparison between use of commercial versus self-produced concentrate in pig farms**

*E.E. Pérez-Martínez, R. Trueta and C. López*
*FMVZ Universidad Nacional Autónoma de México, Av. Universidad no 3000 Del. Coyoacán, 044510, Mexico; trueta@unam.mx*

In order to increase benefits, farmers have to adopt decisions that are not simple, one of them has to do with self-producing or buying commercial feed depending on their cost per kg of food and its efficiency to increase live weight. The objective of this research is shed light in the economic efficiency of pigs foodstuff, since it represents 70-80% of total cost in Latin America. 208 farms from 8 states of Mexico were selected from the National cattle registry, and interviewed in 2012. The selection was randomly made in each of the following stratums: 50 to 100 sows, 101 to 200, 201 to 500 and 501 or more sows, after applying information filters 92 were analyzed. To compare the costs of feed, farms were divided into two groups according to the origin of the feed: self-produced (milling and mixing) or commercial. The variables compared were: average cost of kg of feed throughout the entire production cycle (fc), and the cost of production per kg live pig sold (fc/kg) excluding all inputs except feed. Costs are in Mexican pesos of 2012. Comparison of means of the two groups was performed using ANOVA with SPSS19® software. In farms using commercial feed, the fc of $5.54 is cheaper (P=0.017) than the cost of self-produce feed ($5.94). In the case of fc/kg, the latter have an average fc/kg of $17.06 that is cheaper by $1.3 pesos to farms using commercial feed. However, no significant difference was found between the groups (P=0.43). The results show that the fc of self-produced feed is significantly more expensive; Their fc/kg is cheaper but no statistical significant difference was found for the two groups. The food efficiency being responsible for the cheaper live kg sold due to a better feed conversion and because diets are formulated based on the specific requirements of each farm. The above results contribute to informed decision-making in both the farm and public policy in favor of the self-produced feed in farms.

**Replacing corn by crude glycerol on *in vitro* fermentation kinetic of complete diets for dairy cows**

*M. Bruni*[1], *M. Carriquiry*[1], *J. Galindo*[2] *and P. Chilibroste*[1]
[1]*Faculty of Agronomy. University of the Republic, Department of Animal Production and pastures., Ruta 3 Km 363, 60000 Paysandú, Uruguay,* [2]*Institute of Animal Science, Carretera Central, Km 471/2, San José de las Lajas, La Habana, Cuba, 32700, Cuba; mbruni@fagro.edu.uy*

The study was conducted to assess the effect of the substitution of 50% (CD50) or 100% (CD100) of corn by crude glycerol (76.5% of glycerol, ALUR Uruguay) in complete diets (CD) on the kinetics of *in vitro* gas production (GP). The CD was composed of forage, corn silage and concentrate (40:30:30 DM basis) with a chemical composition of 920, 150, 560, 260, and 14 g/kg MS of OM, CP, NDF, ADF and EE respectively. The *in vitro* fermentation characteristics were determined using the gas production technique, with a pressure transducer, manual reading of GP and frequent releases of the gas. Samples were incubated in triplicate in three different runs. The ruminal inoculum was collected from two lactating dairy cows fed with the CD diet. Data was fitted to the nonlinear model GP(t)=b(1-e-c(t-L)) to determine the kinetics of gas production where GP is gas production (ml/g DM) at the time t, (b) the potential GP which occurs at the fractional rate (c), and L is lag time (h). Kinetics parameters were analyzed with a mixed model that included the effect of treatment as fixed effect and run as a random effect. Data are expressed as $\text{lsmeans}_{(s.e)}$. There were no differences among treatments for b $[258.9_{(25.2)}$, $277.2_{(38.6)}$ and $264.3_{(27.7)}$ ml/g DM ], c $[0.045_{(0.006)}$, $0.043_{(0.004)}$ and $0.042_{(0.005)}$ $h^{-1}]$ or L $[0.95_{(0.48)}$,$0.87_{(0.66)}$ and $0.86_{(0.71)}$ h ] for CD, CD50 and CD100, respectively. Replacement of corn by crude glycerol did not alter PG profiles *in vitro*. Thus, it can be concluded that corn can be substituted for this by-product without affecting the activity of ruminal microorganisms. Further research is needed to evaluate the effects of glycerol inclusion in the dairy cow diet in milk production and composition, metabolism and reproduction.



**Transition to sustainable pathways of livestock production requires new methodological approaches**

*P.V. Baret*[1], *F. Vanwindekens*[1], *T. Lebacq*[1,2] *and D. Stilmant*[2]
[1]*Université de Louvain, Earth & Life Institute, Croix du Sud 2 L7.05.14, 1348 Louvain-la-Neuve, Belgium,* [2]*Centre wallon de recherches agronomiques, Rue de Serpont 100, 6800 Libramont, Belgium; philippe.baret@uclouvain.be*

Livestock systems of the 21st century face new challenges as they will have to address production, environmental and socio-economical issues. A way to tackle this increase in complexity is to combine research based knowledge with breeder's knowledge in a transdisciplinary framework. As interdisciplinary works are still a limited share of scientific production, the everyday practice of farmers is, by design, a combination of disciplinary skills. Moreover, the importance of interactions and trade-offs in farming systems is poorly documented in technical approaches. New methodologies are in development to combine qualitative and quantitative data and to make sense of this complexity. For example, we developed fuzzy cognitive maps methods in order to convert speeches of the farmers on their practices in an object – cognitive maps – that can be statistically processed. On the other hand, we use methods inspired by the transition theory to describe and discuss the diversity of pathways in dairy systems. in both situations, the rationales are to keep most of the information visible and to build syntheses not only on technical and ex-ante typology but also on socio-economical and anthropological dimensions. The diversity is higher than initially expected whatever the considered scale from farm to the chain value. To make sense of this diversity is both a scientific objective and a societal horizon.

**Effect of high N efficiency rations on dairy cows metabolic profile and milk yield**

*L. Migliorati, L. Boselli, A. Dal Prà, F. Petrera and F. Abeni*
*Consiglio per la Ricerca Sperimentazione agricoltura, FLC, Via Porcellasco 7, 26100 Cremona, Italy; luciano.migliorati@entecra.it*

Aim of this study was to assess the effect of dietary crude protein (CP) optimization on dairy cows metabolic profile and milk yield by adopting feeding techniques based on the reduction of CP content and on high diet N efficiency. A Northern Italy commercial farm placed in in Grana Padano area was used. In six control per year, for two consecutive years, the herd was monitored for metabolic profile by a random sampling of fifteen Italian Friesian cows (5 in early, 5 in mid and 5 in late lactation) to compare two diets with different CP content provided in two consecutive years. In the 1$^{st}$ year cows were fed a 16% CP diet on DM basis (C) and 2$^{nd}$ year CP was 5% reduced by to 15.2% (T). The experimental design was a cohort study. Blood samples were drawn from jugular vein just before daily distribution of total mixed ration. Plasma metabolites related to energy and lipid (glucose, cholesterol, triglycerides, NEFA, and BHB), nitrogen (urea, creatinine, total protein, albumin, globulin), mineral (Ca, P, Mg, Na, K and Cl) and enzymatic activities related to hepatobiliary damage and functionality (AST, ALT, GGT) were determined by an automated analyser. Data were statistically processed by MIXED model with year (= dietary treatment) and stage of lactation as fixed effect and cow as random effect. The parameters affected by diet were plasma urea, total protein, with lower values in T than C ($P<0.05$), probably as a consequence of a reduction in total protein supply; plasma Ca and Cl, with higher values in T than C ($P<0.05$). Stage of lactation affected plasma triglycerides, NEFA, BHB, creatinine, total protein, globulin, Mg, ALT, GGT. No difference was found between two years in milk yield. Reduction of CP content in the diet, balancing for cows RUP and RDP MP requirements will have positive effects on nitrogen metabolic profile and on profitability of dairy cows in our experimental condition. This work was funded by European Commission area Environment (LIFE+AQUA Project).



**Tree and shrub leaves as potential additives for nutrient efficient ruminant diets**

*A.S. Chaudhry and M.R. Virk*
*Newcastle University, Agriculture, Food & Rural Development, Newcastle upon Tyne, NE1 7RU, United Kingdom; abdul.chaudhry@ncl.ac.uk*

Ruminant animals can utilise fibrous feeds to enhance nutrient use efficiency in different situations. However, the utilisation of these feeds is limited due to their imbalance nutrient supplies. So ruminant animals not only lose their performance but also excrete undesirable rumen fermentation products such as methane into the environment. Using tree and shrub components as feed additives may not only compensate for the deficiencies but also enhance the nutrient use efficiency of forage consuming ruminants. This 2×4 factorial study statistically compared the potential of tree and shrub leaves as feed additives by analysing replicated leaf samples from 4 types of trees (*Morus alba*=MA; *Magnifera indica*=MI; *Szegium cumunii*=SC; and *Cassia fistula*=CS) and shrubs (*Adhatoda vasica*=AV; *Dodonea viscose*=DV; *Acacia ampliceps*=AA; and *Atriplex lentiformis*=AL). Grass nuts (GN) were used as a tried and tested standard control in this lab. Trees and shrubs differed significantly for most chemical contents, secondary metabolites, *in vitro* rumen degradability (IVD), total gas and methane ($P<0.05$). Amongst tree samples, M. alba contained the highest CP of 21% and lowest NDF of 29% whereas S. cumunii contained the lowest CP of 8% and highest NDF of 48%. Amongst shrubs, A vasica contained significantly more CP but less NDF than other shrubs and even trees and grass nuts (($P<0.05$). The IVD and rumen fermentation patterns over time of all tree and shrub leaves compared well with grass nuts. However, the extent of IVD and rumen fermentation of leaves differed with the tree or shrub type and duration of incubations ($P<0.001$). In fact, IVD of MA, CF, AV and AL were closer to grass nuts but greater than those of SC, MI, DV and AA at most incubations times. It appeared that tree and shrub components can be used as sustainable additives in ruminant diets to optimise rumen function and nutrient use efficiency by reducing undesirable wastes such as methane from forage consuming ruminants.

**Effect of biological treatments on sugarcane bagasse as a rabbit feed**

*M.M. El-Adawy, A.Z. Salem, B.E. Borhami and N. Alaa El-Din*
*Alexandria University, Animal Production, Aflaton Street, Alexandria, Egypt; dr.eladawy@yahoo.com*

The present study aimed to investigate the effects of substitution of 50% of barseem hay by Sugarcane bagasse (SCB) treated with lactic acid bacteria (*Lactobacillus acidophilus*), ZAD (commercial product contain rumen bacteria)(Z) and L+Z in complete pelleted diet on growth performance, nutrients digestibility, some blood constituents and economical efficiency of growing New Zealand White (NZW) rabbits. Fifty 5-weeks-old rabbits were divided between five experimental groups. Rabbit's body weights were affected by feeding biologically treated SCB diets. The group fed SCB treated with LAC recorded 10% increment body weight compared with those fed control diet, followed by group fed L+Z diet. Feeding biologically treated SCB lead to a depression in the feed consumption and improved the feed conversion ratio (FC) during the experimental weeks. The best FC ratio was recorded in group fed LAC diet. Digestibility coefficients of dry matter, organic matter, crude protein, crude fiber, nitrogen free extract were increased in groups fed biologically SCB treated diets, the highest digestibility coefficients were obtained in group fed LAC diet. Most of the rabbit blood parameters in all groups insignificantly differed except cholesterol which significantly decreased in group fed LAC diet compared with the other experimental groups. Feeding LAC diet lead to an increment in the economical efficiency by 33.9% compared with control group. Conclusively, it could be concluded that the replacement of 50% of berseem hay by biologically treated SCB with *L. acidophilus* bacteria or the combination between *L. acidophilus* & ZAD in growing rabbit diets was recommended due to its positive effects on growth performance, digestibility and economical efficiency without any negative effects on growing rabbits.

---



**Smart farming for Europe – value creation through precision livestock farming**

*D. Berckmans*
*Faculty of Bioscience Engineering, Depart. of Biosystems, Division M3-BIORES: Measure, Model & Manage Bioresponses, KU Leuven, Kasteelpark Arenberg 30, 3001 Heverlee, Belgium; daniel.berckmans@biw.kuleuven.be*

The worldwide demand for meat and animal products might increase by 40% in the next 15 years. A question is how to achieve high-quality, sustainable and safe meat production that can meet this demand. At the same time, livestock production is currently facing serious problems such as animal health in relation to food safety and human health. Europe wants improved animal welfare and has made a significant investment in it. At the same time, the environmental impact of the livestock sector is far from being solved. Finally we must ask how the farmer, who is the central figure in this process, will make a living from more sustainable livestock production. One tool that might provide real opportunities is Precision Livestock Farming (PLF). PLF systems aim to offer a real time monitoring and managing system for the farmer. This is fundamentally different from all approaches that aim to offer a monitoring tool without improving the life of the animal under consideration. The idea of PLF is to provide a real-time warning when something goes wrong so that immediate action can be taken by the farmer. Continuous, fully automatic monitoring and improvement of animal health, welfare, yields and the environmental impact should become possible. Several examples are given of PLF systems that are operational today in about 60 compartments for pigs and broilers all over Europe. We give details of which variables these systems measure in real time in a fully automated way. Moreover, we show how in the running EU-PLF project we analyse how these data can generate added value for the farmer. PLF systems can replace the ears and the eyes of the farmer and work 24 h a day and 7 days a week. The challenge now is to show how the farmer gets an advantage from these systems as we start to see in the EU-PLF project. Collaboration between 'animal people' and technical people is needed to make these systems become real support systems for farmers.

## Detecting lameness in sows using acceleration data from eartags

*C. Scheel, I. Traulsen and J. Krieter*
*Institute of Animal Breeding and Husbandry, Christian-Albrechts-University, Olshausenstr. 40, 24098 Kiel, Germany; cscheel@tierzucht.uni-kiel.de*

The aim of this work is to detect lameness early in group housed sows. Using time series of acceleration data sampled from eartags, an expected behavior of an individual sow is to be recognized and extracted from such a series. Deviations thereof are to be measured and utilized to predict a beginning lameness reliably. Data is acquired from a system deployed by MKW-Electronics at the Futterkamp agriculture research farm. Since May 2012, about 200 sows are continuously fitted with eartags measuring acceleration. For a sample of 14 sows, 7 of which were diagnosed lame on the last day of their sample periods of 14 days, a collection of features from their total acceleration time series was computed. These included measures of general activity such as day-wise variance, average variation and average squared variation. Wavelet analysis was used to obtain representations of the signals on the various scales. Further features derived from above measures were computed from each scale of the wavelet coefficient representations, a total of 60. It was assumed that feature values were drawn from the normal distributions given by sows' individual data D. A feature was said to be on (or abnormal) for a day d, if its value v given the data D previous to d lied outside $\pm 1.8\sigma$ so the probability of observing it was smaller than ~0.08, i.e. $p(v|D)<0.08$. To collect statistics for the count 'features per sow in state on', let $(mean,med,min,max)_d$ denote the tuple for a day d containing these statistics. For the lame sows substantially more features deviated significantly: They yielded $(14.6,11,1,41)_{13}$ and $(18.4,15,6,50)_{14}$, while $(5.4,3,0,12)_{13}$ and $(8.3,8,1,20)_{14}$ were found for the healthy sows. Further, the $13^{th}$ day of the lame sows suggested predictive value. Among the wavelet based features, simple means of the wavelet coefficients showed the least discriminative value. Currently an autocorrelation model on finer timescales and more sophisticated features are being developed.



## Usage of tri-axial acceleration of the hind leg for recognizing sheep behavior

*M. Radeski and V. Ilieski*
*Faculty of Veterinary Medicine, Animal Welfare Center, st. Lazar Pop Trajkov 5/7, 1000 Skopje, Macedonia; miro@fvm.ukim.edu.mk*

The leg position and its acceleration could indicate strong distinction between sheep's posture and gait types, indicating sheep behavior and welfare. Tri-axial accelerometer simultaneously records acceleration, measuring the analogue signal in each of its three axes (x, y, z) and converting it to gravity units in predetermined time interval. The objective of this study was to analyze the position and locomotion of the sheep's hind leg using attached accelerometer and to interpret gathered data for posture and gait type determination. Six sheep, 3 rams and 3 ewes, divided by gender, were used in this study. The accelerometer was attached on the lateral side of the metatarsal region on the left hind leg, measuring the acceleration of the three axes, on fast mode logging interval (0.03 sec.). The relative frequencies of the acceleration values of the three axes were calculated in epochs of 3 seconds and categorized in 8 acceleration categories (from -4 to 6 g). The mean acceleration of the x axis for standing position was 0.937±0.035 g, compared to lying where the meanof x axis was -0.09±0.05 g, ($P<0.01$). The relative frequencies of acceleration values of x-axis and sum vector showed significant differences in the category 0-1 g for walking (x-axis, $\bar{x}$=65.56±8.37% and sum vector, $\bar{x}$=37.74±5.92%), trotting ($\bar{x}$=35.87±7.07%; $\bar{x}$=16.41±6.16%) and running ($\bar{x}$=22.24±4.86%; $\bar{x}$=4.91±3.03%). The value category 3-4 g for walking ($\bar{x}$=3.39±1.5%; $\bar{x}$=4.97±2.15%), trotting ($\bar{x}$=9.99±4.15%; $\bar{x}$=16.31±4.95%) and running ($\bar{x}$=19.65±4.05%; $\bar{x}$=30.09±2.19%). Accelerometers provide detailed real time analysis of the leg position in sheep's natural environment, enabling their use in behavioral research and sheep locomotion, as well as a tool for assessing the individual and herd behavior and welfare. Likewise, accelerometers could be used as a diagnostic tool for early detection of pathological processes on the leg, like lameness, which is a topic for further research.

Session 05 Theatre 4

**Effect of cow traffic on an implemented automatic 3D vision monitor for dairy cow locomotion**

*T. Van Hertem[1], M. Steensels[1], S. Viazzi[1], C. Bahr[1], C.E.B. Romanini[1], C. Lokhorst[2], A. Schlageter Tello[2], E. Maltz[3], I. Halachmi[3] and D. Berckmans[1]*
*[1]KU Leuven, M3-BIORES: Measure, Model & Manage Bioresponses, Kasteelpark Arenberg 30, 3000 Leuven, Belgium, [2]Wageningen UR, Livestock Research, P.O. Box 65, 8200 AB Lelystad, the Netherlands, [3]Agricultural Research Organization (ARO), Institute of Agricultural Engineering, P.O. Box 6, 50250 Bet Dagan, Israel; claudia.bahr@biw.kuleuven.be*

The aim of this study was to quantify the effect of cow traffic on an implemented 3D vision system for automatic monitoring of dairy cow locomotion in a commercial dairy farm. Data were gathered from a Belgian commercial dairy farm with a 40-stand rotary milking parlour. This resulted in a forced cow traffic twice a day when all Holstein cows passed the video recording system through a corridor. The video recordings were made with a 3D depth camera, positioned in top-down perspective. The entire monitoring process including video recording, filtering and analysis and cow identification was automated. To investigate how many video recordings could finally be used for monitoring dairy cow locomotion, cow video data were obtained from 289 consecutive milking sessions. Per session, 220±11 cows were identified on average by the RFID antenna, and 192±18 videos were recorded (87.4±6.8%) by the camera. After linking the cow identification to the recorded videos, 174±16 cow-videos were available (79.1±6.4%) for analysis. After all video processing, an average of 107±24 number of recorded cow-videos (48.9±11.2%) per session could be used for analysis. Cow traffic in the corridor where the recordings were made, had a big influence on the performance of the system. Heavy cow traffic reduced the number of recordings, the number of identified cows in each video, and more videos were filtered out due to incorrect cow segmentation in the videos. This study is part of the EU Marie Curie Initial Training Network BioBusiness (FP7-PEOPLE-ITN-2008) and part of the Industrial Research Fund (IOFHB/13/0136).

---

Session 05 Theatre 5

**Development of kinect based self calibrating system for movement analysis in dairy cows**

*J. Salau, J.H. Haas and W. Junge*
*Christian-Albrechts-University Kiel, Institute of Animal Breeding and Husbandry, Olshausenstraße 40, 24098 Kiel, Germany; jsalau@tierzucht.uni-kiel.de*

Lameness as well as body condition loss are problems of animal health, productivity, fertility, and animal welfare. Automatically analyzing dairy cows' movement and body characteristics without the need to examine individual animals separately, reduces stress for the animals and delivers accurate and objective measurements. In this study a system based on Microsoft Kinect 3D cameras is going to be developed to assess dairy cow's body characteristics using Computer Vision. As a prototype, a gate with pass line height ~2 m and passage width 2.4 m was constructed and equipped with 6 Kinects. To observe several steps of passing cows, 2 Kinects were mounted in both upper corners having a combined horizontal field of view of 110°. Therefore, they are able to record the approaching as well as the leaving cow. Additionally, from both sides the udder, ankle joints, and lower rump are monitored by Kinects in ~0.8 m height. A firmly installed future construction will provide guidance to keep the freely walking cows on the center line, but provisionally a Holstein Friesian cow was led through the gate by ropes for test recordings at Kiel University's research farm Karkendamm in August 2013. Software to simultaneously operate multiple Kinects has been written in C++ and a data format suitable for analysis has been developed. Synchronization of the recordings from the 6 cameras has successfully been managed, and algorithms for calibration of the system and sorting the images according to 'cow' or 'no cow' have been implemented. Segmentation of 'cow' images, automatically marking the claws and analyzing the animal's moving direction have already been developed. Reasonable next steps are the analysis of trajectories of claws and joints, the description of posture, and the extraction of body traits to monitor changes in body condition. The almost complete 3D information about the animal makes the measurement of volumes and circumferences thinkable.

**Precision dairy monitoring: what have we learned?**

*J.M. Bewley*
*University of Kentucky, Animal and Food Sciences, 407 WP Garrigus Building, 40546-0215, USA; jbewley@uky.edu*

Technologies are changing the shape of the dairy industry across the globe. In fact, many of the technologies applied to the dairy industry are variations of base technologies used in larger industries such as the automobile or personal electronic industries. Undoubtedly, these technologies will continue to change the way that dairy animals are managed. This technological shift provides reasons for optimism for improvements in both cow and farmer well-being moving forward. Many industry changes are setting the stage for the rapid introduction of new technologies in the dairy industry. Dairy operations today are characterized by narrower profit margins than in the past, largely because of reduced governmental involvement in regulating agricultural commodity prices. The resulting competition growth has intensified the drive for efficiency resulting in increased emphasis on business and financial management. Furthermore, the decision making landscape for a dairy manager has changed dramatically with increased emphasis on consumer protection, continuous quality assurance, natural foods, pathogen-free food, zoonotic disease transmission, reduction of the use of medical treatments, and increased concern for the care of animals. Lastly, powers of human observation limit dairy producers' ability to identify sick or lame cows or cows in heat. Precision dairy management may help remedy some of these problems. Precision Dairy Management is the use of automated, mechanized technologies toward refinement of dairy management processes, procedures, or information collection. Precision dairy management technologies provide tremendous opportunities for improvements in individual animal management on dairy farms. Although the technological 'gadgets' may drive innovation, social and economic factors dictate technology adoption success. These factors, along with practical experiences working with multiple technologies will be discussed in this presentation.

Session 05 Theatre 7

**Characterization of dairy farms with and without sensor technology**

*W. Steeneveld, H. Hogeveen and A.G.J.M. Oude Lansink*
*Business Economics group, Hollandseweg 1, 6706 KN Wageningen, the Netherlands; wilma.steeneveld@wur.nl*

To improve management on dairy herds, sensor systems have been developed that can measure physiological, behavioural, and production indicators on individual cows. Recently, these systems are starting to get used on dairy farms. An overview of sensor systems used on dairy farms as well as reasons to invest or not to invest in sensor systems is not available. The first objective of this study is to give an overview of the sensors currently used in the Netherlands. The second objective is to investigate the reasons of investing and not investing in sensor technology. The third objective is to characterise farms with and without sensor technology (on farm size, milk production level, age of the farmer and whether there is a successor available). An online survey was conducted among 1,693 dairy farms which were all clients of one specific accounting firm. In total, 532 farms (32%) responded and for these farms additional farm information was provided by the accounting firm. After data editing, 14 farms were deleted from the dataset because of incorrect filling out of the survey. From the 518 farms, 202 farms (39%) invested in sensor technology (121 farms had an automatic milking system and 81 had a conventional milking system). The sensor technologies most commonly used on farms with a conventional milking system were activity meters for the dairy cows (70%), electrical conductivity of the milk (35%) and activity meters for the young stock (28%), In total, 316 (61%) had not invested in sensor technology. The most important reasons of not investing in sensor technology was that the farms prefer to invest their money in other things for the farm (48%), uncertainty about the profitability of the investment (38%) and poor integration with other farm systems and software (13%). The results of the characterization of the farms with and without sensor technology (with a discriminant analysis) are not available yet, but will be presented at the conference.

## Session 05 Theatre 8

**Economic modelling to evaluate the benefits of precision livestock farming technologies**

*C. Kamphuis*[1] *and H. Hogeveen*[1,2]

[1]*Wageningen University, Hollandseweg 1, 6706 KN Wageningen, the Netherlands,* [2]*Utrecht University, Yalelaan 7, 3584CL Utrecht, the Netherlands; claudia.kamphuis@wur.nl*

Precision livestock farming (PLF) technology is an emerging research field that develops management tools aimed at continuous automatic monitoring of animal production, which includes real-time monitoring of growth, health and welfare. PLF intends to support farmers in making daily management decisions with extra 'senses', and to make farmers less dependent on human labour. Many PLF concepts have been developed in recent years, but the uptake of most of these technologies on commercial farms has been slow. Reasons for this slow uptake include that these PLF technologies generate substantial amounts of data but that this data is not converted into useful information for decision management. For example, measuring an animal's weight continuously is nice to have, but as long as the herd manager cannot translate this data into information useful for decision-making, weight data as such has limited value for the farmer. Another reason is that investments in PLF technologies can be significant, whereas the (economic) benefits of the investment are unknown. Insight in the on-farm economics of PLF is, therefore, important. This study's objective is to develop a tool that models the value creation of PLF technologies on farms. For the dairy, pig fattening and broiler industry, essential social (e.g. labour conditions and labour hours) and economic parameters (e.g. feed conversion, growth) that impact costs and benefits on-farm have been identified. Twenty global key-suppliers of PLF technologies have been approached to get insight in their views which of these socio-economic parameters are affected by their PLF technology and to what extent. The acquired knowledge will be used to model the costs and (socio-economic) benefits of PLF technologies. The value creation tool will assist, ultimately, in the development of PLF technologies that add value to on-farm decision-making processes.

---

## Session 05 Theatre 9

**Developing SmartFarming entrepreneurship – first year results EU-PLF**

*H. Lehr*[1], *J. Van Den Bossche*[2], *D. Roses*[3] *and M. Mergeay*[4]

[1]*Syntesa Partners & Associates, technologielaan 3, 3053 Haasrode, Belgium,* [2]*SO Kwadraat, Technologielaan 3, 3053 Haasrode, Belgium,* [3]*Abrox, Technologielaan 3, 3053 Haasrode, Belgium,* [4]*M&M Corporation, Technologielaan 3, 3053 Haasrode, Belgium; maurice.mergeay@mm-corporation.com*

One of the main challenges in modern livestock farming is the lack of well-trained and market-savvy high-tech innovators. The EU-PLF project aims to help by developing a blueprint for a Smart Farming service sector that is driven by high-tech entrepreneurs in conjunction with market leaders. The blueprint will look at service and business models, value creation of Smart Farming technologies and will in particular investigate the relationship between high-tech start-ups and established market players in the innovation process in the Smart Farming sector. The blueprint will be validated by early stage companies or start-ups that are going to be trained in the blueprint and coached using the SO Kwadraat coaching methodology. This methodology is based on an individual coaching by an experienced high-tech entrepreneur. A competition will be held between these entrepreneurs and the best will receive funding to demonstrate their technology on farm. Within the EU-PLF project we call this the SME Drive. In this contribution we will look at the livestock service sector and its analysis from the BrightAnimal project, lay out the foundations for the SME Drive, motivate the methodology that we have chosen for selection of teams and report the results from the selection process.

**Practical problems associated with large scale deployment of PLF technologies on commercial farms**

*T. Banhazi[1], E. Vranken[2], D. Berckmans[3], L. Rooijakkers[2] and D. Berckmans[4]*
*[1]USQ, NCEA, West st, Toowoomba QLD 4350, Australia, [2]FANCOM B.V, Research Department, P.O. Box 7131, 5980 AC Panningen, the Netherlands, [3]SoundTalks, Research Department, Kapeldreef 60, 3001 Heverlee, Belgium, [4]KU Leuven, M3-BIORES: Measure, Model & Manage Bioresponses, Kasteelpark Arenberg 30, 3000 Leuven, Belgium; thomas.banhazi@usq.edu.au*

Precision Livestock Farming (PLF) technologies are being described as usually hi-tech management tools aimed at continuously and automatically monitoring different aspects of animal production including production efficiency, environmental sustainability of the farming operation and the health and welfare of animals. Quite a few PLF technologies have been developed by different research organizations in recent years, but the large scale filed trials of these technologies never been attempted before. Thus, the main objectives of the Seven Framework funded EU-PLF and Smart-pigs projects are to (1) deploy selected PLF technologies on commercial farms and (2) to evaluate the benefits derived by producers from the information provided by these technologies. In addition, the EU-PLF project is aimed at developing a 'blue-print' for future SMEs so they are aware of the challenges associated with managing these technologies on farms. Both project started with the deployment of large number of technologies on farms that was both inspiring and of course challenging. The four suppliers of PLF technologies within these two projects have been approached to obtain an exhaustive list of 'problematic issues' encountered on farms during the deployment phase of the projects. The article will list and discuss the practical difficulties associated with deploying these technologies on farms and the solutions that were invented to overcome these challenging problems. The authors hope that the acquired knowledge will be used by future technology developers to avoid making the same mistakes again.

---



**Preliminary comparison of SSN and linear models used for evaluation of meat content in young pigs**

*M. Szyndler-Nędza, K. Bartocha and M. Tyra*
*National Research Institute of Animal Production, Department of Animal Genetics and Breeding, ul. Sarego 2, 31-047 Kraków, Poland; magdalena.szyndler@izoo.krakow.pl*

The aim of the study was to develop new linear models for estimating carcass meat percentage in the present pig population and to evaluate SSN as an alternative estimator of meatiness on live animals. The study included 656 pigs of different breeds. These animals were evaluated in the Polish Pig Testing Station during the period from 2008 to 2013. On the day of slaughter, live ultrasonic measurements of P2 and P4 backfat thickness and of longissimus muscle in point P4M were taken. Thereafter animals were slaughtered and the right sides were divided into primal cuts and subjected to a detailed dissection. Linear models for predicting meat percentage were developed based on the following sets of input variables: Model 1: P2, P4, P4M, breed, Model 2: optimal model. A two-layer ANN architecture was used to predict carcass meat content with a breed indicator, P2, P4 and P4M measurements, explanatory and contextual variables as inputs (Model 3: SSN23). A total of 200 training cycles were conducted and in each the data set was split into subsets for training (70% of samples), validation (15%) and testing (15%). The architecture contained 3-14 hidden neurons and was trained using the Levenberg-Marquardt algorithm. After training, 20 top networks ranked by the smallest generalization error on the test set were chosen as a representative sample of high performing models. Prediction of carcass meat content based on the model 1 and Model 2 was estimated with error MAE=2.26 and MAE=2.17, respectively, and the correlations for these equations were R=0.717 and R=0.738, respectively. The best neural networks already with 3 neurons in the hidden layer, estimated meat content with a similar magnitude of error ($MAE_{TE}$=2.21 and R=0.722). Increasing the number of neurons in the hidden layer did not improve the training results of these networks. The work was funded by the National Science Centre N N311 082240.

**Lameness detection in gestating sows using positioning and acceleration data**

*I. Traulsen[1], S. Breitenberger[2], K. Müller[3] and J. Krieter[1]*
*[1]Institute of Animal Breeding and Husbandry, Christian-Albrechts-University, Olshausenstr. 40, 24098 Kiel, Germany, [2]Linz Centre of Mechatronics GmbH, Altenberger Str. 69, 4040 Linz, Austria, [3]Chamber of Agriculture Schleswig-Holstein, Am Gutshof 1, 24327 Blekendorf, Germany; jkrieter@tierzucht.uni-kiel.de*

Automatic lameness detection in sows is an important issue in group housing of sows. The aim of the present study was to use data of a positioning system including acceleration measurements to detect lameness in grouped housed sows. Data were acquired at the Futterkamp research farm from 05/2012 until 04/2013. In the gestation unit, 212 grouped housed sows (296 parities, parity 2 to 11, gestation day 10 to 130) were equipped with an ear sensor to sample position and acceleration per sow and second. Three activity indices were calculated per sow and day: path length walked of a sow during the day (path), number of squares (25×25 cm) visited during the day (squares) and variance of the acceleration measurement during the day (variance). Additionally, data on lameness treatments of the sows were used. To determine the influence of a lameness event, all indices were analyzed in a linear random regression model. Test-day, parity class and day before treatment had a significant influence on all activity indices ($P<0.05$). Indices path and squares increased with increasing parity while variance slightly decreased. Healthy sows showed an average activity index around 1 (path=1.00 (s.e.=0.012), squares=1.00 (s.e.=0.017), variance=0.96 (s.e.=0.045)). Three days before treatment indices were significantly lower (e.g. $path_1=0.89$ (s.e.=0.039), $path_2=0.99$ (s.e.=0.040), $path_3=0.97$ (s.e.=0.043). A trend over a longer time period (up to 14 days) could not be observed. Gestation day, included as a linear quadratic covariate, influenced only path significantly. However, all activity indices decreased with increasing gestation day reflecting that sows closer to farrowing moved less. In conclusion, all activity indices were influenced by a lameness event while most differences were found for path length.



**Effect of individualized feeding strategy on milk production of Holstein cows**

*C. Gaillard, M. Vestergaard, M.R. Weisbjerg and J. Sehested*
*Aarhus University, Foulum, 8830 Tjele, Denmark; charlotte.gaillard@agrsci.dk*

Earlier studies indicated a positive effect on peak yield and lactation milk yield of an individualized feeding strategy, applied in early lactation, when ration energy concentration was reduced when cows entered deposition phase. The objective of the present study was to determine the effect of a high energy diet from calving until live weight nadir followed by standard diet as compared to a standard diet throughout the period. It was hypothesized that the high energy diet would lead to an increased milk yield until diet shift, and that a higher milk yield would be maintained also after diet shift. Sixty-two Holstein cows (30% $1^{st}$ parity), managed for 16 months lactations in a robot milking system, were allocated to normal (NOR) or experimental (EXP) feeding strategies at calving. Diets were based on the same feeds, partially mixed rations *ad libitum* with supplementary concentrate fed restrictively in the milking robot. Cows on NOR were fed 60:40 forage:concentrate ratio throughout the period. Cows on EXP were fed a higher energy density by 50:50 forage:concentrate ratio until they reached ≥42 DIM and a live weight gain ≥0 kg/d, and were then shifted to the NOR diet. Data were evaluated using a linear mixed-model. From days 0 to 42, the multiparous EXP tended to produce more milk per day (lsmean ±SEM) than the multiparous NOR cows (40.0 vs 36.2 kg, ±1.4; $P=0.063$). For the primiparous cows, milk production was not influenced by treatment ($P=0.17$). From days 79 to 98, after the shift in diet, EXP had similar milk production as NOR cows for both parities. In conclusion, the energy-enriched EXP diet tended to increase milk production for the multiparous cows during the first period but this effect was not carried over to the following period with NOR feeding.

**Monitoring the nest building and parturition progress using acceleration data**

*C. Cornou and A.R. Kristensen*
*University of Copenhagen, Faculty of Health and Medical Sciences, Department of Large Animal Sciences, Grønnegårdsvej 2, 1870 Frederiksberg C, Denmark; cec@sund.ku.dk*

A method to monitor nest building and farrowing activities using activity types classified from acceleration data is suggested. The progress of farrowing (start, end and birth of each piglet) is analyzed for 18 sows housed in crate, of which half was provided with straw (S), and half received no straw (NS). A nest building indicator (NBI) is defined, based on classified series of activity measurements. A nest building indicator (NBI) is defined, and consists of three parameters: (1) the duration of the NBI corresponds to the number of hours where sows are more than 50% of their time performing behavior classified as Active, allowing for a maximum of one hour with Active behavior below 50%; (2) the intensity of the NBI is defined as the proportion of activity during the NBI. Both the duration (14.1 to 17.0 h for group NS and S) and the peak of intensity (around 9 h before farrowing) of the NBI are very similar to results of previous studies describing the nest building behavior using video analysis; (3) the last hour of the NBI can be used as an indicator of imminent farrowing since it occurs in average, during the last 15 min before the onset of parturition. Shifts of activities (from Active, Lying laterally and Lying Active behavior) are studied at the approach, during and after farrowing. An increase of Lying Active behavior is observed during the last 2 to 4 hours prior parturition. Moreover, the average number of these shifts during parturition is significantly higher than during the 6 hours after the birth of the last piglet, especially for the group NS. This result could be used as an indicator of the farrowing end. The suggested method appears reliable to monitor nest building and the progress of parturition, and is less time consuming than video analyses. A better monitoring of these phases, can help to give more attention to sows and piglets around parturition, which can result in a better welfare and a reduction of piglet mortality.

---



**Finding cattle hidden in the bushes and woods by using GPS**

*A.H. Herlin*
*Swedish University of Agricultural Sciences, Dept. of Biosystems and Technology, P.O. Box 103, 23053 Alnarp, Sweden; anders.herlin@slu.se*

Surveillance of grazing animals in semi-natural pastures can be time consuming because the area may be difficult to overview because of terrain and vegetation. By using some animals in the flock with GPS devices that are connected to the mobile network, provision of positions at some pre-defined time-points per day will facilitate the supervision of the herd as the animals will likely to be easier. On a large area (>200 ha) with mainly forest and semi-forested natural grassland in a coastal area in south eastern part of Sweden, 5 animals in a herd of about 20 were equipped with GPS-devices that each were programmed to give two positions per day. The animals spent 171 days on pasture from early May until late October. Potentially, the GPS units should have delivered 1222 positions, of which 535 positions really were collected from the GPS units giving a efficiency rate of 43.4%. Four units that hang in the collar under the neck had a much lower efficiency rate (35.2%) than the one unit that was positioned on top of the neck of the animal, which retrieved about 77.4% of the potential GPS positions. The most important factors to be considered before adopting this technology are: the GPS unit's efficiency may vary by model and brand, the battery capacity, that determines how many positions that can be produced and thus the time that the GPS-unit is in operation, the GPS-unit needs to be positioned on top of the neck (by using a counterweight in the neck collar), the GSM-network in the current pasture may vary depending on terrain, vegetation and the mobile operator and the units need to be robust enough to withstand wind, weather and other stress when animals graze in the shrubs. All these factors need to be taken into consideration before deciding to use GPS-technology to monitor and find grazing animals in semi-natural pastures. In many cases, GPS-technology could be a valuable tool for the farmers who have more mosaic pastures and be an efficient way to reduce the time spent to find the animals.

**Development of an on-farm SowSIS (sow stance information system): hurdles to tackle**

*J. Maselyne[1,2], L. Pluym[2,3], J. Vangeyte[2], B. Ampe[4], S. Millet[4], F. Tuyttens[4], K.C. Mertens[2] and A. Van Nuffel[2]*
*[1]KU Leuven, MeBioS, Kasteelpark Arenberg 30 bus 2456, 3001 Heverlee, Belgium, [2]ILVO, Technology and Food Science Unit, Burg. van Gansberghelaan 115 bus 1, 9820 Merelbeke, Belgium, [3]UGent, Porcine Health, Salisburylaan 133, 9820 Merelbeke, Belgium, [4]ILVO, Animal Sciences Unit, Scheldeweg 68, 9090 Melle, Belgium; jarissa.maselyne@ilvo.vlaanderen.be*

Apart from health and welfare concerns, lameness increases labour and treatment costs and leads to premature culling and euthanasia of sows. The compulsory group housing of gestating sows since 2013 makes visual observations of individual lameness cases harder and is expected to increase lameness incidence. Recently, a SowSIS (Sow Stance Information System) was developed and validated for objective lameness detection in sows at ILVO (Merelbeke-Belgium). This system automatically performs force measurements and additionally generates visual stance variables using a transportable device with 4 load cells (one for each leg of the sow). An adapted version of SowSIS has now been implemented in 4 electronic sow feeders (ESF). Besides some technical hurdles, a main problem with the implementation was seen. The ESF used was with a trough at an angle of 45° versus the sow's position. Not only did this result in an uneven balance of the sows (might be corrected for), the data also showed that sows often placed their hind legs on the same load cell (the left hind load cell for a trough at the right side of the sow). During some measurements a lot of movements of the sows were recorded as well. This raises the question: do all designs of ESF's allow the sow to eat in a comfortable position? This could be food for some interesting research. Implementation of the SowSIS thus needs to be re-though to be able to measure the 4 legs separately for all sows. Different options are now being explored. The system will allow sows to be measured daily without interfering with their daily routine. Time series analyses will be used to further develop this system as early warning system for lameness.

---



**Automated recognition of activity states in dairy cows by combining multiple sensors**

*J. Behmann[1], K. Hendriksen[2], U. Müller[3], S. Walzog[1], H. Sauerwein[3], W. Büscher[2] and L. Plümer[1]*
*[1]Institute of Geodesy and Geoinformation, Department of Geoinformation, University of Bonn, Meckenheimer Allee 172, 53115 Bonn, Germany, [2]Institute of Agricultural Engineering, Department of Livestock Technology, University of Bonn, Nußallee 5, 53115 Bonn, Germany, [3] Institute of Animal Science, Department of Physiology & Hygiene, University of Bonn, Katzenburgweg 7-9, 53115 Bonn, Germany; sauerwein@uni-bonn.de*

Activity patterns of dairy cattle have received increased interest in recent years. They promise insights into health state and well-being. However, positioning alone is not sufficient for activity monitoring with currently available sensors. Fusion of positioning with signals from additional sensors promises a comprehensive monitoring of sequences of single activity states. The interpretation of fused signals, however, is challenging. In an experiment, a Local Positioning System was combined with a heart rate sensor in order to estimate five activity states. During the experiment the sensor signals of 12 cows (more than 150 h in total) were recorded and annotated. For signal interpretation, we used a combination of a Support Vector Machine (SVM), a state-of-the-art classification method, and a Conditional Random Field (CRF). SVMs distinguish single states, whereas CRFs label state sequences considering specified constraints. The different durations of states such as 'lying' and 'lying down' are important prior knowledge in our context. Conditional Random Fields, however, do not have the concept of duration. Therefore, we augmented the concept of CRFs by a memory which allows considering duration distributions of the different activity states. In the results, the application of the CRF to the SVM result caused a modest increase in accuracy (5%) but a major improvement in the determination of long sequences. The length of the longest common subsequence increases on average from 3,481 periods to 6,207 periods. The robust detection of long lying sequences allowed for the unaffected extraction of the resting pulse rate.

## Dynamic precision feeding of growing pigs using a new automatic feeder linked to a weighing station

*P. Massabie, M. Marcon and N. Quiniou*
*IFIP-Institut du Porc, BP35, 35650 Le Rheu, France; nathalie.quiniou@ifip.asso.fr*

A feeder prototype has been developed by IFIP in association with an equipment manufacturer, to perform precision feeding for group-housed growing pigs. For 96 pigs, the system consists in one weighing station (W) connected with five automatic feeders (AF). The W station is placed on force sensors and open when at least one AF is free. It allows for weighing pigs individually, with an 0.1 kg accuracy, and sorting them toward (1) the AF (pigs whose daily cumulated feed intake is presently below the daily allowance), (2) a pen for specific care or departure for the slaughterhouse (after selection by the farmer), (3) the original pen for the others. In case 1, the pig enters the free AF and its ear RFID tag is recognized by an antenna. Then, both back and front doors close automatically, and successive doses of feed fall in the through. At the end of the meal, the pig leaves the AF through a corridor, shared with the other AF, and joined other pigs by pushing a no-return door. Each AF is connected to a computer that calculates the daily amino acids and energy requirements per pig, based on sex and body weight. The quality and quantity of feed to deliver is achieved by mixing two diets differing by their amino acids to energy ratio. The minimum delivery is 30 g per dose per diet. The objective of this poster is to present the results obtained from the first batch used to evaluate the system. All pigs were restricted with a daily increase in feed allowance (+27 g/d) up to a plateau (gilts: 2.4, castrates: 2.7 kg/d). A two-phase feeding strategy (48 pigs) or a multiphase feeding one (48 pigs) were used. Data of the W station were compared with manual weighing at three stages. Weight controls of dietary doses delivered were performed every 2 weeks. The growth performance and carcass characteristics at slaughter (average and variability) were measured.



## Development of automated calf cough detection: preliminary results

*J. Vandermeulen[1], E. Tullo[2], I. Fontana[2], D. Johnston[3], B. Earley[3], M. Hemeryck[4], C. Bahr[1], M. Guarino[2] and D. Berckmans[1]*
*[1]KU Leuven, M3-BIORES: Measure, Model & Manage Bioresponses, Kasteelpark Arenberg 30, 3001 Heverlee, Belgium, [2]Università degli Studi di Milano, Department of Health, Animal Science and Food Safety, via Celoria 10, 20133 Milan, Italy, [3]TEAGASC, Anim & Grassland Res & Innovat Ctr, Anim & Biosci Res Dept, Dunsany, Meath, Ireland, [4]SoundTalks NV, Kapeldreef 60, 3001 Heverlee, Belgium; joris.vandermeulen@biw.kuleuven.be*

Detecting calf diseases, such as Bovine Respiratory Disease, earlier could decrease their severity. The goal in this ongoing study is to develop an early warning system for respiratory disease in calf houses using sound recordings. The study objective is to develop a fully automated detection of calf coughs and distinguish them from all other sounds in the calf house. The data were acquired at the TEAGASC research farm, Ireland using recording equipment provided by SoundTalks. The calves (80 Holstein-Friesian or Jersey male calves) were assigned to 4 calf houses and artificially reared for 84 days. The sound was continuously recorded over the duration of the study (24 hours a day) using one microphone in each house. To develop the automated calf cough detection, 15 randomly chosen sound intervals of 5 minutes were analysed. Labelling of calf coughs was done off-scene by audio-visual scoring of sound data and provided 100 individual coughs. Including 100 coughs a total of 2,523 candidate sounds were found. Detecting only 100 coughs from 2,523 sounds required the generation of a high specificity algorithm. Currently the developed algorithm gives 59% sensitivity and 94.7% specificity which resulted in 59 true positives and 132 false positives. Less than 50% of the found sounds are calf coughs. Further research will therefore concentrate on improving sensitivity and specificity and on acquiring more labelled coughs. The authors gratefully acknowledge the European Community for financial participation in Collaborative Project EU-PLF KBBE.2012.1.1-02-311825 under the 7th Framework Programme.

**An exploratory study of the relation of pH data from pens boluses with othe sensor data**

*R.M. De Mol[1], Y.A.H.M. Van Roosmalen[1], M.-H. Troost[2], A. Sterk[3], R. Jorritsma[4] and P.H. Hogewerf[1]*
*[1]Wageningen UR Livestock Research, P.O. Box 65, 8200 AB Lelystad, the Netherlands, [2]Rovecom, Elbe 2, 7908 HB Hoogeveen, the Netherlands, [3]Agrifirm, Landgoedlaan 20, 7325 AW Apeldoorn, the Netherlands, [4]Utrecht University, Yalelaan 7, 3584 CL Utrecht, the Netherlands; rudi.demol@wur.nl*

In the Dutch Smart Dairy Farming project, automated measurements of the pH in the rumen of dairy cows were used to analyze processes during the transition period. The pH data can be used to monitor the developments during the end of the dry period and the start of the lactation. If these pH values are related with other sensor data (like activity and feed intake), then monitoring these other data might results in possibilities to detect metabolic problems. On two practical dairy farms (farm 1: 300 cows, automatic milking system; farm 2: conventional milking, 110 cows), in total 70 cows were fitted with Smaxtec pH bolus a few weeks before calving. Valid measurements were available per bolus over two months. Other sensors were used to measure milk yield, activity, rumination, concentrates intake, visits to concentrate feeder (and milk robot on farm 1) and body weight. This resulted in a data set to be used to explore the relation between pH value and other traits. The first results show: (1) no clear relation between pH and concentrates intake; (2) great fluctuations in pH on the day of calving (caused by changes in feeding?); (3) that an increase in rumination might be correlated with a decrease in pH The pH data were aggregated by logistic regression; making it possible to convert the pH data of one day in an average level and a slope describing the variance within a day. More results will be included in the presentation.



**Automatically monitoring dairy cows'body condition with a time-of-flight camera**

*J. Salau[1], J.H. Haas[1], A. Weber[1], W. Junge[1], U. Bauer[2], J. Harms[2], S. Bieletzki[3], O. Suhr[3] and K. Schönrock[3]*
*[1]Kiel University, Institute of Animal Breeding&Husbandry, Olshausenstraße 40, 24098 Kiel, Germany, [2]Bavarian Research Center for Agriculture, Institute for Agricultural Engineering&Animal Husbandry, Prof.-Dürrwaechter-Platz 2, 85586 Poing, Germany, [3]GEA Farm Technologies, Siemensstraße 25, 59199 Bönen, Germany; jsalau@tierzucht.uni-kiel.de*

Enduring negative energy balance strongly affects health, fertility, and performance, thus the body condition should be monitored systematically and accurately. The subcutaneous fat bounded by skin and fascia trunci profunda located at the gluteus medius muscle, respectively the longissimus dorsi muscle is called backfat. Measuring backfat thickness (BFT) with ultrasound is highly reflective of the body condition. This study introduces a system for estimating BFT using a Time-Of-Flight camera. 3D data is provided from analyzing the phase shift between outgoing and reentering infrared signal. The cows' lower back was recorded from top view (2.55 m height) in a feeding station. Software for automated camera setup, calibration, animal identification, image sorting (error rate 0.2%), segmentation, determination of the region of interest (error rate 1.5%), and information extraction was developed. Profiles of the digital cow surface were taken between hips or ischeal tuberosities/tail and hip. In total 13 camera traits were calculated and compared to BFT. Data was collected from 96 primiparous and multiparous HF cows from July 2011 to May 2012 at dairy research farm Karkendamm, Kiel University. BFT was measured with a portable ultrasound generator once a week. All traits significantly depend on the animal and showed large effect sizes $\eta^2$. The precision in measuring was comparable between BFT and camera traits but strongly varied between cows. BFT estimation using the cow as random or fixed effect, lactation week and observation month as fixed effects, and a linear regression on 2 camera traits gave normally distributed residuals with homogeneous variance and correlation 0.96 among BFT values and estimator.

**Creation of universal MIR calibrations by standardization of milk spectra: example of fatty acids**

*C. Grelet[1], J.A. Fernandez Pierna[1], H. Soyeurt[2], F. Dehareng[1], N. Gengler[2] and P. Dardenne[1]*
*[1]Walloon Agricultural Research Center (CRA-W), 24 Chaussée de Namur, 5030 Gembloux, Belgium, [2]University of Liège, Gembloux Agro-Bio Tech, Passage des Déportés 2, 5030 Gembloux, Belgium; c.grelet@cra.wallonie.be*

Due to the interest of fatty acids (FA) for the nutritional quality of milk fat and their links to the physiological status of cows, there is an increasing interest to analyse the FA profile of cow milk by mid-infrared spectroscopy (MIR). Indeed, this technology is worldwide used for milk analysis and allows fast and cheap analysis. To build robust MIR equations, a large spectral dataset including samples covering the maximum chemical variability and FA reference values is needed; however this kind of dataset is difficult to obtain mainly due to the cost of reference analysis. Therefore, the sharing of data is a potential solution, but different brands and types of MIR apparatus exist, and therefore different spectral responses and formats, involving difficulties to directly merge all datasets. The objective of this study was to standardize the spectra coming from different apparatus on the same format through a Piece-wise Direct Standardization (PDS) in order to group them into a common database and then to construct robust MIR FA equations. For this, milk samples were collected in Belgium, United Kingdom, Luxembourg, Ireland, France and Germany during 8 years, and measured in different MIR instruments (Foss and Bentley). The reference FA concentrations were measured by gas chromatography. This allowed to group 1,804 samples covering a large FA and spectral variability as well as different geographical areas, breeds, diets and seasons. Calibrations were performed using the chemometric tool Partial Least Squares (PLS) and the first results showed $R^2$ ranging from 0.67 to 0.99 and a ratio of standard error of cross-validation to standard deviation (RPD) ranging from 1.73 and 9.74. The perspective of this work is to allow the use of unique (universal) equations in different countries and different MIR apparatus.

---



**Water-soluble vitamins and reproduction in sows: beyond prolificacy**

*J.J. Matte*
*Agriculture and Agri-Food Canada, Dairy and Swine Research and Development Centre, 2000 College Street, J1M 0C8, Canada; jacques.matte@agr.gc.ca*

Information on vitamin requirements in reproducing sows is often lacking and has become outdated during recent decades, a period during which the reproductive performance was greatly influenced by genetic improvements and changes in husbandry practices. This enhanced sow productivity, although efficient in terms of prolificacy, has brought some pitfalls such as heterogeneity of birth weight and piglet development within litter with carry-over effects later in life in terms of growth performance and quality of carcass and meat. Nursing piglets are entirely dependent for their vitamin provision on the transfer (in utero, colostrum and milk) from the dam and this dependency lasts for approximately half the life cycle (conception to slaughter) of a pig. An adequate maternal transfer of vitamins is therefore critical, especially for high prolific sows, but this has likely not paralleled the increase of prolificacy during the last decades. Perinatal interventions to sows and piglets could be necessary to alleviate litter heterogeneity and 'imprint' this effect for the whole life of pigs. This approach is in line with a recent concept of 'lactocrine' or 'neonatal' programming in piglets. The amount of information available for each vitamin in reproducing sow is heterogeneous and then, empiricism has been and is still present as reflected by large variations in recommendations from private or public sources. Vitamin requirements for sows need to be updated beyond the prevention of deficiencies (practically absent nowadays), i.e. towards the best possible reproduction efficiency. This is a challenge for the future because the different criteria of reproductive performance are actually evolving from prolificacy alone to other aspects such as survival and disease resistance in piglets. The roles of some vitamins in immunology of reproduction, antioxidative capacity and immune competence may become key contributors to improvement of reproductive performance according to these new criteria.

**High dietary vitamin E and selenium improves oxidative status of finisher lambs during heat stress**

*S.S. Chauhan[1,2], P. Celi[2,3], E.N. Ponnampalam[4], D.L. Hopkins[5], B.J. Leury[2] and F.R. Dunshea[2]*
*[1]Department of Animal Husbandry, H.P., Shimla 171 005, India, [2]The University of Melbourne, Melbourne School of Land and Environment, Parkville, VIC 3010, Australia, [3]University of Sydney, Faculty of Veterinary Science, Narellan, NSW 2567, Australia, [4]Department of Environment & Primary Industries, Werribee, VIC 3030, Australia, [5]Centre for Red Meat and Sheep Development, P.O. Box 129, Cowra, NSW 2794, Australia; chauhans@student.unimelb.edu.au*

A study was conducted to elucidate the role of dietary vitamin E (Vit E) and selenium (Se) on the oxidative status of lambs exposed to hot conditions during finishing. Forty-eight lambs (crossbred; 42±2 kg BWt, 7 m old), were allocated to three doses of Vit E and Se with standard finisher pellets. The doses of Vit E and Se for control, moderate, supranutritional diets were 27.6, 130, 227.5 IU/kg DM as α- tocopherol acetate and 0.16, 0.66, 1.16 mg Se as SelPlex™ /kg DM, respectively. After 4 weeks feeding in the individual pens, including 1 week of adaptation, lambs were moved to metabolism cages for 1 week and allocated to one of 2 heat regimes: thermoneutral (TN) (18- 21 °C and 40-50% relative humidity (RH)) or heat stress (HS) (28-40 °C and 30-40% RH) conditions. HS elevated ($P<0.001$) both respiration rate and rectal temperature at 13:00 h and 17:00 h, however a decline ($P=0.08$) in respiration rate and a significant reduction ($P=0.05$) in rectal temperature was observed in lambs fed Supranutritional levels of Vit E and Se (SVitESe). Plasma reactive oxygen metabolites (ROMs) concentrations were increased ($P<0.001$) under HS, however were reduced by SVitESe (117 CARRU) as compared to control diets (128 CARRU). Plasma advanced oxidation protein product levels were increased ($P<0.001$) during HS, however, lambs fed SVitESe had 35% lower ($P=0.005$) than maximum levels in controls. Taken together, our results indicate that high levels of dietary Vit E and Se are required for maintaining oxidative status of lambs finishing during hot conditions.



**Effects of flavonoids dietary supplementation on egg yolk antioxidant capacity and cholesterol level**

*P. Simitzis, M. Goliomytis, K. Papalexi, N. Veneti, M. Charismiadou, A. Kominakis and S. Deligeorgis*
*Agricultural University of Athens, Animal Breeding and Husbandry, 75 Iera Odos, 11855 Athens, Greece; pansimitzis@aua.gr*

An experiment was conducted to examine the effects of supplementing laying hen feed with different levels of hesperidin or naringin, bioflavonoids that are abundant and inexpensive by-products of citrus cultivation, on the yolk antioxidant capacity and cholesterol level. Seventy-two laying hens, approximately 12 months old, were assigned into 6 experimental groups of twelve hens each. One of the groups served as control (C) and was given a commercial basal diet, without bioflavonoid supplementation, whereas the other five groups were given the same diet further supplemented with hesperidin at low (750 mg/kg of feed) (H1) or high (1,500 mg/kg) (H2) concentration or naringin at low (750 mg/kg) (N1) or high (1,500 mg/kg) (N2) concentration or α-tocopheryl acetate (200 mg/kg) (E). Measurements of yolk antioxidant capacity were performed on 8 eggs from each dietary group, at 0, 4, 7, 28 and 63 days after the beginning of the experiment. Yolk cholesterol level was determined on the final day (63rd) of the experimental period. Oxidative stability of egg yolk, expressed as ng MDA/g yolk, was significantly improved in the hesperidin and naringin groups even from the first four days of the supplementation period ($P<0.001$). However, no flavonoids effect on yolk cholesterol level (mg/g) was observed. Antioxidant properties of flavonoids seem to be a promising natural agent for improving the health status and the shelf life of laying hens' egg. This research project was implemented within the framework of the Project 'Thalis – The effects of antioxidant's dietary supplementation on animal product quality', MIS 380231, Funding Body: Hellenic State and European Union.

## Session 06 Theatre 4

**Diurnal plasma zinc fluctuations in highly prolific sows**

*M.M.J. Van Riet[1,2], G.P.J. Janssens[1], B.F.A. Laurenssen[3], K.C.M. Langendries[1,2], B. Ampe[2], E. Nalon[2,4], D. Maes[4], F.A.M. Tuyttens[1,2] and S. Millet[2]*

*[1]Ghent University, Heidestraat 19, Merelbeke, Belgium, [2]Inst. for Agricultural and Fisheries Research, Scheldeweg 68, Melle, Belgium, [3]Wageningen University, De Elst 1, Wageningen, the Netherlands, [4]Ghent University, Salisburylaan 133, Merelbeke, Belgium; miriam.vanriet@ilvo.vlaanderen.be*

Plasma zinc (Zn) is widely used to determine Zn status in animals. However, diurnal plasma Zn fluctuations, observed in humans, may influence interpretation. The aim of this study was to determine plasma Zn fluctuations throughout the day in gestating sows. The experiment included 10 highly prolific sows in mid-gestation (day 44±7). The sows weighed 207±38 kg at the start of the experiment and were fed 2.6 kg of a gestation diet provided in two equal portions (7:45 am & 14:45 pm). The dietary Zn content, originating from the ingredients and ZnO, was 129 mg/kg. All sows had *ad libitum* access to water. The sows were housed individually in farrowing crates. Blood samples were collected from the jugular vein via an ear cannula which was flushed with a saline solution containing heparin after each sampling. Blood samples (n=24 per sow) were taken during a 24 h period before feeding in the morning, at morning feeding, every 30 minutes for 4 hours after feeding, every hour until the next feeding, again every 30 minutes for 4 hours after afternoon feeding, and 4 times during the night. Plasma was analysed for Zn content. Plasma Zn decreased throughout the day ($P<0.01$), indicating that plasma Zn shows a circadian rhythm. The highest values were measured at 7:45 am and the lowest at 17:15 pm. No postprandial effect was observed ($P=0.10$), neither a parity effect ($P=0.64$). However, the variation between measurements at consecutive time points for a sow was extremely high. These variations may question the use of plasma Zn for Zn status in individual sows, while it may be useful to compare groups. The observed diurnal plasma Zn fluctuations should be taken into account if plasma Zn is considered as marker to determine Zn status.

---

## Session 06 Theatre 5

**Copper and zinc recommendations in pig diets, a review**

*P. Bikker, R.M.A. Goselink, J. Van Baal and A.W. Jongbloed*

*Wageningen UR Livestock Research, P.O. Box 338, 6700 AH Wageningen, the Netherlands; paul.bikker@wur.nl*

Copper and zinc are essential trace elements for pigs. Deficiencies may result in impaired feed intake and growth, parakeratosis, anaemia, and a reduced immune response. Published dietary recommendations are poorly documented by dose-response studies and practical diets generally include these trace elements in levels close to legal limits. Since absorption of the digestive tract is in accordance with the requirements of the animal, the majority of dietary copper and zinc is excreted in the manure and contributing to the accumulation in the soil and in ground and surface waters. Therefore the national authorities, the feed industry and research organisations joint forces in a renewed interest into copper and zinc requirements of pigs. The final aim would be a reduction in dietary inclusion levels without loss in animal performance and health. We reviewed recent literature and conducted several studies into responses of growing pigs to incremental dietary copper and zinc levels in order to establish response curves and derive requirements of growing pigs. We propose that 80 mg, gradually decreasing to 60 mg zinc and 12 mg copper per kg of diet is adequate for growing pigs from 12 kg body weight to slaughter. Because of the consistent beneficial effect of microbial phytase on the availability and absorption of zinc, but not of copper, this effect should be included in feed optimisation. The growth-promoting effects of pharmacological levels of dietary copper (EU max. 170 mg/kg) will be addressed, including consequences of a reduction in level or duration of the high copper supply, and the effect after withdrawal. Finally, we propose the development of a factorial method to determine requirements of reproductive sows. Especially absorption from the digestive tract, maternal retention in gestation and mobilisation in lactation are not well known. Therefore it is recommended to further investigate and quantify these factors to better substantiate recommendations for these animals.

**Standardized total tract digestibility of calcium in growing pigs**

*J.C. González-Vega[1], C.L. Walk[2] and H.H. Stein[1]*
*[1]University of Illinois, Urbana, 61801, USA, [2]AB Vista, Marlborough, Wiltshire, United Kingdom; jcgonzal@illinois.edu*

A series of experiments were conducted to determine the digestibility of Ca in Ca-containing feed ingredients fed to growing pigs. In Exp. 1, 48 pigs (16.7±2.5 kg) were used to determine the true total tract digestibility (TTTD) of Ca in canola meal without and with phytase. Endogenous losses of Ca without and with phytase were determined using the regression procedure, but values were not influenced by phytase (160 and 189 mg per kg DMI, respectively). The TTTD of Ca in canola meal with phytase (70.3%) was greater ($P<0.05$) than in canola meal without phytase (46.6%). In Exp. 2, 104 pigs (17.7±2.5 kg) were used to determine the standardized total tract digestibility (STTD) of Ca in monocalcium phosphate (MCP), dicalcium phosphate (DCP), and calcium carbonate without and with phytase. Basal endogenous losses of Ca (120 mg per kg DMI) were determined using a Ca free diet. Values for STTD of Ca in MCP (85.9 and 86.3%) and DCP (77.8 and 78.9%) were not influenced by phytase, but for calcium carbonate, the STTD of Ca with phytase (73.1%) was greater ($P<0.05$) than the STTD of Ca without phytase (60.4%). In Exp. 3, 70 pigs (19.4±1.0 kg) were used to determine the STTD of Ca in fish meal in a corn-based diet without added fat (89.0%) and with added fat (88.1%), but dietary fat did not influence the STTD of Ca. In Exp. 4, 9 pigs (23.8±1.3 kg) were used to determine the place in the intestinal tract where Ca is absorbed. Pigs were surgically prepared with a T-cannula in the duodenum and a second T-cannula in the ileum. Duodenal, ileal, and fecal samples were collected. Results indicated that Ca absorption starts anterior to the duodenal cannula and that no net absorption takes place in the large intestine. The duodenal endogenous flux of Ca (1,030 mg per kg DMI) was greater ($P<0.05$) than the ileal and the total tract endogenous flux of Ca (420 and 670 mg per kg DMI, respectively). In conclusion, based on these data it is possible to formulate diets using STTD of Ca in Ca-containing feed ingredients.

---



**Influence of copper sulphate and tribasic copper chloride on growth performance of weaning piglets**

*D. Kampf[1] and F. Paboeuf[2]*
*[1]Orffa Additives BV, Vierlinghstraat 51, 4251 LC Werkendam, the Netherlands, [2]Orffa France SA, 29 Rue Bassano, 75008 Paris, France; kampf@orffa.com*

Copper sulphate (CS) has a strong antimicrobial effect and is able to improve the growth of piglets. CS is highly soluble and there is a risk the dissociated copper (Cu) will react with other feed ingredients. Cu sources such as tribasic copper chloride (TBCC) dissolve slowly and the Cu remains bound in its original structure for longer thus supplying more intestinal efficient Cu. A feeding study was conducted at Invivo NSA research facility CRZA, Montfaucon (France) to examine the effect of Cu at excessive dosages either from CS or TBCC on the growth performance of weaning piglets. 128 piglets weaned at 28 days of age (Ø 8.5 kg body weight) were blocked on the basis of initial weight, sex and maternal origin and allotted to 32 pens (16 pens of 4 piglets each per treatment). 2 feeding phases were used (1: day 1-14, 2: day 15-42) and pelleted feed (based on wheat, corn, barley, soy and rapeseed meal) was offered *ad libitum*. Treatments were assigned as CS (basal diet + 155 mg/kg Cu from copper sulphate) and as TBCC (basal diet + 155 mg/kg Cu from tribasic copper chloride). Animals were individually weighted at day 1, 4, 7, 14 and 42. Feed intake was recorded daily from day 1-4, at day 7 and afterwards weekly until trial end. Data evaluation was carried out using GLM procedure of SAS. Feed intake up to day 28 did not vary between treatments, from day 28-42 a significant higher feed intake was observed by feeding TBCC (1296 vs 1198 g; $P=0.04$) which remained as a tendency (892 vs 856 g; $P=0.08$) throughout the trial. Daily gain was up to day 28 similar between treatments and after clearly improved by TBCC (720 vs 678 g; $P=0.02$). In addition daily gain was also distinctly enhanced by TBCC over the entire trial period (599 vs 578 g; $P=0.04$). Feed conversion was not influenced by one of the Cu sources. Consequently, the replacement of CS with TBCC resulted (under good sanitary conditions with no diarrhoea and low mortality in an elevated growth rate indicating a superior efficiency of TBCC.

**Effect of zinc oxide on growth and health of *Escherichia coli* F4-challenged susceptible weaning pigs**

*P. Trevisi[1], S. Durosoy[2], Y. Gherpelli[3], M. Colombo[1] and P. Bosi[1]*
*[1]University of Bologna, Via Fanin 50, 42123 Bologna, Italy, [2]Animine, 335 Chemin du noyer, 74330 Sillingy, France, [3]Istituto Zooprofilattico Sperimentale Bruno Ubertini, Via Pitagora 2, 42100 Reggio Emilia, Italy; paolo.trevisi@unibo.it*

Colibacillosis by *Escherichia coli* F4 (ETEC) depends on inherited presence of receptors for adhesion to pig intestine. The effects of a potentiated zinc oxide (ZnO) form (HiZox®, Animine), already tested under *ex vivo* and *in vivo* conditions, were measured on ETEC-challenged piglets vs the pharmacological dosage of the regular ZnO. Thirty-six 24-days-old pigs were selected to be ETEC-susceptible by the gene marker Mucin 4, assigned to three different diets obtained from the same basal batch, but varied with 150 mg (basal+ZnO, C); 300 mg (basal+HiZox®, HZ); 2,500 mg/kg Zn (basal+ZnO, PC), orally challenged (broth with $10^7$ cfu) on d7, and slaughtered on d14 or d15. Data were analyzed as factor design (diet, litter) and orthogonal contrasts calculated. For C, HZ and PC: 3, 4 and 3 pigs died, respectively. HZ and PC improved overall growth ($P<0.05$), reduced fecal score post-challenge ($P<0.05$) and days pigs had fecal score >2 ($P=0.078$). Feed intake, total *E. coli* shedding (before and 3 day post challenge), and ETEC excretion post challenge were not affected. Total IgA activity in blood serum obtained at slaughtering tended to be higher in HZ and PC, than in C ($P=0.09$), while F4-specific IgA were not affected. Morphometry measures in the jejunum mucosa were not changed by the diet. HiZox® can substitute the pharmaceutical ZnO use at 1/10 dosage in the feeding of weaned piglets challenged with ETEC.



**Two aspects of zinc in piglet feeding: an essential nutrient or an environmental factor?**

*H.D. Poulsen and K. Blaabjerg*
*Aarhus University, Animal Science, Blichers Alle 20, 8830 Tjele, Denmark; hannedamgaard.poulsen@agrsci.dk*

Sufficient zinc (Zn) supply is vital for appropriate activity of a lot of enzymes critical for growth, gastrointestinal health, immunity, skin diseases etc. Extra Zn supply has been shown to alleviate diarrhea and stimulate growth in undernourished children, and high dietary Zn supply has for many years been used to effectively alleviate post-weaning diarrhea in piglets. Zinc oxide (ZnO) is the most studied Zn source, and feeding of high Zn levels is mainly used for two weeks after weaning. The other aspect of inclusion of high Zn concentrations in diets for weaned piglets is related to the increasing environmental concern because Zn may affect soil fertility and the life of soil organisms and may also stimulate to the occurrence of resistance to Zn in the gut microflora. Despite the extensive use of high dietary Zn, the mechanisms behind the alleviating effect on diarrhea have not been fully uncovered. This presentation summarizes the current understanding of Zn as a nutrient in piglet feeding and at the same time addresses the environmental concern related to the use of high Zn concentrations in the feeding of weaned piglets. Many hypotheses related to e.g. effects on the gastrointestinal bacterial population have been proposed reflecting a pharmacological approach of ZnO. Other theories are related to the fact that high levels of ZnO have to be applied because the bioavailability of Zn is expected to be much lower in ZnO compared to e.g. Zn sulphate or chelated Zn sources. Finally, an alternative hypothesis is related to Zn as a nutrient where the daily Zn supply is the core issue. This hypothesis is related to the observed correlation between serum Zn status, diarrhea alleviation, and growth improvements in weaned piglets fed high Zn concentrations. An understanding of the mode of action of Zn in piglets after weaning including the transport of Zn at cellular level across the epithelium is anticipated to improve the underlying roles of Zn in preventing diarrhea in newly weaned piglets. Such knowledge will also found the base for balancing the two sides of Zn, pig health and productivity on one side and environmental aspects on the other side.

**Nutritional modulation of the antioxidant system of the body**

*P. Surai*
*Feed-Food Ltd., Dongola road 53, KA7 3 BN Scotland, United Kingdom; psurai@feedfood.co.uk*

During evolution living organisms have developed specific antioxidant protective mechanisms to deal with free radicals. These mechanisms are described by the general term 'antioxidant system'. It is diverse and responsible for the protection of cells from the actions of free radicals. This system includes: natural fat-soluble antioxidants (vitamins A, E, carotenoids, ubiquinones, etc.); water-soluble antioxidants (ascorbic acid, uric acid, etc.); antioxidant enzymes (AO): glutathione peroxidase (GSH-Px), catalase (CAT) and superoxide dismutase (SOD); as well as thiol redox system. The protective antioxidant compounds are located in organelles, subcellular compartments or the extracellular space enabling maximum cellular protection to occur. Some antioxidants are synthesized by the body (AO enzymes, glutathione, ubiquinol, etc.) while others (vitamin E, carotenoids, flavonoids, etc.) as well as trace minerals (Se, Zn, Mn, Cu) must be supplied with the diet. The antioxidant system of the living cell includes three major levels of defence. The first level of defence is responsible for prevention of free radical formation by removing precursors of free radicals or by inactivating catalysts and consists of three AO enzymes: SOD, GSH-Px and CAT plus metal-binding proteins. The second level of defence consists of chain-breaking antioxidants such as vitamin E, ubiquinol, carotenoids, vitamin A, ascorbic acid, uric acid and some other antioxidants. Glutathione and thioredoxin systems also belong to the second level of antioxidant defence. The third level of defence is based on systems that eliminate damaged molecules or repair them and includes lipolytic, proteolytic and other enzymes. Dietary supplementation of natural antioxidants (vitamins E, C, carotenoids, phenolic compounds, etc.) as well as trace minerals is considered to be an important element of modulation of the antioxidant system of the body and maintenance of productive and reproductive performance of farm animals in stress conditions of commercial meat, egg and milk production. It will be reviewed in the paper.

---



**Effect of additional consumption of chromium on reproductive performance of the Gilts**

*M. Covarrubias, J.A. Romo, R. Barajas, H.R. Güémez, J.M. Romo and J.M. Uriarte*
*Universidad Autónoma De Sinaloa, Facultad De Medicina Veterinaria Y Zootecnia, Gral. Angel Flores S/N Poniente Col. Centro, 80000, Culiacán, Sinaloa, Mexico; jumanul@uas.edu.mx*

One hundred pre-puberty gilts 75±8 days old weighting 25.85±4.74 kg, were used to determine the effect of the addition to chromium methionine in the diet from 75 days-age until first service on the reproductive performance of gilts. Agreement with a completely randomized design, gilts were randomly assigned to treatments as follows: (1) Feeding corn- soybean meal-based diets from day 75 until parturition (Control; n=50); or (2) Control plus 0.4 mg of organic chromium/kg of diet (CR; n=50). Organic chromium was offered as chromium methionine-premix MiCroplex® (Zinpro, Co. Eden Prairie, MN). Additional chromium delayed (P=0.03) age to first estrus 211 vs 202 day-ages respect to Control. The presence of second estrus was later ($P<0.05$) in gilts fed CR in relationship to Control (135 vs 130 day-ages). However age to first service was similar (P=0.57) between treatments (248 vs 254 days). CR increased body weight (P=0.03) both at first estrus (126.4 vs 120.6 kg) as the second estrus (135.1 vs 130.0 kg). First service body weight was higher (P=0.03) for CR supplemented gilts (142.3 vs 138.3 kg), and the back fat thickness was greater ($P<0.01$) with mean values of 12.8 and 11.2 mm for CR and Control, respectively. The farrowing rate was similar between treatments (P=0.51) with mean of 66 and 72% for Control and CR, respectively. Total born piglets number was unaffected by treatments (P=0.69) with mean of 11.18 and 10.89 piglets by Control and CR treatments, respectively. Alive born piglets number was similar in both treatments (P=0.64). These results suggest that despite additional chromium consumption increases body weight and back fat thickness; these changes are not reflected as improvement on performance during the first reproductive cycle of gilts.

**Vitamin A and E circulate in different cattle plasma fractions**

*L. Hymøller and S.K. Jensen*
*Aarhus University, Animal Science, Blichersallé 20, 8830 Tjele, Denmark; lone.hymoller@agrsci.dk*

Analysis of fat soluble vitamins in plasma fractions of protein, chylomicrons, and lipoproteins can give information about their plasma transport and physiological function. Vitamin A (A) and vitamin E (E) are known to circulate in different plasma fractions; hence they were used to verify a simple plasma fractionation method for cattle plasma. Blood was collected in Na-EDTA coated tubes from 3 Holstein bull calves, which for 3 weeks prior to sampling were fed 3 kg/d of concentrate. The concentrate contained 3 µg/kg A and 120 mg/kg E. Plasma was isolated by centrifugation at 1,500×*g* for 10 min. Chylomicrons were floated by ultracentrifugation of 6 ml plasma at 100,000×*g* for 1 h in a Beckman-Coulter 70.1 fixed angle rotor. Fractions of lipoproteins (LPF) and protein were separated by adding 1.5 ml OptiPrep™ (Axis-Shield, Norway), layering Hepes buffered saline 0.85%, pH 7.4 on top, and applying ultracentrifugation at 300,000×*g* for 18 h. Fractions were harvested top-down with Pasteur pipette and analysed for A and E by HPLC. Average amounts of A and E found in each plasma fraction, in % of the total amount of A and E in all fractions, were: Protein: A (54%) and E (7%). Chylomicrons: A (13%) and E (15%). LPF1 (VLDL): A (1%) and E (2%). LPF2 (LDL/HDL): A (5%) and E (62%). LPF3 (HDL): A (27%) and E (15%). A is transported in chylomicrons from its site of uptake in the intestines but circulates in plasma bound to specific binding proteins, which are responsible for its biological action in the body. Likewise E is transported from the intestines to the liver in chylomicrons. Binding proteins specific to E is known to exist, however E is found in the LDL/HDL fraction of plasma. This indicates that cattle plasma was separated into relevant plasma fractions using the presented ultracentrifugation method. In conclusion, by fractionating cattle plasma by ultracentrifugation it was shown that A mainly circulated in heavier plasma fractions i.e. protein and HDL, whereas most E was found in fractions representing LDL/HDL.

---



**In different worlds: transport of dietary vitamin D2 and vitamin D3 in cattle plasma fractions**

*L. Hymoller and S.K. Jensen*
*Aarhus University, Animal Science, Blichersallé 20, 8800 Tjele, Denmark; lone.hymoller@agrsci.dk*

It is often stated that vitamin $D_2$ ($D_2$) and $D_3$ ($D_3$) have similar physiological effects but studies in cattle showed, that $D_2$ is less efficient than $D_3$ at securing vitamin D status, measured as liver derived 25-(OH)-$D_2$ (25OH$D_2$) and 25-(OH)-$D_3$ (25OH$D_3$). The aim was to investigate if binding to different plasma fractions could explain the physiological inefficiency of $D_2$ compared to $D_3$ in cattle. 3 Holstein bull calves were fed 75 µg/d $D_3$ and had ad lib access to hay, naturally containing $D_2$. Blood was collected weekly for 11 weeks and plasma isolated by centrifugation. Plasma was fractionated by ultracentrifugation and fractions of protein, chylomicron, and lipoprotein (LPF) analysed for content of 25OH$D_2$, 25OH$D_3$, $D_2$, and $D_3$. Contents of 25OH$D_2$ vs 25OH$D_3$ and of $D_2$ vs $D_3$ in each fraction were compared as average of all samples by Student´s t-test. Average percentages of each metabolite in each fraction were: Protein: 25OH$D_2$ (8%) vs 25OH$D_3$ (75%; $P\leq0.001$) and $D_2$ (10%) vs $D_3$ (16%; P+0.03). Chylomicron: 25OH$D_2$ (39%) vs 25OH$D_3$ (17%; $P\leq0.01$) and $D_2$ (0%) vs $D_3$ (6%; $P\leq0.01$). LPF1 (VLDL): 25OH$D_2$ (0%) vs 25OH$D_3$ (0%) and $D_2$ (52%) vs $D_3$ (33%; P+0.07). LPF2 (LDL/HDL): 25OH$D_2$ (9%) vs 25OH$D_3$ (2%; P+0.07) and $D_2$ (4%) vs $D_3$ (31%; $P\leq0.01$). LPF3 (HDL): 25OH$D_2$ (34%) vs 25OH$D_3$ (6%; $P\leq0.001$) and $D_2$ (0%) vs $D_3$ (15%; $P\leq0.01$). $D_2$ and $D_3$ were only transported evenly distributed in protein, whereas the % of $D_2$ and $D_3$ found in other fractions differed. If this means that $D_2$ is transported in more volatile plasma fractions than $D_3$, then $D_2$ could be more prone to degradation and excretion, contributing to its lower physiological efficiency. 25OH$D_3$ was mainly transported in protein whereas 25OH$D_2$ was transported in chylomicron and LDL/HDL. Associating with designated binding proteins is vital to the function of 25OH$D_3$. Hence transportation in other plasma fractions than protein could explain a compromised physiological function of 25OH$D_2$ compared to 25OH$D_3$. In conclusion, $D_2$ and 25OH$D_2$ are transported in different plasma fractions than $D_3$ and 25OH$D_3$, respectively, which could offer an explanation to a lower physiological efficiency of $D_2$ than $D_3$ in cattle.

**Encapsulation method to protect unsaturated fatty acids from rumen biohydrogenation *in vitro***

*S. Abo El-Nor[1], M. Strabel[2], A. Cieślak[2], R.M. Gawad[1], H. Khattab[3], P. Zmora[2], M. El-Nashaar[1], H. El-Sayed[3] and S. Kholif[1]*
*[1]National Research Center, el-Behouth Street, Dokki, 12622, Giza, Egypt, [2]Poznan university of life science, Department of Animal Nutrition and Feed Management, Poznan, Poznan, Poland, [3]Ain Shams University, animal production, shoubra el-kheima, kalyoubia, Egypt; nor1957@hotmail.com*

The alginate/carrageenan calcium beads of LO were evaluated *in vitro* to verify the ability of these products to protect PUFA and other unsaturated fatty acids from biohydrogenation by ruminal microbes. Biohydrogenation process was evaluated in-vitro using batch culture system (BCS). Linseed products evaluated were: linseed beads containing 15% oil in dry matter, linseed beads containing 20% of oil in dry matter. The treatments were: control (C) without supplements, LB1 (control + 4% of linseed beads containing 15% oil), LB2 (control + 4% of linseed beads containing 20% oil), LO (control + 4% of LO). Each product was supplemented (4% dry matter) to the substrate that comprises of a mixture of meadow hay and barley meal in the ratio of 60:40 and incubated using BCS for 48 h. All samples (feeds and rumen fluid after incubation) were analyzed for fatty acids profile. The results indicated the final beads content from LO was about 87% and 86% for LOB1 loaded with 15 vol% LO and LOB2 loaded with 20 vol% oil, respectively. The encapsulation process didn't have a significant effect on PUFA fraction ($P>0.01$). After incubation in batch culture system linseed beads decreased ($P<0.01$) total rumen saturated and monounsaturated fatty acids content. Omega 3 and omega 6 fatty acids contents increased statistically ($P<0.01$) by LB1 and LB2 and numerically by LO treatment. The omega 3 values were 2.5, 17.9, 18.5 and 9.7 mg/100 mg FAME, and omega 6 values were 8.7, 25.2, 26.9 and 17.6 mg/100 mg FAME (fatty acid methyl esters), respectively for control, LB1, LB2 and LO treatments. In conclusion, new encapsulation method has the potential to protect LO from rumen biohydrogenation *in vitro*, however further *in vivo* experiments are required.

---

Session 06 Poster 15

**Layer fertility and chick hatchability as influenced by dietary fatty acids**

*A.M. Jooste, O.S. Olubowale, J.P.C. Greyling and F.H. De Witt*
*University of the Free State, Animal, Wildlife and Grassland Sciences, P.O. Box 339, Bloemfontein, 9300, South Africa; joosteam@ufs.ac.za*

This study was conducted to determine the effects of dietary fatty acids (FA) on fertility of laying hens and chick hatchability. Five isoenergetic (12.4 MJ ME/kg DM) and isonitrogenous (170 g CP/kg DM) diets were formulated using different lipid sources at a 3% inclusion level to alter the dietary FA profile. The control diet was formulated using a composite of linseed- and fish oil (50:50), while fish oil was used in the polyunsaturated n-3 (PUFA n-3) treatment. Sunflower oil was used in the polyunsaturated n-6 (PUFA n-6) diet, whereas in the mono-unsaturated n-9 (MUFA n-9) diet, high oleic acid (HO) sunflower oil was used. Lastly, tallow was used in the saturated fatty acid (SFA) treatment. One hundred and twenty five hens (n=25/treatment) and 50 cockerels (n=10/treatment) of the Hy-Line Silver-Brown genotype were randomly allocated to the 5 dietary treatments at 20 weeks of age. From 69 weeks of age, hens were inseminated with 0.06 ml. undiluted semen from cockerels within the same dietary treatment. A total of 588 eggs/treatment were individually marked (date & hen number) over a period of 49 days and incubated. Eggs were candled on D7 and D14 to determine embryonic mortalities, with a 24 h window for hatching being provided (D21 + 24 h). Although the PUFA (n-3) treatment resulted in the lowest ($P<0.05$) egg weights (59.3 g) and fertility (84.6%) it recorded the highest ($P<0.05$) hatchability (76%). Contrary, the PUFA (n-6) treatment recorded the lowest hatchability (58.2%) of all treatments, despite its high fertility (89.6%). Results indicate that the dietary fatty acid profile, in particular the n-3 and n-6 concentrations, requires consideration in the formulation of breeder diets to limit embryonic mortalities during incubation.

**Cockerel semen quality as influenced by dietary fatty acid profile**

*A.M. Jooste, O.S. Olubowale, J.P.C. Greyling and F.H. De Witt*
*University of the Free State, Animal, Wildlife and Grassland Sciences, P.O. Box 339, Bloemfontein, 9300, South Africa; joosteam@ufs.ac.za*

The effect of dietary fatty acid profile on the semen quality of Hy-Line Silver-Brown cockerels was investigated. Five isoenergetic (12.4 MJ ME/kg DM) and isonitrogenous (170 g CP/kg DM) diets were formulated using different lipid sources at a constant inclusion level of 3%. The control diet consisted of a blend of linseed- and fish oil (50:50 ratio), while the remainder of the treatments consisted of fish oil (PUFA n-3), sunflower oil (PUFA n-6), high oleic acid (HO) sunflower oil (MUFA n-9) and tallow (SFA). Fifty cockerels, 20 weeks of age, were randomly allocated to the 5 dietary treatments (n=10/treatment), and fed the respective diets for 15 weeks prior to onset of the study at week 35 of age. This ensured 6 complete spermatogenesis cycles; including a training period for ejaculation by means of the massage technique. Semen was collected weekly for a period of 11 weeks and evaluated for semen quality and sperm morphology individually. At week 46 of age, the semen fatty acid methyl esters (FAME) were determined. Data were analysed using the LSM procedures. The PUFA n-3 treatment resulted in the lowest ($P<0.05$) semen volume (0.35 vs 0.42 ml), total sperm output ($120\times10^9$ vs $1.48\times10^9$/ml), sperm motility (48.1 vs 55.1%) and ejaculation rate (79.2 vs 94.2%), compared to the SFA treatment with the highest values for these traits. Differences for sperm morphology were mainly non-significant and no clear trends on sperm viability could be established. It was evident that the semen FAME could be altered, according to the dietary FAME with a consequent influence on cockerel semen quality. Dietary fatty acid profile must be considered when formulating diets for breeding cockerels to ensure optimal semen quality and fertility.

---



**Effect of selenium and vitamin E on the biomarkers of oxidative stress in lactating cows**

*M.S.V. Salles[1], T.S. Samóra[1], A.M.M.P.D. Libera[2], L.C. Roma Junior[1], L. El Faro[1], F.N. Souza[2], M.G. Blagitz[2], C.F. Batista[2] and F.A. Salles[1]*
*[1]APTA, Department of Agriculture of Sao Paulo State, Av Bandeirantes 2419, 14030-670, Brazil, [2]FMVZ, São Paulo University, Av. Orlando M. Paiva, n. 87, 05508-270, Brazil; marcia.saladini@gmail.com*

Antioxidant is an important area of research because immune cells are particularly sensitive to oxidative stress. In fact, antioxidants and some trace minerals have important role in immune function and may affect health. The aim was to study the effect of sunflower oil (SFO) and/or the antioxidants vitamin E and selenium supplementation on the biomarkers of oxidative stress in lactating dairy cows. Twenty-eight cows were divided in four treatments (randomized block design: early and mid lactation), as follows: C (control diet); A (3,5 mg/ kg dry matter (DM) of organic selenium + 2,000 IU of vitamin E/cow per day); O (4% of SFO in DM diet); OA (4% of SFO in DM diet + 3,5 mg/ kg DM of organic selenium + 2,000 IU of vitamin E/ cow per day. Cows were fed with 0.50 of concentrate, 0.42 of corn silage and 0.08 of coast-cross hay (DM). Blood samples were collected at 0, 28, 54 and 80 days after supplementation. The addition of antioxidants in the diet increased ($P=0.002$) the average levels of glutathione peroxidase during the experimental period (76.24, 85.76, 67.45 and 85.12 U/g Hb, SEM=3.88 to C, A, O and OA, respectively). At 54 days reduced glutathione tended to increase ($P=0.07$) in cows fed with A treatment (median 3.59, 4.83, 4.18, 4.02 U/l Hb to C, A, O and OA respectively). At 54 days antioxidant activity tended to decreased ($P=0.05$) in cows fed with AO treatment (median 1.13, 1.18, 1.15, 1.11 mmol/l to C, A, O and OA, respectively). The levels of malondialdehyde were not changed among treatments. The addition of selenium and vitamin E in the diet reduced the oxidative stress of lactating dairy cows. Financial support:FAPESP.

## Session 06 Poster 18

**Effect of sodium selenite on cadmium chloride induced renal toxicity in male Sprague-Dawley rats**

*F. Jabeen[1] and A.S. Chaudhry[2]*
*[1]GC University Faisalabad, Zoology, Wildlife & Fisheries, Allama Iqbal Road, 38000 Faisalabad, Pakistan, [2]Newcastle University, AFRD, Agriculture Building, NE1 7RU Newcastle Upon Tyne, United Kingdom; abdul.chaudhry@newcastle.ac.uk*

Cadmium is an environmental pollutant and is known for various mechanisms of toxicity in organisms under different experimental conditions. Selenium is an essential micronutrient with important biochemical functions in organisms. This study was designed to investigate the protective effect of sodium selenite (Se) on the cadmium chloride (Cd) induced renal toxicity in male Sprague-Dawley rats. Following approval by the Ethics Committee, twenty male Sprague Dawley rats (28 days old) were housed at the animal house. The rats were acclimatized to their housing and feeding for 2 weeks before they were weighed and distributed into 4 groups of 5 rats each with similar initial body weight per group. A same commercial diet and fresh water was available *ad libitum* to these rats. The rats were given subcutaneous doses of 1 mg/kg body weight (BW) of either normal saline (control) or Cd or Se or Cd plus Se on alternate days for 4 weeks. All rats were then weighed at weekly intervals for 4 weeks before their sacrifice on 29$^{th}$ day. To see the effect of sodium selenite on cadmium chloride induced renal toxicity body and kidney weights, comet assay, Terminal deoxynucleotidyl transferase (TdT) dUTP Nick-End Labeling (TUNEL) assay and histology of kidney were carried out. Cells were scored by comet assay-IV software. The data was analysed by Minitab16 software to determine the treatment effects on different parameters. The Cd group showed decrease in body weight, increased DNA damage, apoptosis and altered histology as compared with the control group. The Se group also showed DNA damage, apoptosis and altered histology but it was less severe than the Cd group. In the Cd+Se group, protective effects of Se on Cd-induced changes were observed. While the Se was able to curtail the toxic effect of Cd, the Cd or Se alone were genotoxic and cytotoxic for rats receiving a sub-lethal dose of 1 mg/kg BW.

---

## Session 06 Poster 19

**Vitamin C prophylaxis in hybrid catfish (*Clarias macrocephalus* × *Clarias gariepinus*) subjected to thermal**

*S. Boonanuntanasarn*
*Suranaree University of Technology, Institute of Agricultural Technology, 111 University Avenue, Tambon Suranaree, 30000, Thailand; surinton@sut.ac.th*

A sudden temperature change could create thermal stress and result in adverse effects on the physiological process and health of fish. The present study investigated the effects of a high level of supplementary vitamin C provided to hybrid catfish (*Clarias macrocephalus* × *Clarias gariepinus*) to minimize the thermal stress represented by hematological parameters and non-specific immunity. Dietary vitamin C was fed to hybrid catfish at 250, 500, 1000, 1,500, and 2,000 mg/kg for an 8-week experimental period. The hepatic vitamin C concentration in the fish was similar to that of all tested feeds used throughout this experiment. The survival rate and feed conversion ratios were similar in all groups. During the first four weeks, the increased dietary vitamin C significantly enhanced the specific growth rate and weight gain of hybrid catfish in the linear response ($P<0.05$), suggesting that elevated level of vitamin C during this growth period resulted in maximum growth responses. An increase in dietary vitamin C (1000-1,500 mg/kg) would be essential for juvenile fish to achieve a maximum growth rate. To assess the thermal stress, the hybrid catfish were transferred to water having a temperature of 19 °C for 24 h during week 4 (the first thermal stress) and week 8 (the second thermal stress). The increment of vitamin C (1000-1,500 mg/kg) quadratically increased the total protein and total immunoglobulin ($P<0.05$) at the second thermal stress. Under this stress, increase in dietary vitamin C linearly increased lysozyme activity ($P<0.05$). The alternative complement activity was slightly higher with the higher vitamin C level. Taken together, long term supplementation of high vitamin C level (1000-15,000 mg/kg) would be beneficial to the immunocompetence of hybrid catfish. Thus, vitamin C porphylaxis would be a valuable method to maintain the physiological and immunological process for hybrid catfish production, especially in a situation of thermal stress.

**Environmental resistance against bovine babesiosis – is it conferred by Mediterranean foliage?**

*A. Shabtay, H. Eitam, R. Agmon, M. Yishay, A. Orlov and Z. Henkin*
*Agricultural Research Organization, Ruminant Science, P.O. Box 1021, 30095 Ramat Yishay, Israel; shabtay@volcani.agri.gov.il*

Brushland and grassland are the main pasture types in Israel. It was previously shown that during the hot and dry season, Mediterranean brushlands foliage comprises ca. 60% of cows' diet. This foliage contains high amounts of polyphenols, and possess anti-parasitic traits. We hypothesized that brushland foliage might thus confer environmental resistance to bovine babesiosis, one of the most economical diseases in temperate regions. Blood samples from cows from brushland and grassland herds were collected to assay osmotic fragility, plasma polyphenols and antioxidant capacity. A nested PCR was used to detect *Babesia* sp. DNA in these samples. Ticks were collected from the animals for classification. In a controlled experiment cows were supplemented with oak (*Quercus calliprinos*) leaves for 5 weeks. Blood was sampled before and at the end of the experiment for the analysis of osmotic fragility, polyphenols determination and fluorescence assay (FACS and microscopy analysis). A significant increase in polyphenolic compounds in plasma of brushland cows suggests that polyphenols are absorbed to the blood. Based on the osmotic fragility results it is hypothesized that these metabolites interact with erythrocytes and turn them more resistant to osmotic pressure. As babesiosis is characterized by osmotic and oxidative stresses, these results imply that dietary brushland foliage may be effective in reducing the incidence of babesiosis. Indeed, PCR results indicate that brushland cows were 50% less infected with *Babesia* species as compared with grassland cows, in spite of similar proportion of *Boophilus* in the two habitats. Fluorescence analysis confirmed the interaction between oak derived polyphenols and erythrocytes. This was associated with decreased osmotic fragility. These results indicate that brushland foliage compounds may inhibit invasion of *Babesia* species into bovine erythrocytes, presumably by serving as a physical barrier.

---



**The effect of protected choline on metabolism and productivity of highly productive cows**

*M.G. Chabaev, S.I. Tjutjunik, R.V. Nekrasov, N.I. Anisova, N.G. Pervov, V.N. Romanov and A.M. Gadjiev*
*All-Russian Institute of Animal Husbandry, Department of Animal Feeding, Moscow region, Podolsk district, 142132, Dubrovitcy, Russian Federation; nek_roman@mail.ru*

This study contributes to the development of modern standards for choline supplementation in the diet of highly productive cows. The cows of Black and White Holstein breed (early lactation) were divided into 4 groups: control and 3 experimental (n=9) based on physiological and productive analogy between animals. We used protected form of choline at 0.8 (group 2), 1.0 (group 3) and 1.2 g (group 4) per kg of milk. The diet for control and experimental cows was balanced according to the Russian nutritional standards. The results of this study indicated beneficial effects of protected choline on productivity of cows. The average daily milk production (4% fat) was higher on 1.7, 2.2 and 1.9 kg in the experimental groups 2, 3 and 4 versus controls ($P\leq0.01$), respectively. Similarly, choline content in the milk increased by 1.8, 2.1 and 2.2-fold, respectively. The consumption of fodder per kg of produced milk decreased on 5.5-6.9%. Physiological evaluation showed improved digestibility of fodder's nutrients including dry matter on 2.5-3.1%, crude protein on 2.2-2.5%, crude fat on 2.0-2.4%, crude fiber on 1.7-2.2% and non-nitrogen extracted substances on 3.1-3.5% due to the addition of choline. Increased digestibility was associated with improved microbial processes in the rumen. We found that number of bacteria increased on 9.5-28.6% including protozoa (24.1-48.2% increase) as compared to controls. The concentration of volatile fatty acids in the rumen fluid was higher on 6.4-11.3% in experimental cows. The biochemical blood tests showed improved anabolic processes and functional activity of liver in the cows fed with choline that in turn, resulted in increased milk production. We recommend supplying 1 g protected choline per kg of milk to normalize digestive and metabolic processes and increase milk production.

**Effect of dietary chromium-nicotinate on the portion of fatty parts and chemical composition of fat in Large White breed**

*O. Bucko, A. Lehotayova, P. Juhas, K. Vavrisinova and O. Debreceni*
*Slovak University of Agriculture in Nitra, Department of Animal Husbandry, Tr. A. Hlinku 2, 94976 Nitra, Slovak Republic; ondrej.bucko@uniag.sk*

The aim of the experiment was to consider the effect of addition of chromium-nicotinate on the portion of fatty parts in the carcass body and to determinate the changes in the chemical composition of fat. The experiment included 34 fattening pigs of Large White breed. The pigs were divided in a control group of 19 pigs (9 barrows and 10 gilts) and experimental group consisted of 15 pigs (7 barrows and 8 gilts). Both control and experimental group of pigs were fed the same diet which consisted of three feed mixtures applied at the different growth phases. The diet of experimental group was supplemented with 750 μg/kg chromium-nicotinate in the form of inactivated yeast *Saccharomyces cerevisiae*. The yeast was fermented on the substrate originated from the natural resources with a higher content of trivalent chromium. The samples for the determination of chemical composition of fat were analysed from the backfat and kidney fat by FTIR method using the Nicolet 6700. We compared selected fatty parts of the carcass body: weight of the jowl, weight of the belly, weight of the backfat, the fat from the thigh, weight of kidney fat and backfat thickness. The results showed that these fatty parts had higher values in the control group of pigs. It was found out a statistically significant difference between the experimental and control groups at $P<0.05$ in the parameters the weight of backfat and backfat thickness. The results of the chemical composition of fat showed that the control group of pigs contained a higher content of stearic acid in the fat from backfat ($P<0.05$) and in the kidney fat ($P<0.01$). It can be concluded that the addition of chromium-nicotinate in the diet of pigs caused a lower portion of fatty parts in the carcass body which was shown in the weight of backfat and backat thickness. There was a statistically significant difference in the content of stearic acis but no differences in other fatty acids. This work was supported by projects VEGA 1/2717/12 a ECACB Plus – ITMS: 26220120032.



**Performance and meat quality of light lambs fed concentrates with different sources of magnesium**

*M. Blanco[1], G. Ripoll[1], I. García[2], R. Bernal[2] and M. Joy[1]*
*[1]CITA, Unidad de Tecnología en Producción y Sanidad Animal, Avda. de Montañana, 50059 Zaragoza, Spain, [2]Magnesitas Navarras S.A., Av. Roncesesvalles s/n, 31630 Zubiri, Spain; mblanco@aragon.es*

The aim of this study was to compare the performance, carcass and meat quality of Rasa Aragonesa lambs (n=56) fed a commercial concentrate (87.8% DM, 17.7% CP, 2.7% EE) with different sources of magnesium (Mg) until slaughter at 22 kg LW. Lambs were randomly assigned to four treatments at weaning (LW=12.6±1.53 kg). Magnesium was incorporated in the concentrate as 100% caustic MgO (C), caustic semicalcined MgO and MgCO3 (Mg2), caustic semicalcined MgO and calcined dolomite (Mg3) or caustic semicalcined MgO and Mg(OH)2 (Mg4). The source of Mg did not affect weight gains (293±53.2 g/d; $P>0.05$), daily concentrate intake (0.95±0.021 kg/d; $P>0.05$) and slaughter age (78±10.4 days; $P>0.05$). Serum concentration of sodium, phosphorus, Mg and calcium during the fattening period was not affected by the source of Mg in the concentrate ($P>0.05$). Carcass quality was not affected by the source of Mg except for fat lightness and yellowness indexes. Mg2 lambs had greater lightness than Mg3 and Mg4 lambs (71.4, 68.7 and 68.0, respectively; $P<0.05$); C and Mg2 lambs had lower yellowness than Mg3 and Mg4 lambs (10.0, 9.9, 11.3, 11.5, respectively; $P<0.05$). Meat pH and colour were not affected by the source of Mg ($P>0.05$) but lipid oxidation was affected by the interaction between the source of Mg and the time of oxygen exposure ($P<0.01$). When meat was exposed to oxygen for 0, 2 and 5 days, lipid oxidation was similar among sources of Mg but when it was exposed for 7 days lipid oxidation of Mg3 and Mg4 lambs was greater than that of C and Mg2 lambs (1.4, 1.3, 1.0 and 1.1 mg malonaldehide/kg meat, respectively; $P<0.05$). In conclusion, the source of Mg in the concentrate did not have relevant effects on performance, carcass and meat quality. However, Mg2 lambs had meat with greater lightness and lower yellowness and decreased the lipid oxidation improving the shelf life.

**Effect of the source of magnesium fed during pregnancy and lactation on performance of sheep**

*M. Blanco[1], P. Amezketa[2], M. Arandigoyen[2] and M. Joy[1]*
*[1]CITA de Aragon, Avda. Montañana, 930, 50059-Zaragoza, Spain, [2]Magnesitas Navarras S.A., Av. Roncesvalles s/n, 31630 Zubiri, Spain; mjoy@aragon.es*

The aim of this study was to assess the effect of the inclusion of different source of magnesium (Mg) in concentrate fed during pregnancy and lactation on the performance of Rasa Aragonesa ewes (n=116) and their lambs. The concentrates fed to the ewes were commercial concentrates for pregnancy (88.5% DM, 18.0% CP, 30.7% FND and 13.4% FAD) and lactation (88.8% DM, 19.5% CP, 32.5% FND and 13.3% FAD). Magnesium was incorporated in the concentrate as caustic MgO (C), caustic semicalcined MgO and $MgCO_3$ (Mg2), caustic semicalcined MgO and calcined dolomite (Mg3) or caustic semicalcined MgO and $Mg(OH)_2$ (Mg4). At mating, ewes were randomly distributed in four treatments. During pregnancy, the source of Mg had no effect on live weight (LW; 66 kg) and body condition score (BCS; 3.7), however, affected sodium (Na), potassium (K), calcium (Ca) and Mg concentrations in plasma. During lactation, the source of Mg affected LW ($P<0.001$) and BCS ($P<0.05$). Ewes of Mg2 treatment lost 4 kg LW, Mg3 and Mg4 ewes lost 2 kg and C ewes maintained LW. After the 45 days of lactation, Mg3 and Mg4 ewes had greater BCS than Mg2 and C ewes (3.2, 3.2, 2.9 and 2.8, respectively; $P<0.001$). Treatment affected Na, K, Ca and P concentrations in plasma. Regarding lambing, the source of Mg did not affect birth rate (111 lambings), the number of lambs per lambing (160 lambs, 58.6% simple and 41.4% twins), lamb birth weight (3.6 kg) or mortality (4.4%). However, the source of Mg affected the lambs' weight gains during lactation as Mg4 lambs had greater weight gains than C, Mg2 and Mg3 lambs (259, 227, 220 and 222 g/d, respectively; $P<0.05$). Consequently, Mg4 lambs were heavier at weaning than C, Mg2 and Mg3 lambs (14.5, 13.1, 13.0 and 12.6 kg, respectively; $P<0.05$). In conclusion, the source of Mg in the concentrate had relevant effects only in lactation, with greater LW and BCS of ewes and daily growth lambs in Mg4 treatment.

---



**Source of Se supplementation affects milk and blood serum Se concentrations in dairy cattle**

*L. Vandaele[1], B. Ampe[1], S. Wittocx[2], L. Segers[2], M. Rovers[2], A. Van Der Aa[3], S. De Smet[4], G. Du Laing[5] and S. De Campeneere[1]*
*[1]Institute for Agricultural and Fischeries Research (ILVO), Animal Sciences, Scheldeweg 68, 9090, Belgium, [2]Orffa Additives BV, Vierlinghstraat 51, 4251 LC Werkendam, the Netherlands, [3]Excentials BV, Vierlinghstraat 51, 4251 LC Werkendam, the Netherlands, [4]Ghent University, Faculty of Bioscience Engineering, Proefhoevestraat 10, 9090 Melle, Belgium, [5]Ghent University, Faculty of Bioscience Engineering, Coupure Links, 9000 Gent, Belgium; leen.vandaele@ilvo.vlaanderen.be*

The aim of the present study was to evaluate three different selenium (Se) sources: sodium selenite (NaSe), selenium-yeast (SeYeast) and L-selenomethionine (SeMet) in their potential to affect blood and milk Se levels. A feeding trial was set-up with 26 high producing Holstein Friesian cows. After a two week pre-treatment period without Se supplementation, cows were divided in four homogenous groups and received either no supplementation (Ctrl) or 0.3 mg per kg dry matter intake (DMI) of either NaSe, SeYeast or SeMet for the next 7 weeks. Cows were given maize and prewilted grass silage *ad libitum*, a mixture of soybean-meal and rapeseed-meal to equalize energy and protein intake and a balanced concentrate. Milk and blood serum samples were taken during the pre-treatment period (week 0) and at week 3 and 7 after the start of supplementation. Blood serum Se as well as milk Se concentration was analysed. Se source only marginally affected milk production and composition. The mean blood serum Se was around 30µg/l at week 0 across all groups and 39 and 23µg/l for Ctrl, 63 and 57µg/l for NaSe; 71 and 71 µg/l for SeYeast; 68 and 69µg/l for SeMet at week 3 and week 7, respectively. Milk Se on average 16 µg/kg at week 0 more rapidly increased at week 3 for SeMet (61 µg/kg) than SeYeast (45 µg/kg) and NaSe (26 µg/kg). At week 7 milk Se was significantly different between SeMet (75 µg/kg); NaSe (46 µg/kg) and the Ctrl (21 µg/kg), whereas Se Yeast (63 µg/kg) was not significantly different from SeMet or NaSe. In conclusion, the type of Se source clearly affects the increase in milk Se content after supplementation.

**Attapulgite (Ultrafed®) protective action against *E. coli* and *C. perfringes* toxigenic strains in pigs**

*I. Skoufos[1], I. Verginadis[2], I. Simos[2], A. Tzora[1], A. Mentis[3], A. Tsinas[1], G. Magklaras[1], N. Theofilou[4] and S.P. Karkabounas[2]*
*[1]TEI of Epirus, Division of Animal Production, Kostaki, 47100, Arta, Greece, [2] University of Ioannina, School of Medicine, University of Ioannina, 45110, Ioannina, Greece, [3]Hellenic Pasteur Insitute, 127 Vasilissis Sofias, 11521, Athens, Greece, [4]Geohellas S.A., 8A Pentelis, 17564, Athens, Greece; gmag@teiep.gr*

The protective action of a commercial product based on attapulgite (Ultrafed®) on pig intestinal epithelium cells was investigated, in the presence of bacterial toxigenic strains of *Escherichia coli* and *Clostridium perfringens*. Pig intestinal epithelium cells, cell line IPEC-J2, were grown in culture medium DMEM, in order to determine their growth curve. Then, attapulgite was added in the cell culture in concentrations ranging from 0.0125 to 0.1 mg/ml, in order to specify the concentration for the highest survival rate of the cell line, in comparison to its commercial form (Ultrafed®). The concentration of 0.0125 mg/ml was chosen and then the *E. coli* 0157:H7 strain was added in the cell line, in the presence and absence of attapulgite, at decimal dilutions of 10, $10^2$, $10^4$ and $10^6$. Incubation took place for 48 h, in order to determine the survival rate of the intestinal epithelium cells and the level of protection that attapulgite offers at various concentrations of *E. coli*. The above experiment was repeated with the inclusion of *C. perfringens* type A. In increased concentrations of *E. coli* and *C. perfringes* (10/ml to $10^6$/ml), attapulgite appears to have a protective action in comparison to the control group (incubate the IPEC-J2 cells and *E. coli* or *C. perfringes*), increasing by 21% the number of surviving cells in the presence of *E. coli* (10/ml) and 13% at the highest concentration ($10^6$/ml). Similarly, the number of surviving cells in the presence of *C. perfringens* increased 13% at the lowest concentration (10/ml) and 9% at the highest ($10^6$/ml). In conclusion, attapulgite exhibits a significant protective *in vitro* action for pig intestinal cells against intestinal toxigenic bacterial strains.

---



**Feeding seasons impact on FA profile in cows' milk and lipid fraction in serum**

*P. Lipińska[1], A. Jóźwik[1], N. Strzałkowska[1], E. Poławska[1], T. Nałęcz–tarwacka[2], W. Grzybek[1], M. Wiszniewska[1], B. Pyzel[1], E. Bagnicka[1] and J. Horbańczuk[1]*
*[1]Institute of Genetics and Animal Breedning PAS, Jastrzębiec, Postępu 36A, 05-552 Magdalenka, Poland, [2]Warsaw University of Life Science, Ciszewskiego 8, 02-786 Warsaw, Poland; lipinskapaulina87@gmail.com*

The aim of study concerned the feeding season impact on FA profile, level of SCC in bovine milk and lipid fraction in serum. FA profile and lipid fraction was determinated in samples (n=80) which were collected from Montbeliarde (MB) and Holstein-Friesian (HF). All samples have been gathered evenly in summer (S) and winter (W) to represent seasons of feeding. In both seasons cows which have been participating in research were fed with TMR. As an addition in summer season cows had unlimited access to the pasture. The results gained from experiment concluded that there is a significant effect of on FA profile in milk of both breeds. Higher content of functional FA were found in milk of cows which were grazing during summer. Assay of FA profile in samples in both seasons gave the following results: level of SFA was higher in HFs' milk-(W) 69.63% (S) 61.69%, than in MBs' milk-(W) 63.07% (S) 59.69%, including short chain SFA-HF (W) 7.74% (S) 3.87% and MB (W) 6.21% (S) 2.21%. Level of MUFA was higher in MBs' milk (W) 31.54% (S) 31.93%, than in HFs' milk (W) 27.25% (S) 29.16%. Content of PUFA was lower in MBs' milk (W) 3.25% (S) 3.51% than in HFs' milk (W) 3.81% (S) 3.89%. Lipid fraction of serum showed no significant differences regardless of the feeding season. Obtained results presents that the content of nutrients in bovine milk is determined by both genetic and environmental factors. The beneficial effect of using the pasture is conjugated with higher level of functional components in milk. Application of fodder in summer feeding is promising alternative compared to TMR usage only. BIOFOOD–innovative, functional products of Animals origin no. POIG.01.01.02-014-090/09 co-financed by the European Union from the European Regional Development Fund within the Innovative Economy Operational Programme 2007-2013.

**Influence of feeding and managment on gut microbiota and barrier function**

*B.B. Jensen*
*Aarhus University, Dept of Animal Science, Blichers Allé 20, 8830 Tjele, Denmark; bentborg.jensen@agrsci.dk*

The diverse collection of bacteria colonising the gastrointestinal tract of farm animal, plays an essential role not only for the well-being of the animal, but also for animal nutrition and performance and for the quality of animal products. For the last decades, antibiotics have been used to promote animal growth and health. There are, however, increasing concerns in relation to the development of antibiotic-resistant bacterial strains and the potential of these to impact human health. Minerals as zinc and copper are not feasible alternatives because their excretion is a possible treat to the environment. Consequently, there is a need to develop feeding and management strategies to serve as means for controlling problems associated with animal production. Probiotics has been intensively studied as alternative means to improve animal health. However, there is still a need to clarify the probiotic effectiveness in animals. Use of probiotics in animal production is problematic as maintenance of viability is key to their beneficial activity, but difficult to achieve with commonly used feed processing technologies. One specific context where probiotics organisms may be reliably delivered is the systems utilising fermented liquid feeds. Liquid feed may be fermented by wield lactic acid bacteria or by using specific isolates as starters; the later systems has advantages in terms of reproducibility and speed of fermentation. Low protein diets with AA supplementation improve gut health in piglets. Prebiotics seems to be able to improve performance and enteric health, although the mechanisms are not completely understood. Organic acids in right concentrations decrease survival of pathogenic bacteria and increase growth performance. Feed processing is of great importance for maintaining a healthy gastrointestinal ecosystem in slaughter pigs. Use of coarsely ground meal feed has been shown to reduce the occurrence of pathogenic enterobacteria significantly. However such feeding strategies impair growth performance.

---



**Priming of porcine γδ T-cells with condensed tannins: effects of tannin structure and polymer size**

*A.R. Williams[1], K. Reichwald[1], C. Fryganas[2], I. Mueller-Harvey[2] and S.M. Thamsborg[1]*
*[1]University of Copenhagen, Department of Veterinary Disease Biology, Dyrlægevej 100, 1870 Frederiksberg C, Denmark, [2]University of Reading, Chemistry and Biochemistry Laboratory, 1 Earley Gate, Reading RG6 6AT, United Kingdom; arw@sund.ku.dk*

Diets containing bioactive plant secondary metabolites such as condensed tannins (CT) have been proposed as a possible method of promoting natural health and immunity in livestock, without the use of synthetic antimicrobial drugs. Here, we investigated whether CT could effectively prime porcine γδ T-cells *in vitro*, and also whether priming effects were dependent on CT polymer size and monomer composition, namely the ratio of procyanidin (PC) to prodelphinidin (PD) units. CT were purified from three different sources: hazelnut skin (PC-rich), blackcurrant leaves and white clover flowers (both PD rich). Ratios of PC, PD and mean degree of polymerization were determined by HPLC. Blood was collected by venipuncture from healthy pigs into lithium heparin tubes, and mononuclear cells (PBMC) were isolated on histopaque. PBMC were cultured at $10^6$ cells/ml with each of the extracts for 48 hours. Concavilin A was used a positive control. Priming was quantified by upregulation of CD25 (IL-2 receptor). Cells were stained with antibodies against γδ TCR and CD25 and acquired by flow cytometry. Percentages of $\gamma\delta TCR^+CD25^+$ cells and CD25 mean fluorescence intensities (MFI) in the γδ T-cell population were compared by ANOVA. All CT-containing extracts increased both the percentage of $\gamma\delta TCR^+CD25^+$ cells ($P<0.001$) and CD25 MFI ($P<0.01$). There was no effect of monomer composition, with both PC and PD type CT molecules able to induce effective priming. Within PD-molecules, larger molecules induced a more pronounced up-regulation. CT are able to effectively prime porcine γδ T-cells *in vitro*, and the response was observed in both PC and PD-type tannins. Further experiments are ongoing to determine how this may influence the host immune response to infection with pathogens.

### Early-life fructooligosaccharide supplementation changes later pig immune and growth response

*E. Apper[1], C. Meymerit[2], J.C. Bodin[3] and F. Respondek[1]*
*[1]Tereos Syral, Scientific and Regulatory Affairs, Z.I. Portuaire, 67390 Marckolsheim, France, [2]Interface Elevage, Route de Pau, 64410 Vignes, France, [3]Jefo Europe, 2 Rue Claude Chappe, 44470 Carquefou, France; emmanuelle.apper-bossard@tereos.com*

Early-life nutrition may influence gut immunity and metabolism at short and longer term and impact pig growth at later physiological stages by imprinting. Short-chain fructo-oligosaccharides (scFOS) are soluble fibres defined as prebiotics. Thus, scFOS may exert long-time effects on immune response, growth performance and carcass characteristics of pigs. 129 sows in 2 replicates received a control or a scFOS diet (10 g/d of maltodextrin or scFOS, Profeed® respectively) from the d 109 of gestation to the weaning of piglets. After weaning (d 21), piglets of each litter were divided into two groups to receive the CTRL or the scFOS diet (1.2 g/d of maltodextrin or scFOS) during the first-age period. At 53 and 74 d of age, piglets were challenged by Influenza vaccination. Colostrum was sampled 1 h after the birth of the last piglet to determine nutritional and immune composition. Piglet blood serum was sampled one d prior the vaccination, at d 74 and 95 to measure concentrations of Influenza-IgG. Growth performances of piglets were recorded from birth to slaughtering. Sow supplementation with scFOS decreased colostrum fat content ($P<0.05$), and increased IgG content for older sows (interaction diet × parity, $P<0.05$) without modifying performance of suckling piglets ($P>0.10$). It also reduced the age of piglets at slaughtering ($P<0.001$) and tended to increase fat percentage in muscle ($P=0.07$). Piglet scFOS supplementation modulated immune response against Influenza ($P=0.05$), decreased age at slaughtering ($P<0.001$) and muscle fat content ($P=0.001$), leading to a higher technical added value for the breeder ($P<0.05$). Early-life scFOS supplementation leads to long-term effects on piglet immunity and growth. Mechanisms remain unclear but microbiota and epigenetic changes may be involved.

---



### Impact of antibiotherapy on growing rabbits assessed by a whole-blood transcriptomic approach

*V. Jacquier[1,2,3], S. Combes[3], M. Moroldo[2], I.P. Oswald[1], C. Rogel-Gaillard[2] and T. Gidenne[3]*
*[1]INRA, UMR1331 TOXALIM, Toulouse, France, [2]INRA, UMR1313 GABI, Jouy-en-Josas, France, [3]INRA, UMR1388 GENPHYSE, Castanet-Tolosan, France; vincent.jacquier@toulouse.inra.fr*

In previous studies, rabbits fed antibiotics have a decreased caecal activity and a modified caecal microbiota. Our aim was to study whether antibiotic supplementation could also influence blood parameters assessed by transcriptome analysis of gene expression profiles. We compared control animals (group C) to animals fed a diet supplemented with tiamulin and apramycin antibiotics (group AB) between 16 and 70 days of age (d). Growth performances were recorded at 29, 45, 60 and 70 d. The health status was monitored daily. These phenotypic data were analyzed with a general linear model. Transcriptomic studies of total blood were carried out by comparing the two groups at 29 and 45 d, using a 60K microarray enriched with well-annotated immunity-related genes. Statistical analyses were performed using Limma package in R. Daily gain (49.6 vs 46.8 g) and body weight (2,662 vs 2,532 g at 70 d) were higher ($P<0.01$) in the AB group compared to the C group, respectively. Feed intake and gain-to-feed ratio did not differ between groups. The AB diet improved morbidity (3.4% for group AB vs 16.0% for group C, $P=0.01$) but no difference was observed during experimentation between groups for mortality (11% in average) and the health risk index (19% in average). The transcriptomic study showed a differential expression of 10 genes at weaning (29 d), but no difference between groups at 45 d. By comparing the two ages for each group, we detected 2,539 differentially expressed (DE) genes in the group C, whereas no differential expression was found in the group AB. The DE genes classified by the IPA system were mainly related to cancer, gastrointestinal disease, and cell death and survival functions. Our results suggested a leveling of the gene expression in blood by antibiotics that needs to be further correlated with data on growth and health.

**Exposure to antibiotics affects intestinal microbiota and immune development in broilers**

*D. Schokker[1], A.J.M. Jansman[1], J.M.J. Rebel[2], F.M. De Bree[2], A. Bossers[2], G. Veninga[3] and M.A. Smits[1,2]*
*[1]Wageningen Livestock Research, Edelhertweg 15, 8219PH Lelystad, the Netherlands, [2]Central Veterinary Institute, Edelhertweg 15, 8219PH Lelystad, the Netherlands, [3]Cobb Europe, Koorstraat 2, Boxmeer, the Netherlands; dirkjan.schokker@wur.nl*

Gut microbial colonization and immune competence development are affected by early-life environment and interventions, such as treatment with antibiotics. The interplay between microbiota in the intestinal tract and the host is important for the development of immune competence. In humans, it has been shown that antibiotic treatment in early life can lead to a higher risk for developing immune system disorders such as allergy and asthma. The impact of early life antibiotic treatment on gut microbial colonization and immune development in broilers has not been studied extensively so far. In the present study, in a treatment group antibiotics (amoxicillin) were administrated via the drinking water (67 mg amoxicillin/l) to 1-day-old chicks for a period of 24 hours. A control group was included not receiving antibiotics. Birds were housed in a floor pen system. At three time-points (day 1 (prior to antibiotic administration), 5, and 14) birds were sacrificed and jejunal digesta were collected to analyse the composition of resident microbiota and corresponding jejunal tissue samples were taken to perform genome-wide gene expression profiling. We provide evidence that early 1-day antibiotic treatment affects both the microbiota composition in the intestinal tract as well as the intestinal gene expression in broilers over a period of at least two weeks. Functional analysis of the changes in gene expression showed a significant effect on the expression of immune related processes. The antibiotic treated birds showed less activity in immune related processes compared to the controls, this effect was mainly observed at day 5.



**Feed restriction reduces IgA levels and modifies the ileal cytokine expressions in growing rabbits**

*C. Knudsen[1], S. Combes[1], C. Briens[2], J. Duperray[3], G. Rebours[4], J.M. Salaün[5], A. Travel[6], D. Weissman[7], I. Oswald[8] and T. Gidenne[1]*
*[1]INRA, UMR GenPhySE, Chemin de Borde Rouge, Castanet Tolosan, France, [2]CCPA, ZA du Bois de Teillay, Janzé, France, [3]In vivo NSA, Talhoüet, Vannes, France, [4]TECHNA, Route de St Etienne de Montluc, Couëron, France, [5]SANDERS, Centre des affaires Odyssée, Bruz, France, [6]ITAVI, Centre INRA de Tours, Nouzilly, France, [7]INZO, Chemin de l'église, Chierry, France, [8]INRA, UMR Toxalim, Chemin de Tournefeuille, Toulouse, France; christelle.knudsen@toulouse.inra.fr*

Postweaning short-term restriction strategies are commonly used in rabbit breeding to reduce mortality and morbidity. However, little is known about the implications of the immune system in that reduction. This work studied the consequences of feed restriction and dietary digestible energy (DE) concentration on the local immune response, according to a 2×2 factorial design: 320 animals were alloted at weaning (35 days of age) in four groups, with two diets differing in DE (10.13 vs 9.08 MJ DE/kg) and two intake level (*ad libitum* 'AL' or restricted at 75% of AL). Ten animals per group and per age were sacrificed at 50 and 63 days of age. Feaces, blood and ileum were collected. Fecal and plasmatic IgA levels were determined by ELISA and ileal cytokine expressions were measured by RT-qPCR. Feacal IgA levels were reduced by 58% with feed restriction ($P<0.001$) and increased by 47% with a high energy diet ($P<0.01$), regardless of the age. Plasmatic IgA levels were only affected by feed restriction at 63 days of age (-48%, $P<0.01$). Cytokine expressions were similar for both ages and diets, but were affected by feed restriction with a higher expression of IL-1β and IL-2 (respectively +30%, $P<0.05$ and +77%, $P=0.07$) and a 15% lower expression of TNF-α ($P=0.08$) compared to AL animals. Thus, feed restriction and, to a lesser extent, the dietary energy level modulate gut immunity.

**Effects of maternal nutrition on immune competence and microbiota composition of piglets**

*A. De Greeff[1], S. Vastenhouw[1], P. Bikker[2], A. Bossers[1], F. De Bree[1], D. Schokker[2], P.J. Roubos[3], P. Ramaekers[3], M. Smits[1,2] and J.M. Rebel[1]*

*[1]Central Veterinary Institute, Wageningen UR, P.O. Box 65, Lelystad, the Netherlands, [2]Wageningen Livestock Research, Wageningen UR, P.O. Box 65, Lelystad, the Netherlands, [3]Nutreco, Research & Development, Veerstraat 38, Boxmeer, the Netherlands; annemarie.rebel@wur.nl*

Maturation and programming of the intestinal immune system in piglets is initiated by early life microbial colonization. A significant contribution to microbial colonization of piglets comes from the sow: via vertical transmission of vaginal flora during birth and transmission of mucosal immune memory and flora by feaces, colostrum and milk. In this study we determine the effect of a maternal nutritional intervention with an antibiotic on early microbial colonization of piglets. We used antibiotic treatment as a harsh intervention to investigate the hypothesis that the microbial composition in sows, may have an effect on the early microbial colonization of piglets. Two groups of gestating sows (n=15) were either treated with the antibiotic amoxicillin during the last week before farrowing, or left untreated. On day 21 pigs were weaned and transferred to pens in a nursery facility. Immediately after birth of their first piglet a vaginal swabs was taken; fecal samples were collected on days, -28, -7, 0 (farrowing), 1, 7 and 28 (weaning). Fecal samples of piglets were collected after rectal stimulation on days 1, 7, 28, 32 and 49. On the same days, one pig per litter was sacrificed to harvest ingesta, intestinal scrapings and blood. The effect of treatment on microbial colonization of the vagina and faeces of sows and the jejunum of piglets was determined and correlated with each other. Furthermore, the effect of treatment on immune competence in the intestine of piglets was determined by studying gene expression profiles of jejunal scrapings of piglets to investigate the correlation between microbiota and intestinal immune development.

---



**Intermittent suckling improves galactose absorption in weanling pigs**

*D.L. Turpin[1], P. Langendijk[2], T.-Y. Chen[2], T. Thymann[3] and J.R. Pluske[1]*

*[1]Murdoch University, 90 South Street, 6150 Murdoch, Australia, [2]SARDI, Roseworthy Campus, 5371 Roseworthy, Australia, [3]University of Copenhagen, Nørregade 10, 1165 København, Denmark; d.turpin@murdoch.edu.au*

The abrupt changes at weaning cause a reduction in feed intake and a temporary decrease in gastrointestinal (GIT) function. Intermittent suckling (IS) is the removal of the sow from her piglets for a period of time each day in lactation, and can stimulate feed intake before and after weaning. We hypothesised that piglets separated from the sow for 8 h per day in the 7 d before weaning would have better post-weaning GIT absorptive capacity compared to piglets in a conventional weaning regime. Gilt litters (n=64) were allocated to 1 of 3 weaning regimes: conventional weaning (CW28) (n=22; piglets had continuous access to the sow until weaning at d 28); IS, starting at d 21 (IS21) (n=21); and IS starting at d 28 (IS28) (n=21). For IS treatments, litters were separated from the sow from 0800 to 1600 in the 7 d before weaning (d 28 and d 35 for IS21 and IS28, respectively). Creep feed was provided from d 10. Two piglets were selected per litter and oral sugar absorption tests (SAT) using either 10% mannitol (MAN; 5 ml/kg BW) or 10% galactose (GAL; 5 ml/kg BW) were performed. The MAN SAT was performed 3 d before, and 4 d after, weaning. The GAL SAT was performed 4 d after weaning. Plasma MAN and GAL concentrations were determined from blood samples 20 min after administration. GLM procedures of SPSS were used to compare values across treatments. The absorption of MAN was lower (P=0.001) after weaning between treatments compared with pre-weaning values, likely indicating loss of mucosal surface area. There was a trend for a difference in GAL absorption between treatments (P=0.120; 109, 291 and 452 nmol/ml for CW28, IS21 and IS28 respectively), suggesting that IS improves GAL GIT absorptive capacity and particularly if weaning age is increased.

**Visible inflammation of the rumen wall correlates with caecal lipopolysaccharide concentrations**

*C.A. McCartney[1], R.C. Cernat[1], H.H.C. Koh-Tan[2], E.M. Strachan[3], T.J. Snelling[1], C.D. Harvey[3], N.N. Jonsson[2] and R.J. Wallace[1]*
*[1]Rowett Institute of Nutrition and Health, University of Aberdeen, Bucksburn, Aberdeen, AB21 9SB, United Kingdom, [2]School of Veterinary Medicine, University of Glasgow, Glasgow, G61 1QH, United Kingdom, [3]Harbro Ltd., Turriff, Aberdeenshire, AB53 4PA, United Kingdom; christine.mccartney@abdn.ac.uk*

Sub-acute ruminal acidosis (SARA) induction studies, which involve feeding high proportions of concentrate in the diet, have observed an increase of ruminal soluble lipopolysaccharide (LPS) concentration in cattle. However, little is known about the role of hindgut LPS, and the concentrations of soluble LPS in ruminants under farm conditions. In this study, soluble LPS concentrations from the rumen and caecum were determined at slaughter from 98 commercially reared beef cattle. Scoring systems were used to assess the condition of the rumen wall both pre- and post-cooking at the abattoir (80-90 °C, 10 min). Pre-cooking scores were taken for: pinkness/inflammation, bare/necrotic areas and papillae shape. A score was also taken for any blackness of the rumen wall which is evident post-cooking. Ruminal fluid and caecal content were analysed for soluble LPS concentration. A 10-fold higher concentration of soluble LPS occurred in the caecum in comparison to the rumen ($0.85\times10^6$ and $0.09\times10^6$ EU/ml, respectively). LPS concentrations exhibited high variation, even within farms. ANOVA analysis showed a significant relationship between caecal LPS concentrations and pinkness/inflammation of the rumen wall (P=0.042), and ruminal LPS and rumen pre-cooking damage ($P<0.001$). There were no other significant results. The high concentration of soluble LPS in the caecum and its relationship with pinkness of the rumen wall suggests that LPS originating from the caecum may contribute to the pathological signs of SARA in the rumen wall. Future studies should therefore investigate further the role of the hindgut on the pathogenesis of SARA.



**Perturbation of gut homeostasis in pigs by a high dietary concentration of zinc**

*A.J.M. Jansman[1], M. Hulst[1], D. Schokker[1], A. Bossers[2], R. Gerritsen[3], J.M. Rebel[2] and M.A. Smits[1,2]*
*[1]Wageningen Livestock Research, P.O. Box 65, 8200 AB Lelystad, the Netherlands, [2]Central Veterinary Institute, P.O. Box 65, 8200 AB Lelystad, the Netherlands, [3]Agrifirm Group, Landgoedlaan 20, 7325 AW Apeldoorn, the Netherlands; mari.smits@wur.nl*

Recent studies have begun to identify key signaling pathways of the cross-species homeostatic regulation between gut microbiota and its host. To gain more knowledge about the interplay between microbiota, diet-components and the mucosa of the porcine gut, we used a high zinc concentration (2,500 mg Zn as ZnO per kg diet) from day 14-23 post weaning as a 'model intervention' to induce perturbation of gut homeostasis. Results were compared with a control treatment using a diet with a low concentration of zinc. We used genome-wide gene expression profiling and community-scale analysis of gut microbiota on intestinal mucosal tissue and digesta samples, respectively, to identify zinc-induced shifts in gut homeostasis. The interventions had significant effects on the expression of several immune related genes and on processes known to be regulated by zinc ions. To discriminate between direct effects of zinc on intestinal epithelial cells and indirect effects via gut microbiota and their metabolic products, we also exposed *in vitro* cultured porcine intestinal epithelial cells to zinc and investigated their transcriptional response. In this system the expression of several cell surface-associated zinc transporters and an intracellular heavy metal transporter (MT1A) was up-regulated. MT1A is known to regulate the cytokine IL6 which is a potent inducer of the acute phase response. Zinc oxide also induced the expression of a set of oxidative stress genes. Marked differences and similarities were found between the *in vivo* and *in vitro* induced transcriptional responses. These studies provide insight in the interplay between microbiota, diet components and the intestinal mucosa of the porcine gut and may pave the way towards nutrition-based interventions to improve pig health.

**Effect of benzoic acid and calcium chloride on microbiota, mineral balance and bone in growing pigs**

*J.V. Nørgaard[1], O. Højberg[1], K.U. Sørensen[1], J. Eriksen[1], J.M.S. Medina[2] and H.D. Poulsen[1]*
*[1]Aarhus University, Blichers Alle 20, Foulum, 8830 Tjele, Denmark, [2]Aarhus University Hospital, Nørrebrogade 44, 8000 Aarhus C, Denmark; jan.noergaard@agrsci.dk*

Dietary manipulation of urinary pH can reduce nitrogen emission from pig production. The objective was to study the long-term effects of feeding benzoic acid (BA) and calcium chloride (CaCl2) to pigs on the mineral balance and the microbial fermentation pattern in the gastrointestinal tract. Four diets containing the combinations of 0 or 10 g/kg BA and 0 or 20 g/kg CaCl2 were fed to 24 pigs in a factorial design. Calcium carbonate (CaCO3) was added to the diets without CaCl2 to provide equimolar levels of Ca. The pigs were fed the diets from 36 kg body weight (BW) until slaughter at 113 kg BW, and they were housed in balance cages for 12 d from 60 to 66 kg BW. Data were analysed using mixed linear models. Supplementation of BA and/or CaCl2 had only minor effect on accumulation of organic acids (acetate, propionate, butyrate and lactate) throughout the gastrointestinal tract, although CaCl2 reduced supplementation ($P<0.01$) colon acetate and propionate concentrations. In the urine, pH was reduced by both BA ($P=0.04$) and CaCl2 ($P<0.01$). The combined BA and CaCl2 supplementation resulted in the lowest retentions of both P ($P=0.03$) and Ca ($P<0.01$). Metacarpal III bones were lighter ($P=0.04$) and shorter ($P=0.02$) with CaCl2, but BA supplementation had no effect. The combined CaCl2 and BA supplementation reduced the P concentration in the bones ($P<0.05$), whereas the Ca concentration was not affected. Total mineral density in the bones measured by CT-scanning was reduced ($P<0.01$) in the medial and distal sections by BA supplementation and reduced ($P<0.01$) in the proximal, medial and distal sections by CaCl2 supplemantation. In conclusion, long-term dietary supplementation of combined BA and CaCl2 affected the nutrient balance of P and Ca, as well as it reduced mineral concentration and total mineral density of metacarpal III bones.

---



**Analysing the back of dairy cows in 3D imaging to better assess body condition**

*A. Fischer[1,2,3], T. Luginbühl[4], L. Delattre[4], J.M. Delouard[4] and P. Faverdin[2,3]*
*[1]Institut de l'Elevage, Le Rheu, 35650, France, [2]Agrocampus Ouest, UMR1348 PEGASE, Rennes, 35000, France, [3]INRA, UMR1348 PEGASE, St-Gilles, 35590, France, [4]3D Ouest, 22300, Lannion, France; amelie.fischer@rennes.inra.fr*

Body condition is an important trait in dairy cow management, usually measured with the body condition score (BCS), which is subjective, and not very sensitive. The aim of this work was to develop and to validate a method, NEC3D, estimating BCS with 3D pictures of dairy cows back, from the pins to the hooks, which is commonly used as reference area in the BCS scoring systems. A 57 cows 3D-shapes dataset with large BCS variability (0.5 to 4.75 on a 0-5 scale), transformed with a principal component analysis, was built for calibration. The principal components were performed on BCS with multiple linear regressions. Four anatomical points had to be identified manually to normalise the pictures. Influence of two different ways of points' identification and of the picture resolution on method quality was analysed. External validation was evaluated on two additional datasets: one with cows used for calibration, but with a different stage in milking (valididem) and one with cows not used for calibration (validdiff). To fully qualify the method, the reproducibility was estimated with 6 cows using 8 3D-shapes of each cow obtained the same day. Both ways of points' identification had quite good results in terms of calibration ($R^2=1$) and differed slightly on validation quality (RMSE=0.34 vs 0.32 for validdiff). Nec3d was 2.8 times more reproducible than usual BCS ($\sigma=0.1$ vs 0.28). The lowest resolution implied a loss of reproducibility, but did not increase the error of prediction. A simplified acquisition system, implying low resolution, could therefore be developed. The error of prediction was similar for valididem and validdiff, indicating that the NEC3D is not less efficient for cows not used in the calibration set. Assessing body condition thanks to 3D shapes appears to be a promising tool which can improve phenotyping of this trait.

**Risk and prediction of aerobic-induced silage bale deterioration**

*K.H. Jungbluth[1], G. Jia[2], M. Li[2], Q. Cheng[2], J. Lin[2], Y. Sun[2], C.H. Maack[1] and W. Büscher[1]*
*[1]Institute of Agricultural Engineering, University of Bonn, Livestock Technology, Nussallee 5, 53115 Bonn, Germany, [2]China Agricultural University, Department of Information and Electrical Engineering, Qing hua dong lu 17, 100083 Beijing, China, P.R.; kjungblu@uni-bonn.de*

Conservation of grass in silage bales has been developed as an alternative to clamp silos. Engineering progress enhanced throughput and density in baling technique. The effect of reheating at the opened silo is avoided by the closed cover of every single bale. Many users are convinced of lower feedlosses and better silage quality in bales compared to clamp silos. Previous investigations have proved high silage quality in bales. Bulk density which is an important factor in bales, leads to mechanical stable shape and reduces film- and transport costs. A compact pressed bale minimises losses in case of small damages at the film surface and improves silage quality. Silage reheating is responsible for energy and nutrition losses in preserved fodder, potentially leading to deterioration of silage and endangering animals` health. It is effected by microorganisms, the fermenting products and physical parameters. The cooperation project pursues the qualitative and quantitative measurement and evaluation of the physical parameters influencing aerobic-induced reheating of silage bales. A test bench was used to measure physical parameters like density and temperature. It works with a penetrometer measuring the bale´s resistance against a cone and the temperature inside the bale. To get data about the dynamic of reheating under controlled conditions the same crop pressed into bales is ensiled in tons and glasses and observed by sensing elements and thermography. The fermenting products and the aerobic stability of the silage are tested in standard laboratory analysis. Model based prediction of reheating is one objective in the project. In the presentation the multi-sensor experimental equipment will be demonstrated. Statistical results will be illustrated by two- and three-dimensional mapping.

---



**Interrelationships between live weight, body measurements, BCS and energy balance of dairy cows**

*M. Ledinek[1] and L. Gruber[2]*
*[1]Univ. of Natural Resources and Life Science, Gregor Mendel Straße 33, 1180 Vienna, Austria, [2]AREC Raumberg-Gumpenstein, Raumberg 38, 8952 Irdning, Austria; maria.ledinek@boku.ac.at*

This paper characterizes changes of energy balance, live weight, body measurements and BCS during lactation. Live weight is predicted and the influence of stage of lactation is discussed. Data were derived from Simmental (FV100), Holstein (HF100) and crossbred dairy cows of AREC Raumberg-Gumpenstein. Parameters were measured during 11 experimental periods (63 cows each) during a whole year. Ration was forage-based with concentrate supplementation to meet individual requirements. Breed failed to influence energy balance and its development during lactation. Live weight and BCS decreased significantly with increasing genetic potential for milk performance from FV100 (730 kg, 3.57 points) to HF100 (613 kg, 2.76 points). FV100 differed significantly from all HF groups. A genetic proportion of 12.5% Simmental in Holstein cows caused a significantly higher live weight and tendentially more BCS compared to HF100. Changes of parameters during lactation were significant. In contrast to BCS, live weight started to increase while energy balance was negative. In dry period correlation coefficients between live weight and most body measurements were higher, especially the relationship to BCS. The two models for live weight prediction with three body measurements as regression variates (belly girth, heart girth, BCS or chest depth) had smallest RMSE (17.0 and 18.7 kg). Regression coefficients changed during lactation. In conclusion breeds differ especially in BCS and live weight, depending on genetic proportion and their genetic potential of milk production. BCS describes mobilisation and recovering of body reserves better than live weight. During dry period bigger belly girth, heart girth, body width and chest depth depend more on fatness than on actual size of bones compared to lactation. Furthermore the physiological stage seems to change the influence of body measurements on live weight during lactation.

**Effect of condensed tannins from legumes on nitrogen balance and ruminal fermentation in dairy cows**

*A. Grosse Brinkhaus[1,2], G. Bee[2], M. Kreuzer[1] and F. Dohme-Meier[2]*
*[1]ETH Zurich, Universitätsstrasse 2, 8092 Zurich, Switzerland, [2]Agroscope, Institute of Livestock Sciences, Route de la Tioleyre 4, 1725 Posieux, Switzerland; anja.grosse-brinkhaus@agroscope.admin.ch*

Condensed tannins (CT) in legumes are able to form complexes with proteins, which might affect ruminal fermentation processes and their intestinal availability. In a replicated 3×3 Latin square arrangement, 6 ruminally cannulated Holstein cows (36±20 d in milk) were randomly assigned to 3 treatments. All cows received a diet consisting of maize-silage, hay and concentrate. Approximately 20% of the diet were replaced by pellets of either lucerne (L), sainfoin (S; 223 g CT/kg dry matter (DM)) or birdsfoot trefoil (BT; 30 g CT /kg DM). Each of 3 consecutive experimental periods consisted of 21-d of adaptation and 7-d of collection where feed intake was recorded daily and faeces and urine were collected quantitatively. On d 2 and 5 of each collection period, ruminal fluid and blood were sampled at 0700 and 1700 h. Milk, fat and protein yield were similar (P>0.05) in all treatment groups. Daily intake of total DM and N did not (P>0.05) differ among treatments. While faecal N excretion was similar (P>0.05), urinary N excretion (P<0.05) and the proportion of urinary N of total N intake were lower (P<0.05) in S compared to L cows. Regarding these traits, BT cows did not differ (P>0.05) from S and L cows. In the evening, ruminal ammonia and blood urea concentration in S cows were 29 and 31% lower (P<0.1) than in BT and L cows, respectively. Concomitantly, lower (P<0.05) urea concentration in urine and milk were observed in S compared to BT cows with intermediate concentrations in L cows. Ruminal populations of fibrolytic bacteria Ruminococcus flavefaciens and Prevotella ssp were 26 and 27%, respectively, lower (P<0.05) in S compared to L cows, whereas BT cows did not differ from S and L cows. In conclusion, the results indicate that feeding S, other than BT, decreases ruminal crude protein degradation and metabolic stress by lowering the ammonia load.



**Using on-farm milk progesterone levels to define new fertility traits for dairy cows**

*D. Sorg, M. Wensch-Dorendorf, K. Schöpke, G. Martin and H.H. Swalve*
*Martin-Luther-University Halle-Wittenberg, Institute of Agricultural and Nutritional Sciences, Theodor-Lieser-Str. 11, 06120 Halle, Germany; diana.sorg@landw.uni-halle.de*

Common fertility traits in dairy cows (e.g. days open) are strongly influenced by management decisions. In times of declining dairy cow fertility, there is the need to define new traits reflecting the true biological fertility of the cow, such as post partum commencement of luteal activity (CLA). Our goal was to use progesterone (P4) levels in milk, measured under field conditions with an automated on-farm ELISA device, to define traits and analyse their suitability for breeding purposes in terms of genetic parameters. We used different methods (visual inspection of the P4 profile, computer algorithm) and thresholds to define CLA (week of the first P4 rise above the threshold) and PLA (proportion of luteal activity = proportion of samples above the threshold). 1,446 P4 profiles (herds 1-4) and 296 profiles (herd 5) were obtained by measuring once (herds 1-4) or twice per week (herd 5) from week 3 to 9 pp. CLA and PLA least square means differed between methods (P<0.0001). When omitting every 2$^{nd}$ sample in herd 5 (i.e. simulating once weekly measurements) mean CLA and PLA shifted (P<0.0001) from 4.8 to 5.1 weeks and from 0.33 to 0.46, respectively. A single-trait animal model with the fixed effects herd-year-season and lactation class yielded heritabilities ($h^2$) from 0.039 to 0.076 for the CLAs and from 0.129 to 0.181 for the PLAs (VCE and PEST software). In a two-trait animal model with the same effects, $h^2$ were similar, but PLAs had a high genetic correlation ($r_g$) with their according CLAs (-1.0 to -0.999). Interestingly, the highest $h^2$ of 0.245 and 0.247 were found for the animal's mean P4 value. Its rg with PLA was 0.946. New fertility parameters need to be carefully defined and standardized in order to measure exactly the same trait. Regarding the low $h^2$ of different CLAs, we suggest that the PLA or simply the animal's mean P4 could be used in the future.

**A multivariate approach to identify variables affecting the carbon footprint of dairy cattle farms**

*M.G. Serra[1], A.S. Atzori[1], M. Cellesi[1], G. Zanirato[2] and A. Cannas[1]*
*[1]University of Sassari, Dipartimento di Agraria, viale Italia 39, 07100 Sassari, Italy, [2]Filiera AQ srl, via Cadriano 27, 40127 Bologna, Italy; gabriserra@uniss.it*

The use of multivariate discriminant analysis was tested, to identify technical variables useful to distinguish among dairy cattle farms with high or low carbon footprint (CF). A detailed survey on 285 dairy cattle farms from Southern Italy was carried out and their CF was estimated using a modified TIER 2 of IPCC (2006). Farm mean CF resulted in 1.66±0.04 kg of $CO_2$eq/kg of fat and protein corrected milk (FPCM). Milk yield (range: 1,170-11,100 kg of FPCM/year per cow) explained most of the variability. Based on the pattern of the relationship between milk yield (MY) and CF, the farms were divided in 2 subsets, one below (LMY) and the other above (HMY) 5,000 kg/year of MY per cow. Within each group, CF was regressed on MY and the 89 farms included within ±SD/3 were discarded, to consider only those with a large distance from the regression predicted CF. Thus, farms with high CF (HCF; above the regression line) were separeted from those with low CF (LCF; below the regression line). By using 87 farm variables, a linear discriminant analysis (LDA) was then performed, separately for LMY and HMY farms, to identify LCF and HCF farms. The LDA assigned 2 of the 139 HCF farms to the wrong group (LCF for HMP), while all LMP farms (57) were correctly assigned by the LDA to HCF or LCF groups. A 2$^{nd}$ LDA was then performed, by using the 29 (HMP farms) and 33 (LMP farms) most important farm characteristic variables, selected on the basis of the coefficients assigned by the 1$^{st}$ LDA. In the 2$^{nd}$ LDA, 21.6% of HMP and 8.8% of LMP farms were not assigned correctly to the HCF or LCF groups, respectively. The most important discriminating variables were associated to monthly values of air temperature, dietary characteristics and herd profile. However, even the variables with lowest LDA coefficients were necessary to allow the LDA an appropriate CF estimation.



**Increase the efficiency of dairy cattle an effective tool to reduce enteric $CH_4$ emissions by Brazil**

*A.S. Cardoso, F.O. Alari and A.C. Ruggieri*
*Sao Paulo State University, Animal Science, Av. José Adriano Arrobas Martins, 801 Jaboticabal-SP, 14888-300, Brazil; abmael2@gmail.com*

Livestock are identified as major emitter of greenhouse gases (GHG), especially enteric $CH_4$. Use of vaccines, changes in animal diets and other techniques are discussed as option for reducing these emissions. However, the reduction in the herd by increasing production efficiency may represent a more effective reduction in greenhouse gas emissions. The aim of this study was to calculate how much you can reduce greenhouse gas emissions by dairy cattle in Brazil by increasing the efficiency of production. The enteric $CH_4$ emission was calculated using the IPCC factors (2006) of the current dairy herd of cows with 32.09 million and an average production of 1382 liters of milk per cow per lactation vs a herd that needed to produce the same amount of milk with the highest yield from milk producers, being 9,000 l/cow/lactation and a herd of 4.99 million cows. In this evaluation we assumed that the milk national consumption remained constant. Currently, dairy cows represent 20.61% of the total emission caused by agriculture in Brazil (13,133 Gg $CH_4$ in 2010), and this value can be reduced by increasing the efficiency of the cows, as defined, to 455.4 Gg $CH_4$ per year based on the year 2010. An increase of the efficiency up to a much higher average milk yield as mentioned would result in a reduction of 17.15% of the total emission and the percentage of the dairy herd of the total emission would be reduced to 3.47%. This represents a reduction of 2,251.7 Gg of $CH_4$ launched to the atmosphere each year. Preliminary data indicate that investing in the efficiency of dairy cattle production in Brazil may be the most effective means of reducing GHG emissions.

**Average milk yield per feeding day during extended lactations on four Danish dairy farms**

*J.O. Lehmann, L. Mogensen and T. Kristensen*
*Aarhus University – Foulum, Institute of Agroecology, Blichers Allé 20, 8830 Tjele, Denmark; jespero.lehmann@agrsci.dk*

An on-going project at Aarhus University investigates the use of extended lactations on four private farms in Denmark. The four farms have 95-160 cows and vary considerably by breed, level of milk yield and management system. They supplied 10,849 energy corrected milk (ECM) yield recordings (Jan/2007–May/2013) from 1,041 completed lactations (i.e. a new calving was recorded) with a calving interval between 13-21 months. Milk yield was recorded either six or 11 times per year. Data was split in 12 subsets by herd and parity group (1, 2 and 3+), and each subset was separately analysed with a fourth order Legendre Polynomial using a linear mixed model in R. A drying off date was not known for 468 lactations, and these were either assumed dried off seven weeks before the coming calving or if the daily kg ECM yield dropped below 15% of maximum yield. Average kg ECM yield per feeding day, which is the sum of lactating days and dry days, ranged from 18.0-34.2 when evaluated for each of 48 groups of cows (4 herds, 3 parity groups and 4 calving interval groups). Likewise, the standard deviation of these averages ranged from 0.7-4.8. Average kg ECM yield per feeding day was 30.2±4.2, 28.4±4.0, 20.9±2.8 and 20.2±2.5 for each of the four herds. The corresponding values for a 305-day lactation (without dry period) were 36.3±5.2, 34.0±4.9, 24.9±3.7 and 23.4±2.9 respectively. Average kg ECM yield per feeding day was 22.0±4.8, 26.1±5.5 and 25.7±5.2 across herds for cows in first parity, second parity and third parity or higher respectively. Average kg ECM yield per feeding day was 24.1±5.6, 23.7±5.2, 24.7±5.4 and 25.3±5.4 for cows with a calving interval between 13-15 months, 15-17 months, 17-19 months and 19-21 months respectively. These preliminary results suggest that increasing the calving interval beyond 12 months and thus use extended lactations were possible on these farms with similar yields per feeding day across calving interval groups.

---



**Shortening of the dry period did not prevent hypocalcemia at calving**

*C. Kronqvist*
*Swedish University of Agricultural Sciences, Department of Animal Nutrition and Management, Kungsängen Research Centre, 75323 Uppsala, Sweden; cecilia.kronqvist@slu.se*

Hypocalcemia and milk fever is common in dairy cows and occurs at the onset of lactation, when the calcium requirements for milk production increase. The aim of this study was to investigate if shortening of the dry period, which earlier has been shown to ameliorate metabolic problems and also decrease the colostrum production after calving, could have an impact on the extent of hypocalcemia after calving. A blood sample was taken in the week before calving and 18 to 30 hours after calving on 30 cows, of which 14 had had a normal dry period of 8 weeks and 16 had had a shortened dry period of 4 weeks. The cows were entering their second (n=20), third (n=5) or fourth lactation (n=5). Calcium level in plasma was analyzed. The cows were milked twice daily, and total production of colostrum during the first 3 milkings after calving was measured in the cows. Data were analyzed in a mixed model, with cow as a random factor and sampling time (before or after), dry period length (normal or short), lactation number (1 to 3), breed (Swedish Red and Swedish Holstein) and the interaction between dry period length and sampling time included. For milk yield, time and the interactions with time was not included in the model. The total production of colostrum during the first 3 milkings differed between the groups, with 21.1 L from the cows with normal dry period and 14.2 L from the cows with the shortened dry period (P=0.004). However, the extent of hypocalcemia after calving was similar between the groups, with calcium levels of 1.73 mM in cows with normal dry period and 1.80 mM in cows with shortened dry period (P=0.42). The conclusion was that shortening of the dry period decreased the production of colostrum. However, this did not result in a decrease in the extent of hypocalcemia after calving.

**Reducing concentrate supplementation in an alpine low-input system: response of two dairy cow types**

*M. Horn[1], A. Steinwidder[2], J. Gasteiner[2], M. Vestergaard[3] and W. Zollitsch[1]*
*[1]BOKU-University of Natural Resources and Life Sciences Vienna, Gregor-Mendel-Straße 33, 1180, Austria, [2]AREC Raumberg-Gumpenstein, Trautenfels 15, 8951, Austria, [3]Aarhus University, Blichers Allé 20, 8830 Tjele, Denmark; marco.horn@boku.ac.at*

The objective of the present study was to investigate the impact of concentrate supplementation level (SL) on feed intake, productivity, body condition, reproductive performance and metabolic parameters of two different dairy cow types (CT) at the onset of lactation (first 49 days in milk). The two CT compared were conventional Brown Swiss (BS, n=19 lactations, Ø body weight=619 kg) and a special strain of Holstein Friesian (HFL, n=21 lactations, Ø body weight=557 kg). The latter was for decades selected under low-input conditions and was primarily bred for lifetime performance and fitness. Both CT were assigned to either low (L) or high (H) SL. The dataset was analysed using a mixed model. Daily concentrate intakes were 2.9 and 5.6 kg for groups L and H, resp. ($P \leq 0.001$). Total dry matter intake was higher in group H ($P \leq 0.001$), while the effect of CT or the interaction between CT and SL were not significant (15.8, 18.3, 15.5 and 16.6 kg/cow/day for BS(L), BS(H), HFL(L) and HFL(H), resp.). Neither CT nor SL significantly influenced daily energy-corrected-milk yield which was 23.4, 26.2, 25.5 and 25.0 kg/cow/day for BS(L), BS(H), HFL(L) and HFL(H), resp. Compared to BS, HFL tended to have a lower BCS ($P=0.060$). However, no negative effect of reducing concentrate supplementation was observed neither for BCS (3.17, 3.01, 2.99, 2.83 for BS(L), BS(H), HFL(L) and HFL(H), resp.), nor for submission rates until 49 days in milk (50, 36, 63, 36% for BS(L), BS(H), HFL(L) and HFL(H), resp.). Blood metabolites appeared to be driven by CT, rather than by SL. The results indicate that the reduction of concentrate supplementation does not necessarily have negative impacts on BCS, reproductive performance or metabolic parameters at the onset of lactation.



**The main determinants of economic efficiency in suckler cow herds**

*M. Michaličková[1,2], Z. Krupová[3], E. Krupa[3], L. Vostrý[1,3] and P. Polák[2]*
*[1]Czech University of Life Science, Kamýcká 129, 165 21 Prague 6, Czech Republic, [2]NAFC – Research Institute for Animal Production Nitra, Hlohovecká 2, 951 41 Lužianky, Slovak Republic, [3]Institute of Animal Science, P.O. Box 1, 104 01 Prague, Czech Republic; polak@vuzv.sk*

The profitability without including subsidies and its determinants for suckler cow herds during the years 2008 to 2012 were analysed. Influence of the basic production, reproduction and economic parameters on the economic result per cow and year were quantified using the linear regression model. Profit was not reached over the whole time period in the analysed farms. The best economic result (-470.01 €/cow/year) achieved in 2008 was determined mainly by the lower level of costs per feeding day (FD; -48%) along with the higher value of average daily gain of calves (+18%) compared to the rest of the analyzed period. Based on the regression analysis, costs per FD (of cows and calves) and cows' fertility were found as the main factors of economic result in suckler cow herds. Contrary, average daily gain of calves, calving interval and age at first calving were of no significant importance on profit in analysed herds. Value of profit declined by 407 €/cow/year when value of cost per FD of suckler cow increased by 1 €. A similar situation was found for cost per FD of calves. Increasing of these cost by 1 € declined the profitability by 266 €/cow/year. Positive and statistically significant effect of the fertility on the profitability value was found in the applied regression model. Increase of the number of live born calves per 100 suckler cows by 1 calf improved the profit by 19 €/cow/year. This is related to the specific system of suckler cattle farming, where the calf is the only product and revenues are dependent on the suckler cows' fertility and on the mortality of calves. Improvement of the significant indicators influenced the profitability indirectly through the higher number of live born calves and the cheaper rearing of heifers for replacement.

**Comparing agreement in methods of lameness detection in Holstein Friesian dairy cattle**

*A.V. Sawran[1], G.E. Pollott[1], R. Weller[2] and T. Pfau[2]*
*[1]The Royal Veterinary College, PPH, Royal College Street, Camden, NW10TU, London, United Kingdom, [2]The Royal Veterinary College, Hawkshead Campus, Hawkshead Lane, South Mimms, AL97TA, Hertfordshire, United Kingdom; asawran@rvc.ac.uk*

This research aims to ensure prolonged soundness of dairy cows by examining the agreement between methods of lameness detection. Lameness is a considerable welfare and economic issue and numbers of UK milk producers are decreasing. With no let-up in demand for milk, it is important to determine the most efficient, cost effective means of lameness detection; ensuring viability in a farm setting and little disruption to the daily routine. This is relevant when considering that human-cow interaction may be decreased as larger dairy herds become commonplace. Comparison of subjective scoring data (lesion and gait scoring) with force plate data and distal limb radiography may help determine links between ground reaction forces and clinical lameness, making for a successful detection system. To compare agreement between visual lameness score (VLS) and force plate gait analysis, 38 lactating Holstein-Friesian cows were walked over 5 AMTI 3-axis force plates twice daily over 10 months. Focus was on the change in the centre of pressure during impact phase of the stride (CCOP), stance time (ST) and load rate (LR). Data were compared to VLS as well as hindlimb lesion score (LS) and hindlimb radiographs, with scores based on the severity of bone changes (RS). It was hypothesised that cows with high VLS or LS would have significantly different RS, CCOP, ST and LR values to those with low VLS or LS. Data were analysed using ASReml. There were significant correlations (P<0.05) between RS-LR and significant inverse relationships between RS-ST and RS-CCOP. This suggests animals with radiographic abnormalities take longer to load affected limbs and bear weight for less time than on radiographically normal limbs. It follows that CCOP is smaller in radiographically abnormal limbs due to unwillingness to bear weight normally on the affected digit.

---



**Proinflammatory cytokines expression in the blood of goats infected with small ruminant lentiviruse**

*J. Jarczak[1], J. Kaba[2], M. Czopowicz[2], J. Krzyżewski[1], L. Zwierzchowski[1] and E. Bagnicka[1]*
*[1]Institute of Genetics and Animal Breeding, Jastrzębiec, Postępu 36a Street, 05-552 Magdalenka, Poland, [2]Warsaw University of Life Sciences, Faculty of Veterinary Medicine, Nowoursynowska Street, 166, 02-787 Warszawa, Poland; j.jarczak@ighz.pl*

The aim of study was to evaluate the impact of caprine arthritis-encephalitis infection on the expression of immune system genes in goats' blood leukocytes. Twenty six dairy goats were divided into two groups: control – free from virus infection (n=13) and experimental – infected with CAE virus (n=13). The blood samples were taken four times, on 7, 30, 120 and 240 day of lactation. Total RNA was isolated and expressions of IL1α, IL1β, INFα, INFβ, INFγ, TNF, IL-6, IL-10, IL-16 and IL-18 genes were measured with Real-Time PCR using protein zeta gene (YWHAZ) as a reference. The expression of the IL-2, IL-4 and IL-12 genes were not found indicating that they are not constitutively expressed in the blood and do not participate in the defense of the body against CAE virus infection. The expression of TNF-α was 2 fold higher in the blood of animals infected with CAEV (P=0.05). Expression level of IL1α, IL1β and INFα genes was lower in the group of infected animals. Expression level of IL-10 IL-16 and IL-18 did not differ within the groups. The high, positive correlations within the interferon's family were also found. Furthermore, IL-16 showed also high, but negative correlation with all studied interferons. Obtained results can prove that TNF-α is the only one gene from the group of cytokines, involved in the response against the CAE virus in the blood. The presence of CAE virus in the body can influence the level of transcripts of selected cytokines causing their decreased expression.

**Identification of beta-lactoglobulin and kappa-casein genotypes using PCR-RFLP in Holstein cattle**

*Y. Gedik, M.A. Yildiz and O. Kavuncu*
*Ankara University Agriculture Faculty, Animal Science Biometry & Genetics Major, 06110 Diskapi, Ankara, Turkey; ygedik@agri.ankara.edu.tr*

Milk protein polymorphisms have been studied intensively because of associations between milk protein genotypes and economically important traits in dairy cattle. The objective of the present master thesis was to estimate the gene and genotype frequencies by the determined β-lactoglobulin and κ-casein genotypes. Blood samples from 167 Holstein cattle (78 Bala and 89 Ceylanpınar populations) were genotyped for β-lactoglobulin and κ-casein using by PCR-RFLP method. The isolation of DNA from blood samples was performed with salting-out procedures. The amplification of β-lactoglobulin and κ-casein genes was obtained by PCR. PCR products were digested with restriction enzymes HphI and HindIII to determine the genotypes for respectively β-lactoglobulin and κ-casein. The digestion products were properly resolved in the ethidium bromide-stained 2% agarose gels electrophoresis. Counting the number of genes was used to estimate gene and genotypic frequencies of β-lactoglobulin and κ-casein. The $\chi 2$ test was used to check whether the populations were in Hardy-Weinberg equilibrium or not. β-lactoglobulin A gene frequencies were calculated 0.51±0.04 for Bala and 0.34±0.03 for Ceylanpınar populations. $\chi^2$ tests revealed that both populations were in Hardy-Weinberg equilibrium. κ-casein A gene frequencies were calculated 0.80±0.03 for Bala and 0.84±0.02 for Ceylanpınar populations. $\chi^2$ tests revealed that Bala populations were in Hardy-Weinberg equilibrium while Ceylanpınar was not.



**Effect of propylene glycol on the liver transcriptome profile of dairy cows before calving**

*M. Ostrowska, P. Lisowski, B. Zelazowska and L. Zwierzchowski*
*Institute of Genetics and Animal Breeding PAS, Jastrzebiec, st. Postepu 36A, 05-552 Magdalenka, Poland; malgorzata.ostrowska85@gmail.com*

During periparturition period, cows require adequate nutritional intake to cover all energy requirements associated with the calving, start of milk and to avoid metabolic disorders such as postpartum retention or ketosis. Improperly conducted farming, causes huge economic losses resulting from the cost of treatment and decreased milk yield of cows. In order to eliminate the negative factors, a large number of feed additives (propylene glycol PG, or glycerol) are used as a source of glucose for the organism. The effect of PG on the expression of genes in the cow's liver, on the 7$^{th}$ day before parturition was studied. Cows (12) were divided into 2 groups (6 each): control W-450, (fed 450 ml of water) and PG-450 (450 ml of PG), from 21$^{st}$ day before up to 28$^{th}$ day after calving, were used for this study. Analyses were carried out by one colour microarray Affymetrix Bovine Gene 1.1 ST Array Strip. The functional interpretation of microarray data was performed using the algorithms: DAVID, IPA, GO and KEGG. Comparison of the liver transcriptome of the PG-450 vs W-450 cows, on the 7$^{th}$ day before calving identified 367 genes differently expressed between both groups, logFC>0.5 and $P<0.05$. Functional analysis of genes has enabled the identification of important biochemical pathways and gene networks such as: glycosaminoglycans biosynthesis: XYLT2, DSE, DNA replication, recombination and repair: ITGB3BP, RHOBTB3, TFIIH, nucleic acids metabolism (ENPP1) and vascular remodelling (VEGFA). The obtained results indicate that the administration of the glucogenic supplement (PG-450) into cows before calving causes cascading changes in expression of many genes, including genes of glycosaminoglycans biosynthesis, metabolism of nucleic acids, which are undoubtedly related to the adaptation of the cow's liver to calving and milk production. This study was financially supported within NCN grants: N N311 075 039, 2012/05/B/NZ9/03425

**Four methods for describing dairy cow milk yield during extended lactations on private farms**

*J.O. Lehmann, L. Mogensen and T. Kristensen*
*Aarhus University, Foulum, Institute of Agroecology, Blichers Allé 20, 8830 Tjele, Denmark; jespero.lehmann@agrsci.dk*

The use of extended lactations in high-yielding dairy production may help mitigate greenhouse gas (GHG) emissions per kg milk produced. This is a result of fewer replacement heifers and fewer dry days per cow per year, and thus a reduced total herd feed use. Total cow milk yield and the level of yield towards the end of the lactation are crucial factors for quantifying the performance of each cow during an extended lactation, and thus also for the effect on GHG emissions. An on-going project at Aarhus University investigates the use of extended lactations on four private farms in Denmark. Each farm has 95-160 cows and vary considerably by breed, level of milk yield and management system. A data set from these farms with 10,849 energy corrected milk yield (ECM) recordings (Jan/2007-May/2013) from 1,041 completed lactations with a calving interval between 13-21 months was split into 12 subsets by herd and parity (1, 2 and 3+). Mean number of observations per lactation ranged from 7.1-12.4. A range of equations exist in the literature that models milk yield over time for traditional lactations of less than 12 months, and three of these, Wood's, Wilmink's and Dijkstra's, were fitted to data using nonlinear mixed models in R. No convergence could be achieved for one subset with Wilmink's and two subsets with Dijkstra's. In addition, a fourth order Legendre Polynomial (LegP) was fitted using a linear mixed model, and the objective was to rank the ability of these four methods to describe ECM yield during extended lactations. Models were evaluated with Akaike's Information Criterion (AIC), Bayesian Information Criterion (BIC), Root Mean Squared Prediction Error (RMSPE), RMSPE for the last 25% observations and the numerical difference between these two RMSPE values. The LegP ranked number one for all five criteria as it achieved the best fit for 7, 6, 11, 12 and 8 out of 12 subsets by criteria respectively. RMSPE of the LegP ranged from 2.10-3.15.

---



**Investigation of the inverdale (FecXI) mutation in Turkish Awassi sheep**

*Y. Gedik, H. Meydan, M.A. Yildiz and O. Kavuncu*
*Ankara University Agriculture Faculty, Animal Science Biometry & Genetics Major, 06110 Diskapi, Ankara, Turkey; ygedik@agri.ankara.edu.tr*

Since 1980 there has been increasing interest in the identification and utilisation of major genes for prolificacy in sheep. Mutations that increase ovulation rate have been discovered in the 'BMPR-1B', 'BMP15' and 'GDF9' genes, and the others are known to exist from the expressed inheritance patterns although the mutations have not yet been located. In the case of BMP15, four different mutations have been discovered but each produce the same phenotype. The modes of inheritance of the BMP15 gene includes X-linked over dominant genes with infertility in homozygous females. In this study 172 prolific Awassi sheep were tested for the presence of the $FecX^I$ mutation of BMP15. None of the sheep sampled carried the Inverdale $FecX^I$ mutation in BMP15 gene.

## Grassland for ruminant husbandry, international perspectives and globalisation

*F. Taube*
*University Kiel, Germany, Crop and Plant Breeding Department, Herrmann-Rodewald-Straße 9, 24118 Kiel, Germany; ftaube@gfo.uni-kiel.de*

The concept of sustainable intensification has recently been developed to raise productivity (as distinct from increasing volume of production) while reducing environmental impacts. This means increasing yields per unit of inputs (including nutrients, water, energy capital and land) as well as per unit of undesirable outputs (such as greenhouse gas emissions, water pollution or loss of biodiversity). It is thus helpful to understand 'intensification' as referring to 'environmental factor productivity' or 'eco-efficiency'. Worldwide, grassland is the most important agroecosystem delivering ecosystem services ranging from feed supply for ruminants and soil carbon storage to habitats of biodiversity. However, worldwide, grassland is under threat due to intensified land use and land-use changes from grass to arable. In this presentation, we (1) highlight ecosystem services of selected grassland biomes abroad of Europe, (2) show evidence of sustainable as well as non-sustainable intensification options in these grassland biomes linked to European dairy systems by exports of agricultural commodities (e.g. soy) and (3) derive research strategies for north-west European grassland research and management to match sustainable intensification strategies for the grassland-based dairy industry.



## Sustainability of grazing: effect of grazing on economy, environment and society

*A. Van Den Pol-Van Dasselaar*
*Wageningen UR Livestock Research, P.O. Box 65, 8200 AB Lelystad, the Netherlands; agnes.vandenpol@wur.nl*

In northwest Europe, grazing of ruminants is a matter of public concern. Main reasons for this are animal welfare, biodiversity and the positive image of grazing. However, grazing also has negative effects, e.g. on nitrate leaching and on gross grass yield. Furthermore, the effects of grazing are dependent on the farm situation, i.e. farm characteristics and farm management. The aim of this paper is to provide an overview on the sustainability of grazing, from (1) a social point of view, (2) an ecological point of view and (3) an economic point of view. Recent research results will be used to illustrate the different aspects of sustainability of grazing. A European-wide questionnaire on the importance of grasslands for sustainability, issued by the project Multisward (FP7-244983), clearly showed that the importance of grazing is strongly recognised by all stakeholder groups. They in fact ranked grazing the highest out of a number of functions of grasslands. The working group 'Grazing' of the European Grassland Federation has shown advantages and disadvantages of grazing. Furthermore, the Working Group showed that the number of grazing dairy cattle in Northwest Europe is decreasing and that society is concerned about this trend. The farm situation is a crucial factor for the impact of grazing. Certain farm conditions are economically seen less favourable for grazing: large herds, automatic milking and high milk yields per cow. These conditions have been studied with the whole farm model DairyWise and the environmental and economic consequences will be presented. Results from the project Autograssmilk (FP7-SME-314879) provide further insight into the possibilities and constraints of the combination of automatic milking and grazing. Finally, the Dutch program 'Grass and Grazing' studies competitiveness of future milk produced from grazing from a socio-economic and ecological perspective. The impact of grazing systems on the various aspects of sustainability is explored using modelling techniques and empirical data from experiments. Based on all results, we conclude that grazing is not a black and white story.

**The ideal dairy cow for pasture-based production systems**

*P. Dillon, D.P. Berry and P. French*
*Teagasc, Animal & Grassland Research and Innovation Centre, Fermoy, Co. Cork, Ireland; pat.dillon@teagasc.ie*

Grazing production systems are often characterised as generally lower input cost but also lower output per animal or per unit area compared to confinement production systems. Animal characteristics pertinent to profitable grazing production systems are: (1) propensity for high grass dry matter intake; (2) ability to achieve high output per unit area from grazed grass; (3) extremely good fertility and longevity (i.e. both male and female); (4) healthy; (5) easy care and docile, and arguably more importantly in the future; and (6) be robust to fluctuations in feed quality and quantity without compromising output. Exploitable genetic variation clearly exists in all of these traits. The Irish dairy cow breeding objective, the economic breeding index (EBI), is simultaneously selecting on all six characteristics. Selection on the EBI has resulted in genetic gain in all the economically important traits and the impact has been demonstrated as increased profit under research and commercial herd conditions. The level of crossbreeding, guided by sire selection on EBI, is also increasing in Ireland resulting in further improvements in profit. Crossbreeding is expected to intensify further with the availability of sexed semen; recent advancements in semen sexing technologies have reduced its impact on conception rate. The benefit of crossbreeding however will only materialise if elite genetics of both parental breeds, on EBI, are available. Emerging technologies are, and will continue to advance genetic gain further but must only be used where the fundamentals of a successful and sustainable breeding strategy (i.e. pertinent genetic evaluations and breeding objective) are already in place and validated.



**Dairy system sustainability in relation to access to grazing: a case study**

*V. Decruyenaere[1], S. Herremans[1,2], M. Visser[2], A. Grignard[1], D. Jamar[1], S. Hennart[1] and D. Stilmant[1]*
*[1]Walloon agricultural Research Centre, 8, rue de Liroux, B-5030 Gembloux, Belgium, [2]Université Libre de Bruxelles, Landscape Ecology and Plant Production Systems Unit, 50 Avenue Franklin Roosevelt, 1050 Brussels, Belgium; d.stilmant@cra.wallonie.be*

Extension of the zero-grazing management scheme questions sustainability of dairy production. The aim of this work was to compare, during two seasons, the technico-economical and environmental performances of two experimental dairy herds with similar genetic potential. The first herd has full access to grazed grasslands during summer time while the second remains in the cowshed. Management scheme didn't significantly impact dairy production, neither in quantity nor in quality. It only impacts the distribution of the production across the year. Indeed, from May till July grazing cows produced, on average, significantly more milk than zero-grazing ones (22.7 vs 19.7 kg/d/cow; P=0.02), while the opposite was true from November till January (17.8 vs 20.4 kg/d/cow; P=0.03). From economical point of view, results underlined an increase of all the production costs (feeding, bedding and reproduction costs), with an average increase of 30%, across both years, when shifting from a grazing to a zero-grazing management scheme (22.8±4.0 and 29.7±4.5 € per 100 kg of standardised milk). From environmental point of view, the zero-grazing system had better nitrogen balances with, on average, across both years, 112 against 131 kg N/ha. In term of climate footprint, the grazing system led to lower greenhouse gases emissions than the zero-grazing system due to the high dependency of the zero-grazing system on fertiliser and feedstuff inputs. This result is of value whatever the functional unit mobilised with 8,550±595 and 10,700±208 kg $CO_2$ eq/ha and 1,140±144 and 1,350±76 kg $CO_2$ eq/ton of milk, for grazing and zero-grazing systems, respectively. Based on these results, grazing based systems appeared to be more sustainable than zero-grazing ones.

**Carbon footprint of a typical grass-based beef production system in Chile**

*A. Hänsch and R. Larraín*
*Pontificia Universidad Católica de Chile, Facultad de Agronomía e Ingeniería Forestal, Ciencias Animales, Vicuña Mackenna 4860, Macul, Chile; larrain@uc.cl*

The objectives of this study were to develop a model to estimate carbon footprint for beef producers in Chile, following protocol PAS2050 from 2008, and use it to estimate carbon footprint in a hypothetical farm using average production conditions at the VIII region of Chile (approximate location 36°26'38.60"S, 71°55'14.48"O). Carbon footprint was estimated including calf production and rearing on a pasture based system, and finishing in feedlot for 120 days. Emissions were accounted until the finished animals leaved the farm to the slaughterhouse (from cradle to farm gate) at 18 months of age. All feed was produced inside the farm and the herd was from a British beef breed with a slaughter weight of 475 kg for steers. The breeding herd and stockers were maintained on a pasture-based system and the finishing period was under confinement with corn silage and corn grain as main ingredients. The standard unit for the estimations was 1 kg of carcass, using 56% carcass yield for finished animals and 50% for cull cows. The estimated carbon footprint was 17.33 kg $CO_2$e/kg carcass. The distribution of the emissions among the different production phases was 73.2% for cow-calf system, 19.8% for stockers and 7.0% for the finishing period. The distribution of $CO_2$e emissions from the farm was 74.5% from $CH_4$ from enteric fermentation and manure management, 20.8% from energy use, 2.7% from $NO_2$ from manure management and soil management and 2.0% from use of urea and limestone. The carbon footprint was within the range of values estimated by other authors in similar systems in other countries (16 to 27 kg $CO_2$e/kg carcass) and below the estimated footprint for brazilian beef under more extensive conditions (28 kg $CO_2$e/kg carcass without considering land use change). We concluded that carbon footprint of beef under common production conditions of south-central Chile are in the lower range as observed in similar systems in other countries.



**Pasture-based dairy production systems in the United States**

*S.P. Washburn*
*North Carolina State University, Animal Science, Box 7621, Raleigh, NC 27695-7621, USA; swashbur@ncsu.edu*

Pasture-based systems nearly disappeared in the last half of the 20$^{th}$ century, replaced by non-pasture systems with larger herds of mostly Holstein cows being fed total mixed rations (TMR). Non-pasture systems dominate US production but pasture-based dairy systems including organic have grown in the past 25 years. Rising costs for energy, concentrates, and large investments in equipment and facilities needed in confinement systems contribute to renewed interest in pasture. Pasture-based systems can vary from where pasture is used as the primary source of nutrients to systems in which pasture is only used as supplemental forage for cattle primarily fed a TMR. In the US, use of pasture varies because of diverse environments with many species of annual and perennial forages and also because of availability of relatively economical supplemental concentrates and by-product feeds. Because of such diverse systems, production per cow also varies based on feeding regimens and differing genetic strategies including use of crossbreeding. Dairy graziers may match pasture resources with nutritional needs such that seasonal breeding and calving may be advantageous, thereby necessitating emphasis on cow fertility. Seasonal breeding and calving is not as widespread in the US as in areas where pasture is usually the major source of nutrients but calving for US pasture-based herds tends to be late winter or spring in cooler regions versus fall or early winter in warmer climates. Production per cow may often be less in pasture-based systems but such systems can be competitive because of lower costs for feeding, health, cow turnover, and investments in equipment and facilities. Pasture-based systems may also benefit from emerging markets. We have documented differences in milk composition and flavor characteristics from cows consuming pasture compared to those fed a TMR. Organic dairy markets are growing and increased awareness of favorable fatty acid composition may lead to further market opportunities for dairy products from pasture-based cows.

**Grass based pasture dairy systems in the Mid-West: an overview**

*V.J. Haugen*
*University of Wisconsin-Extension, Department of Agriculture and Life Sciences, Cooperative Extension, 225 North Beaumont Rd, Suite 240, Prairie du Chien, WI 53821, USA; vance.haugen@ces.uwex.edu*

Grass based dairy systems have always been a part of US agriculture. Economic difficulties in the 1980's and new electric fence technology gave rise to an alternative path for dairy producers to achieve economic success and caused large numbers of dairy farmers to explore grass based dairy farming as a sole or hybrid system with confinement dairy farming. This style of dairy farming has helped numerous farmers compete in the difficult changes in dairy polices here in the Midwest United states. An overview of the production methods for successful grass based dairying will be presented along with economic comparisons for conventional, grass based and organic dairy systems.



**Grassland use in South America and cow-calf system under pastoral conditions in Uruguay as a case**

*G. Quintans and G. Banchero*
*National Institute for Agricultural Research, Beef and wool, Ruta 8 km 281, 33000, Uruguay; gquintans@tyt.inia.org.uy*

In South America there is an important area of natural grassland known as 'Pampa Biome', occupying more than 70 million of ha distributed through Argentina, Uruguay and south of Brazil. In this area, the most important agricultural activity is livestock production, with emphasis on beef cattle. Most of the beef production systems are developed under extensive conditions, on native pastures and open sky. Uruguay with only 176,000 km$^2$ contributes to this with 12 million ha of native pastures and 11 million beef cattle. Around 75% of farmers carry out cow-calves production systems. Native pastures are the main component of the beef cow´s diet, with high variation in productivity during the year and between years, since they are highly dependent on the weather conditions. Consequently, the pattern of food intake is not constant and farmers have to cope with this variability to maintain a moderate to high level of productivity. For that, different management decisions should be taken, related to i- the animals: managing of the stocking rate, strategically chosen periods of supplementation with concentrates, application of suckling control technologies like temporary calf removal or early weaning, or management related to ii- the pastures: incorporation of improved forage through sowing legumes or fertilization of native pastures. It is important that most of the native grass species are of summer growth, so winter is the season with the lowest productivity, in which the pasture daily growth rate is between 0 and 5 kg of dry matter (DM) compared to 4 to 12 in summer. The level of crude protein depends on the season of the year: 70-80 g/kg of DM in winter and 100-120 g/kg of DM in spring. The average digestibility of the DM is low (40-50%). The height of the sward, that predicts forage availability, is being used by farmers to manage the stocking rate in each paddock. Also, nutrients requirements are taken into account when food resources must be assigned within different animal categories.

**Enhancing the provision of ecosystem services by cattle grazing on a ski station**

*I. Casasús, A. Sanz and J.A. Rodríguez-Sánchez*
*CITA Aragón, Avda. Montañana 930, 50059 Zaragoza, Spain; icasasus@aragon.es*

Ecological intensification of farming systems implies the combination of technically efficient production and the provision of positive impacts on the environment. Livestock can be used as a tool for landscape management, sometimes even addressing the needs of other economic activities using the same territory. This is the case of ski stations located on alpine pastures used by livestock in the grazing season. A study was conducted in a Pyrenean ski resort (Northern Spain), with the objective of analysing current use of pastures by cattle, providing management recommendations if needed, and determining the willingness of farmers to adopt them. The 297-ha ski station was grazed by a communal herd of 314 cows during the summer and autumn. The patterns of space use were studied throughout the grazing season and entered into a Geographic Information System, including spatial information about pastoral value, altitude, slope, exposure and distance to tracks, infrastructures, water and salt areas. Cattle grazed on 64% of the resort, preference for the different sites and vegetation types depending on pasture quality or physical aspects. Cows preferred areas of higher pastoral value, lower slope and altitude, and closer to salt areas and infrastructures than the rejected ones. Some vegetation communities were preferred while others were little grazed or avoided, with stocking rates varying through the grazing period. Recommendations were made for a more adequate livestock distribution, suggesting modifications of temporal and spatial management or provision of infrastructures (fences, salt points) to ensure a proper use of each pasture type, avoiding non-grazed stubble that would compromise the stability of the snowpack in the winter. Farmers considered that the ski station was beneficial for the economy of their valley and that reciprocally it profited from livestock grazing. Thus, they were prone to implement the suggested management correction measures in order to strengthen the synergies between both activities.

---



**Comparison of methods for estimating herbage intake in grazing dairy cows**

*A.L.F. Hellwing, P. Lund, M.R. Weisbjerg, F.W. Oudshoorn, L. Munksgaard and T. Kristensen*
*Aarhus University, AU Foulum, P.O. Box 50, 8830 Tjele, Denmark; annelouise.hellwing@agrsci.dk*

Estimation of herbage intake is a challenge under both practical and experimental conditions. The aim of this study was to compare methods for estimating herbage intake. Twenty cows were grazing seven hours daily (8:00-15:00) on spring and on autumn pastures. The experiment lasted 16 days in each season, and the herbage intake was estimated twice during each season. In order to generate variation the cows were allocated to low or high stocking rate. Cows were fed a mixed ration based on maize silage in the barn during night. Herbage intake was estimated with nine different methods: (1) animal performance; (2) intake capacity; (3) content of $^{13}C$ in faeces and diet; (4) $^{13}C$ in faeces and diet, and with assumption of $^{13}C$ discrimination in the digestive tract; (5) $^{13}C$ in faeces and diet, with assumption of $^{13}C$ discrimination and in combination with *in vitro* organic matter digestibility (IVOM); (6) simultaneous use of two internal markers (ingestible neutral detergent fibre (INDF) and acid detergent lignin (ADL)); (7) titanium oxide in combination with IVOM; (8) titanium oxide in combination with INDF; and (9) titanium oxide in combination with ADL. Furthermore, grazing time of the individual cows was recorded. The cows spent in average 50% and 40% of the time on grazing activities during spring and autumn, respectively. Estimated average herbage dry matter intake varied from 2.2 kg DM for method 6 to 7.6 kg DM for methods 3 and 4. There was a low correlation between the different methods, but observed grazing time had a high correlation to several of the methods. It was concluded that methods based on animal performance, $^{13}C$ including discrimination factor, $^{13}C$ including discrimination factor and in combination with IVOM, and titanium oxide in combination IVOM gave realistic estimates of the herbage intake. Furthermore, it was concluded that the best method to apply under farm conditions would be the traditional animal performance method.

**Nutritional management of transhumant sheep and goats at the region of Thessaly-Greece**

*A. Siasiou[1], V. Michas[2], I. Mitsopoulos[2], V. Bampidis[2], S. Kiritsi[2], V. Skapetas[2] and V. Lagka[2]*
*[1]Democritus University, Department of Agricultural Development, Pantazidou 93, 58200 Orestiada, Greece, [2]Alexander Technological Educational Institute, Department of Agricultural Technology, Sindos Thessaloniki, 57400 Sindos, Greece; lagka@ap.teithe.gr*

Transhumance is a semi-extensive production system linked with the annual movement of livestock in order to cope with the seasonality of climate conditions and to exploit efficiently the natural resources. The transhumant sheep and goat farming sector in Greece is still active in the most of its regions. The aim of this paper is to examine the feeding practices of the system. For this purpose data about nutritional management from a representative sample of transhumant herders from the Region of Thessaly were collected. Consequently, data were analyzed according to supplementary feed components, dietary levels in dry matter (DY), metabolized energy (ME) and digestible crude protein (DCP) as well as nutritional requirements of the animals analogous to their productive stage and productivity. Rations are calculated by linear programming. The results of the survey show elevate pasture, forbs and browse as the primary source of nutrients for transhumant sheep and goats, especially during spring and summer. Furthermore, they highlight the effort of producers to cover more efficiently nutritional needs of animals, mainly during late pregnancy and the lactation period and less during the mating period, by providing conserved forage, supplementary of grazing, crop residues, and/or concentrates. These concentrated feeds are used to balance the ration that fulfills daily needs according to the productive stage of the animals. This paper is part of the project 'The dynamics of the transhumant sheep and goat farming system in Greece. Influences on biodiversity' which is co-funded by the European Union (European Social Fund) through the Action 'THALIS'. The authors also acknowledge the invaluable contribution of Dr. Zaphiris Abas who died in a car accident on 28 December 2013.

---



**Different rotational stocking management on *Brachiaria brizantha* structural characteristics**

*M.L.P. Lima, E.G. Ribeiro, F.F. Simili, L.C. Roma Junior, J.M.C. Silveira, A. Giacomini and A.E. Versesi Filho*
*Department of Agriculture, São Paulo State Government, Rod Carlos Tonani, km 94, Sertãozinho SP, 14160-970, Brazil; lucia.plima@hotmail.com*

The present study evaluates the structural characteristics of *Brachiaria brizantha* (Hochst. ex A. Rich.) swards, all year long, in Central Brazil. Three stocking systems were tested, in which cattle were allowed to graze: (1) cycles of 28 days ($T_{28d}$); or (2) when the sward reached a pre-grazing height of 30 cm ($T_{0.3m}$) and irrigated when the sward reached a pre-grazing height of 30 cm (T-irrig$_{0.3m}$). The variables canopy height (CH), light interception (LI) and leaf area index (LAI) were measured in the pre-grazing period using the AccuPAR LP80 system (DECAGON Devices). Eight grazing cycles with sward recovery period followed by 2 days of grazing were evaluated using a completely randomized block design and split plot design, with 2 repetitions per stocking system and 4 paddocks per system. The interaction between grazing system and season affected CH (P=0.0205), LI (P=0.0389) and LAI (P=0.0487). During the fall, winter and spring, the $T_{0.3m}$ rotational system presented lower plant height, without achieving the objective of 30 cm of pre-grazing height. The CH results were 28.9, 21.0 and 27.9 cm, during fall; 20.9, 19.1 and 21.6 cm, in winter; 22.7, 19.7 and 22.2 cm, in spring; and, 36.1, 30.4 and 31.2 cm, in summer for the $T_{28d}$,$T_{0.3m}$ and T-irrig$_{0.3m}$ stocking systems, respectively. The LI was higher ($P<0.001$) forT-irrig$_{0.3m}$ during winter. The LI results were 92, 84 and 93% during fall; 85, 85 and 91%, in winter; 86, 84 and 81%, in spring; and, 94, 94 and 94%, in summer for $T_{28d}$, $T_{0.3m}$ and T-irrig$_{0.3m}$ stocking systems, respectively. LAI ($P<0.001$) was lower for $T_{28d}$, and $T_{0.3m}$ during winter and for $T_{0.3m}$ during the fall. respectively. The LAI results were 5.9, 4.3 and 5.7, during fall; 3.7, 3.8 and 4.6, in winter; 4.4, 4.1 and 4.5, in spring; and, 7.2, 6.7 and 7.2 in summer, for $T_{28d}$, $T_{0.3m}$ and T-irrig$_{0.3m}$, respectively. It is concluded that the most appropriate stocking system is the $T_{28d}$ in summer and fall.

**Effect of canola oil inclusion in the diet of dairy cows on fatty acids composition of milk**

*K.C. Welter[1], C.M.M.R. Martins[2], M.M. Martins[1], J.C.S. Souza[1], A.S.V.P. Palma[1] and A. Saran Netto[1]*
*[1]Faculdade de Zootecnia e Engenharia de Alimentos/USP, Departamento de Zootecnia, Rua Duque de Caxias Norte, 225, 13365-900, Brazil, [2]Faculdade de Medicina Veterinária e Zootecnia/USP, Departamento de Nutrição e Produção Animal, Rua Duque de Caxias Norte, 225, 13635-900, Brazil; saranetto@usp.br*

The objective of this study was to evaluate the effect of inclusion of canola oil in the diet of dairy cows on fatty acids (FA) composition of milk. Eighteen lactating Holstein cows were distributed in 6 contemporary 3×3 Latin Square design, 3 periods and 3 treatments: T1=control diet, T2=inclusion of 3% of canola oil and T3=inclusion of 6% of canola oil in the diet (based on dry matter). The identification and quantification of milk fatty acids were made by gas chromatography. The results were analyzed using the MIXED procedure of SAS (2001), with significance level $P \leq 0.05$. The inclusion of canola oil (0, 3% and 6%) increased (g/100 g FA) linearly the concentration of (1) Rumenic (P=0.032), (2) Long Chain ($P<0.0001$), (3) Monounsaturated ($P<0.0001$) and (4) Polyunsaturated FA (P=0.013) in fat acid composition, which increased (5) hypocholesterolemic/hypercholesterolemic (h/H) ($P<0.0001$) ratio of milk. In another way, the inclusion of canola oil reduced the concentration of (6) Short Chain FA ($P<0.0001$), (7) Medium Chain ($P<0.0001$) and (8) Saturated ($P<0.0001$) in milk, according to the following equations: (1) $Y = 0.49(0.06) + 0.01709(0.005881)X$; (2) $Y = 40.24(1.49) + 2.8067(0.2342)X$ ; (3) $Y = 33.32(1.05) + 2.0978(0.1564)X$; (4) $Y = 2.29(0.1516) + 0.05201(0.01983)X$; (5) $Y = 0.57(0.04) + 0.08550(0.004564)X$; (6) $Y = 10.90(0.47) - 0.7778(0.07139)X$; (7) $Y = 49.65(1.11) - 2.2816(0.1551)X$; (8) $Y = 63.97(1.11) - 2.1958(0.1677)X$. In conclusion, the inclusion of canola oil in the diet of dairy cows increases the long chain fatty acids concentration, rumênico fatty acid and (h/H) ratio, resulting in healthier milk for human consumption.



**A typology of transhumant sheep and goat farms in Greece**

*A. Ragkos[1], A. Siasiou[2], G. Galidaki[3] and V. Lagka[1]*
*[1]Alexander Technological Educational Institute of Thessaloniki, Department of Agricultural Technology, Sindos, 57400 Thessaloniki, Greece, [2]Democritus University of Thrace, Department of Agricultural Development, Pantazidou 193, 58200 Orestiada, Greece, [3]Aristotle University of Thessaloniki, Department of Forestry and Natural Environment, University Campus, Thessaloniki, Greece; annasiasiou@yahoo.gr*

The transhumant sheep and goat system constitutes an example of extensive livestock farming. Such farms can be found in all Greek Regions; nonetheless, apart from their common feature of displacement towards mountainous pasturelands for 5 to 7 months in summer, they exhibit a broad range of characteristics, which renders their common examination rather difficult. The purpose of this study is to construct a typology of transhumant sheep and goat farms in Greece. The dataset was formulated with data from the Greek Payment and Control Agency (OPEKEPE) and from all local public veterinary services, in charge of issuing movement licenses. After refining the list with consecutive personal communications, the final dataset comprised 2,957 sheep and goat farms. Data were analyzed by means of the two-step cluster analysis (TSCA) in order to detect homogenous groups of farms within each Region under a set of criteria including farm size, distance of movements, classification of rangelands used, origins of transhumant farmers, means of transportation of animals, restrictions from the terrain etc. The results reveal groups with common characteristics, each one of them requiring different approaches in order to maintain their viability and increase their contribution in local economy and employment by taking advantage of farm and rural policy measures. This paper is part of the project 'The dynamics of the transhumant sheep and goat farming system in Greece. Influences on biodiversity' which is co-funded by the European Union (European Social Fund) through the Action 'THALIS'. The authors also acknowledge the invaluable contribution of Dr. Zaphiris Abas who died in a car accident on 28 December 2013

**Factors affecting profitability in the backgrounding of beef calves: a South African perspective**

*P.J. Fourie*
*Central University of Technology, Free Sate, Agriculture, Private Bag X20539, 9300, South Africa; pfourie@cut.ac.za*

To increase profitability of beef systems, backgrounding prepare weaned calves for finishing on high energy rations to promote rapid weight gain in a feedlot. Backgrounding operations may be pasture or dry-lot based or some combination thereof. The primary objective of this study was to investigate factors leading to poor average daily gain (ADG), high morbidity and mortality rate and the increased costs of gain. The secondary objective was to study the management practices followed in the backgrounding of beef calves concerning purchasing, adaptation, processing, raising, health management and marketing strategies. Forty questionnaires were administered to farmers, small and large feedlots doing backgrounding of beef calves. Data was generated by using the SUM equation and the means, minimum and maximum were generated by using PROC MEANS in SAS. On-farm observations were employed in collecting data and discussions with other farmers and experts doing backgrounding of beef calves. According to this study, the ADG for summer differed significantly ($P<0.05$) from that of winter as the ADG during summer was 22.2% higher than that of winter. However, the study also showed that the feed intake in summer differed significantly ($P<0.05$) from that of winter with summer feed intake being 13% higher than winter feed intake. The production costs per calf in this study were R300.50±158.60 for feeding costs, R138.10±90.80 for remedies, R56.40±22.10 for processing and R37.50±24.30 for transport costs. It was also evident that parainflueza 3 was the infectious disease that mostly led to morbidity and mortality. With protozoal diseases, gall-sickness and red-water was the major cause of mortality. Mortality as a result of nutritional disorder including bloat and acidosis was reported by 37.5% of the respondents.



**Milk production, pasture intake and grazing behaviour of dairy cows in winter grazing**

*M. Ruiz-Albarrán[1], O.A. Balocchi[2] and R.G. Pulido[1]*
*[1]Universidad Austral de Chile, Ciencia Animal, Campus Isla Teja, P.O. Box 567. Valdivia, Chile, [2]Universidad Austral de Chile, Producción Animal, Campus Isla Teja, P.O. Box 567, Valdivia, Chile; rpulido@uach.cl*

The aim of this research was to determine the influence of pasture allowance offered to autumn calving dairy cows during winter on pasture utilization, milk production, and grazing behavior. The trial was conducted in Vista Alegre Experimental Station of the University Austral of Chile, for 63 days using thirty-two multiparous Holstein-Friesian dairy cows, during winter of 2011. Prior to experimental period, milk production averaged 20.2±1.7 kg/day, live weight was 503±19 kg and 103±6 days in milk. The experimental treatments were two pasture allowances (PA) offered above ground level; low PA (17 kg DM cow/day) vs high PA (25 kg DM cow/day). Al the cows were supplemented with grass silage (GS) offered at 6.25 kg DM cow/day, 3.6 kg DM cow/day of high-moisture corn and 1.0 kg DM cow/day of concentrate. Pre-grazing mass ($P>0.05$) averaged 1.370 (SD 74.5) and 1.370 (SD 36.0) kg DM cow/ha for the high and low PA, respectively. Decreasing PA influenced milk production (incremented 25% milk production per ha), milk protein (20 kg/ ha) and milk fat (17 kg/ha) ($P<0.05$), without no effect on milk production per cow ($P>0.05$). Increasing PA increased grazing time from 195 to 245 min/day (for low and high PA, respectively), but with no changes on pasture or total intake ($P>0.05$). In conclusion, pasture allowance is a significant factor influencing milk production and milk solids per hectare of autumn calving cows grazing on restricted pasture mass during winter.

**Effect of dietary supplemental pomegranate extract on cryptosporidiosis in neonatal calves**

*S. Weyl-Feinstein[1,2], A. Markovics[3], M. Yshay[2], R. Agmon[2], I. Itzhaki[1], A. Orlov[2] and A. Shabtay[2]*
*[1]Faculty of Sciences, Department of Evolutionary and Environmental Biology, University of Haifa, 31905 Haifa, Israel, [2]Institute of Animal Science, Newe Ya'ar Research Center, Agricultural Research Organization, Department of Ruminant Science & Genetics, P.O. Box 1021, Ramat Yishay 30095, Israel, [3]kimron Veterinary Institute, Department of parasitology, P.O. Box 12, 50250 Bet-Dagan, Israel; sarah@agri.gov.il*

A study was conducted in order to evaluate the anti-diarrheal and anti- parasitical effects of a concentrated Punica granatum extract (CPE) supplementation in milk, on intestinal morbidity in neonatal calves. Two field experiments were performed (n=81): The first experiment was conducted with 41 calves, where at the age of 3-14 days, treatment group (n=21) received CPE at 3.75% of daily milk ration containing 10% of polyphenols. The second experiment (n=40) tested the effect of a smaller dose (0.6% of daily allocation) of a higher CPE concentrated (12.3% polyphenols). Between 5-13 days, daily fecal samples were collected and fecal oocyst count (FOC) of *Cryptosporidium parvum* was quantified. Diarrhea prevalence and frequency of use in antibiotic therapy were recorded. Treatment calves who were supplemented with 3.75% of extract (H-CPE exp.), showed a reduction of daily average FOC ($P=0.004$), diarrhea intensity ($P<0.001$) and morbidity duration ($P<0.001$) in compare with control. In L-CPE exp., where 0.6% was added in the milk, there was no effect on FOC nor on ADG measured at weaning. However, compare with control, treatment group was characterized by a shorter duration of diarrhea ($P=0.02$), increased total weight gain at the age of fourteen days in male calves ($P=0.0015$), and reduced number of calves who were treated with antibiotic therapy ($P=0.03$). These results suggest that feeding supplemental CPE to calves ration diminishes diarrhea intensity, reduces *Cryptosporidium* oocyst shedding, improve weight gain at the peak of cryptosporidiosis and decreases veterinary therapy frequency.

---



**Ruminal fermentation characteristics on dairy cows grazing during autumn**

*M. Ruiz-Albarrán[1], M. Cerda[1], M. Noro[1], O.A. Balocchi[2] and R.G. Pulido[1]*
*[1]Facultad de Ciencias Veterinarias, Ciencia Animal, Campus Isla Teja, P.O. Box 567, Valdivia, Chile, [2]Facultad de Ciencias Agrarias, Producción Animal, Campus Isla Teja, P.O. Box 567, Valdivia, Chile; rpulido@uach.cl*

The aim of this research was to determine the influence of pasture allowance (PA) and type of silage (TS) on ruminal pH and $N\text{-}NH_3$ and VFA concentration in rumen of dairy cows grazing. The trial was conducted in the University Austral of Chile, for 56 days during autumn of 2011. Four Holstein-Friesian ruminally cannulated dairy cows (milk averaged 174 days in milk, 632 kg/cow BW, and 17.0 kg milk). The cows were randomly assigned to four dietary treatments resulting from the combination of two PA (17 vs 25 kg DM cow/day) and two types of silage (TS), offered at the level of 6.25 kg DM cow/day (grass, GS vs maize, MS). Also, all the cows received 3.6 kg of concentrate. The Fistulated cows were allocated to a 4×4 Latin square design, containing four periods and four dietary treatments, in a 2×2 factorial arrangement. The pastures offered containing on average 28.0% of CP and 2.75 Mcal/kg DM of ME. The DM and ME contents of the grass and maize were similar (29.8 of DM and 2.74 Mcal of ME/kg DM. Ruminal pH (6.3 and 6.4), ruminal $N\text{-}NH_3$ (11.0 and 11.4 mmol/l), total VFA (103 and 104 mmol/l for 17 and 25 PA, respectively) were not affected by PA. Supplementation with MS decreased $NH_3\text{-}N$ concentration ($P<0.05$) (12.1 vs 9.8 mmol/l for GS and MS, respectively). Molar proportion of propionate was higher ($P<0.05$) for cows receiving 25 of PA than 17 kg of PA (16.8 and 18.8, respectively). A positive interaction was found for the two factors on VFA production ($P<0.05$), where a greater concentration was reported for cows grazing 25 kg DM cow/day of PA and receiving grass silage (112±5.9 mmol/l). In conclusion, dairy cows grazing 25 of PA and maize silage appeared to be more efficient on ruminal characteristics of fermentation.

**Evaluation of warm season grasses grown in a highly restricted soil of La Pampa, Argentina**

*S.S. Paredes[1], N.P. Stritzler[1,2], B.C. Lentz[1,2], A.A. Bono[1], R.R. Zapata[1,2], H.J. Petruzzi[1,2] and C.M. Rabotnikof[2]*
*[1]INTA, CR La Pampa-San Luis, Spinetto 785, L-6300, Argentina, [2]UNLPam, Facultad de Agronomía, RN 35 Km 334, L-6300, Argentina; stritzler.nestor@inta.gob.ar*

The weak structure of the soils of the semiarid Pampas in Argentina, and their trend to suffer erosion processes, make important to establish beef production systems based on permanent pastures. The warm season grasses (WSG) might be an important alternative for a better use of soils of this region. The objective of this study was to evaluate the amount and nutritive value of deferred forage produced by three different WSG in mono- or polyphytic pastures in a weakly structured soil. The experiment was carried out in La Pampa, Argentina (37° 41' 35" S; 63° 31' 58" W) in a Petrocalcic calciustoll soil, with carbonates from 10 cm depth and petrocalcic horizon at 50 cm depth, in plots of 300 cm long by 200 cm wide. Each plot had four rows at 50 cm apart and plants within the row spaced by 30 cm. The experimental design was in completely randomized blocks, with three replications per treatment. Three different WSG were evaluated: Eragrostis superba (ES); Panicum coloratum (PC) and Panicum virgatum (PV) in monophytic plots or combinations by two or the three species. The treatments (T), then, were: T1. ES; T2. PC; T3. PV; T4. ES-PC; T5. ES-PV; T6. PC-PV; T7. ES-PC-PV. The aerial biomass of all plots was cut simultaneously during the winter of 2012, and the material collected was used to determine total dry matter, dry matter *in vitro* digestibility (DMD) and crude protein (CP). The results were, for total standing DM (in kg/ha), DMD (%) and CP (%), respectively: T1: 1267; 46.5 and 2.96; T2: 2012; 43.1 and 2.62; T3: 4652; 27.2 and 1.72; T4: 5878; 41.5 and 2.47; T5: 8108; 31.6 and 1.71; T6: 6177; 39.8 and 2.27; T7: 8217; 42.3 and 2.42. The polyphytic treatments had higher total DM ($P<0.05$) than the monophytic ones with no clear trends on the nutritive value of the forage. However, the inclusion of PV increased the total DM and reduced both DMD and CP.

---



**The effect of feeding methods on performance, feeding behavior and oxidative stress in dairy calves**

*M. Boga[1], U. Kilic[2], M. Görgülü[3], S. Yurtseven[4] and S. Göncü[5]*
*[1]University of Nigde, Bor Vocational School, University of Nigde, Bor Vocational School, 51700 Nigde, Turkey, [2]University of Ondokuz Mayis, Department of Animal Science, University of Ondokuz Mayis, Faculty of Agriculture, 55139 Samsun, Turkey, [3]University of Cukurova, Department of Animal Science, University of Cukurova, Faculty of Agriculture, 01130 Adana, Turkey, [4]University of Harran, Department of Animal Science, University of Harran, Faculty of Agriculture, 63100 Şanlıurfa, Turkey, [5]University of Cukurova, Department of Animal Science, University of Cukurova, Faculty of Agriculture, 01130 Adana, Turkey; mboga@nigde.edu.tr*

This study was performed to evaluate the effects of different feeding regimes on growth parameters, feeding behavior and the oxidative stress parameters in preweaning calves. In the study, a total of 45 Holstein calves were used. The calves were divided into three groups. The first group was fed with a calf starter and alfalfa hay as free choice, the second group was fed with total mixed ration (TMR), and the third group was fed calf starter only as pellet during the pre-weaning period. Behavioral observations were recorded six times per day and these observations were made for a period of 1 hour starting at 08.00, 11.00, 14.00, 17.00, 20.00, and 23.00 with 5 min intervals for 8 week to determine the feeding behavior. The recorded activities include eating, ruminating, drinking, licking objects, resting, and others. These activities of calves were not changed as a percentage in all feeding systems. At the end of the experiment, the calves fed TMR had lower live weight gain and feed consumption($P<0.05$) than the others. Furthermore, treatments did not cause a change in oxidative stress parameters such as lipid hydroperoxide (LOOH) concentrations, total antioxidant response (TAR), total antioxidant status (TOS). In conclusion, different concentrate based starter did not cause any behavioral changes during preweaning period.

**Nutritional mitigation alternatives for GHG emissions for Latxa dairy sheep production system**

*C. Pineda-Quiroga, N. Mandaluniz, A. García-Rodríguez, E. Ugarte and R. Ruiz*
*NEIKER, Animal Production, P.O. Box 46, 01080, Spain; eugarte@neiker.net*

Dairy sheep production in the Basque Country has been traditionally based on the Latxa breed with different feeding practices during the productive cycle. The assessment of the diets is critical to evaluate and improve its efficiency and reduce GHG emissions. The objective of this study was: to asses GHG emissions of diets provided during different production stages (prelambing-P, lactating indoors-LI and spring lactation with grazing-LG), and to propose nutritional mitigation alternatives. To asses GHG emissions, feed samples were collected in 4 commercial flocks. As mitigation alternatives, 3 fat-rich feedstuffs (rapeseed and sunflower cake and palm fat) were evaluated, using 2 levels of fat inclusion (3% and 5%), combined with 2 forages of different quality (high and low). GHG emissions of the commercial diets (potential gas production-PGP, organic matter digestibility-IVOMD, volatile fatty acids-VFA and methane-M) as well as the GHG mitigation alternatives were monitored *in vitro* by gas production technique. According to the results, grass based diets yielded higher PGP (ml/g Incubated organic matter) ($524^{a}$ LG vs $268^{b}$ LI and $310^{b}$ P), total VFA (mmol) ($5.47^{a}$ LG vs $4.45^{b}$ P, $5.01^{ab}$ LI) and M emissions ($1.91^{a}$ LG vs $1.46^{b}$ LI and $1.45^{b}$ P), and lower percentage IVOMD ($52^{a}$ vs $66^{b}$ LI and $59^{b}$ P). According to the mitigation alternatives, a significant reduction in PGP was observed in both fat-rich feedstuffs incubated with high quality forage ($258^{a}$ in 3% vs $308^{b}$ in 5%). Methane emissions (mmol/day) were reduced when high quality forage was incubated at 3% fat level ($1.13^{a}$ in 3% vs $1.37^{b}$ in 5%) and when low quality forage was incubated at 5% fat level ($1.13^{a}$ in 5% vs $1.15^{b}$ in 3%). In conclusion, grass-based diets produce higher GHG emissions and the use of supplementary fat-rich feedstuffs in combination with good quality forages appear to be an alternative for mitigation of grass-based systems and to improve their sustainability.

---



**Browsing frequency and milk fatty acids of goats foraging on open grassland and dense shrubland**

*M. Renna, G. Iussig, A. Gorlier, M. Lonati, C. Lussiana, L.M. Battaglini and G. Lombardi*
*University of Torino, Department of Agricultural, Forest and Food Sciences, via Leonardo da Vinci 44, 10095 Grugliasco (TO), Italy; manuela.renna@unito.it*

In the southwestern Alps, the presence of a large number of small semi-natural grasslands interspersed within forests has supported small dairy goat farms over years thanks to the preference of these browsing animals for herbaceous species, shrubs and tree leaves. A study was carried out to evaluate browsing frequency (BR) and fatty acid (FA) profile of milk obtained from Camosciata delle Alpi goats foraging on paddocks laid out on a mountain open grassland (OG) and a dense shrubland (DS). The paddocks were exploited at the same stocking density by the same flock, at one-month interval from each other during the summer season. Forty-five plots were randomly selected inside each paddock and used to assess their botanical composition and goat BR of the most abundant plant species. Individual milk from seven goats was sampled in two consecutive days in each paddock and analysed for the FA profile. A general linear model for repeated measure design was used to compare the FA profile of milk between OG and DS paddocks. The goats selected mostly forbs and woody species leaves in DS [*Veratrum album* L., 96% of BR; *Senecio fuchsii* Gmelin, 88%; *Sorbus aucuparia* L., 85%; *Sorbus aria* (L.) Crantz, 85%; *Salix caprea* L., 80%] and eutrophic species in OG [(*V. album*, 100% of BR; *Lolium perenne* L., 80%; *Poa annua* L., 69%; *Dactylis glomerata* L., 67%)]. Fatty acid analysis showed that DS milk had significantly higher percentages of conjugated linoleic acid (0.65 vs 0.52 g/100 g fatty acid methyl esters; $P<0.05$), trans FA (6.40 vs 5.42; $P<0.05$) and omega 3 FA (1.07 vs 0.82; $P<0.05$), and a lower omega 6/omega 3 FA ratio (3.01 vs 4.59; $P<0.001$), than OG milk. The results show that the exploitation of different forages by goats may result in an increase of dairy products diversity, consequently improving the productive ecosystem function.

**Prediction of voluntary intake by the short term intake technique in warm-season grasses**

*F.M. Ingentron[1,2], S. Degiovanangelo[2], J. Melión[2], B.C. Lentz[2,3], C.M. Rabotnikof[2], N.P. Stritzler[2,3], N. Balzer[2] and H.M. Arelovich[1,4,5]*

*[1]CONICET, Rivadavia 1917, C-1033AAJ CABA, Argentina, [2]UNLPam, Facultad de Agronomía, Animal Production, RN 35 Km 334, L-6300 Santa Rosa, La Pampa, Argentina, [3]INTA, CR La Pampa-San Luis, Spinetto 785, L-6300 Santa Rosa, La Pampa, Argentina, [4]CIC, Calle 526, 1900 La Plata, Argentina, [5]UNS, Departamento de Agronomía, San Andrés 800, 8000 Bahía Blanca, Argentina; stritzler.nestor@inta.gob.ar*

This study was designed to estimate the precision of the short term intake technique (STIR) as a predictor of dry matter voluntary intake (DMVI) by rams. Three different warm-season grasses at vegetative stage: fingergrass (*Digitaria eriantha* ssp. *eriantha* cv. Irene), kleingrass (*Panicum coloratum* cv. Verde) and switchgrass (*Panicum virgatum* cv. Alamo) were used as forage sources, in three different experiments. In each one, 6 Pampinta rams of 56 kg of mean liveweight and 10 months old, were individually fed indoors, *ad libitum* and with free access to water during 14 days, being the last 7 days for measurement of DMVI. Simultaneously, another group of 6 rams of the same age and liveweight were used to measure STIR. These animals were fed on a lucerne hay diet, at maintenance level. The whole ration was offered once a day, at 9:00 am. After a 4-hours starving period, the STIR of one of the forage sources at a time (fingergrass, kleingrass or switchgrass) was calculated, by substracting the amount of DM refused from the DM offered, after 4 minutes of active consumption measured with a chronometer by one observer for each ram. The DMVI across all forages varied between 55.11 and 83.98 g DM/ kg BW-0.75. The STIR ranged from 10.08 to 19.34 g DM/50 kg BW/minute. The correlation coefficient between both variables including data of all three warm-season grasses fed resulted 0.7045. It can be concluded that the STIR technique is a reliable predictor of DMVI for warm-season grasses at vegetative stage.

---



**Grazing behaviour of small ruminants on diverse karst land vegetation**

*D. Bojkovski, I. Stuhec, D. Kompan and M. Zupan*

*University of Ljubljana, Biotechnical faculty, Department of Animal Science, Groblje 3, 1230 Domzale, Slovenia; danijela.bojkovski@bf.uni-lj.si*

Limestone and dolomite rocks are the main components of Slovenian Karst land. Natural conditions for intensive agriculture are not favourable and costs for cultivation are too high. The area as such is consequently exposed to the process of abandonment and bush encroachment. Small ruminants are most appropriate livestock to cultivate this type of land. To obtain more knowledge of the grazing behaviour of small ruminants, an experiment was designed on pasture in the mountain Karst region. The animals which were grazing in the fenced area, were divided in six plots from which three plots were covered with grass, herbs and legumes (i.e. grassy plot, GP). In the other three plots, the area was also overgrown with hazel, beech trees and bushes (i.e. woody plot, WP). The autochtonous sheep breed Istrian Pramenka and crossbred goats (Saanen × Alpine goat) were included in the experiment. A group of 50 animals were rotated between the six plots with 10 randomly selected sheep and 10 goats being observed. The results indicated that botanically diverse paddocks together with climate conditions affected the behaviour. Both species spent more than 65% of their time grazing. Sheep spent more time grazing in GP ($P=0.05$), while goats in WP ($P<0.0001$). Grazing time was affected by the temperature. With a lower temperature (lower than 15.4 °C versus above 15.4 °C) we saw an increasing activity. On the second day of observation, the animals grazed more compared to the first day (both species $P<0.0001$), most likely due to a lower forage availability. In sheep, drinking and salt consumption were higher in the WP, suggesting that salt triggered additional water consumption. Goats drank and consumed more salt on the second day ($P<0.0001$). Studying more natural behaviour of ruminants may provide a firm knowledge on thev ethological traits of small ruminants. Understanding of their grazing behaviour is necessary for optimizing the management in marginal areas, such as the Karst region in Slovenia.

**The effect of gradual weaning on leukocyte relative gene expression levels in dairy calves**

*D. Johnston[1], D.A. Kenny[1], R. Rubio Contreras[1], S.M. Waters[1], M. McCabe[1], A. Kelly[2], M. McGee[1] and B. Earley[1]*
*[1]Teagasc, Animal & Grassland Research and Innovation Centre, Dunsany, Co. Meath, Ireland, [2]University College Dublin (UCD), Belfield, Dublin 4, Ireland; dayle.johnston@teagasc.ie*

The study objective was to examine effect of gradual weaning on the immune response in dairy calves. Eight Holstein-Friesian (H-F) ((mean age ± s.d.) 23±7 d, (mean weight ± s.d.) 46±6 kg) and eight Jersey (J) bull calves (37±8 d, 34±5 kg) were group housed on sawdust bedded floored pens from day (d) -56 to d 14 of the study. Calves were fed using an electronic feeding system (Foster-Tecknik SA 2000, Engen, Germany). The pre-weaning, weaning and post-weaning periods were defined as days -56 to -14, -13 to 0 (milk feeding ceased), and 1 to 14 respectively. During the pre-weaning period calves were offered 6 l milk replacer (MR) and *ad libitum* concentrate (C). Weaning was initiated when calves were consuming 1 kg of C per day for three consecutive days. During the weaning phase MR was gradually reduced from 6 l to 0 l over a 14 day period (d -13 to d 0). Relative to weaning (d 0), on d -14, 1, and 8, blood samples were collected via jugular venipuncture for gene expression profiling. Relative gene expression levels were higher ($P<0.05$) in J calves compared with H-F for CXCL8 (J; 10.6 v H-F; 6.6±1.52) and GRα (J; 2.6 v H-F; 1.3±0.34). Gene expression differences ($P>0.05$) were not observed between the two breeds for Fas, TLR4 and TNFα. An effect of time was observed for Fas with increased relative gene expression between d -14 (1.5±0.08) and d 1 (1.9±0.20) and decreased expression between d 1 and d 8 (1.7±0.10). An immune response to gradual weaning was observed in H-F and J calves as the expression pattern of the pro-apoptotic gene, Fas, was changed over time. The differences between the breeds in relative gene expression suggest that J calves may have a more sensitive immune system, demonstrated by increased levels of the pro-inflammatory cytokine, CXCL8, and of the glucocorticoid receptor, GRα.

---



**Monitoring the physiological and behavioral stress response of calves following pre-marketing mixing**

*S. Weyl-Feinstein[1,2], A. Orlov[2], M. Yshay[2], R. Agmon[2], V. Sibony[2], M. Steensels[2], I. Itzhaki[1] and A. Shabtay[2]*
*[1]Faculty of Sciences, Department of Evolutionary and Environmental Biology, University of Haifa, 31905 Haifa, Israel, [2]Institute of Animal Science, Newe Ya'ar Research Center, Agricultural Research Organization, Department of Ruminant Science, P.O. Box 1021, Ramat Yishay 30095, Israel; sarah@agri.gov.il*

Beef cattle individuals who are not familiar with each other are routinely mixed prior to marketing. This practice constitutes a stressful event that affects animal welfare and meat quality. We questioned whether mixing 34 days before marketing, may still induce stress response that could affect productivity and pH of meat. 22 calves were raised in triplets from the age of 7.3±0.07 months until the age of 15.7±0.07 months. They were mixed 34 d prior to marketing to form two groups (n1=13, n2=9). The rumination and activity of the calves were monitored continuously -20 d to +33 d in relation to mixing, and bull calves were weighed prior to mixing, 3 and 33 days post-mixing (DPM). Blood concentrations of non-esterified fatty acid (NEFA), and anti-oxidative capacity were measured pre- and post-mixing (PM), and after slaughter, the pH of meat was measured. The results demonstrated that mixing had significantly decreased the calves' weight gain by 6.2±2 kg 3 DPM. Daily rumination was decreased by 2 fold 24H PM ($P<0.0001$). Calves' daily number of steps was 3 folds higher, and did not restored to PM values even 25 DPM ($P<0.0001$). Plasma NEFA concentration ascended from 120±13 to 428±40 μmol/l, 24H PM ($P<0.0001$), and the serum anti-oxidative capacity 24H PM significantly decreased. Finally, the mean pH value of LTL in meat was 6.5±0.02, higher than required for proper meat aging. Precise farming management enables monitoring of physiological, behavioral and meat quality outcome, and may thus serve as predictive tool for the recovery period needed from mixing to marketing. Further investigations are required in order to assess additional impacts of mixing on meat quality.

## Session 10 — Theatre 3

### Investigating the use of rumination sensors during the peripartum period in dairy cows

*D.N. Liboreiro, K.S. Machado, M.I. Endres and R.C. Chebel*
*University of Minnesota, 1364 Eckles Avenue, St. Paul, MN 55108, USA; miendres@umn.edu*

The objective of this study was to evaluate the association between rumination and peripartum disorders in dairy cows housed in freestalls in the USA. Nulliparous and parous Holstein cows (n=296) were fitted with rumination time monitors (HR Tags, SCR Engineers Ltd., Israel) from d -17 to 17 relative to calving. Weekly blood samples (0 to 20 d relative to calving) were used for determination of beta-hydroxybutyrate (BHB) concentration and incidence of ketosis (BHB>1,400 mmol/l). Blood sampled on d 0, 1, and 2 relative to calving were used for determination of total Ca concentration and incidence of sub-clinical hypocalcemia (Ca<8.5 mg/dl). Cows were examined for retained placenta (RP) and metritis by study personnel. Statistical analysis was done by ANOVA for repeated measures using the MIXED procedure. Nulliparous cows had lower rumination times than parous cows (P<0.01). There was a tendency for RP to be associated with rumination time (P=0.08); from d -4 to 10 relative to calving, animals with RP had reduced rumination time (P<0.01). Although there was no association between sub-clinical hypocalcemia and rumination time, the interaction between sub-clinical hypocalcemia and days relative to calving was associated with rumination time (P<0.01); on d -16, -13, -11, and 0 relative to calving, animals with sub-clinical hypocalcemia had reduced rumination time. Concentration of Ca was correlated with rumination time (r=0.15; P=0.02). Similarly, ketosis was not associated with rumination time but the interaction between ketosis and days relative to calving was associated with rumination time (P<0.01); from d 6 to 17 relative to calving, animals with ketosis had reduced rumination time. Concentration of BHB was correlated with rumination time (r=0.16; P<0.01). In summary, peripartum disorders were associated with altered rumination time during the peripartum period. It is therefore suggested that rumination sensors can be a valuable tool to predict cows at risk for transition disorders.

---

## Session 10 — Theatre 4

### Monitoring stress behavior in grazing beef cows

*R. Gabrieli*
*Ministry of Agriculture and Rural Development, Extension Service, P.O. Box 30, Beit Dagan, 50250, Israel; ragav@shaham.moag.gov.il*

Commercial beef breeding is carried out in vast pastures. Hierarchy within the herd determines priority of access to limited resources, resulting in stress to the lower ranking cows. Identifying these cows may enable the breeder to improve grouping management and minimize stress. This will lead in turn to improved production and lower replacement costs. Newly developed wireless pedometric system, was modified to serve as a standalone station in a 300 ha. pasture in northern Israel. Tags were installed on 30 cows randomly selected. Comparative analysis of individual hourly activity we believed, may convey a cow social status within the herd. In order to test this hypothesis, we conducted observations near the food trough during feeding time. Cows that reached the trough immediately, got up to two rejections, and stabilized a place to eat, were ranked 'dominant'. Cows that were rejected continuously or didn't approach the trough at all were ranked 'inferior'. Ranking was validated by measuring cortisol level in hair samples. Data analysis showed that peak and low activity hours were identical between the two groups (06:00-07:00; 15:00-16:00, and 03:00-04:00 respectively). Higher than average activity hours were also identical (10). Hourly activity data was variable within all cows, expressed in high values of CV (coefficient of variance, $77\%^{+}_{-}7\%$), meaning that all cows expressed continuously changing activity. Differences between max. and min. hourly activity were significantly higher in the 'inferior' compared to the 'dominant' cows ($327^{+}_{-}45$ and $254^{+}_{-}$ 36 respectively) expressed in higher fluctuations in the activity graphs. Data were analyzed using 'jmp' software. Continuous activity pattern, reflects the measure of freedom each cow has, to choose her own behavior within a grazing herd. Assuming that establishing stable daily activity routine, expressed in a less fluctuated graph, is beneficial for each cow, detecting those that are unable to do so and grouping them apart, will minimize social stress and improve production.

**Using real time cow information for daily grazing management**

*A.H. Ipema, G. Holshof and R.M. De Mol*
*Wageningen UR Livestock Research, P.O. Box 65, 8200 AB Lelystad, the Netherlands; bert.ipema@wur.nl*

Over the last decade, the number of grazing dairy cows in north western European countries has decreased sharply. The reduction in cow pasture grazing is even more likely to occur when AM (automatic milking) is in place. This decrease in grazing is the main reason for starting the project AUTOGRASSMILK funded from the European Union's Seventh Framework Programme under grant agreement no. SME-2012-2-314879. One of the objectives of the project is the optimisation and integration of AM with cow grazing using new technologies. During the grazing season 2013 a herd consisting of 55 HF/FH cows, milked with an AM system, was subjected to strip grazing on the research station Dairy Campus in Leeuwarden, The Netherlands. For grazing 24 ha of permanent pasture was available. The herd had 12 hours daily access to grazing. During the 12-h night period cows were kept indoors and fed with a fixed amount of TMR. Drinking water was only available indoors. At 06:00 h cows got access to a first strip grass. After this strip has been grazed, cows were expected to return to the barn for milking. At 12:00 h cows that not returned back to the barn were fetched. After cows were milked they got access to the second strip for the rest of the daily grazing period. At the end of this grazing period (18:00 h) cows that were still in the pasture were fetched. All individual cow data regarding milk production (yield and frequency) and activity (lying and standing, steps) were recorded during 17 weeks with commercially available technologies. The average yield of the herd was 25.2 kg milk/cow/day with in average 2.4 milkings/cow/day. During the 12-h night period in which the cows were in the barn the number of steps was in average 457/cow/day; during the 12-h day period with access to pasture the number of steps was in average 2,437/cow/day. Possibilities for using these AM and activity data for daily management decisions under grazing circumstances will be discussed in more detail at individual cow level as well as at herd level.



**The potential of using sensor data to predict the moment of calving for dairy cows**

*C.J. Rutten[1], W. Steeneveld[2], C. Kamphuis[2] and H. Hogeveen[1,2]*
*[1]Utrecht University, Department of Farm Animal Health, Faculty of Veterinary Medicine, Yalelaan 7, 3584 CL Utrecht, the Netherlands, [2]Wageningen University, Business Economics Group, Hollandseweg 1, 6706 KN Wageningen, the Netherlands; c.j.rutten@uu.nl*

On dairy farms, management around calving is important for the health of dairy cows and the survival of the new-born calves. Although an expected calving date is known, farmers will have to check their cows regularly to estimate at which moment they might calve to ensure that assistance is available when needed. A prediction that is more precise than the expected calving date could help farmers to improve detecting a cow that starts calving. In the Dutch research project Smart Dairy Farming, a total of 450 cows on 2 farms were equipped with the Agis SensOor, which measures rumination activity, activity and temperature on an hourly basis. Data were collected over a one year period during which the exact moment of 300 calvings were recorded using camera images of the calving pen taken every 5 minutes. The start of parturition and the time that elapsed untill the calve was born were recorded. In a preliminary study, sensor data from 14 days before to 7 days after calving were plotted and judged manually? on the existence of patterns that could be associated with the moment of calving. In the hours before calving rumination activity decreases, while activity increases. The measured temperature did not show a distinguishable pattern in the hours before calving. With sensors it seems to be possible to measure changes in activity and rumination activity in the hours before calving. This means that there is potential to develop detection algorithms that can be used to predict the moment of calving for dairy cows. Further analyses will include filtering of the data and statistical analyses. This research was supported by the Dutch research program Smart Dairy Farming.

## Repeatability of rumination time in individual dairy cows

*P. Løvendahl[1], A. Fogh[2] and M.V. Byskov[2]*
*[1]Aarhus University, Dept. Molecular Biology and Genetics, Center for Quantitative Genetics and Genomics, Blichers allé 20, 8830 Tjele, Denmark, [2]Knowledge Centre for Agriculture, Agro Food Park 15, 8200 Aarhus N, Denmark; peter.lovendahl@agrsci.dk*

Rumination time may be related to feed intake, digestibility, feed efficiency, methane emission, estrus and metabolic health status of dairy cows. Sensors recording rumination time are commercially available in a version including an activity sensor (RuminAct™), and are already installed in several dairy herds. Data on rumination time is displayed in minutes per 2 h intervals. A thorough evaluation of rumination time as indicator trait has been initiated focusing on data quality and repeatability using data from an experimental herd. For this purpose, data on daily rumination time from 131 Holstein and 51 Jersey cows was collected between 0 to 400 DIM from cows in parity 1-3. Rumination time recorded in 2 h intervals were summarized over 24 h initiated at midnight. In total 19% of recordings of daily rumination time were excluded due to one or more missing values at 2 h level, due to heat or due to very low rumination time during 2 or 24 h rumination time. Data was analyzed by the MIXED procedure in SAS, using breed, parity and DIM as fixed effects and cow number as random effect. Estimates of repeatability were calculated based on variance components for the entire period of 0 to 400 DIM or for intervals of 50 days during the 0 to 400 DIM. Mean rumination time increased with parity and DIM. Holstein cows ruminated for 410±120.3 min per day, whereas Jersey cows ruminated for 368±106.5 min per day. The repeatability of rumination time was 0.77 during the entire period of 0 to 400 DIM. Repeatability varied from 0.76 to 0.86 within the periods of 50 DIM, with the lowest repeatability at the beginning and at end of lactation and the highest repeatability in mid-lactation. In conclusion, rumination time seems to be a highly repeatable trait, and further studies will scrutinize the possible relationships to feed intake.

---



## Dairy farm evaluation of rumen pH bolus data: identifying the benefits

*T.T.F. Mottram[1,2]*
*[1]eCow Ltd, King St Business Centre, Exeter, EX1 1BH, United Kingdom, [2]Royal Agricultural University, School of Agriculture, Cirencester, GL7 6SJ, United Kingdom; toby.mottram@rau.ac.uk*

Ruminal pH is an very important parameter for nutritional status particularly of dairy cows and it is claimed by Atkinson that using rumenocentesis 25% of cows have rumen pH values below 5.5 pH. Since 2005, boluses measuring pH continuously and using wireless telemetry have been used for research purposes, mainly in fistulated cows. This paper reports the use of 120 rumen pH telemetry boluses on 30 farms in South West England in 2013/14. The farms were selected to represent a range of farm types from continuous grazing, through mixed grazing and concentrate feeding in a robotic milker to TMR fed continuous housed cows milked three times a day. The data were collected by a nutritionist visiting the farm regularly with a handset to download data, analysing feed and talking to the farmer about events that affected rumen pH. The pH data were recorded in the reticulum which has a pH level approximately 0.25 pH units above that in the ventral sac from which rumenocentesis is conducted. Fewer than 5% of recordings were below 5.75 indicating that SARA was not common in this sample of cows. The variety of responses to the rumen data will be presented as narrative case studies, they include one farmer saving 70 p per cow per day by removing a minor food ingredient from the diet that raising mean rumen pH and increasing the amount of night time feeding without affecting milk yield. In a grazing situation one farmer changed his fence moving routine which optimised rumen pH and raised milk yields. Several farms detected irregularities in rumen pH probably caused by changes in feed offered to the cows by different staff. Optimal pH values in different feeding systems will be discussed but the most important parameters to create pH targets for dairy farmers appear to be the daily range of pH, the mean daily pH, the number of feeds per day and detection of management changes.

**Biopara-milk a whole cow simulation model – assessment of rumen pH predictions**

*V. Ambriz-Vilchis[1,2], N. Jessop[1], R. Fawcett[1], D. Shaw[3] and A. Macrae[2]*
*[1]Bioparametrics Ltd., SRUC Building, West Mains Road, EH9 3JG, Edinburgh, United Kingdom, [2]DHHPS R(D)SVS and The Roslin Institute, The University of Edinburgh, EBVC, EH25 9RG Roslin, United Kingdom, [3]The Roslin Institute and R(D)SVS, The University of Edinburgh, EBVC, EH25 9RG, Roslin, United Kingdom; v.ambriz-vilchis@sms.ed.ac.uk*

The use of new technologies to measure physiological, behavioural and production parameters can improve management strategies and performance. An example of this is the use of boluses to measure rumen pH flux. Also mathematical modelling is helpful tool to describe the complexity of the rumen and to predict multiple responses of the rumen to diets. Biopara-Milk® is a whole cow model, it simulates the digestive system, predicts performance and circadian pH. The aim of this study was to compare Biopara-Milk pH predictions against those obtained with the rumen boluses in lactating dairy cows. Fourteen multiparous dairy cows were offered a TMR diet with concentrate fed to yield. Cows were administrated a bolus to measure pH. Model input data: detail information on the feed (chemical composition and degradation kinetics) and the animals (BW, BCS, lactation potential, milk composition) were used to run Biopara-Milk®. Correlation coefficient, regression analysis, the limits of agreement method and mixed effect model were performed to assess the relationship between pH obtained with the boluses and data from Biopara-Milk, all statistical analyses were performed using R. Average pH values per hour were obtained with both methods correlation coefficient(r), and coefficient of determination (R2) between pH data were acceptable ($r=0.69$, $R2=0.47$ $n=218$, $P<0.001$). The Limits of Agreement method showed that disagreements between the two methods were evenly distributed across the range. Estimates obtained with Biopara-Milk® were only 0.02 points of pH lower than those obtained with the boluses. The simulation exercise shown the capabilities of Biopara-Milk® to predict pH flux in dairy cows. Future work will enable the use of Biopara-Milk® as a diagnostic tool.

---



**Feeding concentrate in early lactation based on rumination time**

*M.V. Byskov[1], M.R. Weisbjerg[2], B. Markussen[3], O. Aaes[1] and P. Nørgaard[4]*
*[1]Knowledge Centre for Agriculture, Agro Food Park 15, 8200 Aarhus N, Denmark, [2]Aarhus University, Dept. of Animal Science, Blichers Allé 20, 8830 Tjele, Denmark, [3]University of Copenhagen, Laboratory of Applied Statistics, Universitetsparken 5, 2100 København, Denmark, [4]University of Copenhagen, Dept. of Veterinary Clinical and Animal Sciences, Grønnegårdsvej 3, 1870 Frederiksberg C, Denmark; pen@sund.ku.dk*

Precision feeding of dairy cows facilitates optimization of milk production. Accordingly, the objective was to study the effect on milk production when stepping up concentrate at 3 rates in early lactation according to individual daily rumination time (RT). Data was collected in 3 commercial dairy herds with Holstein cows, where daily RT was recorded by rumination sensors (Qwes HR™). Cows were fed a partially mixed ration and concentrate in the milking robot. Concentrate was stepped up over the first 28 and 17 DIM for primiparous and multiparous cows. Cows were assigned to either an experimental group (EXP) or a control group (CON) immediately after calving. In addition, all cows in the EXP and CON were assigned to either a high, medial or low rumination group according to their individually RT at 4 to 7 DIM. Cows in the EXP assigned to the high ($E_H$), medial ($E_M$) or low ($E_L$) rumination group were stepped up to 6, 4 or 3 kg concentrate during the experimental period. Concentrate was stepped up to 4 kg during the experimental period for all cows in the CON, regardless of whether the cows were assigned to the high ($C_H$), medial ($C_M$) or low ($C_L$) rumination group. In total, 40 and 41 primiparous cows and 66 and 66 multiparous cows on the EXP and CON finished the trial. Primiparous cows in the EXP showed higher ECM yield compared with primiparous cows in the CON (26.1 vs 25.6 kg/day). The same applied for primiparous cows in the $E_L$ compared to $C_L$ (25.6 vs 25.1 kg/day). No effect on milk production was found for multiparous cows. In conclusion, adjusting concentrate allocation rate in early lactation based in RT shows potential effect in ECM yield for primiparous cows.

**Ability to estimate feed intake from presence at feeding trough and chewing activity**

*C. Pahl[1], A. Haeussermann[1], K. Mahlkow-Nerge[2], A. Grothmann[3] and E. Hartung[1]*
*[1]Christian-Albrechts-University Kiel, Institute of Agricultural Engineering, Max-Eyth-Str. 6, 24118 Kiel, Germany, [2]Chamber of Agriculture Schleswig-Holstein, Futterkamp, 24237 Blekendorf, Germany, [3]Agroscope Reckenholz-Tänikon ART, Tänikon, 8356 Ettenhausen, Switzerland; ahaeussermann@ilv.uni-kiel.de*

Monitoring feed intake of dairy cows can be a useful tool for improvement of health and feeding strategies on farms. Weighing troughs are a very accurate solution for this purpose but they are too costly for widespread utilization. In the last decade, several sensor systems for monitoring behavior and localization of animals were developed and tested. In this connection, the aim of this study was to evaluate whether records on feeding time or chewing activity or a combination of both comprise enough information to allow an estimate on feed intake with a sufficient accuracy. Feed intake and feeding time per cow were recorded by 36 weighing troughs. Presence at the trough was synonymous with feeding time. Simultaneously, chewing activity of seven cows was recorded by MSR-ART sensors. The latter comprised a halter, a pressure sensor able to recognize jaw movements, and a logger for data storing. Data sets of 45 measuring days were available for statistical analysis, which included analysis of temporal and quantitative agreement of the three considered variables. Average daily feed intake of the seven cows ranged from 19.6 kg to 23.3 kg. The average feeding time per day varied between 232 min and 331 min, the average chewing time between 228 min and 347 min. First analyses indicate that feed intake and feeding and chewing time were highly correlated in individual cows. Premises for a good prediction of feed intake are accurate data records and a sufficient variation of the input variables in the calibration set. The ongoing evaluation focuses on the influences of individual cows and daily patterns on the ability to accurately estimate feed intake. Final results will be presented at the conference.

---



**Precision farming and models for improving cattle performance and economics**

*I. Halachmi*
*The Institute of Agricultural Engineering, Agricultural Research Organization (A.R.O.), The Volcani Centre, P.O. Box 6, 50250 Bet Dagan, Israel; halachmi@volcani.agri.gov.il*

The feed intake is the single most costly parameter in intensive livestock operation. Over 64% of the total expenses in a livestock farm can be associated with feed cost. Therefore, knowing the cow, individual feed intake has a potential economic value. An accurate cow individual feed intake can support (1) decision regarding breeding and culling strategy towards feed efficiency and (2) concentrate allocation in milking robots and computerized feed dispensers. In research institutions, cow individual feed intake can be measured directly by using weighting scales locating below the feed troughs. Such a system is rather expensive therefore in commercial farms, models are needed. In order to predict cow individual feed intake, the ARO has developed an equation based on milk yield changes, body weight changes, age, feed diet, season, days in milking and time in the feed lane. Correlation between time in the feeding trough, other variables and food intake in each visit to the feeding trough, as measured in the ARO research farm will be presented in the conference.

**Facilitation of assessment of technical measures in implementing the Broiler Directive (2007/43/EC)**

*A. Butterworth*[1], *G. Richards*[1] *and E. Vranken*[2]
[1]*University of Bristol, Clinical Veterinary Science, Langford, N Somerset, BS40 5DU, United Kingdom,* [2]*FANCOM B.V., Research Department, P.O. Box 7131, 5980 AC Panningen, Netherlands Antilles; andy.butterworth@bris.ac.uk*

The Broiler Directive (2007/43/EC) is unique amongst current EU Directives which address Animal Welfare, in that it uses outcome data collected at abattoirs and on farm to monitor on farm broiler welfare and vary the maximum permitted stocking density on farm. In this paper we describe the process by which manually assessed animal outcome measures for broiler chickens have started to be used alongside the use automated on farm measurements of climate, feed intake, animal growth and camera and sound based automated precision livestock farming methods (eYeNamic) in our pilot studies. We describe how the data collected from this process (both human assessor based and automated farm measures) has enabled the start of a process of 'joint validation' which it is anticipated will lead to advances in automated measurement of environmental and animal parameters, some of which may be potentially fed into the statutory requirement for ongoing assessment under the Broiler Directive 2007/43/EC. The pilot study has identified the key components of the 'baseline standard' for animal based measures being assessed against automated on farm measures, and this information is summarised in the paper. For example, food pad dermatitis, hock burn, walking ability (Gait Score), avoidance distance touch tests and response to the stockman are identified as some of the measures of medium to high priority in terms of high potential for automated measurement, and also relating well to the requirements both of the Broiler Directive, but also as management information of commercial use to broiler production companies. On the other hand, breast lesions, cellulitis, emaciation, joint lesions, scratches and wing fractures were identified as being difficult or impossible to assess on farm, but suitable for measurement in the slaughterhouse.



**Monitoring the hatching time of individual chicks and its effect on chick quality**

*Q. Tong*[1], *T. Demmers*[1], *C.E.B. Romanini*[2], *V. Exadaktylos*[2], *H. Bergoug*[3], *N. Roulston*[4], *D. Berckmans*[2], *M. Guinebretière*[3], *N. Eterradossi*[3], *R. Verhelst*[4] *and I.M. McGonnell*[1]
[1] *Royal Veterinary College, Hawkshead Lane, North Mymms, Hatfield, AL9 7TA Hertfordshire, United Kingdom,* [2]*KU Leuven, Division M3-BIORES: Measure, Model & Manage Bioresponses, Kasteelpark Arenberg 30 – Box 2456, 3001 Leuven, Belgium,* [3]*Anses, Ploufragan-Plouzané Laboratory, BP 53 Ploufragan, 22440, France,* [4]*Petersime N.V., Research and Development, Centrumstraat 125, 9870 Zulte (Olsene), Belgium; tdemmers@rvc.ac.uk*

Monitoring the hatch process demands continuous assessment of the number of hatched chicks, which is to date an interrupt event as opening incubator door is required. Therefore, a real time means to monitor the hatching time of individual chick is of interest to both research and industry. An alternative non-invasive method is presented measuring the eggshell temperature using small accurate temperature sensors. Continuous recordings of eggshell temperature (Tegg) of the focal eggs were analysed and the temperature profiles of plotted Tegg showed a temperature drop around 2-6 °C when chick hatched. Therefore it was possible to obtain the hatching time of individual egg in real-time during incubation based on this registered temperature drop. Furthermore linear regression analysis showed a positive correlation between hatching time ($r=0.32$; $P=0.001$) and chick weight at take-off which indicated that early hatched chicks started to lost weight during holding period.

**The use of vocalisation sounds to assess responses of broiler chicken to environmental variables**

*I. Fontana[1], E. Tullo[1] and A. Butterworth[2]*
*[1]Faculty of Veterinary Medicine, University of Milan, VESPA, Via G. Celoria 10, 20133 Milan, Italy, [2]School of Veterinary Sciences, University of Bristol, Division of Food Animal Science, Langford House, Langford, BS40 5DU Bristol, United Kingdom; ilaria.fontana@unimi.it*

The vocalisation sounds of broiler chicken, have been previously studied to, and in this study we describe the monitoring of broiler chicken vocalisation under normal farm conditions, with sound recorded and assessed at regular intervals throughtout the entire life of the bird from day 1 to day 37 to assess whether there are recognisable, and even predictable, vocalisation patterns, based on frequency and bandwidth analysis, seen in birds at different ages and stages of growth within commercial broiler production timescales. Two experimental trials were carried out in a 'conventional' indoor reared broiler farm, and the audio recording procedures lasted for 37 days. The recordings were made at regular intervals and in the same position of the equipment, inside the broiler house, during each period of data collection. The recordings were made using (Marantz PMD 661 MK II) and collected automatically and without the presence of human operators to provide sound recordings which represented situations without disturbance of the animals beyond that of that created by the farmer. One hour duration digital file were cut into short files of 10 minutes duration, and these sound recordings were analyzed and labelled using analysis software: Adobe Audition CS6. Analysing the sounds recorded, with audio software (Pratt), it was noted that the sounds and the related frequencies, changed in relation to the increasing of the age and the weight of the broilers. Statistical analysis showed a significant correlation ($P<0.001$) between the frequencies of the vocalisations and the age and behaviour of the birds. This method, based on the identification of specific frequencies of the sounds emitted, compared with their age and weight has the potential to be used system to evaluate the health and welfare status of the animals at farm level.



**Three clinical field trials with the pig cough monitor: an overview**

*M. Hemeryck[1], G. Finger[2], M. Genzow[3] and D. Berckmans[1]*
*[1]SoundTalks, Kapeldreef 60, 3001 Heverlee, Belgium, [2]Tierärztliche Praxis Lindhaus, Ebbinghoff 28, 48624 Schöppingen, Germany, [3]Boehringer Ingelheim, Binger Straße 173, 55218 Ingelheim am Rhein, Germany; martijn.hemeryck@soundtalks.com*

We have participated in 3 clinical field trials together with a pharmaceutical industry partner. The trials demonstrate the use of the SoundTalks Pig Cough Monitor (PCM) as a tool for monitoring animal respiratory health, welfare and farm efficiency on commercial pig fattening farms. The 3 trials were based on a commercial pig fattening farm in North-West Germany. They covered a time range of 3 fattening rounds, each round taking up about 3 months. Depending on the trial, serum drawings and oral fluid samplings were employed beside the PCM to detect a number of pathogens: Porcine reproductive and respiratory syndrome virus (PRRSV), Swine Influenza Virus (SIV) and *Mycoplasma hyopneumoniae*. Diagnostics were analyzed for these pathogens and compared with the PCM cough index (CI; average coughs per hour per day). Common to all trials was a baseline CI between 10 and 20. The 1$^{st}$ trial showed that any change in normal respiratory behavior was reflected in CI. The 2$^{nd}$ trial showed that *M. hyopneumoniae* had a severe effect on CI, co-infection with PRRSV aggravated symptoms. The 3$^{rd}$ trial showed paroxysmal cough for 8 days, picked up by CI, serum and oral fluid sample analyses. The PCM proved to be an objective and unbiased measurement of respiratory health. Any deviation from normal respiratory pattern could be observed as an increase in CI. Time between seroconversion of *M. hyopneumoniae* to clinically relevant cough in the 1$^{st}$ trial appeared to be short and less than previously reported. Infection with *M. hyopneumoniae* in the 2$^{nd}$ trial resulted in a distinct increase in cough, co-infection with PRRSV worsens the symptoms. The 3$^{rd}$ trial demonstrated the use of oral fluid samples in addition to results from serum. Paroxysmal cough was clearly apparent in CI and in line with findings from literature.

**Detecting health problems of individual pigs based on their drinking behaviour**

*I. Adriaens[1], J. Maselyne[1,2], T. Huybrechts[1], B. De Ketelaere[1], S. Millet[3], J. Vangeyte[2], A. Van Nuffel[2] and W. Saeys[1]*
*[1]KU Leuven, MeBioS, Kasteelpark Arenberg 30 bus 2456, 3001 Heverlee, Belgium, [2]ILVO, Technology and Food Science Unit, Burg. van Gansberghelaan 115 bus 1, 9820 Merelbeke, Belgium, [3]ILVO, Animal Sciences Unit, Scheldeweg 68, 9090 Melle, Belgium; jarissa.maselyne@ilvo.vlaanderen.be*

Automatic detection of health, welfare and productivity problems on individual animals enables a fast intervention of the farmer, reducing the risk of economic losses, excessive use of antibiotics and suffering of the animals. Since animals change their behaviour in response to problems or stress, it is hypothesized that drinking behaviour of pigs can be a valid indicator of health or welfare problems. A system was designed to register the drinking behaviour automatically. A High Frequency Radio Frequency Identification (RFID) system was placed around 4 nipple drinkers and 55 pigs were equipped with a tag on their ears. Validation of the RFID system was done by comparing drinking bouts constructed from the RFID data with visual observations. Too long and too short registrations were deleted. On average, 97% of the drinking bouts observed were also registered by the RFID system. This corresponds to 99.2% of the total duration of drinking observed. However, the RFID system overestimated the number and duration of the drinking bouts with 9 and 18%, respectively. This can be corrected for by using flow meter data. To detect altered drinking behaviour as a proxy for individual pigs' problems, Synergistic Control (SGC) is used on the parameters extracted from the RFID data. SGC is a combination of Engineering Process control (EPC) and Statistical Process Control (SPC). Problems can be detected when drinking behaviour differs too much compared with the in-control variation of the process. This concept shows potential for early detection of problems.



**Continuous surveillance of pigs in a pen using learned based segmentation in computer vision**

*M. Nilsson[1], A.H. Herlin[2], K. Åström[1], H. Ardö[1] and C. Bergsten[2]*
*[1]Lund University, Centre for Mathematical Sciences, P.O. Box 118, 22100 Lund, Sweden, [2]Swedish University of Agricultural Sciences, Department of Biosystems and Technology, P.O. Box 103, 23053 Alnarp, Sweden; anders.herlin@slu.se*

This work investigates the feasibility to extract ratio of pigs located at the pig pen dunging area from video. Since pigs generally locate themselves in the wet dunging area if the ambient temperature is too high in order to avoid heat stress, as wetting the body surface is the major path to dissipate the heat by evaporation. Thus, the ratio of pigs in the dunging area and resting area could be used as an indicator for controlling the climate in the pig environment. The computer vision methodology utilises a learning based segmentation approach with several features. This is done in order to overcome some of the limitations found in our setup using gray-scale information only in this difficult imaging environment, which includes shadows and challenging lighting conditions. Additionally, the method is able to produce probabilities per pixel rather than a hard decision. In order to test practical conditions, we filmed a pig pen containing ten young animals from a top view perspective by an Axis M3006 camera with a resolution of 640×480 in three ten minute sessions under different light conditions. The results indicate that a learning based method improves on grayscale methods in the task to reliably find the ratio which could be an important feature to use in order to control climate from observed pig behaviour. This feature could further be incorporated in identifying boxes where individual pigs may deviate from normal behaviour or location in the box which may indicate inferior health or acute illness.

**Real-time analyses of BHB in milk can monitor ketosis and it's impact on reproduction in dairy cows**

*J.Y. Blom and C. Ridder*
*Lattec I/S, 69 Slangerupgade, 3400 Hillerød, Denmark; jyb@lattec.com*

Traditionally, cow-side tests to monitor subclinical and clinical ketosis are limited to one sample in the postpartum period. This report presents the first results of ketosis detection and reproductive performance in dairy herds utilizing Herd Navigator™, where milk samples are automatically analyzed for β-HydroxyButyrate (BHB) at least once daily in the post-partum period. Apart from ketosis, the system also monitors reproduction, mastitis and milk urea levels in real-time. Farm 1 (278 cows, DK) and 3 (126 cows, CA) were milked in DeLaval VMS and cows in farm 2 (151 cows, NL) were milked in a 2×10 parlour barn. β-Hydroxybutyrate (BHB) and Progesterone (P4) were measured on a daily basis in Herd Navigator™, and data processed in the system's biomodels (1,2). BHB was measured 4-60 DFC and P4 from 20 days before end of the voluntary waiting period. All data analyses were performed in Matlab. The number of ketosis alarms, and hence ketosis events varied from farm to farm, ranging from 3 to 38 alarms per 100 calvings. Early (≤10 DFC) and later alarms (>10 DFC) did not differ among herds. Incidence rates for Post Partum Anoestrus varied from 4 to 23 alarms per 100 calvings, and closely related to the incidence rate for ketosis. The analysis of reproductive performance revealed that ketosis, cystic ovaries and post partum Anoestrus highly influences the conception rates and length of the breeding period, irrespective of length of the voluntary waiting period. With the use of Herd Navigator™, inferior cow performance can easily be monitored on a real-time basis, and factors contributing to improper ketosis and reproductive management can be identified and corrected to improve farm performance.

---

Session 11 Theatre 8

**Potential for assessing the pregnancy status of dairy cows by mid-infrared analysis of milk**

*A. Lainé, H. Bel Mabrouk, L.-M. Dale, C. Bastin and N. Gengler*
*University of Liège, Gembloux Agro Bio-Tech, Passage des déportés, 2, 5030, Belgium; aurelie.laine@ulg.ac.be*

In dairy cattle, in opposition to other species, performances recording schemes allow to provide advisory tools that integrate information across the whole population. Mid-infrared (MIR) analysis of milk provides a spectrum for each individual cow's milk sample. The MIR spectrum represents the whole milk composition and can be used to assess the status of the animal (e.g. health, pregnancy, feeding). The main objective of the European project OptiMIR (INTERREG IVB North West Europe Program) is to develop innovative advisory tools based on the MIR data collected by milk recording organization. One of the first objectives is to develop a tool to assess the pregnancy status of cows. The tool is based on the comparison of the observed spectrum with an expected spectrum obtained from a set of spectra with a known status of the cow, here being open. Development was done using Walloon milk recording data. A training dataset (342,832 spectral data from 66,174 cows) was used to obtain residual spectra (i.e. difference between observed and expected spectra). Based on the fact that the pregnancy status of all cows was known, predictive discriminant function was constructed on 2,154 residual spectra randomly selected from the initial dataset. The discriminant function was then applied on the rest of the dataset (12,160 residual spectra) for validation. When considering the period from 21 to 50 days after an insemination, the error rate was about 1.9% with a specificity of 82.4% and a sensibility of 99.8%. These results showed a high potential for using directly the MIR spectrum of milk to detect a change in the pregnancy status of dairy cows. This methodology can also be applied to predict other types of physiological status changes (e.g. udder health related) and can be used on other types of biomarker data (i.e. collected from on-farm sensors). Similarly, integration of on-farm information on expected pregnancy status could improve the presented off-farm tool.

## Session 11 Theatre 9

### Progesterone profiles to evaluate simultaneous use of multiple oestrus detection technologies

*C. Kamphuis[1], K. Huijps[2] and H. Hogeveen[1,3]*
*[1]Wageningen University, Hollandseweg 1, 6706 KN Wageningen, the Netherlands, [2]CRV, Wassenaarweg 20, 6843NW Arnhem, the Netherlands, [3]Utrecht University, Yalelaan 7, 3584CL Utrecht, the Netherlands; claudia.kamphuis@wur.nl*

Automated oestrus detection technologies are increasingly popular in the dairy industry. Generally speaking, farmers invest in one technology assuming that this system finds most, if not all, cows in oestrus. It is, however, known that these technologies do not find all cows in oestrus. There are suggestions that automated oestrus detection may improve when sensor data are combined, e.g. joining activity with rumination data. So far, the option to combine different technologies has not been studied for the obvious reason that there are no commercial farms that implement technologies from multiple providers. The Smart Dairy Farming (SDF) project, a Dutch initiative, joins technology providers, knowledge institutions and dairy farmers to improve longevity of dairy cows by developing innovative tools that aid in animal health, reproduction and feeding strategies. The SDF project offers the unique opportunity to study whether combining multiple sensing technologies can improve automated oestrus detection. To do this, progesterone profiles will be created by daily measuring progesterone in milk from 40 cows, during a 24-day period, at two farms. One automated heat detection technology is implemented on both farms, and each farm has a second, different, technology running simultaneously. Generated heat alerts and records of farmers' observations will be compared with progesterone profiles. Data will be used to provide insight in the following issues: do heat detection technologies alert for cows in oestrus, when do they alert for oestrus events, how do farmers use the information from the heat detection technologies, and whether the exact timing of true oestrus may improve by combining heat alerts. Finally, possible explanations will be discussed for those oestrus events that remain undetected by both oestrus detection systems and farmer's observations.

---

## Session 12 Theatre 1

### Hoof lesion detection of dairy cows with manual and automatic locomotion scores

*A. Schlageter-Tello[1], T. Van Hertem[2], S. Viazzi[2], E.A.M. Bokkers[3], P.W.G. Koerkamp[4], C.E.B. Romanini[2], M. Steensels[2], C. Bahr[2], I. Halachmi[5], D. Berckmans[2] and C. Lokhorst[1]*
*[1]Wageningen University, Livestock Research, Edelhertweg 15, 8219 PH, Lelystad, the Netherlands, [2]KU Leuven, M3-Biores, Kasteelpark Arenberg 20, 3001 Leuven, Belgium, [3]Wageningen University, Animal Production Systems Group, De Elst 1, 6708 WD Wageningen, the Netherlands, [4]Wageningen University, Farm Technology Group, Droevendaalsesteeg 1, 6708 PB Wageningen, the Netherlands, [5]Agricultural Research Organization, P.O. Box 6, 50250 Bet-Dagan, Israel; andres.schlagetertello@wur.nl*

Hoof lesions are an important problem in dairy farms and are mostly detected using manual locomotion scoring (MLS) systems. Recently different automatic locomotion scoring (ALS) systems have been developed. The objective of this study was to determine the capability of a MLS system and an automatic 3D video based locomotion scoring system to detect hoof lesions. The experiment was performed on a dairy farm with 250 milking cows. Hoof lesion assessment, including severity, was performed while cows were hoof trimmed. Before hoof trimming MLS and ALS were performed and expressed in a five-level locomotion score. A cow was lame when MLS or ALS was $\geq 3$. Sensitivity was defined as the percentage of lame cows having hoof lesions. Specificity was defined as the percentage of non-lame cows without lesions. Sensitivity and specificity were calculated with cows having one (severe) hoof lesion as a reference. Lameness prevalence was 33% (MLS) and 36% (ALS). Prevalence of hoof lesions was 84%, whereas for severe hoof lesions it was 55%. MLS had a sensitivity of 35% and a specificity of 79% for cows with one hoof lesions and a sensitivity of 49% and a specificity of 83% for one severe hoof lesion. ALS had a sensitivity of 38% and a specificity of 73% for cows having at least one hoof lesions and a sensitivity of 51% and a specificity of 77% for one severe hoof lesion. In conclusion, both manual and automatic locomotion scores presented poor to moderate capability for detecting cows having at least one (severe) hoof lesion.

## Early detection of metabolic disorders in dairy cows by using sensor data

*R.M. De Mol[1], J. Van Dijk[1], M.-H. Troost[2], A. Sterk[3], R. Jorritsma[4] and P.H. Hogewerf[1]*
*[1]Wageningen UR Livestock Research, P.O. Box 65, 8200 AB Lelystad, the Netherlands, [2]Rovecom, Elbe 2, 7908 HB Hoogeveen, the Netherlands, [3]Agrifirm, Landgoedlaan 20, 7325 AW Apeldoorn, the Netherlands, [4]Utrecht University, Yalelaan 7, 3584 CL Utrecht, the Netherlands; rudi.demol@wur.nl*

The transition period is a crucial period for the dairy cow. A negative energy balance results in an increased risk for metabolic disorders like milk fever, ketosis and left displaced abomasum. Detection of these metabolic diseases may be improved by using sensors. In the Dutch Smart Dairy Farming project, these possibilities have been examined. A detection model has been developed and tested. On a practical dairy farm (300 cows, automatic milking system), sensors have been installed for automated measurements of the milk yield, milk composition (fat, protein), visits to the milking robot (rewarded and unrewarded), concentrates intake, visits to the concentrates feeder (rewarded and unrewarded), activity, rumination activity and body weight. It is known from literature that most of these variables are influenced by metabolic disorders. Sensor measurements were aggregated to daily level. In the detection model there were three types of alerts: (1) level alert: the value was outside a confidence interval (based on moving average and standard deviation on preceding values); or (2) trend alert: the change in successive values was outside a confidence interval; or (3) index alert: given on the day of calving in several situations, e.g. when the weight loss was compared with the average weight loss. An alert for metabolic disorder was generated when the number of alerted variables exceeded a certain level. These metabolic alerts were compared with the reference data to estimate the model performance: first for the data set that was also used for model development and secondly for an earlier data set that was only used for testing. The results from the detection model (sensitivity and specificity) will be included in the presentation.



## Behaviour and performance based health detection in a robotic dairy farm

*M. Steensels[1,2], C. Bahr[2], D. Berckmans[2], A. Antler[1], E. Maltz[1] and I. Halachmi[1]*
*[1]Agricultural Research Organization (ARO) – the Volcani Center, Institute of Agricultural Engineering, P.O. Box 6, 50250 Bet Dagan, Israel, [2]KU Leuven, M3-BIORES, Kasteelpark Arenberg 30, bus 2456, 3001 Heverlee, Belgium; halachmi@volcani.agri.gov.il*

Correct separation of ill cows from the herd is important, especially in robotic dairy farms, where searching for an ill cow can cause disturbances in the routine of other cows. Nowadays, many sensors are available to help the farmer to monitor his cows. Therefore the aim of this study was to apply a behaviour and performance based health detection model on post-calving cows in a robotic dairy farm to detect ill cows. The study was conducted in an Israeli robotic dairy farm with 250 Israeli-Holstein cows. All cows were equipped with a rumination and activity monitoring system, that measured rumination time and activity in intervals of two hours. Milk yield, visits to the milking robot and body weight were recorded in the milking robot. For data analysis, a daily sum was calculated for activity, rumination time, milk yield and visits to the milking robot. For body weight, a daily mean was calculated. A tree based model was developed based on a calibration dataset of historical data of the last year. The resulting model was validated on new data of post-calving. The decision generates for each new cow input a probability of being ill. For classification, the cut-off threshold was set at 0.5. The model was applied once a week on the day that the veterinarian performing the routine post-calving health check, that included testing all post-calving cows for ketosis and metritis. The veterinarian diagnosis served as a binary reference of the model (healthy-ill). The validation dataset consisted of 66 cows. The tree based model had a sensitivity of 66% and a specificity of 79%. A pair-wise t-test showed that the model output was not significantly different from the diagnosis of the veterinarian. Further analysis should reveal how to improve the sensitivity of the model.

## Session 12 Theatre 4

**Monitoring the body temperature of cows and calves with a video-based infrared thermography camera**

*G. Hoffmann and M. Schmidt*
*Leibniz-Institute for Agricultural Engineering Potsdam-Bornim (ATB), Engineering for Livestock management, Max-Eyth-Allee 100, 14469 Potsdam, Germany; ghoffmann@atb-potsdam.de*

In this study a video-based infrared camera (IRC) was investigated as a tool to monitor the body temperature of cows and calves. Therefore, the body surface temperatures were measured contactless using videos from an IRC fixed at a certain place in the automatic milking system (cows) or calf feeder (calves). The body surface temperatures were analyzed retrospectively at bounded body regions (eye, back of the ear, shoulder, vulva) and at two larger areas called head- (before forehead) and body area (behind forehead). The rectal temperature served as reference temperature and was measured with a digital thermometer at the corresponding time point. Altogether, 10 milking cows (Holstein-Friesians, 3 to 9 years of age) and 9 calves (Holstein-Friesians, 8 to 35 weeks old) were examined. Significant differences ($P<0.01$) were found between the measured IRC temperatures among the bounded body regions, i.e. eye (mean: 37.0 °C), back of the ear (35.6 °C), shoulder (34.9 °C) and vulva (37.2 °C), except between eye and vulva ($P=0.99$). The range between the minimum and maximum temperature at the 4 above mentioned regions was large, and many outliers were found. However, the regions eye and back of the ear proved to be suitable for temperature monitoring in cattle. Further examinations showed that the maximum temperatures measured by IRC at the head and body area increased with an increase of the rectal temperature in cows and calves. Ongoing investigations are taking place now in order to define algorithms and reference values. The advance of the IRC is that more than one picture per animal can be analyzed in a short period of time in contrast to single picture cameras. Therefore, this system shows potential as an indicator tool for continuous temperature measurements in cattle.

## Session 13 Theatre 1

**Roles of amino acids in the regulation of food intake by animals**

*G. Wu*
*Texas A&M University, Animal Science, 2471 TAMU, College Station, Texas, 77843-4410, USA; g-wu@tamu.edu*

Food intake by animals is affected by the taste, quantity, and quality of dietary protein and amino acids (AA). For example, food consumption by animals is depressed in response to: (1) a severe deficiency of, or substantial increase in, dietary protein or an individual AA; and (2) an imbalance of dietary AA. Interestingly, high intake of dietary protein contributes to satiety to a greater extent than an isocaloric amount of fats or carbohydrate, further supporting a crucial role for AA in the control of food consumption. Thus, the amount and balance of supplemental AA can either stimulate or suppress food intake, depending on individual AA, composition of AA and other nutrients in the basal diet, as well as endocrine status and developmental stage. The underlying mechanisms are complex and involve hormonal, neuronal, and metabolic signals generated from the digestive system, central nervous system, and other organs. At the stomach level, a variety of neurotransmitters, neuromodulators, and other peptides [including ghrelin and bombesin-related peptides (e.g. gastrin-releasing peptide and neuromedin B produced by gastric myenteric neurons) are utilized to control gastric emptying. These substances also act to relay signals from the mechanoreceptors on the gastric wall to the brain by vagal and spinal sensory nerves. In the small intestine, cholecystokinin and products of its endoproteolytic cleavage, as well as glucagon-like pepetide-1, oxyntomodulin and peptide YY, serve as primary satiation signals to inhibit food intake. In contrast, ghrelin stimulates food intake by animals, and physiological levels of glutamate increase gastric emptying and intestinal motility. At the brain level, both the direct blood pathway and the indirect neuromediated (mainly vagus-mediated) pathway contribute to the effects of dietary AA on food intake. Knowledge about the regulation of food intake by AA is expected to improve nutrition, health and well-being of animals.

**Modulation of redox state in pigs differing in feed efficiency as revealed by a proteomic analysis**

*F. Gondret[1], S. Tacher[1] and H. Gilbert[2]*
*[1]INRA, UMR1348 Pegase, Domaine de la Prise, 35590 Saint Gilles, France, [2]INRA, UMR1388 GenPhySE, cedex, 31326 Castanet-Tolosan, France; florence.gondret@rennes.inra.fr*

Improving feed efficiency of livestock is an important goal for sustainability and profitability. The physiological bases of feed efficiency are multifaceted. The functionality of the liver, a highly specialized organ with a wide range of functions including protein synthesis, energy metabolism and detoxification may be involved in variations of feed efficiency. In this study, two lines of pigs from a divergent selection experiment on residual feed intake (RFI) were compared. Two groups (RFI-: the most efficient; RFI+: the less efficient) were fed *ad libitum* during the growing-finishing period. A third group composed of RFI+ pigs was feed-restricted to the level of spontaneous feed intake of RFI- pigs. At slaughter (110 kg BW), the liver was excised. Soluble hepatic proteins (n=4 pigs per group) were separated by two-dimensional gel electrophoresis (2-DE) and spots having a differential abundance between groups were submitted to MALDI-TOF/TOF identification. A total of 50 protein spots had a differential abundance between the pigs, among which 35 proteins were identified. Members of the peroxiredoxin family (PRDX2, 4, 5 and 6), the omega form of the glutathione-S transferase (GSTO1) involved in the gluthatione derivative biosynthetic process, and the superoxide dismutase (SOD)-1 responsible for destroying free superoxide radicals, had a greater abundance in RFI- than in RFI+ pigs. Feed-restriction in the RFI+ pigs resulted in an increased abundance of the peroxiredoxins and the mu form of the glutathione-S transferase. Activity levels of key enzymes in the anti-oxidant or oxidative pathways in the liver were also measured. Notably, the activities of the oxidative enzymes hydroxyacyl-CoA dehydrogenase and citrate synthase were lower in RFI- pigs than in RFI+ pigs; they did not vary with feed restriction. In conclusion, selection for feed efficiency may have modified the redox state of the animals.

Session 13 Theatre 3

**The control of feed intake by metabolic oxidation in dairy cows**

*B. Kuhla, M. Derno, S. Börner, C. Schäff, E. Albrecht, M. Röntgen and H.M. Hammon*
*Leibniz Institute for Farm Animals Biology, Wilhelm-Stahl-Alle 2, 18196 Dummerstorf, Germany; b.kuhla@fbn-dummerstorf.de*

High-yielding dairy cows enter a negative energy balance after calving because feed intake is too low to meet the energy requirements for maintenance and milk production. At this time, cows mobilize their fat reserves – a process that is thought to interfere with the increase of feed intake. The control of feed intake results from the integration of humoral and nerval signals at feed intake regulatory centers in the brain. Metabolic oxidation in peripheral organs, primarily in the liver, may activate nerval signaling which is integrated and translated into anorexic responses by neurons of the hypothalamus (Ht). To study the relationship between metabolic oxidation and feed intake, cows were kept in respiration chambers with simultaneous recording of gas exchange and feed intake. All feed intake events were strongly cross-correlated with carbohydrate and fat oxidation in late-lactating cows. Transition cows with higher fat mobilization capacity (H cows) exhibited the greater fat oxidation, accompanied by the greater expression of hepatic enzymes involved in various fatty acid oxidation pathways and oxidative phosphorylation, each before and after parturition. H cows revealed greater plasma acyl ghrelin concentrations and the higher acyl:total ghrelin ratio, before and after parturition. The acyl:total ghrelin ratio correlated with several aspects of fat metabolism, but not with feed intake, implying that the ghrelin system plays a prominent role in fat allocation and oxidation rather than exerting orexigenic effects. Also, H cows had less activated orexigenic neurons in the arcuate nucleus of the Ht as indicated by the lower percentage of cFOS activated agouti-related protein (AgRP) neurons. Differential AgRP neuron activation was not associated with feed intake but with decreased oxygen consumption, increased respiratory quotient, and reduced NEFA concentrations, suggesting that AgRP activation plays a pivotal role in the regulation of metabolic oxidation during early lactation.

**Influence of roughage intake on free fatty acid receptors mRNA abundance in bovine adipose tissues**

*P. Friedrichs[1], L. Locher[2], S. Dänicke[3], B. Kuhla[4], H. Sauerwein[1] and M. Mielenz[1,4]*
*[1]Institute of Animal Science, University of Bonn, Katzenburgweg 7-9, 53115 Bonn, Germany, [2]Clinic for Cattle, University of Veterinary Medicine, Bischofsholer Damm 15, 30173 Hannover, Germany, [3]Institute of Animal Nutrition, Friedrich-Loeffler-Institute, Bundesallee 50, 38116 Braunschweig, Germany, [4]Institute of Nutritional Physiology, Leibniz Institute for Farm Animal Biology, Wilhelm-Stahl-Allee 2, 18196 Dummerstorf, Germany; pfri@uni-bonn.de*

Herein we studied if different proportions of roughage (R) in the diet will affect the serum concentrations of acetate (C2), propionate (C3) or butyrate (C4) and the expression of their sensing receptors FFA2 or 3 in subcutaneous (SC) and retroperitoneal (RP) adipose tissue (AT) of dairy cows during the transition period. Twenty pluriparous German Holstein cows were allocated to a low-R group (LR, n=10) receiving a diet with 40% R or a high-R group (HR, n=10) at 70% R (on DM basis) from d 1 to 21 postpartum. Serum samples and biopsies of SCAT (tail head) and RPAT were obtained at d -21, 1 and 21 relative to parturition. FFA2/3 mRNA abundances were quantified by qPCR. Serum SCFA were derivatized and measured on a GC-FID. The statistical analyses were performed using SPSS 21.0 ($P<0.05$). All investigated serum SCFA concentrations were higher at d 21 compared to d -21 and 1. On d 21 the LR group, had 1.7-fold higher C3, 1.2-fold lower C2 serum concentrations ($P<0.10$) and 2.5-fold lower FFA2 mRNA abundance in RPAT than the HR group. In RPAT FFA2 mRNA abundance decreased from d -21 to d 1. FFA3 mRNA abundance was lower on d -21 compared to d 1 in both AT. Differences between both AT were limited to FFA3 mRNA abundance that was higher in RPAT than in SCAT at d -21. As expected, C3 serum concentrations were higher in LR than in HR. The lower expression of FFA2 mRNA in RPAT of animals from the LR group points to a higher lipolysis rate. Different changes over time between both receptors mRNAs might indicate differential nutrient sensing in AT during the transition period.

---



**Neuropeptides linking the control of appetite with reproductive function in domestic animals**

*C.A. Lents*
*USDA, Agricultural Research Service, U. S. Meat Animal Research Center, Reproduction Research Unit, State Spur 18D, P.O. Box 166, Clay Center, NE 68933-0166, USA; clay.lents@ars.usda.gov*

The occurrence of puberty and maintenance of normal reproductive cycles are regulated by secretion of gonadotropin hormones from the pituitary gland, which is dependent upon the pulsatile release of gonadotropin-releasing hormone (GnRH) from the hypothalamus. It is well established that secretion of gonadotropin hormones is metabolically gated. It is generally accepted that metabolic signals relay information about energy balance and the degree of body fat to key neuronal pathways in the hypothalamus to alter the secretory pattern of luteinizing hormone (LH). Neuropeptides such as neuropeptide Y, kisspeptin, and alpha melanocyte-stimulating hormone appear to be central to nutritional regulation of LH release. RFamide-related peptide 3 is proposed as a hypophysiotropic hormone that suppresses LH secretion in mammals. RFamide peptides can play a role in appetite regulation and could, therefore, potentially function to interface nutrition with LH secretion. More recently, nesfatin-1 has been shown to participate in appetite regulation and energy homeostasis and may also have a role in conferring metabolic regulation on the reproductive neuroendocrine secretory axis. The challenge is to use this knowledge to devise management practices for different species that impinge upon these mechanisms to increase production efficiency of domestic animals. The USDA is an equal opportunity provider and employer. Partial support from CSREES (2005-35203-16852) and NIFA (2011-67015-30059).

### Associations of feed efficiency with fertility and sexual maturity in young beef bulls

*A.B.P. Fontoura[1], Y.R. Montanholi[2], M.D. Amorim[3] and S.P. Miller[4]*
*[1]Federal University of Pará, BR 316, Km 61, 68740-970, Castanhal, Brazil, [2]Dalhousie University, 58 River Road, B2N 5E3, Truro, Canada, [3]Western College of Veterinary Medicine, 52 Campus Drive, S7N 5B4, Saskatoon, Canada, [4]AgResearch, Puddle Alley, P.O. Box 50034, 9053 Mosgiel, New Zealand; abpfontoura@gmail.com*

Genetic improvement programs on beef cattle have been focused in feed efficiency, which is measured through residual feed intake (RFI). Recently, there are evidences supporting an antagonist relationship between fertility and sexual maturity with feed efficiency. Thus, the objectives were to obtain measures of fertility and sexual maturity and evaluate the relationships with RFI determinations and grouping sizes. Crossbred bulls (34) were housed and handled together for 112 d performance evaluation, which was followed by semen quality assessment, scrotum circumference measure, scrotum thermography, testis ultrasonography; as well as biometrics and histomorphometry of the testis. The RFI was calculated including body weight and gain ($RFI_{koch}$) and also by including ultrasound for body composition ($RFI_{us}$). More feed efficient bulls have lower sperm motility attributes (progressive motility; 47.3 vs 59.9%); higher abundance of sperm pathologies (tail; 4.3 vs 1.8%); lower testis echogenicity (175 vs 198 pixels); larger seminiferous tubules (mature size; 67.2 vs 31.6 $mm^2$) and; higher variation in the surface temperature at the top of the scrotum than less feed efficient bulls. No differences between feed efficiency groups were observed for organ or body size. The $RFI_{koch}$ model resulted in a difference of 50% higher backfat thickness in less feed efficient bulls. The $RFI_{us}$ was more sensitive for differences in fertility traits in comparison to the $RFI_{Koch}$. In conclusion, more feed efficient bulls collect features of incomplete sexual maturity. Moreover, RFI accounting for body composition should be recommended in the evaluation of fertility measures. Investigations to account for fertility in the selection for improved feed efficiency are warranted.

---



### The effect of the MC4R gene on feed intake and growth in growing finishing pigs

*A. Van Den Broeke[1], M. Aluwé[1], F. Tuyttens[1], S. Janssens[2], A. Coussé[2], N. Buys[2] and S. Millet[1]*
*[1]The Institute for Agricultural and Fisheries Research (ILVO), Animal Sciences Unit, Scheldeweg 68, 9090 Melle, Belgium, [2]KU Leuven, Livestock Genetics, Kasteelpark Arenberg 30, 3001 Leuven, Belgium; alice.vandenbroeke@ilvo.vlaanderen.be*

The Asp298Asn polymorphism of the MC4R gene is known to have an effect on economically important traits like growth rate and backfat thickness. The effect on feed intake, on the other hand, has not yet been studied well in pigs. This intervention study with a 2×2 design (genotype × gender) was set-up to assess main effects and possible interactions between gender and MC4R genotype on daily feed intake (DFI), average daily gain (DG), average daily lean meat gain, feed conversion ratio (FCR) and carcass parameters in pigs between 25 and 115 kg. Eleven rounds of 4 pens (one per homozygous genotype × gender combination) of 6 pigs were evaluated. The pigs had *ad libitum* access to water and feed (three phase diet). During the total growth period, AA pigs had a higher DFI than GG pigs whereas gender had no significant effect. A significant interaction between gender and genotype was observed for DG during the total growth period: AA boars had a significant higher DG than GG boars, while AA and GG gilts did not differ significantly. However, when evaluating DG per growth phase, higher DG of AA pigs was observed in both sexes during the second and third phase of growth and the interaction effect was not significant. Lean meat gain was higher in boars compared to gilts, independent of genotype. Similarly, the boars showed a lower FCR compared to the gilts, without effect of genotype. GG pigs had a higher meat percentage than AA pigs whereas gender had no effect. In conclusion, these results indicate that the higher DG of the AA pigs compared to the GG pigs observed in other studies, is probably due to a higher DFI. However, over the total period of growth, the effect of genotype on daily gain could be observed inboars but not in gilts. The higher feed intake resulted mainly in extra fat disposition.

## ω-3 fatty acids modulate expression of orexigenic peptides and leptin signaling in hypothalamus

*N. Mohan[1], H. Ma[2], K. Chiang[2] and J. Kang[2]*
*[1]National Research Centre on Pig, Indian Council of Agricultural Research, Rani, Guwahati, 781131, India, [2]Massachusetts General Hospital, Harvard Medical School, Department of Medicine, 149, 13th Streeet, Charlestown, MA, 02129, USA; mohannh.icar@nic.in*

Impaired leptin and insulin signaling in the hypothalamus, resultant aberrant energy homeostasis (EH) and increased adiposity are the pathophysiological basis for obesity development. The role of ω-3 FA in the hypothalamic signaling,appetite regulation, EH and obesity development remains to be established. In this study, we used the fat-1 transgenic mice, which are capable of producing ω-3 FA from ω-6 FA and thereby rich in endogenous ω-3 FA in their tissues and organs, to investigate the effect of increased tissue content of ω-3 FA on the expression of EH-related neuropeptides, the signaling of insulin and leptin in the hypothalamus, and adiposity. The hypothalamic neuropeptides expression and phosphorylated STAT3 and PI3 kinase levels in fat-1 transgenic (n=6) and wild type (WT) mouse (n=6) were analyzed using realtime PCR and western blotting respectively. Our results showed that fat-1 mice had lower hypothalamic mRNA levels of orexigenic peptides (NPY, AgRP and orexin), NPY receptor, leptin receptor 2, and orexin receptor than wild type (WT) littermates. On the other hand, anorexigenic peptides (CART and POMC) mRNA levels were higher in hypothalamus of WT mice than fat-1 mice. Reduced phosphorylated STAT3 and PI3 kinase levels in the hypothalamus were also observed in fat-1 mice. Furthermore, the expressions of lipogenic genes (FAS, SCD and SREBP-1c) were reduced in the visceral adipose tissue of fat-1 mice. Feed intake and body adiposity of fat-1 mice were also lower than those of WT littermates. Our results provide evidence that ω-3 FA through multiple concurrent targeting reduces orexigenic signaling in hypothalamus and adiposity, suggesting that ω-3 FA supplementation might be a novel strategy for the management of obesity. The study suggests a role for ω-3 FA in appetite control in animals.

---



## Persistence and competition in the human gut microbiome

*H.B. Nielsen and D. Plichta*
*Technical University of Denmark, Center for Biological Sequence Analysis, Kemitorvet, Building 208, 2800 Lyngby, Denmark; plichta@cbs.dtu.dk*

The genetic diversity of the human gut microbiome includes numerous species, phages and clonal heterogeneity and extends far beyond what is covered by reference genomes. Here we present the first microbiome-wide analysis of dependency-associations that link phages and clonal heterogeneity to their microbial hosts. The dependency-associations reveal specific CRISPR-phage relationships and longitudinal samplings show that the occurrence of some associated genetic elements predicts persistence probabilities of the microbes that carry these. In addition, microbial persistence probabilities reveal interactions between gut microbes and together with transcriptional observations these indicate inter-species competition in the human gut microbiome. These observations are made possible by an exhaustive co-abundance segregation of the gut metagenome into 7,381 highly correlated co-abundance gene groups (CAGs) including 741 metagenomic-species (MGS) across 396 human fecal samples.

**Use of full genome sequence information in fine mapping of QTL**

*G. Sahana*
*Center for Quantitative genetics and Genomics, Aarhus University, 8830 Tjele, Denmark; goutam.sahana@agrsci.dk*

Genome-wide association studies primarily make use of markers that are intended to represent causal variation indirectly, whereas, in principle, next-generation sequencing (NGS) can directly identify the causal variants. This is perhaps the fundamental advantage of sequencing approaches over standardized genotyping panels. NGS is rapidly becoming the primary focus of efforts to characterize the genetic bases of quantitative traits. We have conducted association studies using NGS data for all traits in the breeding goal in dairy cattle. SNP chip data for nearly 12,000 bulls from three cattle breeds were imputed to full sequence level variants using a multi-breed reference of 531 bulls. We were able to identify the causal variants for two embryonic lethal conditions and in general located QTLs more precisely. Even with full genome sequence variants from three breeds, failed to identify QTNs for traits like fertility and mastitis. Failure to identify causal variants affecting quantitative traits could be due to (1) high linkage disequilibrium, (2) multiple linked QTL, (3) different casual polymorphisms segregating in different breeds, evident from the haplotype frequencies and effects, (4) causal variants were not in the analyzed panel: removed in the quality checking, at low frequencies (rare variants) or structural variants, (5) no direct functional validation available. High coverage targeted re-sequencing and/or exome sequencing may help. Exome sequencing will be able to discover not only rare causal variants, but also protein-coding risk variants. However, the difficulty in tracking down association signals to causal polymorphisms has led to assume that a sizable amount of causal variants must have subtle regulatory effects. Our inability to identify QTNs greatly limits their importance in applications across cattle populations. However, as the SNPs identified from NGS-GWAS are causal variants or in high LD with the QTNs, it will be of interest to assess their predictive ability of genetic merits of individual across breeds.

---



**Systems genetics analysis of RNA-Seq data to unravel the genetics of obesity using a porcine model**

*L.J.A. Kogelman[1], D.V. Zhernakova[2], H.-J. Westra[2], S. Cirera[1], M. Fredholm[1], L. Franke[2] and H.N. Kadarmideen[1]*
*[1]Univ of Copenhagen, Dep. of Veterinary Clinical and Animal Sciences, Grønnegårdsvej 7, 1870 Frederiksberg C, Denmark, [2]Univ Medical Center Groningen, Dep of Genetics, A. Deusinglaan 1, 9713 AV Groningen, the Netherlands; lkog@sund.ku.dk*

Obesity is a complex health problem associated with many diseases, e.g. type 2 diabetes and cardiovascular diseases. Microarray gene expression data have been extensively used to detect differentially expressed (DE) genes and expression quantitative trait loci (eQTL) in the past years. However, RNA-Sequencing (RNA-Seq) data have the larger scope and potential of revealing novel genes, transcripts and transcriptional regulations in complex traits. The aim of this study was to elucidate biological pathways and potential biomarkers for obesity in a porcine model, by systems genetics approaches using RNA-Seq data. Previously, we created an F2 pig population which was deeply phenotyped (e.g. weight, confirmation, DXA scanning and slaughter characteristics) and genotyped (Illumina 60K SNP Chip). Based on the selection index theory, we created an Obesity Index (OI), combining the genetic values of nine obesity-related phenotypes into one aggregate genetic phenotype for obesity. Based on the OI, 36 animals were selected for RNA-Seq. Analysis of the RNA-Seq data included detection of DE genes and network construction, followed by pathway detection. To find the downstream effects of the genetic variants associated with obesity we performed eQTL mapping. After adequate quality control and alignment, we identified 198 DE genes, which could after network construction (using GeneNetwork) be divided into two sub-networks, representing resp. immune related processes and developmental related processes. Furthermore, we revealed 761 cis-eQTLs of which several could be linked to obesity, e.g. the PEX10 gene. In conclusion, systems genetics analysis of RNA-Seq data elucidated biologically relevant pathways and potential genetic biomarkers affecting obesity and obesity-related diseases.

**Systems biology and RNA-Seq analyses to detect biomarkers for metabolic syndrome in a porcine model**

*P. Mantas, L.J.A. Kogelman, S. Cirera, M. Fredholm and H.N. Kadarmideen*
*Univ of Copenhagen, Dep. of Veterinary Clinical and Animal Sciences, Grønnegårdsvej 7, 1870 Frederiksberg C, Denmark; qhf673@alumni.ku.dk*

Metabolic syndrome (MS) is a set of risk factors for type 2 diabetes and cardiovascular diseases. The welfare and economic implications increase the need for improving diagnosis and treatment of MS. Next generation sequencing of mRNA (RNA-Seq) has been shown to have great potential for unraveling the underlying mechanisms of complex diseases. The present study aims at identifying molecular pathways and potential biomarkers for MS traits using systems biology analyses on RNA-Seq in a porcine model. Previously, we created an F2 pig population genetically divergent for obesity and obesity-related traits. This population has been phenotyped for about 27 different metabolic parameters, related to metabolic disorders such as glucose and cholesterol. Based on their degree of obesity (obesity index values), a subset of 36 animals has been selected for paired-end RNA-Seq of subcutaneous adipose tissue. Explanatory statistical genetic analysis of the 27 serum parameters classified the animals into high and low level groups for the analysis of extreme phenotypes. After the quality control and normalization of RNAseq data, systems biology analysis were conducted to detect differentially expressed (DE) genes in Bioconductor, build co-expression networks using WGCNA methods, and perform downstream analysis of those results using e.g. pathway analysis. Preliminary results of the DE analysis showed highly significant DE genes in 20 out of the 27 phenotypic traits. Functional annotation revealed genes related to metabolic syndrome, e.g. an upregulation of the B2M gene in animals with elevated LDL levels, which is associated with LDL-associated diseases as atherosclerosis and peripheral arterial disease. WGCNA based analyses are ongoing. In conclusion, the preliminary results show the systems biology methods applied to this RNAseq data in porcine model reveals key novel genes and pathways associated with the metabolic syndrome.

---



**Analysis of the ovine homolog of the T-locus as a candidate gene for short tail phenotype**

*C. Weimann, K. Fleck and G. Erhardt*
*Justus-Liebig-University, Department of Animal Breeding and Genetics, Ludwigstrasse 21B, 35390 Giessen, Germany; christina.weimann@agrar.uni-giessen.de*

Between sheep breeds phenotypic variation of tail length can be observed worldwide. Evolutionary mechanisms explaining the variation of tail length in mammals are still unknown. Previous studies in sheep estimated heritability for tail length of 0.77 and suggest that inheritance may be through a small number of genes. One strong candidate gene for short tail phenotype in sheep is the T-gene (brachyury). Sequence variations in the T-gene are associated with short tail phenotype in mice, cats and dogs. The T-gene encodes a member of the T-box family of transcription factors, whereas T is expressed only in early development in primitive streak, prenotochordal mesoderm, and notochord and plays a central role in mesoderm determination and notochord formation. The objective of this study is to characterize the T-homolog gene in sheep in order to identify sequence variations which are responsible for tail length. In the first step the sequence of the sheep T-homolog was localized on ovine chromosome 8 via BLAST. Sequencing the coding regions of the ovine T-homolog showed high conservation of the sequences between cat, dog, mouse and sheep. The analysis of the sequences of 26 animals belonging to four different sheep breeds (Merinoland sheep, German Grey Heath, Romanov sheep and Rhoensheep) resulted in the detection of several polymorphisms within the coding regions of the gene. These polymorphisms did not correspond to the polymorphisms of the gene described in cat, dog and mouse.

**Sequence-based prediction for a complex trait in *Drosophila* reveals sex-differentiated epistasis**

*H. Simianer[1], U. Ober[1], W. Huang[2], M. Magwire[2], M. Schlather[3] and T.F.C. Mackay[2]*
*[1]Georg-August-University, Department of Animal Sciences, Albrecht-Thaer-Weg 3, 37075 Goettingen, Germany, [2]North Carolina State University, Department of Biological Sciences, 3550 Thomas Hall, Raleigh, NC 27695-7614, USA, [3]University of Mannheim, Institute for Mathematics, A5, 6 B 117, 68131 Mannheim, Germany; hsimian@gwdg.de*

When applied to the complex trait ‚chill coma resistance', a sequence-based additive GBLUP model was found to fail when trying to predict phenotypes in a set of 176 fully sequenced inbred lines of freeze 2 of the *Drosophila melanogaster* Genetics Reference Panel. Thus, we extended the approach to include selected pairwise epistatic interactions in genomic prediction and validated its accuracy in a leave-one-out cross-validation (LOOCV). Accuracy was evaluated by the ability to predict the left out observation in each replicate. SNPs with additive effects and SNP pairs with epistatic interaction were identified via GWAS in the training set for all 176 subsets of the LOOCV. While for the simple GWAS 1.87 mio SNPs (MAF>0.05) were considered, the epistatic GWAS tested all $2.26\times10^{11}$ pairwise combinations of 0.67 mio SNPs with MAF>0.15. The best prediction of the trait expressed in male and female animals, respectively, was obtained with different underlying genetic architectures. While in males a single major gene and many (~3000) pairwise epistatic interactions yielded the best prediction, the optimal model to predict the trait in females encompassed few (4-5) additive major genes and just ~30 pairwise epistatic interactions. The optimal model yielded an accuracy of approximately 0.4 in both sexes, which is remarkably high given the limited size of the training set. Based on the results obtained for the model species D. melanogaster it should be reconsidered, whether a sex-specific genetic architecture might be prevailing for complex traits performed by males and females (such as growth-related traits) in farm animals as well. In this case, sex-differentiated models of genomic prediction may lead to more accurate predictions.

---



**Sequence-based partitioning of genomic variance using prior biological information**

*P. Sørensen, S.M. Edwards, P.M. Sarup and D.A. Sorensen*
*Aarhus University, Molecular Biology and Genetics, Blichers Alle 1, 8839 Tjele, Denmark; peter.sorensen2@agrsci.dk*

Complex traits have a polygenic pattern of inheritance, and result from many DNA-sequence variants in functional elements distributed throughout the genome. Statistical models that simultaneously fit all genetic variants account for a large fraction of the heritable component of the trait, although this depends on the population structure. Evidence collected across genome-wide association studies reveals patterns that provide insight into the genetic architecture of complex traits. Although many genetic variants with small or moderate effects contribute to the overall genetic variation it appears that the sequence variants associated with trait variation are enriched for genes that are connected in biological pathways. Another important finding is that multiple independently associated variants are located in the same genes and that the associated variants are enriched for likely functional effects on genes such as altered amino-acid structure of proteins and expression levels of nearby genes. These biological findings provide valuable insights that can be used to improve statistical models for associating genetic variation with phenotypes of interest. We have developed a statistical modeling approach that evaluates the collective action of multiple genetic variants in biological pathways or other genomic features (e.g. coding, non-coding, regulatory element, genes) on the trait phenotype. The genomic feature models provide novel insight into the genetic architecture of complex traits and diseases by identifying the genomic classification schemes that explain the highest proportions of genetic (co)variances or provide better model fits or predictive ability.

**Prediction of sequence-based genomic breeding values in Fleckvieh cattle**

*M. Erbe[1], H. Pausch[2], R. Emmerling[3], C. Edel[3], K.-U. Götz[3], R. Fries[2], T.H.E. Meuwissen[4] and H. Simianer[1]*
*[1]Georg-August-University Göttingen, Department of Animal Sciences, Göttingen, Germany, [2]Technische Universität München, Chair of Animal Breeding, Freising, Germany, [3]Bavarian State Research Centre for Agriculture, Institute of Animal Breeding, Grub, Germany, [4]Norwegian University of Life Sciences, Department of Animal and Aquacultural Sciences, Ås, Norway; merbe@gwdg.de*

Prediction of genomic breeding values (GBV) based on genotypes from SNP chips has become a routine technique in many dairy cattle breeding programs. The aim of this study was to assess possible advantages in terms of accuracy of prediction when having all biallelic variants from the full sequence available. A data set of 6,778 Fleckvieh bulls was used which were either genotyped with 50K or 777K SNPs or were fully sequenced. All bulls that were not sequenced themselves were imputed up to the full sequence. After imputation, variant calling revealed around 21 M biallelic variants (SNPs + Indels) from which around 12 M were available for further analyses after basic filtering. As a first approach we used GBLUP models for which the genomic relationship matrix was built based on (1) all variants available after filtering, (2) SNPs from the HD SNP set, (3) 1 M rare variants (minor allele frequency (MAF)<0.01), (4) 1 M common variants (MAF>0.40) or (5) 3 M SNPs located in genes. Progeny-based daughter yield deviations (DYD) for six different traits were used as quasi-phenotypes. Data were split into training and validation set corresponding to a forward prediction scheme and accuracy of prediction was defined as cor(DYD,GBV) within the validation set. Across all traits, cor(DYD,GBV) was very similar for most of the scenarios (e.g. 0.59-0.61 for SCS) with no clear advantage for (1) or any other scenario. An exception was observed for the rare variants scenario which yielded the worst results (e.g. 0.42 for SCS). These first results suggest that GBLUP cannot benefit much from the availability of sequence data and further methods will be tested that make better use of it.

---



**Sequence differentiation of five European dairy cattle breeds**

*B. Guldbrandtsen[1] and D. Boichard[2]*
*[1]Aarhus University, Center for Quantitative Genetics and Genomics, Department of Molecular Biology and Genetics, Research Center Foulum, 8830 Tjele, Denmark, [2]INRA, UMR 1313 Génétique Animale et Biologie Intégrative, Domaine de Vilvert, Bat 211, 78352 Jouy en Josas, France; bernt.guldbrandtsen@agrsci.dk*

Modern dairy cattle breeds are the product of founder effects at their definition and closure and subsequent gradual differention by genetic drift and more recently strong artificial selection. Based on whole genome sequencing of animals belonging to 5 modern dairy breeds, Holstein, Jersey, Danish Red, Normande and Montbeliarde we investigate levels of genomic differentiation. Differentiation is assessed by Weir and Cockerham's estimator of Fst. Bi-allelic SNP variants in proteins were annotated and classified with respect to their predicted functional effects. Classes were: synonymous, non-synonymous tolerated, non-synonymous deleterious and extreme (stop gained or lost or out of frame frameshift mutations). In seven short chromosomal regions (on chromosomes 7, 17, 22 and X) the breeds exhibited nearly complete differentiation. The most differentiated positions mostly fall outside identified or predicted gene sequences. Most of the genes close to highly differentiated sites represent uncharacterized proteins. A substantial proportion of the non-synonymous variants exhibited strong differentiation when using the differentiation at synonymous variants as a reference. Most of the highly differentiated variants are in uncharacterized proteins. However, one is in a known coat color gene (BTA18, MC1R), another is in 'bread cancinoma apmplified sequence 3' (BTA19). There is molecular evidence for that selection has played an important role in the differentiation of breeds. That many differentiated genes are uncharacterized suggests the pathways contributing to breed differentiation are very poorly understood.

### The effect of read coverage in whole genome sequencing data

*C.F. Baes[1,2], P. Von Rohr[1], M.A. Dolezal[3,4], E. Fritz-Waters[5], J.E. Koltes[5], B. Bapst[1], C. Flury[2], H. Signer-Hasler[2], C. Stricker[6], R. Fernando[5], F. Schmitz-Hsu[7], D.J. Garrick[5], J.M. Reecy[5] and B. Gredler[1]*
*[1]Qualitas AG, Chamerstrasse 56, 6300 Zug, Switzerland, [2]Bern University of Applied Sciences, Länggasse 85, 3052 Zollikofen, Switzerland, [3]University of Veterinary Medicine Vienna, Veterinärplatz 1, 1210 Wien, Austria, [4]Università degli Studi di Milano, Via Festa del Perdono 7, 20122 Milano, Italy, [5]Iowa State University, Ames, 50011 Ames, USA, [6]agn Genetics, Börtjistrasse 8, 7260 Davos, Switzerland, [7]Swissgenetics, Meielenfeldweg 12, 3052 Zollikofen, Switzerland; christine.baes@qualitasag.ch*

Average base coverage is commonly used to describe next-generation sequencing data. True base coverage, however, may differ substantially across individual chromosomes. False positive SNP are often located near gaps in the reference genome due to erronous alignment; coverage in these regions is commonly higher than the average genome coverage. We confirm that higher-than-average coverage (after the exclusion of sequencing artefacts) does not result in better sensitivity and specificity for SNP detection, but rather results in false positive SNP calls. Risk for this type of error is negatively correlated with the quality and completeness of the reference genome. Whole genome sequence data of 65 key ancestors of Swiss dairy populations were available for investigation. Over 24 billion reads were mapped to the reference genome UMD31 (96.8%), with an average coverage of 12×. Various variant calling software was used to identify variants. The average number of SNVs found per 1 Mb window and autosome ranged from 4,500 (BTA22) to 6,500 (BTA23). Regions of very high and very low SNV identification were observed in all callers (e.g. SNV counts for BTA23 ranged from 500 to >10,000 depending on window). These regions were correlated negatively with gaps in the reference. False-positive SNV calls negatively impact GWAS results and downstream analyses. We show that that more information is available when coverage across 1 Mb windows is investigated compared with average genome coverage.

---



### Imputation from 60K SNP data to whole-genome level in Danish Landrace and Duroc pigs

*P.S. Sarup[1], B.G. Guldbrandtsen[1], B.N. Nielsen[2], P.S. Sørensen[1] and J.J. Jensen[1]*
*[1]Aarhus University, Center for Quantitative Genetics and Genomics, Department of Molecular Biology and Genetics, Blichers Allé 20, 8830 Tjele, Denmark, [2]Danish Pig Research Centre, Axelborg, Axeltorv 3, 1609 København V, Denmark; pernille.sarup@biology.au.dk*

Using the Illumina platform we sequenced 30 Danish Landrace and 80 Duroc pigs from the Danish genetic evaluation system with an average coverage of 11 X. The individuals were selected for sequencing to maximize their genetic contributions to a high density (60K) SNP chip genotyped population. Sequencing data were aligned using BWA (Burrows-Wheeler Alignment Tool), and following that we used tools from GATK (the Genome Analysis Toolkit) to realign around indels, recalibrate quality scores and call variants. SNPs were filtered allowing only SNPs with read depths between 3 and 30. This ensured a high quality of the called variants, by avoiding noise from poorly sequenced or duplicated regions. The high quality of the SNPs was confirmed by a high concordance between SNPs called from the sequences and the SNP chip. We partitioned the genetic variation found within and between the breeds using Wright's F statistics and evaluated the potential for imputation of the genotype of the chip typed individuals to full sequence information using Beagle. The accuracy of imputation primarily depended on the allele frequencies of the imputed alleles with accuracies increasing with increasing allele frequencies.

**Genome variability analysis through whole genome sequencing of a highly inbred strain of Iberian pig**

*A.I. Fernández[1], A. Fernández[1], M. Saura[1], M. Pérez-Enciso[2] and B. Villanueva[1]*
*[1]INIA, Mejora Genética Animal, Ctra. Coruña Km7.5 Madrid, 28040, Spain, [2]CRAG, ICREA, Barcelona, 08193, Spain; avila@inia.es*

The Iberian porcine breed presents several particularities that make it of great interest for genome analysis. The Guadyerbas Iberian strain represents one of the most ancient surviving strains, and is the sole representative of the black hairless variety. However, it has undergone a strong reduction in census size that has led to high inbreeding and loss of genetic variability. Genetic diversity studies using pedigree information, neutral markers and high density SNPs from the Porcine SNP60 BeadChip data have been conducted on this strain. Furthermore, one of the first sequenced porcine genomes has been that of a Guadyerbas pig. The aim of the present study was to analyse the genome variability of Guadyerbas strain through the analysis of the whole genome sequence of four Guadyerbas animals. DNA samples were sequenced with Illumina Hi-Seq 2000 platform, generating 1,474 millions of paired-end reads. Bioinformatics analyses were conducted using CLC Genomics Workbench software, employing the quality-based variant detection method. The mean coverage per animal ranged from 10 to 12×. A mean of 5.5 millions of variants were identified per sample. Variants were classified as multinucleotide variants (1 million) and single nucleotide variants (4.5 million), of which a 30% appeared in heterozygosis in any of the samples, and consequently were considered as actual polymorphisms (≈1 SNP each 2,000 bp). In order to evaluate the accuracy of the SNP calling procedure, a comparison between the genotypes identified by sequencing and those obtained with the Porcine SNP60 BeachChip was conducted. Around 86% of the SNPs contained in the porcine chip were detected by whole genome sequencing, and a concordance of 98% was found between the genotypes assigned by both technologies. These data provides us with new tools to evaluate genome diversity as well as to identify potential causal mutations of the adaptive traits in this breed.

---



**Analysis of putatively deleterious alleles in German Fleckvieh by whole genome-resequencing**

*S. Jansen, C. Wurmser, H. Pausch and R. Fries*
*Chair of Animal Breeding, Technische Universitaet Muenchen, Liesel-Beckmann-Straße 1, 85354 Freising, Germany; sandra.jansen@tierzucht.tum.de*

Genome-wide sequencing of key ancestors allows for a thorough assessment of population's genomic variants, among them variants leading to premature stop codons, that are highly relevant in searching of the genetic causes of diseases. Analysis of 214 resequenced cattle genomes of five different breeds revealed 1038 premature stop codons caused by single nucleotide polymorphisms (SNPs). 797 occurred within 112 German Fleckvieh animals (BFV). False-positive stop codons due to annotation and alignment errors and variants revealing low quality were excluded. Olfactory and nondescript genes and genes belonging to the major histocompatibility complex were not considered. The remaining variants consist of 91 positions in 86 different genes. More than 15% of the confidently predicted premature stop codons truncate the resulting protein sequence by more than 80%, most likely severely influencing the protein function and structure. Confidently predicted premature stop mutations are particularly overrepresented among rare alleles, with more than half of them present at low frequency (AF<0.05), suggesting negative selection. Each BFV animal carries an average of 16 high quality stop mutations among them four in a homozygous state. 40 genes harbouring premature stop codons revealed homozygotes in at least one animal. The occurrence of homozygous nonsense SNPs indicate that these mutations may have a less deleterious effect and are possibly causal for complex diseases. However, 51 of the 91 high-quality stop mutations did not occur in a homozygous state in 112 BFV animals. More than 70% of the stop mutations with missing homozygotes have a frequency lower than 0.05. At this low frequency, ruling out homozygotes requires large samples. For the validation of disease phenotypes, animals resulting from matings of carrier individuals need to be inspected.

## Session 14 Poster 14

**Optimizing NGS data analysis pipelines – comparison of alignment and variant calling tools**

*M. Mielczarek[1], J. Szyda[1] and B. Guldbrandtsen[2]*
*[1]Wroclaw University of Environmental and Life Sciences, Department of Genetics, Kozuchowska 7 Street, 51-631 Wroclaw, Poland, [2]Aarhus University, Department of Molecular Biology and Genetics, Blichers Allé 20, 8830 Tjele, Denmark; magda.a.mielczarek@gmail.com*

Next-generation sequencing has an increasing importance in the life sciences. Currently, two of the most important tasks with NGS data are alignment to a reference genome and variant calling. Numerous algorithms have been introduced for these purposes. We compared five NGS mapping tools on reads sequenced from the Ilumina platform. Three of them were Burrows-Wheeler Transform based (BWA-MEM, BWA-BT, Bowtie 2) and two hash-table based (SMALT, MOSAIK). Each of them was run in a paired-end alignment mode and with default parameters. BWT-based aligners worked faster than hash based aligners. The mappers mentioned above had a very high alignment percentage, apart from the last one. MOSAIK aligned a lower fraction of reads which had a significant impact on subsequent steps (SNP and INDEL calling). Two variant callers – samtools and GATK – were compared. They apply Bayesian methods which use quality scores to identify novel variants. Samtools was faster than GATK. In general samtools detected slightly more SNPs and substantially more INDELS than GATK. Aligners and callers gave different results for the same dataset, thus attention must be paid to the choice of tools, filtering criteria and the creation of systematic validation methods for the variants found.

---

## Session 15 Theatre 1

**Recent advances in exogenous enzyme application for monogastric nutrition**

*A.J. Cowieson*
*DSM Nutritional Products, Wurmisweg 576, 4303 Kaiseraugst, Switzerland; aaron.cowieson@dsm.com*

Feed enzymes have been used to enhance the profitability of animal production since the mid-1980s. Though more than 3,000 peer-reviewed papers have been published in this field there are persistent areas of obscurity in mechanism of action, dose optimization, enzyme combinations, ingredient variance and some basic unanswered questions in nutrition per se. This review will address 4 broad topics considered of most relevance. Firstly, 'super-dosing' of enzymes remains poorly defined and not universally accepted. Secondly, formulation of diets to metabolisable or digestible energy does not give enough resolution to deliver predictable, consistent, effects of feed enzymes as their energetic consequences are a sum of changes to macro-nutrient digestion that are separately regulated. Thirdly, the nutritional 'quality' of diets is critical to scale enzyme effects and to recommend targeted combinations. Finally, the manner in which enzymes are included in least cost formulation can profoundly influence their value, particularly when focus is given only to minimizing feed cost disregarding animal performance. Feed enzymes are one of the most promising technologies to promote sustainable animal production, but their indiscriminate use is likely to result in inconsistent effects at an animal level. In future, the value of enzymes should be divided across feed cost saving and animal performance metrics as undue focus on feed cost alone can reduce the value proposition and may lead to erroneous formulation approaches. It can be concluded that application know-how is becoming as important as the quality of the products themselves to deliver bespoke and dynamic feed enzyme solutions. Meta-analyses, rapid assays for ingredient quality, health and performance monitoring and fundamental mechanistic understanding are pivotal to avoid indiscriminate enzyme use and to promote systematic decision making.

**Protein degradation in the gastrointestinal tract of broilers**

*A.M.B. Giessing[1], M. Klausen[1], M. Umar-Faruk[2] and K. Pontoppidan[1]*
*[1]Novozymes R&D, Feed Applications, Krogshoejvej 36, 2880 Bagsvaerd, Denmark, [2]DSM Nutritional Products, Research Centre for Animal Nutrition and Health, Boulevard Alsace, 68128 Village-Neuf, France; aegi@novozymes.com*

Protein is an essential part of diets for broilers. However, knowledge about the true digestibility of protein from different feed ingredients is relatively limited as most digestibility trials evaluate apparent ileal nitrogen digestibility. This study relates to the degradation of dietary proteins in the gastrointestinal tract of broilers and the interplay between dietary protein and endogenous protein loss. Intestinal contents were sampled from the terminal ileum of sixteen 21-day old broilers fed either a standard maize-soybean meal diet (534.9 g/kg maize and 392 g/kg SBM) or a similar diet containing 200 g/kg of raw soybeans (520 g/kg maize and 235.9 g/kg SBM). Diets were formulated to be iso-caloric (12.6 MJ ME/kg) and iso-nitrogenous (21.4% CP). Protein was extracted from freeze-dried samples and digested by trypsin to facilitate analysis of peptides by Orbitrap-Velos LC-MS. Peptide data was analyzed using Expressionist software (Genedata) and searched in Mascot against a Uniprot database containing chicken (*Gallus gallus*), soybean (*Glycine max*) and maize (*Zea mays*) protein sequences. As expected inclusion of raw soybeans in the diet reduced apparent ileal nitrogen digestibility from 82 to 67% ($P<0.0001$) and caused an increase in pancreas weight relative to body weight ($P<0.0001$). Independent of the diet approximately 400 unique proteins were identified in ileum samples evenly distributed between chicken, soybean and maize proteins. Abundance of chicken protein was found to be 20-25% of the identified ileal protein pool. In ileum samples from birds fed the standard diet there was more maize than soybean protein, whereas in ileum samples from birds fed the diet with raw soybeans there was more soybean than maize protein. Focusing on pancreatic proteases, we observed a change in the relative distribution of chicken S1A proteases in response to addition of raw soybeans to the diet.

---



**Digestibility of organic matter and NDF in concentrates accessed with enzyme methods using Daisy**

*M. Johansen and M.R. Weisbjerg*
*Aarhus University, AU Foulum, Department of Animal Science, P.O. Box 50, 8830 Tjele, Denmark; marianne.johansen@agrsci.dk*

In the NorFor feed evaluation system iNDF and rate of NDF degradation, revealed from i.a. organic matter (OM) digestibility, are the two major characteristics determining the energy value of the ration. The reference method to determination of iNDF was the in situ method with 288 h incubation in the rumen, and for OM digestibility the reference was an *in vitro* OM digestibility method using enzyme mix (EFOS), which is used in the official Danish feedstuff control. The objective was to investigate whether an estimate for iNDF or OM digestibility could be obtained using a commercial laboratory enzyme in Daisy incubator. In total 69 different concentrate feeds were analysed. The analyses included two different enzyme treatments combined with either drying or no drying between the NDF boiling and the enzyme treatments in a $2\times2$ factorial design. The enzyme treatments included Tricoderma Viride enzyme with or without Gamanase. All samples were pre-extracted with acetone to remove fat before the samples were NDF boiled according to the ANKOM procedure including heat stable amylase. Afterwards half of the samples were dried to determine the NDF content, before the samples were incubated with enzymes in Daisy at 40 °C by 24 h. The results showed that the digestibility of OM achieved was at the same level as the digestibility of OM obtained by the EFOS method, with few exceptions. The addition of Gamanase reduced the under estimation for some feedstuffs as example palm cake, and had in general a small but significant effect (0.48% units, $P=0.003$). The prediction of iNDF was not convincing with the enzyme method, neither when including Gamanase. The drying between the NDF boiling and the enzyme treatments had a significant but numerical small effect on the undigested residues (0.31% units, $P=0.04$). In conclusion the method could estimate OM digestibility, but not iNDF, and the step that enables determination of NDF content in the same process could be an opportunity.

**Exogenous phytase for dairy cows**

*J. Sehested[1], P. Lund[1], J. Depandelaere[1], D. Brask-Pedersen[1], L.K. Skov[2] and L.V. Glitsø[2]*
*[1]Aarhus University, Dep. Animal Science, AU-Foulum, 8830 Tjele, Denmark, [2]Novozymes A/S, R&D, Krogshoejvej 36, 2880 Bagsvaerd, Denmark; jakob.sehested@agrsci.dk*

Studies indicate that dietary phosphorus (P) bound in phytate (IP) is not totally available to dairy cows despite phytase activity in the rumen. We therefore investigated the effect of 4 exogenous phytases in an *in vitro* rumen fluid buffer system. All 4 phytases increased degradation of IP, which increased with incubation time and phytase concentration. The most efficient phytase *in vitro* was then used to study the effect of exogenous phytase on IP degradation in the rumen of 4 Danish Holstein cows with ruminal, duodenal and ileal cannulae in a 4×4 Latin Square design with four dietary concentrations of phytase: none, low, medium, or high corresponding to 23, 2,023, 3,982, and 6,015 phytase units/kg dry matter (DM). The total mixed ration (TMR) contained 3.8 g P/kg dry matter (DM), P in IP was 1.7 g P/kg DM and cows ingested 27 g P/d from IP. Exogenous phytase resulted in a higher ruminal pool of phytase. Ruminal and total tract degradations of IP were higher with exogenous phytase and ruminal degradability of IP was: none 86%, low 94%, medium 95%, and high 96%. In a third experiment we investigated the interaction between diet composition and exogenous phytase on IP degradation. Again 4 multi-fistulated Danish Holstein cows were used in a 4×4 Latin Square design with 4 dietary treatments. Two iso-energetic diets, based on clover-grass silage and rape seed meal were formulated. A high fiber diet was obtained by adding soy hulls, and a high starch diet by adding corn meal and dehulled oats. Diets were mixed with (1,900 FTU/kg DM) or without exogenous phytase. Ruminal IP degradation was increased by the high starch diet and by exogenous phytase and the effects were additive. Exogenous phytase increased ruminal IP degradation from 75 to 80% with the high fiber treatment and from 83% to 88% with the high starch diet. In conclusion, ruminal degradation of phytate can be increased by increased dietary starch and by addition of exogenous phytase.

---



***In vitro* comparison of cellulolytic and hemicellulolytic enzyme activity from game herbivores**

*F.N. Fon*
*University of Zululand, Agriculture, KwaDlangezwa, 3886, South Africa; fonf@unizulu.ac.za*

Ruminants have the advantage of the digestion provided by rumen microflora and yet less than 50% of ingested forages are digested. This study investigated herbivore microbial ecosystems in the wild for microbes or enzymes with high fibrolytic activity as potential feed additives. Fresh faecal samples were collected from wildebeest, impala, buffalo and llama at Tala Game Reserve. Faecal inoculum was cultured in the laboratory for 72 h at 38.5 °C after mixing 198 ml of faecal filtrate with 402 ml of salivary buffer containing 6 g mixture of lucerne and maize stover. Crude protein enzyme (CPZ) was precipitated from cultured filtrate by 60% $(NH_4)_2SO_4$. Exocellulase, endocellulase and xylanase enzymes were assayed using crystalline cellulose, carboxymethyl cellulose and xylan as substrates, respectively. Reducing sugars (glucose or xylose) were measured by Dinitrosalicylic method. Each enzyme assay was replicated three times. Enzyme specific activity (µg of reducing sugar/mg CPZ), enzyme catalytic rate (Vmax (µg reducing sugar/mg CPZ/min)) and efficiency were measured. Enzyme activities were subjected to analysis of variance (ANOVA) using the general linear model of SAS. The wildebeest and impala showed the highest ($P<0.05$) exocellulase (3.5 and 2.8 µg glucose/mg CPZ) and endocellulase specific activity (3.7 and 2.6 mg glucose/mg CPZ). Exocellulases from the wildebeest showed the highest ($P<0.05$) catalytic rate (27.3 µg glucose/mg CPZ/min) and efficiency (504). Xylanase specific activity was highest ($P<0.05$) for impala although its catalytic rate (1.3 µg reducing sugar/mg CPZ/min) and efficiency (1.8) was lower ($P<0.05$) than observed in the wildebeest (3.9 µg reducing sugar/mg CPZ/min and 265, respectively). Both llama and buffalo showed lower exocellulase, endocellulase and xylanase activities. Fibrolytic enzymes from wildebeest and impala microflora showed the highest potential of improving fibre breakdown suggesting further research on these systems as feed additives.

**Influence of fermentation and enzyme addition on digestibility of a rapeseed cake rich diet in pigs**

*G.V. Jakobsen, B.B. Jensen, K.E. Bach Knudsen and N. Canibe*
*Aarhus University, Animal Science, Blichers Alle 20, 8830 Tjele, Denmark; grethe.venaas@agrsci.dk*

The political agenda to reduce the use of fossil fuels has resulted in an increase in the production of biodiesel worldwide. In Europe, rapeseed is the main feedstock for biodiesel production. The co-product, rapeseed cake (RSC), contains protein of high nutritional value, but its use in animal feed is limited due to high content of dietary fibre. An *in vivo* study was conducted with pigs to investigate liquid fermentation and enzyme addition as means of increasing digestibility of a diet with 20% RSC. Eight pigs were cannulated at the distal ileum and fed four liquid diets in a double-Latin square design. Control: non fermented liquid feed standard diet; n-FLF: non fermented liquid feed experimental diet; FLF: fermented liquid feed experimental diet; FLF+enz: fermented liquid feed experimental diet + a mixture of β-glucanase, xylase and pectinase (Novozymes). Faeces and ileum samples were collected. The concentration of total non-starch polysaccharide (T-NSP) in n-FLF, FLF and FLF+enz was 15.5, 15.9 and 12.8% of DM; and that of lactic acid 21, 214 and 272 mmol/kg mixture, respectively. Preliminary results of the three experimental diets showed that fermentation increased apparent total tract digestibility (ATTD) of N (69.0 for n-FLF vs 73.2% for FLF, $P<0.05$); and ATTD of T-NSP (50.4 for n-FLF vs 54.3% for FLF, $P<0.05$), with the same tendency for ATTD of P (31.8 for n-FLF vs 46.1% for FLF). Enzyme addition to the FLF decreased the ATTD of T-NSP (54.3 to 46.0%, $P<0.05$) if the T-NSP values measured in the feed were used for the calculations. However, considering a similar value for the FLF and FLF+enz of 15.9% T-NSP of DM for the digestibility calculations, the FLF+enz showed an increase in the apparent ileal digestibility (AID) of T-NSP and a value of 56.7% for the ATTD. These preliminary data with pigs indicate that fermentation increases the ATTD of nutrients and that addition of carbohydrases to FLF can be a way of further increasing NSP degradation.



**Xylanases increase the ileal digestibility of non-starch polysaccharides in pigs fed wheat DDGS**

*M.B. Pedersen[1,2], S. Dalsgaard[2], K.E.B. Knudsen[1], S. Yu[2] and H.N. Lærke[1]*
*[1]Aarhus University, Blichers Allé 20, 8830 Tjele, Denmark, [2]DuPont Nutrition Biosciences ApS, Edwin Rahrs Vej 38, 8220 Brabrand, Denmark; mads.brogger.pedersen@dupont.com*

The effect of a current commercial xylanase (DAN) and experimental xylanase (EX), and EX in combination with protease (EXP), on the degradation and apparent ileal digestibility (AID) of non-starch polysaccharides (NSP) in wheat DDGS, was studied in 8 ileum-cannulated pigs (initial BW 36.6±2.8 kg) following a double 4×4 Latin Square design. The control and three enzyme diets, each containing 96% DDGS, were supplemented with vitamins, minerals, L-lysine, 500 FTU phytase/kg feed, dust-binder and chromic oxide (3 g/kg). The pigs were fed 3 times daily for 1 week and ileal content collected for 8 h on day 5 and 7. For each pig, weekly samples were pooled and analyzed for composition of total- and soluble NSP, total non-digestible carbohydrates (NDC) and marker. Low molecular weight (LMW)-NDC was determined as the difference between NDC and NSP concentration. Data were analyzed with enzyme treatment and week as fixed effects, and pig as random effect. Enzyme treatments were compared to the control treatment, and considered statistically significant at $P<0.05$. Enzyme addition overall increased AID of total NSP ($P=0.024$) with an AID of 25% for control to 32% with DAN ($P=0.02$) and 31% with EXP ($P=0.03$). Enzyme addition overall increased AID of high molecular weight arabinoxylan ($P<0.001$) from 31% for control to 41% with DAN ($P=0.0001$), 40% with EXP ($P=0.0006$), and 39% with EX ($P=0.0021$). Enzyme addition also increased the concentration of LMW-NDC ($P=0.005$) in ileal content from 6.4% in control to 8.6% for DAN ($P=0.0026$) and to 8.3% for EXP ($P=0.0084$), and the content of LMW-arabinoxylan ($P<0.0001$) from 5.0% in control to 7.0% for DAN ($P=0.0001$), 6.8% for EXP ($P=0.0003$), and 6.2% for EX ($P=0.0129$). Collectively, the results suggest that exogenous enzymes increase the degradation of NSP and arabinoxylan in wheat DDGS, with ranking of DAN>EXP>EX.

## Degradation of soyabean components assessed by microscopic visualisation and viscosity measurements

*J. Ravn[1], H.J. Martens[1], I. Knap[2], D. Pettersson[3] and N.R. Pedersen[3]*
*[1]Copenhagen University, Thorvaldsensvej 40, 1871 Frb, Denmark, [2]DSM, DNP Innovation Animal Nutrition & Health, 4002 Basel, Switzerland, [3]Novozymes A/S, Krogshoejsvej 36, 2880 Bagsværd, Denmark; jonasravn87@gmail.com*

The monogastric diet is generally composed of 70% cereals to 30% protein meal. Maize accounts for 80% of cereals and soyabean meal for 70% of the protein meal used for animal feed. To digest feed, all animals use enzymes produced either by themselves or by naturally present gut microbes. Monogastrics cannot digest 15-25% of cereal feed and up to 50% of SBM because of the indigestible components and/or lack of specific enzymes hydrolysing fiber components. Inefficiently digested feed is costly to the producer and the environment. Supplementing feed with specific exogenous enzymes improves its nutritional value. All plant derived ingredients contain soluble and insoluble fiber found in cell walls, which may act as an anti-nutrient in several ways. 1. In cereals, starch and protein are trapped within the insoluble cell walls. Monogastrics cannot access them due to their inability to produce fiber-degrading enzymes. 2. Soluble fibers dissolve in the monogastric gut, hold water, form viscous gels trapping nutrients and slow down digestion rates and feed passage through the gut. This may cause bacterial overgrowth leading to diarrhea and wet litter, having detrimental effects such as foot health. This study was conducted to examine RONOZYME VP for its ability to decrease viscosity of several cell wall components using a viscosity pressure assay, and to visualise the break down of soybean fiber using microscopy. 2 μm cross sectioned soyabeans were stained with specific histological dyes to visualize the different microstructural cell components and to follow the enzymatic reactions on soyabean meal following incubation with RONOZYME VP. Cell wall degradation was monitored indirectly as a decrease in the staining intensity due to breakdown of the cell wall components.

---



## Carbohydrases influence nutrient degradability along digestive tract and fermentation patterns

*C.B. Blanc and P.C. Cozannet*
*Adisseo France SAS, Rue Marcel Lingot, 03600, France; pierre.cozannet@adisseo.com*

An *in vitro* experiment was carried out to assess the effects of non-starch polysaccharide (NSP)-degrading enzymes on fermentation of different raw materials in the large intestine of pigs. Four raw materials (wheat, wheat bran, hulled and dehulled soybean meal) selected for their various content of carbohydrate (CHO) fractions were hydrolyzed using pepsin and pancreatin in the presence or not of the multi activities enzyme complex (Rovabio Excel™). After hydrolysis, the residues were fermented in presence of pig fecal bacteria. Dry matter digestibility was measured during the hydrolysis phase and gas production, fitted using de Groots model ($Y = c / (1+\exp(-a(\text{time}-b)))$), and short-chain fatty acids (SCFA) production were monitored during the fermentation phase. Dry matter degradability after pepsin and pancreatic incubation was 47.7, 84.6 62.0 and 71.6% for wheat bran, wheat, hulled and dehulled soybean meal, respectively. Enzyme addition significantly increased dry matter degradability of all raw materials except dehulled soybean meal. Improvement averaged 5.4, 4.2 and 4.0% units, respectively. Besides the expected influence of the raw material, no effect of enzyme was observed on wheat bran and dehulled soybean meal. For wheat, SCFA production decreased and kinetic was altered with significant increased of lag time and a reduction of the asymptotic value ($P<0.001$). This effect was presumably associated with a reduction of remaining starch in wheat treated with enzyme. In contrast, for hulled soybean meal, the results showed that enzyme supplementation increased the asymptotic value, corresponding to an improvement of *in vitro* nutrient fermentation ($P<0.001$). SCFA was increased with similar acetic, butyric and propionic acid profil ($P<0.001$). Oligosaccharides released from hulled soybean meal enzyme hydrolysis were considered beneficial for microbiota activity and overall gas production.

**Xylanase supplementation improve nutrient digestibility in broilers fed dried cassava pulp diets**

*S. Khempaka, T. Suriyawong and W. Molee*
*School of Animal Production Technology, Institute of Agricultural Technology, Suranaree University of Technology, Nakhon Ratchasima, 30000, Thailand; khampaka@sut.ac.th*

Cassava pulp, a by-product of starch manufacture, contains a lot of starch but low amounts of protein and high fiber content which limits its use as a feedstuff for broilers. The fiber content of cassava pulp, which is mostly present in the form of insoluble fiber, has been reported to reduce the nutrient digestibility. Therefore, the objective of this study was to determine the effect of xylanase supplementation in diets containing dried cassava pulp (DCP) on nutrient digestibility in broiler chickens. Forty-nine 21-day-old male chickens were segregated into 7 groups and placed in individual cages for 10 days. The seven dietary treatments were control and DCP at levels of 8, 12 and 16%, supplemented with 0.1 and 0.2% xylanase. Excreta were collected on the last 4 days of the experimental period and used to measure nutrient digestibility and utilization. The results showed that feeding 8-16% DCP supplemented with 0.1% xylanase and 8 and 16% DCP supplemented with 0.2% xylanase had no significant differences on DM and OM digestibilities compared to control ($P>0.05$). Feeding 12% DCP supplemented with 0.1% had higher N utilization than the control ($P<0.05$), while the other treatments showed no significant differences from the control ($P>0.05$). The addition of 0.1 and 0.2% xylanase showed similar result on improved CF digestibility ($P>0.05$). In conclusions, DCP can be used in broiler diets by up to 16% when supplemented with 0.1% xylanase enzyme without having detrimental effects on nutrient digestibility and utilization.

---



**Inclusion of ethanol precipitation step to determine *in vitro* organic matter and NSP digestibility**

*H.N. Lærke and K.E. Bach Knudsen*
*Aarhus University, Department of Animal Science, Blichers alle 20, 8830, Denmark; hellen.laerke@agrsci.dk*

The Danish feed evaluation system for pigs is based on *in vitro* determination of ileal (IDOMi) and faecal digestibility of organic matter (EDOM), which includes incubation with digestive enzymes. In the procedure material dissolved by enzymes is removed and soluble dietary fibre (SDF) may be lost. Fibre-degrading feed enzymes can increase the content of SDF, which may lead to losses during the *in vitro* procedure. We included a step with precipitation of SDF with 80% ethanol and studied the effect on EDOMi of 8×3 replicate samples of a rye based diet with and without added commercial xylanase supplied by Danisco Animal Nutrition. In 3×3 replicates we studied the *in vitro* ileal digestibility (EDi) of NSP, arabinoxylan and NSP-glucose and composition of NSP in the undigested residue. Data were analysed in SAS using the mixed procedure to account for uneven variance; xylanase and ethanol treatment were main factors and the interaction xylanase × ethanol included in the model. Data are presented as LSmeans with SEM. Overall, ethanol precipitation decreased EDOMi from 86.0+0.05 to 80.4+0.42% ($P<0.0001$), but induced a much higher variability in the analysis. Without xylanase in the diet IDOMi was reduced 6.5%, and in the xylanase supplemented diet it was reduced 4.7%. Hence, in the classical procedure xylanase increased EDOMi by 0.44% (86.2±0.08 vs 85.8±0.08%, $P=0.0003$) while with ethanol it was 2.29% (81.6±0.60 vs 79.3±0.60%, $P=0.0107$). Ethanol inclusion reduced the values of EDi of NSP from 22.9 to 5.5% ($P<0.001$) across enzyme treatment. Xylanase increased the EDi of NSP by 11.0% without ethanol in the analysis ($P<0.0001$) and by 13.0% ($P=0.0075$) with ethanol. A similar trend was seen for arabinoxylan and NSPglucose. No effect of ethanol precipitation or xylanase treatment was seen on the NSP composition of undigested residue. The study shows that soluble fibre components are lost in the *in vitro* procedure, and effects of feed enzymes are masked by this in the classical procedure.

## Session 15 Poster 12

**Phytase increased phosphorus digestibility and bone ash of weaned pigs fed a phosphorus reduced diet**

*C. Schauerhuber[1], N. Reisinger[1], D. Bachinger[1], R. Berrios[2] and G. Schatzmayr[1]*
*[1]BIOMIN Research Centre, Technopark 1, 3430 Tulln, Austria, [2]BIOMIN Holding GmbH, Industriestrasse 21, 3130 Herzogenburg, Austria; diana.bachinger@biomin.net*

A total of 150 weaned pigs with an average weight of 7.75 kg were group housed in 15 pens, each equipped with slatted floors, cup drinkers and an automated feeding system. After 14 days adaptation to a starter diet with adequate phosphorus (P) supplementation, the pigs were weighed individually and 120 pigs were selected for the experimental period. Pigs were allocated to 3 dietary groups, each with 4 replicates. Groups were: negative control (NC), positive control (PC) and phytase (PHY). All groups were fed a corn-barley-soy basal diet with no added inorganic P (0.12% digP) whereas PC was supplemented with 0.65% monocalcium phosphate to provide 0.12% digP, and PHY with 500 U/kg BIOMIN Phytase 5000 (*Escherichia coli*-derived; 0.13% digP equivalent). The basal diet was pelleted to remove endogenous phytase and crushed to mix in BIOMIN Phytase 5000. For evaluation of live weight performance, pigs were individually weighed on day 14, 21, 28 and at the end of the trial (day 42). For determination of P apparent total tract digestibility (ATTD), feces were collected on day 28. For determination of bone ash and bone P content, 4 pigs of each group were slaughtered on day 42. Added phytase activity was determined to be 589 U/kg. No significant differences were observed in FCR and ADWG by PHY compared to NC and PC, whereas PC performed significantly better than NC for both parameters (FCR: P=0.041 and ADWG: P=0.030). Phosphorus ATTD was significantly improved by PHY compared to NC (P=0.000) and PC (P=0.003). Also bone ash (corrected for dry matter) was significantly increased by PHY compared to NC (P=0.014), whereas bone P content showed no significant differences. In conclusion, the supplementation with BIOMIN Phytase 5000 increased P digestibility and bone mineralization of weaned pigs fed a P reduced diet whereas live weight performance was not negatively affected by P deficiencies.

---

## Session 15 Poster 13

**Effect of the enzymatic preparation based on beta-glucanase on production performance in laying hens**

*H. Belabbas[1] and N. Adili[2]*
*[1]University of Mohamed Boudiaf, Department of Microbiology and Biochimistry, M'sila, 28000, Algeria, [2]University of El-Hadj Lakhdar, Department of veterinary medicine, Institute of veterinary sciences and agricultural sciences, Batna, 05000, Algeria; nezar.adili@yahoo.fr*

The objective of the current study is to test the effect of an enzymatic preparation based on Beta-glucanase, on the improvement of production and the quality of the egg (weight and color) in laying hens. Two diets were tested, the first one was based on corn and the second one contained corn supplemented with an enzymatic preparation based on beta-glucanase (Safizym GP 40, Lesaffre-France). These diets were isocaloric and isonitrogenous, and fed to 320 hens (n=160/group) aged 42 weeks, in floury form during 4 weeks. The results of the experiment show that the addition of beta-glucanase to laying hens has a positive significant effect on the improvement of production rate (07%). The size of the egg increases considerably and its color becomes so clear. In addition to that, in the presence of beta glucanase in the feed, the average egg weight improves (3.33%). The current study confirms that enzymatic preparation based on beta-glucanase has positive effects on the performance of production and eggs quality.

## Session 15 Poster 14

**Effect of enzyme on performance of gestating and lactating sows and litter**

*P.C. Cozannet and F.R. Rouffineau*
*Adisseo France SAS, Rue Marcel Lingot, 03600, France; pierre.cozannet@adisseo.com*

The effect of dietary inclusion of Rovabio Excel AP on lactating sows and litter performances was tested. Rovabio Excel AP is an enzyme formulation containing β-glucanase and xylanase. At the commercial dose of 50 g/ton, the enzyme preparation provided: 1,500 viscometric units/kg of endo-1,3(4)-β-glucanase and 1,100 viscometric units of endo-1,4-β-xylanase. A total of 7 different trials involving from 21 to 142 sows each were gathered and considered in a meta-analysis to evaluate the effects of feeding a non-starch polysaccharide hydrolyzing enzyme throughout lactation period on sow performances: body weight loss, feed intake, backfat condition and litter performances. Each individual trial did not show significant differences between control or enzyme-treated groups. In the meta-analysis, feed intake of the lactating sows was not statistically affected by dietary treatments. However, significant and positive effect of enzyme was observed on body weight at weaning correlated with a significant reduction of the weight loss during the lactation period. The improvement of body weight at the end of lactation was about 3.0 kg. This effect was more pronounced on early parity as the body weight improvement for primiparus, second parity animals and third and older parities animals were 2.84, 1.66 and 0.40 kg, respectively ($P<0.01$). Sows fed diets supplemented with enzyme tended to have greater back-fat thickness at weaning ($P=0.09$) and service ($P<0.05$). No effect was observed on litter. The number and the average weight of the piglets did not differ between treatments. Overall, it can be concluded that supplementation of enzyme in the lactating diets of the sows significantly improved body weight at weaning by reducing the weight loss during the lactating period and partially solved issues with sow hyperprolificity. Further experiments with repetition of numerous reproductive cycle with and without enzyme would be of interest.

---

## Session 16 Theatre 1

**Seasonal weight loss tolerance in farm animals: a proteomics and systems biology approach**

*A.M. Almeida[1,2,3,4]*
*[1]IBET – Instituto de Biologia Experimental e Tecnológica, Av. República Qta. do Marquês, 2780-157 Oeiras, Portugal, [2]ITQB/UNL – Instituto de Tecnologia Quimica e Biologica, Av. República Qta. do Marquês, 2780-157 Oeiras, Portugal, [3]CIISA – Centro Interdisciplinar de Investigação em Sanidade Animal, FMV, Av. Univ. Técnica, 1300-477, Portugal, [4]IICT – Instituto de Investigação Científica Tropical, CVZ-FMV, Av. Uni. Técnica, 1300-477, Portugal; aalmeida@fmv.utl.pt*

Over domestication, farm animals have adapted to constraints typical of their production systems, such as diseases and parasitosis or seasonal weight loss (SWL). SWL is in fact one of the major constraints to animal production in tropical regions. Coupled to production studies, the study of the metabolic changes to food restriction, highlighting energy and protein metabolic saving mechanisms, can be a useful approach to identify the physiological pathways relevant in breed selection and to identify genetic biomarkers that could be used for the selection of breeds and varieties with metabolic pathways more capable of energy and nitrogen retention, thus increasing productivity. Our research group has been conducting studies on several aspects related to weight loss physiology over the past 15 years, focusing primarily on small ruminants after using the rabbit as an experimental model. We have conducted a Systems Biology approach aiming to combine proteomics with transcriptomics and more recently metabolomics. Herein, we will present the results of three major experiments: (1) Experiments with rabbits; (2) Experiments with meat producing sheep; and (3) Experiments with dairy goats. In the three cases comparing breeds of animals susceptible to SWL were compared to breeds of animals that clearly show a certain degree of tolerance to SWL. An overview of the major results and conclusions herein obtained will be described as well as major implications on research directions.

**Proteomics and biomarkers of poultry production**

*P.D. Eckersall and E.L. O'Reilly*
*University of Glasgow, Bearsden Rd, Glasgow, G61 1QH, United Kingdom; david.eckersall@glasgow.ac.uk*

The application of proteomic investigation to poultry production has been limited although there have been reports of the use of this advanced technology to investigate muscle and serum from chicken. There is great potential for application of proteomics in poultry science and can be illustrated by two recent investigations into biomarkers within the chicken plasma and intestinal proteomes. An investigation of the chicken plasma proteome based on 2 dimension electrophoresis and mass spectrometry and which compared plasma from birds with raised acute phase proteins related to gait abnormality, to plasma from birds with a negligible acute phase reaction revealed increases in ovotransferrin, PIT 54, haemopexin, alpha-1 acid glycoprotein and fibrinogen among the high abundance protein of chicken plasma. A quantitative proteomic investigation using difference gel electrophoresis (DiGE) has determined the effect of microbial challenge by used poultry litter, on the intestinal proteome of chickens. The challenge lead to a significant fall in growth rate and to 28 protein spots on the DiGE gel of protein extracted from the proximal jejunum being differently expressed 10 days after the introduction of the used litter. Actin was in a high proportion of the protein spots showing increased expression along with proteins such as calcium binding protein, histone binding protein and apolipoprotein A1. Heat shock proteins and disulfide-isomerase A3 precursor were among protein that were reduced in expression. Detection and characterisation by proteomics of altered expression of specific protein in plasma or in intestinal extract demonstrates systemic or localised responses to the external environment which can contribute to poor growth, health and welfare of chickens and illustrates how this technology can have significant impact in poultry production.



**Proteomic technologies to identify stress and welfare markers in livestock**

*A. Marco-Ramell[1], L. Arroyo[1], D. Valent[1], M. Olivan[2], A. Velarde[3] and A. Bassols[1]*
*[1]Universitat Autònoma de Barcelona, Bioquímica i Biologia Molecular., Facultat de Veterinaria, 08193 Cerdanyola del Vallès, Spain, [2]SERIDA, Apdo 13, 33300 Villaviciosa, Spain, [3]IRTA, Finca Camps i Armet s/n, 17121 Monells, Spain; anna.bassols@uab.cat*

Welfare problems are important for ethical reasons and because they may cause great economic losses. Due to the limitations of stress markers as serum cortisol, we have applied proteomic techniques to the animal stress/welfare problem, with the aim to identify new biomarkers with potential application in animal production and veterinary medicine. Here our last results on pigs will be presented. Two experimental stressful situations have been studied in pigs (high-density housing and individual gestational stalls). On the other hand, the effect of a positive condition addressed to increase the welfare of the animals, i.e. a close human-animal relationship was also tested. Serum was the sample analyzed in the first two conditions. Two different quantitative proteomic approaches, a gel-based (2D-DIGE) and a LC-MS-based (iTRAQ) method, revealed variations in serum protein composition. In both cases, two main homeostatic mechanisms were altered during housing related stress, namely the innate immune system, especially acute phase proteins, and the redox system (increased carbonylation of proteins). The use of iTRAQ allowed the identification of other differentially expressed proteins of the immune system (α-antichymotrypsin-3, TRFL and PG3) and antioxidant enzymes (PRDX2, GSH0 and GPX). Changes in apolipoproteins were also common to both conditions. Finally, these proteomic approaches allowed to identify the presence in serum of intracellular proteins, which may point out to the existence of cell damage. Serum has inherent problems for proteomic approaches due to the presence of high abundant proteins (albumin, immunoglobulins) that can mask changes in other, less abundant proteins. The use of peripheral mononuclear cells (PBMCs) provides an interesting approach to overtake the limitations of serum.

**Farmed fish quality and welfare monitoring using proteomics**

*P.M. Rodrigues[1], N. Richard[1], T. Silva[2], D. Schrama[1], J. Dias[2] and L. Conceição[2]*
*[1]CCMAR-Universidade do Algarve, Campus de Gambelas, 8005-139 Faro, Portugal, [2]SPAROS Lda, CRIA, Universidade do Algarve, 8005-139 Faro, Portugal; pmrodrig@ualg.pt*

Aquaculture large commercial scale has deeply increased, connected to the rapid growth of the average seafood consumption per person in the last fifty years. Proteomics research areas in aquaculture are mainly focused on nutrition, welfare and health management since these have proven to be major constrains to an efficient production in aquaculture systems. In terms of nutrition, the shift from the usage of finite marine-harvested resources to a richer diet in vegetable protein and oil sources is the new trend with reduced growth performance and feed efficiency being the major drawbacks. Scientific knowledge has been more extensively applied over the last two decades and Proteomics has emerged as a powerful tool towards a deep understanding of marine organisms' biology, helping aquaculture to reach a high productivity of a better quality product in a sustainable way. Fish diseases mainly caused by virus, parasites and bacteria are responsible for the main economic losses in aquaculture. This is the case for the Mediterranean fish Gilthead seabream that develops a disease known as winter syndrome, when water temperatures are low, affecting growth. Specifically designed diets were produced to help fish overcome this period. Liver proteomics and metabolomics data clearly shows that the diet can affect gilthead seabream's response to seasonal thermal stress in a positive way. In terms of fish welfare, proteomics can also be an important tool, helping the development of new aquaculture practices that ensure that farmed marine animals can be reared in an environment that optimizes their capacity to cope with unavoidable challenges/stress (high stock densities, handling and pre-slaughter stress), guaranteeing these animals a good state of welfare and health. This work is part of project 21595-INUTR, co-financed by FEDER through PO Algarve 21 in the framework of QREN 2007-2013.

---



**Proteomics as a tool to better understand beef tenderness**

*E. Veiseth-Kent and K. Hollung*
*Nofima AS, P.O. Box 210, 1431 Ås, Norway; eva.veiseth-kent@nofima.no*

Consumers rate meat tenderness as the most important quality trait of beef. However, a great variation in meat tenderness is seen between the same muscles from different carcasses, and also between different muscles, often leading to consumer dissatisfaction when purchasing meat. Thus, reducing this unwanted variation remains an important goal for the meat industry. During the last decade, meat scientists have applied a variety of proteomics approaches in order to both obtain a better understanding of postmortem proteolysis and its connection with meat tenderness, and to identify potential markers for beef tenderness. Through employing both gel-based and LC-MSMS-based proteomics, we have been able to unravel changes that occur during postmortem storage of beef muscle to proteins that were previously thought to be unaltered during storage (e.g. actin and myosin heavy chain). Moreover, the involvement of stress-related proteins (e.g. heat shock proteins) during the conversion of muscle to meat has also been revealed. Although several groups have used proteomics to identify potential protein markers for tenderness, the power of these markers to explain tenderness variation is neither high nor generic to different muscles, breeds or processing conditions, making it unlikely that on-line measurements of protein markers will ever be introduced into meat processing plants. This work has nevertheless contributed towards building a better picture of this complex factors affecting meat tenderness, however applying this knowledge in order to improve meat tenderness is a great challenge. One solution could be to associate the high-resolution phenotypic data resulting from proteomics experiments with genomics approaches and marker-assisted selection. Through this approach it should be possible to associate protein markers for tenderness with genetic markers, and thereby improve meat tenderness through selective breeding.

**Proteomic tools to assess meat authenticity**

*M.A. Sentandreu and E. Sentandreu*
*CSIC, Institute of Agrochemistry and Food Technology, 7, Agustin Escardino street, 46980, Spain; ciesen@iata.csic.es*

Meat authenticity and traceability has become an issue of primary importance in our modern society. This assumption can be easily deducted from recent events dealing with the adulteration of meat products with non-declared species such as the horse meat scandal in 2013, which has evidenced the global claim of consumers to dispose of clear and reliable information about the food they consume. In the case of processed meat products this is going to be especially true, because a simple visual inspection will not allow discriminating between the different ingredients so easily as in the case of fresh meat. In the meat industry, the main areas susceptible to fraud deal with the origin of meats, substitutions of meat ingredients by other animal species or tissues, alterations of the processing methods and/or additions of non-meat components such as water or additives. In this respect, legal authorities must protect citizens against misdescription and fraud by means of robust, reliable and sensitive methodologies. Common strategies that have traditionally been used to assess meat authenticity include chemometric analysis of large dataset obtained by different techniques, immunoassays or DNA analysis, for example. Despite their advantages, these approaches are not exempted from some important limitations, especially in the analysis of highly processed meats where raw material can undergo an important degradation. Advances in mass spectrometry applied to the analysis of peptides and proteins constitutes an interesting and promising alternative to existing methods due to its high discriminating power, robustness and sensitivity. The possibility to develop standardized extraction protocols, together with the higher resistance of peptides to food processing as compared to DNA sequences, would overcome some of the existing limitations in quantitative measurements. Moreover, the use of affordable and routine mass spectrometry equipment would make this technology suitable for control laboratories.

---

Session 16 Theatre 7

**The milk proteome in diary science: from animal welfare to food safety**

*P. Roncada[1,2], C. Piras[1], A. Soggiu[1], I. Alloggio[1] and L. Bonizzi[1]*
*[1]Università degli Studi di Milano, Dipartimento di Scienze Veterinarie e Sanità Pubblica- Università degli Studi di Milano, via Celoria 10, 20133 Milano, Italy, [2]Istituto Sperimentale Italiano Lazzaro Spallanzani, Dipartimento di Scienze Veterinarie e Sanità Pubblica- Università degli Studi di Milano, via Celoria 10, 20133 Milano, Italy; paola.roncada@guest.unimi.it*

Milk is an important biological fluid because of his contents that is made of proteins, lipids, sugars, mineral salts. Milk proteomics is an emerging field of science that contributes to better understanding the role of proteins during first days of lactation; help to elucidate problems during the management of dairy products; and could help to improve food safety because of his high intrinsic power of discovery of bacteria through metaproteomics. Over the last decade, proteomics has been successfully applied to study quality in milk processing, in microbial biotyping to milk safety (especially in mastitis), and in physiological mechanism of nutrition. Different techniques, used alone or in combination, could be applied to milk proteomics, from 'classical approach gel based (one and two dimensional electrophoresis) to gel free (LC-MS/MS) methods; and metagenomic-metaproteomics analysis. Due to relatively simple, not- invasive method of collection, abundance, milk represents a key biological fluids for proteomics based studies, from animal health to dairy process to food safety. Authors are grateful to COST ACTION FA1002 Farm Animal Proteomics for networking provided.

**Influence of weight loss on the wool proteomics profile: a combined iTRAQ and fibre structural study**

*A.M. Almeida[1,2,3,4], J.E. Plowman[5], D.P. Harland[5], A. Thomas[5], T. Kilminster[6], T. Scanlon[6], J. Milton[7], J. Greeff[6], C. Oldham[6] and S. Clerens[5]*
*[1]CIISA – Centro Interdisciplinar de Investigação em Sanidade Animal, FMV, Av. Univ. Técnica, 1300-477 Lisboa, Portugal, [2]ITQB/UNL – Instituto de Tecnologia Química e Biológica da UNL, Av. República, Qta. do Marquês, 2780-157 Oeiras, Portugal, [3]IBET – Instituto de Biologia Experimental e Tecnológica, Av. República, Qta. do Marquês, 2780-157 Oeiras, Portugal, [4]IICT – Instituto de Investigação Científica Tropical, CVZ, Fac. Med. Veterinária, Av. Univ. Técnica, 1300-477 Lisboa, Portugal, [5]AgResearch New Zealand, Lincoln Research Centre, Private Bag 4749, Christchurch 8140, New Zealand, [6]Department of Agriculture and Food WA, 3 Baron-Hay Court, South Perth WA 6151, Australia, [7]University of Western Australia, 35 Stirling Hwy, Crawley WA 6009, Australia; jeff.plowman@agresearch.co.nz*

Seasonal weight loss is the main limitation to animal production worldwide, significantly affecting the productivity of milk, meat and wool farms, particularly in drought-prone areas of the world where most of the large-scale wool production farms are located. Although the effect of nutritional status on wool quality parameters has been extensively studied, little is known on how it affects wool protein composition. Here, a proteomic approach has been applied to study changes in fibre structure and protein composition in wool from merino sheep subjected to experimentally induced weight loss. Results indicate that there is a significant reduction in the fibre diameter of wool from the animals on a restricted diet over a 42-day period. At the same time, significant increases in the expression of the high sulfur protein KAP13.1 and proteins from the high glycine-tyrosine protein KAP6 family in the wools from the animals on the restricted diet were also detected. Such findings have strong implications for the wool industry that favors finer wool.

---

## Session 17 Theatre 1

**Time management in stable routines**

*T. Thuneberg[1], L. Välitalo[1] and A. Teppinen[2]*
*[1]HAMK University of Applied Sciences, Mustialantie 105, 31310 Mustiala, Finland, [2]ProAgria Southern Finland, Seutulantie 1, 04410 Järvenpää, Finland; terhi.thuneberg@hamk.fi*

Daily stable routines include feeding, taking horses to the paddocks and keeping the stable clean. Depending on the stable's activity, there might also be other tasks to be done. There are almost as many ways to execute the routines as there are stables. In Finland, working times at stables have been researched only few times. HAMK University of Applied Sciences implemented a case study together with advisor organization ProAgria at five stables in 2013. Its aim was to find out how much time is used for stable duties daily, and what the reasons are for diversity in workloads. One day was spent at each stable measuring the working times and distances between the facilities. The stables represented purposely different forms of activity, but the basic routines were still the same. The least time consuming way to feed roughages to the horses, in the stable and at the paddock, were to use ('traditional') small square baled hay. Big bales and (slow feeding) hay nets are more time consuming ways to feed haylage. Time to roughage feeding varied 16-42 seconds per horse. Giving the concentrates was fastest with a ladle from a feeding trolley. The average time per horse was 16 seconds. When taking the horses outside to the paddocks, the most significant thing was the distance between the paddocks and the stable. Cleaning the boxes in the traditional way is very labor-intensive manual work. Bedding materials and systems, as well as the cleaning facilities, varied causing quite big variation in the mean cleaning time per one box (1.7-6.4 minutes). The worker, the equipment and the distance between the facilities, such as manure and bedding storage, affected the most to the working time in total. Because of different methods in the stables, the results cannot be compared directly. In any case, the balance in daily working hours may be as much as a couple of hours, which can mean even 0,5 difference in man-years, when observing a boarding stable for 20 horses.

**Shelter use by Icelandic horses in the winter period**

*J.W. Christensen and K. Thodberg*
*Aarhus University, Animal Science, Blichers alle 20, 8830, Denmark; jannewinther.christensen@agrsci.dk*

This study investigates the use of shelters by Icelandic horses during a winter period (Dec 2013-March 2014), and whether shelter use is affected by (1) the number of entrances (1 vs 2) and (2) a partition inside the shelter. Shelter use is further correlated to weather conditions, social status and level of Faecal Cortisol Metabolites (FCM). Thirty-two Icelandic horses participated in the study. The horses were pastured in 8 groups of 4 horses; each containing three 2-years old mares and one older (>10 years) mare. Each group had access to a shelter (30 $m^2$) which, in the first study period (Dec-Jan) had either one or two entrances (n=4 each). In the second study period (Jan-Feb), all shelters had two entrances and half were equipped with a partition inside the shelter. Infrared cameras were placed inside all shelters to enable registration of shelter use by each individual horse. In addition, location on the pasture and activity was scanned two days per week (12 scans/group/day). Faeces for analysis of FCM were collected weekly for the last three weeks of each study period. Temperature loggers were placed inside each shelter and further data on weather conditions (e.g. wind speed, humidity, rain/snow) were available from a database. The horses were scored for body condition twice (Henneke horse body condition scoring (BCS) System, scale 1-9) and were found to have a BCS above average (6.0±0.1, i.e. 'moderately fleshy') in the first study period and slightly above average in the second study period (5.2±0.1). Data on shelter use in relation to treatment, weather conditions, social status and FCM is currently being analysed and will be presented.



**Testing of a wood pellet product as bedding material in horse stables**

*C.U.P. Herholz, S. Wägeli, M. Feuz, C. Häni, T. Kupper, A. Burren and B. Reidy*
*University of Applied Sciences, School of Agricultural, Forest and Food Sciences HAFL, Länggasstrasse 85, 3052 Zollikofen, Switzerland; conny.herholz@bfh.ch*

The bedding material EQ-Bedding® consists of wood pellets supplemented with five different herbs. It has been used for different animal categories (e.g. pigs, chicken, rabbits). The objective of this study was to test EQ-Bedding® as alternative bedding material for horses. EQ-Bedding® was used on two different horse keeping farms in three boxes on each farm (n=6). The experimental setup consisted of a 7-day control period of the horses on their usual bedding (straw n=5, linen shives n=1) followed by a 42 days testing period with EQ-bedding®. Dry matter, the content of major nutrients (nitrogen, phosphorus, potassium) and microbial cell counts of the beddings were measured at different time points during the experimental period. At laboratory scale, water binding capacity was determined. Ammonia generation of EQ-Bedding® compared to straw was measured with wind tunnel experiments. General clinical health status and hoof health were examined by a veterinarian. Lying behavior of the horses was determined by video surveillance during the night. Ease of handling, dust development and absorptive capacity as well as behavior of the horses were continuously documented by the horse keepers with an observation protocol. Laboratory findings, horses' lying behavior and general clinical health status indicated similar results for the examined bedding materials. Water binding capacity, ammonia generation, ease of handling and hoof health were advantageous for EQ-Bedding®, whereas straw was beneficial in regard to occupation of the horses. EQ-Bedding® proved to be a suitable bedding material for horses. Its use may be preferred in urban horse stables or due to its lower ammonia generation for horses suffering from respiratory problems.

**Equine cortical bone: *in vivo* and *ex vivo* properties assessment**

*M.J. Fradinho[1], A.C. Vale[2], N. Bernardes[3], R.M. Caldeira[1], M.F. Vaz[2] and G. Ferreira-Dias[1]*
*[1]CIISA, Faculdade de Medicina Veterinária, ULisboa, Av. Univ. Técnica, 1300-477 Lisboa, Portugal, [2]ICEMS and Departamento de Engenharia Mecânica, IST, ULisboa, Av. Rovisco Pais, 1049-001 Lisboa, Portugal, [3]Faculdade de Medicina Veterinária, ULisboa, Av. Univ. Técnica, 1300-477 Lisboa, Portugal; amjoaofradinho@fmv.ulisboa.pt*

Throughout the horse life, the skeletal system is subjected to considerable loading and strengths. Thus, a better understanding about the structure and mechanical properties of bone tissue is of major importance for the sport horse industry. The aims of this study were: (1) to assess equine cortical bone status *in vivo* by quantitative ultrasonography (QUS); (2) to evaluate some mechanical properties and characterize its regional variation along the third metacarpal bone (MCIII); and (3) to compare the *in vivo* results with the *ex vivo* mechanical tests on the same bone. Eight Lusitano mares (6.4±2.6 years old) were assessed with a QUS device on the mid-section of the right MCIII (dorsal and lateral regions). Samples were collected post-mortem from different sections and regions of those bones, and submitted to compression and bending tests. Values of speed of sound (SOS) on the lateral region were higher than for the dorsal one ($P<0.0001$), but they were correlated ($r=0.803$; $P<0.05$). In compression, the maximum stress ($\sigma max$) was influenced by bone region and bone section, and an interaction was found between these two effects ($P<0.01$). Samples from the dorsal region were stiffer than samples from the lateral region ($P<0.0001$), and lower values of the Young's modulus (E) were obtained on the distal section of the MCIII ($P<0.05$). Also, lower bending resistance was observed in the distal section, when compared with the proximal and the mid-section of the bone ($P<0.01$). *In vivo* SOS measurements on dorsal region presented a good relation with the estimated E from *ex vivo* compression tests ($r^2=0.92$; $P=0.0002$). The results show that QUS could be used as a non-invasive method in the assessment of mechanical properties of the equine cortical bone.

---



**Comparison between equestrian simulator training and training on horses**

*M. Von Lewinski[1], N. Ille[1], C. Aurich[1], R. Erber[1], M. Wulf[1], R. Palme[1], B. Greenwood[2] and J. Aurich[1]*
*[1]Vetmeduni, Veterinärplatz 1, 1210 Vienna, Austria, [2]Racewood, Park Road, Tarporley, United Kingdom; joerg.aurich@vetmeduni.ac.at*

Although the first riding simulator was developed over 25 years ago, such systems have not been widely used in equestrian training. Recently, new simulators have become available which can be programmed for dressage, show jumping, polo or racing. In this study, we have compared cortisol release, heart rate and heart rate variability (HRV) variables SDRR (standard deviation of beat-to-beat interval) and RMSSD (root mean square of successive beat-to-beat differences) in 12 riders (6 male and 6 female) jumping a course of obstacles on a horse and simulating a jumping course on a jumping simulator. Salivary cortisol concentrations from 60 min before to 60 min after simulator training were higher ($P<0.05$) than during training with a horse but did not change acutely with the simulated or real show jumping efforts. Heart rate of the riders increased versus baseline (94±3 and 92±5 beats/min) when jumping with a horse and on a simulator ($P<0.001$) but reached higher values on the horse versus the simulator (175±3 vs 123±5 beats/min, $P<0.01$ between tests, interaction time x test $P<0.001$). Both HRV variables decreased during the simulated jumping course (SDRR from 30±4 to 5±1 ms, RMSSD from 22±4 to 7±2 ms; both $P<0.001$) and when jumping with a horse (SDRR from 69±14 to 28±6 ms, RMSSD from 53±12 to 25±5 ms; both $P<0.001$). From 30 min before to 30 min after the jumping tests, HRV variables were always higher in association with the simulated course versus the course jumped with a horse (SDRR $P<0.05$, RMSSD $P<0.01$). The changes in heart rate indicate that simulator training required less physical effort than training on a horse. Based on differences in HRV, training with a horse was associated with a more pronounced sympathetic dominance than simulator training. Although simulator training mirrored the situation on a horse, the demands on a horse were more complex.

**Field study on saddle pressure distribution in healthy horses**

*M. Glaus[1], S. Witte[2], S. Wägeli[1], A. Burren[1] and C.U.P. Herholz[1]*
*[1]Bern University of Applied Sciences, School of Agricultural, Forest and Food Sciences HAFL, Länggasstrasse 85, 3052 Zollikofen, Switzerland, [2]Vetsuisse Faculty Bern, ISME Equine Clinic, Länggasstrasse 124, 3012 Bern, Switzerland; m.glaus1@gmx.ch*

Objective of the study was to evaluate how saddle pressure is distributed in healthy horses and if a pressure mat is suitable to give reliable information on saddle fit. In addition the influence of breed, saddle type, saddle pad, discipline, conformation of whithers and back as well as the relative weight of the rider, influence on saddle pressure distribution was assessed. Forty randomly chosen horses were measured with the system 'Team Satteltester®'at rest and during extended and collected walk and trot on a straight line and on both reins. The pressures measured over the horse's back were recorded as the percentage distribution front to back, and left to right as well as the component registered over the spine. Additional information was taken from a questionnaire and all data were statistically analyzed by the programs STS 3.42 and NCSS. Heavy or draft breeds showed higher pressures towards the front of the saddle (whithers) than warmbloods. A sunken back conformation resulted in higher pressures in the back half of the saddle (lumbar region) than in horses with a rounded back. The use of 'alternative' saddle pads such as gel pads, foam material pads or yoga mats seemed to be disadvantageous compared to traditional thin textile saddle pads. Western saddles were recorded to have higher percentage pressures at the front of the saddle compared to other saddle types and a treeless saddle had the highest pressure distribution over the spine. No significant differences were detected between different disciplines, withers conformation and relative rider weight. The study showed that saddle pressure measurements under field conditions can give valuable information on saddle pressure distributions for practical purposes. To further assess the influence of different factors higher numbers of horses need to be examined.

---



**Stress response and interaction with the horse of male and female riders in equestrian show jumping**

*N. Ille, J. Aurich, M. Von Lewinski, R. Erber, M. Wulf, R. Palme and C. Aurich*
*University of Veterinary Medicine Vienna, Veterinärplatz 1, 1210 Vienna, Austria; natascha.ille@vetmeduni.ac.at*

Traditionally, horse riding has been restricted to men but today equestrian sports are dominated by women. We hypothesized that male and female riders differ in their stress response, the physical influence on the horse through their seat and the response they evoke in the horse. In Exp 1, cortisol and heart rate variability (HRV) were compared between male (n=7) and female riders (n=7) jumping a course of obstacles and in the horses (n=7) ridden by male and female riders. Saliva for cortisol analysis was collected repeatedly and cardiac beat to beat (RR) intervals were recorded from horses and riders. From the RR intervals, heart rate and HRV variables SDRR (standard deviation of RR interval) and RMSSD (root mean square of successive RR differences) were calculated. In Exp 2, pressure under the saddle was compared between male and female riders (n=5 each). In Exp 1, cortisol in riders did not change and did not differ between male and female riders. Cortisol in horses increased ($P<0.001$) irrespective of the sex of their rider. Heart rate in riders increased from walk to jumping ($P<0.001$) while HRV decreased ($P<0.001$), these changes at no time differed between male and female riders. In horses, heart rate increased as in the riders ($P<0.001$). The horses` SDRR and RMSSD decreased during walk and remained low at trot, canter and jumping ($P<0.001$) but were not affected by the rider. In Exp 2, saddle pressure was highest in the middle and lowest in the caudal part of the saddle ($P<0.001$). In trot ($P<0.05$) and canter ($P<0.01$) pressure was slightly lower in female versus male riders due to weight differences and not a different seat of male and female riders. In conclusion, there were no fundamental differences in the physical effort, stress response and seat in the saddle between male and female riders and in the stress response of horses to men and women.

## Session 17 Poster 8

**Feeding behaviour of free ranging Lusitano mares in an irrigated pasture**

*A.S. Santos[1,2], R.J.B. Bessa[3], M.A.M. Rodrigues[2], M.P. Cotovio[2], E.G. Mena[2] and L.M.M. Ferreira[2]*
*[1]EUVG, Coimbra, 3020-210, Portugal, [2]CECAV-UTAD, Quinta de Prados, 5001-01, Portugal, [3]CIISA – FMV, UTL, Lisboa, 1300-477, Portugal; assantos@utad.pt*

Feeding behaviour of grazing horses is influenced by abiotic factors such as environmental conditions, by biotic factors such as nutritive value and availability of available pasture, feed supplementation, and inherent animal factors such as grazing experience, hunger, sex, age and their physiological state. This study aimed to assess the influence of two feeding levels on the feeding behaviour of Lusitano mares grazing on an irrigated summer pasture. For this purpose, 24 Lusitano mares were used. Animals were divided into two distinct groups: group 100: animals fed the maintenance level (G100), and group 130 (G130): fed above maintenance level. G100 and G130 had access to different grazing areas in order to ensure their feeding level. The study was conducted in mid -August in an irrigated pasture in Azambuja. Time spent by animals in different activities (i.e. selecting and feed prehension, resting periods, etc.) were assessed by recording their activity every 15 min during the light period of the day (from 7:00 am to 20.15 pm) on 2 consecutive days. From a total of 18:15 h of observation, animals spend an average of 10 h in search/selection processes, representing 55.6% of total observation time. G100 animals spend more time in these processes than G130 animals (10:08 vs 9:40 h). Both groups showed similar behavioural routines throughout the day, animals tended to focus feeding search/prehension activities in three periods of the day: from 7:00 to 9:00, 13:00 to 15:00 and from 18:00 to 20:00. This study proved to be of great interest, not only being innovative in the Lusitano breed, but also by obtaining feeding search/prehension data without interacting and/or influencing the animals. These data can be later translated in more appropriate and efficient management practices.

## Session 17 Poster 9

**Effect of antioxidant supplementation to horses on muscle integrity and resistance to training**

*F. Barbe[1], A. Sacy[1], P. Bonhommet[2] and E. Chevaux[1]*
*[1]LALLEMAND, 19 rue des Briquetiers, 31702 Blagnac, France, [2]ECOPHAR, 48 av. Pierre Piffault, 72000 Le Mans, France; echevaux@lallemand.com*

Intense training in horse is a major source of oxidative stress, due to the excess production of free radicals in the cell respiration process. This condition adversely impacts the resistance to training, by decreasing muscle cell membrane permeability and functionality. Various trials have provided evidence that exercise-induced disturbances could be partially prevented: these studies focused on vitamin E, vitamin C, selenium, lipoic acid and trace elements. An alternative approach to prevent oxidative stress is to act at the first line of defense by promoting antioxidant enzymes: superoxide dismutase (SOD), catalase (CAT) and glutathione peroxidase (GPx). SOD is an endogenously produced enzyme able to catalyze the dismutation of superoxide into less harmful hydrogen peroxide, which is then converted in water and oxygen by CAT and GPx. In rodent models, a specific variety of melon (*Cucumis melo* L.), containing high level of SOD, can induce the endogenous expression of the 3 enzymes, therefore stimulating the overall first line of antioxidant defense. In a previous study, supplementation with SOD increased blood resistance to haemolysis and reduced the muscular membrane permeability induced by training. In the present experimental trial, SOD was combined with organic selenium (acting as a cofactor of GPx) and vitamin E and their combined effects were investigated on markers of muscle cell integrity: CPK and SGOT. 38 horses, originating from 9 French clinics, were divided into 3 groups and received the antioxidant combination either as curative treatment for myositis (group 1), as preventive treatment (group 2) or as curative then preventive treatment (group 3). Groups 2 and 3 were kept in training, while training was stopped for group 1. The antioxidant combination was able to maintain CPK and SGOT values at acceptable levels in long term when horses were maintained in training.

## Session 17 Poster 10

**The influence of load to the heart frequency of the sport horses**

*E. Mlyneková, M. Halo and K. Vavrišínová*
*Slovak University of Agriculture, Department of Animal Husbandry, Tr. A. Hlinku 2, 94976 Nitra, Slovak Republic; klara.vavrisinova@uniag.sk*

In order to achieve potential performance while keeping the metabolic balance in the organism of individual animals it is very important to monitor the basic physiological parameters. The primary indicator of the effect of load on the organism is the monitoring the tested individual's heart frequency. In our work we analysed the heart frequency of 11 warmblood horses on a motion load regulator. The test lasted 3 weeks with a gradual increase of load being applied progressively. The load was applied only in a step with gradual increase of time of load in an up-sloping direction. By analysis of the heart frequency of tested horses we did not register statistically significant differences among them. The mean values of heart frequency in tested horses were 61.31 beats per minute. When evaluating the maximum value of heart frequency we detected a twofold increase in comparison with the mean values. Only on two monitored horses we measured the increase to the level of 138 bpm, resp. 147 bpm. However, even these increased values were not statistically significant. The detected values confirm that the tested load did not cause any physiological changes and its length and degree of load were correctly selected for achieving greater degree of trained condition at tested horses.
This work was supported by Grand Agency of the Slovak Ministry of Education, Research. No: 1/0391/12.

---

## Session 17 Poster 11

**Shelter preferences of horses during summer**

*E. Hartmann[1], R. Hopkins[2] and K. Dahlborn[1]*
*[1]Swedish University of Agricultural Sciences, Anatomy. Physiology and Biochemistry, Box 7011, 750 07 Uppsala, Sweden, [2] University of Greenwich at Medway Chatham Maritime, Kent, ME4 4TB, UK, Natural Resources Institute, Chatham Maritime, ME4 4TB Kent, United Kingdom; kristina.dahlborn@slu.se*

Providing horses kept outdoors with shelter for protection from heat and insects during summer is seldom considered, while protecting horses from the extremes of winter is a requirement. We tested horses' daytime shelter seeking behaviour in summer, specifically whether horses would benefit from shade and whether they seek shelter for protection from insects. Eight Warmblood horses were kept individually on paddocks for two days with access to two shelter types: (1) closed shelter with roof and wind nets on three sides; (2) open sided shelter with roof only. Behaviour was recorded at 5 min intervals from 09.00-16.00. Rectal- and skin temperature on the neck and hindquarter was measured three times daily. Mosquitos were caught with Mosquito Magnet® traps and flies with sticky traps placed inside the shelters and outside near the paddocks. Data were analyzed with GLMM (Glimmix, effect of weather on shelter use), GLM (Proc mixed, effect of shelter use on body temperature) and Spearman rank correlation. Ambient temperature ranged from 16-25 °C (THI index 66±4). Five horses preferred the closed shelter and were observed inside 28% of the time (open shelter <1%). Results also indicated that horses showed a similar shelter seeking behaviour during six out of eight days. Horses used the shelter at lower wind speeds and higher solar radiation ($P<0.001$), although there was no effect of shelter use on body temperature ($P>0.05$). Horses were more likely to use the shelter on days with high numbers of flies ($P<0.01$) but no effect of shelter was detected on the frequency of insect defensive behaviour performed (e.g. tail swish, $P=0.119$). Horses make use of shelters during summer, even when temperatures are moderate. This choice therefore mainly depends upon weather conditions and partly on individual preferences. Shelters did not sufficiently protect from insect harassment.

**Behavioural differences in Italian Heavy Draught Horse foals fed different protein levels**

*C. Sartori[1], N. Guzzo[1,2], S. Normando[2], L. Bailoni[2] and R. Mantovani[1]*
*[1]University of Padua, Department of Agronomy, Food, Natural Resources, Animals and Environment, Viale dell'Università 16, 35020 Legnaro (PD), Italy, [2]University of Padua, Department of Comparative Biomedicine and Food Science, Viale dell'Università 16, 35020 Legnaro (PD), Italy; roberto.mantovani@unipd.it*

The present work aimed to evaluate possible behavioral differences in Italian Heavy Draught Horse (IHDH) foals reared in semi-covered stable and fed 2 isoenergetic diets with high (HP; 13.8% CP on DM) or low (LP; 11.1%) protein level. Protein in HP diet was consistent with common levels provided to IHDH foals. The behavior of 21 foals aged 12.4±1.2 mo. and stabled in 4 pens by sex (S) and diet (D) was videorecorded and analyzed to build a suitable ethogram including 28 behaviors in 6 categories: feeding, movement, rest, individual cleaning, affiliate/sexual activities, agonistic interactions. The percentage of the daily time spent in each category and single behaviors was analyzed via GLM including S, D, S×D, individual age, and moment of day (morning, afternoon or night). Dominance matrices were also built to put in model the rank within group, but it was not significant and discarded after preliminary analysis. Foals spent 35% of day in feeding and 35% in resting, whereas social activities involved 24% of time. Males showed greater levels of affiliate/sexual activities (9% of day vs 5%; $P<0.01$), whereas females spent more time in feeding (38 vs 35%; $P<0.001$). About diet, foals fed HP spent more time in moving (23% of day vs 12% in LP foals; $P<0.001$), whereas LP individuals dedicated more time to rest activities (41 vs 30%; $P<0.001$). This is consistent with findings about protein effects in diet (e.g. role of some amino acids on serotonin production). Otherwise, no stereotypes were found, and daily activity was consistent with typical values for draught breeds both in LP and HP foals. A parallel study also showed no differences in growth and meat quality, also supporting that a reduction of CP in foal diet is suitable with the maintenance of performance and welfare.

---

Session 17 Poster 13

**Slow feeding systems for horses: does a net placed over the forage reduce the feed intake?**

*G. Gerster, A. Zollinger, C. Wyss, B. Strickler and I. Bachmann*
*Agroscope – Swiss National Stud Farm, Les Longs Prés 191, 1580 Avenches, Switzerland; gabriela.gerster@agroscope.admin.ch*

In a natural environment, equids spend over 16 hours a day grazing. Stabled horses are traditionally fed 2-3 times a day with limited amount of forage. This may lead to disturbances of the digestive system and/or behavioral problems. To increase the feeding time without increasing the total feed intake, a new system was developed. It consists of covering the forage with a net which makes the feed intake more challenging thus leading to a longer feeding time. The aim of this study was to investigate whether the hay and haylage intake per hour decreases when using such a net. Subjects were six adult mares with an average body weight of 576±55 kg. They were housed together and fed in foraging stalls for 60 min five times a day. Forage was located on a feed bunk and was covered with the net tied to a metal frame. The mesh size of the applied net was 4.5×4.5 cm. The horses' access to the forage was computer managed. Horses were observed during 4 days of feeding hay and haylage respectively, each with and without net. The forage intake was recorded three times a day from 10.30 to 11.00, 13.30 to 14.00 and 16.30 to 17.00. To investigate the impact of the net on the feed intake, the recorded data was statistically analyzed using a Wilcoxon-Rank Sum test as implemented in R. For haylage the difference in feed intake with or without net was not significant. The mean consumption rate for haylage was 1.70 kg DM/h with net and 1.84 kg DM/h without net. Offering hay instead of haylage we found a significant difference of the feed intake for three out of the six horses. Mean consumption rate for hay with and without net was 1.51 kg DM/h and 1.69 kg DM/h respectively. The results indicate that a net with 4.5×4.5 cm mesh size does not sufficiently decrease the feed intake. Therefore, we do not recommend it for application in practice. A net with 3×3 cm mesh size will be tested in the near future and the results will be available for the conference.

**Horse meat composition and properties from adult horses slaughtered in Lithuania**

*R. Sveistiene and V. Razmaite*
*Lithuanian University of Health Sciences, Institute of Animal Science, R. Zebenkos 12, Baisogala, 82317, Lithuania; ruta@lgi.lt*

Despite the fact that Lithuanians have no specialized horse meat production systems, the horse meat is exported. The objective of the study was to determine the influence of age, gender and carcass massiveness on horse meat properties from adult horses slaughtered in Lithuania. Immediately after slaughter in an accredited abattoir, the carcasses were chilled at 2-4 °C for 24 h. With the aim to avoid the damage of the carcasses for export, samples of m. pectoralis profundus were excised from the left of each carcass before their shipment. All horse meat composition and properties were detected according to EU reference methods. The data were subjected to the analysis of variance in general linear (GLM Multivariate) procedure in SPSS 17. The model included the fixed effect of age, gender and massiveness. Differences among means were determined using the LSD test. The differences were regarded as significant when $P<0.05$, but the differences of $0.05 \leq P<0.10$ would be considered as trends. Horse meat was characterized by a favourable polyunsaturated (PUFA)/saturated (SFA) (0.67-0.77) and n-6 PUFA/n-3PUFA (0.63-0.78) ratios of intramuscular fat. The age increase from 4 to 21 years of adult horses did not appear to affect meat composition, rheological properties and fatty acid composition. However, the increase of Warner-Bratzler shear force values was observed with the horse age increase. Meat quality properties were affected by gender. The stallions were characterized by the lowest percentage of intramuscular fat and the highest percentage of protein if compared with the mares and castrates. The mares tended to have a lower content of cholesterol compared with castrated males. Carcass massiveness showed the differences in meat color parameters and protein content. Horse gender and carcass massiveness also tended to show differences in fatty acid composition and some textural properties.

---



**Effect of storage time on the acceptability of foal meat**

*M.V. Sarries[1], M.J. Beriain[1], K. Insausti[1], J.A. Mendizabal[1], M. Ruiz[1], E. Valderrama[1], A. Perez De Muniain[2] and A. Purroy[1]*
*[1]ETSIA, Departamento Produccion Agraria, Campus Arrosadia S/N, Edif. Los Olivos, Universidad Publica De Navarra, 31006, Pamplona, Spain, [2]ASCANA, c/ Ainziburu S/N, 31170, IZA, Spain; vsarries@unavarra.es*

Nowadays there is a low equine meat production system to satisfy the demand of the modern meat consumer which has interest in tasting different meat products. Burguete breed is an autochthonous equine breed from Navarra (Spain). The production system of these native horses is extensive (mountain), linked to the environment, and based on pasture. Its sustainable presence should be addressed through a potential strategy to improve the sustainability of the rural local development. The aim of this work was to study the effect of storage time on the acceptability of foal meat. The longissimus dorsi muscle was removed from carcasses of six 15-months-old Burguete breed foals to examine the lipid stability, instrumental colour, percentage of pigments by spectrophotometry and acceptability of foal meat. The above parameters were measured along the time of storage (days 1, 3, 6 and 9) at 2 °C. The time of storage played an important role in the storage of foal meat. The highest lipid degradation was found on day 9 compared with day 1 (1.93 vs 0.18 mg/ malonaldehyde/kg fresh meat; $P<0.001$). In addition, the L* and a* values were decreased between the both mentioned periods of time (30.10 vs 35.26; $P<0.001$; 7.92 vs 10.24; $P<0.001$, respectively). The metmyoglobin pigment appeared on the superficial foal meat at the day 3, being the predominant pigment (63.2%). The acceptability of foal meat showed positive values in periods of storage which were no longer than the day 3 (7.63/15 cm) (0 cm = like very much; 15 cm = dislike very much). To conclude, the maximum time of storage of foal meat should not be exceeded more than the $3^{rd}$ day in order to avoid the degradation process of meat which affected the organoleptic characteristics and acceptability of meat.

**Challenge programme: ethics teaching in animal science**

*H.G.J. Gremmen*
*Wageningen University, Department of Philosophy, Hollandseweg 1, 6706 KN Wageningen, the Netherlands; bart.gremmen@wur.nl*

In the past decades societal debates on animal production have become more numerous and intense. Many of the debated issues have an important ethical component. Sustainability, biodiversity and animal welfare are just a few of the relevant societal values involved in the attempts of policymakers, industry and NGO's to steer animal production in an ethical acceptable direction. In this session we focus on the role animal scientist's play in this process as experts providing new knowledge and technology. How can they learn to prepare themselves on their possible contribution and roles? Ethics teaching in academic programmes is an important key in this process of preparation. In some cases professional ethicists have developed ethics courses, which they teach in animal science programmes. However, apart from the fact that in a lot of cases it is not possible to hire professional ethicists, the results could also be counterproductive: necessary hands-on experience with real and actual cases is often lacking. In those cases animal scientists could try to integrate ethical material in their regular courses. Questions in this session are: what are the ethical issues on the agenda? How to find and provide material form the ethical background literature? How to connect the ethical issues to the regular material of the course? How to assure a gradual build-up of ethical competence during the full length of the educational trajectory? In this session we will start with a short introduction of a few examples of the integration of ethics in animal science curricula.



**An industry perspective on putting global animal welfare principles into action**

*S. Vaughn*
*Zoetis, Veterinary Medicine Research and Development, 333 Portage Street, Kalamazoo, Michigan 49007, USA; sherry.e.vaughn@zoetis.com*

The mission of the animal health industry is to research, develop, and deliver medicines and vaccines both for companion animals and livestock. Research and innovation are of vital importance to develop new solutions and improve available animal health products to keep animals healthy, to safeguard both animals and people against diseases and to secure a safe and sustainable food supply. Ethical and regulatory responsibilities are taken into account when demonstrating the safety and efficacy of products before they are used to treat animals. The animal health industry supports a rigorous, scientific product licensing system based exclusively on safety, quality and efficacy. Animal based research remains vital to the discovery, innovation, development, and licensing processes which lead to the development and improvement of animal health products. Medicines have to be safe, of high quality and effective: for animals, for those administering the medicine, for the environment, and for consumers of animal products. In a global animal health research and development environment, how are animal welfare principles put into action? How is corporate culture created to demonstrate a commitment to the humane use of animals in research, development, and manufacturing? How are the principles of animal use for research and testing (refinement, reduction, replacement-the 3Rs) integrated into the culture of an animal health company? What standards are used in countries with evolving animal welfare regulations? Are there factors that might influence variations in the quality of animal care globally? What might be expected of you in regard to animal welfare, as an animal scientist, in today's evolving world?

## Session 20 Theatre 1

**Towards sustainable cows, good herd practices, and quality dairy products in the USA**

*A. De Vries*
*University of Florida, Department of Animal Sciences, 2250 Shealy Drive, Gainesville, FL 32611, USA; devries@ufl.edu*

The efficiency and environmental sustainability of the dairy industry in the USA has increased greatly in the past 70 years. Milk production per cow per year has quadrupled since 1940 due to increased knowledge about management, nutrition, reproduction, herd health and genetics. The emphasis on milk production in genetic selection programs initially led to reductions in functional traits such as fertility, but inclusion of functional traits since 1994 has reversed these trends in the last 10 years. A large emphasis on transition cow management has reduced health problems after calving, but the transition period remains a sensitive time. Longevity of dairy animals has remained at approximately 5 years, including 2 years until first calving. This average longevity is the result of calving heifers that push out the worst cows in a national dairy herd of constant size. Milk quality has improved in the last decade, due to better management and motivated in part by quality premiums. Culled dairy cows account for approximately 15% of dairy farm revenue, and are approximately 15% of all processed cattle in the USA. A new trend is the emphasis on reducing the environmental footprint of dairy production. Half of the carbon footprint of retail milk is due to milk production on the farm. Compared to 1944, the same amount of milk is produced with 89% fewer animals, 65% less water, 90% less land, and a 63% lower carbon footprint. These gains are mostly realized by a much greater milk production per cow which dilutes maintenance costs. Although fluid milk production is decreasing per capita, the total volume of milk produced increases annually by approximately 2%. Drought is a major constraint in some parts of the US with some concern that the pumping of water from aquifers may not be sustainable. Rapid genetic progress through the implementation of genomics is emerging, which will results in a better cow that is producing a lot of quality milk while staying naturally healthy.

---

## Session 20 Theatre 2

**Optimization of production efficiency and environmental impact within the Austrian cattle production**

*B. Fuerst-Waltl[1], F. Steininger[2], L. Gruber[3], K. Zottl[4], W. Zollitsch[1], C. Fuerst[2] and C. Egger-Danner[2]*
*[1]Univ. of Nat. Res. and Life Sci. (BOKU), Gregor-Mendel-Strasse 33, 1190 Vienna, Austria, [2]ZuchtData, Dresdner Straße 89/19, 1200 Vienna, Austria, [3]Agricultural Research and Education Centre, Raumberg 38, 8952 Irdning, Austria, [4]LKV Niederösterreich, Pater Werner Deibl-Str. 4, 3910 Zwettl, Austria; birgit.fuerst-waltl@boku.ac.at*

Under the condition of limited resources and increasing human population growing competition for high-quality, plant-based sources of energy and protein will increase worldwide. Disruptions in global inventories due to climate change may also encourage more emphasis on production efficiency. The presented project 'Efficient Cow' has the aim to evaluate the possibilities of genetically improving production efficiency in cattle breeding under Austrian circumstances. A key question is the definition of adequate efficiency traits. Presumably efficiency is a combination of already existing traits: milk, beef and functional traits and traits aiming at feeding efficiency and health. In addition to routine performance recording, new traits like body weight and other body measurements, health traits with emphasis on metabolism and feet and legs, and data about feed quality and feed intake are collected on about 170 farms within one year. Additionally, 2,000 Fleckvieh cows and 1000 Brown Swiss cows of these farms will be genotyped within the EU-project Gene2Farm. The farms are representing different production and feeding systems. Genetic parameters for newly defined efficiency traits and genetic correlations to other traits within the total merit index will be estimated. Further, the environmental effect of different milk production levels is modelled. One expected output is a recommendation for low-cost routine recording of traits associated with production efficiency, either as direct or as auxiliary traits. A higher efficiency is economically important and contributes to a reduction of nutrient losses and greenhouse gas emissions.

**The robustness of the dairy cow herd seen through the diversity of cows adaptive responses**

*E. Ollion[1,2], S. Ingrand[1], S. Colette-Leurent[3], J.M. Trommenschlager[4] and F. Blanc[2]*
*[1]Inra, UMR 1273 Métafort, Theix, 63122 Saint Genès-Champanelle, France, [2]VetagroSup, UMR 123 Herbivores, Av. de l'Europe, 63370 Lempdes, France, [3]Inra, UE 326 Domaine expérimental du Pin-au-Haras, Borculo, 61310 Exmes, France, [4]Inra, UR0055 ASTER, Av. L Buffet, 88500 Mirecourt, France; emilie.ollion@vetagro-sup.fr*

The ability of the dairy cow herd to maintain itself and its performances in a disturbed environment is qualified as robustness and relies on both the farmer practices and on individual capacities of cows to adapt. We hypothesized that adaptive responses of dairy cows facing disturbances are related to the trade-offs (TO) expressed in a restricted environment between milk production (MP), reproduction and survival. Our analyses were performed using 493 lactations of cows from two Inra experimental units. The method consisted of three main steps: (1) to characterize the trade-offs profiles for each cow, using a clustering method (2) to describe their adaptive response profiles when facing disturbances, (3) to analyze the links between both profiles. The TO profiles were based on indicators of MP and body condition score (BCS) dynamics and reproduction performances over the 12 first weeks postpartum. Three main profiles were identified for cows who successfully reproduced (350 cows): P1 with priority given to MP rather than BCS (156 cows); P2 characterized by a short drop of BCS, but a delayed reproduction and the highest lactation yield (72 cows); P3 associated with the maintaining of BCS, low MP and good reproduction performances (122 cows). Three adaptive response profiles were highlighted using the same cows when experiencing a disturbance, as described by Ollion *et al*. The analysis of the links between the two kinds of profiles showed that 82% of the P1 cows maintained MP close to the predicted level but lost BCS during disturbances. Cows from P2 TO profile were equally spread over the three types of adaptive response profiles. Finally, cows expressing P3 TO profile were stable in BCS but their MP fluctuated during disturb events.

---



**Genetic variation and consistency among feed efficiency traits in Holstein and Jersey cows**

*S. Van Vliet, J. Lassen and P. Løvendahl*
*Aarhus University, Molecular Biology and Genetics, Research Centre Foulum, 8830 Tjele, Denmark; sophie.vanvliet@agrsci.dk*

Breeding for improved feed efficiency requires well recorded phenotypes of feed intake, milk yield and other traits involved in the feed efficiency complex. These are only available from research stations and therefore only for small numbers of recorded animals. Most often cows are kept in similar ways in research herds and commercial farms so that the herd includes cows of the first 3 parities, with about 40% from first parity. However, recording is rather intense so that traits are recorded on daily basis, and for practical reasons expressed as weekly averages. This experiment used 458 Holstein and 216 Jersey first parity cows housed at the Danish Cattle Research Centre (Foulum), fed total mixed rations ad lib and supplemented with up to 3 kg concentrates per day during AMS milkings. The lactation was further divided into 3 sections of 4 week around key stages: (1) weeks 1-4; (2) 6-9; (3) 18-21. For dry matter intake the within section heritability and repeatability was for Holsteins: (1) 0.35/0.76; (2) 0.42/0.78; (3) 0.43/0.73 and for Jerseys: (1) 0.15/0.72; (2) 0.28/0.80; (3) 0.53/0.82. The estimates for Jerseys had large standard errors that may be the main reason for the estimates fluctuating more than in the Holsteins. The study also included residual feed intake, as dry matter intake adjusted for ECM yield and metabolic live weight. The within section heritability and repeatability was for RFI in Holsteins: (1) 0.30/0.71; (2) 0.17/0.70; (3) 0.35/0.63 and for RFI in Jerseys: (1) 0.19/0.67; (2) 0.16/0.63; (3) 0.20/0.83. The results show that feed intake and RFI are heritable traits in dairy cows, and that the traits are highly repeatable within shorter time periods, and also stable and consistent attributes of the individual cow over the timespan of complete lactations.

## Meeting challenges for organic dairy production in North Carolina

*S.P. Washburn and K.A.E. Mullen*
*North Carolina State University, Animal Science, Box 7621, Raleigh, NC 27695-7621, USA; swashbur@ncsu.edu*

Organic dairy production in NC began in 2007 when a few herds became certified to sell organic milk. There were economic challenges the first few years as drought conditions forced those producers to import expensive organic feedstuffs but all farms have managed to survive. Udder health was also a concern for those producers because of hot summers and parasitic flies as potential vectors of mastitis-causing bacteria. Milk sample cultures from 7 organic and 7 conventional dairy farms in NC in 2010 documented that overall udder health and prevalence of mastitis-causing pathogens were similar between management systems. Further studies were conducted to examine herbal preparations as alternatives to antibiotics for managing udder health. Herbal products used at dry off did not negatively affect subsequent milk production and generally were not inferior to conventional dry cow therapy in preventing new infections. Thyme oil, an ingredient in one herbal product, was tested in milk culture and documented to have significant antibacterial activity. In a 4-year study at our pasture-based research site, cows were managed using organic practices (n=250) vs conventional practices (n=252) except that all cows received conventional concentrate supplements. Production, incidences of mastitis, and most reproduction measures were similar between management groups. However, organically managed cows had lower SCS (2.78 vs 3.04, SE=0.16, P=0.02) but tended to have longer calving intervals (371 vs 364 d, SE=3 d, P=0.06) than conventional cows. Though organic dairying is still new to the area and likely not yet optimally managed, our studies provide evidence that organic milk production systems have potential viability in the region.



## Nutritional trajectory allows to increase the efficiency of beef cows whatever their body condition

*A. Delatorre[1], F. Blanc[2], P. D'Hour[3] and J. Agabriel[1]*
*[1]INRA, UMR1213, F-63122 Saint-Genès-Champanelle, France, [2]Clermont Université, VetAgro Sup, UMR1213, F-63000 Clermont-Ferrand, France, [3]INRA, UE 1296, 63820 Laqueuille, France; anne.delatorre@clermont.inra.fr*

Feed costs represent the largest operating costs in beef production, one way to increase profitability is to improve the use and efficiency of energy intake. The objectives of this work were to (1) characterize the energetic efficiency of cows experiencing 2 nutritional trajectories after calving and (2) to determine the effect of body condition score at calving ($BCS_c$, scale 0-5) on this efficiency. This latter was estimated by the residual energy expenditure ($E_{resid}$) which accounts for the net energy available for other expenditures than milk and tissue growth; $E_{resid}$ (in MJ/d/kg $BW^{0.75}$ in NE for lactation)= Eintake – (Emilk + Eretained in tissues). $E_{resid}$ was calculated in 40 cows differing in $BCS_c$ (moderate, M vs fat, F) experiencing 2 energy trajectories during the first 120 days post-partum period (P1): Control level (FC (n=9) and MC (n=9) vs Low level (FL (n=9) vs ML (n=10). During Period 2 (P2, 120-196 days post-partum) all cows were turned out to pasture. BW, body condition and milk production were measured in P1 and P2. Body lipids reserves of each cow were assessed 3 times over the experiment by measuring subcutaneous adipose cells diameter. Milk production was similar between the 4 groups of cows. Over P1, FL and ML cows lost 25 and 43 kg respectively compared to FC and MC cows. $E_{resid}$ was 25% lower in L than in C cows (P<0.05) whatever their BCS. Over P2, BW and BCS gains were similar in FL and ML cows. At the end of P2, FL and ML cows weighed 20 and 10 kg less than FC and MC cows. Over P1+P2, the difference in $E_{resid}$ between the 2 nutritional trajectories was 23% in both F and M cows (P<0.05). These results showed that lean cows submitted to an underfeeding/refeeding trajectory present similar abilities to withstand nutritional perturbations than fatter ones. $E_{resid}$ changes could be interpreted as an indicator of cows' efficiency within a nutritional trajectory.

**Health traits and their role for sustainability improvement of dairy production**

*K.F. Stock, J. Wiebelitz and F. Reinhardt*
*Vereinigte Informationssysteme Tierhaltung w. V., Heideweg 1, 27283 Verden, Germany; friederike.katharina.stock@vit.de*

The potential of improving animal health and welfare by breeding implies considerable benefits for overall sustainability of dairy production, and genetic correlations between health traits and longevity could serve as quantifiers for the effects of targeted breeding measures. Standardized health records from on-farm documentation systems of 104 German dairy farms were available for this study, with information on about 130,000 lactations of dairy cows. Genetic parameters were estimated in linear repeatability animal models with REML for quasi-continuous health traits, including mastitis, reproduction disturbances, metabolic disorders, and claw diseases. Heritabilities were 0.04-0.13 for claw diseases and 0.02-0.09 for the other diseases. Univariate BLUP EBV for health traits were used for correlation analyses with EBV for longevity from the routine national genetic evaluation for dairy cattle (functional herd life, fHL) and from test runs with refined trait definition (6 survival periods, with separate consideration of survival in days 0-150 and after day 150 in parities 1-3). Reliabilities of EBV for health traits were rather low for most of the 4,334 Holstein bulls, so only the 239 bulls with >50 daughters were considered for the correlation analyses. Moderately positive correlations of on average 0.3 were found between EBV for health traits and fHL. For health traits with specific risk periods within lactations (mastitis, displaced abomasum, ovary cycle disturbances), up to >2-fold closer correlations with corresponding than non-corresponding survival periods indicated advantages of refined trait definition. Selection based on EBV for health traits is the most specific approach to improving sustainability by increase of disease resistance, but direct health information is available for only parts of the population. Combined use and optimized modelling of new phenotypes has the potential to increase breeding progress with regard to health and longevity of the dairy cow.

---

Session 20 Theatre 8

**Impact of growth and age at first calving on production and reproduction traits of Holstein cattle**

*L. Krpalkova[1], V.E. Cabrera[2], M. Stipkova[1], L. Stadnik[3] and P. Crump[2]*
*[1]Institute of Animal Science, Prague, Přátelství 815, 104 00 Praha 10, Czech Republic, [2]University of Wisconsin, 1675 Observatory Dr., Madison, WI 53716, USA, [3]Czech University of Life Sciences Prague, Kamýcká 129, 165 21, Prague 6 – Suchdol, Czech Republic; lenka.krpalkova@centrum.cz*

The objective of this study was to evaluate the effect of body condition score (BCS), body weight (BW), average daily weight gain (ADG) and age at the first calving (AFC) of Holstein heifers on production and reproduction parameters in the 3 subsequent lactations. The data set consisted of 780 Holstein heifers calved at 2 dairy farms in the Czech Republic from 2007 to 2011. Their BW and BCS were measured at monthly intervals during the rearing period (5 to 18 mo of age). The highest milk yield in the first lactation was found in the group with medium ADG (5 to 14 mo of age, 0.949 to 0.850 kg/d ADG). The highest average milk yield over lifetime performance was detected in heifers with the highest total ADG (≥0.950 kg/d). Difference in milk yield between the evaluated groups of highest ADG (in total and postpubertal growth ≥0.950 kg/d and in prepubertal growth ≥0.970) and the lowest ADG (≤0.849 kg/d) was approximately 1,000 kg/305 d per cow. The highest milk yield in the first lactation was found in the group with the highest AFC≥751, for which fat and protein content in the milk was not reduced. Postpubertal growth (11 to 14 mo of age) had the highest impact on AFC. The group with lowest AFC≤699 d showed a negative impact on milk yield but only in the first 100 d of the first parity. Highest ADG was detrimental to reproduction parameters in the first lactation. Highest BW at 14 mo (≥420 kg) led to lower AFC. Groups according to BCS at 14 mo showed no differences in AFC or milk yield in the first lactation or lifetime average production per lactation. We concluded that low AFC≤699 d did not show a negative impact on subsequent production and reproduction parameters. Therefore, a shorter rearing period is recommended for dairy herds with suitable management.

## Use of milk somatic cell count from the colostral phase as an early indicator for health

*R. Schafberg[1], K.F. Stock[2] and H.H. Swalve[1]*
*[1]Martin-Luther-University Halle-Wittenberg, Institute of Agricultural and Nutritional Sciences, Theodor-Lieser-Str. 11, 06120 Halle, Germany, [2]Vereinigte Informationssysteme Tierhaltung w.V. (vit), Heideweg 1, 27283 Verden/Aller, Germany; renate.schafberg@landw.uni-halle.de*

During a three-year period (2010 to 2012) within the project BHNP (Breed for Health NeoPartus) data on various indicators for health status were collected for 15,273 calvings of Holstein cows from seven large herds in Germany. Within our project, it was attempted to measure cell counts during the early colostral phase, i.e. within the first three days after parturition. However, due to the field study nature of the project, samples obtained varied with respect to the day collected post-partum. In total, 8,447 samples with an average collection day of 5.3 p.p. and a range between 0 to 20 days p.p. were obtained. Noting that cell counts from the colostral phase in general are on a higher level than lactation cell counts, it was found that cows with cell counts above 400,000 show a drastic and highly significant increase in the probability for a later treatment for udder health. For early cell counts below 100,000, between 100,000 and 400,000, and above 400,000, probabilities for a later treatment within the first 100 days of lactation increased from 20.32%, to 23.97%, and to 37.55% based on a mixed model analysis accounting for effects of herd-year-season and parity-age-at-calving. This finding was not due to an auto-correlation as herd managers were unaware of the results from the colostral phase samples. Probabilities for later treatment also corresponded to data for culling within the lactation. Estimates of heritabilities for SCSel (Somatic cell count in early lactation) were 0.23 when data was restricted to days in milk 0 to 4 only (n=4,006 lactations) and 0.10 for the full data set (n=7,483). In summary, the results confirm the use of somatic cell counts from the colostral phase as a potential management tool and as a new trait for genetic selection.



## Genetic evaluation of peripartal problems in Bavarian Fleckvieh and Brown Swiss

*A.H. Haberland, D. Krogmeier, E. Zeiler, E.C.G. Pimentel and K.-U. Goetz*
*Bavarian State Research Center for Agriculture, Institute of Animal Breeding, Prof.-Duerrwaechter-Platz 1, 85586 Poing, Germany; anne.haberland@lfl.bayern.de*

In addition to productivity, modern dairy cattle breeding programs also have been focusing on health traits to increase the profitability of dairy herds. Data collection is a first step towards an optimized breeding goal and can be used to provide dairy farmers with a useful feedback about the health status of their herds. In Bavaria, a herd monitoring program has been implemented on the basis of veterinary diagnoses and surveillance by farmers. It enables to trace the condition of single cows and offers an overall picture of the herd. Initial evaluations with the data collected by farmers involved peripartal problems in Fleckvieh (FV) and Brown Swiss (BS) cows. In this study, we report heritabilities and genetic correlations among retained placenta (RP), downer cow syndrome (DCS) and calving ease within first (CE_1) and later (CE_2+) lactations. The incidence of RP within the first lactation was twice as high in BS (5.4%) as in FV (2.7%). In contrast, BS cows showed a higher proportion of easy calvings (83%) than FV cows (74%). For both breeds, the incidence of DCS increased from 0.2% in the first lactation to about 2.8% in the following lactations. Using a linear sire model, heritability estimates for RP, DCS, CE_1 and CE_2+ were 0.014, 0.043, 0.059 and 0.002 for FV and 0.018, 0.050, 0.018 and 0.014 for BS, respectively. Genetic correlations ranged from low to moderate. Moderate correlations were found between RP and CE_1 (0.45 for FV), and between RP and CE_2 (0.36 for FV). A threshold sire model yielded higher heritabilities for RP and DCS, but slightly lower heritabilities for CE_1 and CE_2+. Further evaluations within the scope of the project will involve udder health, calf diseases, reproductive and metabolic disorders. Ultimately, the routine data collection initiated by the project will be used to incorporate breeding values for health traits into the joint German-Austrian breeding program.

**Estimates of genetic parameters for milk production in Tunisian Holstein cows**

*A. Hamrouni[1], M. Djemali[1] and S. Bedhiaf[2]*
*[1]Institut National Agronomique de Tunis, Production Animale, 43 Avenue Charles Nicolle 1082- Cité Mahrajène. Tunisie, 1082 Cité Mahrajène, Tunisia, [2]INRAT, Rue hédi Karray 2049 Ariana. Tunisie, 2049 ariana, Tunisia; abirturki@yahoo.fr*

The objective of this study was, therefore, to estimate (co)variance components of milk, fat, and protein yields in the first 3 lactations with a multivariate model. Genetic parameters of milk, fat, and protein yields were estimated in the first 3 lactations for registered Tunisian Holsteins. Data originated from the Tunisian Genetic Improvement Center. They contained 48,244, 32,940 and 18,034 test-day production records collected on 7,852, 5,258 and 2,878 first, second and third parity cows respectively. (Co)variance components were estimated using residual maximum likelihood (REML) as applied in SAS 9.0. Estimates of 305 days heritabilities pooled over three lactations were 0.26, 0.22 and 0.24 for milk, fat and protein yield, respectively. Heritabilities for 305-d milk, fat and protein yields in the first three lactations were moderate (0.14 to 0.20) and in the same range of parameters estimated in management systems with low to medium production levels. Genetic correlations among 305-d yield traits ranged from 0.25 to 0.44. The lowest genetic correlation was observed between the first and third lactation, potentially due to the limited expression of genetic potential of superior cows in later lactations. It was concluded that these genetic and parameters could be used for the genetic evaluation of dairy cattle in Tunisia.

---

Session 20 Theatre 12

**Temperature and humidity influence milk yield and quality in Scottish dairy cows**

*D.L. Hill and E. Wall*
*SRUC, Animal & Veterinary Sciences, Roslin Institute Building, Easter Bush, Midlothian EH25 9RG, United Kingdom; davina.hill@sruc.ac.uk*

A better understanding of the response of livestock to current and future weather patterns is essential to enable farming to adapt to a changing climate. We investigated the effects of recent weather patterns on milk yield and milk fat content in an experimental dairy herd in Scotland over 21 years. Holstein Friesian cows were either maintained indoors with summer grazing or were continuously housed. Milk yield was measured on a daily basis and fat content (%) was sampled once a week. We calculated a Temperature Humidity Index (THI) from daily measurements of external air temperature and humidity from an onsite weather station. Data from ~1,400 cows were analysed using Restricted Maximum Likelihood. We found that THI influenced milk yield, and the shape and magnitude of the effect depended on whether animals were inside or outside on the test day. Animals produced more milk when they were indoors than outdoors, largely due to differences in diet. The milk yield of cows outdoors was lower at the extremes of THI than at intermediate values, declining more steeply at higher THI values than at lower ones. The highest yields were obtained when the weekly maximum THI, measured at 9am, was ~62 units. Cows inside, sheltered from extremes of THI, increased milk yield as THI increased. Milk contained a higher concentration of fat when animals were outside than inside, and fat content was lower at extremes of THI than at intermediate values in both environments. These results show that milk yield and quality are impacted by extremes of THI under conditions currently experienced in Scotland. Potential decreases in productivity due to predicted increases in THI over the 21$^{st}$ century may be offset through changes in farm management practices or selective breeding from animals that are more resilient to climate change.

## Effect of rearing period of heifers and herd level of milk yield on performance and profitability

*L. Krpalkova[1], V.E. Cabrera[2], J. Kvapilik[1], J. Burdych[1] and P. Crump[2]*
*[1]Institute of Animal Science, Přátelství 815, 104 00 Praha 10, Czech Republic, [2]University of Wisconsin, 1675 Observatory Dr., Madison, WI 53716, USA; lenka.krpalkova@centrum.cz*

The objective of this study was to evaluate the effects of variable intensity in rearing dairy heifers on 33 commercial dairy herds included 23,008 cows and 18,139 heifers as analysis in age at first calving (AFC), average daily weight gain (ADG), and milk yield level on reproduction traits and profitability. Data were collected in 1 yr period (2011) in the Czech Republic. The results show that those herds with more intensive rearing periods had lower conception rates among heifers. The differences in those conception rates between the group of greatest ADG (≥0.800 kg/d) and the group of least ADG (≤0.699 kg/d) were approximately 10 percentage points. All the evaluated reproduction traits differed between AFC groups. Conception at first and overall services (cows) was greatest in herds with AFC≥800 d. Shortest days open (105 d) and calving interval (396 d) were found in the middle AFC group (799 to 750 d). The highest number of completed lactations (2.67) was observed in the group with latest AFC (≥800 d). The earliest AFC group (≤749 d) was characterized by the highest depreciation costs per cow at CZK 8,275 (US$ 414), and the highest culling rate for cows of 41%. The most profitable rearing approach was reflected in the middle AFC (799 to 750 d) and middle ADG (0.799 to 0.700 kg) groups. Most MY (≥8,500 kg) occurred with earliest AFC of 780 d. Higher levels of MY led to lower conception rates in cows, but the most MY group also had the shortest days open of 106 d, and calving interval of 386 d. The same MY group had the highest cow depreciation costs, total profit, and level of profitability without subsidies of 2.67%. We conclude that achieving low AFC will not always be the most profitable approach, which will depend upon farm-specific herd management. The group of farms having the most MY achieved the highest profit despite having greater fertility problems.

Session 20 Poster 14

## Replacement of soybean meal and straw against alfalfa hay in TMR of high yielding dairy cows

*H. Scholz[1] and T. Engelhard[2]*
*[1]Anhalt University of applied Sciences, LOEL, Strenzfelder Allee 28, 06406 Bernburg, Germany, [2]State Institute for Agriculture, Forestry and Horticulture Saxony-Anhalt, Lindenstrasse 19, 39606 Iden, Germany; h.scholz@loel.hs-anhalt.de*

Alfalfa is an excellent forage for high yielding dairy cows. The price of Soybeans has been term trend in the recent years and many people in the European Union prefer non-genetically modified food. A protein replacement for soybean or rapeseed meal can be achieved to a certain extent by roughage, provided that the effects of the feed must be observed. For instance, additional health aspects are taken into account through the use of alfalfa hay or alfalfa dried products over the use of straw (mycotoxins). An economic evaluation of the exchange can then prove a certain excellence. The investigation was carried out in the State Institute for Agriculture, Forestry and Horticulture Saxony-Anhalt (Iden). 74 Holstein Friesian cows were divided into two feeding strategies (straw and Soybean meal vs alfalfa hay). The Total Mixed Ration (TMR) was similar between 2 groups: 5-6 kg corn silage, 3.5-4.5 kg Grass silage, 2.0 kg Sugar beet pellets, 4.0-5.0 kg cereal meal and 4.0-4.3 kg raps meal per day. Soybean meal with 0.8-0.9 kg + 1.0-1.2 kg straw were substituted with 2.0-2.2 kg alfalfa hay. TMR had a nutrient content of 160 g crude protein per kg dry matter with 7.0 MJ NEL per kg DM. Feed intake in the first 60 days of lactation of the alfalfa group was average 20.7±2.7 kg DM per day and showed no significant differences to the group with straw and soybean meal (20.4±2.7 kg DM/d). Milk yield and milk composition was similar between the two groups. The amount of milk-urea was in the alfalfa group (187 mg/l) significant reduced in compare to the group with straw and soybean meal (200 mg/l). Metabolic status of the cows was similar between the two feeding strategies (BHB, NEFA, GLDH, ASAT). The investigation shows the adequate protein replacement of soybean meal and straw towards alfalfa hay without compromising on milk yield and milk composition as well as animal health.

**Quantitative behavior in dairy cows under the conditions of automatically and conventionally milking**

*H. Scholz, B. Füllner and A. Harzke*
*Anhalt university of Applied Sciences, Strenzfelder Allee 28, 06406 Bernburg, Germany; h.scholz@loel.hs-anhalt.de*

Automatically milking systems increased in the last years in the European Union. Behavior of lactating dairy cows in the barn can be used to evaluate the effects of changing in the procedures of milking like Forced Cow Traffic (AMS and VMS) and the design of management in the waiting area (AMR24). Conventional milking systems (6 farms) were used in this investigation as 'standard'. High producing dairy cows should be lying for a minimum of 12 hours a day. In this investigation were used 18-25 dairy cows in each farm for measurement of quantitative behavior over 24 hours. The behavior of cows was observed directly with Time-Sampling-method (5 minutes). In the farm with AMR24 data collection of behavior was every month over a period for 1 year. In the system of AMR24 were found an average lying time of 702±107 minutes per day and in the conventional milking systems of 652±139 minutes per day. Cows under the condition of AMS/VMS show an average lying time per day of 593±149 minutes, while between free cow traffic and forced cow traffic no significant difference. Under the condition of AMR24 and most of other farms delivery of feed were once a day. Between milking systems there are significant differences in feeding behavior (AMS/VMS: 279±65 minutes; AMR24: 307±65 minutes and Conventional 247±72 minutes per day). Ruminating time in the investigation was 501±78 minutes per day while ruminating during lying was 69% and ruminating during standing was 31%. Times to spend standing (included milking time) were highest in AMS/VMS (560±145 minutes) than AMR 24 (426±95 minutes) and Conventional (537±149 minutes per day). It can be concluded that the milking system show a significant influence of quantitative behavior of dairy cows.

---



**Comparison of high-yielding dairy cows fed concentrates vs those fed no concentrates over ten years**

*P. Kunz[1], M. Buergisser[1], A. Liesegang[2] and M. Furger[3]*
*[1]Bern University of Applied Sciences, Laenggasse 85, 3052 Zollikofen, Switzerland, [2]University of Zurich, Winterthurerstrasse 260, 8057 Zürich, Switzerland, [3]Agricultural Education Centre Plantahof, Kantonsstrasse 17, 7302 Landquart, Switzerland; peterkunz@bfh.ch*

Milk prices in Switzerland have been falling for years. As a result, farmers seek to reduce production costs by feeding lower amounts of the most expensive feed component, i.e. concentrates, or cease feeding concentrates altogether. In order to clarify how high-performance dairy cows respond to a lack of concentrates in their rations, the high-performance herd (75 Brown Swiss cows) at the experimental farm of the Plantahof Agricultural Education and Advisory Centre was divided into two groups: one herd (forage herd =FH) has not received concentrates since 2003, while the other herd (concentrate herd =CH) received the same roughage feed components as the FH and an additional 1,500 kg concentrates per cow and lactation. The ration was composed of hay, dried grass and maize silage in winter, and hay, grass and maize silage in summer. Both herds were of equal genetic value, as the same bulls were used to sire progeny. The different feeding regimes resulted in differences between the two trial herds: over the past six years, feed intake in CH cows has been higher (25.6±3.3 kg DM/day) than in FH cows (21.5±2.3 kg DM/day). Similarly, over the past nine years milk yields of CH cows (10,323±731 kg/lactation) has exceeded that of FH cows (8,279±341 kg/lactation). During the first 100 days of lactation in the year 2012, the concentrations of glucose, insulin and free fatty acids in the blood were similar in each of the 15 cows of both herds. The FH cows tended to have higher blood concentrations of β-hydroxybutyrate and leptin and lower concentrations of triiodothyronine. All measured blood hormones and metabolites remained within the physiological range. This long-term study shows that high-yielding cows are able to adapt to a ration without concentrates without enhanced risk of metabolic diseases.

**Performance in primiparous cows from two breeds submitted to different rearing strategies**

*A. Sanz, J.A. Rodríguez-Sánchez and I. Casasús*
*CITA de Aragón, Avda. Montañana 930, 50059 Zaragoza, Spain; asanz@aragon.es*

An experiment was conducted to analyse the influence of breed (Pirenaica vs Parda de Montaña) and nutrition level during the rearing period (from weaning at 6 months to first mating at 15 months) on the performance of primiparous cows calving at two years. Twenty-five autumn-born-heifers were assigned to two nutrition levels (High vs Low). At 15.2 months heifers were treated with an intravaginal progesterone device (PRID, CEVA, Spain) and Ovsynch protocol, being inseminated 14 days later. Blood samples were collected weekly during rearing and postpartum periods for progesterone analysis (Ridgeway Science, UK). Productive parameters were controlled from heifers' birth until weaning of their first calves (30 months). As expected, nutrition level influenced heifer average daily gain (ADG) during the rearing phase (0.814 vs 0.624 kg, in High and Low, $P<0.001$) and therefore their weight at 15 months (452 vs 399 kg, in High and Low, $P<0.05$). No difference was found in weight at onset of puberty (323 kg, 56% of adult weight in these breeds) due to feeding level, but Pirenaica heifers reached puberty 50 days later than Parda breed (326 vs 276 days, $P<0.05$). These divergence did not traduced in significant differences in the conception weight or age (426 vs 466 kg; 497 vs 503 days; in Pirenaica and Parda), nor fertility rate (93%). Primiparous cows weight, age and body condition at first calving were not affected by breed or rearing nutrition level. However, Pirenaica cows showed lower calf birth weight (32 vs 38, $P<0.01$), slightly higher ADG during it first lactation (0.109 vs -0.024 kg, $P=0.06$) and greater body condition at 120 days-weaning (2.8 vs 2.6, in a 1-5 point scale, $P<0.01$). No difference was found in calf gain (0.651 kg/d) or postpartum anoestrous period (62.9 days) due to breed or rearing nutrition level. Regardless of their different precocity and aptitude, these preliminary results confirm the feasibility of advancing the first service from 21 to 15 months of age in both breeds.

---



**Energy-protein supplement affects somatic cell count in cows milk in different stages of lactation**

*A.M. Brzozowska[1], J. Oprządek[1] and P. Micek[2]*
*[1]Institute of Genetics and Animal Breeding of the Polish Academy of Sciences, Department of Animal Improvement, Jastrzębiec, Postępu 36A, 05-552 Magdalenka, Poland, [2]University of Agriculture in Kraków, Department of Animal Nutrition and Feed Management, Mickiewicza 24/28, 30-059 Kraków, Poland; j.oprzadek@ighz.pl*

The aim of the study was to determine the effect of dietary energy-protein supplement (E-PS) in the ration of dairy cows on somatic cell count (SCC) in relation to stage of lactation. Energy-protein supplement, stage of lactation and their interaction were included as fixed effects in the model, whereas individual effect of the animal and date of sampling were random effects. SCC was the lowest at the beginning of lactation and then gradually increased. Daily milk yield of cows over 300 days of lactation (DL) supplemented with E-PS was significantly higher than in cows which were not supplemented, however their milk was also characterized by a higher number of somatic cells and a lower fat and protein content. Both the interaction between E-PS and stage of lactation ($P=0.026$) as well as E-PS alone ($P=0.003$) had a significant impact on SCC. The increase of SCC during supplementation was found after calving (up to 99 DL) and over 300 DL. High SCC at the beginning and at the end of lactation can be explained by their increased number in a smaller amount of milk (perinatal and dry period), whereas low SCC during peak of lactation may be the effect of 'diluting' the number of somatic cells in milk. Energy-protein supplement had a significant impact on milk yield, its composition as well as SCC of cows, however different stages of lactation determine using E-PS as the results were dependent on them. A study was carried out within the grant 2011/01/B/NZ9/00587.

**Effects of pre partum supplementation on milk yield of crossbred dairy cows in Tanzania**

*J. Madsen[1], K.A. Gillah[2] and G.C. Kifaro[3]*
*[1]University of Copenhagen, Department of Large Animal Sciences, Groennegaardsvej 2, 1870 Frederiksberg C., Denmark, [2]Government of Tanzania, Ministry of Livestock Development and Fisheries, Dar es Salaam, P.O. Box 9152, Tanzania, [3]Sokoine University of Agriculture, Department of Animal Science and Production, Morogoro, P.O. Box 3004, Tanzania; jom@sund.ku.dk*

Supplementation in the late stage of pregnancy may be crucial as replenishment of body reserves to be used in the following lactation may to a large extent determine the level of initial milk production. However, supplementation in the late pregnancy feeding is seldom practiced in tropical Africa. The present study was conducted to evaluate the effect of feeding a formulated concentrate to crossbred dairy cows during the late stage of pregnancy on milk production, reproduction and milk quality. Forty eight pregnant crossbred (*Bos taurus* × *Bos indicus*) cows with seven months of pregnancy were used in the study. The cows were equally divided into three treatments. The first treatment (HMR-PPP) was fed 4.0 kg of home-made ration (HMR) per cow per day for an average of 56 days pre partum and the same amount during 24 weeks postpartum. The second group of cows (HMR-PP) was given 4 kg/day HMR only after calving, and the third group (MB-PP) was a control group fed 4 kg/day of maize bran for 24 weeks postpartum which supplied 91.9 g/kg CP. The control group simulated the farm's concentrate feeding practice The present experiment shows that milk production can be increased and that it is economical to substitute or supplement the commonly used maize bran with better concentrate. Supplementation of a smaller amount of concentrate distributed both pre- and post-partum increased milk production significantly more than supplementing all the concentrate post-partum. In most situations the prices on concentrate is low compared to the price on milk and supplementation can be advised. When the prices are high it is recommended to distribute the allocation both pre-and post-partum.

---



**Influence of inbreeding on semen parameters in Swiss dairy bulls**

*A. Kreis[1], J. Kneubühler[2], E. Barras[3], A. Bigler[4], M. Rust[5], U. Witschi[2], H. Jörg[1], F. Schmitz-Hsu[2] and A. Burren[1]*
*[1]Bern University of Applied Sciences, School of Agricultural, Forest and Food Sciences, Länggasse 85, 3052 Zollikofen, Switzerland, [2]Swissgenetics, Meielenfeldweg 12, 3052 Zollikofen, Switzerland, [3]Holstein Association of Switzerland, Route de Grangeneuve 27, 1725 Posieux, Switzerland, [4]Swissherdbook, Schützenstrasse 10, 3052 Zollikofen, Switzerland, [5]Braunvieh Schweiz, Chamerstrasse 56, 6300 Zug, Switzerland; alexander.burren@bfh.ch*

In the present study, the relation between the inbreeding coefficient and the semen parameters volume and concentration of ejaculation was studied in Swiss dairy bulls. For the study, the company Swissgenetics and the Swiss dairy cattle breeding associations provided semen and pedigree data of a total of 2,622 Swiss dairy bulls of the breeds Brown Swiss BS (925), Holstein HO (351), Original Braunvieh OB (107), Red-Factor-Carrier RFC (25), Red Holstein RH (668), Swiss Fleckvieh SF (296) and Simmental SI (250). The semen of bulls was collected between 2000 and 2012. For the calculation of the inbreeding coefficient, the CFC software was used. The relation between the coefficient of inbreeding and semen parameters has been investigated by a linear mixed model. For this purpose, the software R and the packages nlme and car were used. The mean pedigree completeness of the first 6 bull generations was between 98.1% and 100%. The mean inbreeding coefficient was 5.01%, 4.32%, 3.61%, 2.95%, 2.84%, 2.46% and 2.22% for the BS, HO, SI, OB, RFC, RH and SF breed, respectively. The regression coefficients and the standard errors of the two sperm parameters ejaculate volume and concentration were -0.1378±0.25 ml and -16.94±5.8 million/ml per percent inbreeding increase. Both characteristics are significantly influenced by the inbreeding coefficient (volume: $P=0.0000$; concentration: $P=0.0035$). In a follow-up study, we will analyse the influence of the inbreeding coefficient on the non-return rate of a bull.

**Relationships between muscle and fat thickness and muscle development in Pinzgau heifers**

*P. Polák and J. Tomka*
*National Agricultural and Food Centre, Research Institute for Animal Productio, Department for Animal Welfare, Breeding and Product Quality, Hlohovecká 2, Lužianky, 951 41, Slovak Republic; polak@vuzv.sk*

Economical conditions of last decades influenced traditional breeders of Pinzgau cattle to change production from beef – milk semi intensive production to beef production in cow-calf units. Muscle development of Pinzgau cattle as rustical dual purpose breed is not sufficient to compete with specialised beef breeds. The aim of the investigation was to analyse relationship between muscle thickness measured by ultrasound and muscle development as information of exterior classification. There were 70 heifers of Pinzgau breed in age between 17 and 26 months used in investigation. Muscle and fat thickness were measured by Aloka PS 2, with probe UST-5044-3,5 MHz, 172 mm on last lumbal vertebra (muscle – MLLV, fat – FLLV) and on ischium (muscle – MI, fat – FI). Body size, body length, body weight, muscle development of shoulder, back and rump were obtained by linear evaluation of exterior used for beef breeds in Slovakia, done by certified person. MLLV had significant high correlation coefficient with muscle development on back rp=0.52 and MI with muscle development on rump rp=0.32. Interesting was also that fat layer on both positions had significant medium positive correlations with classification for body size (FLLV 0.41; FI 0.43) but low negative correlations with classification for body weight (FLLV -0.11; FI -0.34). High value of correlation coefficients between MLLV and muscle development on back and rump shows that ultrasonographical measurement can have potential to be one of the objective selection criteria for improving muscle development of rear part in Pinzgau cattle.



**Influence of dietary composition on the carbon, nitrogen and hydrogen ratios of Grana Padano cheese**

*L. Migliorati[1], L. Boselli[1], M. Capelletti[1], G. Pirlo[1] and F. Camin[2]*
*[1]Consiglio Ricerca Agricoltura, FLC, Via Porcellasco 7, 26100 Cremona, Italy, [2]Ricerca Innovazione, Qualità Alimentare e Nutrizione, Via E. Mach 1, 38010 San Michele D'Adige (TN), Italy; luciano.migliorati@entecra.it*

Most of the farms producing milk for Grana Padano cheese use maize silage as the main forage; however, possible alternatives can be other cereal silages, such barley. Animal products' isotopic content is the expression of what do they fed. Ratio between $\delta^{13}C$ and $\delta^{12}C$ in cheese or milk is essentially influenced by $C_4$ to $C_3$ plants ratio in dairy cows diet. This study aimed at evaluating the effect of a partial (trial 1) a total (trial 2) substitution of $C_4$ plant (maize silage) with $C_3$ plant (barley silage) in dairy cow diet on the alteration of stable isotope ratio in Grana Padano cheese. Animals were randomly divided in two groups (40 cows per pen). In trial 1 a maize silage based diet (MSI100) and a maize silage:barley silage 50:50 ratio diet (MS50) were compared while in trial 2 a maize silage based diet (MSII100) and a barley silage based one (MS0) with no maize silage supplementation, were tested. Experimental design was a cross – over 2×2 with two periods of four weeks each, the first two for adaption and the other two for sampling. Data were subjected to the one-way analysis of variance (ANOVA) using the post-hoc Bonferroni test to verify the existence of differences between dietary treatments. The $\delta^{13}C$ values of the overall diet and cheese casein were significantly correlated with the percentage of maize diet: each 10% increase in the maize content corresponded to a shift of 0.7‰ in the $\delta^{13}C$ of cheese casein. On the contrary $\delta D$ and $\delta^{15}N$ values of casein were only slightly influenced by the amount of maize included. A threshold value of -22.6‰ for $\delta^{13}C$ in cheese casein, above which it is not possible to exclude the presence of maize in the diet, was found. Stable isotope ratio analysis of Grana Padano has revealed that variations in cow diet isotope ratios, especially the $\delta^{13}C/\delta^{12}C$ one, could significantly affect the same ratios in milk derivatives.

**The association between BoLA – DRB3.2 alleles and somatic cell contains in Polish Holstein cows**

*J.M. Oprzadek, A. Brzozowska, A. Pawlik, G. Sender and M. Łukaszewicz*
*Institute of Genetics and Animal Breeding, Animal Improvement, Jastrzebiec, 05-552, Poland; j.oprzadek@ighz.pl*

The aim of this study was to evaluate the suitability of the polymorphism of BoLA-DRB3.2 gene for defining a phenotypic value of somatic cell count of Polish Holstein-Friesian cattle. The material consisted of 808 Polish Holstein cows. Animals originated from 190 fathers. PCR-RFLP alleles were identified according to procedure described by Van Eijk. Restriction patterns were obtained with BstYI, HaeIII and RsaI enzymes. For statistical analysis 17 alleles were selected, which frequency in the herds was equal to or higher than 2%.SCC changes during the lactation, therefore the influence of environmental factors on this trait can be eliminated by analyzing the cumulative values of the traits for the whole lactations or – which is more successful – by using a model for daily yields. There was a significant relationship between the occurrence of alleles of BoLA-DRB3.2 gene in the genotype and SCC. In case of estimating the effects for all lactations together the effect of substitution was observed for 11 alleles. The effects of four alleles (*12, *23, *24, *ndb) were statistically significant in the analysis of all lactations together and their effects have been repeated in at least two lactations in the analysis of each lactation separately. The most numerous relationships were found for allele *24, which influences were statistically significant in lactations I to III in the analysis of all lactations together. Supported by Polish State Committee of Scientific Research Grant: N31103731/0685.

---



**The effect of a second grazing period on carcass traits of indigenous Cika and Simmental young bulls**

*M. Simčič, M. Čepon and S. Žgur*
*University of Ljubljana, Biotechnical Faculty, Department of Animal Science, Groblje 3, 1230 Domžale, Slovenia; mojca.simcic@bf.uni-lj.si*

Carcass traits of Cika and Simmental young bulls either semi-intensively fattened indoors (S-INT) or finished indoors after a previous (second) grazing period was studied. Bulls from the S-INT group were characterised by higher average daily gain compared to bulls previously grazed and finished indoors. Only, fatness score and pelvic/kidney fat proportion were higher in Cika carcases compared to Simmental, while chest depth was higher in Simmental. Bulls which have been finished indoors after a previous grazing period had significantly lighter carcasses, lower dressing percentage, lower EUROP conformation and fatness score but heavier empty reticulo-rumen proportion compared to bulls fattened indoors with S-INT diet. Breed × diet interaction was significant for carcass weight, conformation, fatness, chest depth and empty reticulo-rumen, while slaughter weight affected only carcass weight, conformation and carcass length. Observed difference in the lean meat proportion between Cika and Simmental bulls as well as the difference between indoors fattened bulls and indoors finished bulls after grazing period were not significant. Breed significantly affected fat and bones proportion. Carcass of Cika bulls contained more fat and less bone proportion compared to Simmental bulls. However, diet did not affect the tissue proportion in the carcasses. Cika young bulls had more red and more yellow meat compared to Simmental. Bulls fattened indoors had darker meat than bulls finished indoors after a second grazing period. Thus, Cika young bulls should be put on the pasture for the second grazing period younger to prevent hostile behaviour among animals in the herd and losing energy due to early sexual maturity. Consequently, a second grazing period could be efficiently set up in the growing-fattening scheme and apparently would not affect the quality of carcass traits. This way beef production would positively affect consumer's interest.

## Session 20 Poster 25

**Using pro- or synbiotic in colostrum and milk on immunoblobulins' uptake, growth and health of calf**

*F. Moslemipoor and F. Moslemipur*
*University of Gonbad Kavoos, Animal Production, Shahid Fallahi St., Gonbad Kavoos, Golestan Province, Iran, 49717-99151, Iran; farid.moslemipur@gmail.com*

In the study, effects of probiotic and synbiotic addition to colostrum and milk on passive transfer of immunoglobulins, and growth and health parameters in calves were investigated. Sixteen newborn male Holstein calves were assigned to four treatment groups and received the treatments for 48 days in a completely randomized design. Treatments were included as (1) colostrum and milk without any additive (control), (2) colostrum and milk with 2.5 g probiotic/day (the recommended level), (3) colostrum and milk with 5 g probiotic/day, and (4) colostrum and milk with synbiotic (2.5 g probiotic + 25 g prebiotic/day). Feeding of colostrum was performed immediately after the birth and then three times a day. Blood collections were done in days 0, 1, 8, 16 and 48. Results showed that serum Immunoglobulin G concentrations were significantly greater in groups received probiotic than control and synbiotic groups ($P<0.001$), where the greatest was observed in 2.5 g probiotic group. There were no significant differences among groups in serum total protein concentrations. Starter intake in control group was significantly greater than other groups. Adding of probiotic significantly increased the weight gain over the study than control and prebiotic groups ($P<0.05$). Feed conversion ratio was improved by adding of probiotic to colostrum and milk ($P<0.05$). No significant differences were observed in fecal and health scores among groups. In conclusion, probiotic addition alone can increase immunoglobulin's uptake from colostrum into calf and also improve growth performance and the recommended level (2.5 g/day) is sufficient to achieve the beneficial effects.

---

## Session 20 Poster 26

**Pathogens of the milks produced by industerial or traditional dairy farms of Daland (North of Iran)**

*F. Vakili Saleh and F. Moslemipur*
*University of Gonbad Kavoos, Animal Production, Shahid Fallahi St., Gonbad Kavoos, Golestan Province, Iran, 49717-99151, Iran; farid.moslemipur@gmail.com*

To determine kind of pathogenic factors in the milk of Daland's dairy cattles, 4 industerial and 4 traditional herds in this region were selected. Three cows from each herd that seems to be infected to mastitis organoleptically were selected and two morning milk samples were collected from each. At first, samples incubated in general culture to observe colonies, then separative tests were performed to identify *Staphylococcus aureus* and *Escherichia coli*. Antiobiogram test was performed to determine sensetivity to oxytetracycline, chlortetracycline and gentamycine. Results of microbal culture showed the incidence of mastitis in many of suspected cow. *S. aureus* infection rate was 75% in industerial herds' samples and 100% in traditional ones. *E. coli* infection rate was 25% for industerial herds' samples and 100% for the another. Results of antibiogram test showed that all pathogens were resistant to chlortetracycline, resistant to semi-sensitive to oxytetracycline and sensitive to gentamycine. In general, results demonstereted the high overlap of organoleptic method with bacteriologic method and *S. aureus* and *E. coli* are mediated in mastitis in this region. Furthemore, pathogens in this region are relatively resistant to tetracycline and gentamycine is suitable replacement.

**Genome-wide association of milk fatty acids in Chinese and Danish Holstein populations**

*X. Li[1,2], A.J. Buitenhuis[1], C. Li[2], D. Sun[2], Q. Zhang[2], M.S. Lund[1] and G. Su[1]*
*[1]Molecular Biology and Genetics, Department of Molecular Biology and Genetics, Aarhus University, 8800, Denmark, [2]College of Animal Science and Technology, China Agricultural University, Beijing, China, P.R.; xiujin.li@agrsci.dk*

The fatty acid synthesis is a complicated process which is involved in many genes and several biological pathways. The objective of this study was to carried out a genome-wide association analysis using HD chip to detect QTL regions for fatty acids in Chinese and Danish Holstein populations. The dataset consisted of 784 Chinese and 371 Danish Holstein. Sixteen fatty acid traits were measured independently in each country. 371 Danish Holstein were genotyped using bovineHD beadchip while 784 Chinese cows were genotyped using the Illumina BovineSNP50 BeadChip, and then imputed to HD marker data using Beagle with 96 HD genotyped Chinese bulls as reference population. After quality control, there were 497,835 SNPs available. A linear mixed model (LMM) with fixed effects of herd, parity and a single SNP, the effect of days in milk modeled using lactation curve and random polygenic effect was implemented separately in each population. A significant threshold –log10 (P-value) = 7.00 (according to Bonferroni correction) was used to detected significant SNPs. For C14:1 and C14 index, significant SNPs for each population and common for both populations were detected. For C18 and C18 index, significant SNPs were detected only in Chinese Holstein. For C16:1, significant SNPs were detected only in Danish Holstein. For the other eleven traits, No significant SNP was detected in any population.



**Effect of extended lactation on chemical composition and sensory quality of milk**

*G.M. Maciel, U. Kidmose, L.B. Larsen and N.A. Poulsen*
*Aarhus University, Department of Food Science, Blichers Alle 20, 8830 Tjele, Denmark; Guilherme.deMouraMaciel@agrsci.dk*

From a health and animal welfare point of view, about 65% of the health problems in dairy production occur between calving and early lactation and extended lactation can be an attractive strategy to reduce by one third the susceptibility of animals to the critical period, and increase their longevity. However, cows in extended lactation may have their udder epithelium compromised; resulting in a more leaky blood milk barrier and increased levels of blood constituents in the milk, including proteolytic enzymes. The aim of this study was to evaluate whether an extended lactation strategy would potentially compromise technological and sensory qualities of milk by affecting milk composition significantly. In total, milk samples were collected from 70 Holstein cows enrolled in 18 months extended lactation at the Cattle Research Centre, Aarhus University, Denmark. From each cow, milk samples were collected once within each of the following lactation windows (mid lactation: week 20-25; late lactation 1: week 40-45; and late lactation 2: week 55-60). The milk was characterized for its overall milk composition, conductivity, pH, chloride content, fat globule size, as well as determination of proteolysis degree, and relationship with technological milk properties such as coagulation and cheese yield. In addition, a triangle test was performed by a trained sensory panel on pasteurized milk on pooled samples originating from six mid-lactating and six late 2-lactating cows, respectively. Furthermore, a sensory profiling was carried out on samples from individual cows, as well as on pooled samples from six mid-lactating and six late 2 lactating cows, respectively. Significant differences between the pooled mid-lactation and late-lactation samples were observed in the sensory triangle test. These results from the sensory profiling will be associated with the chemical fingerprints of the milk.

## Session 20 Poster 29

### Effects of body condition score at calving on metabolic state of beef cows grazing native pasture

*P. Soca[1], M. Carriquiry[1], M. Claramunt[2], G. Ruprechter[2] and A. Meikle[2]*
*[1]Universidad de la República, Facultad de Agronomía, Producción Animal y Pasturas, Ruta 3 km 363 Paysandú, 60000, Uruguay, [2]Universidad de la República Facultad de Veterinaria, Producción Animal, Lasplaces 1550, Montevideo. Uruguay., 11600, Uruguay; psoca@fagro.edu.uy*

The objective of this experiment was to analyze the effect of body condition score (BCS) at calving on betahdroxybutirate (BHB), urea, adiponectin, insulin and IGF-I levels of primiparous beef cows grazing native pasture. Primiparous beef (n=56) cows in anestrous were classified by BCS at calving (low≤3.5 and moderate≥4; 1-8 visual scale). All cows had a suckling-restriction treatment of 12 days Days (0=initiation of the treatment) at 55±10 days postpartum (DPP. Repeated variables were analyzed using a time repeated measure analysis and lineal model respectively. BHB and urea concentrations were reduced after the suckling treatment was initiated, and BHB levels increased again after the treatment was finished. The BHB decrease observed during the restriction treatment was associated with a immediate increase in IGF-I concentrations, which is probably due to the reduction in milk production (decreases of aprox. 30% in energy requirements). Both IGF-I and insulin concentrations were greater in moderate BCS cows. Adiponectin concentration was affected by BCS at calving as levels were maintained in moderate BCS cows but decreased in low BCS cows. As adiponectin is known to sensitize tissues to insulin, the decrease observed in low BCS suggests a metabolic strategy oriented to limit anabolic process by periphery tissue. Increased pregnancy rates were observed in moderate BCS cows (0.84 vs 0.46, P<0.01).These results confirmed BCS at calving relevance as link between energetic nutrition- metabolic and reproduction processes in primiparous beef cows grazing native pasture.

---

## Session 20 Poster 30

### Growth performance of feedlot-finished beef cattle of different genetic groups

*L.D.C. Vieira[1], A. Berndt[2], H. Borba[1], R.R. Tullio[2], R.T. Nassu[2], F.B. Ferrari[1], P. Meo Filho[2] and M.M. Alencar[2]*
*[1]FCAV/Unesp, Jaboticabal/SP, 14884900, Brazil, [2]Embrapa Southeast Livestock, Sao Carlos/SP, 13560970, Brazil; alexandre.berndt@embrapa.br*

Crossbreeding is frequently used in order to increase livestock production efficiency and meat quality, but in tropical regions *Bos taurus* breeds can show adaptation problems. So crossbreeding between adapted and non-adapted breeds is an alternative. Ninety-six steers and heifers (offspring of Bonsmara (BO), Brangus (BR) or Canchim (CA) and Nellore (NE), ½ Angus + ½ Nellore (TA) or ½ Senepol + ½ Nellore (SN)) were evaluated. Animals were fed in individual pens for 100 days, receiving diets with 55% corn silage and 45% concentrate, 13.1% crude protein, 71% of total digestible nutrients, 37.5% neutral detergent fiber and 3.2% ether extract. Animals were offered the diet twice a day and were weighed every 28 days. Statistical data treatment was performed using XLSTAT statistical program with a significance level, P-value, at 0.05. There were no significant differences (P>0.05) between sire breeds. Steers were significantly heavier (P<0.05) at slaughter (530.8 vs 499.2 kg), gained more weight per day (1.77 vs 1.59 kg/d) and had higher consumption (14.8 vs 13.8 kg DM/d) than heifers. Animals offspring of SN cows were slaughtered lighter (496.6 kg) than animals offspring of TA cows (520.6 kg) and pure Nellore cows (527.4 kg). Animals BO × TA gained more weight per day (1.80 kg) than animals BR × SN (1.47 kg). Steers offspring of NE cows were more efficient (8.0 vs 9.4 kg DM/kg gain) than heifers. Steers offspring of TA cows had higher consumption (1,421 vs 1,252 kg DM) than steers offspring of NE cows. In order to produce more meat, steers perform better, independently of sire breed, however they were not more efficient than heifers.

## Session 20 Poster 31

**Effect of herd composition on sustainable beef production systems in commual systems in Namibia**

*I.B. Groenewald, P. Shikomba and G.N. Hangara*
*University of the Free State, Centre for Sustainable Agriculture, P O Box 339, 9300 Bloemfontein, South Africa; groenei@ufs.ac.za*

The study on the effect of herd composition on beef cattle production was conducted in five constituencies of Omusati Region. A sample of 100 communal cattle farmers were interviewed using a comprehensive questionnaire on overall cattle herds' composition and it's dynamics. Herd size and its structure, reproductive characteristics, mortality, marketing strategies were studied in details. The study results were interpreted using descriptive statistical methods. The study results shows that the farmers in Omusati Region are not keeping the correct herd composition due to many reasons such as high mortality rates in some cattle group, poor off-take, and keeping of old and unproductive cattle in their herds. The imbalanced herd compositions will have a long term negative impact on the efficiency of beef production in northern communal areas. Keeping of old cows result in longer ICP and poor weaning rate due to high calve mortality (26%), which severely affecting sustainable productivity of beef herds. The goals of growing herd size will have detrimental effect on available grazing areas which are already under pressure due to high number of cattle at the rate of 58 cattle per farmer on average. Social issues which are common in communal areas farmers need to be factor onto all program which aim at improve sustainability of beef farming. This is because the study found that they have influence on the current herd compositions. For example the reasons for high number of oxen is a results of keeping oxen for immediate sources of income when need arise, draft power, and prestige amongst others. Correct cattle herd compositions have a high potential to bring about improvement in land degradation, marketing and over application of good management practice for sustainable beef productions system.

---

## Session 20 Poster 32

**Performance, carcass and meat quality of forage-fed steers as an alternative to concentrate-based be**

*M. Blanco, M. Joy, P. Albertí, G. Ripoll, B. Panea and I. Casasús*
*CITA, Unidad de Tecnología en Producción y Sanidad Animal, Avda. Montañana 930, 50059 Zaragoza, Spain; mblanco@aragon.es*

The performance and carcass quality under different management systems were studied to search alternatives to indoors intensive beef production. During the winter housing period, 8 bulls were *ad libitum* fed concentrate+straw (C) while 16 steers were fed a total mixed ration. From April, steers grazed on mountain meadows+1.8 kg DM corn/d. Half the steers grazed until slaughter at 500 kg (G-supp) while the remaining steers were finished indoors (54 d) on a total mixed ration (TMR). In the housing period, C bulls had greater weight gains than the steers (1.772 vs1.221 kg/d, P<0.001). In the finishing period, TMR steers had greater weight gains than G-supp steers (1.371 vs 0.942 kg/d, P<0.01). At slaughter (499±9.2 LW; P>0.05), C bulls were younger than G-supp and TMR steers (442, 569, 539 d; P<0.001). TMR steers had worse conformed carcasses, greater fat and lower edible meat proportions than G-supp steers and C bulls (P<0.01). The management strategy did not affect meat pH but C bulls had lower shear force than G-supp and TMR steers (66, 83 and 79 N/cm2, respectively; P<0.05). Meat from C bulls had lower yellowness (b*; P<0.001) and Chroma (C*; P<0.01) than that from both groups of steers. Steaks of TMR steers had greater intramuscular fat content than those of G-supp steers (P<0.05) whereas that of C bulls was intermediate. Steaks of G-supp steers had greater PUFA n-3 than those of C bulls and TMR steers (P<0.001), and greater PUFA n-6 than TMR steers (P<0.001). Thus, G-supp and TMR steers had lower n-6:n-3 ratio than C bulls (5.23, 6.13 and 20.07, respectively; P<0.001). Consumers scored higher the taste, tenderness and overall impression of the steaks from C bulls than those from both groups of steers (P<0.05). Consequently, forage-feeding can be an alternative to concentrate-fed cattle, but the proportion of corn in diets of forage-fed cattle should be increased to improve meat tenderness and guarantee the consumers acceptability.

**Change in milk fatty acid profile of dairy cows fed with hydrolysed sugarcane and sunflower product**

*A.C.N. Alves[1], F.F. Simili[2], M.L.P. Lima[2], L.C. Roma Junior[2], J.M.B. Ezequiel[3] and C.C.P. Paz[2]*
*[1]Clínica Veterinária Jardim Botânico, Ribeirão Preto, SP, 14021-636, Brazil, [2]Instituto de Zootecnia SAA/ APTA, Department of Agriculture, São Paulo State Government, C. Postal 63, Sertãozinho, SP, 14160-970, Brazil, [3]UNESP, Departamento f Animal Science, Via de acesso Prof. Paulo Donato Castellane, Jaboticabal, SP, 14884-900, Brazil; flaviasimili@gmail.com*

This study assesses the inclusion of lipids source to dairy cattle diet, by adding sunflower seeds, oil or meal, and the resulting changes on milk fatty acid profile. The concentrate diet consisted of sunflower seeds (SS), sunflower meal (SM) and sunflower oil (SO) associated with sugar cane (SC) and hydrolyzed sugarcane (HSC). The sugar cane was hydrolysed by adding 1% $Ca(OH)_2$. The four experimental diets were: sunflower seeds and sugarcane (SS-SC); sunflower seeds and hydrolysed sugarcane (SS-HSC); sunflower meal and hydrolysed sugarcane (SM-HSC); sunflower oil and hydrolysed sugarcane (SO-HSC). Twenty four multiparous crossbred dairy cows averaging 70 milking days were assigned to a completely randomized design. Milk fatty acid profile was determined by gas chromatography. The diets did not change the milk short chain (P>0.05) but increased the long chain fatty acid profile (P<0.05). The SO-HSC diet increased the milk CLA, compared to other diets. The results were 0.036, 0.056, 0.068 and 0.070 g/100 g milk for C18:2 T11 and 0.67, 0.85, 0.79 and 1.42 g/100 g milk to C18:2 C9 for SS-SC, SS-HSC, SM-HSC and SO-HSC, respectively. The biggest difference was observed between SS-SC and SO-HSC diets probably because the calcium added to HSC protected the lipid by creating a fat bypass. The hydrolyzed sugarcane combined with sunflower oil in the diet fed to dairy cows changed the long chain fatty acid in the milk, thus improving its nutritional quality.



**Comparative study of the productive traits of Simmental cattle of different origin**

*V.I. Seltsov and A.A. Sermyagin*
*All-Russian Research Institute of Animal Breeding, Laboratory for Breeding and Selection of Animals, Dubrovicy 60, Moscow region, 142132 Podolsk, Russian Federation; alex_sermyagin85@mail.ru*

The aim of this experimental study was to compare the productive traits of the daughters of sires of the Russian Simmental (RS) and Austrian Fleckvieh (AF) from 2008 to 2011. The growth and conformation of calves and the productive traits of fist-calving cows (n=110) were studied. The daughters of AF sires were characterized the higher body weight (+30 kg,P<0.01) at the age of 18 month: 409±7 vs 379±6 kg for RS daughter. For the heifers derived from AF sires had the higher values of chest width by 8.0% (P<0.05), chest depth by 2.5%, withers height by 2.0%, body length by 1.7% and thickness of the skin for point of elbow and last rib by 14.0 and 7.4%. The daughters of AF sires had the higher frequency of dystocia by 1.7% and gestation length by 4 days (F=8.4; P<0.01). The value of calving easy for measurements of newborn calves depended from the depth of the head (r=0.30; P<0.05), the heart girth and the width between the points of shoulder and hip. The daughters of AF sires had the increased 305-d milk yield comparing to RS cows with the average values of 4,700±78 and 4,452±92 kg (+248 kg; F=4.1; P<0.05), respectively. For the chemical composition of milk the daughters of AF sires comparing to RS analogues had lower fat content (-0.12%, P<0.05), lower protein content (-0.13%, P<0.05) and lower solids (-0.43%, P<0.01). The daughters of AF sires had more depth and girth of the udder. The study of the technological milk properties showed the higher milk density (31.2°A), higher thermostability (79.5%) and higher value of resazurin-rennet test (quality class 2) in daughters of AF sires. The rates of hoof growth and wear were more balanced for daughters of RS sires than AF analogues: 118 vs 124% (F=10.1; P<0.01-0.001). Thus, the AF sires used on the RS population permit increase milk production and improve the farm traits. However, for the hoof hardness and chemical compositions of milk this can have negatively effect.

**Feedlot performance of the Drakensberger in comparison with other cattle breeds: a meta-analysis**

*L.J. Erasmus, M. Niemand and W.A. Van Niekerk*
*University of Pretoria, Department of Animal and Wildlife Sciences, Private Bag X20, 0028 Hatfield, South Africa; willem.vanniekerk@up.ac.za*

Despite economic beef production from indigenous cattle breeds, there are negative perceptions resulting in Drakensbergers not being favoured as feedlot animals. Drakensbergers are perceived to have lower growth performance and are more prone to health problems when compared to other breeds. The objective of this study was to compare the Drakensberger with all other feedlot animals using historic growth performance and health data from feedlots and Phase C performance tests by means of a meta-analysis. After separate analyses on different feedlots and test centre, a meta-analysis were performed which incorporated combined results from all the feedlots and test centres respectively. Effects of breed, gender, season, year and feedlot on average daily gain (ADG), feed conversion ratio (FCR), as well as morbidity ratios were tested. The final meta-analyses included five feedlots, consisting of 497,798 head of cattle; and four test centres, which represented data from 6,139 head of cattle. The feedlot analysis, which included 48,600 Drakensberger cattle and 449,198 head of cattle from other breeds and crossbreeds, revealed a difference in mean ADG of 20 grams in favour of other breeds (P<0.01). Other breed types had a higher mean ADG than Drakensbergers in the test centres (1.613 kg vs 1.737 kg P<0.01) but FCR did not differ (P>0.05). Although differences in growth performance between Drakensbergers and other cattle breeds were observed, they are most likely not economically or biologically significant. Although trends revealed higher respiratory-related disease occurrence in Drakensbergers, the health data at most feedlots and test centres was either not accurate or not available and therefore not representative of the actual health status of cattle in the feedlot industry, implicating that more studies are urgently needed. Results suggest that there is no justification to discriminate against the Drakensberger breed based on feedlot growth performance.

---



**Quality of musculus longissimus lumborum et thoracis from slaughter cows in Slovakia**

*M. Gondeková, I. Bahelka, P. Polák and D. Peškovičová*
*NAFC-Research Institute for Animal Production Nitra, Institute of Animal Husbandry Systems, Breeding and Product Quality, Hlohovecká 2, 95141 Lužianky, Slovak Republic; peskovic@cvzv.sk*

The aim of the study was analysis of nutritional, physical-technological and organoleptic quality of cow's meat. Cows of various breeds and breed combinations were kept on the farms in different regions of Slovakia and slaughtered in 2006-2012. Meat samples were randomly taken from 323 cows. The place of sampling was musculus longissimus lumborum et thoracis (8/9 rib; approx. 800 g). Samples were taken 24 h pm from the right side of the carcass. According to the age of cows at slaughtering the group was divided into two categories. Category A consisted of the cows under 4 years (n=79), category B were cows over 4 years (n=244). The analysis of the basic chemical composition of meat, measurement of colour and $pH_{48}$ were performed within 48 h pm. Degree of marbling using a 10 point scale (1-very abundant, 10-practically devoid) was dezermined. 7 days after slaughter a consumer test for determination of sensory traits of meat was performed. The samples were sliced into 2.0 cm thick steaks and grilled for four minutes. Sensory traits (odour, taste, tenderness, juiciness) were evaluated by 5 point scale (1-the worst, 5-the best). Overall, the sensory properties were evaluated by 841 randomly selected consumers. Comparing two age categories of cows, the content of intramuscular fat was expectedly determined higher in older cows. Significant results were noticed in total water (75.69 vs 74.79 g/100 g, P<0.05) and protein content (20.07 vs 20.69 g/100 g, P<0.001). The statistically significant differences in physical-technological parameters were found in colour lightness L (30.38 vs 29.44, P<0.05) and in shear force (13.9 vs 18.99 kg, P<0.001). Older cows had a tendency toward darker meat color and had a higher value of shear force. The consumer evaluation of the organoleptic qualities didn´t find any statistical significancy with regard to the age of cows, but more favourable results in the group of younger cows were determined.

**Assessing placental function in our livestock species to ensure adequate fetal development**

*K.A. Vonnahme*
*North Dakota State University, Animal Sciences, Hultz Hall Room 181 Dept 7630, P.O. Box 6050, Fargo, ND 58108-6050, USA; kim.vonnahme@ndsu.edu*

Fetal survival is dependent upon proper establishment of the placenta. Inadequate nutrition and other stressors can alter placental function; however it appears that the placenta has the potential to adapt in order for fetal growth to be optimal. In environments where fetal growth can be compromised, potential therapeutics may augment placental function and delivery of nutrients to improve offspring performance during postnatal life. Supplementation of specific nutrients, including protein, as well as hormone supplements, such as indolamines during times of inadequate nutrition may augment placental function. Assessment of placental vascular reactivity increases our understanding of mechanisms that aid in nutrient transport. Moreover, understanding how the maternal environment impacts uterine and umbilical blood flows and other uteroplacental hemodynamic parameters can assist in development of management methods and therapeutics for proper fetal growth.

---



**Effect of nutrient restriction during pregnancy in heifers on maternal and offspring's metabolism**

*S. Wiedemann[1], U. Köhler[2], S. Spiegler[2], F.J. Schwarz[2] and M. Kaske[3]*
*[1]Institute of Animal Breeding and Husbandry, CAU, Olshausenstr. 40, 24098 Kiel, Germany, [2]Physiology & Animal Nutrition, TUM, Weihenstephan, 85354 Freising, Germany, [3]Vetsuisse Faculty, University of Zurich, Department for Farm Animals, Winterthurer Str. 260, 8057 Zurich, Switzerland; swiedemann@tierzucht.uni-kiel.de*

The plane of maternal gestational nutrition affects health and performance of the offspring in many mammalian species. This study analysed the impact of dam's nutrition during early and late pregnancy (1) on metabolism and placental indices of growing heifers and (2) on growth and key blood metabolites of their calves during the first year of life. Red-Holstein heifers inseminated with sexed semen of one bull received either recommended energy and crude protein (standard diet) throughout pregnancy (C; n=7), 60% of standard diet for the first 2 pregnancy months followed by 100% of standard diet (ER; n=7) or 100% of standard diet for the first 7 months and 60% of standard diet for the last 2 months of pregnancy (LR; n=7). Offspring was fed 30 kg of milk replacer during the first 8 weeks of life and thereafter a total mixed ration. Blood metabolites and body weights of dams and their offspring were assessed regularly. Placenta weights and areas of placentomes were examined directly after birth. In the calves, hyperglycemic clamps and hyperinsulinemic euglycemic clamps were performed at an age of 8 weeks and 6 months to assess the pancreatic and peripheral insulin response, respectively. Restriction periods resulted in reduced weight gains of the ER- and LR-dams, respectively. The impact of the severe alteration in the dams' nutrient supply on blood metabolites and placental development were lower than expected from results of other species. Birth weight and weight gain during the first year of life of the offspring were not influenced by the 40% reduction of dams' standard diet. Results of the blood analyses and clamps revealed no strong long-term influence of maternal nutrient restriction on regulation of calves' metabolism.

**Foetal programming and long-term implications for metabolic and endocrine function**

*M.O. Nielsen, P. Khanal, L. Johnsen and A.H. Kongsted*
*Facuty of Health and Medical Sciences, University of Copenhagen, Department of Veterinary Clinical and Animal Sciences, Grønnegårdsvej 3, 1st floor, 1870 Frederiksberg C, Denmark; mette.olaf.nielsen@sund.ku.dk*

Some 20 years ago, observations from human epidemiological research revolutionized the scientific view of the importance of foetal life development for body functions in postnatal life. Until then, it was believed that the genome received from the parents at conception in mammals would define the phenotype at birth and the ability of an individual to cope with the postnatal environment. Since then, it has been clearly demonstrated that malnourishment during fetal life can predispose to a wide range of developmental disorders later in life. This is due to the phenomenon termed foetal programming (FP), which alters the way the genome becomes phenotypically expressed. In our Copenhagen sheep model, we have demonstrated that FP induced by late gestation malnutrition in the form of either under- or overnutrition interferes with postnatal growth and development and functions of a wide range of metabolic and endocrine systems, which are key to farm animal productivity and economic efficiency. Birth appears to be a critical set point for when permanent programming effects are induced in sheep. Phenotypical manifestations of FP are not clearly observable at birth, but become increasingly manifested as the animal approaches adulthood. The economic consequences of FP in farm animal production have not been properly assessed, but the accumulating evidence suggest that the consequences of FP for farm animal productivity are so significant that we need to incorporate the foetal life history into future decision making tools and management strategies targeting replacement stock and the reproducing female. Especially since FP traits can be passed on to the next generations by epigenetic mechanisms.

---

Session 21 Theatre 4

**Consequences of nutrition and growth during gestation for beef production**

*P.L. Greenwood and D.L. Robinson*
*NSW Department of Primary Industries, University of New England, Armidale NSW 2351, Australia; paul.greenwood@dpi.nsw.gov.au*

The embryonic, fetal and neonatal periods are when most developmental processes occur and the systems that support life independent of the dam are established. There is increasing interest in how to manage breeding females and their offspring to either minimize the consequences of adverse environmental effects or to enhance productivity and efficiency. This has arisen at least partly from studies in humans and animal models that have demonstrated consequences of prenatal growth and development for health in later life. This paper reports results of our Australian Beef Cooperative Research Centre studies on factors including chronic, severe nutritional restriction during pregnancy and/or lactation within pasture-based systems on production characteristics of offspring. They are discussed within the context of other research on effects of fetal programming for beef production and enterprise profitability. Fetal programming and related maternal effects can explain substantial amounts of variation for growth-related production characteristics such as body weight, feed intake, carcass weight, muscle weights, meat yield, and fat and bone weights at any given age. These effects are less evident when assessed at the same body and carcass weights. Some influences of maternal and early-life factors are evident for efficiency traits, but fewer affect beef quality characteristics in later life, explaining only small amounts of variation in these traits. Maternal nutrition during pregnancy can also have carry-over effects during the postnatal period until weaning, particularly on lactational performance, which needs to be considered when designing experiments aimed at understanding influences resulting specifically from altered prenatal development. Furthermore, the environment into which the offspring are reared and grown to market weights may also interact with influences imparted earlier in life to affect productive outcomes.

**Low weight at birth increases the chance of the gilt reaching first farrowing**

*F. Thorup*
*Danish Pig Research Centre, Feeding & Reproduction, Axeltorv 3, 1609 Copenhagen, Denmark; ft@lf.dk*

This project investigated if birth weight, litter size at nursing or dam parity affect the chance of the weaned female piglet farrowing a litter. In one herd producing DanAvl LY gilts for breeding, all female piglets were weighed at birth and raised in litters of 9 or 13 piglets, by removing male piglets after farrowing. Parity of the dam was registered. After sale to production herds, data were retrieved on delivered gilts, including date of mating and farrowing. Results were statistically analyzed using a khi-test with yates correction. From 1096 weaned gilt piglets, farrowing results for 56% were retrieved. Farrowing results for some gilts were lost due to farms not wishing to participate in the study and to gilts losing their ear tags. It was presumed that all weaned female piglets had the same risk of being lost irrespective of factors in early life. Piglets with a birth weight below 1.5 kg were grouped as small piglets. They had a 6% higher chance of farrowing than large piglets, weighing above 1.45 kg (P=0.05). Nursing in a normal litter size of 13 piglets gave a non-significant higher chance of farrowing a litter when compared to piglets nursing in a litter of 9 piglets. The parity of the dam ($1^{st}$ vs $2^{nd}$ to $9^{th}$ parity) did not affect the chance of farrowing. Small piglets at birth show a slow growth rate, when compared to larger piglets. It may be speculated that restricted growth rate has a positive effect on the development of leg quality, which in turn may affect the piglet's chance of reaching maturity, mating and completing gestation. It is concluded that small piglets at birth have significantly higher chance of producing a litter if they are weaned than piglets that are large at birth. Dam parity and litter size while nursing had no significant effect.

---



**Impact of early life nutrition on survival and production performance in multiple-born lambs**

*S. McCoard, Q. Sciascia, F. Sales and D. Van Der Linden*
*AgResearch Limited, Animal Nutrition & Health Group, Private Bag 11008, Palmerston North 4474, New Zealand; sue.mccoard@agresearch.co.nz*

Increasing the number of lambs born in sheep has the potential to increase farmer profits by enhancing productivity per ewe. However, competition between littermates in mid-late gestation leads to lower birth weight, increased mortality and can influence future meat and milk production compared to their singleton counterparts. Nutrition in early life (prenatal and neonatal) can influence survival and future performance, a phenomenon is known as nutritional programming. Despite the impact on survival and production, few strategies are available to farmers to address issues of multiple-bearing ewes on-farm. Understanding how nutrition influences the development and growth of key tissues (e.g. placenta, skeletal muscle, adipose tissue, gastrointestinal tract and mammary gland) and identifying critical time windows for intervention will contribute to future strategies to enhance the performance of multiple-born lambs. Supplementation with specific nutrients such as amino acids is an example of a potential intervention strategy. It is well established that amino acids are not only building blocks for proteins, but that they also play key roles in the regulation of many metabolic pathways that influence tissue development. Therefore, the strategic use of amino acid supplementation may provide an opportunity to improve the survival and production performance of multiple-born lambs through modulation of nutritional programming in early life.

**Maternal undernutrition during gestation does not affect growth and physiology of rabbit offsprings**

*G.K. Symeon[1], Z. Abas[1], M. Goliomytis[2], I. Bizelis[2], G. Papadomichelakis[2], O. Pagonopoulou[1], S.G. Deligeorgis[2] and S.E. Chadio[2]*

*[1]Democritus University of Thrace, Department of Physiology, Medical School, 6th km Alexandroupoli-Makri, 68100, Alexandroupoli, Greece, [2]Agricultural University of Athens, Department of Animal Science and Aquaculture, 75, Iera odos, 11855, Athens, Greece; gsymewn@yahoo.gr*

It is now well established that a stimulus or insult during embryonic development establishes a permanent physiological response (fetal programming). Moreover, maternal nutritional status has been recognized as a prominent cause of programming. Thirty rabbit does were randomly divided in three treatment groups: Control, early undernourished (fed to 50% of maintenance requirements during days 7-19 of gestation) and late undernourished (fed to 50% of maintenance requirements during days 20-28 of gestation). During gestation, body weight change, feed intake, serum glucose, corticosterone, insulin and leptin were recorded. At parturition, kindling performance of the does was also recorded. At weaning, 32 rabbit offsprings per group were randomly selected end fattened till 70 days of age. During fattening, body weight, feed intake, serum glucose, insulin and leptin were recorded. At 70 days of age 16 rabbits per group were randomly selected and slaughtered for the assessment of carcass and meat quality parameters. All data were analyzed using GLM procedures fitting the treatment as the fixed factor as well as time, sex and their interactions, when applicable. For the offsprings data especially, the effect of the doe was included in the model as a random factor. The dietary treatments affected maternal weights during gestation. At kindling, control does produced litters with higher proportions of stillborn kits while birth weight of kits was not different among groups. Growth and feed intake of the offsprings was also not different among groups. Maternal undernutrition did not affect carcass and meat quality of L. lumborum muscle as well as the physiological characteristics of the offsprings.

---



**Impact of sow backfat thickness around weaning on nutritional status and litter performance**

*J. Alvarez-Rodriguez[1], G. Sayez[1], A. Sanz[2] and D. Babot[1]*

*[1]Universitat de Lleida, Av Rovira Roure 191, 25198 Lleida, Spain, [2]CITA de Aragón, Av Montañana 930, 50059 Zaragoza, Spain; asanz@aragon.es*

This study assessed retrospectively the effects of sow backfat thickness (BFT) at weaning (<13 vs ≥13 mm), BFT change during the weaning to oestrus interval (WOI, 6 days) (-1 vs +1 mm) and parity (2$^{nd}$, 3$^{rd}$-5$^{th}$, 6$^{th}$-7$^{th}$), and their single interactions, on energy metabolites (blood glucose and lactate) at -5, 0 and +5 days (d) relative to oestrus; plasma progesterone (P4) at 5 d post-mating; and litter birth weight, uniformity and vigour at <24 h old (n=54 sows, 796 piglets, 674 alive). During WOI, sows were fed 3 kg/d of a lactation diet (13.1 MJ ME/kg, 0.91% Lys). After mating, they were gradually fed 1 to 3 kg/d of a gestation diet (9.1 MJ ME/kg, 0.60% Lys). Glucemia was greater at oestrus than in the rest of period (4.5 vs 3.7±0.1 mmol/l, P<0.05) but did not differ among groups (P>0.05). Blood lactate decreased from d -5 to oestrus (41.8 vs 27.9±1.3 mmol/l, P<0.05) and remained steady at d +5. The oldest sows (6$^{th}$-7$^{th}$ parity) showed greater blood lactate when they loosed BFT during the WOI than when gained (35.9 vs 29.9 mmol/l, P<0.05), but it did not differ in the rest of parities (P>0.05). The proportion of sows showing low blood P4 (<20 ng/ml) at +5 d post-mating was greater in sows gaining BFT during WOI (50.0 vs 7.1%, P<0.01). There were no differences among groups in total born, stillborn or mummified piglets (P>0.05). Litter birth weight was greater in sows with BFT≥13 mm at previous weaning coupled with positive BFT change during WOI than in those loosing BFT before mating (21.2 vs 18.0 kg±1.1 kg, P<0.05). The 2$^{nd}$ parity sows showed best litter uniformity than older sows (23.2 vs 30.7±1.7% coefficient of variation of litter birth weight). Piglet latency to right itself was lower when sows had <13 mm BFT at weaning and gained BFT during WOI than in the rest of groups (5.1 vs 6.4±0.5 seconds, P<0.05). In conclusion, BFT level and its dynamics before mating may have carry-over effects on litter early performance.

**Recovery from intrauterine growth restriction in piglets defined by their headshape: a pilot study**

*C. Amdi, J. Hales, A.T. Nguyen and C.F. Hansen*
*University of Copenhagen, Department of Large Animal Sciences, Grønnegårdsvej 2, Frb C, Denmark; ca@sund.ku.dk*

Piglets that have suffered from intrauterine growth restriction (IUGR) have a dolphin-like headshape (due to brain sparing) at birth but this headshape is not observed at weaning, suggesting that they either die or lose the dolphin-like headshape. We have previously shown that piglets given an IUGR score at birth based on their headshape ingest insufficient amounts of colostrum, have an increased risk of dying in early lactation, have lower liver glycogen and glucose levels at 24 hours. The aim of this study was to investigate when piglets lose their dolphin-like headshape and if it is related to body weight (BW). In addition the relationship between BW gain and headshape was investigated. Data from 370 piglets were used in this pilot study to see at what stage IUGR piglets lose their dolphin-like headshape and look like 'normal' piglets as part of a larger trial investigating different sow diets. Piglets were classified as either 'normal', 'mildly IUGR' (m-IUGR) or 'severe IUGR' (s-IUGR) based on the head morphology at birth and re-scored after one and two weeks. Data was analysed in proc mixed in SAS. Results showed that at birth 218 piglets were deemed 'normal' (average BW of 1.6 kg), 138 piglets were m-IUGR (average BW of 1.2 kg) and 14 piglets were s-IUGR (average BW of 0.8 kg). After one week all the piglets given an s-IUGR score were still alive and were re-scored as m-IUGR. In total 314 piglets were re-scored as normal with an average BW of 2.7 kg and 56 piglets given an m-IUGR score with an average BW of 1.7 kg of these 42 piglets had also been scored m-IUGR the previous week. After two weeks all piglets were re-scored as normal with an average BW of 3.9 kg. Results showed that IUGR score at birth influenced gain from 0 to 14 days ($P<0.001$). The results suggest that if an IUGR piglet can reach approx. 2 kg BW the dolphin-like headshape seems to disappear, however, daily gain at least up to d 14 is still influenced by their headshape at birth.

---



**Correlations between ovum pick–up *in vitro* production traits and pregnancy rates in Gyr cows**

*C.R. Quirino[1], W.H.O. Vega[1], A. Pacheco[1], A.S. Azevedo[1], C.S. Oliveira[2] and R.V. Serapião[3]*
*[1]State University of North Fluminense, LRMGA, Rua Alberto Lamego 2000, Pq. Califórnia, Campos dos Goytacazes, 28013-602 – Rio de Janeiro, Brazil, [2]EMBRAPA, Rua Eugênio do Nascimento, 610 – Dom Bosco – Juiz de For a, MG, 36038-330, Brazil, [3]PESAGRO, BR 465, Km 47 – Bairro Ecologia, Seropédica, RJ, 23.890-000, Brazil; crq@uenf.br*

The growth of Gyr breed in Brazil in terms of genetic gain for milk, along with the conditions of the market, led to the use of Ovum Pick–up *in vitro* production (OPU – IVP) as leader biotechnology for multiplication of genetic material. The aim of this work was to study the phenotypic correlations between the OPU – IVP linked characteristics and pregnancy rates registered in an embryo transfer program using Gyr cows as oocyte donors. Data collected from 211 OPU sessions and 298 embryo transfers during the years 2012 and 2013 were analyzed. Statistical analysis was performed. Estimates for simple Pearson correlations were calculated. Correlations were accomplished for: Number and ratio of viable cumulus-oocyte complexes (Nvcoc and Pvcoc); number and proportion of cleaved embryos on day 4 of culture (Ncleavd4 and Pcleavd4); number and proportion of transferable embryos on day 7$^{th}$ of culture (Ntembd7 and Ptembd7); number and proportion of pregnancies at day 30$^{th}$ after transfer (NPrD30 and PPrD30) and number and proportion of pregnancies at day 60 after transfer (NPrD60 and PPrD60). Moderate to high correlations were found for all numerical characteristics, suggesting these as the most suitable parameters for selection of oocyte donors in Gyr programs. The case Nvcoc is proposed as a selection trait due to positive correlations with percentage traits and pregnancy rates at the 30$^{th}$ and 60$^{th}$ day.

## Genomic selection to reduce boar taint in Danish pigs

A.B. Strathe[1,2], T. Mark[2], I.H. Velander[1] and H.N. Kadarmideen[2]

[1]Pig Research Centre, Department of Breeding and Genetics, Axeltorv 3, 1609 Copenhagen V, Denmark, [2]University of Copenhagen, Department of Clinical Veterinary and Animal Sciences, Groennegaardsvej 3, 1870 Frederiksberg C, Denmark; strathe@sund.ku.dk

Producer and animal welfare organizations within EU have signed a voluntary declaration to stop castration by 2018. Hence, developing a genetic solution to boar taint (BT) is of key interest. The aim of this work was to develop genomic selection and phenotyping scheme against BT for all 3 Danish pig breeds, and to present preliminary results from the ongoing geno- and phenotyping. A performance test for BT has been developed, where the main chemical compounds and a human-nose score (HNS) of boar taint were recorded on either live male breeding candidates or slaughter boars. Using the Danish Duroc as an example, concentrations of androstenone (AND) and skatole (SKA) in the back fat was currently available for approximately 700 boars while 2,192 boars were recorded for HNS. Furthermore, 396 boars with BT phenotypes had genotype information (60K SNP chip). Heritabilities of HNS, log(AND) and log(SKA), estimated using a multi-trait single-step GBLUP model, were 0.12, 0.55 and 0.17, respectively. The most notable genetic correlation was between HNS and log(AND), being 0.55. Genetic correlations between log(AND) and litter size and semen traits suggest currently that selection against BT will have minimal impact on fertility. The association analysis with 33028 SNPs revealed that log(AND) levels in fat tissue were significantly affected by 31 SNPs on pig chromosomes SSC1 and SSC6. On SSC6, a larger region of 10 Mb was shown to be associated with log(AND) covering several candidate genes potentially involved in the synthesis and metabolism of androgens. In summary, the project uses genomics for aiding a future meat production of meat from entire males, being free of boar taint so that consumers will continue to appreciate an odorless meatball.

---



## Model comparison based on genomic predictions of litter size and piglet mortality

X. Guo[1], O.F. Christensen[1], T. Ostersen[2], D.A. Sorensen[1], Y. Wang[3], M.S. Lund[1] and G. Su[1]

[1]Aarhus University, Dept. of Molecular Biology and Genetics, Blichers Allé 20, 8830 Tjele, Denmark, [2]Pig Research Centre, Danish Agricultural and Food Council, Axeltorv 3, 1609 Copenhagen, Denmark, [3]China Agricultural University, Dept. of Animal Science and Technology, Yuanmingyuan West Road No. 2, 100193 Beijing, China, P.R.; xiangyu.guo@agrsci.dk

Prediction of breeding values (BV) based on dense markers across the genome has become an important component of pig breeding programs. This study compared traditional BLUP, genomic BLUP (GBLUP), and single-step models for genomic predictions of litter size and piglet mortality in the Danish Landrace and Yorkshire populations. The data consisted of 778,095 records of 309,362 Landrace and 472,001 records of 190,760 Yorkshire sows. There were 332,795 Landrace and 207,255 Yorkshire animals in the pedigree, among which 3,445 (1,366 boars and 2,079 sows) Landrace and 3,372 (1,241 boars and 2,131 sows) Yorkshire individuals were genotyped. To validate prediction reliability, the data were divided into a training dataset (80%, born before 2012-04-01) and a validating dataset (20%). The traits were total number of piglets born (TNB), litter size at five days after birth (LS5) and mortality rate before day 5 (Mort). BV were predicted using a traditional BLUP model with pedigree-based relationship matrix, a GBLUP model with marker-based relationship matrix, and a single-step model with a combined relationship matrix constructed from marker and pedigree information. The reliabilities of estimated breeding values (EBV) were computed as the squared correlation between EBV and corrected phenotypic value ($y_c$), divided by the heritability of $y_c$. In general, the single-step model predicted EBV more accurately than the GBLUP model, and the latter in turn better than the traditional BLUP. The results indicate that genomic information can increase the reliabilities of predictions of litter size and piglet mortality, and that the single-step model is a useful tool for practical genomic prediction.

**Genomic evaluation of both purebred and crossbred performances**

*O.F. Christensen[1], P. Madsen[1], B. Nielsen[2] and G. Su[1]*
*[1]Aarhus University, Molecular Biology and Genetics, Blichers alle 20, P.O. Box 50, 8830 Tjele, Denmark, [2]Danish Agriculture and Food Council, Pig Research Center, Axeltorv 3, 1609 Copenhagen V, Denmark; olef.christensen@agrsci.dk*

Genomic selection has offered a new paradigm for livestock breeding within purebred populations, but it also offers greater opportunities for incorporating phenotypic information from crossbred individuals and selection for crossbred performance. Wei and Van der Werf presented a model for genetic evaluation using information from both purebred and crossbred animals for a two-breed crossbreeding system. The model provides breeding values for both purebred and crossbred performances. However, an extension of that model to incorporate genomic data has not been made yet. Here, a single-step method for genomic evaluation of both purebred and crossbred performances is presented. The model is a reformulation and extension of the model by Wei and Van der Werf, and it contains two combined breed-specific partial relationship matrices that incorporate both pedigree and marker information. The method allows phenotypic information from crossbred animals to be incorporated into genomic evaluation in a coherent manner for such crossbreeding systems.



**Benefits of distributing males and females among phenotyping candidates in genomic selection**

*T.O. Okeno[1], M. Henryon[2,3] and A.C. Sørensen[1]*
*[1]Aarhus University, Center for Quantitative Genetics and Genomics, Department of Molecular Biology and Genetics, P.O. Box 50, 8830 Tjele, Denmark, [2]University of Western Australia, School of Animal Biology, 35 Stirling Highway, Crawley WA 6009, Australia, [3]Danish Agriculture and Food Council, Pig Research Centre, Axeltorv 3, 1609, Copenhagen V, Denmark; tobias.okenootieno@agrsci.dk*

Distribution of males and females among genotyping candidates in genomic selection was found to influence genetic gains and inbreeding rate. No findings have been reported on phenotyping candidates, since most studies assumed 100% phenotyping. We demonstrated that phenotyping 80% high ranking candidates before genotyping realized most genetic gains achievable with 100% phenotyping. In that study, however, the potential benefits of allocating phenotypings to males and females was not investigated. The aim of this study will be to determine the effect of male-female ratio among phenotyping candidates on genetic gains and inbreeding rate. In parallel to earlier studies on genotyping strategies, we hypothesize that at low phenotyping proportions, it is important to phenotype the most intensively selected sex, while at higher proportions, candidates of the other sex become more important. Five hundred piglets per generation from 10 boars randomly mated to 100 sows will be simulated. Genetic architecture of the founder population generated will reflect the linkage disequilibrium observed in Danish pigs. All candidates will be assumed genotyped after phenotyping at proportions of 20-100%. Candidates will be allocated for phenotyping at random, estimated breeding value and true breeding value. At each proportion males-females' ratios will be 100:0, 75:25, 50:50, 25:75 and 0:100. Selection will be for a single trait with heritability $h^2$=0.4. Each scenario will be run for 12 generations and replicated 100 times. Genetic gains and inbreeding rate will be estimated in the last 5 generations. This study will provide guidelines on the distribution of males and females among the phenotyping candidates in genomic selection.

**Crossbred reference can improve response to genomic selection in crossbreds**

*H. Esfandyari[1,2], A.C. Sørensen[2] and P. Bijma[1]*
*[1]Animal Breeding and Genomics Centre, Wageningen University, Wageningen, 6708 GA, the Netherlands, [2]Center for Quantitative Genetics and Genomics, Department of Molecular Biology and Genetics, Aarhus University, Denmark, 8830, Denmark; hadi.esfandyari@agrsci.dk*

Genomic selection (GS) can be used to select purebreds for crossbred performance (CP). However, performance of purebred parents can be a poor predictor of performance of their crossbred descendants. Training on crossbred data for GS accounts for genetic differences between purebred and crossbred animals and can be beneficial to improve CP, but requires collection of phenotypes and genotypes at the crossbred level. In addition, estimating LD-marker effects from crossbred data can be complicated by the fact that crossbreds represent a mixture of the specific marker-QTL associations that exist within the breeds that contributed to the cross, which can differ between breeds because of differences in the extent and even direction of LD. To address these problems, the objective was to compare crossbred response in a two-way crossbreeding program using a variety of training populations. A genome of 1 M with 1000 marker was simulated. After 2,000 generations of random mating to establish a historical population, 2 lines with effective size 100 were isolated and mated for another 500 generations to create 2 pure breeds that are distantly related. These breeds were used to generate separate, combined and $F_1$ training data sets of individuals with marker genotypes and phenotypes for a trait controlled by 100 segregating QTL and narrow-sense heritability of 0.30 and dominance variation of 0.1. In general, training on crossbreds led to higher cumulative response to selection in crossbreds than using purebreds as training. When training was on crossbreds without genotypes, using genotype probabilities and ignoring parental allele origin led to better performance in crossbreds than considering allele origin in estimation of marker effects. In conclusion, GS can be applied to select purebreds for CP by using a dominance model and training on crossbreds' genotype probabilities.

---



**Using milk spectral data for large-scale phenotypes linked to mitigation and efficiency**

*H. Soyeurt[1], A. Vanlierde[2], M.-L. Vanrobays[1], P.B. Kandel[1], E. Froidmont[2], F. Dehareng[2], Y. Beckers[1], P. Dardenne[2] and N. Gengler[1]*
*[1]University of Liège, Gembloux Agro-Bio Tech, Passage des Déportés 2, 5030 Gembloux, Belgium, [2]Walloon Research Centre, Rue de Liroux 9, 5030 Gembloux, Belgium; hsoyeurt@ulg.ac.be*

Even if producing milk efficiently has always been a major concern for producers, the direct environmental impact of their cows is becoming a novel one. Traits linked to this issue were identified as methane emission ($CH_4$), dry matter intake (DMI) and feed efficiency (FE); however they are available on a small scale. Researches showed that $CH_4$ could be predicted from milk mid-infrared (MIR) spectra, allowing large-scale recording at low cost. The main objective of this study was to show, using a modelling approach, that DMI and FE could be derived from milk MIR spectra. For that, knowledge of body weight (BW) is required; however it was unknown in this study. Derived procedure was based on milk yield and composition, MIR $CH_4$, and modelled standard animal requirements, allowing the prediction of expected BW. An external validation was conducted based on 91 actual records. 95% confidence limit for the difference ranged between -0.66 and 18.84 kg for BW, from -0.02 to 0.26 kg/day for DMI, and from -0.02 to 0.002 kg of fat corrected milk/kg DM for FE. Root mean square errors were 39.66 kg, 0.56 kg/d, and 0.03 kg/DM for the 3 studied traits. P-value for the t-test was not significant for BW and DMI. This suggests the possibility to obtain expected BW and therefore DMI from MIR spectra. Single trait animal test-day models used 1,291,850 records to assess the variability of studied traits. Significant variations were observed for the lactation stage, parity, genetics, and age. These findings were in agreement with the literature except for early lactation. This suggests in conclusion that the MIR information gave similar results for DMI and $CH_4$ for the major part of lactation. The use of this novel method to predict expected BW offers new possibilities interesting for the development of genomic and genetic tools.

## Session 23 Theatre 7

### A genome-wide association study of caudal and intercalary supernumerary teats in Red Holstein cattle

*H. Joerg[1], C. Meili[1], A. Burren[1], E. Bangerter[1] and A. Bigler[2]*
*[1]Bern University of Applied Sciences, School of Agricultural, Forest and Food Sciences, Länggasse 85, 3052 Zollikofen, Switzerland, [2]swissherdbook, Schützenstrasse 10, 3052 Zollikofen, Switzerland; hannes.joerg@bfh.ch*

Supernumerary teats represent a common abnormality of the bovine udder. They negatively affect machine milking ability and are undesirable at dairy expositions. The goal in dairy breeding is udder clearance. Supernumerary teats are positioned at the rear udder (caudal), between the normal front and rear teats (intercalary) or as appendix of normal teats (ramal). In this study, we performed a genome-wide association study based on 50,733 SNPs and proportion of the occurrence of supernumerary teats of at least twenty progenies of 1,079 Red Holstein bulls. We partitioned the genetic variance onto all the 29 autosomes based on genomic relationship matrix. Additionally, the procedure egscore which uses principal component analysis was used to correct for population stratification in the dataset. We performed allelic genome-wide association analyses for the fractions of caudal und intercalary supernumerary teats. We identified a locus showing strong evidence of association with the occurrence of caudal supernumerary teats in Red Holstein breed. 6 SNPs on chromosome 20 had a P-value of $<4\times10^{-9}$. The chromosome 20 explained 24% of the genetic variation of caudal supernumerary teats in Red Holstein breed. The detected QTL region is not known as QTL region for udder clearance. Our study unraveled the categorical trait udder clearance due to the separation of the caudal and intercalary supernumerary teats.

## Session 23 Theatre 8

### Genomic prediction using a G-matrix weighted with SNP variances from Bayesian mixture models

*G. Su, O.F. Christensen, L. Janss and M.S. Lund*
*Aarhus University, Department of Molecular Biology and Genetics, Center for Quantitative Genetics and Genomics, AU-Foulum, 8800 Tjele, Denmark; guosheng.su@agrsci.dk*

Bayesian mixture models often predict more accurate breeding values than GBLUP, but GBLUP may be preferred for routine evaluations because of low computational demand. The objective of this study is to achieve the benefits of both models by using results from Bayesian models as weights on SNP markers in construction of a G-matrix. The data comprised 5,221 progeny-tested Nordic Holstein bulls genotyped with the Illumina Bovine SNP50 BeadChip. Weighting factors in this investigation were squared posterior SNP effect and posterior SNP variance estimated from a Bayesian mixture model with four normal distributions, based on data sets which were 0, 1, 3, or 5 years before genomic prediction. In building the G-matrix, either weights were assigned to each marker (single-marker weighting) or to groups of 5-150 markers (group-marker weighting). The analysis was carried out for milk, fat, protein, fertility and mastitis. Deregressed proofs were used as response variables for genomic prediction. Averaged over the five traits, the Bayesian model led to 1.7% higher reliability of genomic estimated breeding value (GEBV) than the GBLUP model with an original G-matrix ($GBLUP_0$). When using posterior variances as weights on a G-matrix, the reliabilities of GEBV were similar to those obtained from the Bayesian model. Using squared posterior SNP effect as weights gave only small increase of GEBV reliability in some group-marker weighting scenarios, compared to $GBLUP_0$. In addition, group-marker weighting performed better than single-marker weighting in terms of reducing bias of GEBV. Finally, weights derived from a data set having a lag up to 3 years did not reduce reliability of GEBV. The results indicate that the posterior SNP variances estimated from a Bayesian mixture model are good alternative weighting factors, and common weights on groups of 30 markers is a good strategy when using markers of the 50K chip.

**Sire-based genomic prediction of heterosis in White Leghorn crossbreds**

*E.N. Amuzu-Aweh[1,2], H. Bovenhuis[1], D.J. De Koning[2] and P. Bijma[1]*
*[1]Wageningen University, Animal Breeding and Genomics Centre, P.O. Box 338, 6700 AH Wageningen, the Netherlands, [2]Swedish University of Agricultural Sciences, Department of Animal Breeding and Genetics, P.O. Box 7023, 750 07 Uppsala, Sweden; esinam.amuzu@wur.nl*

Reliable methods for predicting heterosis will improve the efficiency of crossbreeding schemes by reducing their dependency on tedious and expensive field-tests of multiple pure-line combinations. Heterosis has been predicted at the population level, but this concept could be extended by predicting heterosis for individual sires. This would quantify genetic variation between sires from the same pure-line, and can increase heterosis in crossbred populations. We aimed to derive the theoretical expectation for heterosis due to dominance in crossbred offspring of individual sires, then predict heterosis for these offspring. We used 60K SNP genotypes from 3,427 sires, allele frequencies from 9 pure-lines, and egg production records from 16 crosses between those lines, representing ~210,000 crossbred hens. We derived the expected heterosis due to dominance for offspring of a specific sire, which is a function of the within- and between-line difference in allele frequency between the sire and the dam-line that they are mated to. Next, we estimated heterosis by regressing offspring performance on this function of allele frequencies. Results show that it is possible to predict heterosis at the sire level, thus distinguishing between sires within the same pure-line whose offspring will show different levels of heterosis. However, the within-line variance in heterosis among sires was low (0.72%) compared to between-line variance (99%). This study was based on an average genome-wide function of allele frequencies. In a preliminary analysis where we used a subset of SNPs with a putative effect on heterosis, we observed a larger variation in heterosis among sires. We conclude that based on a dominance model, it is possible to identify specific sires whose offspring are expected to have relatively higher levels of heterosis than others.



**Identification of SNP of two candidate genes associated with reproductive trait in Egyptian buffalo**

*S. Ahmed[1], H. Abd El-Kader[1], O. Abd Elmoneim[1] and A. El Beltagy[2]*
*[1]National Research Center, Cell Biology Department, Genetic Engineering Division, El Bohoth St., 12311, Dokki, Giza, Egypt, [2]Animal Production Research Institute, Animal Biotechnology, Nady El Seid St., 12126, Dokki, Giza, Egypt; selnahta@yahoo.com*

Improvement of reproductive efficiency has been applied through crossbreeding and improved management techniques rather than direct selection. Selection accuracy measures the correlation between the predicted value and true breeding value. Accordingly, genetic prediction of fertility traits could be potentially affecting the improvement of reproductive traits in buffalo. The aim of our study was to determine the single nucleotide polymorphism (SNP) of insulin-like growth factor1 (IGF-1) and insulin-like growth factor1-receptor (IGF1R) associated with calving interval period which is the most important criterion of fertility. Genomic DNA was extracted from whole blood of unrelated Egyptian buffalo females. The animals were divided into high and low fertility groups according to the calving interval period (CI). The PCR products of investigated genes were subjected to single-strand conformation polymorphism analysis. The polymorphisms of genes were sequenced to determine the single nucleotide polymorphism (SNPs).The results illustrated that the IGF-1 SNP gene locus showed two genotypes. The genotypes frequencies of IGF-I$^1$ and IGF-I$^2$ were 82.1 and 17.9%, respectively. The IGF1R SNP gene locus showed four genotypes. The genotypes frequencies were 48.4% for IGF1R$^1$, 30.5% for IGF1R$^2$, 14.7% for IGF1R$^3$ and 6.4% for IGF1R$^4$. The correlations between fertility and the molecular results of genes studied revealed that IGF-1 gene locus genotypes had no impact on fertility, while IGF1R gene locus genotypes predicted the phenotype of the high fertility for IGF1R$^2$ genotype with approximately 82.7% accuracy. The conclusion of this initial study is the SNP IGF1R$^2$ could be an interesting candidate molecular marker for genetic improvement of fertility in Egyptian buffalo.

**Effect of EGF gene polymorphisms on fattening and slaughter traits in pigs**

*A. Mucha[1], M. Tyra[1], K. Ropka-Molik[2], K. Piórkowska[2] and M. Oczkowicz[2]*
*[1]National Research Institute of Animal Production, Department of Animal Genetics and Breeding, ul. Sarego 2, 31-047 Kraków, Poland, [2]National Research Institute of Animal Production, Labolatory of Genomics, ul. Sarego 2, 31-457 Kraków, Poland; aurelia.mucha@izoo.krakow.pl*

Epidermal growth factor (EGF) belongs to a group of peptide growth factors. It stimulates cell proliferation, differentiation and survival, and also promotes the growth of epithelial, endothelial and fibroblast cells. The aim of the experiment was to use the DNA mutations in the EGF gene to determine associations between the genotype and fattening and slaughter traits in Polish Landrace gilt. Animals were kept at Pig Performance Testing Station under the same housing and feeding conditions. The animals were slaughtered when reach the weight of 100 kg. The polymorphisms in EGF gene were genotyped by PCR methods. In the analysed group of gilts, the highest frequency was found for the AA genotype (0.61) and the lowest for the BB genotype (0.39) of the EGF gene. It was concluded from the preliminary results of the study that gilts of BB genotype were characterized by the highest daily weight gain and those of AB genotype by the lowest ($P<0.05$). Gilts of BB genotype also had the highest average backfat thickness from 5 measurements ($P<0.05$ with AA genotype). For the other traits, such as age and body weight at slaughter and meat weight of primal cuts, animals with BB genotype were characterized by the lowest parameters compared to gilts with AA and AB genotypes ($P<0.01$ and $P<0.05$). The present results therefore suggest that A allele of the EGF gene has a positive effect on fattening and slaughter traits.

---

Session 24 Theatre 1

**Meat inspection: a key tool to assess health and welfare at farm level**

*A. Velarde[1], M. Oliván[2], A. Bassols[3] and A. Dalmau[1]*
*[1]IRTA, Animal Welfare Subprogram, Veïnat de Sies s/n, 17111, Monells, Spain, [2]SERIDA, Apdo 13, 33300 Villaviciosa, Spain, [3]UAB, Departament de Bioquímica i Biologia Molecular, Facultat de Veterinària, Campus UAB, 08193 Cerdanyola del Vallès, Spain; antonio.velarde@irta.cat*

Good health is an important principle of animal welfare and includes abscence of injuries and diseases. The main elements of the current pig meat inspection are ante-mortem examination of animals and post-mortem examination of carcasses and organs. The Welfare Quality® project has developed an integrated and standardised welfare assessment that includes valuable information regarding the health and welfare status of the animals on the farm of origin and transport. Skin damage and ear and tail wounds are assessed by visual inspection after scalding, and not only helps to determine number of marks on the carcass, but also may recognize the source (fighting, rough handling, overcrowding or poor facilities design) according to the anatomical location and damage type. Post-mortem inspection enables also detection of macroscopic lesions caused by some pathogens e.g. pleurisy and pneumonia in the lungs, pericarditis in the heart and white spots in the liver. There have been several occasions within the EU where outbreaks of epidemic diseases have first been detected during meat inspection. However, palpation/incisions used in current post-mortem inspection should be performed separately from the slaughterline because of the risk of microbial cross-contamination. New technologies like the 'omics' approach aim to identify animal-based biomarkers for health and welfare. These measurements may be taken from different fluids and tissues in the carcass. At the slaughterline, animals from several farms can be sampled on the same day, reducing the risk of disease transmission. However, to use the slaughterline records to improve health and welfare, a feedback system to the farm should exist.

### Monitoring animal welfare problems in fallen stocks

*J. Baumgartner, M.T. Magenschab, K. Haas and C. Hofer-Kasztler*
*Vetmeduni Vienna, Veterinaerplatz 1, 1210 Vienna, Austria; johannes.baumgartner@vetmeduni.ac.at*

Fallen stocks are cadavers of farm animals not intended for human consumption. In 2012 a total of 355,000 fallen stocks was collected by certified animal waste processing plants (AWPP) in Austria. Based on the livestock, the proportion of fallen animals was 11% for young cattle, 2% for cattle >1 year and 5% for pigs >50 kg. The causes of death of fallen stocks remain unclear within the standard animal waste processing procedure. However, there are indications of unnecessary pain and prolonged suffering in a considerable number of fallen animals due to improper medical treatment and/or the lack of euthanasia at the right time. The aim of our study was to develop a feasible monitoring system for sorting out cadavers suspicious for animal welfare problems within the routine animal waste processing procedure for a concluding inspection by an official veterinarian. In a first step, the content of 100 collection trucks (n=1070 cattle, 987 pigs) delivered to an Austrian AWPP was evaluated. We found 134 fallen cattle (13%) and 203 fallen pigs (21%) with one or more predefined pathological signs indicative of severe pain and prolonged suffering. Decubitus ulcers were most frequent in cattle (75%) and bite wounds in pigs (62%). Based on these results, we developed a simple checklist with the five most welfare relevant pathological findings. Additionally 16 drivers of collecting trucks were trained to identify fallen animals with these findings and to ear tag them at the farm of origin. In the testing phase the drivers pre-selected 11.2% of fallen cattle (n=170) and 6.6% of fallen pigs (n=562), which proved to be justified in 95%. We conclude that our monitoring concept for pre-selection of suspicious fallen stocks produces reliable results with low additional workload. In combination with intensive training of farmers in proper treatment of moribund animals, it may help to reduce the number of fallen stock with obvious signs of unnecessary suffering.

---



### Boar carcass skin lesions reflect their behaviour on farm

*D.L. Teixeira[1], A. Hanlon[2], N. O'Connell[3] and L.A. Boyle[1]*
*[1]Teagasc Moorepark, Pig Development Department, Fermoy, 00000, Ireland, [2]University College Dublin, Dublin, Ireland, [3]Queens University Belfast, Belfast, United Kingdom; dayane.teixeira@teagasc.ie*

The aim was to investigate if aggressive and mounting (sexual) behaviours performed by boars are associated with skin lesions scored at the abattoir. A total of 70 entire boars were compared to 71 gilts, housed in 5 single-sex groups per gender (14 pigs/group) in fully slatted pens. On days -14 and -1 pigs were weighed. Posture, aggression and mounting behaviours were recorded in 3×2 hour periods on days -13, -9, -7 and -2 prior to slaughter. Carcass cold weights were obtained from the abattoir. Carcass skin lesions were assessed as per the Welfare Quality® protocol and bruises were classified as fighting, mounting and handling-type. Data were analysed by Mixed models, Kruskall-Wallis and Spearman correlations. On Day -1 boars were heavier than gilts (100.7 vs 99.0 kg; s.e.m. 0.59; P<0.05) but gender did not affect carcass cold weight (77.3±0.4 kg). Boars performed more aggressions (1.8 vs 1.0/pig; s.e.m. 0.22; P<0.05) and mounts (0.4 vs 0.005/pig; s.e.m. 0.02; P<0.05) than gilts. In general, lying, sitting and standing were similar in both genders (P>0.05). Boars had higher carcass skin lesion scores (1.9 vs 1.3; s.e.m. 0.10; P<0.05) and more fighting-type bruises (4.5 vs 2.3; s.e.m. 0.35; P<0.05) than gilts. In both genders, there were positive correlations between aggression and carcass skin lesions score (actor: r=0.264, P<0.01; recipient: r=0.302, P<0.001) and fighting-type bruises (actor: r=0.442, P<0.001; recipient: r=0.297, P<0.001, respectively). Recipient of mounting behaviour was also associated to carcass skin lesion score (actor: r=0.157, P<0.08; recipient: r=0.187, P<0.05) but not to mounting-type bruises (r=-0.034, P>0.05). The results reinforce the importance of on-line monitoring of carcass skin lesion in the routine inspection procedures as a complementary tool to identify critical points along the slaughter chain and as indicator of animal welfare in the farm.

**Effect of mixing boars prior to slaughter on behaviour and skin lesions**

*N. Van Staaveren[1,2], D. Teixeira[2], A. Hanlon[1], N. O'Connell[3] and L. Boyle[2]*
*[1]UCD, Belfield, Dulbin, Ireland, [2]Teagasc, Moorepark, Fermoy, Ireland, [3]QUB, Belfast, Northern-Ireland, United Kingdom; nienke.vanstaaveren@teagasc.ie*

The effect of mixing boars prior to slaughter on behaviour and its association with skin lesions was studied. Pigs from 3 single sex pens were randomly allocated to one of 3 treatments (n=20/group; 6 focal pigs/group) on day of slaughter over 5 replicates: boars unmixed (MUM); boars mixed (MM) and boars mixed with gilts (MF). Frequency of aggressive and sexual behaviours during holding on farm and at lairage was recorded. Skin lesions (LS) were scored according to severity (0-5) on farm, at lairage (focal pigs only) and on the carcass. Loins were assessed for bruising (LB) according to severity (0-2). Effect of treatment and time on behaviour and lesions was analysed by SAS V9.3 PROC MIXED. Correlations between behaviour and LS/LB were calculated. MM performed more aggressions than MUM (50.4±10.72 vs 20.3±9.55; $P<0.05$) and tended to perform more aggressions than MF (37.2±10.77; $P=0.06$). MM performed more mounting than MF and MUM (30.9±9.99 vs 11.4±3.76 and 9.8±3.74 respectively; $P<0.01$). MUM pigs tended to show a smaller increase in LS than MF and MM pigs and MM pigs showed a greater increase than MF pigs ($P=0.08$). No significant effect of treatment was found on carcass LS/LB ($P>0.05$). Pigs involved in more fights after mixing had higher LS at lairage ($r=0.23$; $P<0.05$) but not on carcass ($P>0.05$). Mounting was not associated with LS/LB ($P>0.05$). LS scored on farm was correlated with LS scored at lairage ($r=0.45$, $P<0.001$) and on carcass ($r=0.21$, $P<0.01$). LS at lairage and on carcass tended to be correlated ($r=0.19$, $P=0.07$). Mixing boars prior to slaughter stimulates mounting and aggression and increases LS particularly when mixed with other boars. The increase in aggression and mounting is not reflected in higher LS/LB at the carcass. Carcass LS is correlated with that scored on farm suggesting that such information recorded at meat inspection could be used by farmers to inform their herd health and welfare management plans.

---



**Welfare (findings at slaughter) consequences following a reduction in antibiotic use**

*N. Dupont and H. Stege*
*University of Copenhagen, Grønnegårdsvej 2, 1870 Frb C, Denmark; nhd@sund.ku.dk*

During 2010 the Danish authorities introduced legislation subjecting pig herds with high antibiotic (AB) usage to fines and regulation. The following year overall use of AB dropped by ~25%. Hence, the aim of this project was to determine if this decrease had any effects on animal welfare in finisher, measured as prevalence of pathological findings and variation of lean mean percent (LMP), as variations in LMP tends to be correlated with variations in daily weight gain. Herds with an AB consumption of >3.5 kg active compound in the year before June 2010 and a reduction in AB consumption of >10% the following year were randomly selected from the national database, Vetstat. Organic and outdoor herds, herds that had suffered severe disease outbreaks, performed eradication programs or made any other major changes during the study period were excluded. AB consumption was calculated as gram active compound AB/pen places and as average number of daily doses given per 100 animals per day (ADD/100 animals/day). Data on number of animals produced, pathological findings and LMP at slaughter were collected. Multi-level models for binary outcome were performed (significance level 95%) (PROC GLIMMIX, SAS Enterprise Guide 4.3). In average the 65 included finisher herds sent 11201 finishers for slaughter (std.dev 5,569). AB consumption decreased with 55% in the year following June 2010 regardless of calculation method (g/pen place: from 12.1 to 4.9; ADD/100 animals/day: 5.8 to 2.3). In the year following June 2010 a significant increase was observed in the proportion of finishers at slaughter with abscesses (2.9% to 4.4%; $P<0.001$) and osteomyelitis (0.3% to 0.5%; $P<0.001$). Standard deviation of LMP within each herd did not change significantly before and after June 2010 (2.4; 2.5; $P=0.68$). The 52% and 67% significant increase in the proportion of pigs with abscesses and osteomyelitis respectively, suggests that a reduction in AB consumption may affect animal welfare.

**Feed intake cannot be used as predictor of gastric ulcers in lactating sows**

*T.S. Bruun*[1], *J. Vinther*[1] *and C.F. Hansen*[2]

[1]*Danish Pig Research Centre, Danish Agriculture & Food Council, Axeltorv 3, 1609 Copenhagen, Denmark,* [2]*University of Copenhagen, Department of Large Animal Science, Faculty of Health Sciences, Groennegaardsvej 2, 1870 Frederiksberg, Denmark; tch@lf.dk*

In the scientific literature only limited information is available on gastric ulcers in sows, and currently it is not possible under commercial conditions to identify sows with gastric ulcers. The aim of this study was to determine whether occurrence of gastric ulcers differed between sows with low and high average daily feed intake in lactation, and whether feed intake could be used as a predictor of stomach health in sows. One herd using liquid feeding systems allowed daily recordings of feed allowance for the individual sow. 25% of the sows (n=81) having the lowest average daily feed intake (LOW) and highest (HIGH) average daily feed intake (n=85),respectively, were selected. LOW sows all consumed less than 5.9 Danish feeding units for sows ($FU_{sow}$) per day and HIGH sows all ate more than 6.9 $FU_{sow}$ per day. Stomachs from sows culled day 0 to 5 post-weaning were collected and a pathologist inspected the esophageal part of the stomachs visually and palpated the esophagus. The pathological changes were described by an index from 0 to 10. Index 0: no pathological changes. Index 1-3: degree of parakeratosis of the pars esophagea (PE). Index 4-5: degree of erosion of the PE. Index 6-8: degree of ulcers and/or scars in PE. Index 9-10: stenosis of the esophageal lumen (10 mm (index 9) or 3 mm in diameter (index 10)). Data was analyzed using Fisher's Exact Test by PROC FREQ in SAS. Odds-ratio was calculated for the risk of having a gastric ulcer index of either 6-10 or 8-10 for LOW versus HIGH, respectively. The preliminary results showed that the probability of having gastric ulcers with an index from 6 to 10 (P=0.64) or 8-10 (P=1.00) was identical for LOW and HIGH sows. Based on the present data, feed intake cannot be used as a tool for prediction of gastric ulcers in lactating sows. However, this finding will be validated in 2 additional sow herds.

---



**Prevalence of gastric ulcers in culled sows is representative of the herd**

*J. Vinther and T.S. Bruun*

*Danish Pig Research Centre, Danish Agriculture & Food Council, Axeltorv 3, 1609 Copenhagen, Denmark; tch@lf.dk*

True prevalence of gastric ulcers in sows has scarcely been investigated. Culled sows have often been used for a pathological inspection of the esophageal part of the stomachs, but it has remained uncertain whether culled sows were representatives of the herd. The aim of this study was to investigate whether prevalence of gastric ulcers in culled sows was representative of the herd. Herds were not randomly selected. From 8 herds that were closing all sows were culled in 2nd half of 2012 and beginning of 2013. Four farmers used home-mixed feed and 4 farmers used compound feed. At slaughter, information about cause of culling (ordinary culling (OC) or culling due to closing herd (CC)) was available. More than 220 stomachs per herd were collected. A pathologist inspected the esophageal part of the stomach visually and palpated the esophagus. The pathological changes were described by an index from zero to ten. Index 0: no pathological changes. Index 1-3: degree of parakeratosis of pars esophagea (PE). Index 4-5: degree of erosion of PE. Index 6-8: degree of ulcers and/or scars in PE. Index 9-10: stenosis of the esophageal lumen (10 mm (index 9) or 3 mm in diameter (index 10)). Data was analyzed by logistic regression using PROC GLIMMIX in SAS. The cause of culling (OC or CC) was included as a fixed effect, and herd was included as a random effect. Results showed that the prevalence of gastric ulcers and scars (index 6-10) was 76.6 and 81.2% for CC and OC, respectively (P=0.102). The prevalence of more severe gastric ulcers and scars (index 8-10) was 25.3 and 29.0% for CC and OC, respectively (P=0.197). From a practical point of view, the prevalence of gastric ulcers in culled sows can be considered to be representative of the prevalence of gastric ulcers in the herd, and therefore pathological inspection of stomachs of culled sows is recommended.

### Effect of amount of straw provided to growing/finishing pigs on gastric ulceration at slaughter

*M.S. Herskin[1], H.E. Jensen[2], A. Jespersen[2], B. Forkman[3], M.B. Jensen[1], N. Canibe[1] and L.J. Pedersen[1]*
*[1]Aarhus University, Dept. of Animal Science, AU-FOULUM, Tjele, Denmark, [2]University of Copenhagen, Dept. Veterinary Disease Biology, KU-SUND, Frederiksberg, Denmark, [3]University of Copenhagen, Dept. Large Animal Science, KU-SUND, Frederiksberg, Denmark; mettes.herskin@agrsci.dk*

Gastric ulceration is an important health and welfare issue in pig production, the economic losses of which are substantial. We examined the effect of permanent access to straw on gastric lesions in pigs housed in pens (18 pigs per pen and 0.7 $m^2$/pig) with partly slatted flooring and provided with 10 vs 500 or 1000 g straw/pig/day from 30 kg live weight to slaughter. A total of 45 LYD pigs, three from each of 15 different pens, were used. The pigs were fed a standard commercial dry feed for ad lib intake. Stomachs were collected following euthanization. The size of the non-glandular part (pars oesophagea) was measured longitudinally direction (L; along the inside of cuvatura minor) and transversally ($T_{1-3}$) at three points. Lesions were characterized and the stomachs were categorized as either ulcerated (score 7, 8 or 9) or non-ulcerated (score 1-6). Data were analysed in PROC GLIMIXED of SAS after pooling the results from pigs provided with 500 and 1000 g straw (considered permanent access). Among the pigs with access to 10 g straw (n=18), 33% of the stomachs were categorized as ulcerated, whereas only 7% of pigs with permanent access to straw (n=27) was categorised as ulcerated ($F_{1,30}$=4.3; P=0.048), leading to an odds ratio of gastric ulceration of 6.1 (95% CI: 1.1-38.0). The weighed area of the pars oesophagea (($LxT_1$+$LxT_2$+$LxT_3$)/3 for each pig) did not differ between the treatments. Permanent access to straw led to reduced occurrence of gastric ulceration compared to pigs given 10 g straw/day. The present findings may be attributed to nutritional as well as environmental effects of the straw, and suggest that provision of straw may be a relevant strategy to limit the prevalence of gastric ulceration in pig production.

---



### Economics of higher health and welfare pig production

*B. Vosough Ahmadi, C. Correia-Gomes, E.M. Baxter, R.B. D'Eath, G.J. Gunn and A.W. Stott*
*Scotland's Rural College (SRUC), West Mains Road, EH9 3JG, Edinburgh, United Kingdom; bouda.v.ahmadi@sruc.ac.uk*

Efficient livestock production is threatened by factors that interfere with production, such as diseases, management, housing and behaviour problems which affect health, performance and growth and/or welfare of the animals. However, producers' profit is directly linked not only to efficient production but also to the characteristics of the product demanded by consumers such as high animal welfare standards, food-safety, organic attributes, etc. The required adjustments of the production systems to fulfil consumer demand are often costly and require major changes. However, in certain scenarios win-win situations for health/welfare and economic aspects of the livestock systems could be achieved. In this paper, we bring together three studies to assess the relative contribution of alternative trade-offs and synergies available in this area. Also the magnitude and importance of these problems are discussed. Understanding the overall herd health score (HHS) of breeding and finishing herds as well as having an insight into the economic consequences of health/welfare issues was the focus of the first study. Based on a systematic literature review, consulting with veterinarians and an online survey, HHS systems for breeding and finishing herds were developed. The financial consequences of two diseases were then estimated. In the second and third studies, analytical models were used to establish acceptable trade-offs between profit and welfare within three farrowing systems and within three possible housing and management alternatives with respect to tail-docking. The following outcomes were demonstrated and will be discussed in the paper: (1) the potential risk of alternative prevention, control and management; (2) possibility of a win-win situation for welfare and profit in non-crate farrowing systems; and (3) possibility of improving welfare in relation to tail-docking and the changes required in pig housing and management before tail docking can be stopped.

**Dynamic monitoring of weight data at the pen vs at the individual level**

*D.B. Jensen[1], N. Toft[2], A.R. Kristensen[1] and C. Cornou[1]*
*[1]Faculty of Health and Medical Sciences, University of Copenhagen, Department of Large Animal Sciences, Groennegaardsvej 21870, Frederiksberg C, Denmark, [2]National Veterinary Institute, Technical University of Denmark, Section for Epidemiology, Bülowsvej 27, 1870 Frederiksberg C, Denmark; daj@sund.ku.dk*

The PigIT project, led by the University of Copenhagen, aims at improving welfare and productivity in growing pigs using ICT methods. Automatically and manually recorded data are currently being collected in a production herd. One of the first steps of the project is to make use of the manually recorded weight data from finisher pigs. Data are collected at insertion and at the exit of the first pigs in the pen, and in few pens, the weight is recorded weekly. Dynamic linear models are fitted on the weight data, at the pig level (univariate), at the double pen level using averaged weight (univariate) and using individual pig values as parameters in a hierarchical (multivariate) model including section, double pen, and individual level. Variance components of the different models are estimated using the Expectation Maximization algorithm. The difference of information obtained at the individual vs pen level is thereafter assessed. Whereas weight data is usually monitored after a batch is being sent to the slaughter house, this method provides weekly updating of the data. Perspectives of application include dynamic monitoring of weight data in relation to events such as diarrhoea, tail biting and fouling in order to assess whether it is possible to detect deviations of patterns before or during the occurrence of these events.

---



**Relationship between meat quality and stress biomarkers in pigs**

*M. Oliván[1], J. González[2], A. Bassols[3], R. Carreras[2], Y. Potes[4], E. Mainau[3], L. Arroyo[3], R. Peña[3], A. Coto-Montes[4], K. Hollung[5], X. Manteca[3] and A. Velarde[2]*
*[1]SERIDA, Apdo 13, 33300 Villaviciosa, Spain, [2]IRTA, Finca Camps i Armet s/n, 17121 Monells, Spain, [3]UAB, Cerdanyola del Vallès, 08193 Barcelona, Spain, [4]Universidad de Oviedo, C/ Julián Clavería s/n, 33006 Oviedo, Spain, [5]Nofima, P.O. Box 210, 1431 Aas, Norway; antonio.velarde@irta.cat*

This study was performed within the ANEMOMA project, which addresses the demand for improved knowledge on animal welfare and its relationship with meat quality in pigs, in order to develop an animal-based approach to animal welfare assessment. The specific objective of this work was to identify biomarkers to assess the stress response to two different rearing managements (indoor vs outdoor) and two types of transport procedures (short and smooth driving 'smooth' vs long and rough driving 'rough') to slaughtering. Rectal temperature was measured at farm and at the end of the trip. At slaughter, blood samples were taken for biochemical analysis. Carcass and meat quality traits (lesions, pH, EC, colour) were recorded and muscle samples were taken for 2D-DIGE proteomic analysis. Multivariate analysis was applied to study the complex relationships between meat quality and physiological/biochemical/proteomic variables. The biplot obtained by PCA showed that PC1 and PC2 explained 32% of the variability. PC1 distinguished in the right side pale and exudative meat (higher L* and EC), obtained from animals coming from the rough transport treatment, with more skin lesions, higher muscle temperature post mortem and increased levels of glucose, triglycerides and creatine kinase (CK) in blood at slaughter, which indicates higher ante mortem stress. This response to stress also affected to some of the proteins related to the post mortem muscle metabolism. In the left side of the plot, there were meat samples with normal pH decline, obtained from animals subjected to the smooth transport procedure. PC2 and PC3 contributed to discriminate rearing treatments, with animals from outdoor showing higher skin lesions, CK levels and meat a* value.

**Effect of different dietary protein sources on proudictive performance of Farafra sheep**

*H.H. Abd El-Rahman, Y.A.A. El-Nomeary, A.A. Abedo, M.I. Mohamed and F.M. Salman*
*National Research center, Animal production Department, Al-Behose ST. Dokki Cairo Egypte, 2226, Egypt; elnomeary@yahoo.com*

Theses study was conducted to evaluate the effects of using different sources of protein on feed intake, nutrient digestibility, average daily gain (ADG) and some ruminal parameters of sheep. Amino acid assay and the solubility of tested protein sources were done. Economic feed efficiency was also determined. Twenty Farafra male lambs having live body weight on an average 41.39±1.66 kg were divided randomly into four groups (Fife animals each) lambs of each group on one of four iso-colaric and iso-nitogens diets of different protein sources being soybean meal (SBM), black seed meal (BSM), cotton seed meal (CSM) and sesame seed meal (SSM). The feeding trial extended for 65 days followed by digestibility and N balance trials. The results showed that average daily weight gain was higher ($P<0.05$) in lambs feed diets CSM (235 g) and BSM (211 g), respectively, as compared to diets SSM (206 g) and SBM (180 g). No significant difference was found in dry matter intake (DMI) among tested diets. Better ($P<0.05$) feed conversion (g DMI) were found in lambs feed on diets CSM (6.21) and BSM (6.76) as compared to diets SSM (7.12) and SBM (7.54). Economic feed efficiency was the best on diet CSM (98.75%) as compared to diets BSM (55.12%), SSM (36.58%) and SBM (0.0%). No significant differences were observed in nitrogen intake of the four of tested groups, however, nitrogen balance values of sheep given CSM diet were highest ($P<0.05$) compared with other groups. No significant differences were observed TDN values among of tested rations. It was concluded that cotton seed meal was better protein sources followed by black seed meal for sheep performance compared with soybean meal and sesame male.

---



**Influence of GH genotype and feeding regime on reproductive parameters in Serra da Estrela ewes**

*M.R. Marques, J.P. Barbas, J.M. Ribeiro, A.T. Belo, T. Cunha, P. Mesquita, R.M.L.N. Pereira and C.C. Belo*
*INIAV, Fonte Boa, 2005-048 Vale de Santarém, Portugal; rosario.marques@iniav.pt*

Growth hormone (GH) is one of the main hormones involved in the onset of puberty and gestation. It is known to influence reproductive parameters in sheep, and is involved in the homeorhetic control of the metabolism by regulating homeostatic signals. The objective of this work is to investigate how GH2-Z gene copy genotypes [AA (R9R/S63S), AB (R9C/S63S), and AE (R9R/S63G)], and feeding regime (group R – growth rate restricted to 79 g/day; or group N – growth rate of 106 g/day during pre-pubertal phase) influences reproductive parameters of Serra da Estrela primiparous ewes. Ovarian activity was assessed through plasma progesterone (P4) measurement by radioimmunoassay. Blood samples of 90 lambs born in September were collected once every 10 days, from 5 month of age till forty days after the first oestrous detection. Onset of ovarian cyclicity was achieved when P4 levels reached $\geq$0.5 ng/ml and ewes became cyclic every 17 days. The onset of puberty was detected using vasectomised rams to identify oestrous behavior. Oestrous synchronization and ovulation induction were performed before artificial insemination (AI) with refrigerated semen. Ewes' cyclicity was reached on average at 251.4 days of age. This average was not affected by ewes' GH genotype. A delay in the onset of cyclicity was identified in ewes from AB genotype submitted to R feeding regime (group R=261.9 vs group n=245.9 days, $P<0.05$). Neither genotype nor feeding regime affected the age at puberty (273 days). Fertility, fecundity and prolificacy after AI were respectively 40.0, 55.6 and 138.9%. Global results (AI+ natural breeding) for these reproductive parameters were 88.1, 98.8 and 112.2%. Fertility was higher ($P<0.05$) for AB (96.4%) and AE (95.2%) genotypes than for AA genotype (77.1%). No differences were detected in fertility between R (88.2%) and N (91.3%) feeding regimes. In conclusion, ewes GH genotypes influence global fertility results.

### Use of non-traditional feed ingredients in goats diet

*H.M. El-Sayed[1], A.A. Abedo[2] and N.E. El-Bordeny[1]*
*[1]Fac. of Agric., Ain-Shams Univ., Animal Production Dept., Hadayek Shobra, Kaluobia, 68, Egypt, [2]National Research Centre, Animal Production Dept., Dokki, Giza, 12622, Egypt; abedoaa@yahoo.com*

This work was done to study effect of use date seed, olive pulp, rady cill, poultry litter and an-aerobically digested manure as non-traditional feed ingredients on goats performance. Twenty-four kids weighed 18.7 kg in average were divided into four groups and fed four different experimental diets; D1 as a control diet (100% concentrate feed mixture), and from D2 to D4 contained 35% date seeds, 27% olive pulp from total concentrate and contained 30% rady cill, poultry litter and an-aerobically digested manure in D2, D3 and D4, respectively. Animal were fed in the previous mixture to cover the maintenance requirements and were allowed to graze natural vegetation mixture of Atriplex and Acasia. The results indicated that feed intake, water consumption and nutrients digestibility were not significant different between different groups. Nitrogen intake and nitrogen retention recorded ($P \leq 0.05$) highest values for diet contained rady cill (D2). Also D2 recorded the highest value of ruminal total volatile fatty acids, while D1 recorded the highest value of ruminal ammonia. Blood parameters included total protein, albumin, globulin, urea, creatinine, ALT and AST were within the normal range, and the highest ($P \leq 0.05$) values of total protein, albumin and cholesterol were noticed with diet contained rady cill.



### The effect of feeding level on goats milk fatty acids and on insulin and leptin concentrations

*E. Tsiplakou[1], S. Chadio[2] and G. Zervas[1]*
*[1]Agricultural University of Athens, Nutritional Physiology and Feeding, Iera Odos 75, 11855, Greece, [2]Agricultural University of Athens, Anatomy and Physiology, Iera Odos 75, 11855, Greece; eltsiplakou@aua.gr*

The effect of long term under/over feeding in ruminants has focused on milk chemical composition but not on milk fatty acids (FA) profile. Thus the objective of this study was to determine the effects of long-term under/overfeeding on goats milk chemical composition and FA profile and on blood plasma insulin and leptin concentrations. Twenty-four 3 years old dairy goats of local breed were used for the experiment. Three months post partum the goats were divided into three homogenous sub-groups (n=8). Each group fed the same ration, but in quantities which covered 70% (underfeeding), 100% (control) and 130% (overfeeding) of their energy and crude protein requirements. The results showed that the milk of underfed goats had significantly higher fat content compared to overfed, while the lactose content was significantly lower compared to that of control and overfed goats. As the milk FA profile concerns the underfeeding reduced significantly the concentrations of $C_{16:0}$ and long chain FA (LCFA) and increased that of $C_{18:0}$ and polyunsaturated FA (PUFA) in goats milk fat compared to the controls. The overfeeding increased significantly the cis-9, trans 11 $C_{18:2}$ (CLA) milk fat content compared to the control and increased significantly the medium chain FA (MCFA) compared to underfeeding. The concentrations of non esterified FA (NEFA) and $C_{18:1}$ in blood plasma was significantly higher in underfed, compared to overfed, goats while the opposite was observed for the $C_{18:2n6c}$ and $C_{18:3n6}$ FA concentrations. No differences were observed in blood plasma leptin concentration between the dietary treatments, while the concentration of insulin was significantly higher in overfed compared to underfed goats. In conclusion, the long term under/overfeeding affected the goats milk chemical composition, milk and blood plasma FA profile and insulin blood plasma concentration.

**Effect of *Rosmarinus officinalis* on growth, blood factors and immune response of newborn goat kids**

*B. Shokrollahi[1], F. Amini[1], S. Fakour[2] and F. Kheirollahi[3]*

*[1]Sanandaj branch, Islamic Azad University, Department of Animal Science, Kurdistan, Sanandaj-6616935391, Iran, [2]Sanandaj branch, Islamic Azad University, Department of Clinical Sciences, Kurdistan, Sanandaj-6616935391, Iran, [3]Razi University, Kermanshah, Daneshgahe razi, 73441-81746- Kermanshah, Iran; borhansh@yahoo.com*

This study was aimed to evaluate the effect of different levels of rosemary (Rosmarinus officinalis) extract on growth rate, hematology and with cell -mediate immune response in Markhoz newborn goat kids. Twenty four newborn goat kids (aged 7±3 days) were randomly allotted to four groups with six replicates. The groups included control (received raw milk), group T1, T2 and T3 which received enriched-milk with 100, 200 and 400 mg aqueous rosemary extract /kg of live body weight/day for 42 days, respectively. Body weights of kids were measured weekly until the end of the experiment. On day 42, about 10 ml blood samples were collected from each kid through the jugular vein. Cell-mediate immune response was assessed through the double skin thickness after intradermal injection of phytohematoglutinin (PHA) at day 21 and 42. Overall ADG and total gain were highest (3583.3 gr) in group T2 compared to other groups. Significant differences in globulin ($P<0.05$), and WBC (White blood cells) ($P<0.001$) were observed among groups. Globulin (2.65 gr/dl) was decreased significantly in T3 kids and non-significantly diminished in T1 (2.825 gr/dl) and T2 (2.825 gr/dl) kids in comparison with the control group. WBC level was highest in T2 ($24.075\times10^6$/µl) and also T1 ($20.975\times10^6$/µl) animals. There were not any significant differences in hemoglobin (Hb), packed cell volume (PCV), red blood cells (RBC), lymphocytes and neutrophils among groups. There were significant differences in increase of skin thickness to intradermal injection of PHA among groups at 42-day ($P<0.01$) and group T3 had the highest response to PHA injection but there were not the same significant differences among groups at 21-day. In conclusion, the results indicated that enriched-milk with aqueous rosemary extract can have a positive effect on growth rate and immunity of newborn goat kids.

---



**Relations among birth weight and some body measurements at birth in Saanen kids**

*H. Onder[1], U. Sen[1], S. Ocak[2], O. Gulboy[1], C. Tirink[1] and H.S. Abaci[1]*

*[1]Ondokuz Mayis University Faculty of Agriculture, Animal Science, Samsun, 55139, Turkey, [2]Zirve University, Middle East Sustainable Livestock, Biotechnology and Agro-Ecology Research and Development Centre, Gaziantep, 27260, Turkey; honder@omu.edu.tr*

This study was carried out at the commercial dairy goat farm in Bafra province of Samsun, Turkey. Number of single born female was 14, single born male was 10, multiple born female was 18, and multiple born male was 20, which constituted the material of this study. All data of the kids collected after kidding when they set on feet. The effect of sex and birth type on birth weight and body measurements; height at withers, height at sacrum, body length and chest girth were measured. Result showed that sex and interaction term between sex and birth type has no effect on interested traits. Birth type has significant effect on all interested traits except height at sacrum. Single born kids yield higher values than females as expected. The highest Pearson correlation coefficient was obtained between birth weight and chest girth with value of 0.912. All data showed Johnson SU distribution which is closely related to Normal distribution. So, linear regression was used for model building with stepwise variable selection approach. The best model was obtained as birth weight=-6.532+0.238(chest girth)+0.055(body length) where coefficient of determination was 0.839. The result of study indicated that there was high relation among birth weight, chest girth and body length, except for height at withers and height at sacrum in Saanen kids.

## Concentrations of NEFA, lactate and glucose in lambs are different to cattle at slaughter

*S.M. Stewart, P. McGilchrist, G.E. Gardner and D.W. Pethick*
*Murdoch University, School of Veterinary and Life Sciences, South Street, Murdoch, WA, 6150, Australia; s.stewart@murdoch.edu.au*

During the pre-slaughter period, sheep may experience acute stress and extended periods of feed deprivation. Studies in slaughter cattle have shown a moderate impact on plasma non-esterified fatty acid (NEFA) concentration in cattle but a larger acute stress immediately prior to slaughter, demonstrated by marked increases in plasma glucose and lactate. Sheep undergo similar pre-slaughter management to cattle in Australia, thus we hypothesise that lambs will have a similar metabolic profile at slaughter. Blood was collected at exsanguination from 1,536 lambs (mean age at slaughter 298±57 days) from the Sheep Co-Operative Research Council Information Nucleus Flock. The lambs were managed on two research stations (Katanning, Western Australia and Armidale, New South Wales) and were assigned to ten slaughter groups based on carcass weight targets for the desired market. Lambs were yarded the day prior to slaughter, held in curfew on-farm for 2 hours, weighed and then transported to commercial abattoirs where they were held in lairage for an average of 17.9±1.79 hours. Blood was collected at exsanguination and the plasma was analysed for non-esterified fatty acids, glucose and lactate. Plasma metabolite concentrations differed markedly between slaughter groups ($P<0.05$). Across all of these groups the mean plasma NEFA, glucose and lactate concentrations were 1.57±0.45 mmol/l, 4.05±0.56 mmol/l and 2.94±1.74 mmol/l. Contrary to our hypothesis, NEFA concentrations were markedly higher at slaughter in lambs compared with cattle. Alternatively lactate and glucose concentrations were lower than concentrations in slaughter cattle. This suggests that different mechanisms may exist between sheep and cattle in terms of their metabolic response to feed deprivation and acute stress during the pre-slaughter period.



## Validation of indicators to assess consciousness in sheep

*M.T.W. Verhoeven[1,2], M.A. Gerritzen[2], L.J. Hellebrekers[3] and B. Kemp[1]*
*[1]Wageningen University, P.O. Box 338, 6700 AH Wageningen, the Netherlands, [2]Wageningen UR Livestock Research, P.O. Box 65, 8200 AB Lelystad, the Netherlands, [3]Utrecht University, P.O. Box 80.163, 3508 TD Utrecht, the Netherlands; merel.verhoeven@wur.nl*

The aim of this study (part A and B) was to validate indicators used to assess consciousness in sheep at slaughter: eye lid-, threat- and withdrawal reflex responses as well as breathing pattern and EEG. In part A, ewes (n=10, 35±4 kg) were anaesthetised twice, six days apart, for ±30 min after which they regained consciousness. Propofol (anaesthetic agent) was used to achieve different levels of (un)consciousness. In part B, ewes (n=21, 38±4 kg) were exsanguinated by neck cut, severing both jugular veins and carotid arteries. In part A, sheep lost and regained consciousness after propofol infusion (T=0) at 43±6 s and 30±5 min, based on visual appraisal of the EEG. Threat- and withdrawal reflex responses were absent after 2±1 and 3±1 min and returned at 15±5 and 30±7 min, respectively. The eye lid reflex response was absent after 5±2 and returned at 11±6 min, but did not disappear in 8 out of 20 observations. Based on the EEG, however, sheep were considered unconscious. In part B, sheep lost consciousness after the neck cut (T=0) at 15±4 s, based on visual appraisal of the EEG. The EEG was considered flat, indicating brain death, at 27±8 s. Non-rhythmic breathing was observed in all sheep directly following the neck cut. Seven sheep showed a threat reflex response prior to loss of consciousness. No withdrawal reflex responses were seen after the neck cut, during exsanguination. Eye lid reflex responses, however, were observed until 74±17 s. The findings show substantial variation between sheep in timing of loss and regain of reflexes in relation to the EEG. Variation in the withdrawal- and eye lid reflex response are difficult to interpret in the assessment of consciousness. The challenge is to further develop a way to use EEG in slaughter practice or find the most accurate way of using (multiple) indicators to assess consciousness.

**Effect of breed and litter size on the display of maternal and offspring postnatal behavior in sheep**

*P. Simitzis, K. Galani, P. Koutsouli and I. Bizelis*
*Agricultural University of Athens, Animal Breeding and Husbandry, 75 Iera Odos, 11855 Athens, Greece; pansimitzis@aua.gr*

A preliminary study was conducted to examine the effects of genotype and litter size on the expression of ewe maternal and lambs neonatal behavior in 3 Greek breeds. Animal observations were carried out on 7 Orini Epirus (4 singles and 3 twins), 7 Karagouniko (3 singles and 4 twins) and 7 Chios multiparous ewes (4 singles and 3 twins) during an hour before parturition and the first hour after the birth of each lamb. Overall labour was not significantly different among the 3 breeds, although was shorter for first-born twin lambs compared to singletons ($P<0.05$). Chios ewes devoted significantly more time grooming their lambs in the immediate postnatal period compared to the Karagouniko ewes ($P<0.05$). As it was found, second-born twin lambs received less grooming attention than singles and first-born twins ($P<0.05$). In general, the birth of lamb stimulated intensive grooming attention for the first 30 min, followed by a gradual decline in the next 30 min period ($P<0.05$). Latency for lamb to stand, walk and reach the udder was not influenced by litter size and breed, although Chios tended to reach the udder later than the other lambs ($P=0.120$). The time standing during the first hour after birth was significantly higher in Karagouniko compared to Chios lambs ($P<0.05$). This study demonstrates that significant genotype differences exist among the examined breeds in several aspects of both maternal and neonate behavior. Ewe behavioural patterns associated with the birth and care of the neonatal lambs are essential for the survival and growth of the offspring. Further experimentation is therefore warranted to reach to reliable conclusions and elucidate the mechanisms that control maternal and neonatal behaviour in Greek sheep breeds.

---



**Sire selection for muscling improves the lightness and redness of lamb meat**

*H.B. Calnan[1,2], R.H. Jacob[1], D.W. Pethick[1,2] and G.E. Gardner[1,2]*
*[1]Cooperative Research Centre for Sheep Industry Innovation, Armidale, New South Wales, 2350, Australia, [2]Murdoch University, Murdoch, Western Australia, 6150, Australia; h.calnan@murdoch.edu.au*

Meat colour was measured in 7732 lambs produced at 8 sites across Australia over a 5 year period (2007-2011) and slaughtered in 125 groups as part of the Sheep Cooperative Research Centre's information nucleus flock experiment. Lambs were the progeny of sires of different types (merino, maternal and terminal) selected for a diverse range in Australian Sheep Breeding Values for post weaning eye muscle depth (PEMD). 24 hours post slaughter the m. longissimus was cut at the 12$^{th}$ rib, a probe was used to measure pH and the meat surface was allowed to bloom/oxygenate for 30 minutes before fresh colour measures of lightness ($L^*$), redness ($a^*$) and yellowness ($b^*$) were captured using a Minolta colorimeter. These measures were analysed in a multivariate analysis before least square means were produced for $L^*$, $a^*$ and $b^*$ using a linear mixed effects model in SAS. The base model included fixed effects for site, year of birth, slaughter group, sex and dam breed within sire type as well as random effects for sire and dam by year. In a second analysis sire PEMD estimates were included in the model as a covariate. Increasing sire PEMD across a range of -2 to 4 was associated with an increase ($P<0.01$) in the predicted means for $L^*$, $a^*$ and $b^*$ by 1.15, 0.38 and 0.55 units respectively. When pH measured 24 hours post-mortem was accounted for in the model, the impact of sire PEMD estimates on $L^*$, $a^*$ and $b^*$ were halved. This suggests that selection for sires based on PEMD may impact meat colour via changes in muscle metabolism that affect the extent of glycolysis and pH decline post slaughter. Our findings suggest that using sires with high PEMD will improve the lightness and redness of the meat produced.

## Effect of age on sensory scores of Australian Merino sheep meat

*L. Pannier[1], G.E. Gardner[1], A.J. Ball[2] and D.W. Pethick[1]*
*[1]Murdoch University, School of Veterinary and Life Sciences, Murdoch, 6150 WA, Australia, [2]University of New England, Meat & Livestock Australia, Armidale, 2351 NSW, Australia; l.pannier@murdoch.edu.au*

In addition to nutritional attributes and lean meat, eating quality is a key driver influencing the consumer demand for sheep meat. It is generally believed that meat from older animals reduces the overall consumer acceptance of meat products, however there are limited sensory studies available to define these differences within untrained consumers, or to identify factors that affect them. A lamb versus hogget comparison study was conducted, which was part of a larger sensory experiment within the Information Nucleus program of the CRC for Sheep Industry Innovation. This study tested genetic and non-genetic factors, and objective meat quality traits on sensory scores. We hypothesised that hogget meat will have lower sensory scores compared to lamb meat. Sensory scores were generated on the longissimus thoracis et lumborum (loin) and semimembranosus (topside) muscle from 189 Merino lambs, average age 355 days, and 209 Merino hoggets, average age 685 days. Five day aged grilled steaks were tasted and scored (1-100 score) by untrained consumers for tenderness, juiciness, flavour and overall liking. The difference in magnitude of sensory scores of the topside was greater than for the loin. Lamb topside had 8.4, 7.0, 3.3 and 5.8 more sensory scores for tenderness, overall liking, juiciness and flavour than hogget topside samples, whereas the lamb loin had only 4.7, 2.7, 1.6 and 2.2 more sensory scores for tenderness, overall liking, juiciness and flavour than hogget loin samples ($P<0.01$). Within each age group loins were more acceptable than topsides. In support of our hypothesis, older sheep had reduced sensory scores however these differences were minimal for the loin. These preliminary results highlight the better eating quality of lambs but show an acceptable eating quality of hogget meat, particularly for the loin, which opens the possibility of developing a high quality hogget product.

---



## A dual X-ray absorptiometer for estimating body composition in lamb abattoirs

*G.E. Gardner[1], R. Glendenning[2], R. Coatsworth[2] and A. Williams[1]*
*[1]Murdoch University, School of Veterinary and Life Sciences, South Street, Murdoch, 6150, Australia, [2]Scott Technology Ltd, 630 Kaikorai Road, Dunedin, New Zealand; g.gardner@murdoch.edu.au*

Scott Technology Ltd. have developed a robotic boning system for use in lamb abattoirs. This system operates at chain speed and makes use of 2D x-ray images taken of the carcass to identify cutting lines. The opportunity exists to replace this x-ray system with a dual energy x-ray absorptiometry (DEXA) system which meets robotic image requirements while also enabling the determination of body composition. This study tested a prototype DEXA design that we have developed with the hypothesis that this system could determine body composition at chain speed. 12 Merino lamb carcasses with weights ranging between 14 and 16.3 kg were scanned using the prototype DEXA system to derive an estimate of fat composition. The carcasses were then scanned using Computed Tomography (CT) to determine fat, lean and bone weights which were then expressed as a percentage of carcase weight. Composition of these carcases varied between 9-15.5% for CT fat, 68-72.5% for CT lean, and 16.2 to 20.7% for CT bone. The DEXA estimate of fat composition was then used to predict CT fat, lean, and bone percentage. Our findings supported the hypothesis as the DEXA system was able to predict carcase composition with a precision (R2) of 0.88, 0.34, and 0.60 for predicting the percentage of CT fat, CT lean and CT bone in the carcase. These results demonstrate the capacity of this prototype DEXA system to estimate body composition and lean meat yield in lamb carcasses at abattoir chain speed.

**Paper based point of care testing for detection of *Staphylococcus* in milk from sheep and cow**

*L. Chapaval[1], E. Carrilho[2], M.D. Rabelo[1], L.F. Zafalon[1], J. Oiano[1] and A.R.A. Nogueira[1]*
*[1]Embrapa Southeast Livestock, Animal Health, Molecular Microbiology, Rodivia Washington Luiz, km 234, P.O. Box 339, São Carlos, SP – 13560-970, Brazil, [2]Universidade de São Paulo, Departamento de Química e Física Molecular, Avenida Trabalhador São Carlense, 400 P.O. Box 780, São Carlos, SP – 13560-970, Brazil; lea.chapaval@embrapa.br*

Tests point-of- care test (POCT) of bacterial infectious agent's offers substantial benefits for the diagnosis of diseases, especially for reducing the time required to obtain results. Microfluidic biosensors based on lateral flow devices offer low cost and are highly sensitive to POCT. For the development of a multiplex immuno-specific biosensor for the detection of *Staphylococcus aureus* and two coagulase-negative Staphylococci (SCNs) from the milk of cows and sheep, functionalized gold nanoparticles conjugated antibodies will be used as antibacterial agents for signaling. The project is at baseline and, until the present time, these are the activities executed. Twenty milk samples were collected, from cows and sheeps, for the California Mastitis Test (CMT). In animals showed positive for CMT, 5 ml of milk collected aseptically, was sent for bacteriology. For detection and identification of *Staphylococcus* species, the procedures described Koneman (2007) were used. Confirmation of the identity of the microorganisms mostly found was made by PCR, using primers specific for *S. aureus* and *Staphylococcus epidermidis*. Growth curves were constructed for knowledge of physiological behavior of microorganisms to perform the extraction of different amounts of DNA for quantification through microchips. Cell quantification was estimated by spectrophotometer and plate count. This has been done to evaluate the sensitivity of detection of POCT. This project intends to establish the advantages of the POCT because it does not require any preprocessing of the sample and will be capable of detecting whole bacterial cells, will be detected visually, and provide a tool for milk producers.

---



**Ultrasound measurements of longissimus dorsi and subcutaneous fat in Teleorman Black head lambs**

*E. Ghita, C. Lazar, R. Pelms, M. Gras, T. Mihalcea and M. Ropota*
*National Research Development Institute for Animal Biology and Nutrition, Animal Biology, Calea Bucuresti nr. 1, Balotesti, Ilfov, 077015, Romania; elena.ghita@ibna.ro*

Romania is among the European countries with the highest stock of sheep and with a high potential of sheep meat export. In order to produce carcasses that are competitive on the European market, the Romanian sheep farmers must increase the proportion of meat and to decrease the proportion of bones and fat in the carcass. The purpose of our experiment was to evaluate the properties of longissimus dorsi muscle (depth, area and perimeter) and subcutaneous fat thickness on live animals using the ultrasound method on Teleorman Black Head lambs, a local breed. The studies were conducted on weaned lambs, 48 females and 25 males, aged 100 days, with the average body weight of 30.05 and 33.22 kg, respectively. The ultrasound measurements were performed with an Echo blaster 64 using LV 7.5 65/64 probe, supplied by TELEMED ultrasound medical systems. The ultrasound images were recorded using Echo Wave II software version 1.32/2009. The first measurement point was 5 cm from the spine, at the $12^{th}$ rib; the second measuring point was between $3^{rd}$ and $4^{th}$ lumbar vertebrae. The average values of the fat thickness, depth, area and perimeter of LD muscle were: 2.41 mm, 21.88 mm, 8.93 sq cm and 124.22 mm. The results show differences between the two measuring points, but they were not statistically significant ($P>0.05$). The measurements were also analysed by sex and significant ($P<0.05$) differences were noticed only for the depth of the muscle at the $12^{th}$ rib and for fat thickness between $3^{rd}$ and $4^{th}$ lumbar vertebrae. There were significant phenotypic correlations between the body weight and the ultrasound measurements, the correlation coefficients ranging between 0.348 and 0.782. These studies of ultrasound measurements on live animals will continue and the results will be used to select animals for meat production.

**Determining body condition score and milk somatic cell count of Turkish Saanen goats**

*S. Atasever, U. Sen and H. Önder*
*Ondokuz Mayis Üniversity, Department of Animal Science, Faculty of Agriculture, Samsun, 55139, Turkey; ugur.sen@omu.edu.tr*

The aim of the study was to evaluate body condition score (BCS) and somatic cell count (SCC) of milk in Turkish Saanen goats. Data were obtained from 95 goats during the morning milking. While a 1 to 5 scale was used to assess BCS, direct microscopy was performed to analyze SCC. Effect of stage of lactation (SL) and number of kids at birth (NK) were evaluated in three subgroups. While SL was not effected the parameters, goats with single kid had significantly higher BCS ($P<0.05$) than goats with twins. Correlation coefficient between evaluated parameters was estimated as weak ($r=0.08$). In conclusion, milk SCC and BCS levels of Turkish Saanen goats were assumed in acceptable thresholds.



**Prediction of pelvic area through body measurements in Dorper ewes**

*I.M. Van Rooyen and P.J. Fourie*
*Central University of Technology, Free State, Agriculture, 1 Park Road Bloemfontein, Free State, South Africa, Central University of Technology, Free State (CUT) Private Bag X20539, Bloemfontein, 9300, South Africa, South Africa; irooyen@cut.ac.za*

Birth stress or dystocia is associated with perinatal mortality, low lifetime rearing and amongst other a small pelvis area in ewes. It would thus make sense to include pelvic area as a criterion to select breeding ewes. The main objective of this study was to investigate and quantify the correlations between a number of easy to measure external linear body parameters (body height, shoulder height, chest depth, front quarter width, hindquarter width, rump length, etc.) and pelvic measurements (height, width and area) in Dorper ewes. The secondary objective was to investigate if the selection according to breed standards resulted in indirect selection for ewes with different pelvic areas. In this study, the pelvises, body weight and certain linear body measurements of 332 young Dorper ewes (±12 months old; 48.0±5.9 kg) were taken. Analyses of variance were conducted to compare means of different parameters using GLM procedure of SAS. Product moment correlations between the variables were also conducted. A stepwise regression was carried out to determine the individual effect of body measurements on pelvic area. An F to enter level of 0.10 was used to determine the significance of the partial contribution of each effect. Pelvic area was included as a covariate. The overall mean pelvic area of yearling Dorper ewes was $35.44\pm4.9$ $cm^2$. Stud ewes recorded significantly ($P<0.05$) higher pelvic areas ($37.38\pm4.3$ $cm^2$) than commercial ewes ($33.92\pm3.8$ $cm^2$). Results indicate no significant correlation between all the body measurements taken in this study and pelvic area. This indicates the need to measure the pelvic area of ewes directly. The pelvic meter and techniques specially developed for this study proved to be very useful to measure the pelvises of sheep, being very accurate and relatively easy to use in ewes.

## Session 25 Poster 17

**Effects of condensed tannins on weight gain and intake of boar goats in savannas**

*S.P. Dludla[1], L.E. Dziba[2], P.F. Scogings[1] and F.F. Fon[1]*
*[1]University of Zululand, The Department of Agriculture, Private bag x1001, 3886 KwaDlangezwa, South Africa, [2]The Council for Scientific and Industrial Research (CSIR), Department of Natural Science, P.O. Box 395, 0001 Pretoria, South Africa; spdludla@gmail.com*

Plant secondary compounds including tannins occur widely among plant species consumed by herbivores. Condensed tannins have both beneficial and detrimental effects on intake, digestion and growth of ruminants, depending on numerous factors. Domestic goats (Capra hircus) spend most of their time browsing compared to sheep and cattle. However, little is known about the effects of condensed tannins (CT) on goats' body weight. It was hypothesized that goats fed low CT concentration diets will lose less body weight and have higher feed intake than goats fed high CT concentration diets. Twenty four, 13-month old female Boer goats were randomly assigned to four groups of six goats and housed in individual pens ($1\times3$ $m^2$). All four groups were fed a standard diet but with different CT concentrations of respectively 0, 3, 5 and 10% CT and lucerne hay was provided to stimulate rumen functioning. All goats had free access to water through water nipples installed at the back of the pens. Experiments were conducted for four weeks after one week of conditioning. Goats were weighed daily every morning to monitor daily body weight gain. One-way ANOVA was used for statistical analysis. There was an increase ($P<0.05$) in the body weight of goats fed diets low in CT. It was observed that an increase in the condensed tannin concentration reduced ($P<0.05$) pellet intake. As the covariates, there was a significant effect of the pellet intake ($P<0.05$) and lucerne intake ($P<0.05$) on the body weight gain of goats. This suggests that an effect on body weight was the function of feed intake. It was concluded that different CT concentration affects body weight gain and feed intake. Digestibility studies are suggested for further research.

---

## Session 25 Poster 18

**The effect of a PMSG treatment on parturition time in Saanen does**

*U. Sen and H. Onder*
*Ondokuz Mayis University, Faculty of Agriculture, Animal Science, Samsun, 55139, Turkey; ugur.sen@omu.edu.tr*

The aim of this study was to determine the effect of a PMSG treatment with intravaginal progesterone application on parturition time in Saanen does. Time of estrus of 3-5 years old Saanen does was synchronized as follows; intravaginal sponges (30 mg flugestone acetate) for 11 days followed by an IM injection of 500 IU PMSG (group PP; n=43) or intravaginal sponges (group P n=44) and natural estrus (group control; n=65, C). During the kidding time, does were monitored at an hourly interval during all day. Parturition time was divided into 8 stages; Stage 1; 00:00-03:00, Stage 2; 03:00-06:00, Stage 3; 06:00-09:00, Stage 4; 09:00-12:00, Stage 5; 12:00-15:00, Stage 6; 15:00-18:00, Stage 7; 18:00-21:00, Stage 8; 21:00-00:00. Time of birth was recorded, which was defined as the time when the first kid had fully emerged from the vagina. Does in C group had lower parturition rate at stage 3 compare to does in PP and P groups ($P<0.05$), also does in PP group had lower parturition rate at stage 6 compare to does in PP and P groups ($P<0.05$). Births in all does showed a unimodal distribution and most part of parturition of does in PP, P and C occurred between 06:00-18:00 daylight hours (83.7, 86.4 and 81.5%, respectively), which is significantly different from other parturition times ($P<0.05$). These observations suggest that estrus synchronization may effects parturition time and kidding is concentrated during daylight hours in Saanen does.

## Faecal NIR to predict chemical composition of diets consumed by ewes

*N. Núñez-Sánchez, A.L. Martínez Marín, M. Pérez Hernández and D. Carrion*
*Universidad de Córdoba, Departamento Producción Animal, Carretera Madrid-Cádiz km 396, 14014, Córdoba, Spain; domingo.carrion@genusplc.com*

There is a great interest in an inexpensive and fast method to accurately predict the chemical composition of the diet consumed by an animal with free access to several feed ingredients. The aim of this study was to use Near Infrared Reflectance Spectroscopy (NIRS) to directly predict chemical composition and forage content of diets consumed by ewes from the analysis of their faeces. Twenty four ewes (46 kg BW) at maintenance state placed in individual cages were used. Each animal was randomly allocated to four diets fed in four successive 28 days periods. A total of 16 diets were fed (six ewes per diet). Diets were designed to have different chemical composition. Daily rations of 800 g (as-fed) were formulated using alfalfa (0-100% DM) or straw (0-80% DM) as forage base, and different mixtures of maize, peas and lupines, offered as whole grains, and pelleted sunflower meal. Composition range (DM basis) was: Ash, 3.2-10.5%, crude protein, 8.8-17.5%; neutral detergent fibre, 18.1-72.9%. The rations were completely consumed. Faeces samples were collected at the end of each feeding period, directly from the rectum. A total of 96 faeces samples were dried and ground. The samples were scanned in small ring cups in a FOSS-NIRSystems 6500 SY-II scanning monochromator. Faecal Modified Partial Least Squares calibration equations, selected in terms of standard error of cross validation (SECV) and coefficient of determination ($r^2$), showed good predictive values for the prediction of alfalfa (3.11 and 0.99), straw (3.53 and 0.99), ash (0.44 and 0.95), crude protein (0.95 and 0.89) and neutral detergent fibre (3.01 and 0.96) of the diets. These results support the viability of Faecal NIRS as a fast and reliable analytical method which allows accurately predict the forage and chemical composition of diets consumed by ewes having free access to a variety of feed ingredients.

---



## Influence of dietary extruded linseed on goat cheese fatty acid profile

*M. Renna, C. Lussiana, A. Mimosi and R. Fortina*
*University of Torino, Department of Agricultural, Forest and Food Sciences, via Leonardo da Vinci 44, 10095 Grugliasco (TO), Italy; manuela.renna@unito.it*

An experiment was conducted to evaluate the effect of dietary extruded linseed (EL) on the fatty acid (FA) profile of goat cheese. Twenty-six Saanen goats in mid lactation were divided into two balanced groups and fed for seventy days a 40:60 forage:concentrate total mixed ration supplemented (EL group) or not (C group, control) with 3 g/100 g diet of rumen unprotected EL. Separate bulk milk tanks (one per treatment) were used to produce twenty fresh and twenty 30-days ripened cheeses, which were analysed for the FA composition by gas chromatography. Data were subjected to a two-way analysis of variance to evaluate the effect of diet, ripening, and their interaction on cheese FA profile. Samples of EL cheeses showed significantly higher levels of caprylic (+7.3%; $P<0.05$), capric (+11.1%; $P<0.01$) and stearic acids (+20.6%; $P<0.001$), total trans-octadecenoic acids (+17.8%; $P<0.05$), the majority of non-conjugated octadecadienoic acids, and α-linolenic acid (+36.8%; $P<0.001$) than C samples. Hypercholesterolemic saturated FA (sum of lauric, myristic, and palmitic acids; -4.6%; $P<0.05$), total conjugated linoleic acid isomers (-14.9%; $P<0.05$), linoleic acid (-6.0%; $P<0.05$), and the omega 6/omega 3 FA ratio (-17.2%; $P<0.001$) were significantly lower in EL than C samples. Total polyunsaturated and very long-chain omega 3 FA were not significantly different between EL and C samples. Ripening did not exert significant modifications of the cheese FA profile, and the interaction between diet and ripening was not statistically significant for any of the detected parameters. We concluded that extruded linseed supplementation is a reliable feeding strategy to lower the hypercholesterolemic saturated FA and the omega 6/omega 3 FA ratio in goat dairy products.

**Effect of glycerol on *in vitro* fermentation parameters and digestibility**

*E.H.C.B. Van Cleef[1], J.M.B. Ezequiel[1], A.P. D'Aurea[1], M.T.C. Almeida[1], H.L. Perez[1], J.R. Paschoaloto[1], F.O. Scarpino-Van Cleef[1] and F.B. Costa[2]*
*[1]UNESP, Department of Animal Science, Jaboticabal, SP, 14884-900, Brazil, [2]TECTRON, Toledo, PR, 85908-220, Brazil; ericvancleef@gmail.com*

Two studies were conducted to evaluate effects of glycerol on *in vitro* fermentation and digestibility. Ruminal fluid was collected from 2 ruminally canulated sheep fed diet composed of 40% corn silage, 30% corn, 5.5% soybean hulls, 23% soybean meal, 0.55% urea, and 1% mineral premix. Buffered ruminal fluid (150 ml) was placed into 250 ml plastic flasks equipped with special caps with tubing, which conducted the fermentative gas to a storage graduated container. Diets containing 0, 10, 20, or 30% dry glycerol product containing 68% food-grade glycerol were added at 1 g DM/flask. Gas production, pH and DM disappearance were determined after 24 h of incubation at 39 °C. In Study 2, IVDMD was measured. Twenty-five replicates of each diet were weighed into ANKOM F57 bags, and incubated in an ANKOM DAISY$^{II}$ fermenter for 48 h, at 39 °C. After that, 40-ml of 6N HCl and 8 g of pepsin was added to each digestion vessel, and incubated for another 24 h. Both studies were arranged as completely randomized design with each flask or bag being the experimental units. Data were analyzed using the GLM procedure of SAS, and orthogonal contrasts were used to determine linear, quadratic, and cubic effects of glycerol. The addition of glycerol increased *in vitro* gas production (linear, P=0.008; cubic, P=0.003) in up to 14% for treatment with 20% glycerol, and 7% for treatment with 30% glycerol. The pH was linearly increased (P=0.013) from 5.82 to 5.97 (control diet and 30% glycerol diet, respectively). Differences in dry matter ruminal disappearance were negligible among treatments, presenting an average of 66.7%. The glycerol product linearly increased diets IVDMD in up to 4.4% (P=0.004). Feeding up to 30% glycerol product to fermentation systems showed it is a good alternative energy source, increasing microorganisms' activity and improving the IVDMD.

---



**Estimation of some meat quality parameters using artificial neural networks**

*U. Sen, O. Gulboy and H. Onder*
*Ondokuz Mayis University, Faculty of Agriculture, Animal Science, Samsun, 55139, Turkey; honder@omu.edu.tr*

The aim of the study was to estimate some meat quality parameters; drip loss (DL), cooking loss (CL), shear force (SF) and color characteristics (L, a, b) with explanatory variables of intramuscular fat and pH values of longissimus dorsi (LD) muscle from 64 lambs, which were slaughtered at 150 days of age, by using Artificial Neural Networks (ANN) model with multi-layer perceptron. The results of this study showed that DL, CL and SF values of LD muscle were reliably estimated with coefficient of determination values of 0.81, 0.94 and 0.84, respectively by using pH and intramuscular fat values with ANN models. However L, a and b color characteristics were estimated insignificantly with coefficient of determination values of 0.28, 0.55 and 0.31, respectively. As conclude, intramuscular fat and pH values may use for estimation of DL, CL and SF, except for color characteristics, with ANN models.

**Deuterium enrichment in body water of sheep after intravenous or intraruminal administration of D2O**

*C.C. Metges[1], S. Goers[1], H.M. Hammon[1], U. Agarwal[2] and B.J. Bequette[2]*
*[1]Leibniz Institute for Farm Animal Biology (FBN), Institute of Nutritional Physiology, Wilhelm-Stahl-Allee 2, 18196 Dummerstorf, Germany, [2]University of Maryland, Department of Animal and Avian Sciences, 4147 Animal Sciences Center, College Park, Maryland 20742-2311, USA; metges@fbn-dummerstorf.de*

The $D_2O$ method to measure fractional gluconeogenesis (GNG) in monogastric species is considered the gold standard because it includes all gluconeogenic precursors. In ruminants, we investigated whether the route of $D_2O$ administration affects equilibration of deuterium with protons from water in various body water pools. Four sheep (23.5±1 kg BW), equipped with a rumen fistula and a jugular vein catheter, were fed a pelleted ration (35 g/kg BW×d; 9 MJ ME/d) at 2-h intervals. Two boli of 7 g $D_2O$ (99.2 atom% (AP) D/kg BW) were given at 8.00 and 12.00 h via the rumen (IR) or jugular vein (IV) in a cross-over design separated by 2 wk. Plasma was sampled prior to and at 1-h intervals for 11 h after the first bolus. Rumen fluid and urine were collected prior to and at 3, 6, 9 and 11 h. D enrichments were measured by isotope ratio mass spectrometry. Paired t-test was used to evaluate route effect. Plasma D enrichments did not differ with route of $D_2O$ administration. A plateau was reached 2 h after the first bolus (IR: 0.69; IV: 0.71 AP excess (APE); P>0.1) with a further increase to a second plateau at 1400 h (IR: 1.46; IV: 1.48 APE; P>0.1). Urine D enrichment 3 h after the initial IR dose tended (P=0.09) to be lower than with the IV route (IR: 0.47; IV: 0.78 APE). Both routes lead to a similar maximum urinary enrichment 9 to 11 h after the first bolus (IR: 1.47; IV: 1.45 APE; P>0.1). Rumen fluid D enrichment attained a plateau 6 h after the first bolus (IR: 1.51; IV: 1.43 APE; P>0.1). For measurement of fractional GNG using the $D_2O$ method, the D labelling of body water pools was similar with the IR and IV routes. Further measurement of D labelling of gluconeogenic precursors and glucose will verify whether both routes of administration yield similar estimates of GNG in ruminants.

---



**Lamb birth weight and beta-hydroxybutyrate concentration in ewes fed diet containing crude glycerin**

*D.M. Polizel[1], R.S. Gentil[1], E.M. Ferreira[1], R.A. Souza[1], A.P.A. Freire[1], F.L.M. Silva[1], M.C.A. Sucupira[2] and I. Susin[1]*
*[1]ESALQ/University of São Paulo (USP), Av. Padua Dias, 11, Piracicaba, SP, Brazil, [2]FMVZ/USP, Av. Prof. Orlando Marques de Paiva, 87, São Paulo, SP, Brazil; ivasusin@usp.br*

Crude glycerin is a co-product from biodiesel production and glycerol is its main component. Glycerol is a glyconeogenic substrate for ruminant and can improve energy balance of animals during high nutritional requirement periods. The objective in this trial was to determine the effects of partial replacement of corn by crude glycerin (CG) on lamb birth weight, beta-hydroxybutyrate (BHB) concentration, body weight (BW) and body condition score (BCS) in ewes. One hundred and eighteen, 90 days pregnant, Santa Ines ewes were fed either a control diet (no CG added) or a 10% CG (83.6% glycerol, DM basis) diet, corresponding to G0 and G10, respectively. Diets were isonitrogenous (13.0±0.3% CP, DM basis) and composed by 70% concentrate and 30% raw sugar cane bagasse. At parturition, ewes were weighted and BCS was recorded subjectively using a scale ranged from 1 to 5. To determine BHB concentration, 32 ewes were used. Blood samples were obtained at -14, -7, 0, 7, 14 and 28 days relative to lambing, 2 hours after feeding. Data were analyzed using the MIXED procedure of SAS and BHB concentration was determined as repeated measures over time. The LSMEANS option was used to obtain the means. Crude glycerin did not affect (P>0.05) lamb's birth weight (3.86 vs 3.87 kg for G0 and G10, respectively), ewe's body weight (66.4 vs 65.9 kg) and body condition score (2.9 vs 2.8) at parturition. There was an interaction between sampling day and diet. Ewes fed the diet with CG had decreased (P=0.02) BHB concentration at lambing (0.405 vs 0.290 mmol/l) with no difference in the remaining days. Crude glycerin can replace 10% of corn in the diet of pregnant ewes with no detrimental effect on lamb's birth weight and ewes BCS or BW. Adding 10% of CG may decrease ewe's fat mobilization at parturition.

**Feeding silage from olive cake improves milk fatty acid composition in sheep milk**

*O. Tzamaloukas, M. Orford, D. Miltiadou, M. Hadjipanayiotou and C. Papachristoforou*
*Cyprus University of Technology, Department of Agricultural Sciences, Biotechnology and Food Science, P.O. Box 50329, 3603 Limassol, Cyprus; ouranios.tzamaloukas@cut.ac.cy*

The present study examined the effects of feeding dairy ewes a diet supplemented with ensiled olive cake on milk fatty acid (FA) content. A mixture of skins, pulp, woody endocarp and seeds obtained after extraction of oil (crude olive cake) was ensiled according to a method developed earlier. Sixty second-parity Chios ewes in their mid-lactation were selected at random to form three balanced groups. Animals were housed indoors and allocated to the following feeding regimes: (1) no inclusion of ensiled olive cake (NO-group), (2) inclusion of 500 g/day/ewe (OC-group) and (3) inclusion of 1000 g /day/ewe (OOC-group). Total rations offered were iso-nitrogenous and iso-energetic to meet 1.1× maintenance energy and milk production requirements. Milk samples were collected from each ewe during the morning milking after 3 and 4 weeks on the different diets. Each sample was analysed for FA content by applying a modified GC-MS method. The effect of the inclusion of olive cake silage was tested using a general linear model for repeated measurements. Mono-unsaturated FAs, and particularly oleic acid, was significantly increased ($P<0.001$) in milk from both OC and OOC animals. Expressed in g/100 g of fat, oleic acid was 20.2, 25.5, 26.5 in NO, OC and OOC group, respectively, at week 3. Poly-unsaturated FAs were also increased significantly when ensiled olive cake was included in the diet. Conjugated linoleic acid showed a similar trend with significantly higher ($P<0.001$) concentration in OC and OOC milk (0.45, 0.60 and 0.62 g/100 g milk fat from NO, OC and OOC groups, respectively, at week 3). Similar results were obtained during week 4. Overall, the inclusion of olive cake silage in diets significantly increased beneficial unsaturated FAs in ovine milk.

---



**Covariance components for milk traits using multi-trait regression model for Alpine goat in Croatia**

*M. Špehar, D. Mulc, D. Jurković and Z. Barać*
*Croatian Agricultural Agency, Ilica 101, 10000 Zagreb, Croatia; mspehar@hpa.hr*

The objective of this study was to estimate covariance components for daily milk yield (DMY), fat (FC), and protein (PC) content for Alpine goat in Croatia. Data included 267,773 test-day records for 14,820 does recorded from 1998 to 2014. Pedigree file included 20,468 animals. Test-day records were modelled using multi-trait fixed regression repeatability test-day model. Fixed class effects in the model were: parity, litter size as a number of born lambs, year and month of kidding. Days in milk was treated as covariate. The effect was fitted using the Legendre polynomial of order 4 and nested within parity and number of lambs born. Direct additive genetic effect, flock-test-day, and permanent environment effect within lactations were included in the model as random effects. Covariance components were estimated using Residual Maximum Likelihood as implemented in the VCE-6 program. Heritability estimates were 0.33±0.02, 0.18±0.01, and 0.27±0.01, for DMY, FC, and PC, respectively. Genetic correlations were −0.26±0.04 between DMY and FC, −0.27±0.04 between DMY and PC, and 0.59±0.04 between FC and PC. Random flock-test-day determined between 25 and 36% of phenotypic variance. Permanent environment effect obtained another 19% of phenotypic variation for DMY, while lower proportion (4 and 6%) was explained for FC and PC. Results indicate possibility of using multi-trait test-day model for genetic evaluation of the Alpine goat instead of currently used single test-day model due to improved accuracy of evaluation by accounting genetic correlations among traits.

**Fatty acid composition in intramuscular and subcutaneous fat of light carcass lambs**

*M. Margetín[1,2], M. Oravcová[2], J. Tomka[2], P. Polák[2], D. Peškovičová[2] and Z. Horečná[1]*
*[1]Slovak University of Agriculture Nitra, Tr. Andreja Hlinku 2, 949 76 Nitra, Slovak Republic, [2]National Agricultural and Food Centre, Hlohovecká 2, 951 41 Lužianky, Slovak Republic; margetin@cvzv.sk*

Fatty acids profile (totally 69 FAs) of intramuscular fat (IMF; Musculus longissimus lumborum et thoracis) and subcutaneous fat (SCF; root of the tail) defined in 80 light carcass lambs of both sexes (52 ram-lambs; 28 ewe-lambs) reared on ewes` milk and on milk replacer was determined by gas chromatography. Differences among individual FAs or FAs groups in dependence on the fat type we detected by means of multifactorial ANOVA. The average empty live weight of lambs before slaughter was 18.1±2.09 kg and average age of lambs 58±6.0 days. Content of LA (8.30 g/100 g FAME), GLA (0.075), ALA (0.522), AA (2.52), EPA (0.290), DPA (0.496) and DHA (0.174 g/100 g FAME) was in IMF significantly higher ($P<0.001$) than in SCF (5.35, 0.039, 0.402, 0.180, 0.015, 0.107, 0.026 g/100 g FAME). Conversely, the content of SFA (C12:0, C14:0, C16:0) and also C18:1 trans9, C18:1 cis9, tVA and RA was in SCF significantly higher than in IMF ($P<0.001$). We found highly significant effect of fat type on MUFA, PUFA, cis-MUFA, n-6 and n-3 PUFA, essential FAs and ratio of n-6/n-3 and LA/ALA ($P<0.001$). The content of PUFA, n-6 and n-3 FAs was in IMF two times higher (14.5, 11.1, 1,72) than in SCF (7.7, 5.6 and 0.61 g/100 g FAME; $P<0.001$). The content of CLA in IMF (0.489 g/100 g FAME) and SCF (0.514 g/100 g FAME) was similar ($P>0.05$). The values of atherogenic index and thrombogenic index had been significantly higher in SCF (1.15 and 1.70) as in IMF (0.87 and 1.31; $P<0.001$). From dietary point of view composition of IMF is better than in SCF. This work was supported by the Slovak Research and Development agency under contract No. APVV-0458-10.



**Plasma antioxidant stability of Churra ewes supplemented with grape pomace**

*C. Guerra-Rivas[1], B. Gallardo[1], A.R. Mantecón[2] and T. Manso[1]*
*[1]Universidad de Valladolid, Departamento de Ciencias Agroforestales, Avd. Madrid s/n, 34004 Palencia, Spain, [2]Instituto de Ganadería de Montaña (CSIC-ULE), Finca Marzanas, 24346 Grulleros (León), Spain; tmanso@agro.uva.es*

The effects of natural sources of polyphenols, such as by-products from wine industry, to prevent lipoperoxidation *in vivo* and preserve animal health and product quality has a great interest. The aim of this study was to evaluate the chemical composition, including secondary metabolites, of grape pomace from red wine and its effect on plasma oxidation stability. Eight healthy non-pregnant, non-lactating Churra ewes housed in individual pens were divided, on the basis of LBW, into two dietary treatments (n=4): control group (CTL, without grape pomace) and GP group (GP, with 7.5% of GP from red wine). Ewes were fed at maintenance level (45 g DM per kg LW0,75) with a 50:50 lucerne:concentrate TMR and the diet was offered at 8:00 h and 17:00 h. After 20 days of adaptation period to diets, ewes were blood sampled by jugular venipuncture in the morning at 0, 3, 6 and 9 h after feeding. Lipid peroxidation was analysed in plasma samples using the thiobarbituric acid-reactive substances (TBARS) method. Data were subjected to ANOVA using the GLM procedure of SAS. Dry matter content of grape pomace was 41.1% and NDF, ADF, CP and EE content (g/kg DM) was 376, 317, 119 and 73 respectively. The extractable polyphenols, hidrolyzable polyphenols, condensated tannins and total anthocyanins content of grape pomace (g/kg DM) was 44.1, 8.16, 44.5 and 3.47 respectively. Plasma lipid oxidation (TBARS) was not significantly influenced by GP supplementation ($P>0.05$) but was lower in GP treatment than CTL treatment at 3 h (1.84 vs 1.45 μM MDA), 6 h (1.96 vs 1.47 μM MDA) and 9 h (2.31 vs 1.74 μM MDA) after feeding. On the basis of these results, we conclude that grape pomace could be an attractive antioxidant for ruminant nutrition and products quality. Further experiments are necessary to verify the GP antioxidant capacity.

**Grape pomace in early lactating ewes: milk performance and oxidative stability of lambs meat**

*C. Guerra-Rivas[1], B. Gallardo[1], P. Lavín[2], A.R. Mantecón[2], C. Vieira[3] and T. Manso[1]*
*[1]Universidad de Valladolid, Ciencias Agroforestales, Avd. Madrid s/n, 34004 Palencia, Spain, [2]Instituto de ganaderia de montaña, Grulleros (León), 24346, Spain, [3]ITACYL, Estación Tecnológica de la Carne, 37770 Guijuelo, Salamanca, Spain; tmanso@agro.uva.es*

The aim of this study was to evaluate the effects of different levels of grape pomace and other commonly used antioxidant (vitamin E) in n-3 PUFA diets fed to early lactating ewes on milk performance and oxidative stability (TBARS) of plasma lipids and suckling lambs meat. After lambing, forty eight Churra ewes were fed daily with a TMR containing a 2.7% (as-fed basis) of linseed oil and lucerne and concentrate at a 40:60 ratio. Ewes were assigned to four dietary treatments: CTL (without grape pomace), VIT E (with 500 mg per kg TMR of vitamin E), 50 GP (50 g per kg TMR of grape pomace) and 100 GP (100 g per kg TMR of grape pomace). Intake and milk yield and composition were recorded weekly during the first month of lactation. At 11 kg LW, suckling lambs were slaughtered and samples were taken from M. longissimus. TBARS were measured in samples of plasma of each ewe and in suckling lambs meat 7 days after slaughter. Data of intake and TBARS were subjected to ANOVA using the GLM procedure and milk production and composition using the MIXED procedure of SAS. Intake, milk yield and milk composition (fat and protein percentage) were not different ($P>0.05$) between treatments. TBARS values in plasma of ewes were reduced by 24% in VIT E and 80 GP and 12% in 40 GP group when they are compared to the CTL diet, however, differences between treatments were not statistically significant ($P>0.05$). With reference to oxidative stability of suckling lambs meat, VIT E, 50 GP and 100 GP treatments decreased TBARS values ($P<0.05$) and no statistical differences ($P>0.05$) were found between VIT E and treatments with grape pomace. It can be concluded that grape pomace could be a suitable way to improve the oxidative stability of suckling lambs meat without detrimental effects on milk performance.



**Kinetics of lithium chloride excretion in conditioned averted dairy goats**

*C.L. Manuelian, E. Albanell, M. Rovai, A.A.K. Salama, G. Caja and R. Guitart*
*Universitat Autònoma de Barcelona, Animal and Food Science, Group of Ruminant Research (G2R), Campus universitari, Edifici V, 08193 Bellaterra, Spain; gerardo.caja@uab.es*

Lithium chloride (LiCl) activates the vomiting centre inducing conditioned food aversion (CTA) to paired foods. CTA treated livestock is used to graze weeds and green covers on crops (i.e. vineyards) in many countries, although LiCl is not approved in the EU. There is no information on LiCl excretion at the effective dose for inducing CTA in small ruminants, which is a key for regulation. We studied the Li elimination in 6 dairy goats at late lactation after dosing 200 mg LiCl/kg BW. Goats were in metabolic cages where intake, feces, urine and milk were measured, and blood sampled, for168 h. Li was analized by graphite furnace Atomic Absorption Spectrometry and data corrected by basal concentration. Withdrawal period (Wp) was calculated according to the European Medicines Agency recommendations (95% tolerance limit). Plasmatic Li reached a peak at 4 h (14.5±0.8 mg Li/l plasma) with an individual biological half-life of 40.3±3.8 h. All Li administered was fully recovered at 96 h, mainly by urine (92.4±4.4%); however, the estimated Wp calculated for faeces was 11 d. After LiCl administration, goats decreased water intake and feed intake, which were fully recovered after 1 and 2 d, respectively. Nevertheless, milk yield did not vary. These results support the moderate gastrointestinal discomfort reported by after dosing LiCl. In conclusion, Li was fully eliminated in dairy goats following a similar pattern to monogastrics, although for a complete elimination, it took longer in goats. Additionally, the concentration of Li in milk was very low and estimated not enough to induce CTA in suckling kids. The previously reported benefits of grazing green covers in crops with CTA induced small ruminants and the complete elimination results observed in the present study, let us to suggest to the EU regulatory agencies to consider the authorization of LiCl for this practice in livestock.

**B and T-cell epitopes prediction for *Brucella melitensis* OMP25 antigen**

*S. Yousefi, M. Tahmoorespour, M.H. Sekhavati, T. Abbassi Daloii and A.A. Khabiri*
*Ferdowsi University of Mashhad, Animal Science, Mashhad, Khorasan Razavi, 9177948974, Iran; soheil.yousefi01@yahoo.com*

Brucellosis is a disease affecting various domestic which is caused by a bacterium *Brucella*. Du to serious economic and medical consequences ofthis disease various efforts have been made to prevent the infection through the use of recombinant vaccines based on *Brucella* outer membrane protein (OMP) antigens. The objective of this study was to obtain the bioinformatics characteristics of OMP25 antigen as a major *Brucella* OMP antigenes. Up to this, online prediction software applications were used such as: B and T-cells epitopes prediction, secondary and tertiary structures, 3D structure, antigenicity ability and enzymatic degradation sites. The bioinformatics approached used in the present study was confirmed by the results of four different experimental epitope predictions. Bioinformatics analysis identified B-cell epitopes locations at 26-44, 59-79, 88-112, 146-166 and 175-202amino acid residual, whereas T-cell epitopeswas determined at 1-10, 14-22, 122-132, 154-162 and 206-213 positions. All final B and T-cell predicted epitopes had antigenicity ability expect of 1-10 and 14-22 residuals. Finally, a common B and T-cell epitope was identified (154-162 amino acids) for OMP25 antigen. Bioinformatics analysis showed that this region had proper epitope characterization and so it seems to be useful for developing of vaccine based epitope design. *In vitro* synthesis of determined peptides and experimental validation are essential for using predicted epitope as an effective vaccine against *Brucella* pathogen, in this regards our laboratory has already initiated research in this direction.

---



**Effect of the vitamin E supplementation prior to slaughter on plasma metabolites in light lambs**

*L. Gonzalez-Calvo[1], M. Blanco[1], F. Molino[1], J.H. Calvo[2] and M. Joy[1]*
*[1]CITA de Aragon, Avda. Montañana, 930, 50059-Zaragoza, Spain, [2]ARAID Aragon, Avda. Montañana, 930, 50059-Zaragoza, Spain; mjoy@aragon.es*

Dietary vitamin E (VE) supplementation has been recommended to increase meat shelf life. Alpha-tocopherol, the most potent form of vitamin E, is a major free-radical-trapping antioxidant in plasma and tissues that attenuates the oxidative stress and decreases the formation of low-density lipoproteins (LDLs). However, it is expensive which requires accurate feeding to reduce the period of α-tocopherol supplementation. Grazing is a cheap option to increase α-tocopherol content in the muscle and reduce the oxidation processes. Single reared male lambs (n=54) of Rasa Aragonesa breed were weaned at 45 days of age and fed a basal concentrate (C; 30 mg α-tocopheryl acetate/kg of concentrate) and a supplemented concentrate (500 mg α-tocopheryl acetate/kg of basal concentrate) for 0 (C), 10 (VE10d), 20 (VE20d) and 30 (VE30d) days before slaughter (23±1.4 kg; 75±1.3 days). Additionally, 8 unweaned lambs were continuously stocked on alfalfa pasture supplemented with the basal concentrate (ALF). Jugular blood samples were collected at before slaughter into EDTA vacuum tubes to determine α-tocopherol, cholesterol, high density lipoproteins (HDL), LDL and triglycerides (TG) concentrations. Plasmatic α-tocopherol concentration increased with VE supplementation ($P<0.05$), regardless the length of the feeding period. ALF lambs had intermediate plasmatic α-tocopherol concentration between C and VE supplemented lambs. Cholesterol, HDL and LDL were lower in VE30d than in C lambs ($P<0.01$). ALF lambs had greater cholesterol, HDL, LDL and TG concentrations than the concentrate-fed lambs, because lambs suckled milk. In summary, VE supplementation increased α-tocopherol plasmatic concentration but 30 days were needed to decrease plasmatic LDL, HDL and cholesterol concentrations. In unweaned light lambs, grazing increased α-tocopherol concentration in plasma, but milk intake had a greater effect than forage intake or VE supplementation on plasmatic metabolites.

## Session 25 Poster 33

**Influence of birth weight and feeding program on growth performance of Baladi kids fed milk replacer**

*R. El Balaa[1], P. Aad[2] and S. Abi Saab[3]*
*[1]Balamand University, IUT-IF, Koura, P.O. Box 100 Tripoli, Lebanon, [2]NDU, Zouk, Zouk, Lebanon, [3]USEK, Kaslik, Kaslik, Lebanon; rodrigue.elbalaa@balamand.edu.lb*

The purpose of this study is to assess the influence of birth weight and different feeding programs on the growth performance of Baladi goat kids, in an intensive rearing system. Twenty eight Baladi goat kids were used in the experiment, they were divided into three groups, the first one (GI) was considered as the control group, containing eleven animals of heavy birth weight (HBW), the second group (GII) containing eleven individuals of light birth weight (LEW) and the third group (GIII) was formed by six individuals of (HBW). Two feeding programs were used: Program A based on Providing GI and GII a constant concentration of milk replacer of 230 g/l during the experiment, while program B consisted on feeding milk replacer of increasing concentration (from 180 g/l to 230 g/l j during the first 10 days, and a constant concentration of 230 g/l until the end of the experiment to GIII. All animals were slaughtered at an age of 50 days. The body measurements recorded during the experiment allowed the elaboration of the following formula to estimate kids' body weight based on the body length and heart girth: BW (kg) = 0.198 × BL (cm) + 0.141 × HG (cm) – 6.674 GII showed the lowest Feed Conversion Rate (FCR) 1.8±0.6, which means that it is the most economical rearing system. Slaughter weight showed significant difference ($P<0.05$) between GI and GIII, it was higher for GI with 8.2±1 kg and 6.2±1.5 kg for GII. Carcass weights showed significant difference ($P<0.05$) between GI and GII, it was higher for GI with 3.8±0.6 kg and 2.9±0.8 kg for GII The most economical rearing system using milk replacer turned out to be that of light birth weight (LBW) kids fed a constant concentration of 230 g/l. Although the success of this study opens the door for an interesting breeding system; however, further research on its economic feasibility will follow to assert the optimal application of this breeding method.

---

## Session 25 Poster 34

**Construction of two-promoter vector for co-expression of Omp25 and Omp31 antigens of *B. melitensis***

*M.H. Sekhavati, M. Tahmoorespour, T. Abbassi-Daloii, S. Yousefi, A.A. Khabiri, R. Akbari and M. Azghandi*
*Ferdowsi University of Mashhad, Animal Science, Mashhad, Khorasan Razavi, 9177948974, Iran; soheil.yousefi01@yahoo.com*

Brucellosis, also known as undulant fever or Mediterranean fever, is caused by *Brucella* bacteria and is still one of the major, highly transmissible and zoonotic diseases affecting people and animals. Economic and medical consequences of brucellosis have led to efforts to prevent the infection through the use of vaccines. It is demonstrated that Omp25 and Omp31 are main antigens that causing the immune response. Moreover, these antigens were found to be the most exposed outer membrane proteins (OMPs). Also, there are several reports which have shown that DNA vaccine and recombinant protein vaccine of Omp25 and Omp31 are protective against the virulent challenge of *Brucella* species in mice. Therefore, these antigens are the best candidates for vaccine development against sheep brucellosis. In this study, in order to co-expression of these antigens the vector with two-promoter was planned. So, two pair primers that contain T7 promoter, lac operator, ribosome binding site and multiple restriction sites were designed. These megaprimers were used to amplify pET-32a(+) vector as template using SOE-PCR protocol. The PCR products contain two junk sequences that considered to substituting with Omp25 and Omp31 using enzyme digestion. To verify the reaction, the construct cloned and then sequenced. The Omp25 and Omp31 genes were cloned separately and then sequenced. Their sequences were submitted in NCBI with accession number KJ193850 and KJ193851, respectively. Thereafter these genes cloned in two-promoter construct. Colony PCR and enzyme digestion were performed to verify the accuracy of final construct with about 1500 bp that contain these antigens. The co-expression of Omp25 and Omp31 is on process.

**Stage of lactation affects the fatty acid profile of Chios sheep milk**

*O. Tzamaloukas, M. Orford, D. Miltiadou and C. Papachristoforou*
*Cyprus University of Technology, Department of Agricultural Sciences, Biotechnology and Food Science, 30 Arch. Kyprianou str, 3036 Lemesos, Cyprus; c.papachristoforou@cut.ac.cy*

The present study examined possible variation in fatty acid (FA) composition of milk, in association with stage of lactation in the Chios sheep breed. Fifteen second parity Chios ewes were used from the start until week 20 of lactation. Animals were housed indoors and group fed the same diet to meet 1.1× maintenance energy (0.401 MJ/kg weight$^{0.73}$) and milk production requirements. From each ewe, milk samples collected during the morning milking (weeks 1, 4, 8, 12, 16 and 20 of lactation), were analysed for fat, protein, lactose and solids non-fat using established ISO methods, and for FA content, by applying a modified GC-MS method. The effect of the stage of lactation on milk constituents was tested using a general linear model for repeated measurements. Results showed that stage of lactation significantly affected the FA composition of milk ($P<0.001$). De novo synthesized FA, particularly C4:0 to C12:0, were significantly affected showing a similar trend, with an increase during early lactation (peak at week 4) and a decrease thereafter. Stage of lactation also affected most of the saturated and unsaturated long chain FA (≥16 carbons). Expressed in g/100 g of fat, the content of long saturated FA (sum of C16 to C24) was low during early lactation (36.1 and 31.8 in weeks 1 and 4, respectively) and increased thereafter (41.9 and 42.8 in weeks 16 and 20, respectively). Poly-unsaturated FA (PUFA) content was higher in early lactation (4.6 and 5.1 in weeks 1 and 4, respectively), and decreased over time (3.4 and 2.9 in weeks 16 and 20, respectively). Conjugated linoleic acid (CLA) showed a similar trend to other PUFA with high content at the start of lactation (0.61, 0.77 in weeks 1 and 4 respectively) and decreasing thereafter (0.41, 0.24 in weeks 16 and 20, respectively). Depending on FA type, stage of lactation differentially affected the milk FA content of Chios ewes on a standard diet.

---



**Digestibility and microbial protein synthesis of sheep treated with *Bacillus thuringiensis***

*F.C. Campos[1], P.P. Santos[1], P.S. Corrêa[2], R. Monnerat[3], C.M. Macmanus[4], A.L. Abdalla[1] and H. Louvandini[1]*
*[1]CENA/USP, LANA, Av Centenário 303, 13400-970, Piracicaba, São Paulo, Brazil, [2]CENA/USP, CMBL/GRASP, Av Centenário 303, 13400-970, Piracicaba, São Paulo, Brazil, [3]EMBRAPA, CENARGEN, Av. W5 Norte, 70770-917, Brasilia, DF, Brazil, [4]UnB, FAV, Asa Norte, 70910-900, Brasilia, DF, Brazil; louvandini@cena.usp.br*

The use of *Bacillus thuringiensis* (Bt) has been studied because of its possible toxicity in the different life stages of helminths, making it a potential alternative method for the control of nematodes of sheep. Nevertheless, little is known about their mode of action in the metabolism of the ruminant. The objective of the present study was to evaluate the apparent digestibility (AD), microbial protein (MP) synthesis and nitrogen balance (NB) of sheep fed Bt. Eighteen Santa Ines sheep with a mean age of 3 months and body weight 18±3.5 kg were randomly assigned to 2 treatments where one group received $2.5 \times 10^6$ spores Bt/kg BW/day orally and the other did not receive any Bt (control) over 57 days. The animals were allocated to metabolism cages after this period for 7 days. Orts, faeces and urine were measured daily and chemically analyzed to obtain the coefficients of apparent digestibility (CAD) of dry matter (DM), crude protein (CP), neutral detergent fiber (NDF) and acid detergent fiber (ADF). Using liquid chromatography, the pool of urine of the animals was investigated to obtain the values of the synthesized MP and their contribution to microbial nitrogen (MN) by the method of purine derivatives in the urine. The Bt and control treatments did not differ ($P>0.05$) between CAD (%) for DM, CP, NDF and ADF (70.71 vs 71.53 (SE 1.38); 67.86 vs 68.66 (SE 2.75); 68.37 vs 68.04 (SE 1.49) and 57.92 vs 58.52 (SE 2.02), respectively). Likewise, no differences were found in the values of MN absorbed (g/day) (2.99 vs 2.88 (SE 0.02)) and NB (g/day) (18.09 vs 19.01 (SE 0.15)), respectively. These data confirm that the inclusion of $2.5 \times 10^6$ spores/kg BW of Bt in the diet of ruminants did not reflect damage in AD, MP synthesis and NB of the animals.

**Fatty acids of intramuscular fat in light carcass lambs raised in different nutritional conditions**

*M. Margetín[1,2], D. Apolen[2], M. Oravcová[2], K. Vavrišinová[1], O. Bučko[1] and L. Luptáková[1]*
*[1]Slovak University of Agriculture Nitra, Tr. Andreja Hlinku 2, 949 76, Slovak Republic, [2]National Agricultural and Food Centre, Hlohovecká 2, Lužianky, 951 41, Slovak Republic; margetin@cvzv.sk*

The meat quality of 40 light carcass lambs reared on ewes' milk (EM) of two genotypes (EM1 and EM2) and 40 carcass lambs reared on milk replacer (MR) of the same genotypes (MR1, MR2) was assessed on the basis of fatty acids profile (FAs) of intramuscular fat (IMF; MLLT). EM and MR lambs were divided into four identical groups. Female and male lambs were represented in each group in the same proportion. The average empty live weight of EM1, EM2, MR1 and MR2 lambs before slaughter was 17.6, 20.1, 17.8 and 17.6 kg. Profile of fatty acids (totally 69 FAs) was determined by gas chromatography. By means of ANOVA we detected significant differences among individual FAs or FAs groups in dependence on lamb type (EM1, EM2, MR1, MR2) and lamb sex. In EM1 and EM2 lambs the content of tVA (0.955 and 0.888), RA (0.672 and 0.681), EPA (0.352 and 0.673), DPA (0.625 and 0.916) and DHA (0.252 and 0.290 g/100 g FAME) was significantly higher ($P<0.001$) than in MR1 and MR2 lambs (0.112 and 0.133; 0.148 and 0.085; 0.061 and 0.075; 0.179 and 0.265; 0.079 and 0.074 g/100 g FAME). Conversely, the content of SFA (C12:0, C16:0, C17:0, C18:0) in EM1 and EM2 lambs was significantly lower than in MR1 and MR2 lambs ($P<0.001$). The differences in above mentioned FAs between EM1 and EM2 lambs and MR1 and MR2 lambs were statistically insignificant in most cases. Ratio of n-6/n-3 in MR1 (14.56) and MR2 lambs (15.76) was several times higher than in EM1 (3.25) and EM2 lambs (4.15; $P<0.001$). The content of CLA in IMF of EM1 lambs (0.749 g/100 g FAME) and EM2 lambs (0.862 g/100 g FAME) was several times higher than in lambs reared on milk replacer (0.193 in MR1 and 0.152 g/100 g FAME in MR2 lambs; $P<0.001$). We found highly significant effect of lamb type on MUFA, PUFA, cis-MUFA, trans-MUFA, n-6 and n-3 PUFA and ratio of LA/ALA ($P<0.001$).



**Milk quality of indigenous Greek dairy sheep and goats raised under different production systems**

*A. Tzora[1], I. Skoufos[1], C.G. Fthenakis[2], G. Tsangaris[3], G. Bramis[4], A. Karamoutsios[1], A. Gelasakis[4] and G. Arsenos[4]*
*[1]TEI of Epirus, Kostaki, 47100, Arta, Greece, [2]University of Thessaly, 43100 Karditsa, Greece, [3]Biomedical Research Foundation of the Academy of Athens, 11527 Athens, Greece, [4]Aristotle University of Thessaloniki, 54124 Thessaloniki, Greece; tzora@teiep.gr*

The milk chemical composition of indigenous Greek dairy breeds of sheep (Chios, Boutsko, Karagouniko) and goats (Skopelos, indigenous Capra prisca), raised under different production systems, was investigated. A reference flock comprised of animals from the above five breed (n=15 per breed) was created. The criteria for selection of ewes and does was the number (first, second, third) and stage of lactation (early-, mid-, late-). Individual milk samples were collected in three time points (45, 90 and 130 days postpartum). In each sample, the compositional content (fat, protein, lactose, total solids, total solids non fat) and acidity (pH), using Milkoscan, were measured. Average estimates (±SD) of the above parameters across lactation were 6.8±1.84, 5.6±0.83, 4.2±0.86, 17.6±3.12, 10.7±0.94, 6.9±0.18 for Chios breed, 8.5±2.61, 5.9±0.84, 4.6±0.45, 19.8±2.84, 11.4±0.61, 6.9±0.18 for Boutsko breed and 9.7±2.64, 5.6±0.26, 4.5±0.21, 20.5±2.46, 11.0±0.17, 6.9±0.18 for Karagouniko breed. Similar, average estimates (±SD) for Skopelos goats were 4.8±1.34, 3.8±0.44, 4.4±0.37, 13.8±1.69, 9.0±0.45, 6.8±0.12 and 4.6±1.72, 3.7±0.37, 4.4±0.40, 13.5±1.70, 8.9±0.49, 7.0±0.14 for indigenous Capra prisca goats. Significant differences were observed in all milk parameters across breeds at different stages of lactation. No significant differences were observed between sheep and goat breeds. The results provide useful information of milk quality for dairy breeds of sheep and goats raised in different production systems in Greece. This research is implemented through the Operational Program 'Competitiveness and Entrepreneurship, Cooperation, Large Scale Projects, GOSHomics' and is co-financed by the European Union and Greek National Funds.

**Robust Sheep EID: novel electronic tools are needed to aid the enforcement of EU regulation 21/2004**

*R. Mobæk[1], E.N. Sossidou[2], E. Ugarte[3], A. Lauvie[4], G. Molle[5], D. Gavojdian[6], G. Steinheim[1] and Ø. Holand[1]*
*[1]Norwegian University of Life Sciences, P.O. Box 5003, 1432 ÅS, Norway, [2]hellenic Agricultural Organization-Demeter, Veterinary Research Institute, 57001 Thermi, Thessaloniki, Greece, [3]Neiker Tecnalia, Campus Agroalimentario de Arkaute. Apto 46, 01080 Vitoria-Gasteiz, Spain, [4]INRA, Quart Grossetti, 20250 Corte, France, [5]Agris Sardegna, Loc. Bonassai, SS 291 Sassari-Fertili, Sassari, Italy, [6]Romanian Academy for Agricultural and Forestry Sciences, Drumul Resitei, 325400 Carensebes, Romania; ragnhild.mobaek@nmbu.no*

EU regulation 21/2004 requires all sheep born after 31/12/2009 to be tagged with electronic identification (EID). Use of electronic transponders allows automatic reading of individual animal codes directly into a database. Ideally, this legislation facilitates an electronic system that can be used as a tool for monitoring sheep movements. However, how the legislation is put into practice varies among countries. This study presents preliminary results from a survey conducted to understand country-specific adaptations to the regulation in the main sheep-producing Mediterranean countries: Spain, France, Italy and Greece. In each country, the implementation status of the regulation was mapped by interviewing relevant people/institutions within the sheep sector. In addition, various parameters that could be an obstacle to an effective operation of the countries' electronic traceability system were investigated. Obstacles identified in all countries studied were: National databases established, but individual movements not recorded; Loss of ear tags/boluses; No/low availability of cheap and user-friendly EID devices; Parameters inherent in sheep production systems inhibiting true tracing of movements; Farmers' attitudes towards EID and compliance to legislative requirements. The country-specific data reported here may be imperative to suggest improvements for implementing the EID regulation and to develop new electronic tools that can be adopted throughout EU, in order to help farmers in their routine practices.

---



***In vitro* studies on milk properties of the indigenous breed Capra prisca in nitric oxide stimulation**

*I. Skoufos[1], A. Metsios[2], A. Tzora[1], C.G. Fthenakis[3], G. Arsenos[4], G. Tsangaris[5], G. Papadopoulos[1], C. Voidarou[1] and S. Karkabounas[2]*
*[1]TEI of Epirus, Kostaki, 47100, Arta, Greece, [2]University of Ioannina, 45110 Ioannina, Greece, [3]University of Thessaly, 43100 Karditsa, Greece, [4]Aristotle University of Thessaloniki, 54124 Thessaloniki, Greece, [5]Biomedical Research Foundation of the Academy of Athens, 11527 Athens, Greece; jskoufos@teiep.gr*

The study investigated the potential influence of goat milk from the indigenous Greek breed Capra prisca on nitric oxide (NO) production in human B cell line U937 (immortal B lymphocytes). Nitric oxide is a known cytotoxic molecule, contributing to the antimicrobial action of B lymphocytes, a signal mediator, regulating inflammation. Cells from human B lymphocyte line U937 were cultured in DMEM at 37 °C, 5% $CO_2$ and then counted in Newbauer haemocytometer. They were plated at a density of 10,000 cells/ml in 96 well plates. Platelet activating factor (PAF), 10 µl/well from a stock solution of 100 µM, Lipopolysaccharide (LPS) from *Escherichia coli* 2.5 µl/well was obtained from a stock solution of 10 mg/ml and very small volumes of goat milk (1-10 µl/ml B cell line U937 culture medium) were used as agonists and added separately in each of the well plates. Following 24 hours of incubation U937 cell numbers were estimated and NO concentration was measured by colorimetric methods. It was revealed that PAF and LPS addition increased NO concentration by 140% and 120% respectively, compared to the controls (B cell line U937 with 10 µl of PBS). The addition of goat milk stimulated NO production by 210% compared to the controls. Nitric oxide stimulation in the human immortal B cell line under the influence of proper stimulants, such as LPS, PAF, resulted in the release of higher amounts of NO from human B cell line U937, indicating a possible antimicrobial and anti-inflammatory effect associated with the addition of goat milk. This research has been co-financed by the European Union (ESF) and Greek national funds through the Operational Program ARCHIMEDES III.

**Milk composition of ewes fed diets with soybean seeds**

*N.M.B.L. Zeola, A.G. Silva Sobrinho, C.T. Hatsumura, T.H. Borghi, V.T. Santana, F.A. Merlim, C.R. Viegas, J.C. Barbosa, R.S. Romeiro, S.W.B. Santos and A. Fernandes*
*São Paulo State University, College of Agricultural and Veterinarian Sciences, Department of Animal Science, Via de Acesso Prof. Paulo Donato Castellane, s/n., 14884-900, Brazil; americo@fcav.unesp.br*

The objectives of this research were to evaluate the centesimal composition and fatty acids profile of 1/2 Lacaune 1/2 Ile de France ewes fed diets containing different levels of soybean seeds. After lambing, ewes fed diets with and without soybean seeds, all containing 40.0% of sugarcane and 2.2% of mineral mixture, as the treatments: T0 (23.3% ground maize, 24.6% soybean meal and 9.9% citrus pulp), T7 (7.0% soybean seeds, 22.2% ground maize, 19.2% soybean meal and 9.4% citrus pulp) and T14 (14.0% soybean seeds, 21.0% ground maize, 14.0% soybean meal and 8.8% citrus pulp), formulated according to NRC. Ewes were milked third postpartum day to the $59^{th}$ day. The experimental design was completely randomized (three treatments and eight replicates) with comparisons made by Tukey´s test and analysis of variance according SAS. No differences were found in the milk of ewes to the variable total solids (14.87%), solids not fat (9.84%), protein (4.98%), fat (5.03%), lactose (4.11%), casein (4.15%), saturated fatty acids (72.07%), palmitic (26.71%), myristic (12.36%) and capric (10.60%), monounsaturated (21,48%), oleic (17.49%) and polyunsaturated (6.37%), conjugated linoleic acid (0,76%) and the atherogenicity index (3.11), however, a higher proportion of mineral was observed in the milk of ewes fed a diet containing 7% of soybean seeds (5.14%), compared to those fed 14% (4.45%), while for the polyunsaturated linoleic acid difference was found in the milk of sheep fed a diet containing 14% of soybean seeds (5.58%) in relation those fed 0 and 7% (3.82%). The use of a diet with 14% of soybean seeds in the diet of dairy ewes is recommended, providing higher tenor of linolenic acid (polyunsaturated) in the sheep milk, improving the qualities of the lipid and nutritional fractions of this feed for consumers.

---



**Study of mastitis in Valle del Belice dairy sheep using survival analysis approach**

*A.M. Sutera[1], M. Tolone[1], D.O. Maizon[2], M.L. Scatassa[3] and B. Portolano[1]*
*[1]Università degli Studi di Palermo, Scienze Agrarie e Forestali, Viale delle Scienze, 90128 Palermo, Italy, [2]Instituto Nacional de Tecnología Agropecuaria, EEA Anguil, CC 11, 6326 La Pampa, Argentina, [3]IZS della Sicilia 'A. Mirri', Via G. Marinuzzi, 90129 Palermo, Italy; annamariasutera@libero.it*

Mastitis is the most prevalent disease present in livestock species leading to economic loss. The aim of this work was to investigate the risk of having mastitis in Valle del Belice primiparous dairy ewes during the first lactation. All test-day records were collected from 5 flocks and were grouped in two data sets: mastitis due to minor pathogens (MIN) and mastitis due to major pathogens (MAJ). The follow up period of a ewe was the lactation. The traits were defined as the number of days between lambing and the occurrence of the first infection event due to MIN or MAJ. Ewes that were infected by MIN or MAJ before the end of the study were referred to as a failure (event = 1), whereas ewes that did not show presence of MIN or MAJ were considered right censored (event = 0). The analysis was performed with the Survival kit using the Cox frailty proportional hazard model, that contains the baseline hazard function at day d; the time-independent covariates: litter size and age at first lambing; the time-dependent covariates: milk production within flock (MK); somatic cell count within flock (SCC); the time-independent random effect of flock-year-season. For MIN and MAJ about 48% and 85% of the records were right censored, respectively. For the MIN, the effects of MK and SCC were significant ($P<0.001$). Higher risk of being infected (hazard ratio, HR=1.41) was observed for ewes with lowest MK level. Higher HR was found for ewes that had the highest level of SCC (2.23). For MAJ, the effect of SCC was significant ($P<0.001$). Higher HR was found for ewes that had the highest level of SCC (3.96). Results suggest that selection for decreased SCC may be effective to reduce mastitis incidence and the breeding goal should favor ewes with lowest observed SCC.

### Nutraceutical characteristics of meat from lambs fed diets containing mulberry hay

*A.G. Silva Sobrinho, L.G.A. Cirne, V.T. Santana, E.A. Oliveira, F.A. Almeida, V. Endo and N.M.B.L. Zeola*
*São Paulo State University, College of Agricultural and Veterinarian Sciences, Department of Animal Science, Via de Acesso Prof. Paulo Donato Castellane, s/n., 14884-900, Brazil; americo@fcav.unesp.br*

The objective of this experiment was to evaluate the fatty acid profile of meat from sheep fed mulberry hay. A total of twenty four Ile de France sheep weighing average 15 kg and aged approximately 60 days old at the beginning of the experiment, were housed in individual pens. Experimental design was completely randomized. The following diets, with forage:concentrate ratio 50:50, were supplied *ad libitum*: D1: sugarcane + concentrate without mulberry hay; D2: sugarcane + concentrate with 12.5% mulberry hay and D3: sugarcane + concentrate with 25.0% mulberry hay. The sheep were slaughtered when body weight reached 32 kg and the carcasses were kept refrigerated at6 °C during 24 hours. The lipids and methylated esters of fatty acids were extracted from the meat of the longissimus lumborum muscle. The inclusion of mulberry hay in the diet increased ($P<0.05$) the percentage of the saturated fatty acids, pentadecanoic (C15:0), from 0.31 to 0.42; heptadecanoic (C17:0), from 1.15 to 1.62; arachidic (C20:0), from 0.12 to 0.13; the monounsaturated fatty acid, heptadecenoic (C17:1), from 0.70 to 0.92; the polyunsaturated fatty acids, α-linolenic (C18:3ω3), from 0.18 to 0.57; ɣ-linolenic (C18:3ω6), from 0.12 to 0.15 and eicosapentaenoic (C20:5ω3), from 0.08 to 0.17. A quadratic effect ($P<0.05$) was observed for the fatty acids nervonic (C24:1ω9) and conjugated linoleic acid (CLA; C18:2cis9t11), from 0.37 to 0.59, for mulberry hay levels of 16.98 and 12.56%. Polyunsaturated fatty acids are essential to human health, especially in the prevention of atherosclerotic and thrombotic diseases. The IA (0.65%) and IT (1.73%) of lamb meat were not affected ($P>0.05$) and were considered low. Mulberry hay added to the diet of sheep reared in confinement improved the polyunsaturated fatty acids profile of the meat, which are associated with properties beneficial to human health.



### Organic resilient animal farming systems to meet future livestock production challenges

*M. Vaarst*
*Aarhus University, Animal Science, P.O. Box 50, 8830 Tjele, Denmark; mette.vaarst@agrsci.dk*

Organic livestock farming has the potential to be an important driver for a sustainable development of future agricultural and food systems. Sustainable is here understood as a concept comprising social, economic, environmental and institutional aspects. However, severe challenges face this development in the current agrifood systems building on increased industrialization of animal production. To continuously present an alternative to this, it is necessary to strengthen the agro-ecological and systems oriented approaches to livestock farming, on small as well as on large scale, e.g. optimizing the use of free range systems including semi-natural areas, agro-forestry, or taking holistic grazing approaches where appropriate. Another approach could be to combine more farms into a larger farming system, benefitting from the diversification of enterprises and farm elements. Complex well-integrated systems should support the animals to maintain a high health status. In this context, animal health is understood as resilience, which means that they have a high ability to withstand shocks and influences from their surroundings. As such, health is understood as much more than disease freedom, although species-specific high-quality feed and a management system which supports their resilience and welfare will lead to less disease, hence low or no medicine use. Appropriate robust breeds suitable for the local conditions are crucial to enhance animal as well as system resilience. Agro-ecological systems build on context specific human knowledge and experience, and therefore rely on the farmers' ability to observe, assess and judge situations as well as their skills and complex knowledge. This presentation will focus on how integrated and well-designed animal farming systems can contribute specifically to the enhanced resilience and welfare of the animals, and thereby also the humans which are involved in the production. It will mainly focus on a Global North perspective, but also include selected Global South examples.

**What makes organic livestock production sustainable?**

*F.W. Oudshoorn[1], A. Stubsgaard[2] and C.G. Sørensen[1]*
*[1]Aarhus University, Engineering, Blichers Allé 20, 8830 Tjele, Denmark, [2]Knowledge Center for Agriculture, Organic Farming, Agro Food Park 15, 8200 Aarhus, Denmark; fwo@eng.au.dk*

Environmental sustainability has been presented as the principal focus for agricultural production of the future. A few decades ago, local toxic pollution of groundwater, surface water and acidification were the major concerns, whereas now global GHG emissions and limited resources are in focus. There is a consensus on live cycle assessment (LCA) as the prime approach for optimisation, even though the issue of scope and system expansion are being discussed. Concurrent assessments on profitability of the production, measured as remuneration per kg, ha or animal unit has always been set as the condition for production performance. In addition, a common acceptance has been reached between producers, consumers, and society that social and governance issues play a crucial role in the sustainability of agricultural business and organic agriculture in particular. Documenting and visualising social issues (e.g. animal welfare, biodiversity, landscape, product quality) together with environmental and economic issues, pertaining to the society and the consumer, will provide the basis for informed societal future development steps. Organic livestock production can focus and improve on these indicators of sustainability (SI), and hereby advocate for governing structures to support the SI not yet justified by increased organic market sizes. Multiple sustainability assessments of organic livestock production systems (dairy, pig-meat, egg) have been carried out in Denmark showing SI rankings for soil use, animal husbandry, nutrient flows, water use, energy and climate, biodiversity and plant production, working conditions, quality of life, economic viability, and farm management. Visualization by polygons via web interface was used for dissemination and farm development.

---



**Organic livestock farming systems in the Central France: evolution of the performances and drivers**

*P. Veysset[1], M. Benoit[1], J. Belvèze[2], O. Patout[3], J.L. Reuillon[4], E. Morin[2] and M. Vallas[5]*
*[1]INRA, UMRH, Clermont-Theix, 63122 Saint Genès-Champanelle, France, [2]Idele, Chemin de Borde Rouge, BP 42118, 31321 Castanet Tolosan, France, [3]AVEM, ZA du Cap du Crès, BP419, 12104 Millau, France, [4]Idele, 9 Allée Pierre de Fermat, 63170 Aubière, France, [5]Pôle Agriculture Biologique Massif Central, VetAgro Sup, 89 av de l'Europe, BP35, 63370 Lempdes, France; veysset@clermont.inra.fr*

The 'Organic Farming Massif Central' hub and fifteen partners lead a program on sustainability and on the technical and economic operation of OF livestock systems in the Massif Central. This systemic and multi-year study (2008-2011) is based on data from a constant sample over four years, from 56 farms comprising four types of products: cattle and sheep, dairy and meat. Over 4 years, the technical and economic results are quite stable, and at a good level, but with great variability inter-farms. With lower labor productivity, but with a more diversified crop rotation, a good food self-sufficiency and good technical skills, the farms with the highest income get an income more than four times higher than the farms with the lowest income.

**Definition of a global research program for organic farming in Walloon area for 2015-2020 horizon**

*D. Stilmant, D. Jamar, V. Decruyenaere, A. Vankeerberghen, M. Abras and E. Froidmont*
*Walloon agricultural Research Centre, Organic farming and protein autonomy research unit, 9, rue du Liroux, 5030 Gembloux, Belgium; d.stilmant@cra.wallonie.be*

In order to define research priorities for Walloon organic farming sector and for an increase in protein autonomy, during the 2015-2020 period, a three steps approach was performed. (A) Interviews were conducted, on an individual basis, with 90 farmers, 60 of them have livestock productions (dairy, beef, pig or poultry). (B) Ten focus groups were organized with (1) advisory services, (2) administration and farmers labor unions, (3) research institutions. (C) Based on these outputs and on a review of prospective exercises performed in this field, subjects list will be set up and submitted to the different organisms involved in the focus groups for ranking. The priorities that will be identified will support future research programming. Based on the 3 first focus groups, the following subjects were highlighted for livestock productions: (1) Due to grassland proportion in Wallonia (±50%), beef production is a major production sector but main part of organic animals are sold as suckler calves in conventional chains. To face this problem farmers have to learn to fatten their animal with local resources, butchers have to learn to cut organic carcass (different from Belgian Blue Conventional bred) and consumers to appreciate a less lean and tender but more tasty meat. (2) In such context, decision support tools leading to a better use of grasslands and to better local fodder resources associations are also necessary. (3) In order to promote protein autonomy in dairy and monogastric systems, a special focus seems also necessary on cereal/pea/faba associations production and valorisation. (4) Systemic approaches, taking into account soil, plant and animal performances and system externalities are also necessary to justify agro-environmental support of this sector. Final results of this approach could be of interest for EAAP community.



**Challenges and future perspectives of different organic beef cattle farms of Southern Europe**

*A.J. Escribano[1], P. Gaspar[2], F.J. Mesias[3], M. Escribano[1] and A.F. Pulido[3]*
*[1]Faculty of Veterinary Sciences, University of Extremadura (Spain), Animal Production, Avda. de la Universidad s/n, 10071 Caceres, Spain, [2]Faculty of Agriculture, University of Extremadura (Spain), Animal Production, Carretera de Caceres s/n, 06071 Badajoz, Spain, [3]Faculty of Agriculture, University of Extremadura (Spain), Economy, Carretera de Caceres s/n, 06071 Badajoz, Spain; ajescc@gmail.com*

The livestock sector must be redesigned in order to achieve food security sustainably. Several authors have argued that organic farming may contribute positively in this sense. However, the externalities of the organic livestock systems are influenced by several factors. Due to this, the aim of this paper is to shed light on the needs and perspectives of the organic beef cattle farms located in the 'dehesa' ecosystem (SW Spain). For this purpose, we created the 'Feasibility of Success Index' (FSI) by using indicators that took into account the context of the extensive livestock systems located in the Mediterranean basin. Two groups of organic farms were analyzed: farms that sold calves at weaning age as conventional ones (Organic 1); farms that fattened and sold their calves as organic (Organic 2). These groups were compared to the Conventional farms, in order to facilitate the understanding of the situation of the organic farms. For the assessment, the MESMIS approach was used. The results showed that the Group 2 showed an advantageous position (FSI=61.63%). The Organic 1, despite showing a lower external dependence, they also showed low economic and productive results (FSI=56.59%). Finally, the Conventional farms scored the highest for the productive and economic indicators. However, they scored the lowest in the FSI (54.46%). In general terms, all three groups must improve in certain common aspects that are crucial in the current and future context of the beef cattle sector: reducing the dependence on external feedstuff, the dependence on subsidies, the bargaining power, and the business diversification.

**Characteristics of the diets in organic pig production**

*A. Prunier[1], G. Rudolph[2], D. Bochicchio[3], G. Butler[4], S. Dippel[5] and C. Leeb[2]*
*[1]INRA, UMR 1348, PEGASE, 35590 Saint-Gilles, France, [2]Univ. of Natural Resources, BOKU, Gregor-Mendel-Str. 33, 1180 Vienna, Austria, [3]CRA-SUI, Agricultural Research Council, 41018 San Cesario sul Panaro, Modena, Italy, [4] School of Agriculture, Newcastle University, Newcastle upon Tyne, NE1 7RU, United Kingdom, [5]IAWAH, Friedrich-Loeffler-Institut, Doernbergstr. 25/27, 29223 Celle, Germany; armelle.prunier@rennes.inra.fr*

A better knowledge of the rations pigs receive, should help identify weaknesses and hence improve the efficiency of organic pig production. A research project was initiated in 8 EU countries (ProPIG from the ERANET CoreOrganic II) involving 72 farms: 59 with sows (53 farrow-to-finish = FF, 5 with and 1 without weaners), 11 fattening (F) and 2 with weaners and fatteners. Farmers were asked to describe their feeding practices and the nutrient content of feeds used was recorded, either from the manufacturer claim or calculated from ingredients. Four FF farms used a single diet for all pigs. For sows, 46% of the farms fed the same diet. For fatteners, 58% of the farms used a single diet, 38% used two diets and 5% used 3 diets. For weaners, 73% of the farms used a single diet and 27% used two diets. Nutrient feed contents were 13.3±1.0 MJ ME, 141±19 g crude proteins (CP) and 5.0±1.2 g total P (tP) /kg for pregnant sows, 12.8±0.9 MJ ME, 159±19 g CP and 5.2 g±1.2 tP/kg for lactating sows, 12.8±1.0, 175±23 g CP, 5.3 g tP/kg for weaners, 12.7±0.1 MJ ME, 165±23 g CP and 4.7±1.1 g tP/kg for fatteners (means ± sd). Major ingredients were triticale (from18% in weaners to 27% in fatteners, 51% homegrown = HG), barley (from 22% in lactating sows to 28% in weaners, 48% HG), wheat (from 18% in weaners to 27% in fatteners, 23% HG), maize (from 13% in pregnant sows to 16% in fatteners, 52% HG), peas (from 8% in pregnant sows to 12% in fatteners, 38% HG), and fava beans (from 3.9% in fatteners to 10.4% in weaners, 67% HG). Results suggest using specific feeding for different types of pigs may improve feeding efficiency and reduce the environmental impact.

---

Session 26 Theatre 7

**Mussel meal in diets to growing/finishing pigs: influence on performance and carcass quality**

*A. Wallenbeck[1], M. Neil[2], N. Lundeheim[1] and K.H. Andersson[2]*
*[1]Swedish University of Agricultural Sciences (SLU), Department of Animal Breeding and Genetics, Box 7023, 75007 Uppsala, Sweden, [2]SLU, Dep. of Animal Nutrition and Management, Box 7024, 750 07, Sweden; anna.wallenbeck@slu.se*

Mussel meal has potential as an alternative protein source for pig since it has high protein content and a balanced amino acid pattern. Mussels are good filterers of water and an effective tool to clean waters from nitrogen and phosphorus. Using mussel meal as feedstuff in animal production could close the agro-aqua nutrient cycle, bringing nutrients back from the water to the agro-ecosystem. The objective of this study, included in the EU project ICOPP, was to investigate the effects of inclusion of mussel meal in diets for two genotypes of growing/finishing pigs on performance and carcass quality. Sixty-four growing/finishing pigs (32 Yorkshire × Hampshire, 32 Yorkshire × Duroc) was fed either a commercial diet with conventional protein ingredients (C) or a diet containing 5% inclusion of dry mussel meal (M). Nutrient composition was equal in the two diets and both diets were given to the two genotypes. This mussel meal was produced from mussel meat only, i.e. no shells were included. Performance and carcass quality was analyzed with procedure MIXED in SAS. The results are in line with our hypothesis that pigs will perform with maintained high performance and carcass quality when mussel meal replaces conventional protein feed (Daily weight gain: 956 vs 948 g/day, Feed conversion: 27.1 vs 27.9 MJ NE/kg, Lean meat content: 58.8 vs 58.2%, Dressing percentage: 78.1 vs78.6% and Daily lean meat growth: 457 vs 451 g/day inthe M and C treatmet, respectivley, P>0.05 fo all traits). Genotype (Hampshire or Duroc boar) had no influence on performance (P>0.05 for all traits). The daily gain was relatively high in this study, on average 952 g/day from 37 kg live weight to slaughter (107 kg), and we conclude that mussel meal can substitute conventional protein sources in growing/finishing pig diets with maintained production results.

**Performance and behaviour of free-range pigs in relation to feed protein level and forage crop**

*M. Jakobsen and A.G. Kongsted*
*Aarhus University, Agroecology, Blichers allé 20, 8830 Tjele, Denmark; malene.jakobsen@agrsci.dk*

The aim of 100% organic diets based on locally produced feed ingredients is a huge challenge in organic pig production due to a limited production of suitable protein feed sources. The pig is an omnivore and has a unique foraging capacity for seeking food on and below the soil surface. It is hypothesised that foraging in the range area can contribute considerably to the nutritional needs of growing-finishing pigs. Foraging activity, nutrient intake from the range area and pig performance was investigated in 36 pigs foraging on either lucerne or grass and fed either a standard organic mixture (HP: high protein) or a mixture of barley and wheat (LP: low protein) from approximately 60 to 90 kg live weight in three replicates. All pigs had access to 4 $m^2$ per pig per day and were fed according to 80% of energy recommendations. Behavioural observations were carried out 12 times over the entire experimental period of 40 days. For both crops, LP pigs rooted significantly more than HP pigs but the effect of feed protein level was larger in grass (44 vs 19% of all observations) compared to lucerne (28 vs 16% of all observations). Feed protein level had no significant effect on grazing behaviour but pigs on lucerne grazed significantly more than pigs on grass (10 vs 4% of all observations). A significant interaction between forage crop and feed protein level was found on daily weight gain and feed conversion ratio. Compared to the HP pigs, the pigs on LP treatment had 33% lower daily weight gain (589 vs 878 g) and 31% poorer feed efficiency (3.75 vs 2.59 kg feed per kg weight gain) in grass paddocks but only 18% lower daily weight gain (741 vs 900 g) and 14% poorer feed efficiency (2.95 vs 2.54 kg feed per kg weight gain) in lucerne paddocks. The LP pigs foraging on lucerne used 169 g less feed crude protein per kg weight gain than the HP pigs. The results indicate that direct foraging on lucerne can pose an important contribution to protein supply in organic growing pigs.

---



**Anti-parasitic effects of plant secondary metabolites on swine nematodes**

*A.R. Williams[1], M. Peña[2], C. Fryganas[3], H.M. Ropiak[3], A. Ramsay[3], I. Mueller-Harvey[3] and S.M. Thamsborg[1]*
*[1]University of Copenhagen, Department of Veterinary Diease Biology, Dyrlægevej 100, 1870 Frederiksberg C, Denmark, [2]Technical University of Denmark, National Veterinary Institute, Bülowsvej 27, 1870 Frederiksberg C, Denmark, [3]University of Reading, SAPD, P.O. Box 236, Reading RG6 6AT, United Kingdom; arw@sund.ku.dk*

Organic production presents challenges to animal health and productivity. In organic pig production, animals must have access to outdoor pastures which increases exposure to gastrointestinal parasites. Moreover, the routine use of synthetic anti-parasitic drugs is not allowed. Thus, novel parasite-control options are required. We present results from a comprehensive *in vitro* screen of plant secondary metabolites (PSM) from diverse plant sources on the economically important pig parasites *Ascaris chlamydiae* and *Oesophagostomum dentatum*. We focused on two PSM classes commonly found in natural diets – condensed tannins (CT) and sesquiterpene lactones (SL). Different CT-types were purified from various plant sources to reflect their diversity; SL were purified from forage chicory. These PSM were tested in inhibition assays of worm motility and migratory ability. CT had potent activity against *A. suum*, with substantial inhibition of migration of *in vitro* hatched larvae ($EC_{50}$ values ranging from 40 to 120 μg/ml). In contrast, migratory ability of *O. dentatum* larvae was not significantly affected. However, the motility of adult *O. dentatum* was reduced after *in vitro* incubation with CT. The purified chicory extract showed potent inhibition of *A. suum* larval migration (EC50 value of 42 μg/ml) and was also active against larval and adult stages of *O. dentatum*. Electron microscopy demonstrated significant structural damage to the cuticle and digestive tissues in nematodes exposed to PSM. Plants rich in PSM such as CT and SL show promise as natural anthelmintics against two highly prevalent swine parasites. Experiments to determine *in vivo* efficacy and mechanisms of nematocidal action are on-going.

## Effect of genotype, sow rearing system and outdoor access on piglet survival in extensive systems

*H.M. Vermeer[1], G.P. Binnendijk[1] and J. Leenhouwers[2]*
*[1]Wageningen UR Livestock Research, P.O. Box 65, 8200 AB, the Netherlands, [2]TOPIGS Research Center IPG, Schoenaker 6, 6641SZ, Beuningen, the Netherlands; herman.vermeer@wur.nl*

In organic and low input systems, health and survival are at risk in piglets shortly after birth and weaning. Increasing litter sizes, lower birth weights and suboptimal climatic conditions increase the risk for perinatal mortality. Post weaning, the protein provision and housing conditions are often suboptimal. In order to improve survival and health of newborn and weaned piglets, an experiment with a 2×2×3 factorial design was conducted: (1) Sows of two genotypes: fathers of these gilts with high (HPS) and low (LPS) breeding value for piglet survival to affect preweaning survival; (2) Sows born and reared until first mating under conventional (CONV) or organic (ORG) conditions to affect future maternal behaviour; (3) Suckling piglets born in 3.75×2.00 m indoor farrowing pens with straw bedding according to EU organic standards and kept indoors (IN), with an outdoor run (RUN) or with outdoor run including piglet access to pasture (PAST) to affect gut health after weaning. We monitored 96 gilts during their first and second lactation resulting in 192 litters. Sow observations focused on maternal behaviour and activity pre- and post-farrowing, whereas piglet observations focused on performance, health, mortality and weight development. Data were analysed using GLMM in Genstat with genotype, rearing system, outdoor area and parity as main effects in the model. Preliminary results show 16.9 vs 19.4% mortality of liveborn piglets in CONV and ORG, 21.4 vs 16.2% for HPS and LPS, 14.8 vs 20.7 vs 18.6 for IN, RUN and PAST. Daily gain during the first three days after birth was 142 g/d for IN and 123 and 126 g/d for RUN and PAST. After weaning the daily gain was 338 vs 382 vs 359 g/d for IN, RUN and PAST. With growing interest in loose housing in conventional farrowing pens these potential effects of genotype and rearing conditions can also be meaningful for piglet survival in conventional systems.

---



## Milk production and fatty acid content in milk on organic farms feeding three levels of herb silage

*M.B. Petersen[1], K. Søegaard[2] and S.K. Jensen[3]*
*[1]AgroTech, Agro Food Park 15, 8200 Aarhus N, Denmark, [2]Aarhus University, Department of Agroecology, Blichers Allé, 20, PO 50, 8830 Tjele, Denmark, [3]Aarhus University, Department of Animal Science, Blichers Allé, 20, PO 50, 8830 Tjele, Denmark; mbp@agrotech.dk*

Herb-rich feed to dairy cows has been shown to increase beneficial n-3 fatty acids (FA) in milk, but most of these studies have been carried out with semi-natural grasslands or alpine pastures. The main objective of the present study was to investigate the effect of substituting traditional clover-grass silage with herb silage composed of 10 herb species from managed swards on milk production, especially milk FA composition, on three Danish commercial organic dairy farms. The three farms differed in size (50, 189, and 235 animals on farm A, B, and C, respectively) and breed (Jersey, Danish Holstein, and Cross breed on farm A, B, and C, respectively), but on all three farms clover-grass silage normally made up >60% of DM. The herb mixture was sown out on all three farms, harvested and conserved in silage bales. During two to four weeks herb silage replaced the normal clover-grass silage. Feed intake was estimated and milk samples from 20 cows on each farm were taken at the start and again at the last day of the feeding trial. Feed and milk samples were analyzed for chemical and FA composition. Herb-silage constituted 27, 21 and 64% of DM on farm A, B, and C, respectively. Milk yield was not affected on farm A and B, but decreased on farm C on the herb-silage diet ($P<0.001$). On all three farms milk protein % increased ($P<0.005$), and milk urea decreased on farm B ($P<0.005$) and farm C ($P<0.005$). Milk fat % was unaffected on all three farms, but milk FA profile was significantly affected on all three farms. Content of n-3 FAs increased by 7, 16 and 5% on farm A, B, and C, respectively ($P<0.005$) and CLA by 11, 14, and 17% on farm A, B, and C, respectively ($P<0.005$). The results showed that it is possible to maintain milk production and alter milk FA profile with 30% of DM as herb-rich silage.

**Evaluation the autochthonous genetic resources and their use in organic sheep production in Serbia**

*M. Savic[1], S. Prodanovic[2], S. Vuckovic[2], B. Dimitrijevic[1], Z. Beckei[1] and M. Vegara[3]*

*[1]Faculty of Veterinary Medicine University of Belgrade, Animal breeding and genetics, Bul. oslobodjenja 18, 11000 Belgrade, Serbia, [2]Faculty of Agriculture, Nemanjina 6, 11000 Belgrade, Serbia, [3]Norwegian University of Life Sciences (NMBU), Department of International Environment and Development Studies, P.O. Box 5003, 1432 Aas, Norway; mij@beotel.net*

Agricultural development of Serbia has been going through a difficult transition process during the last decade. The new Strategy for Agriculture and Rural Development from 2014 to 2024 defines goals and priorities for further development of agriculture. One of the priority topics are research and development related to organic production. In the field of organic farming knowledge transfer is particularly important since this type of farming is an example of an innovation system. In order to develop organic production process, the FVM is taking an active role in knowledge transfer related to organic livestock management. The implementation of organic standards in sheep production has been performed in the hilly mountain regions in Serbia. In the framework of multidisciplinary project synergies between ecological intensification methods and impact on the environment development of sheep husbandry have been evaluated. As well, analyses of the value of genetic resources, various types of Zackel sheep, with special emphasis on quality traits such as health effects and robustness for organic production have been performed. The GI parasitic infections have been recognized as the main health problem in organic sheep production. Therefore, improved methods for prevention and alternative medicine use should be developed further and applied more consistently. Low productive traits of Zackel in comparison with the Wurttemberg breed reared in the same system remain a great challenge for the application of smart breeding technologies. Evaluation of both chemical and sensory lamb meat quality have confirmed adding value to traditional products of regional origin.

---



**The effect of laying hens age on egg quality in organic farming**

*L. Zita[1], Z. Ledvinka[1] and Z. Volek[2]*

*[1]Czech University of Life Sciences Prague, Department of Animal Husbandry, Kamýcká 129, 165 21, Prague 6 – Suchdol, Czech Republic, [2]Institute of Animal Science, Department of Physiology of Nutrition and Quality of Animal Products, Přátelství 815, 104 00, Prague – Uhříněves, Czech Republic; zita@af.czu.cz*

The aim of the present study was to evaluate the effect of the age of laying hens, which were reared in organic farming, on egg quality characteristics. A total of 1,080 eggs from Lohmann Brown hens were collected at the beginning of the first and second laying cycle (from 31$^{st}$ to 33$^{rd}$ week of age and from the 77$^{th}$ to 79$^{th}$ week of age). Egg weight and some quality parameters of yolk, albumen and eggshell were monitored. The values of some parameters of egg quality were statistically processed. The effect of the age of laying hens on egg quality parameters was found. The eggshell weight, strength and colour of the eggshell were significantly affected by the age. The weight of eggs, eggshell, yolk, albumen and proportion of yolk increased significantly with increasing age (61.34 vs 69.78 g; 6.13 vs 6.86 g; 14.43 vs 17.78 g; 40.78 vs 45.18 g; 23.58 vs 25.56%). On the other hand, the proportion of albumen, yolk and albumen index and Haugh units score decreased significantly with the age of hens(66.41 vs 64.58%; 23.58 vs 25.56%; 10.24 vs 8.18%; 86.83 vs 79.46). Colour of yolk was lighter with the age of hens (9.35 vs 8.86; $P \leq 0.001$), as well as the colour of the eggshell (28.48 vs 30.49%; $P \leq 0.001$). The eggshell strength was lower with increasing age (45.80 vs 42.31 N/cm$^2$; $P \leq 0.001$). The conclusion of this observation is that the layer's age is one of the important factors determining the egg quality. The weight of eggs increased with layer's age, which causes that the quality of albumen and eggshell was worse. The quality of eggs may be affected e.g. by interactions of age and breed. This research was funded by 'S' grant of Ministry of Education, Youth and Sports of the Czech Republic.

**Effect of age of buffalo cows on milk yield and composition in low input organic production system**

*B. Barna and G. Holló*
*University of Kaposvár, Guba S. str 40., 7400 Kaposvár, Hungary; hollo.gabriella@sic.hu*

The objective of present study was to evaluate the effect of age of buffalo cows on milk yield and milk composition maintained at organic low input production system (Vókonyatanya, Hungary). The diet of cows (n=40) consist of mainly roughage produced organically on farm and buffalo cows spend grazing period outdoors. Milk samples from morning milking were collected during the whole lactation period according to ICAR guidelines. Milk composition was recorded by using ultrasonic milk analyser (Ecomilk ultra pro, Bulteh 2000), the following data were recorded: fat – solids non fat – and protein content. The data processing was made by SPSS 17.0 using ANOVA and Tukey test. Cows were divided into three classes according to their age (1. group:<6 yr, 2. group:6-10 yr, 3. group:>10 yr). The overall average lactation length was 202.27+16.73 days and the overall average milk yield was 596.68+68.84 kg per lactation. Milk production was lower in the youngest (543.24 kg) and the oldest age group (603.84 kg), than that in 6-10 years old age group (624.44 kg). The lowest milk yield value measured in first age group, it differed significantly from values of other two groups. The average fat, solids non fat and protein percentage were 8.92%, 10.07% and 4.18%. The effect of age on protein percentage and fat percentage showed opposite tendency. The youngest age group had the highest protein value, whilst the highest fat percentage measured in the oldest age group (protein: 4.33 v. 4.00%, fat: 8.14 v. 9.44%). Non solids fat percentage changed in line with protein percentage. In conclusion, age of buffalo cows is an important factor in the evaluation of organic milk production traits.

---



**Insects as a feed ingredient for livestock**

*T. Veldkamp and G. Van Duinkerken*
*Wageningen UR Livestock Research, Animal Nutrition, P.O. Box 65, 8200 AB Lelystad, the Netherlands; teun.veldkamp@wur.nl*

The UN Food and Agricultural Organisation (FAO) estimates that the world will have to produce ca. 70% more food by 2050. Rising standards of living increase the demand for animal-derived proteins, and as a result the global feed demand. The EU is currently dependent for more than 70% on imports of vegetable proteins as protein-rich feed ingredients. An alternative protein strategy, and an improved utilisation of biomass residues and left-overs is very important to further increase the sustainability of EU agro & food production. Insects are an alternative animal protein source, which can sustainably be reared on organic side streams. A feasibility study on the use of insects in pig and poultry diets was performed. From this study it was concluded that Hermetia illucens (Black soldier fly, BSF), Musca domestica (Common housefly, CH) and Tenebrio molitor (Yellow mealworm, YM) are potentially interesting insect species. The crude protein content (on dry matter basis) of the larvae of these species is 35-57, 43-68, and 44-69% for BSF, CH, and YM, respectively. These crude protein levels are similar as for defatted soybean meal with a crude protein content in the range of 49-56% on dry matter basis. The Essential Amino Acid Index of the BSF, CH, and YM is above 1 indicating that these protein sources provided more of the essential amino acids than required for broilers and growing pigs. Beneficial functional properties of insect protein need to be further investigated in order to create an added value for insect protein. Main bottlenecks were identified in the area of law and regulation and the possibilities to increase the scale of insect production at a low cost price. In the EU, it is currently not allowed to use insect protein in pig and poultry diets due to the TSE regulation. Further risk assessment of the use of insects as feed ingredient is required. Maximum price of insects as feed ingredient for pig and poultry diets should be in the range of € 1.00-1.50 per kg to be a competitive alternative protein source.

**Insects industry: a new circular economy to assure food security**

*M. Peters*
*New Generation Nutrition, Nieuwe Kanaal 7F, 6709 PA Wageningen, the Netherlands; marianpeters@ngn.co.nl*

FAO reported on the future prospects of edible insects for feed and food security. Despite the fact that the insect production and supply chain are small and immature, the future looks promising. It is clear the future demands an increasing need for nutritious food. To cover the need for more proteins, insects arise and there is an obvious environmental impact of the existing meat production system. Water and soil pollution, emission of greenhouse gasses, major land-use for feed production and depletion of natural sources like fish meal is a serious concern; while immense organic waste stream find it's way to be destructed. The circular industry design promises a source of ingredients for feed and food. In a system that is based on the upgrading and re-use of organic waste; organic waste will gain a higher economic value when it can be used in de insect industry. In this circular design economic development will benefit from natural solutions present in the ecological system. Insects farmed on organic waste will have a low environmental impact and do not leave a large surplus after rearing. The introduction of insect products can significantly contribute to the exponentially growing demand for sufficient, affordable and sustainable proteins in several efficient ways. New opportunities arise: (1) insects as bio-converter; (2) viable and sustainable alternative designs for a new agricultural sector; (3) opportunity for innovative modern entrepreneurs Major challenges for the foundation of a flowering and viable industry are: (1) scaling; (2) increase market and consumer acceptance; (3) tackle legislative frameworks on a national and international level; (4) new cooperative and funding models are developed. As soon as the supply chain is starting to function and there is a market pull created, the insect industry will make a major contribution to the growing demand for more sustainable, cost-effective, high-quality protein in the feed and pet food industry.

---

Session 27 Theatre 3

**Can greenhouse gas emissions be reduced by inclusion of waste-fed larvae in livestock feed?**

*H.H.E. Van Zanten[1,2], F.H. Van Holsteijn[2], D.G.A.B. Oonincx[3], H. Mollenhorst[2], P. Bikker[1], B.G. Meerburg[1] and I.J.M. De Boer[2]*
*[1]Wageningen UR Livestock Research, P.O. Box 135, 6700 AC Wageningen, the Netherlands, [2]Animal Production Systems Group, Wageningen University, P.O. Box 338, 6700 AH Wageningen, the Netherlands, [3]Laboratory of Entomology, Wageningen University, P.O. Box 8031, 6700 EH, Wageningen, the Netherlands; hannah.vanzanten@wur.nl*

The livestock sector requires alternative protein sources, because of the expected increase in demand for animal products. Insects have a high protein content and, therefore, can contribute to this goal. Moreover, use of insects may reduce the environmental impact of livestock production. Insects have the potential to turn organic waste streams into high quality feed products replacing conventional feed with a high environmental impact. The aim of this study was to explore the potential reduction in greenhouse gas emissions from inclusion of larvae of the common housefly fed on waste in diets for piglets. Data were based on a business model, based on experimental studies, to produce 20 ton dried larvae meal per day. Larvae are fed on a substrate of poultry manure and food waste. After harvesting larvae, the remaining substrate is used for anaerobic digestion, replacing fossil fuels and artificial fertiliser. Larvae meal is included in a diet of piglets and replacing fishmeal and soybean meal. Furthermore, we compensated for the losses in production of the current use of food waste (anaerobic digestion) and manure (fertiliser). Results showed that feed with larvae has a GWP between 0.63 and 0.85 kg $CO_2$-eq per kg of feed. Feed without larvae has a GWP of 0.67 kg $CO_2$-eq per kg. The uncertainty in the GWP of the larvae originates from lacking knowledge around three aspects: (1) the potential of the remaining substrate as fertiliser; (2) energy use for larvae production; (3) emissions from substrate during larvae cultivation. To conclude, inclusion of waste fed larvae in diets of piglets has potential to reduce greenhouse gas emissions. Further research is needed for an accurate estimate of the reduction potential.

**Establishment of mother-young relationships in farm mammals**

*R. Nowak*
*INRA, Physiologie de la Reproduction et des Comportements, 37380 Nouzilly, France; raymond.nowak@tours.inra.fr*

The defining characteristic of mammals is that females nurse and care for their young; without this, the neonate has no chance to survive. This review compares three well-studied species (the rabbit, pig and sheep) that differ in their parental strategies and in the problems that neonates have to overcome to adapt to their post-natal life. As a general trend, mother–young interactions vary according to the maturity of the newborn, and the size of the litter. Neonatal survival relies to a great extent on an environment that is ecologically sound for the developmental stage of the neonate, and on optimum interactions with the mother. The hormonal changes observed at the end of gestation (steroids, prolactin, oxytocin) play a crucial role in the onset of maternal behavior. Adaptive maternal care supposes that the mother provides the basic needs of the neonate: warmth or shelter by building a nest (in pigs and rabbits), nutrients, water and immunological protection through nursing (via colostrum and then milk). The first challenge for the newborn is to get access to the udder and the sensory information detected varies according to species. It may be limited to one cue like in the rabbit (mammary pheromone) or to several like in the pig and the sheep. In some species (sheep) a strong bond develops between mother and young, leading to selective maternal care. On the lamb's side, rapid access to the udder is extremely important, not only as a source of food, but also because suckling has strong rewarding properties in the establishment of a preference for the mother. Such a selective bond does not exist in the other two species. A major risk facing all neonates, other than the birth process itself, is inadequate colostrum intake owing to delayed suckling or competition with siblings, which leads to starvation, hypothermia or even crushing, as has been observed in pigs. In sheep, prolificacy may be a problem not only in terms of milk yield available for the litter but also in terms of the lambs being unable to keep contact with their mother in the flock.

---



**How the cow tamed a continent: the story of how the ability to drink milk set the stage for European civilization**

*A. Curry*
*Journalist, Gipsstrasse 22, 10119 Berlin, Germany; andrew.curry@gmail.com*

It may come as a surprise to some that just 35% of the people in the world today – most of them living in central and northern Europe, or descended from people from those regions – can digest the lactose protein in milk past the age of 7 or 8. For the other two-thirds of the world, a glass of milk leads to severe digestive discomfort. Not so long ago, no one on Earth looked forward to a glass of milk. Recent research suggests drinking raw milk is a surprisingly recent development: About 7,500 years ago, probably somewhere in Hungary, the gene that produced lactase was switched on – and left that way for good. That tiny mutation changed the course of European civilization. For the ability to drink milk to become so common so quickly, it must have had a tremendously strong 'selection coefficient,' the advantage that a gene gives to the people who have it. Based on when it might have emerged and its distribution in populations today, researchers estimate that people with the lactase persistence mutation would have produced between 2 and 19% more offspring than people without it. Over the course of three or four hundred generations, that's a stunning advantage, the equivalent of a minor superpower in evolutionary terms. Because researchers know when it started and can trace its flow via ancient DNA and modern populations, lactase persistence may be a perfect proxy for understanding Europe's deep past. The timing of the mutation's appearance and its rapid spread, it turns out, coincided exactly with the spread of agriculture in the late Stone Age, a development also known as the Neolithic. By the early Bronze Age, around 5,000 years ago, lactase persistence in parts of Europe was close to 95% – and cattle herding became a preoccupation. Cattle bones eventually represented more than two thirds of the animal bones in many archaeological sites in central and northern Europe. Yet there's a catch that makes the rapid spread of lactase persistence a puzzle. If the mutation appeared in a person with no access to cow, sheep or goat milk, the mutation wouldn't make much of a difference in terms of reproduction. Instead, the extremely fast evolution of lactose tolerance required a sort of 'perfect storm' of cultural and genetic factors: The lactase persistence mutation, plus a culture of herding and probably cheese-making. Research into the archaeology, genetics, chemistry and demography of ancient dairying is revealing fascinating things about Europe's past. It's also helping answer a question archaeologists have been arguing over for decades: Was the important shift from hunting and gathering to agriculture simply due to the spread of a new technology among existing populations? Or was there an influx of cow-herding colonists from central Anatolia who out-competed the locals thanks to a combination of genes and technology?

## Session 29 Theatre 3

**Which direction for animal nutrition science?**

*Y. Kyriazakis*
*School of Agriculture, Food and Rural Development, Newcastle University, Newcastle-upon-Tyne, NE1 7RU, United Kingdom; ilias.kyriazakis@newcastle.ac.uk*

Already at the time of completion of my PhD studies in 1989 animal nutrition was considered a mature science, as most questions in animal physiology and metabolism had already been addressed. Consequently a career in livestock nutritional sciences did not constitute an attractive option. Nutritional research was revitalised in the '90s, by the development of molecular techniques that allowed questions to be asked at a different level. Some successes associated with these advances are the contribution of maternal nutrition to the development of the offspring (epigenetics). The recent emergence of the issue of Food Security and the interest in the environmental impact of livestock systems have forced us to revaluate the issue of feed and system efficiency, for which animal nutrition has a clear role to play. However, most research questions associated with these issues are reframing old questions in a different way. For example, what will be the outcome of feeding animals with novel or alternative feed ingredients? This does not constitute a revitalised scientific discipline, but delegates nutrition to 'support' science. In my view, however, nutritional science has yet major contributions to make in the interface between different disciplines, as it is there where novel and challenging scientific questions have yet to be answered. Such an interface has been the focus of my work over the last 20 years. In this presentation I will give a couple of examples to substantiate this assertion. The first comes from the ability of nutrition to affect the immune responses of animals infected with pathogens. As we are running out of options to control pathogens in livestock, enhancing the ability of animals to cope with pathogens through nutrition constitutes an attractive proposition. The second deals with the interaction between genotype and nutrition. Through the advent of genomics, genetic change is now happening at a much faster rate and there is an imperative to understand how to manage it. The question of custom-made nutrition for lightweight pigs that have resulted from genetic selection for increased litter size constitutes a good case in point. On the basis of these I suggest that nutritional science has both a significant contribution to make and be an attractive career option for innovative young animal scientists. In my view the award of the Leroy Award for a 25 year career in a 'mature science' discipline provides support to my proposition.

## Session 31 Theatre 1

**Brasilian beef costs of production in comparison with EU ones**

*S. De Zen, D.V.M. Bedoya and M.S. Santos*
*ESALQ – Universidade de São Paulo, Centre for Advanced Studies on Applied Economics (CEPEA), Caixa Postal 09, 13400-970 Piracicaba, Brazil; sergdzen@usp.br*

The beef world production may be efficient from both economic and social points of view. This paper aims to compare production costs of the Brazilian fed cattle with those in Europe and analyze how this aspect (costs) needs to be allocated according to the economic rationality. In Brazil, six farms were evaluated: they represent three typical production systems (complete cycle, sow/calf and fatteing) and are located in two different areas regarding biomes and agriculture structure: Southeastern and Central-Western. In Europe, farms from France, Germany, Spain, Ukraine and Ireland were considered.

**Competitiveness of Italian beef production**

*K. De Roest and C. Montanari*
*Research Center for Animal Production (CRPA), Viale Timavo 43/2, 42121 Reggio Emila, Italy; k.de.roest@crpa.it*

Beef production in Italy has declined in the last ten years due to a reduction of the domestic demand for beef which in particular in these years of economic crisis suffers price competition of pig and poultry meat. This economic condition is a matter of concern for both cattle farmers and slaughterhouses The national suckler cow herd is relatively small and is not able to produce sufficient calves for the fattening farms. The natural and climatic conditions do not allow for low cost production of young beef calves in Italy. Local high quality beef cattle breeds (Chianina, Marchigiana, Romagnola and Piemontese) raised in Central Italy and Piedmont have conquered however high value niche markets, which enables these breeds to survive. The consumption market is highly segmented in very different price categories connected to a regionally differentiated pattern of demand. The production costs on the beef cattle farms increased year by year, because of the rise in feed costs and the costs of the purchase of young calves ('broutards') from abroad. Also in 2014 beef farms specialized in the fattening of imported bullocks from primarily France still enjoy a significant income support from decoupled CAP premia. The application of the CAP reform for the period 2015-2020 will imply a flat hectare rate of direct income premia and the necessity to comply with the requirements of the 'greening' policy of the CAP. This will put further pressure on the profitability of these farms until 2020. The presentation will contain an economic analysis of the supply chain of beef production in Italy, including supply balance sheets, profitability forecasts for Italian fattening and suckler cow farms until 2020 and an analysis of the slaughter costs of the main slaughterhouses of the country. Results of the round table organised by CRPA on the strengths and weaknesses of beef production in Italy to be held on the 11$^{th}$ of June 2014 at the Agribenchmark Beef Conference in Turin will provide further material for a discussion on the future competitiveness of Italian beef production.

---

Session 31 Theatre 3

**Beef competitiveness: different farmers' strategies to improve their efficiency**

*P. Sarzeaud*
*French Livestock Institute, Livestock farming and society, BP 82225, 35652 Le Rheu cedex, France; patrick.sarzeaud@idele.fr*

The economic efficiency of the beef farming systems is at the heart of the concerns regarding the context of international competition. The huge growth of costs of production leads farmers to develop new strategies to resist and remain. In 2013, from a data base of 570 farms, the French livestock institute chose 20 of them that appeared to be more efficient, economically speaking. Those farms are spread over the country and point up the diversity of the beef activity, gathering fatteners and cowcalf producers as well. Indeed, they represent the variety and the difference of strategies implemented by the farmers. Some could have increased the size of the operation in order to reach some scale economy, but others could in the contrary find their high level of rentability in a drastic management of the inputs used, a third orientation is based on the good valorization of the products. In all, the herd management needs to be completed and efficient.

**Productivity and sustainability of beef production systems**

*L. Mogensen[1], T. Kristensen[1], N.I. Nielsen[2], A. Munk[2], E. Søndergaard[3] and M. Vestergaard[1]*
*[1]Aarhus University, Research Centre Foulum, 8830 Tjele, Denmark, [2]Knowledge Centre for Agriculture, Agro Food Park 15, 8200 Aarhus, Denmark, [3]AgroTech A/S, Agro Food Park 15, 8200 Aarhus, Denmark; lisbeth.mogensen@agrsci.dk*

Beef is produced in rather diverse production and farming systems, ranging from intensive indoor fattening based on bull calves of dairy breeds to extensive pasture based systems based on specialized beef breeds. The aim was to evaluate the effect of production system on productivity and sustainability using a holistic approach. The applied method was based on a set of indicators from a sustainability index developed in 2012. Productivity was described in term of use of feed and area. In the most intensive systems, 7 kg DM of feed from 9 $m^2$ of land was used per kg meat compared with 30 kg DM from 155 $m^2$ in the most extensive system. The high land use in the extensive system has a positive effect on biodiversity mainly due to a high share of grazing and low production intensity. On the other hand, the low feed efficiency has a negative effect on climate with a carbon footprint of 30 kg $CO_2$/kg meat in the extensive system compared with 9 kg $CO_2$/kg meat in the intensive system. Regarding effect on environment, the ammonia emission from barn and storage was higher per kg meat produced in the extensive than the intensive system (112 and 29 g $NH_3$-N per kg meat) due to the high feed use and thereby N excretion per kg meat. In contrast, average treatment index for pesticides per ha was lower in the extensive system (0.3 vs 1.2 respectively) due to lower pesticide use for grass production compared with cereals. Animal welfare was evaluated in terms of behaviour, housing condition and feeding. Finally, economic sustainability was evaluated in term of gross margin for meat to secure a balance between long term income and expenses. According to this work we think that a holistic approach is needed to evaluate different beef production systems to secure a sustainable development that at the same time considers nature, climate, animal welfare and socioeconomic conditions.

---



**Factors affecting performance and economic traits of intensively managed beef cattle in Italy**

*G. Cesaro, M. Berton, L. Gallo and E. Sturaro*
*University of Padova, DAFNAE, Viale dell'Università 16, 35132 Legnaro, Italy; giacomo.cesaro@studenti.unipd.it*

In Italy nearly two thirds of the beef bulls are reared in northern regions by specialized intensive fatteners that import stock calves from other European countries, mainly from France. Usually they feed them until slaughter with total mixed rations based on maize silage and concentrates. The increase of feed and animal purchase costs, the growing requirements for environmental sustainability and the expected detrimental changes in Common Agricultural Policy for intensive systems are currently endangering the viability of this sector. This study aimed to analyse the effects of management factors such as beef breed, body weight (BW) at arrival and diets on performance traits, nutrient excretion and net sale gain (NSG), taken as economic indicator, computed as (beef value at sale – beef value at purchase) and expressed per day of fattening (€/d). Seventeen specialized fattening herds were visited monthly during 1 year to collect performance data and diets formulation and composition. The data were analyzed using ANOVA with herd, breed and classes of management factors as fixed effects. The fattening period for beef cattle lasted on average 229 d, with an average BW at purchase and at sale of 370 and 671 kg, respectively. The average NSG computed per day of fattening was 2.44 €/d, with a large variation due to herd and breed. Feed intake averaged 9.75 kg DM, and the average gain to feed ratio was 7.5. Diets were characterised by an average crude protein and phosphorus content of 141 g/kg DM and 3.78 g/kg DM, respectively, the ratio between NSC and NDF averaged 1.4, and average self-sufficiency was 33% on daily DM intake. During the production cycle the intake, the retention and excretion of nitrogen averaged 0.22, 0.04 and 0.18 kg/d, respectively, and those of phosphorous averaged 0.04, 0.01 and 0.03 kg/d, respectively. Further studies are needed for better defining the profitability of this system by including the ratio and cost of feedstuffs purchased outside the farm.

**Heart rate assessment in beef bulls prior slaughter is associated with feed efficiency**

*Y.R. Montanholi[1], S. Lam[2] and S.P. Miller[3]*
*[1]Dalhousie University, Plant and Animal Sciences, 58 River Road, B2N 5E3 Truro, NS, Canada, [2]University of Guelph, Animal and Poultry Science, 70-50, Stone Road E, N1G 2W1 Guelph, ON, Canada, [3]AgResearch Limited, Invermay Agricultural Centre, Puddle Alley, P.O. Box 50034, 9053 Mosgiel, New Zealand; yuri.montanholi@dal.ca*

Feed efficiency is a trait of singular relevance for the beef industry, with strong ties to economic and environmental aspects. Beef cattle genetic improvement programs have begun to target feed efficiency, measured through residual feed intake (RFI; kg/d). However, there are difficulties in obtaining this phenotype and a shortage of indirect phenotypes applicable in commercial settings; therefore, limiting the selection for improved RFI. The cardiac function, measured through heart rate (HR; beats/min), accounts for a portion of the basal energy expenditure associated with variation in RFI; thus, the HR may be an indicator of RFI. Thirty-two yearling crossbred *Bos taurus* bulls were submitted to a 112 d feed intake and performance evaluation (28 d body weight and ultrasound for body composition) for calculating RFI. Groups of 4 bulls were mounted with continuous heart rate monitors (Polar RS800CX, adapted) on the day before being shipped for slaughter, the HR monitors where removed after stunning. The full HR recording was partitioned and averaged into resting (lower HR period overnight), transportation (HR during the 50 min. transport to abattoir) and pre-stunning (HR at stunning box). Bulls were divided into HIGH (n=16; RFI=0.93) and LOW RFI (n=16; RFI=-0.93) groups, the RFI means were different ($P<0.05$) and HR traits were compared between RFI groups. The three HR traits were different ($P<0.05$) between HIGH and LOW. Resting, transportation and pre-stunning HR were 68.9 vs 77.6, 89.3 vs 96.2 and 109.7 vs 125.2 for LOW vs HIGH respectively. The HR may be a promising assessment of RFI given the capacity of distinguishing groups of bulls with distinct RFI under different circumstances. This application of HR assessment may also have usage for monitoring welfare.

---



**The use of carcass traits collected in the abattoir using video image analysis to improve beef yield**

*M.P. Coffey and K. Moore*
*SRUC, Animal Breeding and Genomics, Easter Bush Campus, Midlothian, EH25 9RG, United Kingdom; mike.coffey@sruc.ac.uk*

Data was collected from 1 large abattoir in the UK for 18 months resulting in over 80,000 records to date. As well as carcass traits on the EUROP scale (carcass weight, fat and conformation), predicted weights of 9 carcass traits was available using video image analysis (VIA). The 9 VIA traits are the yield of individual primal cuts with different retail values e.g. striploin, fillet, flank. These data were analysed for variation at the sire level. The average carcass weight reported at slaughter was 355 kg for steers and 316 kg for heifers. The average weight for each VIA primal cut yield was consistently higher in the steers than the heifers. The average carcass conformation in the steers was between classes R- and R= in the EUROP conformation scale whilst the average carcase conformation in the heifers was class R- in the EUROP conformation scale. The average carcass fat in steers was between classes 3= and 3+ and in heifers between classes 3= and 3+ in the EUROP conformation scale. The average age at slaughter for steers was 762 days and for the heifers was 842 days. When summed according to the published retail value of individual carcass traits, carcasses of a given weight, conformation score or fat score differed in their retail value by £400. This demonstrates that the EUROP grid is a crude way of valuing carcasses and is likely more suited to beef production in an intervention market environment rather than valuing it for retail yield. It also demonstrates that VIA data may be suitable for creating EBVs that would enable producers to breed animals of changed distribution for muscle of high value rather than just changed conformation.

**Eating quality of meat from grazing Holstein bulls and Limousine × Holstein bulls and heifers**

*M. Therkildsen[1] and M. Vestergaard[2]*
*[1]Aarhus University, Department of Food Science, Blichers Allé 20, 8830 Tjele, Denmark, [2]Aarhus University, Depatment of Animal Science, Blichers Allé 20, 8830 Tjele, Denmark; margrethe.therkildsen@agrsci.dk*

We examined the eating quality of meat from spring-born crossbred Limousine × Holstein bulls (CB) and heifers (CH) compared with Holstein bulls (HB) raised over two grazing seasons on high-yielding pastures and slaughtered directly from pasture after the 2nd grazing season at a fixed slaughter age (16.9 months). The meat quality was evaluated by a nine-member trained sensory panel on an unstructured scale from 0 to 15, with 0 representing minor aroma and taste characteristics and tough meat and 15 representing intense aroma and taste characteristics as well as tender meat. The evaluation was done on M. longissimus dorsi (LD) and M. semimembranosus (SM) sampled from 8 HB, 8 CB and 8 CH animals 24 h post mortem and aged at 4 °C for additional 13 days. The filet (LD) was prepared as 20 mm steaks on a frying pan to an internal temperature of 63 °C and the round (SM) was prepared as a roast in an oven (100 °C) to an internal temperature of 63 °C. The panel recognised no variation in the aroma and the taste of SM whereas the LD from HB had more gamy ($P<0.003$) and liver ($P<0.02$) aroma and more gamy ($P<0.004$) and bitter ($P<0.001$) and less meaty ($P<0.001$) taste compared with CB and CH. The texture of both cuts was affected by the sex of the animals, thus the tenderness and bite resistance was inferior in cuts from HB and CB compared with CH ($P<0.02$). The tenderness score of 5.7 and 5.2 for SM and 6.2 and 6.1 for LD from HB and CB, respectively, is expected to be too low to fulfil consumer expectations of tender beef. In conclusion crossbred Limousine × Holstein bulls and heifers may be an alternative to purebred Holstein bull because of the aroma and taste, but the texture of the crossbreed bulls needs to be improved either through changes in the production strategy or in the pre and post mortem handling.

---



**Effect of linseed supplementation on fatty acid composition and lipid oxidation of Maremmana beef**

*G. Conte[1], A. Serra[2], L. Casarosa[2], A. Cappucci[2] and M. Mele[2]*
*[1]University of Pisa, CIRAA, via Vecchia di Marina 6, 56122 Pisa, Italy, [2]University of Pisa, DISAAA-a, via del Borghetto 80, 56124 Pisa, Italy; gconte@agr.unipi.it*

The interest in feeding strategies aimed to increase the level of polyunsaturated fatty acid (PUFA) in animal products is growing. Linseed supplementation has been reported as an useful strategy in order to increase PUFAω-3 content. However, less knowledge is available about the effect of PUFAω-3 enrichment on the shelf life of meat. The aim of this work was to evaluate the intramuscular fat oxidative stability of beef obtained from Maremmana young bulls fed a diet supplemented with extruded linseed. 20 Maremmana young bulls were divided in 2 groups: experimental (E- diet supplemented with extruded linseed during fattening) and Control (C- diet without linseed) group. Longissimus dorsi muscles of each animal were sampled, grinded and stored at 4 °C. Fatty acid composition, cholesterol, peroxide, dienes, thiobarbituric acid reactive substances (TBARs), retinol, carotenoids, and cholesterol oxidation products (COPs) levels were determinated at 0, 2 and 6 days of storage. The linseed supplementation led to a significantly increase (+85%) of PUFA ω-3. E group's beef showed a higher susceptibility to lipid oxidation: TBARs and COPs levels increased by 141% after 2 and 6 days respectively. On the contrary, retinol and carotenoids similarly decreased in beef from both groups. In conclusion, the beef from C group was more stable with regard to oxidation, probably due to the lower levels of pro-oxidant substances and to the proper content of antioxidant substances. In contrast, although the level of antioxidant substances at time 0 was similar between groups, the beef from E group showed a significant increase of oxidation products already after two days. These results suggested that linseed supplementation improves meat quality by increasing PUFAω-3 content, but higher levels of antioxidants than those normally present in the intramuscular fat are needed in order to control the oxidation during the storage.

**Intramuscular fat and dry matter of beef are correlated with untrained consumer scores**

*S.P.F. Bonny[1], I. Legrand[2], R.J. Polkinghorne[3], G.E. Gardner[1], D.W. Pethick[1] and J.F. Hocquette[4]*
*[1]Murdoch University, School of Veterinary and Life Science, 90 South St, Murdoch, Western Australia 6150, Australia, [2]Institut de l'Elevage, Service Qualite des Viandes, MRAL, 87060 Limoges Cedex 2, France, [3]Polkinghornes, 431 Timor Rd, Murrurundi, 2338, NSW, Australia, [4]INRA, VetAgro Sup, UMRH 1213 Thelix, 63122 Saint Genes Champanelle, France; s.bonny@gmail.com*

Supplying beef with consistent eating quality to consumers is vital to the beef industry. We investigated the relationship between eating quality and biochemical measurements, intramuscular fat (IMF), moisture content and heme iron. Six cuts (striploin, outside, rump, tenderloin, oyster blade and the topside) from 18 French cattle were grilled, medium or rare. In total 540 untrained French consumers rated the steaks for tenderness, flavour, juiciness and overall liking, according to MSA protocols. These scores were combined on a weighted basis (0.3, 0.3, 0.1, 0.3 respectively) to make a score called MQ4. The sensory scores were analysed using a mixed linear model with cut, age and country included as fixed effects, and animal ID as a random term. The biochemical measurements were then incorporated one at a time as covariates into the model, as well as their interactions with all fixed effects to assess their association with the sensory scores. IMF (range 0.23-9% wet matter) had a positive, linear, relationship with flavour and overall liking (magnitude of effect 6.08 and 5.94 across the IMF range). Moisture content (range 70-77% wet matter) had a negative, linear, relationship with flavour and overall liking (magnitude of effect 6.35 and 6.15 across the moisture content range). When together in the same model, both remained significant implying that moisture content has a relationship with eating quality beyond its correlation with IMF. All other relationships were not significant. The effect of IMF is smaller than that seen by Pannier *et al.* in lamb striploin (10.1, 9.1), but slightly larger than lamb topside (6.6, 5.4).



**Effect of sustained release nonprotein nitrogen and dietary escape microbial protein on beef heifers**

*S.L. Vandoni, N. Keane, S. Maher, P. Caramona and S. Andrieu*
*Alltech Bioscience Centre, Summerhill Road, Dunboyne, Co. Meath, Ireland; svandoni@alltech.com*

Improving nitrogen (N) efficiency in beef cattle diets contributes to cost-effective production. This study investigates the effects of a sustained release (SR) non-protein nitrogen (NPN) source combined with a dietary escape microbial protein (Rumagen®, Alltech Inv., KY) as partial replacement of vegetable protein on finishing beef heifers' live and dead performance. One hundred and eighty continental heifers were allocated to 2 different finishing diets: Control (n=90; basal diet of corn meal, brewers' grain, straw, rapeseed, molasses, soy hulls) and treated (n=90; basal diet reformulated to include Rumagen®). The reformulated diet was designed to decrease crude protein (CP) level from 121 to 117 g/kg DM. Animals were on treatment for a finishing period of 57 days. Liveweight was measured on days 0, 35 and 57. Average daily gain (ADG) was calculated from liveweight data. Cold carcass weight was determined in the slaughterhouse. Data were analysed by ANOVA and adjusted for initial liveweight and days on farm. Despite a lack of difference between liveweight and cold carcass weight, there was a significant difference ($P<0,05$) in ADG from d35 to 57 (1.34 vs 0.96 kg) and overall ADG (1.51 vs 1.42 g). These data support that SR NPN and dietary escape microbial protein could be efficiently used as replacement of part of the vegetable protein source, aiming to improve performance in finishing beef cattle.

**Carbon footprint of the French-Italian beef production chain**

*L. Boselli[1], A. Gac[2], L. Bava[1], L. Migliorati[3], A. Tamburini[1] and J.B. Dollé[2]*
*[1]Università di Milano, Via Celoria,2, Milano, Italy, [2]Institut de l'Elevage, BP 85225, 35652 Le Rheu, France, [3]Consiglio per la Ricerca e la sperimentazione in Agricoltura, Via Porcellasco,7, 26100 Cremona, Italy; leonardo.boselli@unimi.it*

The fattening of weaners imported from France is the main beef production system in the Northern Italy area. The aim of this work was to estimate the whole carbon footprint (CF) of the French – Italian beef production chain using life cycle assessment. Beef production cycle was analyzed from cradle to the Italian farm exit gate, considering each stage (inputs, cow-calf French farm, finishing Italian farm) and taking into account greenhouse gas emissions at animal scale and those related to feed production and animals. A French case study was used to get information about the main cow-calf system based on grass exporting weaners to Italy. A questionnaire was submitted to 10 Italian fattening breeders located in Veneto or on the boundary between Lombardia and Emilia Romagna. Kilogram of live weight sold (LW) was chosen as functional unit (FU). A common Tier 2 methodology from IPCC, 2006 was applied for the two countries to assess methane and nitrous oxide emissions while on-farm $CO_2$ and off-farm emissions were calculated using specific emission factors reported by literature. Biophysical allocation has been applied to split impacts between the two co-products of the French system: cull cows and weaners. The global CF of the young bulls, born in France and fattened in Italy is 19.6 $kgCO_2$-eq per kg LW. The calf-to-weaners phase explains about 60% of these emissions, while the transport represents 3%. These results are in the range of other studies performed on French or Italian young bulls systems. The environmental efficiency of the different systems has now to be analyzed by including other environmental impacts such as eutrophication which is an important issue in some areas in the North of Italy (Po Valley) or carbon storage.

---



**The effect of suckler cow genotype and stocking rate on dry matter intake and methane emission**

*F. Lively, A. Richmond, A. Wylie and D. Lowe*
*Agri-Food and Bioscience Institute, Sustainable Beef Systems, Large Park, BT26 6DR, United Kingdom; francis.lively@afbini.gov.uk*

Grazed grass is a relatively cheap and nutritious feed but herbage quality can vary with management, potentially impacting on animal intake and methane ($CH_4$) production. However, there remains a dearth of $CH_4$ emission data from suckler cows of different genotypes when grazing pasture. The aim of the current study was to investigate the impact that cow genotype (dairy origin vs beef origin) and grazing stocking rate (high vs low) had on their dry matter intake (DMI) and $CH_4$ emissions. The experiment was designed as a 2×2 factorial i.e. two cow genotypes were grazed on a predominantly perennial ryegrass (Lolium perenne L.) sward at either high (HS) or low (LS) stocking rate at a ratio of 40:60, respectively. The study consisted of forty eight spring calving suckler cows of two genotypes; 24 Limousin × Holstein Friesian cows (LF) (dairy origin) and 24 pure Stabiliser cows (ST) (beef origin) of mean initial body weight (BW) 535 kg (s.d. 88) and 567 kg (s.d. 105), and mean body condition score (BCS) of 2.9 (s.d. 0.91) and 3.0 (s.d. 0.81) respectively. Cows were balanced across the four treatments (LF – HS, LF – LS, ST – HS and ST – LS) for weight, age, parity, condition score and calving date. Methane emission and DMI were estimated using the $SF_6$ and n-alkane techniques, respectively, during a two week period in August 2013. Initial analysis of these data using ANOVA shows that there was no significant difference in the DMI of cows on the high or low stocking rate; however the ST cows had significantly greater DMI values than LF cows. There was no significant difference between stocking rates in terms of $CH_4$ emissions (g/d). However, the ST cows emitted significantly greater amounts of $CH_4$ (g/d) than the LF cows; 274 g/d compared with 245 g/d, respectively. When $CH_4$ emissions were expressed as $CH_4$ (g/d) per kg DMI there were no significant treatment effects, hence suggesting that DMI is the primary driver of $CH_4$ production.

**Competitiveness of extensive beef cattle farms located in the dehesa ecosystem (SW Europe)**

*A.J. Escribano[1], P. Gaspar[2], F.J. Mesías[3], A.F. Pulido[3] and M. Escribano[1]*
*[1]Faculty of Veterinary Sciences. University of Extremadura, Animal Production and Food Science, Avda. de la Universidad s/n, 10071. Caceres, Spain, [2]Faculty of Agriculture. University of Extremadura, Animal Production and Food Science, Carretera de Caceres s/n, 06071. Badajoz, Spain, [3]Faculty of Agriculture. University of Extremadura, Economy, Carretera de Caceres s/n, 06071. Badajoz, Spain; ajescc@gmail.com*

Beef pasture-based systems of Europe, especially those located in disfavored areas, play a key role in the maintenance of the ecosystems and rural population. However, they suffer from low competitiveness. In this sense, the design of improvement measures is needed. Due to this, this work has two objectives: (1) to assess the competitiveness of the beef pasture-based systems of the Spanish rangelands ('dehesas'); (2) to design improvement measures. For this purpose, a Competitiveness Index (CI) was created on the basis of technical and economic indicators. According to this index, farms were classified in three groups (1 = highly competitive but very dependent on subsidies; 2 = medium level of competitiveness; 3 = low competitiveness but low dependence on subsidies). Group 1 showed the highest scores for the CI (61.46%), mainly due to their lower external dependence on feedstuff, their lower production costs, their larger amount of fattened calves sold per cow, and their high profitability rate. Group 2 had intermediate scores for the majority of indicators, as well as for the CI (49.20%). However, Group 3 showed very low results for the economic indicators, they also showed the lowest dependence on subsidies, and the lowest scores for the CI (34.40%). In general terms, the profitability rate, fattened calves sold per cow, and the dependence on subsidies are the issues on which farms' competitiveness mainly rely. In this sense, the increase of the self-sufficiency, the implementation of a fattening period, and the establishment of contracts with the next link of the agri-value chain must be encouraged in the farms analyzed.



**Productivity dynamics of a premium Welsh beef supply-chain during a three year period (2010 to 2012)**

*S.A. Morgan[1], S.A. Huws[1], G.H. Evans[2], T.D.M. Rowe[2] and N.D. Scollan[1]*
*[1]Aberystwyth University, Institute of Biological, Environmental and Rural Science, Aberystwyth, SY23 3EB, United Kingdom, [2]Celtic Pride Ltd, Carmarthen, SA31 3SG, United Kingdom; sgm8@aber.ac.uk*

There are a number of specialist beef schemes in the UK which market specific characteristics, such as area of geographical origin, breed (e.g. Hereford or Aberdeen Angus) or brand identity, in order to differentiate their products. One example is the premium Welsh beef brand 'Celtic Pride Ltd', who strive to deliver fully traceable Welsh beef which is of consistently high eating quality via defined protocols covering on-farm, slaughter and processing criteria. Commercially, animals are typically selected for slaughter primarily on degree of fatness. This study's aim was to describe data trends in cattle numbers, sex, breed, age, carcass weight (CW), conformation and fatness, and to examine the relationships between age and carcass weight. Number of cattle slaughtered has increased ~19% per year from 2,925 in year 1 to 4,149 in year 3, with a total of 10,565 cattle slaughtered during the 3 year period. Heifers accounted for 56% and steers 44% of the total number of cattle, with 73% either Limousin or Charolais. During the three years, 77.4% of the total cattle slaughtered achieved the set criteria for age (14-30 mo), carcass weight (240-390 kg), fatness (2-4 L) and conformation (E-R) to qualify for premium. Average daily carcass gain (ADCG) was calculated and plotted against CW. These parameters were found to have a positive relationship, with ADCG ranging 0.22-0.98 kg/d and CW ranging 189.9-463.2 kg. Mean and standard error values were 0.46±0.001 kg/d and 327.7±0.38 kg for ADCG and CW, respectively. From a physiological perspective, a positive relationship between these two parameters was anticipated. Interestingly, the data suggests that slower growing animals with lower ADCG result in lighter carcasses at slaughter, irrespective of target fatness grades.

**Growth and carcass quality of grazing Holstein bulls and Limousine × Holstein bulls and heifers**

*M. Vestergaard[1], C. Cakmakci[1], T. Kristensen[1], K.F. Jørgensen[2], M. Therkildsen[1] and M. Kargo[1,2]*
*[1]Aarhus University, Foulum, 8830 Tjele, Denmark, [2]Knowledge Centre Agriculture, Cattle, Agro Food Park 15, 8200 Aarhus N, Denmark; mogens.vestergaard@agrsci.dk*

We examined the performance of crossbred Limousine × Holstein bulls (CB) and heifers (CH) compared with Holstein bulls (HB) when utilizing two grazing seasons on high-yielding pastures and a fixed slaughter age (16.9 months). The study included 15 HB, 15 CB and 15 CH purchased in April at 20 d of age (1$^{st}$ winter). Calves were kept indoor in groups of 5 animals until weaning at 3 mo. Average daily gain (ADG) from birth to weaning was 724 g/d and not different between treatment groups. Calves were gradually introduced to a grass-silage based ration from 3 to 4 mo, and were then raised on a mixed ryegrass-white clover pasture from 4 to 7 mo (1$^{st}$ summer). From late October till mid-May (2$^{nd}$ winter), animals were kept in the same groups of 5 animals and were housed in deep litter stalls: Animals had free access to a low energy grass-haylage ration. ADG during 2$^{nd}$ winter was 1,012, 1,052 and 930 g/d for HB, CB and CH, respectively. The 2$^{nd}$ summer, animals were grazing in a rotational paddock system (18 paddocks) in the same groups of 5 animals (9 groups) and generally moved to a new sward every week. Animals were slaughtered directly from pasture in mid-August (3×5 animals, one block) or early September (6×5 animals, two blocks). Before turn-out, during 1$^{st}$ summer, and during 2$^{nd}$ winter, HB and CB had higher ADG than CH ($P<0.02$). During the first 11 wk of 2$^{nd}$ summer, ADG of HB, CB and CH were 1081, 1,357 and 847 g/d (SE 50 g/d, $P<0.001$). LW at slaughter was 534, 575 and 480 kg and ADG from birth to slaughter was 948, 1018 and 841 g/d for HB, CB, and CH, respectively ($P<0.001$). Carcass wt, EUROP conformation (1 (lowest) -15 (highest)), and fatness (1-5) was 272, 315 and 249 kg, 3.0, 7.0 and 5.3, and 1.0, 1.2 and 2.9 for HB, CB and CH, respectively ($P<0.001$). Crossbreeding markedly improved carcass conformation but fatness of pasture-fed bulls was not acceptable.

---



**Growth and Carcass traits of early- and late-maturing suckler bulls, at four carcass weights**

*D. Marren[1,2], M. McGee[2], A.P. Moloney[2], A. Kelly[1] and E.G. O'Riordan[2]*
*[1]University College Dublin, School of Agriculture & Food Science, Belfield, Dublin 4, Ireland, [2]Teagasc, AGRIC, Grange, Dunsany, Co. Meath, Ireland; declan.marren@teagasc.ie*

Intake, growth and carcass characteristics of early- and late-maturing suckler bred bulls were determined at four carcass weights. Sixty spring born (21 March, s.d. 34.3 days), early-maturing (EM – Aberdeen Angus and Hereford sired) and sixty spring-born (24 February, s.d. 45.8 days) late-maturing (LM – Charolais and Limousin sired) weaned bulls were assigned to a 2 Breed type, (B) × 4 Carcass weights, (CW) factorial arrangement of treatments, balanced for breed and initial weight within breed type. The two B were: EM and LM. The four CW were 260, 300, 340, and 380 kg. Bulls were adapted to an *ad libitum* concentrates (C) (870 g/kg rolled barley, 60 g/kg soya bean meal, 50 g/kg molasses and 20 g/kg minerals/vitamins) plus grass silage (GS) (DMD 693 g/kg) *ad libitum* diet. Bulls were slaughtered on reaching the treatment mean live weight to achieve the target CW. Data were analysed using the Mixed model procedure of SAS. The model contained fixed effects of B and CW and their interaction. Initial age was included as a covariate. To achieve the same CW EM were heavier ($P<0.01$), had a lower kill-out rate, higher carcass fat score, a longer duration to slaughter and were older at slaughter ($P<0.001$) compared to LM. C DMI did not differ ($P>0.05$) between the breed types. Kill-out proportion ($P<0.001$), carcass conformation score ($P<0.001$) and C DMI ($P<0.01$) increased with increasing CW with the exception of those at 380 kg carcass, where these variables were lower than for the 340 kg carcass groups. There was a B × CW interaction for GS DMI whereby DMI increased with increasing carcass weight for LM but only up to the 340 kg carcass for EM. Carcass fatness ($P<0.001$), GS DMI ($P<0.01$), age at slaughter ($P<0.001$) and days on the farm increased ($P<0.001$) with increasing CW. Although ADG and DMI were not significantly different between B, EM bulls needed longer feeding to reach a common CW.

**Fattening performances of three cattle genotypes under a tunisian feedlot system**

*S. Maatoug Ouzini[1], W. Gharbi[2] and H. Ammar[2]*
*[1]Institut Supérieur des Etudes Préparatoires en Biologie Géologie de la Soukra, 49 avenue 13 août Choutrana II, 2036 Soukra, Tunisia, [2]Ecole Supérieure d'Agriculture Mograne, Mograne Zaghouan, 1121, Tunisia; soniamaatoug@yahoo.fr*

In Tunisia, more than 11,000 farmers practice bulls' fattening in Bizerte governorate, this region is located in the Northern of the country under a humid climate. About 70% of fatteners purchase their animals at 9 months age's corresponding to the first phase of fattening period. The main objective of this study was to compare three breeds on their performance recording. The three breeds (Blonde d'Aquitaine (60 bulls), Limousine (31 bulls) and Charolaise (35 bulls)) received the same feeding practices during eight months. The feedlot system is the common practiced feeding system in Tunisia. Main results showed that at the beginning of the experience no weights' differences were obtained between the three breeds. At the end of the fattening period, Limousine reached the highest weight with 500 kg compared to Blonde d'Aquitaine and Charolaise with 450 and 480 kg, respectively. These results showed that under Tunisian conditions the Limousine breed has the best performances and has a good adaptation to the prevailed production system.



**Effects of differential supplementation during first winter and at pasture on suckler bulls**

*K. McMenamin, D. Marren, M. McGee, A.P. Moloney, A.K. Kelly and E.G. O'Riordan*
*Teagasc, AGRIC, Grange, Dunsany, Co. Meath, Ireland; kevin.mcmenamin@teagasc.ie*

The aim of this study was to determine if additional live weight gain (LWG) achieved during the winter would be retained subsequently through concentrate supplementation at pasture, and the effects on carcass traits of weaned suckler bulls. Ninety spring-born Charolais and Limousin sired bulls (369 kg) were assigned to a 2 (first winter supplementation level–WSL) × 3 (supplementation level (as % of diet) at grass-SG) factorial arrangement of treatments. The two WSL were: (1) *ad libitum* grass silage (GS) + 3 kg of barley-based concentrate (C) daily (3G); and (2) GS + 6 kg C daily (6G), were imposed over 127 days. Subsequently, bulls were turned out to pasture and rotationally grazed for 98 days. The SG were: (1) Grass only (G0); (2) Grass + 0.2 C (G20); or (3) Grass + 0.4 C (G40). After this period bulls were re-housed for finishing on *ad libitum* C + GS. Finishing lasted 76 days followed by slaughter at an average age of ~19 months. Data were analysed using the mixed models procedure of SAS. The model contained fixed effects of WSL and SG and their interaction. Initial weight was included as a covariate. There were no WSL × SG interactions recorded ($P>0.05$) for the traits presented. Bulls offered the higher WSL (6G) had greater daily LWG during the winter and during the finishing phase, but lower LWG at pasture ($P<0.001$), resulting in a heavier slaughter ($P<0.05$) and carcass weight ($P<0.01$) compared to those offered 3G. Bulls offered G40 had higher ($P<0.001$) LWG at pasture than those offered G0 and G20, which did not differ ($P>0.05$), whereas during the indoor finishing period those offered G0 had higher ($P<0.05$) LWG than G20 and G40, which did not differ ($P>0.05$). However, SG had no effect ($P>0.05$) on slaughter weight, carcass weight, carcass fat and conformation scores. Both supplementation during the first winter and at pasture increased animal live weight.

**Effect of lack of mineral supplementation on bone characteristics in beef calves**

*C. Lazzaroni and D. Biagini*
*University of Torino, Department of Agricultural, Forest and Food Sciences, Largo P. Braccini 2, 10095 Grugliasco, Italy; carla.lazzaroni@unito.it*

The evaluation of the mineral status of bovine skeleton has been the purpose of several investigations during the last twenty years, mainly for physiological studies, but nowadays also the mineral supplementation costs and the environmental impact of some minerals have become relevant. Of these elements, phosphorus is the most critical, both for its cost and for the environmental fallout due to excess in rations, and then in animal manure. The effect of mineral supplementation on the performances of 32 fattening bulls slaughtered at 18 or 24 months of age was studied on two breeds with different growing rate (Friesian, early maturing dairy breed, F; and Limousine, late maturing beef breed, L). During the last rearing period, lasting 106-268 d, feed rations were designed to meet the needs of animals for an increase of 1 kg/d, by adding (HM) or not (LM) a supplement of dicalcium phosphate (1.5% on feed). Initial and final animal live weights, so as carcass weight, were recorded to calculate the productive indices (average daily gain, ADG; and carcass yield, CY), and at slaughter the left fore-shin of each animal was collected to study metacarpus characteristics (weight, length, middle circumference, wall thickness). ADG was higher in F-18 and L-24 (1.09 and 1.07 kg/d) and in F-HM and L-LM (1.04 and 1.00 kg/d) animals. CY was higher, as expected, in L than F (317 vs 244%) and in older animals (334 vs 231%), showing also interactions (diets-breeds and diet-age). Metacarpus was heavier in older animals (520 vs 421 g), longer in F than L (246 vs 223 mm) as well as in older animals (242 vs 224 mm), showing interaction breed-age and highest value in F-24 (262 mm), with a longer circumference in L than F (120 vs 115 mm) and in older animals (122 vs 113 mm), and a thicker wall in older animals (7.3 vs 5.3 mm), showing interaction breed-age and highest value in F-24 (8.3 mm). No negative effects of diet were observed on health and productive parameters.

---



**Evaluation of the Greek buffalo breed for meat quality**

*A. Zotos[1], D. Petridis[1], B. Skapetas[2] and V.A. Bampidis[2]*
*[1]Alexander Technological Educational Institute of Thessaloniki, Department of Food Technology, P.O. Box 141, 57400 Thessaloniki, Greece, [2]Alexander Technological Educational Institute of Thessaloniki, Department of Animal Production, P.O. Box 141, 57400 Thessaloniki, Greece; bampidis@ap.teithe.gr*

A study was conducted with 29 male and 18 female Greek buffaloes (Bubalus bubalis) to determine the chemical composition and fatty acid (FA) profile of the longissimus lumborum et thoracis muscle. The moisture, protein, lipid and ash content of the muscle were 70-76, 19-24, 0.36-4.36 and 0.78-1.18%, respectively. Factor analysis revealed that the increase of moisture content led to a decrease of either lipid or protein content, while ash followed the changes of moisture content. The oleic acid was the most predominant FA (30.26-44.12%), followed by palmitic acid (21.10-30.73%) and stearic acid (14.87-27.09%). From the polyunsaturated FA, the linoleic acid ranged from 1.51 to 10.03%, while the sum of conjugated linoleics (CLA) [C18:2ω-6 (9c,11t), C18:2ω-6 (9t,11c)] from 0.25 to 1.65%, the linolenic acid from 0.08 to 1.54% and the arachidonic acid from 0.27 to 6.75%. According to factor analysis, the increase of lipid content was followed by an increase of stearic acid and a decrease of arachidonic acid and CLA. Oleic and linolenic acids were not influenced by lipid content; however, they affected each other, thus the increase of oleic acid was followed by an increase of linolenic acid. Factor analysis also revealed that arachidonic acid and CLA were the characteristic FA of male buffaloes, showing higher values. This research has been co-financed by the European Union (European Social Fund – ESF) and Greek national funds through the Operational Program 'Education and Lifelong Learning' of the National Strategic Reference Framework (NSRF) – Research Funding Program: ARCHIMEDES III. Investing in knowledge society through the European Social Fund.

**Physicochemical and textural properties of sausage type 'Horiatiko' from Greek buffalo meat**

*D. Petridis[1], A. Zotos[1], B. Skapetas[2] and V.A. Bampidis[2]*
*[1]Alexander Technological Educational Institute of Thessaloniki, Department of Food Technology, P.O. Box 141, 57400 Thessaloniki, Greece, [2]Alexander Technological Educational Institute of Thessaloniki, Department of Animal Production, P.O. Box 141, 57400 Thessaloniki, Greece; bampidis@ap.teithe.gr*

The physicochemical and textural properties of sausage type 'Horiatiko' made from Greek buffalo (Bubalus bubalis) meat (65-70%) were extensively evaluated using varying additions of buffalo and pork fat (0-30%). Fresh and grilled samples enriched with 70% buffalo meat and 30% buffalo fat showed the highest levels of moisture (63.9 and 48.2%, respectively) and protein (13.6 and 30.8%, respectively) content. The increase of buffalo fat and, thereafter, the corresponding decrease in pork fat resulted in an increasing mechanical hardness and chewiness of sausages, reaching maximum values at 70% buffalo meat and 30% buffalo fat (95.3 and 49.6N, respectively, results from Texture Analyzer). A strong relationship between mechanical and sensory attributes was evidently established proving also a reliable panel assessment. Hedonically, the most acceptable sample was the recipe with 70% buffalo meat and 30% pork fat and, the second preferable, the sample with 70% buffalo meat, 15% pork fat and 15% buffalo fat. This research has been co-financed by the European Union (European Social Fund – ESF) and Greek national funds through the Operational Program 'Education and Lifelong Learning' of the National Strategic Reference Framework (NSRF) – Research Funding Program: ARCHIMEDES III. Investing in knowledge society through the European Social Fund.



**Effect of grilling on chemical composition of the longissimus muscle of Purunã bulls**

*F. Zawadzki[1], D.C. Rivaroli[1], R.A.C. Passetti[1], C.A. Fugita[1], C.E. Eiras[1], J.L. Moletta[2], J. Torrent[3] and I.N. Prado[1]*
*[1]State University of Maringá, Animal Science, 5790 Av. Colombo, 87020900, Maringá, Paraná, Brazil, [2]Agronomic Institute of Paraná, Animal Science, s/n Av. Euzébio de Queiroz, 84001970, Ponta Grossa, Paraná, Brazil, [3]Oligo Basics, 7269 Bent Bow Trl, 55317-7562, Chanhassen, Minnesota, USA; fernando.zawadzki@hotmail.com*

The longissimus muscle (LM) from Purunã bulls (¼ Aberdeen Angus + ¼ Caracu + ¼ Charolais + ¼ Canchim) fed with Essential® oils (EOL) and glycerine (GLY) was used to investigate the effect of grilling on the chemical composition. Bulls were distributed in a complete randomized design by a factorial scheme 2×2 and chemical composition were analysed by ANOVA using the GLM procedure to examine treatment (fresh vs grilled procedure). Thirty-two bulls (206±20.0 kg, 8±2 mo old) were allocated in individual pens and distributed into 4 treatments: CON, without GLY or EOL; EOL, with EOL (3 g/animal/day); GLY, with 203 g/kg of GLY on dry matter (DM) and GEO, with GLY and EOL; for 252 d (corn silage, ground corn, and soybean meal were used as-fed). The bulls were slaughtered at a commercial slaughterhouse (468±31.5 kg, 19±2 mo old). LM samples were taken between the $11^{th}$ and $13^{th}$ ribs after chilling (24 h at 4 °C), taken to the laboratory and frozen (-20 °C). LM fresh and grilled were analysed (±3 months). LM grilled samples were sliced into 1 cm-thick steaks 2 h before grilling. Industrial grill was used to samples grilling at 200 °C until reaching 75 °C internal temperature. Total lipids were extracted using method with chloroform/methanol mixture. LM samples were defrosted at 4 °C±1.5, ground, homogenised and analysed in triplicate. The moisture and ash contents of the meat were determined according to (ISO-R1442, 1997), and the crude protein content was obtained following (ISO-R937, 1978). LM grilled increase crude protein ($P<0.01$), and total lipids ($P<0.02$). Whereas, moisture ($P<0.01$) and ash ($P<0.05$) decrease in the LM grilled.

**Effect of grilling on fatty acids composition of the longissimus muscle of Purunã bulls**

*F. Zawadzki[1], D.C. Rivaroli[1], A. Guerrero[1], R.M. Prado[1], M.V. Valero[1], D. Perotto[2], J. Torrent[3], I.N. Prado[1] and J.V. Visentainer[1]*
*[1]State University of Maringá, Animal Science, 5790 Av. Colombo, 87020900, Maringá, Paraná, Brazil, [2]Agronomic Institute of Paraná, Animal Science, 274 Rua Máximo João Kopp, 82630900, Curitiba, Paraná, Brazil, [3]Oligo Basics, 7269 Bent Bow Trl, 55317-7562, Chanhassen, Minnesota, USA; fernando.zawadzki@hotmail.com*

Longissimus muscle (LM) from Purunã bulls (¼ Aberdeen Angus + ¼ Caracu + ¼ Charolais + ¼ Canchim) fed with Essential® oils (EOL) and glycerine (GLY) was used to investigate the effect of grilling on the content of fatty acids (FA). Bulls were distributed in a complete randomized design by a factorial scheme 2×2 and FA composition were analysed by ANOVA using the GLM procedure to examine treatment (fresh vs grilled). Thirty-two bulls (206±20.0 kg, 8 mo old ±2) were allocated in individual pens and distributed into 4 treatments: CON, without GLY or EOL; EOL, with EOL (3 g/animal/day); GLY, with 203 g/kg of GLY on dry matter (DM) and GEO, with GLY and EOL; for 252 d (corn silage, ground corn, and soybean meal were used as-fed). The bulls were slaughtered at commercial slaughterhouse (468±31.5 kg, 19 mo old ±2). LM samples were taken between the 11$^{th}$ and 13$^{th}$ ribs after chilled (24 h, 4 °C) and taken to the laboratory and was frozen (-20 °C). LM fresh and grilled were analysed (±3 months). LM grilled samples were sliced into 1 cm-thick steaks 2 h before grilling. Industrial grill was used to samples grilling at 200 °C until reaching 75 °C internal temperature. Total lipids were extracted using method with chloroform/methanol mixture. Fatty acid methyl esters were prepared by triacylglycerol methylation (ISO-R5509, 1978). Then, the esters were extracted with 2 ml of n-heptane and analysis by chromatography (Thermo-3300). The FAs 18:1 n-9c, 18:2 n-6, 20:4 n-6, 22:6 n-3, ∑-polyunsaturated FA (PUFA), ∑-n-6, and ratio ∑-PUFA:∑-Saturated FA and ∑-n-6:∑-n-3 were higher ($P<0.05$) in the LM grilled. Whereas, the FAs 20:5 n-3, 22:0, 22:5 n-3 and ∑-monounsaturated FA decreased ($P<0.05$) in the LM grilled.



**Testosterone levels in calves of different neuroreflection types by habituation test**

*J. Petrák, O. Debrecéni, P. Juhás and K. Vavrišinová*
*Slovak University of Agriculture, Faculty of Agrobiology and Food Resources, Department of Animal Husbandry, Tr. A. Hlinku 2, 94901 Nitra, Slovak Republic; peter.juhas@uniag.sk*

Ability of adaptation in calves to changed breeding conditions is very interesting for economics in animal husbandry. The dynamics of a reaction to loads can be assessed by excitability of the nervous system. Neuroreflection types in terms of nervous system arousal are divided into highly EHb +, medium EHb° and low excitable EHb- types. One of the hormones that have the neuroprotective effect is testosterone, which controls the activity of Brain-derived neurotrophic factor – BDNF. BDNF is an important factor for growth and connecting synapses in the brain. The aim of this work was to monitor the concentrations of testosterone in relation to excitatory type of calves. The calves were divided into excitatory types by quantity of locomotion in the habituation test. We have measured the intensity of locomotion in the habituation chamber. We tested 13 male Holstein calves. The habituation test was performed in calves with body weights of 120-150 kg. Saliva was collected using a gauze swab. Into the EHb+, EHb°, and EHb- type 4, 5, and 4 pieces were included, respectively. Saliva was sampled immediately before test habituation. The concentration of testosterone in saliva was determined by ELISA. The optical absorbance was measured using Microplate Reader Model DV 990BV4, UniEquip Deutschland. The highest levels of testosterone were found out in EHb°, lower levels were in EHb- and the lowest in EHb+ type. Statistically significant differences in concentrations of testosterone were observed among EHb+ versus EHb° ($P\leq0.01$) and EHb° versus EHb- ($P\leq0.05$). Between EHb+ and EHb- groups we have not found statisticaly significant difference in the testosterone concentrations. The highest level of testosterone in this EHb type ensures its neuroprotective function and potential better ability to cope with animal husbandry conditions. This work was supported by projects VEGA 1/2717/12, ECACB – ITMS 26220120015.

**The horse as key player of rural development in Europe**

*N. Miraglia*
*Molise University, Agriculture, Environment, Feed Sciences, Via De Sanctis, 86100 Campobasso, Italy; miraglia@unimol.it*

Over the last years there has been considerable worldwide changes in rural environment: in the past the total number of resident and raised animals decreased with related problems linked to forest and woodlands fires, to soil erosion, to desertification, etc. Actually the exploitation of marginal lands with equids breeding represents many potential economical advantages coming from different kinds of equids productions, from the improving of sustainable agriculture models and from the increase of tourism demand for the discovery of mountain and hilly areas. Equids are now playing a rising role in animal biodiversity preservation and in the micro economy of rural areas. Horse husbandry contribute to the diversification of agricultural activities and to the utilization and preservation of extensively cultivated and natural areas, as well as development of agritourism on horse farms. Such activities are more and more linked to the maintenance of population in rural areas, to new relationships between citizen and cultural rural life and consequent preservation of traditional socio-cultural life. In EU's rural areas the diversity of the 'equine culture' and the equine-related activities are focused to the leisure and tourism activities, to the preservation of rural socio-cultural life and to the most relevant socio-economic issues. In this context equids play a key role in sustainability referred to the safeguard of geographical area, of the environment and landscape and to the preservation of genetic type and biodiversity that present different characteristics in the European districts, from North to Southern countries.



**New formations and understandings of horse and society in the Nordic countries**

*R. Evans*
*Hogskulen for landbruk og bygdeutvikling, 213 Postvegen, 4353, Klepp Stasjon, Rogaland, Norway; rhys@hlb.no*

As we enter the 21st Century, perceptions of horses and riding have changed across the Nordic region. At the end of the 20th Century it became clear that a transition in emphasis was happening, from work horse to hobby horse. Evidence for this transition could be seen in the changing gender of riders and owners, in the growth of new leisure-based activities, and in the growing numbers of horses across the Nordic nations, after their precipitous decline following the mechanisation of their former working fields. Now that consensus on transition is established. Across the Nordic world we see horses undertaking a vital role as companions in combating the stresses of an increasingly urbanized lifestyle, as offering people new opportunities to experience nature and to reflect on their own relationship with non-human animals, and becoming important assets for rural development in equine tourism, the health and well-being sector, and other parts of the consumption and experience economies. This presentation will summarize the current state of the horse world across the diverse nordic lands and identify key components of the new equine economy and new ways of thinking about horses and how we keep them.

**The equine power in Mediterranean countries: is there a place for it?**

*A.S. Santos[1,2], M.J. Fradinho[3], I. Cervantes[4] and N. Miraglia[5]*
*[1]EUVG, Lordemão, Coimbra, Portugal, [2]CECAV-UTAD, Vila Real, 5001.901, Portugal, [3]CIISA-FMV, ULisboa, Lisboa, Portugal, [4]UCM, Dep. Animal Production, Madrid, Spain, [5]Molise University, Dep. Agriculture, Environment and Feed Sciences, Campobasso, Italy; assantos@utad.pt*

Horses have always been a part of human development. From the earliest times, they were a part of the Mediterranean history and culture. Today's horse still portrays a cultural heritage of extreme importance, in all these countries. Mediterranean regions have unique characteristics that make them specially prone to horse use in its most widen expression. Equines were and are an important piece of rural development. Closely associated with a sustainable type of agriculture, they present themselves as an important biological tool for maintaining the rural landscape and ecological biodiversity of Mediterranean agricultural systems. In some parts of Portugal, Spain and Italy, the terrain orography is so characteristic (and difficult), that its agricultural use is still closely dependent on horsepower. Like in other European countries, sports and leisure figures on the use of the horse have been increasing. The number of equestrian practitioners almost tripled over the last ten years and other uses like the therapeutic riding became an important tool in what concerns children and adult treatment and social integration. Another developing sector is the equestrian tourism. The southern Europe countries provides beautiful landscapes, either in mountain or plain areas, witch visit can be enriched when using equestrian activities. In the context of the sustainable use and alternative ways for profitability of some local breeds, production of milk and meat could be a solution for the preservation of these animals. In current times, looking at the horizon and taking into account all the challenges that the world faces (economic, climatic and demographic), Mediterranean countries can look at its equines as, like in the past, a loyal and important fellow traveller in writing our future history.

---



**The horse as a key player of local development in the United States in the 3rd millennium**

*C.J. Stowe*
*University of Kentucky, Agricultural Economics, 307 C. E. Barnhart, Lexington, KY 40546, USA; jill.stowe@uky.edu*

The horse has played an important role in the settlement and development of the United States, from the Native Americans and Spanish explorers to the settlement of the West. Horses offered new advantages in battle, provided early transportation, and were crucial to success in ranching and farming. Although their role has since changed from a necessity to a luxury good, horses will continue to play a critical role in the development of the United States, and in particular, in the measured development of local areas in at least three ways. First, equine-related agritourism can provide tremendous opportunities for visitors to experience and interact with local communities in a unique way. Second, in select areas of the U.S., equine economic clusters will continue to be critical centers of knowledge growth and outreach in the equine industry, supporting equine-related research, business, and education. Finally, the horse will be a critical component of a sometimes overlooked part of local development: land preservation.

**The horse as key player of local development in Brazil: from past to future**

*A.A.O. Gobesso*
*School of Veterinary Medicine and Animal Science/São Paulo University, Animal Nutrition and Production, Avenida Duque de Caxias Norte, 225, 13630-900 Pirassununga/São Paulo, Brazil; gobesso.fmvz@usp.br*

Brazil is a very young country with continental dimensions, suitability agricultural and livestock production. Thus, horses, donkeys and mules, which didn't exist in the continent, were of great importance during the colonial period to reach the most distant locations and were instrumental in the economic cycles of gold, sugar cane, rubber, coffee and thus becoming part of the population's life. As the country kept on growing it started to import Pure blood animals of different breeds and possessing today the condition to export world widely. The largest group of horses, estimated at three million animals, having until today, important role in the extensive production of beef cattle, prevalent in the country. As a challenge for the future, is necessary to study the aspects linked to feeding and management in tropical conditions, including facilities and equipment, to improve the quality and longevity of the horses. Develop mechanisms for control of endemic diseases and thus increase the export and realization of international sports competitions. In the same sense, it is crucial to stimulate the development for the use of horses in rural tourism and leisure activities. To this end, a study was developed to evaluate the size of the horse complex agribusiness in Brazil, which proved the economic importance of this sector for the country. Was also created the Horse Commercial Chamber, to discuss among the breeders associations what are the goals to be achieved for evolution of national production, and severe measures are being adapted to control the most important diseases affecting livestock. Potentially, Brazil has the largest herd of horses at the world in tropical climate and great potential for development of new technologies connected with the breeding and use of equines.

---

Session 32 Theatre 6

**Role of the horse industry in the regional development in France**

*G. Bigot[1], X. Dornier[2], E. Perret[1], C. Vial[3] and N. Turpin[1]*
*[1]Irstea, UMR Métafort, 9 avenue Blaise Pascal, CS 20085, 63178 Aubière, France, [2]Institut Français du Cheval et de l'Equitation, Observatoire Economique et social, route de Troche, BP 3, 19231 Arnac Pompadour cedex, France, [3]Institut Français du Cheval et de l'Equitation, Inra UMR 1110 MOISA, 2 place Pierre Viala, 34060 Montpellier, France; genevieve.bigot@irstea.fr*

In Europe, the horse industry is seldom included in regional development policies. Stakeholders explain this dismiss by the lack of visibility about this sector. In fact, the main difficulties to gather available data can be related to the diversity of horse uses (races, sport and leisure, meat…) and to the contribution of this industry to various sectors such as agriculture, sports and tourism. The aim of this paper is (1) to present data available in national bases in France, and (2) how these data precise the place of the horse industry in the different French regions. The national number of horses is estimated at 1 million animals with significant regional differences. For example, Basse-Normandie is the first French region with a total of 77,000 horses and gathered near 50% of the national livestock for races but only 6% of sport and leisure horses, while Auvergne is the first area for meat production thanks to 5,000 draught mares. The horse industry is well involved in the agricultural sector of these regions, as they both gathered 20% of farms with horses and 20% of horses registered by the agricultural census. Studies in French regions as Languedoc-Roussillon show that the horse industry is also developed in suburban districts where private owners and equestrian structures use small plots left available by the increase of urbanization. These regularly updated data are used in various French contexts as public investments and national CAP implementation. While national and regional policies depend more and more on European regulations, the European horse industry remains invisible despite its obvious role in rural development, probably because of a lack of a Community procedure for the organization of data.

**New strategies in genetic evaluations and tools to preserve the diversity in horse breeds. A review**

*I. Cervantes*
*Departamento de Producción Animal, Universidad Complutense de Madrid, 28040 Madrid, Spain; icervantes@vet.ucm.es*

The aim of the horse breeding programs is to attain more competitive animals to obtain good results in high level sport competitions. The performance of horse traits is affected by environmental factors such as the subjectivity (judges scores), or the rider effect that increase the difficulty to assess the best genetic model to make the genetic evaluation. Moreover, the discontinuous nature of some traits as the rank, has promoted the use of different transformations in the genetic models. But there are other approaches that use the rank without transformations and try to correct the breeding value for the competition level (e.g. Thurstonian models). Crossvalidation can help to choose the best model. Finally, the genomic selection could increase the genetic response and diminish the generation interval in this specie. On the other hand, in small populations the preservation of the maximum genetic diversity is one of the main objectives. This ensures more genetic variation for extreme traits, which could be useful in case the population needs to adapt to a new environment/production system. Both the genealogies and the molecular information are used to monitor the genetic variability. There are several parameters to measure the genetic diversity and many mate strategies to decrease the inbreeding depression effect. In this sense, the effective population size is one of the most important issues given its usefulness as a measure of the long-term performance of the population regarding both diversity and inbreeding. When the selection and the preservation of genetic variability are both important more sophisticated strategies should be applied.

---



**Mare's milk in cosmetics: preliminary tests**

*D. Gavrić[1], M. Kovačič[1], M. Pliberšek[1], P. Keršmanc[1], K. Žmitek[1,2] and K. Potočnik[3]*
*[1]Higher School of Applied Sciences, Department of cosmetics and Institute of cosmetics, Gerbičeva 51a, 1000 Ljubljana, Slovenia, [2]Nutrition Institute, Ljubljana, Tržaška 40, 1000 Ljubljana, Slovenia, [3]University of Ljubljana, Biotechnical Faculty, Department of Animal Sciences, Groblje 3, 1230 Domžale, Slovenia; Klemen.Potocnik@bf.uni-lj.si*

The equine industry has an important impact in the economy like races, sport, leisure use, therapeutic riding and use of horses in tourism. Mare's milk (MM) production hasn't played an important role in EU horse industry. Beside systematic (consuming) use also topic use of MM seems to be very interesting. The beneficial effects of MM on the skin have been recognized for centuries and it is believed Cleopatra have bathed in it to keep her skin young and beautiful. MM naturally contains a range of important nutrients and its unique active ingredients that could support the regeneration and regulation of the skin functions, and also prevent premature ageing. It is said to be beneficial for the treatment of irritated and sensitive skin. But those potential beneficial effects on skin are still to be researched and confirmed, as there is lack of scientific data regarding its dermal applications. In this study we checked the effect of topical use of MM on the skin. MM products vs placebo were tested on three volunteers that reported dry and sensitive skin. The volunteers followed prescribed skin care regimen for two months with test products used on one and placebo on the other side of the face twice a day. Topical use ofMM products resulted in improved elasticity and hydration of the skin, follicular openings size reduction, effective brightening of pigment stains and more pronounced wrinkle reduction compared to placebo products. Due to identified potential of MM for topical use larger study would be performed. Results from this study supported the idea that MM production for cosmetics is a challenge to improve the economic situation of endangered horse breeding. That could be very important in self preservation of that breeds.

## Session 32 Theatre 9

**A new promising approach for donkey (*Equus asinus*) sperm freezing**

*P. Milon[1], I. Couty[1], F. Mea[2], A. Garot[2], Y. Gaude[1], F. Reigner[1] and M. Magistrini[1]*
*[1]Institut National de la Recherche Agronomique, UMR Physiologie de la Reproduction et des Comportements, 37380 Nouzilly, France, [2]Institut Français du Cheval et de l'Equitation, Haras National de Blois, 41043 Blois cedex, France; pierre.milon@tours.inra.fr*

In France, all the 7 donkey breeds are endangered. Actually the most efficient method to keep genetic diversity is sperm cryopreservation of best Jackass in each breed. But today, there is no efficient sperm cryopreservation procedure to ensure adequate fertility. The aim of this study was to improve efficiency of freezing extender for donkey sperm. Recently a study from Italy demonstrated that donkey sperm can be frozen using INRA96 extender supplemented with 2% of egg yolk and 2.5% of glycerol obtaining a fertility rate of 25%. In our laboratory, we previously demonstrated on stallion semen, *in vitro* and *in vivo*, that egg yolk can be replaced by sterilized egg yolk plasma in a ready to use medium for stallion frozen semen (INRA Freeze) composed of INRA96 supplemented with sterilized egg yolk plasma and glycerol. Other authors showed that INRA96 supplemented only with glycerol can be used to freeze stallion semen. The present study was conducted to check the efficiency of three freezing extenders: INRA-Freeze and INRA96 supplemented with 2.5% (IN96-2.5G) or 4% (IN96-4G) glycerol on donkey semen. Semen from 2 Jackasses (Grand Noir du Berry) and 3 ejaculates per jackass were frozen. Then motility parameters using CASA (computer Assisted Sperm Analysis) and membrane integrity (using flow cytometry) were evaluated. The data demonstrate that all the motility parameters are significantly higher when sperm cells are frozen in INRA-Freeze extender compared to IN96-2.5G and IN96-4G. However we did not observed any significant difference on membrane integrity. The next step will be the evaluation *in vivo* of the efficiency of INRA-Freeze extender on donkey stallion semen. The authors thank ANR for financial support ('Investissements d'Avenir', ANR-11-INBS-0003)

---

## Session 33 Theatre 1

**Factors that impact the stress and immune responses of beef cattle and swine**

*J.A. Carroll[1], N.C. Burdick Sanchez[1], T.H. Welsh Jr.[2] and R.D. Randel[3]*
*[1]USDA-ARS, Livestock Issues Research Unit, 1604 East FM 1294, Lubbock, Texas 79403, USA, [2]Texas A&M University, Department of Animal Science, 442 Kleberg Center, College Station, Texas 77843, USA, [3]Texas A&M AgriLife Research, 1710 FM 3053 N., Overton, Texas 75684, USA; jeff.carroll@ars.usda.gov*

Producers are aware that exposing livestock to repeated or extended periods of stress can have detrimental effects on productivity and well-being. Known to a lesser extent are the differing biological responses that occur when livestock are exposed to diverse stressors. Additionally, it is often disregarded that subjecting gestating females to stressful stimuli may elicit epigenetic effects with lasting consequences on the progeny. We previously demonstrated that exposing piglets to a cold environmental temperature (ET) during an immune challenge causes a drop in body temperature that could be fatal. In addition to succumbing to hypothermia, piglets are also often crushed by their dams while seeking warmth, a situation exacerbated in sick pigs during cold stress. In cattle, we discovered that ET can influence the animal's acute phase response following an immune challenge. Specifically, heat stress alters the febrile response during an immune challenge, as well as immune indices that aid in an animal's return to homeostasis. In addition to environmental stressors, we have evaluated practices such as restraint and transportation that can impose stress on gestating females to determine if these events impose any risk to the progeny either in utero or during postnatal life. Specifically, we demonstrated that restraining late-term gestating sows alters the immune system of postnatal pigs when exposed to an immune challenge. In cattle, we demonstrated that exposing gestating cows to repeated transportation alters perinatal and postnatal indices including immune parameters and animal temperament. Our research reveals previously unknown relationships linked to environmental and prenatal stressors that may aid in elucidating the importance of stress regulation in livestock.

**Research overview: the impact of animal transport on the health and welfare of beef cattle**

*B. Earley*
*Teagasc, Animal and Bioscience Research Department, Animal & Grassland Research and Innovation Centre, Teagasc, Grange, Dunsany, Co. Meath, Ireland; bernadette.earley@teagasc.ie*

The transport of livestock can have major implications for their welfare, and there is strong public interest and scientific endeavour aimed at ensuring that the welfare of transported animals is optimal. Physical factors such as noise or vibrations; psychological/emotional factors, such as unfamiliar environment or social regrouping; and climatic factors, such as temperature and humidity, are also involved in the transport process. We have conducted a series of animal transport studies from Ireland to Spain, and from Ireland to Italy, using a roll-on roll-off ferry. The overall objective of the studies was to investigate the physiological, haematological and immunological responses of weanling heifers transported to Spain, and of weanling bulls transported to Italy under EU legislation and to evaluate the implications in terms of animal welfare. During these studies, appropriate physiological, haematological and immunological measurements were made on the animals which quantified the effect of transport (by road and sea) on the degree of stress imposed and the performance of the animals in the post-transport period. Physiological, haematological and immunological parameters (including interferon-$\gamma$ production, cortisol, protein, immunoglobulin, urea, white blood cell numbers and differentials, and haptoglobin) were used to determine the welfare status of animals, before, during and after the respective transport journeys. Age-matched control animals that were blood sampled for the same parameters at times corresponding to the transported animals were retained in Ireland as controls. While transient changes in physiological, haematological and immunological parameters were found in the transported and control animals relative to baseline levels, the levels that were measured were within the normal physiological range for the age and weight of animals that were studied.

---



**Blood transcriptome response to LPS in pigs**

*E. Merlot[1], A. Prunier[1], M. Damon[1], F. Vignoles[2], N. Villa-Vialaneix[3], P. Mormède[2] and E. Terenina[2]*
*[1]INRA, UMR1348 PEGASE, 35590 Saint-Gilles, France, [2]INRA, UMR1388 GenPhySE, 31326 Castanet-Tolosan, France, [3]INRA, UR0875 MIAT, 31326 Castanet-Tolosan, France; elodie.merlot@rennes.inra.fr*

The aim was to describe immune, hormonal and metabolic responses of pigs to a systemic inflammatory challenge using blood transcriptome. Male pigs of 4 different gonadal statuses were used (intact, surgically castrated, immunized against GnRH, and surgically castrated plus immunized against GnRH, 7-9 pigs per group). Blood samples were collected via jugular catheters at 149 d of age, just before (t-1), and 1, 4 and 24 hours (t1 to t24) after a lipopolysaccharide administration (*Escherichia coli* O55:B5 serotype LPS, 15µg/kg, i.v.). Blood lymphocyte/granulocyte (L/G) ratio was measured, and was maximal at t1 ($P<0.001$), back to baseline at t4 ($P>0.1$) and lower than baseline at t24 ($P<0.001$). Total blood mRNA was extracted and the transcriptome of intact pigs was analyzed using the Agilent 60K microarray. Differential expression among time points was observed for 2,148 annotated single genes (Benjamini-Hochberg adjusted P values <0.01). The 5 first functional clusters of genes identified with the David software concerned immunity and inflammation, chemotactism, apoptosis, ion transport, and energy metabolism (209, 196, 185, 127 and 121 genes respectively). Among these genes, 51 were selected to investigate the level of expression in the 4 time points and 4 gonadal statuses by fluidigm © PCR. The L/G ratio significantly influenced 37 of these genes ($P<0.05$). Therefore, for all genes, the effects of time, gonadal status and their interaction were analyzed on the residuals of the L/G covariance analysis. The variation along time in response to LPS was confirmed for 43 genes ($P<0.05$). None of them was affected by the gonadal status. In conclusion, blood transcriptome can be used for investigating LPS responses provided that respective blood cell proportions are known. Furthermore, the innate immune response activated by LPS injection is not sensitive to the sexual status of male pigs.

**Tradeoff between production and survival under stress: role of glucose sparing by the mammary gland**

*N. Silanikove*
*Institute of Animal Science, A.R.O., the Volcani Center, Biology of Lactation Lab., P.O. Box 6, Bet Dagan 50250, Israel; nissim.silanikove@mail.huji.ac.il*

Glucose (Glu) is the most important fuel for energy metabolism in cells. In lactating animals, Glu is the major precursor for lactose production (about 80% of Glu entering into the mammary gland) in addition of being a substrate for synthesis of milk proteins and fat. Glu extraction of 3-4 kg/d from blood into the mammary glands is needed to support the requirements of high producing cows, and account for 60-85% of Glu turnover. Some acute abiotic stresses, such as heat stress, and biotic stress, such as infection by *Escherichia coli*, or exposure to its toxin, lipopolysaccharide induces large increase for Glu demands, which may range between 250 g under heat stress to 1 kg/d to under acute immunological response. The present overview is associated with the physiological mechanisms that coordinate prioritization toward acclimation and survival under heat and inflammatory stresses. The reduction in milk yield under stress can be considered as a biological tradeoff between lactational performance and survival. Glu sparing by the mammary gland under stress by up to 1 kg/d under heat stress and even more under inflammatory stress is the main mechanism that allows supporting the increase demand for energy by immune cells and to maintain essential supply of Glu to tissues of critical importance for survival, such as cerebral and heart cells. Signals induced by milk-born negative feed-back mechanism and inflammatory reactions in mammary gland lumen causes shift of mammary gland cells from aerobic-mitochondrial to aerobic glycolysis (the Warburg effect), and serves as the core machinery for Glu sparing by the mammary gland. Supporting evidences, including key biochemical and regulatory (signal transduction) components that underlies move to aerobic glycolysis, are presented to sustain the concept.

---



**Can biomarkers for oxidative stress early predict Bovine respiratory disease in calves?**

*V. Sibony[1,2], M. Yishay[2], R. Agmon[2], A. Orlov[2], S. Khatib[3], J. Vaya[3] and A. Shabtay[2]*
*[1]Tel-Hai academic College, Kiryat Shmona, 1021, Israel, [2]Newe-Ya'ar research center, agricultural research organization, Beef cattle section, 1021, Ramat Yishay, Israel, [3]MIGAL-Galilee Technological Center, Oxidative stress research laboratory, Kiryat Shmona, 1021, Israel; sibony.vered@gmail.com*

Bovine respiratory disease complex (BRD) represents multifactorial etiology disease model, with major economic impact on the beef cattle industry. BRD is resulted from a stress-related alteration in susceptibility of respiratory tract to colonization by pathogens. Although BRD is the most studied feedlot cattle disease, to date there is no effective tool for its early prediction. Accumulating circumstantial evidence indicate the involvement of oxidative stress in the development of BRD. Therefore, the aim of this study was to examine the relevance of oxidative stress markers – exogenous linoleoyl tyrosine (LT) and plasma antioxidant capacity (FRAP method) in early objective prediction of BRD, in young calves, in response to transportation stress. Blood samples were collected pre-transportation and 1 hour, 4 hours, 1 day, 3 days and 7 days post-transportation (n=32) and the health status of calves was evaluated every morning until the slaughter. Epoxidation of the LT marker lipid component in sick calves increased 1 hour post-transportation ($P<0.0001$), indicating that the LT marker allowed us to distinguish between healthy and sick calves, at least 60 days ahead to the appearance of BRD symptoms ($P=0.0067$). Similarly, healthy calves had significant higher FRAP values already pre-transportation ($P=0.0023$). indicating that the antioxidant capacity pre-transportation may be the key to reducing transportation stress related morbidity. These results show that the susceptibility of calves to BRD can be predicted at early life stages and may even be prevented by antioxidant therapy.

**Organic selenium and chromium supplements in female lambs: transportation stress responses**

*A. Mousaie, R. Vlizadeh and A.A. Naserian*
*Ferdowsi University of Mashhad, Animal science, Azadi square, Mashhad, 91775-1163, Iran; moosaee.amir@gmail.com*

The aim of this study was to investigate the effect of feeding seleno-methionine and chromium-methionine on physiological responses to transportation stress (TS) in 24 Baluchi female lambs of 4 months of age. Experimental treatments included control, seleno-methionine (1.5 mg/kg diet), chromium-methionine (0.8 mg/kg) and seleno-methionine plus chromium-methionine (1.5 and 0.8 mg/kg respectively). TS was applied by transporting animals in a truck in a bumpy road for 30 minutes. Blood samples were collected before and after TS(after unloading animals) and serum analyzed for glucose and cortisol. Paired T-test was used to compare the metabolites differences before with after TS. The results indicated that,transporting animals in inappropriate condition significantly increased serum cortisol and glucose in female lambs. Chromium-fed animals had the lowest blood glucose(P=0.0415). Selenium and chromium fed animals showed lower blood cortisol than control group but it was statistically nonsignificant. Based on our results, it concluded that feeding organic selenium and chromium supplements could ameliorate the adverse effects of TS on Baluchi female lambs.

---



**Tryptophan supplementation as stress modulator in steers**

*G. Marin and R. Larrain*
*Pontificia Universidad Católica de Chile, Facultad de Agronomía e Ingeniería Forestal, Dpto. Ciencias Animales, Vicuña Mackenna 4860, 7820436, Santiago, Chile; larrain@uc.cl*

Tryptophan (TRP) is an essential amino acid that reduced stress response in several species by increasing serotonin and reducing stress-related hormones. There are no reports about the effects of unprotected TRP supplementation on circulating serotonin in ruminants. Previously we observed low rumen degradation of purified L-TRP *in vitro* (<20% at 48 hr), so the goal of the experiment was to test if L-TRP supplementation in steers may increase plasma TRP and serotonin. Four steers were used in a Latin square design with 0, 50, 100 and 150 mg L-TRP/kg BW/d, supplemented for 1 week and followed by a 3 weeks washout period. The animals were fed twice per day to maintenance with TRP given in the morning meal. Blood samples were taken during the afternoon meal (8 h after) at days 0, 2, 5 and 7, to measure TRP, TRP:large neutral amino acids (LNAA) and serotonin (SER). The results were analyzed by ANOVA using repeated measures methodology and day 0 as baseline. Results indicated no effect of diet or diet×day interaction (P>0.05 for TRP, TRP:LNAA and SER). Blood TRP (μmol/l) was 17.66±0.91, 17.94±0.91, 17.24±1.02 and 16.63±0.91, rate TRP:LNAA (%) was 6.40±0.344, 6.38±0.344, 5.74±0.370 and 6.19±0.349 and SER (ng/ml) was 1,482±277.5, 1,647±277.5, 1,238±290.5 and 1,727±290.5 for treatments 0, 50, 100 and 150 mg, respectively. Two steers had symptoms of respiratory distress, one when supplemented with 100 and the other with 150 mg TRP/kg BW. Results indicate that purified TRP was not effective for raising plasma TRP in steers. This and the observed symptoms of respiratory distress suggest that purified TRP was extensively degraded in rumen to produce skatole. This is not consistent with our earlier *in vitro* measurement of TRP degradation. We concluded that it was not possible to raise plasma TRP using purified TRP, and therefore, it was not possible to observe any changes in circulating SER. Funded by FONDECYT.

**Effect of rearing system on Iberian piglet's cortisol concentration at weaning**

*F. Gonzalez, J. Robledo, J.A. Andrada, J.D. Vargas and M. Aparicio*
*School of Veterinay Science, University of Extremadura, Animal production, University Avenue, 10071, Cáceres, Spain; fgonzalej@alumnos.unex.es*

Weaning is a stressful time for piglets and secretion of glucocorticoids is a common endocrine response to overcome this situation. The suckling conditions pre-dispose the physiological maturity of piglets, and this factor affects the secretion of glucocorticoids. Therefore, the wide variety of rearing systems becomes an important factor which must be taken into account to achieve a good adaptation after weaning. The aim of these experiments was to assess the effect of rearing systems on glucocorticoids levels at weaning in two different weaning ages. A total of 360 Iberian piglets were reared on three different systems: Intensive, Traditional and Out-doors (I; T; O) and were assigned at random to be weaned at 28 d or 42 d postpartum. Same number of weaned piglets was studied in each system (n=120) and same number of piglets were weaned at each age (n=60). All animals were weighed weekly from 21 to 63 days old, and salivary was sampled with the same frequency from weaning moment to 63 day old. Saliva was collected between 9:00 and 11:00, and samples were assayed for glucocorticoids using an E.L.I.S.A. cortisol kit following the manufacturer protocol (Diagnostic systems, Texas). Data were analyzed using Spss® and statistical significance were accepted at P<0,05. Comparing the three rearing system, at weaning time there were significant differences in cortisol levels in each weaning age (28 d: CI: 2.27; CT: 1.24 and CC: 1,58 µg/dl; 42 d: CI: 1.44; CT: 1.15 and CC: 0,82 µg/dl). However at the end of the study (63 days old), there were no differences (28 d: CI: 2.78; CT: 1.61 and CC: 1.35 µg/dl; 42 d: CI: 1.01; CT: 0.87 and CC: 1.03 µg/dl). Therefore, the rearing systems have an influence on cortisol levels at weaning but not for the whole post-weaning period, in both weaning ages.

---



**Effect of weaning age on Iberian piglet's cortisol concentration and performance**

*F. Gonzalez, J. Robledo, J.A. Andrada, J.D. Vargas and M. Aparicio*
*School of Veterinay Science, University of Extremadura, Animal production, University Avenue, 10071, Cáceres, Spain; fgonzalej@alumnos.unex.es*

Weaning involves marked physiological changes in piglets. The secretion of glucocorticoids is a common endocrine response to weaning stress. The physical maturity of piglets varies according to their age, and the maturity of immune cells and the corticoids levels changes with age. Therefore, the choice between late or early weaning becomes a important factor in ensuring that animals' physiological balance and productivity levels are restored as soon as possible. In order to assess the corticoid levels in response to different weaning ages, a total of 360 purebred Iberian piglets were studied. Three trials were conducted with 24 litters assigned at random to be weaned at either 28 d (n=12) or 42 d (n=12) postpartum. In each trial and from each weaning age, four groups of 15 piglets were assembled with different sexes and sizes (three litters by group and five animals from each litter, and12 replicates per weaning age). All animals were weighed weekly from 21 to 63 d of age, and salivary samples and food intake was controlled with the same frequency. Saliva was collected between 09:00 and 11:00, and samples were assayed for glucocorticoids using an ELISA cortisol kit. Data were analyzed using Spss® and statistical significance were accepted at P<0.05. At the beginning (P=0.798) and end of the study (P=0.450) there were no significant differences between piglets weights. All groups experienced a similar decline in production rates during the first week post-weaning and similar recovery times the following weeks. At weaning time there was no significant differences between piglets in cortisol levels (P=0.150). However, after weaning, there were differences in food intake and in feed conversion rates (P<0.001), and early weaned piglet had high cortisol levels until the end of the study. Therefore, weaning age has different influence on Iberian piglet´s corticoids levels.

**A metabolic factor that influences the somatic cell count in the milk of lactating cows after intram**

*S. Kushibiki[1], S. Shingu[1], Y. Kiku[2], T. Hayashi[2] and S. Inumaru[2]*
*[1]National Institute of Animal and Grassland Science, Ikenodai-2, Tsukuba, Ibaraki, 305-0901, Japan, [2]National Institute of Animal Health, Kannonndai-2, Tsukuba, Ibaraki, 305-0901, Japan; mendoza@affrc.go.jp*

This study was designed to investigate a metabolic factor that influences the somatic cell count (SCC) in the milk of cows with *Staphylococcus aureus* subclinical mastitis after intramammary injection of the recombinant bovine granulocyte-macrophage colony-stimulating factor (rboGM-CSF, 400 μg/5 ml). Twenty-one cows diagnosed to have subclinical mastitis were used in this study. The infusion of rboGM-CSF into the cistern through a catheter was conducted immediately after the morning milking on day 0. Blood and milk samples of the test quarters were collected prior to the infusion (day 0) and 1 h, 3 h, and 6 h and 1, 2, 3, and 7 days after the injection. Statistical analyses The cows were classified into two groups decrease group (DG) and non-decrease group (ND) based on the SCC on day 7. The results were subjected to an analysis of variance for repeated measures. All data were expressed as means ± standard error of means and analyzed with StatView Software Ver5.0. A value of $P<0.05$ was considered significant. Twelve cows were classified into DG and 9 cows into NG. Plasma insulin concentrations in both groups increased immediately after 1 h, and returned to the pre-infusion level at 3 h. Plasma glucose concentration decreased in the DG group during 1-3 h after the injection. However, plasma glucose concentration in the ND group was not affected by the treatment. At day 0, plasma insulin concentration was higher ($P<0.05$) in the ND group than in the DG group. A decrease in milk SCC due to rboGM-CSF was observed in approximately 60% of the cows, suggesting that insulin resistance was related to the changes in SCC after rboGM-CSF treatment.

---



**Towards sustainable and responsible production of livestock**

*M.R.F. Lee[1], G.B. Martin[2], J.F. Tarlton[1] and M.C. Eisler[1]*
*[1]University of Bristol, Veterinary Sciences, Langford, Bristol BS40 5DU, United Kingdom, [2]University of Western Australia, Inst. of Agriculture, 35 Stirling Highway, Crawley WA 6009, Australia; michaelrf.lee@bristol.ac.uk*

We re-assessed the role of ruminants in meeting food requirements in the context of mixed agricultural systems, and focussed on six major challenges of livestock sustainable intensification globally. (1) Poor health and welfare – parasitism and disease interfere with reproduction, lactation, and growth. Drug resistance drives a decrease in reliance on antimicrobials and antiparasitics. Management and genetics are key for 'One Health' in which human and livestock disease are seen as a continuum. (2) Consumption of human food – a third of the world's cereals is fed to livestock. Ruminants can utilize forages and crop residues that are unsuitable for humans, and can graze land unsuitable for crops. (3) Environmental footprint – manage GHG emissions, and reduce loss of reactive N and P to the environment. The measure of environmental footprint should be the mass of greenhouse gas produced per unit mass of food at a system level. (4) Species and genotypes adapted to the local environment – importing exotic livestock into environments to which they are poorly suited imposes unsustainable strains on animal health and welfare, and on feed and energy resources. Environmentally-adapted native and traditional breeds must be conserved and improved by using genomic selection, epigenetics and 'pan-genomics'. (5) Focus on healthy food – animal products are a rich source of protein, essential amino acids, and micro-nutrients. The developed world over-consumes with negative consequences for health but, for the world's poor, there are advantages of high-quality animal products. (6) Husbandry and management – in the developing world, practices were traditionally optimised for status, dowry, and long-term economic resilience. With risks to food security, priority must go to sustainable output of products of high nutritional value that are also socially acceptable and offer export opportunities.

## Session 34 Theatre 2

**Beyond the production paradigm: feeding the world is a matter of sustainability and equity**

*P. Baret*
*Université de Louvain, Earth and Life Institute, University Square 1, 1348 Louvain-la-Neuve, Belgium; philippe.baret@uclouvain.be*

Change is at the agenda of agricultural systems. New environmental challenges are calling for a reconfiguration of food systems according to multiple terms of reference: food provisioning, expansion of ecosystem services and minimization of environmental impact such as loss of biodiversity and contribution to climate change. In most of the cases, malnutrition is caused by inequity and poor economic conditions and is not directly related to a lack of production. The state of present livestock systems is often determined by path dependency mechanisms and sometimes locked in suboptimal situations. If the objective is to provide affordable food for a majority of humans in a sustainable manner, a lucid appraisal of the motivations of investments in research is a prerequisite. In most of the innovation systems, more than one option exists and, in a competitive research world, allocation of resources is critical. Agroecology is a scientific and practical approach of agriculture addressing the environmental issues and the related socio-economical dimensions. In agroecology, the focus is on the food systems: a systemic and joint approach of food availability, food utilization and access to food. By taking advantage of biological synergies and favouring local resources, it is possible to use fewer resources for food production. A better integration of crop and livestock systems contributes to this objective of sobriety. Reframing research objectives according to this agroecological analytical framework paves the way for new pathways in the development of livestock systems. Use of multi-criteria methods of assessment and acknowledgment of the importance of local knowledge are key elements for making these pathways relevant in a perspective of transition.

---

## Session 35 Theatre 1

**Genetic architecture and genetic correlation of milk production traits in German Holstein population**

*E. Amiri and G. Thaller*
*Animal Breeding and Husbandry, Hermann-Rodewald-Straße 6, 24118 Kiel, Germany; e.amiri79@gmail.com*

The aim is to demonstrate the genetic architecture of milk, fat, protein yield and somatic cell score (SCS) using Illumina BovineSNP50 Beadchip as well as identification of chromosomes or genomic regions influence these traits in German Holstein population. 2,333 German Holstein bulls were genotyped with analyzing of 43,838 known position SNPs distributed on 29 autosomes and XY pseudo-autosomal. Total additive genomic variance was calculated by summation of chromosomal variances and covariance between them or summation of SNP variance and covariance between SNPs for milk, fat, protein yield and SCS traits. Finally, chromosomal genetic correlations were estimated as well as SNP genetic correlation for fat and milk yield on BTA14 and BTA20. Sums of the chromosomal additive genetic variances and covariance between chromosomes were equal with total additive genetic variance as well as sums of SNP variances and covariance between SNPs along the genome. Chromosomal additive genetic variance explained 54.49 to 69.9% of total additive genetic variance. The higher additive genetic variances were on BTA14 for milk and fat yields and BTA6 for protein yield and SCS traits, respectively. Sum of SNPs variance explained 6.3 to 9.6% of total additive genetic variance with higher SNPs additive genetic variance on XY pseudo-autosomes. As a result BTA14 exhibited a strong negative correlation between fat with milk and protein yields. Higher positive correlations between milk, fat and protein yields with SCS were observed for BTA26. Conversely, correlations between traits across SNPs highlight either positive or negative correlation between chromosomal regions in particular traits of interest. This can aid the design of low density chip technology with high correlated SNPs for economical traits in genomic selection.

**Unravelling the genetic background of interdigital hyperplasia of the bovine hoof**

*S. Sammler[1], M. Wensch-Dorendorf[1], K. Schöpke[1], B. Brenig[2] and H.H. Swalve[1]*
*[1]Martin-Luther-University Halle-Wittenberg, Institute of Agricultural and Nutritional Sciences, Theodor-Lieser-Str. 11, 06120 Halle, Germany, [2]Georg-August-Universität Göttingen, Institute of Veterinary Medicine, Burckhardtweg 2, 37077 Göttingen, Germany; svenja.sammler@landw.uni-halle.de*

Interdigital Hyperplasia (IH) is a firm mass located in the interdigital space of the bovine hoof that develops from the interdigital skin or subcutaneous tissue. It has been regarded as a sporadically occurring disease, but can reach a prevalence of over 30% on some farms. A hereditary predisposition is assumed, at least when more than one foot is affected and the animal is young. Estimated heritabilities range around 0.25 to 0.43. Firstly, to fortify the assumption of IH to be heritable, we searched for bull lineages passing on predisposition for IH, secondly, we scanned genome-wide for IH associated SNPs. 1,962 first lactation cows were assessed for sub-clinical and clinical IH status at time of hoof trimming of the rear legs. Cows studied originate from seven large commercial herds. 192 cows (87 IH positive, 105 controls) from herds with the highest IH frequencies (5-23%) were selected for the genome-wide association study with the Illumina BovineSNP50 BeadChip. Stratifying the data into genetic lineages yielded one conspicuous bull line. This bull had six sons with more than five daughters in this data set. Probability for daughters of four of these half brothers to be IH affected was high (10-30%), of a further an intermediate frequency was observed (6%), and only one son had no IH affected daughters. Using a mixed threshold model analysis with the genotype included as a fixed effect, the genome-wide association study revealed 12 SNPs chromosome-wide significantly associated with IH status. Both the existence of IH susceptible bull lineages and the detection of IH associated SNPs show that genetic selection for an improved resistance to IH may be possible. Further analyses are necessary to identify causative genes.

---

Session 35 Theatre 3

**Assessing genomic variation in Swiss dairy cattle using annotated whole genome sequence information**

*C.F. Baes[1,2], M.A. Dolezal[3,4], E. Fritz-Waters[5], J.E. Koltes[5], B. Bapst[1], C. Flury[2], H. Signer-Hasler[2], C. Stricker[6], R. Fernando[5], F. Schmitz-Hsu[7], D.J. Garrick[5], J.M. Reecy[5] and B. Gredler[1]*
*[1]Qualitas AG, Chamerstrasse 56, 6300 Zug, Switzerland, [2]Bern University of Applied Sciences, Länggasse 85, 3052 Zollikofen, Switzerland, [3]University of Veterinary Medicine Vienna, Veterinärplatz 1, 1210 Wien, Austria, [4]Università degli Studi di Milano, Via Festa del Perdono 7, 20122 Milano, Italy, [5]Iowa State University, Ames, 50011 Ames, USA, [6]agn Genetics, Börtjistrasse 8, 7260 Davos, Switzerland, [7]Swissgenetics, Meielenfeldweg 12, 3052 Zollikofen, Switzerland; christine.baes@qualitasag.ch*

Whole genome sequence data of 65 key ancestors of genotyped Swiss dairy populations were available for investigation. More than 24 billion reads were mapped to the reference genome UMD31 (96.8%), resulting in an average coverage of 12× for all animals. Variant identification was done using four publically available variant calling programmes and the results were compared. Concordance analysis with high-density chip data showed an average non-reference sensitivity of 99.5% and average non-reference discrepancy rate of 1.0% for all multi-sample calling results. Depending on variant calling software used, between 16,894,054 and 22,048,382 SNP were identified (multi-sample calling); the proportion of insertions and deletions found ranged between 9.6-13.3% and the proportion of multi-allelic sites ranged from 0.1-3.8%. Annotation with Ensembl's Variant effect predictor showed moderate to large caller differences in the number of variant classes and coding consequences found. As an example, differences in the number of severe consequences varied considerably between callers (missense variants: 60,709-71,922, frameshifts: 1,411-2,883, stop retained: 46-54, stop lost: 38-61, etc.). Our results indicate that it is important to be cognizant of differences in variant calling between different programs, as this may complicate interpretation of genotype determination.

**Across breed QTL detection and genomic prediction in French and Danish dairy cattle breeds**

*I. Van Den Berg[1,2,3], B. Guldbrandtsen[3], C. Hozé[1,2,4], R.F. Brøndum[3], D. Boichard[1,2] and M.S. Lund[3]*
*[1]AgroParisTech, UMR1313 Génétique Animale et Biologie Intégrative, 16 rue Claude Bernard, 75231 Paris, France, [2]INRA, UMR1313 Génétique Animale et Biologie Intégrative, Domaine de Vilvert, 78350 Jouy-en-Josas, France, [3]Aarhus University, Center for Quantitative Genetics and Genomics, Department of Molecular Biology and Genetics, Blichers Allé 20, 8830 Tjele, Denmark, [4]UNCEIA, 149 rue de Bercy, 75012 Paris, France; irene.vanderberg@agrsci.dk*

The objective of this study was to investigate the potential benefits of using sequence data to improve genomic prediction across distantly related breeds. A prerequisite for accurate across breeds predictions is that breeds share a substantial amount of the genetic variance. Therefore the first part studied the proportion of QTL shared across breeds. Assuming that QTL are shared, association between markers and QTL must be conserved across breeds. The second part of the study attempted to assess how close markers need to be to the causal variants to use this common variance in predictions. Sequence data was used to quantify the loss in prediction reliabilities that results from using different sets of genomic markers rather than the true causal variants. Our results suggest that there is a large potential to increase the reliability of genomic predictions using sequence data. A proportion of QTL detected within breed was shared across breed. Prediction of the genomic relationships at causative mutations was more accurate when prediction markers close to the causative mutations were used than when a large number of random markers or markers on the Illumina chips were used. This difference was more pronounced for across breed prediction than for within breed prediction. Genomic prediction across breed will thus be more accurate when a selective number of markers within 1 Kb of the causative mutations is included in the model. Whole-genome sequence data can help to get closer to the causative mutations and therefore improve genomic prediction across breed.

---



**Genetic analysis of uniformity of egg color in purebred and crossbred laying hens**

*H.A. Mulder[1], J. Visscher[2] and J. Fablet[2]*
*[1]Animal Breeding and Genomics Centre, Wageningen University, P.O. Box 338, 6700 AH Wageningen, the Netherlands, [2]Institut de Sélection Animale, Hendrix Genetics, P.O. Box 114, 5830AC Boxmeer, the Netherlands; han.mulder@wur.nl*

Uniformity of eggs is an important aspect for retailers because consumers prefer homogeneous products. One of the aspects is the color of the egg, especially for brown eggs. Existence of genetic variance in the degree of uniformity would enable selection to improve uniformity of egg color. Therefore, the objective of this study was to quantify the genetic variance in uniformity of egg color in purebred and crossbred laying hens and to estimate the genetic correlation between uniformity of egg color in purebred and crossbred laying hens that are laying brown eggs. Egg color records of 167,651 eggs from individually housed purebred laying hens and 85,454 egg color records of crossbred laying hens that were housed in group cages as paternal half-sibs, were analyzed. The sires of the purebred and crossbred laying hens originated from the same line and most sires had both purebred and crossbred offspring. The egg color records of crossbred laying hens were collected per cage, so that it was only known to which cage and sire family the record belonged. Double hierarchical generalized linear sire models were used to estimate genetic variance in the mean of egg color and its residual variance. The genetic variance in residual variance at the exponential scale was 0.077 and 0.071, in respectively purebred and crossbred laying hens. The genetic coefficient of variation at the level of the phenotypic standard deviation was 0.055 and 0.070, in respectively purebred and crossbred laying hens. The genetic correlation between purebred and crossbred uniformity of eggs was 0.71, indicating some re-ranking of sires with respect to uniformity of egg color in purebred and crossbred laying hens. The results indicate good opportunities to improve uniformity of egg color in purebreds and crossbreds by genetic selection, ideally with combined crossbred and purebred selection.

**Everyday multithreading: parallel computing for genomic evaluations in R**

*C. Heuer and G. Thaller*
*Kiel University, Institute of Animal Breeding and Husbandry, Leibnizstrasse 15, 24098 Kiel, Germany; cheuer@tierzucht.uni-kiel.de*

In recent years high dimensional genomics data for thousands of individuals has become available in livestock. With increasing number of data points, computational demands rise and there is need for highly efficient software implementations that make use of modern computer hardware architectures. For specific problems, highly efficient software is available that utilizes multicore processors, graphics processing units and high performance computing clusters. The predominant API's used are CUDA, OpenCL, MPI and openMP. The aim of the study was to incorporate multithreaded functions for genomic evaluations in the user friendly and widely spread R software environment. This is achieved by linking C++ functions that make use of openMP to R. When an R function is being called the number of threads can be specified and the execution will be performed on several available cores. The portfolio of the toolbox that has been developed for R includes frequently used operations like the computation of: covariance matrices, genomic additive and dominance relationship matrices, canonical transformations and general matrix cross products. For genomic evaluations, parallel functions for REML-based mixed model solvers, genomewide association studies as well as a general purpose Gibbs-Sampler for an arbitrary number of random effects, are available. The decrease in computation time depends on the application, the size of the problem and the number of cores available. Matrix cross products, like computing genomic relationships, scale almost linearly with the number of cores. Iterative algorithms like the Gibbs-sampler heavily depend on the problem size in terms of scaling, because only computations inside an iteration can be parallelized. The toolbox uses Eigen for the linear algebra and Rcpp for linking C++ up with R. It offers an easy way of utilizing the computational power present in every workstation PC for genomic evaluations and is embedded in R for ease of use.

---



**Prediction of genomic breeding values for feed efficiency and related traits in pigs**

*D. Do[1], L. Janss[2], A.B. Strathe[1,3], J. Jensen[2] and H.N. Kadarmideen[1]*
*[1]Faculty of Health and Medical Sciences, University of Copenhagen, Grønnegårdsvej 7, 1870 Frederiksberg C, Denmark, [2]Center for Quantitative Genetics and Genomics, Aarhus University, Blichers Alle 20, 8830 Tjele, Denmark, [3]Danish Agriculture & Food Council, Pig Research Centre, Axeltorv 3, 1609 Copenhagen V, Denmark; ddo@sund.ku.dk*

Improvement of feed efficiency is essential in pig breeding and selection for reduced residual feed intake (RFI) is an option. Accuracy of genomic prediction (GP) relies on assumptions of genetic architecture of the trait under selection. This study applied five different Bayesian Power LASSO (BPL) models with different power parameters to investigate genetic architecture of RFI, to predict genomic breeding values, and to partition genetic variances for different SNP groups. Data were 1272 Duroc pigs with both genotypic and phenotypic records for RFI as well as daily feed intake (DFI). The gene mapping confirmed previously identified significant SNPs on 30-31 Mb and 60-61 Mb on chromosome 1 (SSC 1) for RFI and DFI, respectively, and found several novel SNPs on SSC 14 for RFI. Estimated genomic variance varied according to the power parameters. Prediction accuracy for BLP models ranged from 0.33-0.34 and 0.31-0.32 for DFI and RFI, respectively, and approx. 15% higher than that of GBLUP method. However, BPL models are more biased than GBLUP method for both traits and use of different power parameters had no effect on predictive ability of the models. Partitioning of genetic variance showed that SNP groups either by position (intron, exon, downstream, upstream and 5'UTR) or by function (missense and protein-altering) have similar average explained variance per SNP, except 3'UTR SNPs which explain approx. 3 times more variance then SNPs in the other groups. This study supports use of BPL models for GWAS and indicates the potentials to use BPL models in genomic prediction. Further work includes applying other GP methods for RFI and DFI as well as extending these methods to feed efficiency related traits such as feeding behaviour and body composition traits.

## Imputation of genotypes in Danish purebreds and two-way crossbred pigs using low density panels

*T. Xiang[1], O.F. Christensen[1], A. Legarra[2] and T. Ostersen[3]*
*[1]Aarhus University, Center for Quantitative Genetics and Genomics, Department of Molecular Biology and Genetics, Blichers Allé 20, Postboks 50, 8830 Tjele, Denmark, [2]INRA, UR 1388 GenPhySe, 24 Chemin de Borde Rouge – Auzeville CS 52627, 31326 Castanet Tolosan, France, [3]Danish Agriculture & Food Council, Pig Research Centre, Axelborg, Axeltorv 3, 1609 København V, Denmark; tao.xiang@agrsci.dk*

Genotype imputation is commonly used as an initial step of genomic selection. Studies on humans, plants and ruminants suggested that many factors would affect the performance of imputation. However, studies rarely investigated pigs, especially crossbred pigs. In this study, 60K genotypes were available for Danish Landrace and Yorkshire and 7K genotypes were available for crossbred Landrace-Yorkshire. Different scenarios of imputation from 5K SNPs to 7K SNPs on both purebred and crossbred were compared first. Then an imputation scenario from 7K to 60K was implemented on a simulated 60K crossbred dataset. According to results on the former comparisons, it is concluded that genotype imputation on crossbreds performs equally well as in purebreds, when parental breeds are used as the reference panel. When the size of reference is considerably large, it is redundant to use a combined reference to impute the purebred because a within breed reference can already ensure an outstanding imputation accuracy, but in crossbreds, using a combined reference increased the imputation accuracy greatly. Results of imputing from 7K to 60K on simulated crossbred dataset were also in favor of these conclusions. Furthermore, on this simulated dataset imputed 60K crossbred genotypes from 7K SNPs were highly accurate, and this result suggested 60K genotypes could be imputed accurately from 7K SNPs in real genotyped crossbred dataset. The general conclusion is that imputation from 7K to 60K is highly accurate in both purebreds and crossbred pigs.



## Genetic association between leg conformation in young pigs and longevity of Yorkshire sows

*H.T. Le[1,2], E. Norberg[1], K. Nilsson[2] and N. Lundeheim[2]*
*[1]Aarhus University, Molecular Biology and Genetics, Blichers Allé 20, 8830 Tjele, Denmark, [2]Swedish University of Agricultural Sciences, Animal Breeding and Genetics, Gerda Nilssons väg 2, 75007 Uppsala, Sweden; thu.le.hong@slu.se*

Longevity is an important trait in swine production regarding both economic and animal welfare aspects. However, direct selection for improved longevity might be ineffective due to late in life information on this trait. This study aims at studying genetic correlations between leg conformation traits scored in young Yorkshire pigs in nucleus herds and longevity traits of purebred Yorkshire sows in multiplier herds. Two datasets were available: (1) on 97,533 animals (both sexes) with information on movement and overall leg score recorded at performance testing; and (2) on 12,444 sows with information on longevity. The longevity traits were: stayability to survive up to second or third parity (STAY12/STAY13); length of productive life (LPL); number of litters (NoL); lifetime born alive (LBA) and lifetime total born (LTB). The estimated heritabilities ranged from 0.05 to 0.16. Almost all estimated genetic correlations between conformation and longevity traits were significantly favorable (better scores – better longevity). Movement showed higher correlations with longevity (0.36-0.53) compared with overall leg score (0.24-0.33).

## Genetic correlations between morphological factors and test day milk in Valdostana Red Pied breed

*S. Mazza, N. Guzzo and R. Mantovani*
*Dept. of Agronomy Food Natural Resources Animals and Environment, Viale dell'Università, 16, 35020 Legnaro (PD), Italy; serena.mazza@studenti.unipd.it*

The aims of this study were to estimate genetic parameters and correlations between milk yield and morphological factors regarding specific regions of the body in cows belonging to the small indigenous population of Valdostana Red Pied dual purpose cattle. Factor analysis was applied to fleshiness and udder individual type traits (scored linearly 1 to 5 point) for 33,206 first and second lactation cows. Factor1 (F1) reflected 4 muscularity traits, factor 2 (F2) included 3 dimensional udder traits and factor 3 (F3) reflected a good dairyness. Data from 169,008 test-day records of milk yield, fat and protein content belonging to 16,605 cows (first three lactations) were also included in this study. The models considered for the REML single-trait analyses accounted the herd-year-classifier, classes of age at calving and days in milk as fixed effects and the random additive effect of cows for the first dataset; herd test day within lactation, class of gestation, class of age at parity within the lactation number, class of month of parity within the lactation number, permanent environment within and between lactations were taken into account for the repeatability test-day model, together with the additive genetic component. In this model, the shape of the lactation curve was described by $4^{th}$ order Legendre polynomials. These effects together were also considered for the AIREML bi-trait analyses. Heritability estimates were 0.17 for F2, 0.20 for F3 and 0.31 for F1. Regarding the milk yield correlated traits, the values of heritability were 0.13 for fat (kg), 0.17 for protein (kg) and 0.20 for milk (kg). The genetic correlations between milk yield and factors included udder traits were positive and medium-high, whereas the genetic correlations between the antagonistic milk yield trait and factor 1 (fleshiness traits) was about 0.20. These results will help to improve the selection scheme for both aspects of this dual purpose breed.



## Optimizing design of small-sized dairy cattle breeding program with minimal performance recording

*C.M. Kariuki[1,2], A.K. Kahi[3], H. Komen[2] and J.A.M. Van Arendonk[2]*
*[1]Chuka University, Animal Sciences, P.O. Box 109, 60400 Chuka, Kenya, [2]Wageningen University, Animal Breeding and Genomics Centre, P.O. Box 338, 6700 AH Wageningen, the Netherlands, [3]Egerton University, Animal Sciences, P.O. 536, 20115 Egerton, Kenya; charles.kariuki@wur.nl*

Deterministic simulations using selection index theory were performed to determine the best alternative selection strategy to maximize response to selection for a small-sized nucleus dairy cattle improvement program with minimal commercial dams' performance recording. The nucleus was made up of 10 sires and 100 dams. Commercial recorded dams provided information for evaluation. Alternative strategies differed in type and amount of evaluation information. Five strategies were defined i.e. only nucleus records (WN), progeny records in addition to nucleus records (PT), only genomic information (GS), dam performance records in addition to genomic information (GS+DP), and progeny records in addition to genomic information (GS+PT). Alternative PT, GS, GS+DP and GS+PT schemes differed in the number of progenies per sire and size of reference population. The maximum number of progeny records per sire, and size of reference population were 30 and 5,000, respectively. Results showed GS schemes had three advantages when selecting over overlapping generations i.e. (1) they had higher selection intensities as selection is done early in life, (2) the rate of genetic progress was faster due to increased use of young parents, and (3) the generation intervals were greatly reduced. Consequently, GS small-sized nucleus programs out performed WN, PT, GS+DP and GS+PT programs on response to selection. However, PT schemes were superior to all other strategies for accuracy of selection. It was concluded that adoption of genomic selection for situations where establishment of large reference populations is difficult, trade-offs between increased response to selection and high accuracies of EBVs have to be made. A general discussion was done on the genetic improvement of dairy cattle in Kenya.

**Cow identification in heterogeneous dairy herd management software programs**

*K.H. Hermans*
*Ghent University, Department of Reproduction, Obstetrics and Herd Health, Faculty of Veterinary Medicine, Salisburylaan 133, 9820 Merelbeke, Belgium; kristof.hermans@ugent.be*

Data is the biggest and most powerful asset for a farmer. The total amount of cow related data that gets collected per day has grown exponentially. The first and most critical step in the data collection process, is the identification of an entity. We hypothesized that the identification of an entity in commercial dairy herd management programs is not error free. Seven dairy herd management software programs were checked for their allowance of a missing, incorrect or duplicated value towards the following cow identifiers: official ear tag as described by European law, work number as described by European law, cow number, date of birth, gender, identity of mother and father. Four out of 7 programs allowed a null value for the ear tag and in six programs the ear tag could be altered. For the creation of a cow record the cow number had to be filled in six of the seven programs and none of the programs allowed a duplicated record with identical cow number. One can state that the majority of the programs focus on cow number as the main identifier and not on the official ear tag. All of the investigated programs allowed actions which affect the integrity of the cow identifiers. As a consequence these programs are receptive for misidentification, which we defined as decoupling the identity of an entity from its collected data. Hence, there is a high risk that the collected data is not reliable and therefore may lead to serious errors.



**The effect of utilizing sex sorted semen on selection response of sire pathways**

*S. Joezy-Shekalgorabi[1] and A. De Vries[2]*
*[1]Islamic Azad University, Shahr-e-Qods Branch, Department of Animal Science, Shahre Qods, 37515-374, Tehran, Iran, [2]University of Florida, Department of Animal Science, Bldg. 499, Shealy Drive, Gainesville, FL, 32608 Florida, USA; joezy5949@gmail.com*

Selection proportion is the main parameter for prediction selection intensity. The current study has considered the effect of employing sex sorted semen on selection proportion of sire of dam (SD) pathways. Using a deterministic approach, three different strategies were implemented to find out how selection proportion in SD pathway varies as a result of utilizing sex sorted semen on dairy heifers. The strategies were included (1) continuous use of sexed semen only (CS), utilization of sexed semen for both the first and second services with conventional semen afterwards (S2), and (2) utilization of sexed semen for the first service with conventional semen afterwards (S1). To apply the effect of sex sorted semen on sire of dams pathway, it was assumed that for producing each vial of sexed semen we need 3.5 vials of conventional sperm. On the other hand the effect of using such kind of semen on the number of insemination per conception was considered in the model. Results indicated that in all investigated scenarios (combination of various sexed and conventional conception rate) the selection proportion was greater than its related control value (continuous use of conventional semen). Strategy CS led to the largest selection proportion compared with the other strategies. In spite of its positive effect on genetic improvement of dam pathways, when using sex sorted semen, one has to consider its genetic and economic impacts on sire pathways.

**Paternal origins of toro de lidia cattle breed historical castes**

*R. Pelayo[1], L.J. Royo[2], A. Molina[3] and M. Valera[1]*
*[1]University of Seville, Agro-Forest Sciences, Crtra. de Utrera Km. 1, 41014 Seville, Spain, [2]SERIDA, Nutrition, Grasslands and Forages, Carretera de Oviedo s/n, 33300. Villaviciosa (Asturias), Spain, [3]University of Cordoba, Animal Genetic, Campus de Rabanales. Edif. Mendel, planta baja. Carretera Madrid-Córdoba, km 396a, 14071 Cordoba, Spain; rociopega55@hotmail.com*

Five founding lineages or 'castes', (BOE, Real Decreto 60/2001) are defined in Toro de Lidia cattle breed, being most of them currently endangered. Although some works have studied the genetic diversity in this breed, revealing that the level of differentiation between lineages is very large and the maternal and paternal influences are similar to other Mediterranean breeds or Iberian Peninsula breeds, the ancestral genetic origins of the Lidia cattle breed remains controversial. These previous works sometimes do not include all the founding castes. The main objective of this work is to show if these castes have a paternal genetic support, and what its phylogenetic origin is. Total DNA was isolated from 93 blood samples belonging to different Spanish cattle breeds. Individuals from Pérez-Pardal *et al.* and Cortes *et al.* will be also used after genotype standardization totalling 1300 male individuals. DNA samples were genotyped using five bovine Y-chromosome specific microsatellites (INRA189, UMN0103, UMN0307, BM861 and BYM1) and one polymorphism (ZFY10) which allows us to differentiate between both taurine haplogroups. Microsatellites and SNP alleles were combined into haplotypes, identifying a total of 38 haplotypes, 11 of them belonging to haplogroup Y1, and 27 to haplogroup Y2. Haplogroups did not share haplotypes. A median-joining network connecting different haplotypes was constructed using the program NETWORK 4.5.2. In total ten different haplotypes were found in Lidia breed, being five of them exclusive of this breed. Two individual haplotypes merit special attention: Y2H23 exclusive and fixed in Gallardo caste, and Y2H27 only present, and having high frequency, in Cabrera and Navarra Caste.



**Estimation of behavioral genetic parameters in the fighting bull**

*R. Pelayo[1], M. Solé[1], M. Bomba[1], S. Demyda Peyras[2], A. Molina[2] and M. Valera[1]*
*[1]University of Seville, Agro-Forest Sciences, Crtra. de Utrera Km. 1, 41014 Seville, Spain, [2]University of Córdoba, Animal Genetic, Campus de Rabanales. Edif. Mendel, planta baja. Carretera Madrid-Córdoba, km 396a, 14071, Spain; rociopega55@hotmail.com*

Fighting bull is one of the most important Iberian cattle breeds, which is distributed in many countries around the world. This breed has been selected for more than 400 years mainly focused in their behavioural traits, especially those related to aggressiveness. However, there have only been few studies dealing with behavioural traits in this breed. Behavioural traits are more complex to assess since they are often determined in a subjective manner. By this reason, since 2008, all the bulls belonging to the AEGRB, (Spanish Fighting bull breeders association) were assessed using a linear standardised behavioural evaluation formulary (CLEC) that includes 17 linear and 6 environmental variables. All the evaluations were performed by trained observers. The aim of this work was to analyse the main behavioural traits (bravery (B), strength (S), mobility (M) and fixity (F)) in a fighting bull population in order to be included into the breeding program of the AEGRB. Genetic parameters were estimated from 901 animals, aged between 2 and 4 years old, using a multiple trait REML animal model, using B, S, M and F as dependent variables. The model included the herd, the bullfighter category, the body condition score of the animals and the evaluation date as fixed effects. Data were evaluated using VCE software with a coancestry matrix of 3,501 individuals. Heritability values were 0.20±0.10; 0.14±0.07; 0.09±0.07 and 0.10±0.07 for the B, S, M and F respectively. Genetic correlations were all positives and varied from moderate (0.17 between S and F) to high (0.82 between B and S). It is noteworthy that all the genetic correlations between the bravery and the other variables estimated were higher than 0.63. To conclude, all the results obtained in this study indicate that the analysed traits could be used as an efficient tool in the fighting bull breeding program.

## Session 35 Poster 16

**Novel SNPs in caprine SCD gene and its relationship with cholesterol level in southern of Thailand**

*C. Supakorn*
*Walailak University, School of agricultural technology, 222 Thaiburee Thasala Nakhonsithammarat, 80161, Thailand; schina@wu.ac.th*

The present study attempts to study the effect of genetic polymorphisms of stearoyl-coA desaturase (SCD) gene on cholesterol level in meat goat at a commercial goat farm in Thailand. Genetic variability in caprine SCD, analyzing 290 animals belonging to several types of Boer (B), Thai native (TN), Anglo-nubian (AN) and Saanen (SA) breed fractions by single strand conformation polymorphism (SSCP) and DNA sequencing. Four single nucleotide polymorphism were identified in exon 3 (601A->G; Ref. AF422168), exon 5 (878C->T; AF422170), exon 6 (718C->G; Ref. AF422171) and 3'untranslated region (3'UTR) (TGT deletion). Conformational patterns of SCD gene were classified into 2 genotypes in exon 3, exon 5 and 3'UTR and 3 genotypes in exon 5. Five haplotypes were constructed but only the lowest frequency of a haplotype C was not included in dataset. The haplotypes of SCD gene were also significantly correlated with plasma cholesterol (P<0.01) levels. The results suggested a possible use of the SCD locus in gene assisted selection programs for healthy improvement in this meat goat population.

---

## Session 35 Poster 17

**Effect of environmental factors on survival traits from birth to one-year old in Karakul lambs**

*Z. Kiamehr*
*Islamic Azad University, Kashmar branch, Mashhad,Vakil abad,Seyyed razi, Mhran 3, home No:66, 9188764386, Iran; zohal_m_63@yahoo.com*

6,108 records of Karakul lambs bred during 1995-2011 at Karakul sheep breeding center were analyzed for survival features from birth to one-year old. This study aimed to investigate liveability from birth to 90 days and 360 days old. Environmental factors such as birth date (year, month) and sex included in model as fixed effect, as well as birth type and birth weight (linear and second grade) as covariates. Lifereg procedure of SAS was used to analyze the environmental factors affecting longevity. The results showed that birth type, birth month, and sex had a considerable significant effect on survival traits (P<0.01). The effect of age was significant for all traits in this study (P<0.05). Birth weight had significant effect in both linear and quadratic order for all traits (P<0.01). the effect due to inherent traits derives from hormones in two different sexes. the effect of birth month is due to the different climates and the focus of parturition at special dates which leads to easy raising management. the effect of birth type is because of allocating the whole uterine environment to a lamb and its maternal effect. regarding the results of this study with improving the effective factors on survival traits we can improve these traits.

# Session 36 Theatre 1

**Technological advances in fur grading**

*J.L. Christensen*
*Kopenhagen Fur, Langagervej 60, 2600 Glostrup, Denmark; jlc@kopenhagenfur.com*

During the last 20 years, Kopenhagen Fur has developed the grading of mink skins into a high technology process, which in addition to ensuring low production costs, also creates added value to the shareholders through objective information on their individual mink production. As a result of the new quality management programs, as well as the new technology, the Danish fur farmers get valuable information about their skin production on their computer. This information is used to achieve improvements in the production, however it is also used to strengthen the breeding stock through a comprehensive genetic exchange of live animals among fur farmers.

---

# Session 36 Theatre 2

**Response to selection for Orylag® fur production traits estimated by using a control cryopreserved**

*D. Allain[1], G. Auvinet[2] and S. Deretz[2]*
*[1]INRA, URM1388 GenPhySE, CS52627, 3132 Castanet-Tolosan, France, [2]INRA, UE1372 GenESI, Le Magneraud, BP52, 17700 Surgères, France; daniel.allain@toulouse.inra.fr*

An analysis of the response to selection for fur production traits was carried out in a line of orylag fur animals which derived from the Castor rex rabbit by genetic improvement of coat characteristics. It was obtained at INRA in 1990 from a closed original population of 100 Castor rex animals managed in 8 separate mating groups with overlapping generations. Animals were selected on fur bristliness (FB) or absence of residual guard hair in order to obtain a fur only composed of undercoat, hair length (HL), fur compactness (FC) and fur priming (FP). Response to selection was estimated by analysing a mixed model and estimating the genetic effect of the animals in each generation, and also by comparing the generation 2005 of selection with a control population. Embryos from generation 1995 were frozen and thawed in order to be contemporaries of animals born 10 years later. Control group was the offspring (n=87) of generation 1995 (18 sires and 13 does from 7 mating groups issued from 347 embryos) and selected group was the contemporary offspring (n=113) of generation 2005. Probabilities of gene origin were analysed in control group and the whole generation 2005 using pedigree information. The number of founders and effective ancestors were: 36 and 16.8, and 34 and 15.9 in control group and whole selected population at generation 2005 respectively. A significant phenotypic gain was observed in generation 2005: 27.5, 4.4, 6.6 and 35.5% for FB, HL, FC and FP respectively, that may be the cause of the success in the selection process. The genetic trends have been estimated using mixed model methodology with all the information of the selection process (34,084 records; 34,897 animals in pedigree). In agreement with previous results the genetic trends were: 1.11, 1.29, 0.92 and 1.30 $\sigma_g$ after 10 years of selection for FB, HL, FC and FP respectively.

**Low dietary protein provision during fetal life in several generations in mink**

*C.F. Matthiesen and A.-H. Tauson*
*University of Copenhagen, Department of Veterinary Clinical and Animal Sciences, Grønnegårdsvej 3 1st floor, 1870 FRB C, Denmark; cmt@sund.ku.dk*

Low protein (LP) provision during gestation has been shown to induce structural and functional metabolic changes in the offspring, primarily through adaptations aimed towards maximizing the utilization of those nutrients that are available so as to optimize fetal development and survival. Maternal LP provision could be expected to cause a mismatch between prenatal and postnatal life, because growing mink kits usually are provided with high levels of protein and energy. Our objectives were to investigate how LP provision in late gestation affects the reproductive performance, kit survival rate and pre-weaning kit growth of $F_1$, $F_2$ and $F_3$ offspring from dams fed an LP diet in late gestation, raised in small (4-6 kits) or large (8-10 kits) litters until weaning. Sixty-nine dams consisting of 20 control (C), 22 adequately treated (APAP), 13 born by mothers fed an LP diet during late gestation (APLP) and 14 dams born by mothers exposed to LP during late fetal life and fed an LP diet during late gestation (LPLP) were used. C dams were fed an adequate protein diet (AP – 30% of metabolizable energy (ME) from protein) the entire gestation whereas APAP, APLP and LPLP dams were fed an LP diet (19% of ME) in late gestation. The reproductive performance was not affected by gestation diets but the results indicate that dams exposed to LP during fetal life and fed an LP diet during late gestation gave birth to larger litters and had fewer stillborn kits than APAP dams fed an LP diet during late gestation. The birth weight of $F_1$ kits born by mothers fed an LP diet during late gestation was lower ($P<0.05$) than that of controls which confirms previous results. The growth rate of the kits during lactation indicated that kit exposed to LP during fetal life performed better in larger litters than adequately treated ones especially in the transition period from milk to solid feed and around weaning ($P<0.05$).



**Natural or synthetic vitamin E for mink kits**

*S.K. Jensen[1], T.N. Clausen[2] and P.F. Larsen[2]*
*[1]Aarhus University, Department of Animal Sciences, Blichers Alle 20, 8830 Tjelse, Denmark, [2]Copenhagen Fur, Agro Food Park, 8200 Aarhus N, Denmark; sorenkrogh.jensen@agrsci.dk*

Vitamin E is a very important, but also expensive component in mink feed. For commercial use 3 types of vitamin compounds is of interest. Thus, it is of interest to elucidate the relative bioefficiency of the expensive natural vitamin E on alcohol form (RRR-α-tocopherol) and the cheaper natural acetate form (RRR-α-tocopherylacetat) against the cheapest form, synthetic vitamin E (all-rac-α-tocopherylacetat). Twelve groups of 12 mink, 4 week of age were allocated to 4 levels of vitamin E (50, 75, 100 or 150 mg/kg feed) of one of the 3 types of vitamin E, in addition a vitamin E free group served as control. After 3 or 6 weeks on the experimental diets, six mink were sacrificed and plasma, liver, heart, lung and brain was removed, weighed and subsequent analysed for vitamin E content and stereoisomer composition. After 3 and 6 weeks on the experimental diets, α-tocopherol concentration in plasma and tissue was significant higher in mink fed natural vitamin E compared to synthetic vitamin E and dose response curves for the 3 types of vitamin E showed significant differences in non-linear regression curves for plasma and tissues. Likewise the biodiscrimination of the different stereoisomers varied between the different tissues. RRR-α-tocopherol showed the highest bioefficiency and synthetic all-rac-α-tocopherylacetat the lowest bioefficiency.

**The more mink in the same cage, the more bite marks**

*S.W. Hansen*
*Aarhus University, Department of Animal Science, Blichers Allé 20, 8830 Tjele, Denmark; steffenw.hansen@agrsci.dk*

Group housing of mink has been introduced in order to increase cage size, cage complexity and social stimulation in farmed mink. The aim of this study was to compare the number of bite marks in female mink kept in groups of two, three and four female mink in the same cage. Furthermore, the aim was to test if the number of bite marks was reduced in male-female pairs kept in climbing cages (developed for group housing) compared to male-female pairs in standard cages. Bite marks, defined as dark spots on the leather side of the matured mink skin, are shown to be caused by bites of the skin. Each of the five experimental groups: F2 (two females), F3 (three females), F4 (four females), PS and PC (male-female pairs) consisted of 24 cages. Mink in group F3, F4 and PC were housed in climbing cages consisting of a standard cage (bottom cage (0.90×0.30×0:45 m) and a top cage (0.70×0.30×0:45 m) connected through an opening (0.20×0.30 m) between bottom and top cage, a nest box (0.23×0.28×0.20) and a tube and shelf as enrichments. Mink in group F2 and PS were housed in standard cages with access to nest box and same types of enrichments. There was *ad libitum* access to water and fresh food. Seven months old, the mink were killed and pelted and subcutaneous tissue and fat on the leather side of the pelt were removed, and the pelt was examined for bite marks. The score of bite marks was analyzed by Restricted Maxium Likelihood method in the mixed model procedure. Results showed that two, three and four females together had 1.5, 2.5 and more than 5 times as many bite marks as mink kept in male-female pairs. The difference was significant for all combinations of more than two mink. However, there was no difference in the number of bite marks in male-female pairs kept in standard cages and climbing cages. The results show that the more mink in the same cage, the more bite marks and that housing in climbing cages does not reduce the number of bite marks in mink kept in male-female pairs.

---



**Dissected fat deposits for evaluation of total body fat percent in mink**

*V.H. Nielsen[1], D. Mayntz[1,2], A. Sørensen[2], S.K. Jensen[3], B.M. Damgaard[3] and S.H. Møller[3]*
*[1]Aarhus University, Molecular Biology and Genetics, Blichers Alle 20, 8830 Tjele, Denmark, [2]Aarhus University, Bioscience, Ny Munkegade 116, 8000 Aarhus C, Denmark, [3]Aarhus University, Animal Science, Blichers Alle 20, 8830 Tjele, Denmark; vivih.nielsen@agrsci.dk*

Selection for high November weight in mink production is assumed to be associated with increased body fat. However, whole body chemical composition is difficult to determine due to obstacles of grinding the carcass related to the large amount of fat and the fur. Individual body fat deposits are assumed to be a valuable measure of total body fat. Here we study weight of subcutaneous fat in the groin region and perirenal fat of the kidneys as a predictor of total body chemical fat percent and the effect of selection on body chemical fat percent. Groin and perirenal kidney fat was dissected from 5 male and 5 female mink from each of 3 lines (1: selection on *ad libitum* feeding; 2: selection on restricted feeding; 3: a control line) and weighed. Afterwards, the whole body was grinded and total body fat percent was determined. An Analysis of Covariance (ANCOVA) was used for data analysis. Total body fat percent was analysed with line and sex as fixed effects and covariates body weight, body length, subcutaneous groin fat, kidney fat and total fat (groin fat plus kidney fat). The best prediction of total body chemical fat percent was obtained using the sum of groin and kidney fat as covariate. The prediction was better for females than for males. No significant effect of line was found for total body chemical fat percent neither for males nor females.

**Introgression of wool shedding gene pool into the Romane breed sheep**

*D. Allain[1], B. Pena[1], D. Foulquié[2], Y. Bourdillon[3] and D. François[1]*
*[1]INRA, URM1388 GenPhySE, CS52627, 3132 Castanet-Tolosan, France, [2]INRA, UE0321, Domaine de la Fage, Saint Jean et Saint Paul, 12250 Roquefort, France, [3]INRA, UE0332, Domaine de Bourges, La Sapinière, 18390 Osmoy, France; daniel.allain@toulouse.inra.fr*

There is an interest in Europe for the use of breeds that have no wool or shed their wool due to relative value of meat and wool, and increasing shearing costs. In the present paper investigations were made to evaluate (1) genetic variability of wool shedding in the French Romane breed and (2) introgression of gene pool from the Martinik Hair, a hairy sheep known to shed, into the Romane breed through 4 consecutive backcrossing generations. Genetic variability of wool shedding in the Romane breed was observed on a total of 1,259 adult ewes from 15 months to 5 years of age (n=2,984) and 1503 animals measured at both 7 and 19 months of age. Ability to shed during spring varied from 42.7 to 52.8% in lambs and adult ewes respectively but only 1.2% of adult Romane ewes showed total wool shedding. Heritability estimate of wool shedding ability was high (0.69) and a high genetic correlation (0.86±0.09) was observed when wool shedding in lambs and adults were considered as separate traits in the Romane breed. After 4 successive backcrossing generations, ability to shed and shedding extent increased by 75.4% and 130.2% respectively and 79.4% and 150.1% at 7 and 19 months of age respectively in the first introgressed cohort compared to the Romane breed. Thereafter by selection on wool shedding using EBV, a genetic gain of 1 $\sigma_g$ was observed after 2 generations of selection on wool shedding indicating that a rapid progress can be obtain through introgression scheme of wool shedding gene pool in the Romane breed. Our first results about introgression of shedding gene pool from the Martinik Hair breed through 4 successive backcrossing generations seems to be an interesting selection strategy to lead more quickly to a shedding sheep and thus to create genetic progress by improving adaptive traits in the Romane breed which has a high production potential.



**Physical properties of hair fibres obtained from different breeds of pigs**

*N. Mohan[1], S. Debnath[2], R. Mahapatra[1], L. Nayak[2], S. Baruah[1], A. Das[1], S. Banik[1] and M. Tamuli[1]*
*[1]National Research Centre on Pig, Indian Council of Agricultural Research, Rani, Guwahati, Assam, 781131, India, [2]National Institute of Research on Jute and Allied Fibre Technology, Indian Council of Agricultural Research, 12, Regent Park, Kolkata, 700040, India; mohannh.icar@nic.in*

A key by-product of pig farming is the hair or bristle fibres obtained at the time of slaughter. Data on physical properties of pig hair fibres is sparse, restricting its diversified industrial application. A study was conducted to analyse the variations in the tensile properties of the pig hair fibres with respect to breed (Ghungroo, Ninag Megha, Hampshire and Duroc) and location on the body. The mean density of fibres on the body surface of pigs was between 0.079 and 0.154 per $mm^2$. The average length of hair fibre varied from 56.8 mm in the Duroc to 127.1 mm in the Niang Megha breeds and was significantly different between breeds. The average growth rate of fibre was 0.16 mm/day. The fineness of fibres ranged between 29.1 and 124.2 tex. The general stress-strain curve of pig hair fibre showed elastic and plastic zones, with an upward curve in the post-yield zone. The mean tensile strength of the pig hair fibre was 14.05 cN/tex. The average extensibility and Young`s modulus of the pig hair fibre was 31.53% (range 21.1 to 39.5%) and 6.39±0.19 GPa respectively. The maximum load at rupture of the hair fibres varied from 9.2 to 13.8 N. The overall mean work of rupture and flexural rigidity of the hair fibre was $3.61\times10^{-2}$ J/m/tex and 1,059±321 mN $mm^2$/mm respectively in the Hampshire, Duroc, Ghungroo and Niang Megha breeds. The coefficient of static friction of the hair fibres was similar among the different breeds of pigs, irrespective of their location on the animal's body. Some variations in the physical properties of pig hair fibres with respect to breed and location could be observed. In general, the physical and tensile properties of the fibres were similar to other natural fibres of animal origin and they may find use in the manufacture of diverse, environment friendly products.

### Genetic parameters and genetic trends for growth and fur quality trait in sliver blue mink in China

*Z. Liu, X. Xing and F. Yang*
*Institute of Special Economic Animal and Plant Sciences, The Chinese Academy of Agricultural Science, Juye street 4899, Changchun 130112, China, P.R.; liuzongyue7777@sina.com*

Knowledge of genetic parameters is essential for efficient breeding programs. The present study was performed to estimate genetic parameter for growth and fur quality traits in Chinese silver blue mink. Recordings of body weight (BW), body length (BL), guard hair length (GL), underwool length (UL), and the rate of guard hair and underwool lenght (nap) from 1686 silver blue mink spanning six generations were obtained from Da Lian Ming Wei Marten Industry Company. The phenotypes were analyzed using a model with year and sex as fixed effects. Genetic parameters were estimated using a multi-trait animal model with year and sex as fixed effects. The heritability estimate for BW, BL, GL, UL and nap was 0.41, 0.53, 0.53, 0.52, and 0.52. The phenotypic correlation between nap and BW, BL, GL and UL were negative (-0.218, -0.178, -0.074, -0.425). The phenotypic correlation between BW and BL, GL and UL and between BL and GL and UL and between GL and UL was positive (0.928, 0.882, 0.869, 0.806, 0.788, 0.93). The genetic correlation between BL and UL and nap was negative (-0.941, -0.983) as was the genetic correlation between nap and GL and UL (-0.074 and -0.425).The genetic correlation between BW and GL, UL and nap was 0.731, 0.972 and 0981, between BL and GL it was 0.622, between GL and UL and nap it was 0.992 and 0.641 and between UL and nap it was 0.987. The genetic trends were estimated by regressing breeding value on year. The genetic trend for BW and BL was close to zero, and positive. The genetic change for GL, UL, and nap was negative and parallel. Our results herein form a practical basis for designing optimal breeding schemes in Chinese silver blue mink.

---



### Interest of qualified presumption of safety for probiotic developments

*G. Bertin[1,2]*
*[1]Erawan Consulting SARL, 25 rue des Bas, 92600 Asnières-sur-Seine, France, [2]European Probiotic Association (EPA), 18 rue Pasquier, 75008 Paris, France; secretarygeneral@asso-epa.org*

Qualified presumption of safety (QPS) has recently entered EU law with the publication of a new Commission Regulation (EC) No 562/2012. This approach was developed by EFSA (European Food Safety Authority) to provide a generic harmonized pre-assessment of microorganisms or viruses intentionally introduced into feed chain. The concept of 'Qualified presumption of Safety' is based in taking into account 4 pillars: taxonomy, body of knowledge (familiarity), possible pathogenicity and end of use. The three main pillars for assessment are taxonomy (such as genus, species, subspecies or other grouping as homofermentative for *Lactobacillus* etc), familiarity (includes the history of use, scientific literature and so on) and possible pathogenicity (assessment of antimicrobial resistance). The first list published in 2007 is annually reviewed by EFSA in function of the new scientific information published. The actual list concerns a large number of microorganisms but it will be interesting to see how we can use QPS for a non-genera listed.

## Session 37 Theatre 2

**The effect of probiotics on animal health: a focus on host's natural intestinal defenses**

*G. Tabouret*

*INRA, Animal Health, National Veterinary College – UMR1225. 23 chemin des Capelles, 31076 Toulouse, France; g.tabouret@envt.fr*

Probiotics are live microorganisms that, orally administered, confer a realistic health benefit on the host. The gradual reduction of the use of antibiotics as growth promoters has generated renewed interest in the incorporation of such microbial strains in the food in order to maintain the beneficial effect obtained with antibiotics. Microbes added to feed are classified at a regulatory level as zootechnical additives, in the category of gut flora stabilizers for healthy animals and are regulated up to strain level in Europe. The microorganisms used are mainly bacteria (belonging to different types including *Lactobacillus*, *Enterococcus* or *Bacillus*) and fungi such as yeast belonging to the genus *Saccharomyces*. The overall effect on the animal is theoretically revealed by better performances (measured by production parameters) resulting either from a direct nutritional effect or a 'health' effect. This later effect is thought to result from the improvement of host's intestinal natural homeostasis and defenses. Ingested probiotic microorganisms can profoundly modulate composition and activities of microflora whose role is essential in gut homeostasis. An adequate supplementation with probiotics can help in restoring a balanced structure of gut microflora (pathogenic/commensal) which is strongly affected by dietary and management constraints of intensive breeding/production systems. Moreover, probiotics are endowed with an interesting panel of immunomodulatory properties that can be used to significantly reduce gut infectious or metabolic disorders. Here, we will discuss the mechanisms triggered by different probiotic microorganisms to stimulate gut defenses with a specific focus on barrier functions and innate immunity.

## Session 37 Theatre 3

**Feed Additive, can they improve animal welfare?**

*J. Brufau, R. Lizardo and B. Vilà*

*IRTA, Monogastric Nutrition and Animal Welfare, Centre Mas Bover, Ctra. Reus -El Morell Km 3.8, 43120 Constanti, Spain; joaquim.brufau@irta.cat*

The EU regulation on feed additives (1831/2003) has introduced the animal welfare concept. EFSA has published an opinion (2008) in which animal welfare criteria are discussed. The opinion describes how feed additives may be involved in the future of feed manufacturing with respect to animal welfare. However, the animal welfare in animal husbandry has some difficulties to be implemented by farmers and technicians. Recently the EU produced a 'working paper' impact assessment on the European Union Strategy for the Protection and Welfare of animals 2012-2015 which recognize the existence of difficulties on its implementation. Trough the presentation we are going to answer questions like: Zootechnical additive can improve animal performances trough improved gut health? Are these healthy improvements easily measurable? What are the main markers to consider? Can these markers be directly connected with animal performance? How immunity is involved in animal performance? Finally, we would like to conclude that several feed additives may improve animal welfare conditions.

**The effect of live yeast supplementation on digestion of young Beagle dogs**

*E. Štercová[1], E. Auclair[2], D. Kumprechtová[3], G. Bertin[4], J. Novotná[5] and J. Surová[6]*
*[1]University of Veterinary and Pharmaceutical Sciences Brno, Institute of Animal Nutrition, Palackeho 1/3, 612 42 Brno, Czech Republic, [2]Lesaffre Feed Additives, Research and Development, 137 rue Gabriel Péri, 50703 Marcg-en-Baroueul, France, [3]PROBIONIC s.r.o., Skrickova 1626, 664 34 Kurim, Czech Republic, [4]erawan Consulting Sarl, 25, rue des Bas, 92600 Asnières-sur-Seine, France, [5]MediTox s.r.o., Pod Zámkem 279, 281 25 Konárovice, Czech Republic, [6]MediTox s.r.o., Pod Zámkem 279, 281 25 Konárovice, Czech Republic; dana.kumprechtova@gmail.com*

The study evaluated the effect of oral supplementation of live yeast (LY) *Saccharomyces cerevisiae* Sc47(ACTISAF® Sc 47) to 24 Beagle dogs subjected to the Control and LY treatments. LY received $2.9\times10^8$ cfu LY/kg body weight in capsules; Control got empty capsules. The dogs received a diet with 8.5% crude fibre. Body weight was measured weekly, feed consumption daily. Clinical observations were performed once a day during the Acclimation (A) period (D-28 to D-1) and twice a day during the Supplementation (S) period (D0 to D42). Haematology and blood chemistry (urea, creatinine, alkaline phosphatase, alanine amino transferase) were performed on D-28, D-1 and D42. Faecal samples were taken from each animal on D 1, D 8, D 15, D 22, D 29, D 36, D 42 for microbiology, pH, dry matter, lactic acid, ammonia. Nutrient digestibility analysis included NDF, dry matter, ash, crude fibre, crude protein, fat. During the S period the LY dogs showed significantly ($P<0.05$) greater body weight increase than the Control ones. The haematological and blood chemistry parameters were within physiological ranges and were not adversely affected by the LY supplementation. LY had no significant effect on fecal pH values. Contents of volatile fatty acids and lactate were not significantly influenced by the treatment. The LY dogs showed significantly ($P<0.05$) higher digestibility of NDF. In the LY dogs, significantly ($P<0.05$) lower *Escherichia coli* counts were found than in the Controls. Intestinal enterococci were also significantly lower in the LY dogs.

---



**Impact of live yeast on the intestinal transcriptome profile of weaned pigs challenged with ETEC**

*P. Trevisi[1], M. Colombo[1], D. Priori[1], V. Motta[1], R. La Torre[1], R. D'Inca[2] and P. Bosi[1]*
*[1]University of Bologna, Department of Agricultural and Food Science, via Fanin 50, 40127 Bologna, Italy, [2]Lesaffre Feed Additives, 137 Rue Gabriel Péri, 59703 Marcq-en-Baroeul, France; paolo.trevisi@unibo.it*

The ability of live yeasts to modulate the cell signals in response to the infection with *Escherichia coli* F4ac (ETEC) is not deeply studied. This trial aimed to evaluate the effect of *Saccharomyces cerevisiae* CNCM I-4407 supplied at different patterns on the transcriptome profile of the jejunum mucosa of pigs 24 h after the infection with ETEC. Twenty piglets selected to be ETEC-susceptible by the gene marker Mucin 4 were weaned at 24 days of age (d0) and allotted by litter to one of following groups: control (CO), CO+colistin (AB), CO+$5\times10^{10}$ cfu of *S. cerevisiae*/kg feed, from d0 to d8 (PR), CO+$5\times10^{10}$ cfu of *S. cerevisiae*/kg feed from d7 (infection with ETEC) to d8 (CM). On d 7, pigs were orally challenged with ETEC and killed 24 h later. Jejunum mucosa was sampled to confirm the presence of ETEC receptors on villi and to perform the microarray analysis by Affymetrix©Porcine Gene strip. A functional analysis was done by Gene Set Enrichment Analysis. Normalized enrichment score (NES) was calculated for each gene set, and statistical significance was defined when False Discovery Rate % <25 and P-values of NES <0.05. All the pigs were confirmed to be ETEC-susceptible. Based on the gene sets enrichment analysis, the transcriptome profiles of AB and PR groups were similar. In both groups, several gene sets involved in mitosis and in mitochondria development ranked the top, while in CO group gene sets related with cell junction and anion channels were affected. In CM group, gene sets linked with metabolic process, and transcription ranked the top. Moreover, a specific gene set linked with the negative effect on the growth was also affected. In conclusion, a constant supplementation in the feed with the tested strain of yeast, seems to limit the early activation of gene sets related with the impairment of the jejunum mucosa of piglets infected with ETEC.

**New findings on mode of action of *Bacillus toyonensis*: disruption of Quorum sensing**

*M. Castillo[1], G. Jimenez[1], S. Martín-Orúe[2], D. Solà[2] and G. González-Ortiz[2]*
*[1]RUBINUM SA, Av. de la Llana 123, Pol. Ind. La Llana, 08191 Rubí, Spain, [2]SNiBA. Dept. de Ciència Animal i dels Aliments. Facultat de Veterinària, UAB, 08193 Bellaterra, Spain; m.castillo@andersen-rubinum.com*

*Bacillus toyonensis* (formerly classified as *Bacillus cereus* var. *toyoi*) is a gram positive bacterium used in animal nutrition as gut flora stabilizer (Toyocerin®). This probiotic has shown to disrupt Quorum Sensing (QS) systems thus inhibiting the expression of different virulence factors. Two different *in vitro* tests were run. The first one evaluated the invasiveness of *E. coli* (fimbriated (K88) or non-fimbriated (NF)) when grown with their own supernatant, with the pure cultures of B. toyonensis or with co-incubates, to porcine intestinal epithelial cells (IPEC-J2). The second trial evaluated the invasiveness of *E. coli* (K88 and SF) when grown with the ileal and the colonic digesta from weaning piglets supplemented or not with the probiotic. In the first test, the co-incubation of both *E. coli* grown with their own supernatant increased the invasiveness of *E. coli* to IPEC-J2, suggesting an activation of the auto-inducer type-2 (AI-2). However, when they grew with the supernatant from the co-incubation of *E. coli* and *B. toyonensis*, it was not observed such increase in the invasiveness response. In the second trial, the incubation of *E. coli* with the ileal supernatant from the probiotic supplemented animals lead in a reduction in the invasiveness of *E. coli* compared to control animals. These results suggest the ability of *B. toyonensis* to reduce the invasiveness of *E. coli* in the ileal tract by acting on the QS systems and, in particular by degrading acetyl homoserine lactones (AHL) or inhibiting AI-2 signals. The disruption of quorum sensing in pathogenic bacteria has been proposed as a new anti-infective strategy to control bacterial infections. The overall results of the present *in vitro* tests can partially explain the mode of action of *B. toyonensis* against pathogenic bacteria such as *E. coli* and *Salmonella* spp, thus corroborating the beneficial effects in livestock.

---

Session 37 Theatre 7

**Effect of *Saccharomyces cerevisiae* on steers progressively adapting to high concentrate diets**

*G.Z. Ding[1], Y. Chang[2], Q.X. Meng[1] and E. Chevaux[2]*
*[1]China Agricultural University, Animal Science & Technology, Yuanmingyuan West Rd 2#, 100193 Beijing, China, P.R., [2]Lallemand China, RM815, 2# North RDd, East 3rd Ring, 100027 Beijing, China, P.R.; echevaux@lallemand.com*

The objective of the current study was to evaluate the effect of *Saccharomyces cerevisiae* (SC) on rumen fermentation characteristics, kinetics of nutrient degradation and rumen bacteria population in steers progressively adapting more concentrated diets. Ten Simental × Local breed steers fitted with rumen fistulas were assigned to control and treatment groups following a 2-period crossover design. Steers in the two groups were fed the same basal diets but the treatment groups received SC supplementation. Each period was comprised of 4 phases. From the 1st to the 4th phase, steers were fed in a stepwise fashion with incremental levels of concentrate to forage ratios (CTFR) of 30:70, 50:50, 70:30 and 90:10, respectively. Degradation characteristics of Chinese ryegrass hay (CRH) and alfalfa pellet DM and NDF, concentration of ammonia N, VFA, D+L–lactate, xylanase, avicelase, β-glucanase, carboxymethyl cellulose activity and bacteria population were determined. The MIXED procedure was used to analysis the difference among different groups and the model was: $Y=\mu+Di+ Tj+(DT)ij+ Aik +\varepsilon ijk$. Differences between treatments in the same CTFR diet were determined by Student's t-test. SC possesses the capacity to stabilize ruminal pH, but has little effect on TVFA, ammonia N and lactate concentration. Effective degradability and degradation rate of CRH or alfalfa pellet DM and NDF, enzyme fibrolytic activity and the number of total bacteria, rumen fungi, protozoa, fiber-degrading and lactate-utilizing bacteria species tended to be improved by SC. Starch-degrading and lactate-producing bacteria species tended to reduced by SC. SC possesses the potential to reduce the risk of ruminal acidosis and enhance fiber degradation of roughage when animals are fed higher levels of concentrate, both of which are beneficial to beef cattle production.

**_Enterococcus faecium_ as a dry cow treatment improved colostrum quality without affecting gestation**

*A. Tiantong, H.-Y. Peng and C.-J. Chang*
*National Chung Hsing University, Animal Science, 250 Kuo-Kuang Rd., Taichung, 402 Taiwan, R.O.C., Taiwan; attapol_tt@hotmail.com*

*Enterococcus faecium*, along with other probiotic bacteria, is authorized in the EU for use as a probiotic feed additive for many farm animal species. A period of 60 to 80-day milk stasis before parturition is widely practiced in dairy cows nowadays. MMP-2 and MMP-9 participate in degrading gelatin and type IV collagen, the main components of extracellular matrix. We have recently reported that ultrasonicated *E. faecium* SF68 (SF68) was compatible with the milk stasis mammary gland of Holstein cows to serve as a dry cow treatment. It was found enhanced the local innate immune function and promoted the rate of mammary gland involution during the first two-week dry period. In the current study, we further examined the long term effects of intramammary infusion of ultrasonicated SF68 at dry on the last stage of gestation in Holstein cows and the production of colostrum-derived protection factors post-partum. Five Holstein cows received intramammary infusions of untrasonicated SF68 ($5\times10^8$cfu/quarter) at the first day of milk stasis. During the 30 post-infusion days, the total blood leukocyte counts, plasma total protein and immunoglobulins/albumin increased as early as day 3 and until day 30 post-infusion. All infused cows normally delivered live calves. Effects of intramammary infusions of untrasonicated SF68 on plasma levels of progesterone, estradiol, and cortisol are yet to be determined. In another trial, two cows received similar intramammary infusions of ultrasonicated SF68 in two quarters but the contra-lateral quarters received regular antibiotic dry cow treatment. Analyses of samples of the first colostrum showed that colostrum from SF68-infused quarters contained higher ratios of immunoglobulins, BSA, and lactoferrins, as well as MMP-9/MMP-2 with no difference in casein ratio. It appears that intramammary infusion of ultrasonicated SF68 during dry period might have long term effects on colostrum quality.

---



**Effect of antimocrobial substitution with _Bacillus subtilus_ on broiler health and performance**

*P.Y. Aad[1], J. Ibrahim[2], C. Al-Fahel[1], S. Boustany[1] and N. El Ghossein Malouf[1]*
*[1]NDU-Louaize, Sciences, Main St., Zouk Mosbeh, Lebanon, [2]Lebanese University, Agricultural Sciences, Main St., Dekwaneh, Lebanon; paad@ndu.edu.lb*

*Bacillus subtilis* Pb6 (CloSTAT, Kemin), a new generation growth promoter, secretes surfactins within the digestive system, targeting pathogenic bacteria. The objective of this study was to monitor the effect of Clostat on *Escherichia coli* and *Clostridium* intestinal population in 2 Lebanese commercial operations. Two experiments were designed to question the antimicrobial (exp 1; 10,000 birds) and the growth promoting effects (exp 2; 18,360 birds) of CloSTAT as compared to standard antibiotic growth promoters (AGP). Broilers were raised in a parallel barn system, with identical practices except for inclusion of AGP (avilamycine in exp 1 and Maxus in exp 2) or CloSTAT (Treatment, Kemin) in a traditional soya/corn diet. In exp 1, fecal and digestive samples from sacrificed birds were collected on day 10, 20 and 30 coinciding with the growing, transition and fattening rations. Intestinal lesions were scored as the birds were dissected, and digestive content samples saved. Both samples were cultured aerobically on McConkey agar to *E. coli* cfu, and anaerobically on Chromoselect agar (Sigma) to indentify *Clostridium* sp. cfu, using the pour plate method. In exp 2, body weight, feed intake and feed conversion were recorded and monitored weekly. Both CloSTAT and AGP flocks had no significant difference (P>0.10) in body weight, mortality and feed conversion rate in exp 1 and 2. Furthermore, no significant difference (P>0.10) in *Clostridium* and *E. coli* suppression effect between Clostat and AGP in exp 1 at days 10, 20 or 30. Therefore, the introduction of new managerial practices such as phasing out antibiotic growth promoters and replacing with CloSTAT to control bacterial growth and its resistance to antibiotics should become a global goal in modern commercial broiler production. This work was supported by a grant from KEMIN Co, Lebanon.

**Host response and microbiota composition after *Lactobacillus* administration: pig as model for human**

*A. De Greeff[1], D. Schokker[2], F.M. De Bree[1], A. Bossers[1], S.A. Vastenhouw[1], F. Harders[1], M. Kleerebezem[3], J.M. Wells[3], M.A. Smits[1,2,3] and J.M.J. Rebel[1]*
*[1]Central Veterinary Institute, P.O. Box 65, Lelystad, the Netherlands, [2]Wageningen Livestock Research, P.O. Box 65, Lelystad, the Netherlands, [3]Host Microbe Interactomics, Wageningen UR, P.O. Box 338, Wageningen, the Netherlands; annemarie.rebel@wur.nl*

Literature has shown that mouse models do not always reflect the human immune response, in particular for inflammatory processes. It is known that porcine immune responses are more similar to human immune responses. To evaluate the potential of porcine models for probiotic studies, this study evaluates the porcine responses to probiotics for which the molecular responses were previously determined *in vivo* in humans. Three groups of 6 piglets were treated daily with probiotics for a period of 8 consecutive days. The piglets of two groups received each day either $10^{11}$ cfu of *Lactobacillus plantarum* or $10^{11}$ cfu of *Lactobacillus caseii*, the third group received placebo (no bacteria). Probiotics were administered orally prior to the morning meal dissolved in a sports drink. After 8 days of treatment, piglets were sacrificed. At necropsy different locations of the GI tract were sampled for evaluation of epithelial integrity, microbiota composition, and transcriptional host responses. All piglets were clinically healthy during the experiment. No morphological changes in epithelial integrity were observed. Microbiota composition of probiotic groups changed significantly, revealing that the administered lactobacilli were among the most predominant species residing in the jejunum, although lactobacilli were also prominently present in the control animals. Notably *L. plantarum* administration elicited transcriptional modulation of immune related pathways in ileum, which was also the most prominent response-category observed in the human study. More detailed comparative analysis of these initial observations are required to further establish that piglets provide a suitable animal model for human probiotics studies.

---



**Efficacy of a microbial feed additive in weaned piglets**

*B.T. Lund[1], J. Sánchez[2], C. Millan[2], O. Casabuena[2] and M.I. Gracia[2]*
*[1]Chr. Hansen A/S, 10-12 Bøge Allé, 2970, Hørslom, Denmark, [2]Imasde Agroalimentaria, S.L., C/ Nápoles 3, 28224 Pozuelo de Alarcón, Spain; mgracia@e-imasde.com*

A feeding trial was carried out to assess the effect of a microbial feed additive (probiotic, *Bacillus subtilis* DSM 27273) on growth performance of post-weaned piglets. A total of 360 crossbreed (Duroc male × Landrace × Large White female) piglets were sorted according to body weight and allocated to 36 pens (2 weaner rooms) with 12 pens of 10 piglets (5 male and 5 female) per treatment. There were 3 experimental treatments: T1, Negative Control (NC); T2, NC + $4\times10^5$ cfu/g of probiotic and T3, NC + $8\times10^5$ cfu/g of probiotic applied in prestarter (28-42 d) and starter (42-70 d) feeds. Pelleted feeds were *ad libitum* available with no added growth promoter or veterinary antibiotics. Data were analysed as a completely randomised block design by General Linear Model (GLM) procedure of SPSS (v. 19.0). The model includes room (block) and dietary treatment as main effect, and initial body weight was used as a covariate. At the end of the experiment, piglets supplemented with the probiotic at the higher dose (NC + $8\times10^5$ cfu/g) were 3.9% heavier than the Negative Controls, but differences were not significant ($P>0.05$). During the prestarter period, piglets receiving the probiotic at the higher dose (NC + $8\times10^5$ cfu/g) improved feed conversion ratio by 14.7% compared to Negative Controls (1.22 vs 1.43 g/g, $P<0.05$), showing the lower dose intermediate values. For the starter phase, piglets fed with the lower dose of probiotic showed the best feed conversion ratio, compared to Negative Controls (1.43 vs 1.51 g/g, $P<0.05$). For the overall experiment, the supplementation with the probiotic improved feed conversion ratio by 4.0% compared to Negative Control ($1.49^b$, $1.42^a$ and $1.44^a$ g feed/g growth; for T1, T2 and T3, respectively). Under our experimental conditions, it can be concluded that supplementation of piglet diets with the tested probiotic improves feed conversion of piglets for the overall experimental period.

**Effects of kefir as a probiotic source on the performance of young dairy calves**

*M. Gunal and S. Satık*
*Suleyman Demirel University, Animal Science, Cunur, 32260-Isparta, Turkey; mevlutgunal@sdu.edu.tr*

Kefir is a sour, viscous, slightly carbonated and alcoholic milk beverage, which is traditionally fermented with bacteria and yeasts. The influence of kefir on health has been well studied in mice and rats. However, researches on usage of kefir in ruminants are rather limited. The aim of this study was to investigate the effect of kefir as a probiotic on the performance of young dairy calves. Thirty calves were randomly allocated to three treatment groups: Control, commercial probiotic (2 g/d/calf yeast-based and 3 g/d/calf bacteria-based) and kefir (20 ml/d/calf). The experiment was performed in 70 days. Calves were weaned when they started to consume 0.8 kg of calf starter. Starter intake and faecal scoring were measured daily. Faecal samples were collected and analysed on days 14 and 28. Calves were weighted every two weeks. The supplementation of different probiotics did not have any significant effect (P>0.05) throughout the study on live weight and weight gain of the calves as compared to the control group. Similarly, the intake of starter was not affected (P>0.05) by the treatments. On days 14 and 28 the faecal population of lactic acid bacteria (LAB) was not different (P>0.05) between treatments; however the average population of LAB in faeces was greater (P<0.05) with commercial probiotic and kefir groups than the control. The results of this study showed that incorporation of kefir in the diet can allow generating a controlled imbalance in the gastrointestinal tract of calves, but did not affect the calves' growth performance.



**The use of GalliPro® to improve broiler performance in protein-reduced diets**

*D. Harrington[1], E. Scott-Baird[2] and A.B. Kehlet[1]*
*[1]chr. Hansen A/S, Ahn, Bøge Alle 10-12, 2970 Hørshølm, Denmark, [2]ADAS Drayton, Alcester Road, Cv37 9rq Stratford-Upon-Avon, United Kingdom; dkdaha@chr-hansen.com*

Probiotics have been used successfully in commercial poultry production globally. GalliPro, a unique strain of *Bacillus subtilis*, exhibits increased protease activity compared to other commercially available probiotics and may confer benefits of increased nutrient digestibility in addition to other probiotic effects. A study was conducted to evaluate the performance effect of GalliPro (GP) on broilers fed diets with reduced protein levels reared for 42 days. A total of 368 birds (Ross 308) were allocated to 4 treatments and fed standard wheat/barley/soya based rations with reduced digestible protein (DP) levels: T1) standard (S) DP; T2) DP + GP ($8.0\times10^5$ cfu/g feed); T3) -3% DP; T4) -3% DP+GP. At 7 days used litter from a commercial poultry farm was introduced to all pens to provide a low bacterial and coccidial challenge. By day 42, birds in T1 were 2.82 kg and T2 were 3.00 kg, significantly different from each other and heavier than birds in T3 and T4 (2.36 and 2.42 kg, respectively). Birds in T2 had the lowest FCR (1.59) although this was not significantly different from T1 (1.66). FCR in T3 and T4 (1.82 and 1.75, respectively) were not significantly different from each other but were significantly higher than T1 and T2. Mortality did not differ significantly between treatments, although T2-T4 were lower than T1 (2.22, 2.22, 4.11 and 4.33%, respectively). This study indicates that the supplementation of feed with GalliPro may numerically compensate bird performance (FCR) when diets have reduced digestible protein by -3%, most likely by improving diet digestibility. However the compensation is not sufficient to confer equal performance to birds on standard diets. GalliPro does also significantly improve the performance of birds on standard diets.

**Effect of probiotic addition to rations on lactating buffaloes performance**

*T. Morsy, S. Kholif, H. Azzaz and H. Ebeid*
*National Research Center, Dairy Science, Elbehoos st. Dokki, Giza, 12622, Egypt; tareknrc@yahoo.com*

Probiotics have been shown to improve anaerobiosis, stabilise pH and supply nutrients to microbes in rumen microenvironment. In this concern, the yeast *Saccharomyces cerevisiae* (Y) or yeast+propionibacteria (Y+P) were used as ration additives for lactating buffaloes. Their effects on nutrients digestibility, milk yield and composition were studied. Eighteen lactating buffaloes, 2 months after parturition, were divided into three groups of six animals each: (1) control ration consisted of 45% concentrate feed mixture (CFM), 30% corn silage, 15% dried sugar beet pulp and 10% rice straw; (2) control + 30 g yeast/head/day; (3) control + 30 g yeast + 4 g propionibacteria (P169)/head/day. The complete randomized block design, with 12 weeks period was used to study these groups. Results indicated that experimental additives improved ($P>0.05$) nutrients digestibility compared with control. Blood serum urea nitrogen was decreased ($P<0.05$) with probiotic additives. All treatments, significantly increased ($P<0.05$) milk yield, 4% fat corrected milk and all milk components except the solid not fat and ash. The experimental additives were increased ($P<0.05$) unsaturated fatty acids in milk specially conjugated linoleic acids (CLA). In conclusion, yeast with or without propionibacteria supplementation to ration of lactating buffaloes had beneficial effects on milk yield and milk components, in particular, healthy fatty acids (CLA) contents.



**Efficiency of liquide probiotic for suckling piglets**

*R. Leikus, R. Juska and V. Juskiene*
*Lithuanian University of Health Sciences, Institute of Animal Science, R. Zebenkos 12, 82317 Baisogala, Lithuania; violeta@lgi.lt*

The aim of the study was to determine the efficiency of liquide probiotic SCD Bio LivestockTM in the diets of suckling pigs. A trial was carried out at the Lithuanian University of Health Sciences, Institute of Animal Science with piglets of Lithuanian White breed. Piglets were allotted into two groups of 48-50 animals (6 litters) each. The piglets of the control group were fed compound feed. The piglets of the experimental group were offered the same feeds except that liquid probiotic SCD Bio LivestockTM was additionally introduced into the drinking water at a rate of 1:1000 (1 ml/l water). The investigation data showed that the introduction of probiotic SCD Bio Livestock into the drinking water improved the heath of piglets. The number of diarrhoea cases was by 28.4% lower and the length of diarrhoea was 50.7-57.4% ($P=0.015$-0.037) shorter. There was also a tendency towards shorter medical treatment period for piglets (27-28.7%; $P=0.071$-0.078). The usage of the treatment lowered medication costs by 30.3% per piglet. The growth rate of the experimental piglets did not differ significantly from that of control. Inclusion of the probiotic into the drinking water resulted in 3.9-14.3% lower consumption of feeds per kg gain. The piglets from 35 to 60 days of age and during all experiment period consumed less feeds by 6.7 and 11.1%, respectively. Thus, it can be concluded that it is advisable to use liquid probiotic SCD Bio LivestockTM for suckling piglets at a rate of 1 ml/l of drinking water. Use of this liquide probiotic in the diets of suckling pigs leads to higher healthiness, less cases of diarrhoea and other diseases and lower treatment and medication costs.

**Probiotic and prebiotic effects on calf development in the first months of life**

*A. Arne and A. Ilgaza*
*Latvian University of Agriculture Faculty of Veterinary Medicine, Helmana street-8, 3004 Jelgava, Latvia; arne.astra@gmail.com*

Diarrhea is the most common gastrointestinal tract pathology of newborn calves, where for the prevention and treatment antibiotics are commonly used. In recent years there is a special attention on antibiotic therapy replacements and researches for new biotherapeutic products and ways on how to improve efficiency of already known products (such as probiotics). The aim of this study is to determine whether inulin enhances probiotic *Enterococcus faecium* positive effects on calf growth and development. In this research we will present the results which we obtained after using probiotics *E. faecium* ($2\times10^9$ cfu/g) (EF group) and synbiotics, where the latter contained not only the probiotics, but also the prebiotic inulin from specially prepared (12 g, 50% inulin) Jerusalem artischok (*Helianthus tuberosus*) concentrate (EF/IF group) feeding male calves (n=8 in each group) at the age from 4 to 12 weeks (56 days). The collected data is compared with the control group (Co group) results. Once a week we took the measures of general health and live weight gain of calves. After slaughter, the measured weight of carcass, the empty rumen and abomasums are used in the calculation of relative weight. Data shows that gastrointestinal tract disfunction was seen less frequently in EF/IF group animals as in Co and EF group of calves. We assume this is the reason why calves of EF/IF group live weight gain was significantly (P=95%) higher than other animals. It should be noted, that compared to the other groups of calves, a statistically significant (P=95%) proved to be EF/IF animals empty rumen weight and their relative weight. Calves which were fed with probiotics showed better results than the Co group of animals. Overall, we can conclude that, despite the indisputable positive effects of feeding the probiotics *E. faecium*, adding Jerusalem artischok concentrate at intake, which contains inulin, significantly improve animal health, development and it's growth.

---



**The efficiency of a novel synbiotic supplement in a pig diet**

*R.V. Nekrasov[1], M.G. Chabaev[1], N.I. Anisova[1], N.A. Ushakova[2] and L.Z. Kravtsova[3]*
*[1]All-Russian Institute of Animal Husbandry, Department of Animal Feeding, Moscow region, Podolsk district, 142132, Dubrovitcy, Russian Federation, [2]A.N. Severtsov Institute of Ecology and Evolution, Moscow, Lenin prospect, 33, 119071, Russian Federation, [3]OOO 'NTC BIO', Belgorod region, Schebekino, 309292, Russian Federation; nek_roman@mail.ru*

The product of beet pulp solid-state fermentation was used as a major component for a novel synbiotic supplement. It contains phytoparticles-microsorbents, live *Bacillus subtilis* cells (3 strains), *Bacillus licheniformis*, lactic acid bacteria and products of their metabolism, mannan-oligosaccharides of *Saccharomyces cerevisiae* cell walls and phytobiotics as a mix powder of *Echinacea purpurea* and berries of *Silybum marianum*. The efficacy of supplement was tested on 36-days old pigs (n=140) that were randomly divided into control and experimental groups after weaning. The experimental group was fed with enriched fodder containing 0.1% synbiotic supplement by weight per food portion for 40 days. The results of study showed that the introduction of synbiotic supplement in fodder increased average daily gain on 6.8% as compared to controls (489.7 vs 458.6 g) and improved the survival rate in the experimental group by 3%. The amount of feed consumed per every unit of weight gained was 0.785 and 0.720 kg in experimental and control groups, respectively. The biochemical blood analysis confirmed the results of study in terms of higher concentration of total protein and protein index in blood accompanying by improved mineral metabolism. In an experiment designed to study the digestibility of food (n=3), the underweight piglets showed a trend in increased nitrogen balance due to better accessibility of the nutrients in the fodder supplemented with synbiotics. The analysis of large intestine microbiocenosis demonstrated a trend in increased number of lactic acid microorganisms in the experimental animals. Thus, the use of our synbiotics in feeding of pigs at early age is an effective strategy.

## Session 38 Theatre 1

**Prudent use of antimicrobials in farm animal production**

*F.C. Leonard*
*University College Dublin, Veterinary Sciences Centre, Belfield, Dublin 4, Ireland; nola.leonard@ucd.ie*

The increasing threat posed by antimicrobial resistant (AMR) organisms to human and animal health is widely accepted. In the EU in 2011, levels of resistance to the critically important AMs, as designated by the WHO, varied from 50% resistance to ciprofloxacin in *Salmonella* isolated from broilers to 4.5% resistance to cefotaxime in *Escherichia coli* isolates from veal calves. The principal driver of resistance is the use, both prudent and inappropriate, of antimicrobial agents (AMs). Data from the EU show that in 2011 almost 90% of AM sales were in the form of oral formulations such as premixes for herd treatments; oral antimicrobial use may facilitate the transfer of resistance genes between bacteria. A wide variety of factors can influence both the use of AMs and the occurrence of AMR. In addition to AM use, AMR in *E. coli* isolates from calves and poultry and in MRSA from pigs has been linked to farm management factors. A recent EU survey identified the strongest influences on prescribing behaviour as susceptibility test results, the veterinarian's own experience, the risk of AMR developing and ease of administration of the agent. These findings were largely in agreement with those of an Irish study of cattle practitioners, although that study also identified external stressors, such as the expectations of the farmer, as influencing prescribing behaviour. An US study of farmers found that the principal barriers to following proper AM use were lack of time and limited finances. Over 80% of farmers were not concerned that overuse of AMs in animals could lead to AMR in farm workers. Approaches to control of AMR are similar across countries and include legislative controls on AMR monitoring systems and prescribing practices, setting of targets for reduction in use, improving husbandry and disease prevention, education of stakeholders and development of rapid diagnostics and new AMs. Some countries have opted for decoupling of prescription from supply of AMs, banning of prophylactic use and issuing of specific guidelines on administration.

---

## Session 38 Theatre 2

**Improving animal health: prevention is better than cure**

*M. Kluivers-Poodt*
*Wageningen UR Livestock Research, Edelehertweg 15, 8219 PH Lelystad, the Netherlands; marion.kluivers@wur.nl*

Antibiotic use in animal husbandry is a main concern. When developing strategies to reduce antibiotic use, it is important to acknowledge that disease has a strong association with transitions. Transitions encompass the shift from one life phase to another. Animal health is influenced by two major factors that can change during a transition: the environment of the animal (incl. presence of pathogens) and the resilience of the animal. Resilience is the capacity of the animal to adapt to changing circumstances and is influenced by stress. Preventive measures regarding animal health should therefore not only focus on pathogens, but also on the animal itself, by improving resilience. In intensive animal husbandry, animals undergo several transitions, varying from birth to weaning to transportation to a new facility. A transition like weaning not only puts the animal at risk of contact with pathogens, by mixing and regrouping animals, but also induces stress by placing the animal in unfamiliar surroundings with new feeding or drinking systems. In piglets, weaning is considered to be the most critical transition, associated with high incidence of disease and high antibiotic use. And while calving in dairy cattle signals the birth of a new calf and the onset of the new lactation, it is also the opening to the majority of the disease events that occur in the life of a cow. The impact of transitions on animal health is influenced by management practices and husbandry conditions that determine the total load of the animal. Some factors that contribute are well researched, the contribution of others still has to be established. Keeping piglets from a social group together and in the same pen during and after weaning for example, results in better feed intake and a decrease in disease incidence. The challenge is to optimize transition management in existing systems, but also in designing new systems better adapted to the animal's needs. Thereby, disease occurrence and antibiotic use will decrease and animal health and performance will increase.

**Family farm versus mega farm in the USA**

*A. De Vries*
*University of Florida, Department of Animal Sciences, 2250 Shealy Drive, Gainesville, FL 32611, USA; devries@ufl.edu*

The debate about family farm versus mega farm is not very prominent in the USA. The contrast implies that family farms are not large dairy farms, and the underlying connotation of 'family' is that a family farm must be preferred over a non-family farm. The average number of cows on dairy farms in the USA has doubled from 94 in 1997 to 196 in 2013. In 2008, the largest 2.4% of all dairy farms produced 47% of the national milk. These 1,600 farms all had at least 1000 dairy cows each. Almost all large dairy farms (in terms of number of cows) are family owned (compared to a group of investors) and work with mostly hired employees. They are both 'family' and 'mega'. In the American mind, it is admirable that a growing business provides employment to non-family members. Larger farms are generally more efficient in milk production in terms of use of assets and have lower cost of production than smaller farms. Some of these advantages are due to specialization within the employees, for example milk harvesting, calf raising, and health care. It is said that managers of farms with more than 300 dairy cows need to be people managers (including providing training) instead of cow managers. Most large dairy farms are located away for population centers, which is feasible with large open spaces in the middle and western parts of the USA. A public concern sometimes voiced is that the welfare of dairy cattle on large farms is compromised compared to small, family operated dairy farms. However, the data on disease incidence and death rates do not support this. Larger farms typically have better reproductive performance, produce more milk per cow, and with a lower somatic cell count. Smaller farms are more likely to offer pasture to dairy cows, however. The shift from family-operated smaller dairy farms to hired-employee operated larger farms occurred several decades ago and is widely accepted among the public. The current public concern is with animal welfare (but independent of farm size), food safety, and environmental footprint of milk production.

---

Session 39 Theatre 2

**Focus on family farm**

*A. Rozstalnyy*
*FAO, Regional Office for Europe and Central Asia, Benczur ut. 34, 1068 Budapest, Hungary; andriy.rozstalnyy@fao.org*

The UN celebrates the 2014 International Year of Family Farming (IYFF) with aim to raise the profile of family farming and smallholder farming by focusing world attention on its significant role in eradicating hunger and poverty, providing food security and nutrition, improving livelihoods, managing natural resources, protecting the environment, and achieving sustainable development, in particular in rural areas. Family farming includes all family-based agricultural activities, and it is linked to several areas of rural development. Family farming is a means of organizing agricultural, forestry, fisheries, pastoral and aquaculture production which is managed and operated by a family and predominantly reliant on family labour, including both women's and men's. Both in developing and developed countries, family farming is the predominant form of agriculture in the food production sector. There are over 500 million family farms in the world. At national level, there are a number of key factors for a successful development of family farming, such as: agro-ecological conditions and territorial characteristics; policy environment; access to markets; access to land and natural resources; access to technology and extension services; access to finance; demographic, economic and socio-cultural conditions; availability of specialized education, among others. Family farming in Europe and Central Asia has an important socio-economic, environmental and cultural role. Running crop-livestock mixed systems it supports sustainable development as well as a contribution to preservation of traditional food products, contributing both to a balanced diet and the safeguard of the world's agro-biodiversity. FAO undertakes studies of the role of family farming and their contribution to sustainable agricultural production in a number countries of the region. The IYFF gives us a clear opportunity to further highlight the strategic role of family farmers in agricultural and rural development and to strengthen their capacities.

**Future opportunities for dairy farming development: a divers picture**

*M. Scholten*
*Animal Sciences Group, Wageningen University and Research Centre, Houtribweg 39, 8221RA, Lelystad, the Netherlands; martin.scholten@wur.nl*

The abolition of the milk quota system in 2015 will open up new perspectives for European dairy production. The world market for milk will increase, with good prospects for the European dairy sector when it can guarantee top quality in connection to high quantity. This might lead to a 'white revolution' in agriculture in the 21$^{st}$ century, as follow up of the 20$^{th}$ century's 'green revolution' in crops. But the 'white revolution' requires a responsible dairy farming with a credible performance based on customized care for animals, human health and environment. The challenge is to produce more and better milk with less resources. In this context, sustainable and climate smart intensification is most effective. The ecological basic principles of efficiency and adaptation is the key guidance towards a customized intensification given the world wide diversity in cattle husbandry in perspective of regional differences in conditions, traditions and perceptive discourses. There is no unique 'one size fits all' best practice. Both, family scale farming and more industrial farming have opportunities for sustainable intensification. The main general priorities in scientific support of innovations towards a competitive and sustainable dairy farming are: optimizing the use of land for basic feed production (both grass and crops) and manure recycling; excessive manure management (protein and mineral recovery); inclusion of precision farming in herd management based on an integrated breed/feed/ health approach; increased life time production; reduced methane emissions.



**The design of mega dairies in India: optimal facility allocation, how big do you want to go?**

*I. Halachmi*
*Agricultural Research Organization (A.R.O.), The Institute of Agricultural Engineering, The Volcani Centre., P.O. Box 6, Bet Dagan 50250, Israel; halachmi@volcani.agri.gov.il*

Economic and environmental pressures are guiding intensive milk production to large farms located in the Far East. The design and management of large-scale dairy farms require management tools. A combined model – queuing-network, robust 6σ design, and simulation optimization – was developed and applied to dairy farm construction in India. Design criteria were: 10,000 cows in milking, intensive farming, year-round indoors, no grazing, open cowsheds such as in Israel (no freestall, no cubicles), dry manure bedding, while maximized animal welfare, maximizing cow resting time and worker convenience. All design criteria were met. We modeled eight farming aspects: cow traffic, milking parlors, vet treatment, manure handling, cow cooling, feed-center operation, workers' transportation and a problematic junction, and their interrelations. Apparent mutual synchrony of these aspects was well-timed with cow traffic. The algorithms developed herein allow optimal facility allocation. The enterprise adopted the model results. The numerical solution may reflect local farming conditions, but the methodology can be applied elsewhere. In the lecture, the discussion will include the farm itself (stable, cow traffic, etc.) and at least as important is how such a farm fits in its surroundings (effect on infrastructure, supply of feedstuffs, other farmers, manure, social effects, environment, etc.). A second concern is: How big do you want to go. Some experiences so far show that 3,000 animals is a nice upper limit and if you want to go beyond, you will have to use different units instead of a larger farm. A thirdly discussion point will be: the wider aspect (sustainability, health; both animal and human, pros-and-cons of centralised units versus decentralised units, labour management, etc.).

## The Kentucky family dairy farm experience

*J.M. Bewley*
*University of Kentucky, Animal and Food Sciences, 407 WP Garrigus Building, 40546-0215, USA; jbewley@uky.edu*

Often, people around the world, perceive the United States dairy industry to be dominated by large, corporate owned dairy farms. While it is certainly true that large dairy farms produce a large percentage of the milk in the United States, 74% of the 51,000 dairy farms in the United States have fewer than 100 cows. Approximately 97% of United States dairy farms are family-owned. Moreover, the size distribution of dairy farms in the United States is somewhat regional. For example, in Kentucky, the average herd size for 780 dairy farms was 88 cows in 2013. Approximately, Amish and Mennonite families; who have a very different approach to dairy farming own 1/3 of the dairies in the state. These dairy farms remain competitive by following a family-based model for managing labor. Economies of scale and size present competitive advantages for larger dairy farms in a commodity market. However, many different models for economic success may co-exist.



## The Slovenian family dairy farm experiences

*M. Klopčič[1] and A. Kuipers[2]*
*[1]University of Ljubljana, Biotechnical Faculty, Dept. of Animal Science, Groblje 3, 1230 Domžale, Slovenia, [2]Wageningen UR, Expertise Centre for Farm Management and Knowledge Transfer, P.O. Box 35, 6700 AA Wageningen, the Netherlands; marija.klopcic@bf.uni-lj.si*

Farm size varies greatly between countries and within countries between regions. Central and Eastern Europe (CEE) has small dairy farms compared to Western Europe. Exceptions are Slovakia and Czech Republic with relatively large farms and Norway with small farms. Data of the Farm Accountancy Network in Europe indicate that income within a country is only weakly correlated to farm size. Other factors are also important, like land quality, strength of chain, practical know-how and motivation. Future development paths in the dairy sector were studied from the view point of farmers in 4 European countries, on the basis of 1,135 questionnaires. Also group sessions were held to develop a strategic plan for each individual farmer. Expansion of milk volume has the highest priority in all countries studied, but there are surely differences in outlook. Poland is on top with the strategy of expansion. Diversification by including another agricultural branch is common in Lithuania, while diversification in non-agricultural activities scores rather high in Slovenia, but also in The Netherlands. Half of the Netherlands farmers in the sample choose for a wait and see attitude towards the realisation of strategies. In the CEE countries the farmers want to move on faster. Bottlenecks to realise the preferred strategies are land and labour availability. The land structure, especially in Lithuania and Slovenia, is very inefficient. In the study, the attitude toward EU-subsidies, abolition of milk quota, regulations on animal welfare, the (inter)national milk market and client orientation differs considerably between farmers from the East and West. But the Netherlands dairy farmer has a higher self-confidence compared to his colleagues in CEE. On the contrary, the ambition score seems to be somewhat higher for the Polish and Slovenian dairy farmers. In other words, farmers in different parts of Europe plan and react differently dependent on the circumstances and culture. There is no one road to the future. Some examples will be presented of young farmers with small size farms who develop strongly.

**Strategies for improving productivity in small ruminants**

*T. Ådnøy*
*Norwegian University of Life Sciences, Department of Animal and Aquacultural Sciences, Box 5003, 1432 Ås, Norway; tormod.adnoy@nmbu.no*

A simple answer is to record size of the grown animal when measuring production per animal. That gives at least some information on costs of keeping it. Another less feasible means is to use genomic selection if some relationship of genome and productivity could be established. A naïve and theoretical approach is taken to clarify the challenges facing small ruminant production improvement. Productivity may be taken as what product the small ruminant gives us back for a given resource. Depending on the resources available (feed, labor, ..) and the production (meat, milk, fibre, landscaping, ..) this may require different husbandry strategies and genetic constitutions in the domestic animal. One total measure of productivity can be found if costs of resources and income from production is known. But both costs and incomes depend on market prices and are prone to changes depending on availability etc. Some resources may be abundantly available (rangeland in some season of the year, water, air, space, ..). Some resources may be what limits the production (space per farm, cost of winter feed, labor, ..). A ruminant may convert plants that man can not digest to products we need. It can also convert human food resources to the same products. When costs of extra human food resources are lower than the income from the extra gain achieved in small ruminant products it may pay to use it. So what will future market prices be? Will human food scarcity lead to market scenaria that favor small ruminants? Will markets favor ideal biological uses of resources? Can or should the ruminant compete with the pig in converting human food resources to products desired by the current market? An extra cost to consider is the greenhouse gas emission related to production. So we need to say what kind of productivity we want to improve. Then we can look for measures of input resources per output product. Some resource use is proportional to animal size – a simple measure that is often not used.

---



**Seasonal variation in semen quality of Dorper rams using different collection techniques**

*C.M. Malejane, J.P.C. Greyling, M.B. Raito and A.M. Jooste*
*University of the Free State, Animal, Wildlife and Grassland Sciences, P.O. Box 339, Bloemfontein, 9300, Free State, South Africa; greylijp@ufs.ac.za*

The aim of the study was to evaluate the seasonal variation in semen quality of Dorper rams, using different semen collection techniques. Eleven mature Dorper rams (mean body weight of 69.6±9.2 kg and age of 18±4.7 months) were used in the trial. Semen from 6 rams were collected weekly with the aid of the artificial vagina (AV) and weekly from 5 rams using the electro ejaculator (EE) for a total period of 12 months. All ejaculates were macroscopically or microscopically evaluated for semen volume, colour, pH, wave motion, sperm concentration, sperm viability and morphology. Results showed that generally semen may be collected (using both the AV and EE methods) throughout the year, with some variation in semen volume and quality. Overall significant ($P<0.05$) superior semen quality (sperm motility, wave motion) was however recorded with the AV method of collection. Semen of significantly ($P<0.05$) higher quality for sperm motility and sperm abnormalities was generally recorded in summer, autumn and spring (both collection techniques). There was a tendency for the EE to be an inferior method of semen collection in terms of semen colour, sperm motility (32.6±15.8%), wave motion (1.9±0.6), sperm viability or % live sperm (39.0±14.6%) in winter, when compared to the AV method of semen collection. No difference in semen quality was recorded for autumn (natural breeding season) and spring (supplementary breeding season) – both collection techniques. Results show that both AV and EE collection methods are generally able to yield acceptable quality semen from Dorper rams throughout the year (limited seasonality), with the AV overall generally recording the better semen. Winter was recorded to be an unfavourable period in terms of semen quality using the EE, especially when semen is to be used for AI or semen cryopreservation.

**Impact of intensification factors on the economics of dairy sheep**

*Z. Krupová[1], E. Krupa[1] and M. Michaličková[2]*
*[1]Institute of Animal Science, P.O. Box 1, 104 01 Prague, Czech Republic, [2]NAFC-Research Institute for Animal Production Nitra, Hlohovecká 2, 951 41 Lužianky, Slovak Republic; krupova.zuzana@vuzv.cz*

The impact of selected intensification factors on the economics of Improved Valachian breed without considering of subsidies was analyzed using the bio-economic model of the program package ECOWEIGHT. The followed intensification tools were investigated: early weaning of lambs with selling (5 days of age) or with artificial rearing of lambs (to 50 days of age), out-of-seasonal lambing in autumn and production parameters settled in the new breeding goal (130 kg of milk, 1.55 lambs per lambing and higher growth intensity of animals). In the base production system, the average parameters (lambs reared with ewe till weaning, seasonal lambing in winter, and average production parameters for the period 2004-2010) were applied. Compared to the base production system, early weaning followed by selling of lambs was of positive impact on revenues (+4%) and profitability (expressed proportion of profit on the total costs) changed from -16 to -14%/ewe/year. Higher value of sold milk (157 kg/milking period) and lower production of lambs (3.45 kg of lamb/ewe/year) at weaning played the main role in this variant. More effective utilization of costs (profitability ratio of -10%) was found when the early weaning was followed by artificial rearing of lambs. Total revenues changed +19% and costs +10%/ewe/year in this case. For out-off-seasonal lambing, a lower conception rate of ewes (-11 p.p) and production of lambs (11 kg of lamb/ewe/year) was compensated by higher market prices of sheep commodities (+11 to +20%). Application of the parameters given in breeding goal increased the revenues (+35%) and profitability per ewe and year was positive (8%). Early weaning of lambs followed by artificial rearing along with production parameters specified in the breeding goal can be recommended as useful intensification factors in analyzed multipurpose dairy sheep breed. Study was supported by project MZERO0714 of the Czech Republic.

---



**Study of dairy lactation curve of Sicilo-sarde sheep milk traits under low input system**

*N. Ben Mansour[1], S. Bedhiaf[2], M. Khlifi[1], A. Hamrouni[1] and M. Djemali[1]*
*[1]Institut National Agronomique de Tunisie, Tunis, Tunisie, Cité Mahrajène, 1082, Tunisia, [2]Laboratoire PAF. Institut National de la Recherche Agronomique de la Recherche Agronomique de Tunisi, RueHédi Karray, 2049, Tunisia; bedhiaf.sonia@gmail.com*

This study aims to characterize the dairy performances of Sicilo-Sarde breed and adjust their lactation curve. A total of 7504 ewes raised in 6 farms and 5 herds with a total of 35,900 monthly dairy controls were used. To analyze the dairy performances', a linear model including the following factors: farm, herd, year of lambing, duration of nursing, duration of milking, age and season of lambing, was used. The fitting of the lactation curve was performed by the Gauss Newton's algorithm in the NLIN procedure of the SAS program using the Wood's incomplete Gamma function. Main results showed a daily milk production mean of 0.6 liters /day, a total milk production mean of 91.6±44.2 liters corresponding to a pre-weaning period of 102.8±22 days and a milking's duration of 145.42±40.4 days. The linear model showed that the sources of variation with highly significant effects on dairy performances were: farm×herd×year effect, the pre-weaning period and the milking's duration effects. The Wood's function enabled to draw a lactation curve having two phases, an ascending phase and a descending one. The initial production was 0.8 liters, the production reached its maximum in the 43$^{rd}$ day with a mean of 1.64 l and the persistence coefficient was of 96.21%. This study allowed estimating the dairy potential of the Sicilo-Sarde ewe under low input production system.

**Breeding strategies for small ruminant productivity in Turkey**

*S. Ocak[1] and E. Emsen[2]*
*[1]Zirve University, Middle East Sustainable Livestock Biotechnology and Agro-Ecolgy Research and Development Centre, Kizilhisar Campus, 27260 Gaziantep, Turkey, [2]Ataturk University, Department of Animal Science, Faculty of Agriculture, 25240 Erzurum, Turkey; sezenocak1@gmail.com*

Turkey's small ruminants are well adapted to the rangelands and marginal areas of the country. They contribute to the livelihoods of vulnerable and resource-poor farmers who live under extreme conditions in the rural areas and highlands. The number of small ruminants in Turkey fell from almost 51,530 in 1990, to 32,2 in 2012; resulting with a reduction of about 40% in 22 years. The population of Turkey is currently 70 million and expanding with a growth rate of 2.2% per year. An increasing population combined with a decrease in animal population result in nutritional and socio-economic impacts. About four million agricultural enterprises are operating in Turkey; crop production and mixed crop-livestock production account for 96% of this total; the remaining 4% consists of livestock production alone. Small ruminants constitute about 55% of the total number of livestock kept in Turkey livestock enterprises. Typically, Turkey's small ruminant systems consist of small flocks of indigenous animals that are managed using family labour. Three general livestock production systems are used, the sedentary, the transhumant and the nomadic. As a consequence, production levels per head are not high. However, it is possible to increase productivity, by using new techniques, appropriate breeds and production systems. It is important to underline the importance of sustainability in breeding as a main strategy in order to increase the productivity of small ruminants. Before starting a strategic planning for small ruminant breeding we must consider number of factors such as; production system (i.e. nomadic, sedentary, transhumant, mixed farming etc.), the priority order of economic traits (i.e. meat, milk, fibre skins etc.), the national herd structure, socio-economic levels, including level of literacy, product consumption habits and market conditions.

---



**Relationship between technology and the variability of dairy sheep**

*J. Rivas[1], C. De Pablos[2], E. Angón[1], J. Perea[1], R. Dios-Palomares[1], M. Morantes[1] and A. García[1]*
*[1]Universidad de Córdoba, Campus Rabanales, 14014, Córdoba, Spain, [2]Universidad Rey Juan Carlos, Paseo de los Artilleros, 28034, Madrid, Spain; pa1gamaa@uco.es*

Spain has a long tradition of dairy sheep production; each time more the sheep farming is more specialized and technologically advanced. The aim is to study the association between technological packages (TP) and their relationship with the variability of the production in sheep farms of La Mancha. A survey about technology packages (management, feeding, animal health and milk quality, pasture and land use, equipment and housing facilities, and reproduction and breeding program) to a sample of 157 farms was applied. Descriptive statistics, Sperman correlations and multiple regressions step by step analysis were applied. The typical farm was described, the degree of implementation of TP was analyzed and the variation ratio of each TP over the production and the degree of dependence on external inputs were established. The typical farm responds to extensive mixed crop-livestock systems, pastoral (95%), with 888 sheep and 1,124 ha total surface (ST). The mean productivity is 145.3 l of milk and 1.6 lambs per ewe and year. Animal health and the milk quality, the feeding and management processes were using TP with a degree of implementation over 55%. Correlations ($P<0.01$) between the managerial skills and the implementation of other technological packages is evidenced; similar to the correlations between high reproduction and genetic breeding ($r=0.612$), equipment and facilities ($r=0.626$) and animal health and milk quality ($r=0.515$). It was observed that the variability in milk production (35%) and commercial lambs (38%) is explained by the TP of reproduction, the breeding program and managerial skills. Moreover the variability explained over the dependence on external inputs (62%) depends mainly on grazing and the use of land and, to a lesser extent, on food. The TP show synergies one each another and are being implemented in the dynamic context of the company showing multiple interactions towards a dynamic equilibrium of the system.

**Small ruminants productivity improvement by abortions prevention**

*H. Khaled, S. Merdja and A. Bouyoucef*
*LBRA, Veterinary Institute, USD BLIDA, Route de Soumaa, BP 270, Algeria; khaledhamz@yahoo.fr*

Livestock is a main part of agriculture worldwide and, is critical to the employments especially in developing countries. The productivity improvement strategies could be approached from genetic; nutritional and health aspects. In Algeria, sheep predominate the global effective and account for 80% of the total estimated at more than 23 million, with 10 million ewes. Goats are second (13%) comprising 50% of females. these data make the small ruminants' husbandry a real asset for the economy development. One of the most dangerous livestock scourges is represented by reproductive disorders including abortions, which are responsible for major losses to dairy industry. These damages contain foetal mortality, delivery of weak lambs, loss of milk production for lactating females, infertility and infectious risk to humans in contact with these animals. The purpose of our work is to illustrate the role of abortions prevention in the development of livestock. The main results indicate a predominance of abortions in sheep with 56%, in second place goats with 34%. These abortions are enzootic in excess of 5% for 33% of farms investigated, which show the extent of this phenomenon and confirm their negative role as a limiting factor for herds' development. They should be taken with more cautions and this, to contribute in the improvement of livestock health and productivity.



**Do cooperatives influence technical efficiency in extensive sheep production systems?**

*P. Gaspar, A.F. Pulido, A. Rodríguez De Ledesma, A.J. Escribano and M. Escribano*
*Universidad de Extremadura, Producción Animal y Ciencia de los Alimentos, Avda. Adolfo Suárez s/n, 06007 Badajoz, Spain; pgaspar@unex.es*

Over the last few years there has been an important trend among extensive sheep-for-meat producers in South West Spain towards integration in a cooperative system. This process has reached 64.7% of the sheep population in the area and its main goal is the improvement of the productivity and profitability of cooperative farms. Despite this degree of integration, there are several cooperatives operating in the region. They are performing similar strategies but applied differently. Among the various services they provide we could emphasize (1) technical advice to the farmer, (2) subsidies for the improvement of farm fattening facilities, (3) establishment of business partnerships and (4) financial services to members. Moreover, since 2008, CAP subsidies to the sheep sector are linked to farmers' membership to producer associations or cooperatives, promoting specialization and competitiveness of the sector. This paper analyzes the technical efficiency of extensive sheep farms in a cooperative environment. The aim is to identify the impact of cooperative membership in the economic performance of farms using the concept of technical efficiency. For this purpose we have analyzed a total of 121 farms which belong to 3 sheep cooperatives located geographically in an area with similar edaphoclimatic conditions. The methodology used is the data envelopment analysis. This technique creates efficiency indices by comparing the performance of each farm with the best production practices observed which define the efficiency or production frontier. Indices have been obtained for each of the farms analyzed and differences in efficiency depending on the cooperative membership have been observed. In a second step the diverse services provided by cooperatives have been included as exogenous variables, with the aim of explaining the inefficiencies identified at cooperative level.

**Technical-economic structure in extensive sheep farms for decision making**

*A. Rodriguez-Ledesma, A.F. Pulido, P. Gaspar, A.J. Escribano and F.J. Mesias*
*Universidad de Extremadura, Producción Animal y Ciencia de los Alimentos, Avda. de Adolfo Suarez s/n, 06007, Spain; rledesma@unex.es*

In the Spanish sheep primary sector, a strong cooperative structure has been developed in the last few decades, mostly on extensive farms. As a result, a linked business network has been growing in order to fulfill their managing demands, requiring useful decision-making tools. The interrelation cooperative-producer is key to compile all the information needed to obtain practical and easy-to-use tools. Unfortunately, the collected information tends to be scarce, of variable nature and not quite reliable, forcing managers to consider changing, fuzzy scenarios too frequently when it comes to decision making. This study aims to evaluate the relationship between resources, decision nodes, outputs-inputs and the most important information flows in the system. At this point, potential indicators and their use in decision making -for the cooperative management- are analyzed. The cooperative activity takes priority over the the productive activity, in order to ultimately favour the farmer.

---



**The introduction of novel bucks without physical contact can induce reproductive activity in goats**

*L. Gallego-Calvo, J.L. Guzmán and L.A. Zarazaga*
*Universidad de Huelva, Campus de Excelencia Internacional Agroalimentario-ceiA3, Carretera de Palos de La Frontera, s/n, 21819, Huelva, Spain; lourdes_vet88@hotmail.com*

The aim of this work was to determine if the introduction of a novel male without physical contact with females that were previously in contact with familiar males, is enough to induce reproductive activity and compare this response to a classical male effect. Eighty five adult goats were used. They were distributed into two groups according to their previous contact with males: not isolated group (NOTISOL group, n=54) or isolated group (ISOL group, n=31). The ISOL group was isolated from the males contact for a period of 46 days and the NOTISOL group was in permanent contact with three vasectomised adult males. After this period, novel males were introduced in direct contact with females of the ISOL group (classical male effect) and separated by a fence for the NOTISOL group. To ensure that novel males were sexually active at the moment of their introduction, they were treated with melatonin implants inserted at the same time that the females of the ISOL group were isolated from males. Oestrous activity was checked daily with males fitted with a marking harness. The ovulation rate was examined by transrectal ultrasonography after detecting oestrus. Plasma progesterone concentration was tested in blood samples collected weekly by jugular venipuncture. No differences between groups were observed in the percentage of females showing ovarian activity in response to the novel male introduction (84 vs 70%, for ISOL and NOTISOL, respectively, $P>0.05$). However, differences between groups on the oestrous response were observed (74 vs 52%, for ISOL and NOTISOL, respectively, $P<0.05$). The results of the present experiment demonstrate that even only the introduction of active males in a group of females that are in anoestrous and without being isolated from males during a period is able to induce an ovarian response similar to a classical male effect. This work was supported by Grant AGL2012-31733 from C.I.C.Y.T. (Spain).

**The response to the male effect are modified by increasing live weight-body condition score in goats**

*L. Gallego-Calvo, J.L. Guzmán and L.A. Zarazaga*
*Universidad de Huelva, Campus de Excelencia Internacional Agroalimentario-ceiA3, Carretera de Palos de La Frontera, s/n, 21819, Huelva, Spain; lourdes_vet88@hotmail.com*

The aim of this work was to determine in goats if the response to the male effect could be modified by the dynamic effect (gains/losses) of body weight (BW)-body condition score (BCS). Forty adult and non- pregnant goats were used. They were distributed into two groups with 20 animals each according to the initial level of BW and BCS. A group with Low BW and Low BCS (LL group, BW: 36.9+-1.5 kg and BCS: 2.52+-0.04) and other group with High BW and High BCS (HH group, BW: 40.7+-1.3 kg and BCS: 2.89+-0.07). The day 0 was considered the day when the males and females were isolated for a period of 45 days. During the period of isolation between males and females, the female groups were submitted to the experimental levels of nutrition. They were fed 1.9 and 0.4 times over the maintenance requirements for the LL and HH group, respectively. Oestrous activity was checked daily by males fitted with a marking harness. Fecundity was determined by transrectal ultrasonography 45 d after mounting. Weekly BW, BCS and progesterone concentrations were determined. At birth, prolificacy and productivity was recorded. At the moment of the male introduction, LL group showed a BW 40.7+-1.9 kg and BCS 2.74+-0.03 and the HH group a BW 32.8+-0.9 kg and BCS 2.49+-0.04. The gain of BW-BCS modified the ovarian and oestrous response and the productivity to the male introduction, that was higher in the LL group (ovarian response: 100 vs 78%; oestrous response: 95 vs 70% and productivity 1.15+-0.18 vs 0.70+-0.11 kids/present female, for LL and HH group, respectively, $P<0.05$). No other reproductive parameter, fecundity or prolificacy, were modified by the gain/loss of BW-BCS ($P>0.05$). The results of the present experiment demonstrate that the dynamic effect of BW-BCS previous to the male introduction has a critical role on the response to the male effect. This work was supported by Grant AGL2012-31733 from C.I.C.Y.T. (Spain)

---



**Lamb weaning and start of machine milking in dairy ewes: influence on milking efficiency**

*M. Uhrinčať, V. Tančin, L. Mačuhová, J. Margetínová and J. Antonič*
*National Agricultural and Food Centre, Research Institute for Animal Production Nitra, Hlohovecká 2, 951 41 Lužianky, Slovak Republic; uhrincat@cvzv.sk*

The lamb weaning and first milking are the stress factors, which could affect milking efficiency at the beginning of lactation. The aim of trial was to study how these factors influenced Tsigai and Improved Valachian crossbreed ewes (project MLIEKO No. 26220220098). 36 lactating ewes on $3^{rd}$-$9^{th}$ lactation were used. The first of 3 consecutive milkings was performed approximately 12 h from lamb weaning, control milking (CM) followed by two weeks later. Milk samples were collected individually. Differences in the number of somatic cell count (SCC) after log transformation were analysed using One-Way ANOVA and Scheffe's test, relations between order of milking (OM) and parameters of milkability and SCC in first 3 milkings were tested using the REG and GLM procedure (SAS 9.1). There were no significant differences among milkings for total milk yield (TMY) but differences between $1^{st}$ and $2^{nd}$ milking (0.34±0.04; 0.22±0.04 l (LS means±SEM), resp.; $P<0.05$) and between $1^{st}$ and $3^{rd}$ milking (0.34±0.04; 0.19±0.04 l, resp.; $P<0.01$) for machine milk yield was discovered. We also found decrease of milking time in $3^{rd}$ milking against $1^{st}$ milking (75±5; 52±5 s, resp.; $P<0.01$). The changes in milk flow type (MFT) were recorded also. The occurrence of different MFT were changed in $1^{st}$, $2^{nd}$ and $3^{rd}$ milking follows: 25; 33; 31% for bimodal MFT (B), 11; 28; 28% for non-bimodal MFT (N), 31; 22; 19% for plateau (P) and 33; 17; 22% for low plateau (PL). In CM, PL milk flow type did not occur and percentages of B, N and P were 35; 30 and 35%. The average of SCC was 5.39±0.70 at $1^{st}$, 5.66±0.73 at $2^{nd}$, 5.68±0.65 at $3^{rd}$ and 5.26±0.61 log/ml at CM. Significantly negative correlation of SCC with TMY ($r=-0.196$) and OM and lactose ($r=-0.319$) were found out. In conclusion, weaning of lambs and change of milk removal, as possible stress effect, negatively influenced milking efficiency and SCC shortly after weaning only.

**The effect of ewes relocation on milk composition and milk flow kinetic**

*V. Tančin[1,2], L. Jackuliaková[2], M. Uhrinčat'[1] and L. Mačuhová[1]*
*[1]National Agricultural and Food Centre, Research Institute for Animal Production Nitra, Hlohovecká 2, 95141 Lužianky, Slovak Republic, [2]Slovak Agricultural University, Department of Veterinary Science, Tr. A. Hlinku 2, 949 76 Nitra, Slovak Republic; tancin@vuzv.sk*

The effect of relocation and milking in other parlour (treatment) on milk flow kinetic, milkability and milk composition was the aim of this study. 34 ewes of two breeds of ewes Tsigai (14 heads) and Improved Valachian (20 heads) were tested (project MLIEKO 26220220098). Two weeks after lamb weaning the ewes were milked in parallel milking parlour (1×16 stalls) under shelter. On the last evening milking (first experimental milking, EB) before relocation of flock to another parlour, and during next three continuous evening milkings (E0 = second; E1 = third; and E2 = fourth milking of exp.) after relocation the milk flow kinetics were measured using electronic collection jar. On day E0 after morning milking the flock was moved on a pasture and milked in other parlour (1×24-stalls). During E0 we recorded a significant decrease of total milk yield in comparison with EB (0.527±0.04 and 0.647±0.04 l). Significant differences were also recorded in machine milk yield, machine stripping, milking time and in maximum milk flow rate. During E0 there was a higher number of nonbimodal and lower numbers of bimodal flow types. The response of ewe to E0 depended on its response to EB. Ewes with bimodal flow at EB responded more negatively to E0 than ewes with nonbimodal or plateau flow. During E2 there were significantly increased protein content and solids not fat in milk. Thus the treatment significantly influenced the milkability of ewes in a negative way, but more clear response was finding out in ewes with bimodal flow response to machine milking before treatment. We could assume that relocation to other milking conditions caused only short-term changes in milk flow kinetic and milk yield.

---



**Milk partitioning in the udder of Tsigai and Improved Valachian ewes**

*V. Tančin[1,2], L. Mačuhová[1], M. Uhrinčat'[1], J. Antonič[1] and D. Apolen[1]*
*[1]National Agricultural and Food Centre, Research Institute for Animal Production Nitra, Hlohovecká 2, 95141 Lužianky, Slovak Republic, [2]Slovak Agricultural University, Department of Veterinary Science, Tr. A. Hlinku 2, 949 76 Nitra, Slovak Republic; tancin@vuzv.sk*

Milk fractions which are collected during milking (machine milk and stripping milk), residual milk (milk obtained after oxytocin injection) could be used to evaluate machine milking ability of dairy ewes. The volumes of cisternal and alveolar compartment of the udder are different in each breed of ewes. Therefore, the aim of this experiment was to evaluate the milk distribution and basic milk composition in the udder of Tsigai (T) and Improved Valachian (IV) (project MLIEKO 26220220098). The study was performed on 24 ewes (T, n=12; IV, n=12) which were chosen on the base of their milk production (over 0.400 l) from previous control milking. Eight hours from previous milking the volume of cisternal milk was evaluated after Atosiban administration (AT – oxytocin blocking agent) and alveolar milk was obtained after an iv oxytocin injection. The samples of milk were taken two times (after AT and after oxytocin). Dates between two above mentioned breeds of ewes were evaluated by Student t-test. IV had higher total milk yield than T ewes (0.455±0.07 vs 0.363±0.09 l, respectively; $P<0.05$). Milk partitioning in the udder for IV and T differed in cisternal milk (0.298±0.07 l and, 0.199±0.05 respectively; $P<0.001$), but not in alveolar milk (0.157±0.05 and 0.164±0.05, respectively). Ratios between cisternal and alveolar milk were 65:35 and 54:45 for IV and T ewes, respectively. Breeds did not influence milk composition except protein content in alveolar milk in IV and T (5.17±0.37 vs 5.63±0.57., respectively; $P<0.05$). In conclusion, milk distribution in udder of T and IV is first time presented but more data is needed to confirm or explain high difference in milk distribution in udder between two above mentioned breeds.

**Abortion in Bedouin goat: serological diagnosis of Brucellosis, Bluetongue and CAEV**

*A. Kouri, F. Khammar, Z. Amirat and S. Charallah-Cherif*
*University of Science and Technology Houari Boumediene, Faculty of Biology, BP 32. El Alia Bab Ezzouar, 16111 Algiers, Algeria; charalla@yahoo.fr*

Bedouin goat living in dry lands frequently presents pregnancy failures. Height abortion rats restrain economic profitability of goat herds. According to their etiologies, abortions can be classified into two categories either infectious or non infectious. The aim of this study is to determine the implication of some abortive infectious agents in the pregnancy failure in the Bedouin breed. The study was carried out, between March 2011 and December 2012, at Béni-Abbès region in the southwest of Algeria, on 73 females belonging to 18 herds (4 sedentary, 4 semi-sedentary and 10 nomadic systems). Fifty four of these females aborted at different stages of gestation. The other females (19) have kidded. Blood samples taken from jugular vein were centrifuged then serums were frozen at -20 °C. Serological diagnosis of Brucellosis was realized by the Buffered Plate Antigen test and Complement Fixation Test. Bluetongue and Caprine Arthritis Encephalitis Virus were detected by competitive ELISA. Chi-Square and Fisher's exact test were used for statistical analysis. All serums tested were seronegatives for CAEV. Among the whole, 23 females (31.5%) were seropositives for Brucellosis, out of them 2 kidded (8.7%) and 21 aborted (91.3%). Brucellosis affects significantly the incidence of abortion (P=0.02). Forty six females (63%) were seropositives for Bluetongue. Nine of them have kidded (19.6%) and 37 have aborted (80.4%). No significant difference was observed (P=0.1) between the occurrence of abortion and seropositivity for Bluetongue. The Bedouin goat, as a rustic breed, may have developed an adaptive strategy to resist consequences of this disease. Moreover, 17 goats have both Brucellosis and Bluetongue. Twenty one goats were seronegatives for the 3 pathogens agents; in these cases abortion can be attributed to other infectious pathogens or to non infectious causes which should be explored in order to identify pregnancy failure etiologies in Bedouin goat.

---



**Effect of shearing on performance and physiological parameters in lambs under ambient heat stress**

*F. Moslemipur*
*University of Gonbad Kavoos, Animal Production, Shahid Fallahi St., Gonbad Kavoos, Golestan Province, 49717-99151, Iran; farid.moslemipur@gmail.com*

The aim of the study was to investigate effect of shearing on performance and some metabolic and physiologic parameters in fattening lambs under heat stress, 16 male, fattening Dalagh lambs3. (5±0.5 months of age) were divided into two groups and maintained individually for adaptation. One group was completely shorn and another left unshorn. Duration of the study was four consecutive weeks after the shearing (mean ambient temperature 36.22 °C). Weight changes were recorded weekly and feed intake was measured three times a week. Rectal temperature, heart rate and respiratory rate were weekly measured. Blood samples were collected and used for blood metabolites and hormones' determinations. Results showed that shearing has no significant effects on feed intake and weight gain over the study. Feed conversion ratio was improved by the shearing (10.70 vs 13.26 kg/kg). Shearing caused a non-significant decline in rectal temperature (P=0.09). Heart rate was not affected by the shearing, but respiratory rate was significantly decreased by the shearing (P<0.001). Blood glucose, urea and total protein levels were not affected by the shearing. The marked decline in respiratory rate and relative decline in rectal temperature by the shearing indicate a lowered response to heat stress. Furthermore, decrease in thyroxine level is due to the lower body's heat production that generally indicating increase in energy retention.

## Session 40 Poster 17

**Characterisation of production systems and genetic structure of Zulu Sheep populations in R.S.A**

*N.W. Kunene[1], B.S. Mavule[1], E. Lasagna[2] and C.C. Bezuidenhout[3]*
*[1]University of Zululand, Agriculture, Kwadlangezwa, 3886, South Africa, [2]Università degli Studi di Perugia, Biologia Applicata, Giugno 74, 06121, Perugia, Italy, [3] North-West University, Environmental Science and Development Management, Potchefstroom, 2520, Potchefstroom, South Africa; kunenen@unizulu.ac.za*

Zulu or Nguni sheep are a local breed of the KwaZulu-Natal (KZN) province; it serves as a source of meat and income to the rural farmers. The breed adapts well to the hot humid areas of KZN and reared at low input. Zulu sheep numbers are decreasing. Genetic Resource conservationists in South Africa have recognised the significance of planning a conservation program for this breed. Information of production systems, genetic structures of available populations is required for formulating a conservation program. The aim of the study was to describe Zulu sheep production systems and to give a preliminary indication of the genetic structure of four Zulu sheep populations. Information to describe the production system was collected in a form of survey from 96 Zulu sheep farmers in 11 rural areas of KZN. DNA from 119 blood samples of sheep collected in two areas (KwaMthethwa (KM) and Msinga (EM)) and in two research stations (University of Zululand (UZ) and Makhathini research station (MS)) were analysed using 29 microsatellite markers. The survey revealed that founder flocks were fewer than 4 animals bought from other farmers or inherited through patrilineality. About 28% of the flocks occurred in isolation while 43.7% interacted with 2 or fewer neighbouring flocks. The neighbour joining tree showed two main clusters: UZ, KM and MS as one cluster and the EM as another cluster. The STRUCTURE analysis showed that UZ and MS clustered together whereas KM and EM were found in different clusters. The highest $F_{is}$ index in MS indicated a relatively strong presence of inbreeding in this population. This work suggests that an exchange of rams between populations that are genetically distant could be considered for a conservation plan.

---

## Session 40 Poster 18

**Etiology of subclinical mastitis in primiparous goats reared in the region of Epirus, Greece**

*A. Tzora[1], I. Skoufos[1], G. Arsenos[2], I. Giannenas[2], C. Voidarou[1], K. Fotou[1], Z. Kalyva[1] and C.G. Fthenakis[3]*
*[1]TEI of Epirus, Agriculture Technology, Division Animal Production, Kostaki, 47100, Arta, Greece, [2]Aristotle University of Thessaloniki, Faculty of Veterinary Medicine, Thessaloniki, 54124, Thessaloniki, Greece, [3]University of Thessaly, Veterinary Faculty, Karditsa, 43100, Karditsa, Greece; tzora@teiep.gr*

The objective was to investigate the occurrence and etiology of subclinical mastitis in indigenous Greek dairy goats. A total number of 340 primiparous indigenous dairy goats (Capra prisca) were used. Milk samples were taken directly from one udder cistern from individual goats and were subjected to bacteriological analysis. Bacterial growth was observed in 34.57% of total milk samples. From the latter figure 51% concerned only the half-udder that was affected. Bacterial isolates were identified on the basis of colonial morphology, Gram staining and automated phenotypic identification system Vitek 2 (bioMérieux). Highest prevalence was of Coagulase Negative Staphylococci (CNS) 41.07% followed by *Staphylococcus aureus* 23.21%, Enterobacteriaceae 3.57%. The rest isolates were identified as environmental origin. A positive mastitis test reaction (CMT) was revealed in 46.91%. Pathogenic bacteria were isolated from samples which were negative to CMT corresponding to a 28.5% percentage of the total bacterial positive samples. In conclusion, the present study reveals that the CNS is the prevalent etiological agent of the subclinical mastitis in primiparous goats. The diagnosis of the inflammation should be based mainly on the bacteriological analysis. Acknowledgements: This research has been co-financed by the European Union (European Social Fund – ESF) and Greek national funds through the Operational Program 'Education and Lifelong Learning' of the National Strategic Reference Framework (NSRF) – Research Funding Program: ARCHIMEDES III.

**Adoption of technology in sheep farms of La Mancha, Spain**

*J. Rivas[1], C. De Pablos[2], J. Perea[1], C. Barba[1], R. Dios-Palomares[1], M. Morantes[1] and A. García[1]*
*[1]Universidad de Córdoba, Campus Rabanales, 14014, Córdoba, Spain, [2]Universidad Rey Juan Carlos, Paseo de los Artilleros, 28034, Madrid, Spain; pa1gamaa@uco.es*

Recently in Spain the milk production coming from sheep farms shows higher degrees of specialization based on the adoption of technologies. The aim of this research is to examine the pattern of adoption of technologies in sheep farms of La Mancha. Based on previous researches, from 77 questions, only 38 questions were selected by using qualitative and participatory methods; the chosen variables were grouped into six technology packages (TP): management, feeding, animal health and milk quality, pasture and land use, equipment and facilities, and reproduction and breeding program. The survey was applied to a sample of 157 farms. Using descriptive statistics each TP was characterized and the pattern of adoption was determined. The average of technologies adopted was of 18.4±6.0 (48.3%). TP showing higher degrees of implementation are animal health and milk quality (67.8%), feeding (56.0%) and management (55.7%), but their adoption is not sequential or responds to independent events. This research facilitates the identification of a number of technologies that must be implemented from an organizational strategy point of view. Moreover, all technologies are seeking a dynamic balance system that allows firms migrate to more efficient processes without losing their main attributes. As a technological challenge, an andrologic evaluation of ram, gynecologic evaluation of ewes prior to mating, early detection of non-productive animals are proposed; and a better use of productive records for the decision-making; aspects are recommended. The results of this analysis will have an impact on future research that attempts to improve the use of subproducts, forage reserves and improved rangeland management and hygiene control system, taking the quality milk as an strategic asset, so further research is necessary to assess the impact of each technology on the operating of the mixed system in the Mancha region.

---



**Fertility of Libyan Barbary sheep treated with prostaglandin F2α(PGF2α)in different seasons**

*A. Elmarimi, N. Almariol and J. Ahmed*
*Faculty of Veterinary Medicine, Surgery and Theriogenology, Alfornaj area, Tripoli, Libyan Arab Jamahiriya; drelmarimi2005@yahoo.com*

Abstract Bearing in mind the objectives of this study to investigate the Barbary sheep fertility under our climatic conditions and to evaluate the efficiency of the prostaglandin f2α(pgf2α)injections in induction of fertile estrus in different seasons. For these objectives, some experiments were performed during the period from July,2008 to March, 2009. A total number of 300 Libyan Barbary ewes(3-6 years old, weighing 40-60 kg) was used in this experiment. Ewes were kept in privet farms, fed and managed similarly. Ewes were divided to four season groups(summer, autumn, winter and spring groups). Each season group was divided into treated and control. Animals in the treated groups were injected with double injections of 125 μg of prostaglandin f2α intramuscularly(i/m), 11 days apart. While those in control groups were injected with two injections of 1.0 ml of 0.9% NACL saline solution simultaneously with the treated ewes. At the same day (day 11) rams wearing painted sponges on their briskets regions were introduced for natural mating. Treated groups showed shorter estrus response time than control groups in all seasons($P<0.05$). Estrus duration was longer in winter and spring than in summer and autumn ($P<0.001$),but no difference was found between treated and control groups inside seasons. Treated group showed higher pregnancy rate($P<0.001$) in winter season than control group. Percentage of ewes lambed in winter was significantly high ($P<0.001$) among treated ewes than control (80 vs 38%). Lambing rate differed significantly ($P<0.001$) among treated groups in all seasons. Data were collected and calculated statistically using SPSS system for percentages, means, standard deviation(mean ± sd) analysis of variance (anova), Chi square and Dunacan's test were used accordingly. Other values were measured, calculated and analyzed similarly.

# Session 40 Poster 21

**An expert integrative approach for growth rate simulation of sheep**

*S. Zakizadeh[1], M. Jafari[2] and H. Memarian[3]*

*[1]Higher Applied science-Technology Institute, Animal Science Department, Kalantari Highway, 9174994767, Mashad, Iran, [2]Animal Breeding Center of North-East, Mashhad-Sarakh Road, 9179684573, Iran, [3]University of Birjand, Faculty of Natural Resources and Environment, Birjand-Kerman road, P.O. Box 331, Birjand, Iran; sonia_zaki@yahoo.com*

Prediction of highly non-linear behavior of growth rate in animal is of importance in sheep livestock management. In this study, the predictive performance of artificial neural network (ANN) integrated with genetic algorithm (GA) was assessed. GA was used to optimize the parameters and architecture of the ANN. The growth records of 322 Iranian Black sheep were applied in this simulation. Two simulation scenarios were developed using monthly weight records, as well as sex, birth date and ewe parity to predict yearling weight. The scenario S1 was composed of the weight at the 3$^{rd}$ and the 4$^{th}$ month of age (weaning), whereas the S2 included just the birth weight. Correlation coefficient and normalized mean square error (NMSE) were used to evaluate the performance of ANN-GA. The Levenberg Marquat learning rule with the TanH transfer function was used to construct a network for S1, which consisted of 1 hidden layer and 4 neurons. This network was optimized using genetic algorithm with 100 epochs, a population size of 50 and 50 generations. After GA optimization, 8 neurons were suggested for hidden layer. Based on validation analysis, the S1 scenario with 0.29 and 0.87 as the normalized mean square error (NMSE) and correlation coefficient (r), respectively, was chosen as the best scenario, while the S2 produced an NMSE of 0.62 with a correlation coefficient of 0.61. The difference of desired output and actual network output was originated from non-linear relationships during the growth features. Nevertheless, it is concluded that ANN-GA represents a valuable performance for predicting of yearling weight of sheep based on weaning weight.

---

# Session 41 Theatre 1

**Genetic analysis of mentality traits in Rhodesian Ridgebacks**

*K. Lysaker[1], T. Ådnøy[1], P. Arvelius[2] and O. Vangen[1]*

*[1]Norwegian University of Life Sciences, Animal and Aquacultural Sciences, P.O. Box 5003, 1432 Ås, Norway, [2]Swedish University of Agricultural Sciences, Animal Breeding and Genetics, P.O. Box 7027, 75007, Sweden; odd.vangen@nmbu.no*

To improve the behavior of dogs, or the probable mental outcome of a mating, breeders need a tool to assess dog mentality in a standardized way, with a reasonable chance of succeeding in improving the targeted traits. Estimated breeding values (EBV's) is expected to be the best way to accomplish this, and if they are based upon a reliable test with reliable variance components, the genetic gain can be quite fast, given reasonable selection intensities. There have never previously been estimated breeding values for Rhodesian Ridgebacks, and only one other breed has had breeding values estimated for mentality traits based upon the Swedish Dog Mentality Assessment (DMA). This study used factor analysis in order to identify the broader traits for this breed, to see if they differed from the ones used by the Swedish Kennel Club, and estimated heritabilities and predicted breeding values for these traits. Totally 2022 dogs with complete records were included, and the pedigree file included 3,567 animals. Six factors were defined based on principal component analys1s: (1) play; (2) fear/exploration; (3) chase; (4) distance play; (5) defense; and (6) sociability. Heritabilities (and SE) were 0.14 (/0.,04), 0.31 (0.05), 0.10 (0.04), 0.09 (0.03), 0.11 (0.04), 0.10 (0.04), respectively. The heritabilities for the factors were all higher than for the separate variables and/or for the average of those included in the factor. The genetic trends were small, however the largst was for fator 4. For some breeds behavioral problems are so large that EBV's from mentality tests should be included into the breeding goals, similar to what have been done for other health traits through intensive screenings. Mentality tests are important in order to genetically improve behaviour of many dog breeds.

**Estimating dominance effects and inbreeding depression of weaning weight in Pannon White rabbits**

*I. Nagy[1], J. Farkas[2] and Z.S. Szendrő[1]*
*[1]University of Kaposvár, Animal Genetics and Biotechnology, 40 Guba S., 7400 Kaposvár, Hungary,*
*[2]University of Kaposvár, Department of Informatics, 40 Guba S., 7400 Kaposvár, Hungary; nagy.istvan@ke.hu*

Genetic parameters and inbreeding depression were estimated for weaning weight at day 35 of a synthetic Pannon White rabbit population (selected as a closed population at the Experimental Rabbit Farm of the Kaposvár University –Hungary – since 1992). The data file contained 16,533 kindling records of 4,066 rabbit does for the period of 1992-2012. The total number of animals in the pedigree file was 4,913. REML procedure was applied using 4 different single traits repeatability animal model (using VCE 6). All models contained year and month of weaning as fixed effect, individual inbreeding coefficients and complete pedigree equivalents as covariates and animal and permanent environmental effects as random effects. The second and fourth model contained dominance effect while the $3^{rd}$ and $4^{th}$ model also contained number of weaned kits as covariates. The estimated heritability for litter weight was low for all models and ranged between 0.048±0.013 and 0.073±0.012. The estimated permanent environmental effects showed higher relative importance compared to the additive genetic effects and their value varied between 0.079±0.013 and 0.132±0.012. The magnitude of dominance effect was practically zero for the second model but it was 0.0701±0.011 in the fourth model. Analyzing the inbreeding depression 10% increase of the inbreeding coefficient resulted in 0.16 kg estimated decrease based on the first and second models but practically no inbreeding depression was found using models 3 and 4. Based on our results it can be concluded that the models containing number of reared kits performed better compared to the models not having this effect. Also authors believe that including dominance effect is advantageous from the aspect of precision when estimating breeding values. Acknowledgement: This work was supported by the Hungarian Scientific Research Fund (OTKA, Project 106175).



**Heritability for enteric methane emission from Danish Holstein cows using a non-invasive FTIR method**

*J. Lassen and P. Løvendahl*
*Center For Quantitative Genetics and Genomics, Department of Molecular Biology and Genetics, Aarhus University, Blichers Alle, 8830, Denmark; jan.lassen@agrsci.dk*

Enteric methane emission from ruminants contributes substantially to the greenhouse effect. Few studies have focused on the genetic variation in enteric methane emission from dairy cattle. One reason for that is the limited number of methods appropriate for large scale phenotyping needed to generate a sufficient number of animals to accurately estimate additive genetic variance. In Denmark a method that measures gases during milking in robots has been used. A total of 2,104 cows have been phenotyped in commercial AMS farms using a Fourier Transformed Infrared (FTIR) measuring unit that was used to make large scale individual methane emission records. The focus is on using the ratio between $CH_4$ and carbon dioxide ($CO_2$) concentrations in the breath as the phenotype to adjust for the position of the cow's head relative to the measuring device. This ratio reflects the proportion of the metabolisable energy exhaled as methane. The FTIR unit air inlet was mounted in the front part of an AMS close to the feeding bin for 7 days, recording continuously every 5 seconds. All cows were within 365 days from calving when measured. The linear mixed model included fixed effects of herd, month, days in milk, lactation number, and random effects of animal and residual. Variance components were estimated using a REML approach in an animal model design with a pedigree containing 9,661 animals. The heritability of the methane to carbon dioxide ratio was moderate (0.21; SE=0.05). Mean methane/carbon dioxide ratio was 0.089 with a SD of 0.089 ranging from 0.048 to 0.151. In conclusion, the results from this study show that individual cow's methane emission can be accurately measured using FTIR equipment and that the trait is heritable.

## Session 41 — Theatre 4

**Genetic diversity in Slovak Pinzgau cattle based on microsatellites**

*V. Šidlová, N. Moravčíková, I. Pavlík, R. Kasarda, O. Kadlečík and A. Trakovická*
*Slovak University of Agriculture in Nitra, Department of Animal Genetics and Breeding Biology, Tr. A. Hlinku 2, 94976, Slovak Republic; anna.trakovicka@uniag.sk*

Pinzgau cattle belong to the traditional livestock breeds in Slovakia, but significant decrease of the population can be seen in recent years. Currently loss of genetic resources concerns not only the extinction of traditional breeds, but also the loss of genetic diversity within breeds. The aim of the proposed study was to evaluate the genetic diversity of Slovak Pinzgau cattle population based on polymorphism at 8 microsatellite loci (TGLA122, CSSM66, TGLA227, ILST006, CSRM60, ETH3, BM1824, and SPS115). Microsatellites were found highly polymorphic with a mean number of 10.46±0.6609 alleles (ranging from 8 to 14 per locus) and generated a total of 84 alleles across all loci from the 302 individuals analysed, with locus TGLA122 showing the highest number of alleles. High level of polymorphism confirms also the average value of polymorphic informative content (0.7653±0.0182). The suitability of sample size and number of loci analysed were tested using a bootstrap procedure. The data obtained were used to estimate the genetic diversity within population. The mean expected heterozygosity over all loci was 0.7914±0.016 ranging from 0.714 to 0.8598 and the mean observed heterozygosity has reached similar value (0.7984±0.0234) ranging from 0.7053 to 0.8874. The overall average of fixation index was close to zero ($F_{IS}$ = -0.0039±0.0122). The chosen set of microsatellite markers is suitable to describe level of polymorphism and genetic diversity in the whole population.

---

## Session 41 — Theatre 5

**Bayesian approach integrating correlated foreign information into a multivariate genetic evaluation**

*J. Vandenplas[1,2] and N. Gengler[2]*
*[1]National Fund for Scientific Research, Rue d'Egmont, 5, 1000 Brussels, Belgium, [2]University of Liege – Gembloux Agro-Bio Tech, Passage des Déportés, 2, 5030 Gembloux, Belgium; jvandenplas@ulg.ac.be*

Using correlations among traits, two advantages of multivariate genetic evaluations are prediction and improvement of accuracy of estimated breeding values (EBV) for a trait of interest for which phenotypes are unavailable or could be difficult to collect locally. However, accuracy of local evaluations may be low for some traits of interest, while accurate foreign evaluations for involved correlated traits are routinely performed. Therefore, integration of correlated information provided by foreign evaluations into a local multivariate evaluation could improve its accuracy. The aim of this study was to test a Bayesian approach to combine, through a multivariate evaluation, local pedigree and phenotypes for a trait of interest and genetically correlated foreign information, i.e. EBV and reliabilities provided by a foreign evaluation. One population was simulated across 10 generations. Phenotypes for two traits were simulated for each female. Heritabilities of 0.10 and 0.35 were considered for the trait of interest and the correlated trait, respectively. Genetic correlations between traits of 0.10, 0.25, 0.50, 0.75 and 0.90 were considered. Only foreign information for sires were locally integrated. No phenotype was locally considered for the correlated trait. For the trait of interest and for correlations from 0.10 to 0.90, results for 100 replicates showed that average rank correlations among Bayesian EBV and EBV from a bivariate evaluation using phenotypes for both traits were close to 1 for sires and ranging from 0.996 to 0.820 for their progeny. The respective correlations for the local evaluation ranged from 0.995 to 0.696 and from 0.993 to 0.651. Mean squared error, expressed as a percentage of the local mean squared error, was lower than 2.76% for sires and 57.78% for progeny. Thereby, the Bayesian approach has the potential to integrate correlated foreign information into a multivariate genetic evaluation.

**Quantitative and systems genetics analyses of lipid profiles in a pig model for obesity**

*S.D. Frederiksen, S.D. Pant, L.J.A. Kogelman, M. Fredholm and H.N. Kadarmideen*
*Univ of Copenhagen, Dep. of Veterinary Clinical and Animal Sciences, Grønnegårdsvej 7, 1870 Frederiksberg C, Denmark; tpm991@alumni.ku.dk*

The study aims to estimate heritabilities and identify genetic variants and their corresponding pathways for 6 lipid profiles associated with high-density lipoproteins (HDLs). The phenotypic traits in question are the concentration of HDL esterified cholesterol (HDL-CE) and HDL esterified cholesterol/total cholesterol (HDL-CE/CT), and the percentage of HDL free cholesterol (pHDL-CL), HDL triglycerides (pHDL-TG), HDL phospholipids (pHDL-PL) and HDL esterified cholesterol (pHDL-CE). This study used F2 intercross pigs (from crossing Danish production pigs with Göttingen Minipigs) that were measured for several obesity and metabolic phenotypes and genotyped using 60 k porcine Illumina SNPchip. The heritabilities of the lipid traits were estimated using a univariate mixed linear model fitting fixed effects of age, sex, batch and random animal effects in ASReml. HDL-CE, pHDL-TG, pHDL-PL and pHDL-CE have high heritabilities of 0.98, 0.77, 0.99 and 0.41 respectively. HDL-CE/CT has a moderate heritability of 0.37 while pHDL-CL has a low heritability on 0.17. Given this range of heritabilities, our next aim was to detect underlying genetic variants modulating these lipid profiles. This was performed using family-based genome wide association studies (GWAS) using mmscore function of GenABEL software. The genome-wide significance of the estimated associations of SNPs with the lipid profiles were based on the modified bonferroni correction ($P<2.2\times10^{-4}$). HDL-CE and HDL_CE/CT both have 17 significantly associated loci. There were 9, 4, 11 and 11 loci associated with pHDL-CL, pHDL-TG, pHDL-PL and pHDL-CE, respectively. Systems genetic analysis identified several genes in the vicinity of these loci (e.g. FUT4, CLN5, FBXL3 etc.) and gene set enrichment analysis identified metabolic, and olfactory transduction KEGG pathways to be associated with the HDL-CE phenotype. The results indicate significant genomic variation in lipid profiles in our porcine model.

---

Session 41 Theatre 7

**Effect of censoring in a social-genetic model**

*B. Ask, T. Ostersen, M. Henryon and B. Nielsen*
*Pig Research Centre, Breeding & Genetics, Axeltorv 3, 1609 Copenhagen V, Denmark; bas@lf.dk*

The aim of this study was to estimate (1) social-genetic effects in Danish Duroc pigs, and (2) the impact of censoring on genetic parameters, bias, and predictive ability. In commercial-pig breeding, pigs can be removed from pens during performance test, e.g. due to mortality and involuntary culling. This poses problems for social-genetic models because social-genetic effects for individual pigs are influenced by other pigs sharing the same pen during the performance test. It is, therefore, important to determine how to handle censoring in social-genetic models. We did this using data from a performance test at the Danish test station Bøgilgaard. Data included 719 pens, 12-14 boars per pen, and a total of 9,432 pigs. Data were available on average daily gain (ADG), feed conversion (FCR) recorded in the period from 30 kg to 100 kg, and meat percentage (MEAT%). During the test period the mortality and involuntary culling was ~7% and occurred across ~55% of the pens. Single-trait analyses were carried out on three datasets: (1) full data; (2) full data excluding dead and culled animals; and (3) full data excluding pens with dead or culled animals. Three models were applied: (1) a classical-animal model; (2) a social-genetic model; and (3) a social-genetic model with proportional weighing of social effects on the time each pig spent in the pen. Preliminary results show that social-genetic effects were present for ADG and FCR with higher total heritabilities compared to classical heritabilities (0.27 and 0.20). Results also indicate that estimated social-breeding values were more sensitive to the method of handling censoring than classically estimated breeding values. Although, the higher total heritabilities bring promises of higher genetic gains, this potential may be lower if censoring is ignored.

### The WISH network method applied to obesity in an F2 pig popolution

*L.J.A. Kogelman[1], S.D. Pant[1], J. Karjalainen[2], L. Franke[2], M. Fredholm[1] and H.N. Kadarmideen[1]*
*[1]Univ of Copenhagen, Dep. of Veterinary Clinical and Animal Sciences, Grønnegårdsvej 7, 1870 Frederiksberg C, Denmark, [2]Univ Medical Center Groningen, Dep of Genetics, A. Deusinglaan 1, 9713 AV Groningen, the Netherlands; lkog@sund.ku.dk*

High-throughput genotype (HGT) data are extensively used in genome-wide association studies (GWAS) to investigate the biological and genetic background of complex traits, such as obesity. Unfortunately, to date results explaining complete genetic variation are limited. As it has been repeatedly shown that gene × gene interactions may play a key role in complex traits, systems genetics approaches are increasingly used to elucidate more of the currently limited knowledge of complex traits. We previously published the Weighted Interaction SNP Hub (WISH) network method, which uses HGT data to detect modules of highly interconnected SNPs, potentially representing biological pathways associated with the trait under study. Here we used the WISH network method to study obesity in a porcine model for human obesity. We previously created an F2 pig population, which was extensively phenotyped and genotyped. We created an Obesity Index (OI) based on multi-trait selection indexes containing nine obesity-related phenotypes, to perform a GWAS and WISH network analysis, both followed up with pathway analysis. The GWAS revealed 404 SNPs highly associated with the OI, and functional annotation revealed several genes located in or near those SNPs related to obesity (e.g. NPC2). WISH network construction revealed several modules related to obesity, and pathway analysis revealed among others an overrepresentation of purinergic receptor activity, which has been associated before with feeding behavior. In conclusion, this study shows the potential of systems genetics methods that involve network and pathway based approaches to jointly use phenotype and high-throughput genotype data. In particular, the WISH method moves GWAS beyond its scope to detect gene × gene interactions, resulting in the detection of biologically relevant pathways in complex traits.

---



### Combining purebred and crossbred information on nurse capacity and litter size

*B. Nielsen[1], O.F. Christensen[2], B. Ask[1] and I. Velander[1]*
*[1]Pig Research Centre, Breeding & Genetics, Axeltorv 3, 1609 V Copenhagen, Denmark, [2]Aarhus University, Blichers Allé 20, 8830 Tjele, Denmark; bni@lf.dk*

In commercial pig production, sows are mainly two-way crosses of purebred lines. To evaluate how purebred selection affects crossbred performances and how crossbred information might affect purebred selection, an experiment was conducted in three productions herds and 11,247 crossbred sows (Landrace × Yorkshire) were recorded for total number born (TNB), litter size at day 5 (LS5), and nurse capacity (NC). All terminal sires used were Duroc AI boars. The purebred dams and half-sib sisters of the crossbred sows were identified in multiplier herds, resulting in a total of 59,884 Landrace sows and 37,495 Yorkshire sows, and their records of TNB and LS5 were included in the data. Only first-parity sows (gilts) were used to avoid bias due to culling of sows after first litter. A multivariate mixed model approach was applied to develop a so-called reduced animal where genetic effect in crossbreds was traced back to the two purebred lines of Landrace and Yorkshire. The results showed that TNB recorded in purebreds and TNB recorded in crossbreds were two different traits as the genetic variances in crossbreds (0.55 and 0.90) were lower than purebreds (1.42 and 1.02). Also, for LS5 genetic variances were lower in crossbreds (0.34 and 0.62) compared to purebreds (0.69 and 0.72) indicating two different traits. Favorable genetic correlations between TNB recorded in purebreds and TNB in crossbreds were found. They indicated that the genetic transfer from each purebred to the crossbreds was about 50%. For LS5 the genetic transfers from purebreds to crossbreds were 61% and 73% for Landrace and Yorkshire respectively. Unfavorable genetic correlations were obtained between TNB recorded in purebreds and NC recorded in crossbreds, but for LS5 the genetic correlations with NC were non-significant. Preliminary results indicate that genetic gain of NC in purebreds is lower using crossbred recordings compared to purebred recordings. Thus NC needs to be recorded in purebreds.

**Genetic parameters of functional longevity using a multivariate approach in Austrian Fleckvieh**

*C. Pfeiffer[1], B. Fuerst-Waltl[1], V. Ducrocq[2] and C. Fuerst[3]*
*[1]University of Natural Resources and Life Sciences, Vienna, Sustainable Agricultural Systems; Division of Livestsock Sciences, Gregor-Mendel Strasse 33, 1180 Vienna, Austria, [2]INRA, Bovine Genetics and Genomics, UMR 1313, 78352 Jouy-en-Josas, France, [3]ZuchtData EDV-Dienstleistungen GmbH, Dresdner Strasse 89/19, 1220 Vienna, Austria; birgit.fuerst-waltl@boku.ac.at*

A longevity subindex is included in the official Austrian total merit index. It is a combination of functional longevity and five auxiliary traits (hip width, body depth, muscularity, feet and legs, udder). Variance components and EBVs for all these traits are estimated univariately. Genetic correlations between longevity and type traits are computed using the approximate method of Calo *et al.* Through the application of this method residual correlations are neglected. This might cause biased EBVs and reliabilities. To solve this problem an alternative method using a multivariate animal model based on yield deviations (YD) was tested. In total, 74,292 YDs of Austrian Fleckvieh cattle (dual purpose Simmental) of five Austrian regions born between 2002 and 2011 were used. The pedigree included 240,268 animals. Variance components were estimated based on an animal model using ASReml 3.0. The log-likelihood converged well. Heritabilities for longevity (0.08±0.006), muscularity (0.36±0.012) and udder (0.30±0.011) differed from the routinely published values 0.12, 0.28 and 0.24, respectively. Genetic correlations between longevity and hip width (-0.08±0.05), body depth (-0.14±0.05) and muscularity (-0.08±0.04) were lower than the ones used in routine evaluation. Genetic correlations between longevity and feet and legs (0.39±0.05) and udder (0.40±0.04) were slightly higher. Genetic correlations within type traits range from -0.27 (muscularity/udder) to 0.64 (hip width/body depth). Results suggest that the approximate multivariate approach based on YDs is feasible and could be introduced in the routine genetic evaluation.

---



**A new mtDNA haplotype of European brown hares (*Lepus europaeus* Pall.) in southwestern Slovakia**

*L. Ondruska, V. Parkanyi, D. Vasicek, J. Slamecka, M. Bauer, R. Jurcik, K. Vasickova and D. Mertin*
*National Agricultural and Food Centre – Research Institute for Animal Production Nitra, Hlohovecka 2, 95141 Luzianky, Slovak Republic; ondruska@vuzv.sk*

The genetic alterations in mitochondria are common mtDNA, haplotypes that are maternally inherited. Mitochondrial protein coding genes of comprise important components of the oxidative phosphorylation pathway (OXPHOS) and every change in amino acid composition can affect the efficiency of the oxidative phosphorylation. For the analysis of mtDNA we used the samples of liver from 82 adult (male and female) European brown hares of wild population pertain to 10 hunting grounds in the region of south-western Slovakia. The samples were collected from brown hares hunted during winter seasons of 2009-2011. Tissues were stored frozen at -20 °C until DNA extraction. Total genomic DNA was purified by automat Maxwell® 16 Tissue DNA Purification kit AS1030 (Promega). The PCR conditions (PTC-200 DNA Engine Cycler; Bio Rad) consist of initial denaturation at 95 °C for 2 min, and 35 cycles of denaturation at 94 °C for 30 s, annealing at 60 °C for 30 s, primer extension at 72 °C for 30 s, with last extension at 72 °C for 10 min. The amplified segments from each sample were subsequently screened for polymorphism in selected regions of two protein coding genes that have been implicated in changes of germinal, generative and somatic cells through ATPase subunit 6 from and a segment of the mitochondrial NADH dehydrogenase subunit 2. For detection of polymorphism we used 2 restriction endonucleases: NlaIII and Sau96I (8 hrs at 37 °C). We detected in ATP6 82 A/D haplotypes and in NADH2 gene were identified 78 A/C/E haplotypes. Moreover we found a new haplotype in NADH2 in 4 samples in two hunting grounds. This new haplotype was discovered for the first time in the hare population in Slovakia.

**Body composition and muscle structure in F2 offspring of a German Holstein × Charolais cross**

*E. Albrecht, R. Pfuhl, G. Nürnberg, C. Kühn and S. Maak*
*Leibniz Institute for Farm Animal Biology, Wilhelm-Stahl-Allee 2, 18196 Dummerstorf, Germany; elke.albrecht@fbn-dummerstorf.de*

Cattle breeds of either secretion or accretion type like German Holstein (GH) and Charolais (CH) differ in body weight and composition when slaughtered at the same age. These two breeds were the founder breeds of a F2 herd currently under investigation in our institute to identify the genetic and physiological background of phenotypic variability. In total, 247 F2 bulls were analysed for body composition, muscle structure of m. longissimus and the resulting meat quality. Data from two groups (n=40, each) of low (F2-L) and high (F2-H) carcass weight were analysed separately and compared with data from bulls of the founder breeds (n=18, each). The histogram of carcass weights of all F2 bulls showed an intermediate distribution between that of GH and CH. The mean of birth weight and carcass weight of F2-L and F2-H corresponded to the respective values of GH and CH. In contrast, dressing percentage and m. longissimus area differed significantly among all groups, being CH>F2-H>F2-L>GH. Carcass composition of both F2 groups was intermediate between GH and CH. Significantly less fat than in GH was stored in internal fat depots of F2 and CH bulls. Muscle fibre composition of both F2 groups was similar to that in GH. The higher percentage of white fibres in CH corresponded to brighter meat. Correlation analysis between muscle traits and meat quality traits in all F2 bulls revealed that shear force was influenced by fat area percentage, number of marbling flecks, and size of intramuscular adipocytes. Muscle fibre distribution affected colour and shear force value. Muscle fibre distribution was also associated with differences in body composition as observed when analysing groups of F2 bulls according to low or high percentage of white fibres in m. longissimus. This suggests that muscle fibre composition has consequences beyond the muscle itself and offers selection possibilities for improved quality and efficiency in beef production.

---



**Association among leptin and DGAT1 gene polymorphisms and buffalo milk composition measured by US**

*G. Holló[1], B. Barna[1] and I. Anton[2]*
*[1]University of Kaposvár, Guba S. str 40., 7400 Kaposvár, Hungary, [2]Research Institute of Animal Breeding and Nutrition, Gesztenyés str. 1, 2053 Herceghalom, Hungary; hollo.gabriella@sic.hu*

Leptin and DGAT polymorphisms were mostly studied in dairy cattle, but rarely have been reported in buffalo. Our aim was to analyse a possible association between above mentioned polymorphisms and buffalo milk composition data. 44 hair samples were collected for DNA extraction from buffalo cows at the only dairy buffalo organic farm in Hungary. From the same animals altogether 457 milk samples have been collected from morning and evening milking at 14 different occasions during lactation period. Milk composition was examined by using ultrasonic (US) milk analyser (Ecomilk ultra pro, Bulteh 2000). Recorded parameters were: fat, solids not fat (SNF), milk-density, protein, conductivity and freezing point. Genomic DNA was isolated from hair samples and genotypes were determined by qPCR. Statistical analysis were carried out using SPSS 17.0 program, significant differences were calculated by ANOVA using Tukey test. Two genotypes were found in both cases: AA/AA and AA/GC for the DGAT1 gene and CC and TC for the leptin gene. No GC/GC (DGAT1) and TT (leptin) genotypes were found. Genotype frequencies in both cases were the same: 89% and 11%. $\chi$2-probe indicated Hardy-Weinberg Equilibrium in the population. Fat content, SNF, milk density and protein content were higher in TC cows' milk by 0.85%, 0.47%, 1.06 g/cm3 and 0.37% (P>0.02). Concerning DGAT1 gen, AA/AA cows had higher values for SNF (10.47% v. 10.58%), milk density (28.7 g/cm3 v. 29.1 g/cm3) and protein content (4.64% v. 4.73%) compared to other genotypes; however differences were not significant. Preliminary results indicate that examined gene polymorphisms are associated with the composition of buffalo milk and further investigations are planned in this field.

**Genealogical analysis in Slovak Holstein and Red-Holstein populations**

*I. Pavlík, O. Kadlečík and R. Kasarda*
*Slovak University of Agriculture in Nitra, Department of Animal Genetics and Breeding Biology, Tr. A. Hlinku 2, 949 76 Nitra, Slovak Republic; radovan.kasarda@uniag.sk*

The aim of our study was to evaluate basic measures of diversity in Slovak Holstein (H) and Red-Holstein (R) populations by the pedigree analysis. There were 52,082 Holstein and 23,753 Red-Holstein cows included in the analysis. Cows were involved into official animal recording and were born between 1997 and 2012. The computation of basic diversity parameters was realized by software package ENDOG v.4.8. The equivalent number of traced generations was 5.05 in H and 4.77 in R. The average inbreeding coefficient (F) was 1.13% in H and 0.54% in R. The average increase in inbreeding (ΔF) was 0.29% per generation in H and 0.16% in R. The effective number of founders ($f_e$) was 96 in H and 149 in R, respectively; effective number of ancestors ($f_a$) ranged from 69 in H to 90 in R. The effective number of founder genomes ($N_g$) was 37 and 49 in H and R, respectively. Presented results show higher inbreeding level in H than in R including higher increase in inbreeding. On the other hand diversity in R population is more significantly reduced by bottleneck and genetic drift. Due to these facts it is necessary to find optimal mating system to avoid diversity loss problems in the future.

---



**A new statistical approach to identify breed informative single nucleotide polymorphisms in cattle**

*F. Bertolini[1], G. Galimberti[2], D.G. Calò[2], G. Schiavo[1], D. Matassino[3] and L. Fontanesi[1]*
*[1]University of Bologna, Dept. of Agricultural and Food Science and Technology, Viale Fanin 46, 40127, Italy, [2]University of Bologna, Dept. of Statistical Sciences 'Paolo Fortunati', Via delle Belle Arti, 41, 40126, Bologna, Italy, [3]CONSDABI, Piano Cappelle, 82100 Benevento, Italy; francesca.bertolini3@unibo.it*

The identification of the breed or population of origin of animals and of their products has several practical applications that range from management of livestock genetic resources, estimation of hybridization levels and authentication of brand products that are produced using particular breeds or populations. High density single nucleotide polymorphism (SNP) genotyping tools can be used to these purposes. In this work Principal Component Analysis (PCA) and Random Forest (RF) statistical techniques were combined to analyze SNP data obtained from the Illumina 50KBovine BeadChip for 3,712 cattle of four breeds (2,091 Holstein-Friesian, 738 Italian Brown, 476 Italian Simmental and 407 Marchigiana). From this dataset, two groups were created: a training population of 3342 animals and a test population of 370 animals (10% for each breed). PCA was applied on the training population to identify a first panel of 580 informative SNPs. RF was then applied to rank and further reduce to 48 the number of informative SNPs that can be able to assign an animal to the corresponding breed with an Out Of Bag (OOB) error rate of 0.06% in the training population and a 100% of correct assignment rate in the test population. These results indicate that the combination of PCA and RF is a powerful approach that can identify SNPs useful for breed assignment.

**Estimation of genetic parameters and prediction of breeding value for calving ease**

*L. Vostry[1,2], H. Vostra-Vydrova[1,2], B. Hofmanova[2], I. Majzlik[2], Z. Vesela[1] and E. Krupa[1]*
*[1]Institute of Animal Science, Pratelstvi 815, 10401 Prague, Czech Republic, [2]Czech University of Life Sciences Prague, Faculty of Agrobiology, Food and Natural Resources, Kamycka 129, 165 21 Prague, Czech Republic; vostry@af.czu.cz*

The aim of this study was to select a suitable model for genetic parameters of calving ease (CE) and birth weight (BWT). Eight pure breeds of beef cattle were included in the analysis: Aberdeen Angus (26%), Beef Simmental (11%), Blonde d'Aquitaine (4%), Charolais (36%), Gasconne (1%), Hereford (10%), Limousine (9%), Piedmontese (3%). For the correct estimation of genetic parameters for calving ease and body weight three different bivariate models were tested: linear-linear animal model with calving ease classified into four categories (1 – easy, 2, 3, 4 – poor) (L-LM), and bivariate threshold-linear animal model with calving ease classified into four categories (1 – easy, 2, 3, 4 – poor) (T-LM). A total of 81,092 records were obtained from the field database of the Czech Beef Breeders Association. All tested models included fixed effects of herd-mate group (herd × year × season), age of dam, sex and breed of calf, whereas random effects included direct and maternal genetic effects, maternal permanent environment effect and residual error. Direct heritability estimates for CE were with the use of L-LM and T-LM 0.096±0.013 and 0.210±0.024, respectively. Maternal heritability estimates were for L-LM and T-LM 0.060±0.031 and 0.104±0.125, respectively. Genetic correlations of direct CE with direct BWT were 0.46 and 0.50 for tested models, whereas maternal correlations were 0.25 for all tested models. Within traits direct and maternal genetic correlations were found substantial and negative for all tested models (from -0.574 and -0.60 or -0.53 and -0.58), illustrating the importance of including this effect in CE evaluations. The results indicate that all models can be used to estimate genetic parameters of calving ease for beef cattle in the Czech Republic. However, higher suitability for routine evaluation was obtained with T-LM.



**Analysis of inbreeding and genetic diversity of four pig breeds in Czech Republic**

*E. Krupa[1], E. Žáková[1], Z. Krupová[1], L. Vostrý[2] and J. Bauer[1]*
*[1]Institute of Animal Science, Prague, 10400, Czech Republic, [2]Czech University of Life Science, Prague, 16521, Czech Republic; krupova.zuzana@vuzv.cz*

Breeds with a small number of individuals are often endangered by the reduction of genetic diversity and increase of inbreeding. Knowledge of these basic parameters of genetic relatedness in the population is the first prerequisite for maintaining of the breed. The aim of this study was analyze the pedigree data of Czech Large White-paternal line (LW), Duroc (DC), Hampshire (HP) and Pietrain (PN) to determine trends for genetic diversity, and to find the main sources of genetic diversity loss. The total size of the pedigree was 32,913, 13,299, 5,184, and 7,160 animals for LW, DC, HP, PN, respectively. Animals born in the years 2011-2013 (2010-2012 for PN) were selected as reference population. The average pedigree completeness index for one generation back was 99.4%, 97.1%, 98.4% and 94.1 for appropriate breeds. Number of ancestors explaining 100% of gene pool was 125, 157, 8 and 38 animals for LW, DC, HP, PN. For analyzed breeds, the percentage of inbred animals 86%, 76%, 45%, 36%, average inbreeding 0.025, 0.036, 0.035, 0.013 and average co-ancestry 0.033, 0.042, 0.144, 0.033 was found over the past decade. The expected inbreeding under random mating increased during the last 10 years for LW, HP and PN and varied from 0.016 to 0.143. Only for DC expected inbreeding decrease from 0.032 to 0.019. The average relative genetic diversity loss (GD) within last generation interval was 5.42%, 4.25%, 19.54% and 6.08%, respectively. The relative proportion of GD loss due to genetic drift was 80.93, 75.71, 84.64, 71.93 for LW, DC, HP, PN. The effective population size computed on the basis of inbreeding was 43, 21, 20, 107 and based on co-ancestry was 48, 32, 7, 82 for LW, DC, HP, PN. The loss of a genetic diversity over one generation interval was found for HP and PN breed. All breeds are characterized by a high proportion of inbred animals, but the average inbreeding was low. The study was funded by the project QJ1310109.

**Genomic predictions of obesity related phenotypes in a pig model using GBLUP and Bayesian approaches**

*S.D. Pant[1], D.N. Do[1], L.L.G. Janss[2], M. Fredholm[1] and H.N. Kadarmideen[1]*

*[1]University of Copenhagen, Veterinary Clinical and Animal Sciences, Gronnegaardsvej 7, 1870 Frederiksberg C, Denmark, [2]Aarhus University, Center for Quantitative Genetics and Genomics, Department of Molecular Biology and Genetics, Blichers Allé 20, building K23, Room 3026, 8830 Tjele, Denmark; sameer.pant@sund.ku.dk*

Whole genome prediction (WGP) based on GBLUP and Bayesian models (e.g. A, B, C and R) are routinely used in animal breeding but have not been well tested in a genetic mapping population that segregates for QTLs. In our pig model experiment, purebred Duroc and Yorkshire sows were crossed with Göttingen minipig boars to obtain an F2 intercross resource population (n=566) that is known to segregate for QTLs affecting several obesity and metabolic phenotypes. All pigs were genotyped using Illumina Porcine 60 k SNP Beadchip. The objective of this study was to test the differences in predictive accuracy and bias of WGP using Bayesian methods (Bayes Cpi, Bayes R and Bayes Power Lasso) and a polygenic inheritance (GBLUP) model in a resource population with segregating QTLs. Initially, we are applying Bayesian Power LASSO (BPL) model to investigate genetic architecture of obesity phenotypes in order to partition genomic variances attributable to different SNP groups based on their biological and functional role via systems genetics/biology approaches. We apply different methods to group SNPs: (1) functional relevance of SNPs for obesity based on data mining; (2) groups based on positions in the genome, and significance based on previous genome-wide association study in the same data set. Preliminary results from our studies in production pigs indicate that BPL models have higher accuracy, but in some cases more bias, than GBLUP method; although using different power parameters has no effect on predictive ability of the models. Future work involves applying these and other WGP methods to the F2 intercross resource population.



**The response to selection regarding the number of stillborn piglets**

*M. Sviben*

*Freelance consultant, Siget 22B, 10020 Zagreb, Croatia; marijan.sviben@zg.t-com.hr*

The reality of the imagination of the rhythm of selection according to the course of life and the exploitation of swine was established, the cognitions from the field of quantitative genetics were regarded and it was decided to select the sows concerning the number of born piglets at the $1^{st}$ farrowing and the number of piglets alive at the weaning of the $1^{st}$ litter. So, at the Pig Farm in Nova Topola (Bosnia and Herzegovina) the primiparous sows to be selected had to farrow the number of piglets above the average of their generation and to have the number of alive sucking-pigs at the weaning above the generation's mean. From the generation 0 till the generation 19, during 20 years, the sow's prolificacy was increased and the number of lost piglets decreased. After the particular way of selection the composition of losses of piglets changed. It was concluded that sows should be selected with regard to the number of stillborn and the number of piglets died during the suckling period separately. The research was done using the data on Swedish Landrace primiparous in Dubravica, Croatia – and applying the equation $M_{PSP} = M_{PAP} + SR = M_{PAP} + SD \times h^2$, where $SD = M_{SP} - M_{AP}$ and it meant: M: the mean; AP: all parents; SP: selected parents; PAP: progeny of all parents; PSP: of selected parents. $M_{AP}$ was 0.580 stillborn (5.93% of born piglets in mother's litters). 68 (60.71% of all) selected mothers had 88 daughters (60.69% of all). It was established $M_{AP}$=0.662 (6.23% of born piglets), $M_{PSP}$=0.602 (5.56% of born piglets), the selection response SR=-0.060 piglets a litter, since the selection differential was SD=-0.580 and $h_r^2$=0.1034. 44 unselectable sows farrowed 1.477 stillborn piglets a litter. They had 57 daughters with 0.754 stillborn piglets a litter. It could be found: the selection differential SD=0.897, $h_r^2$=0.1026, the selection response SR=0.092 piglets a litter.

**Identification of porcine genome regions regulating collagen content in longissimus muscle**

*A. Fernández[1], A. Martinez-Montes[1], E. Alves[1], N. Ibañez-Escriche[2], J.M. Folch[3], C. Rodríguez[1], L. Silió[1] and A.I. Fernández[1]*
*[1]INIA, Ctra. Coruña Km7.5 Madrid, 28040, Spain, [2]IRTA, Lleida, 25198, Spain, [3]UAB, CRAG, Barcelona, 08193, Spain; afedez@inia.es*

Collagen is one of the major components of intramuscular connective tissue and it is directly involved in meat properties. The aim of the present study was to identify genome regions and genes associated with collagen content in longissimus dorsi from an experimental porcine backcross F1 (Iberian × Landrace) × Landrace through different genomic approaches. A total of 124 backcrossed animals with collagen content measures in longissimus dorsi were available for the analyses. DNA samples from these animals were genotyped with the Porcine SNP60 BeadChip and RNA samples from muscle were hybridized against Affymetrix porcine expression microarray. A linkage QTL scan was conducted using 8,247 evenly spaced SNPs. Four QTLs were identified on SSC4 (138-143 Mb), 6 (146-150 Mb), 7 (0-21 Mb) and 13 (208-210 Mb). Furthermore, powerful candidate genes were identified: the product of PODN gene (146.8 Mb – SSC6), involved in the determination of the collagen fibril morphology; the product of the BMP6 gene (5.1 Mb – SSC7), regulator of the levels of type X collagen; the product of the RUNX1 gene (208.2 Mb – SSC13), that promotes type II and X collagen expression. An additional genetical genome analysis was conducted to identify the genome regions regulating the expression of genes involved in collagen metabolism. Those probes from the expression microarray which expression showed significant correlation with collagen content (492 probes) were further analyzed in an expression GWAS analyses conducted with 38,648 SNPs. The most relevant region corresponded to the positions 43-48 Mb on SSC2 associated with the SOX6 gene expression. Moreover, this gene falls within the associated region and its product participates in the regulation of the collagen II production. Further studies on the selected candidate genes should be performed to identify the causal mutations.

---



**Genetic structure of candidate genes for meat quality parameters in Large White pigs**

*N. Moravčíková and A. Trakovická*
*Slovak University of Agriculture in Nitra, Department of Animal Genetics and Breeding Biology, Tr. A. Hlinku 2, 94976 Nitra, Slovak Republic; nina.moravcikova1@gmail.com*

The aim of this study was to identify and analyse polymorphic variants of two cathepsin genes, cathepsin L (CTSL) and cathepsin S (CTSS), and fat mass and obesity associated gene (FTO) in population of 150 Large White pigs. The cathepsins belong to the enzymes family of lysosomal proteinases, which are in pigs mainly tested in relation to muscle transformation during dry-cured ham processing. The FTO gene is associated with fat deposition traits in pigs. The single nucleotide polymorphisms of porcine CTSL (q.143C>T; FN556187), CTSS (q.171A>G; FN556188) and FTO (g.276T>G; AM931150) genes were analyzed by PCR-RFLP methods. After digestion with restriction enzymes were detected in population alleles with frequency: CTSL/TaqI C 0.93 and T 0.07 (±0.014), CTSS/BseNI G 0.94 and A 0.06 (±0.013), and FTO/TaiI T 0.54 and G 0.46 (±0.029). Predominant in analyzed animals were homozygous $CTSL^{CC}$ (86.67%), $CTSS^{GG}$ (88.67%) and heterozygous $FTO^{TG}$ (66.67%) genotypes. The observed heterozygosity was also transferred to the low or median polymorphic information content 0.12 ($H_e$ 0.13), 0.10 ($H_e$ 0.11) and 0.37 ($H_e$ 0.67) for CTSL, CTSS and FTO genes, respectively. Comparison of loci Ne showed highest effective allele number for FTO gene (1.98). Observed Wright's fixation indexes $F_{IS}$ had negative values (CTSL -0.07, CTSS -0.06, and FTO -0.3419), which represent slight excess of heterozygote compared with Hardy-Weinberg equilibrium expectations. In subsequent study will be used these results in association with meat quality parameters of evaluated animals for analysis of CTSL, CTSS and FTO genes role in economically important traits in pigs.

**Single nucleotide polymorphisms of the bovine ER1 and CYP19 genes in Holstein and Limousine cattle**

*A. Trakovická, T. Minarovič and N. Moravčíková*
*Slovak University of Agriculture in Nitra, Department of Animal Genetics and Breeding Biology, Tr. A. Hlinku 2, 94976 Nitra, Slovak Republic; nina.moravcikova1@gmail.com*

The aim of this study was identification of SNPs in bovine oestrogen receptor (ER1) and cytochrome P450 aromatase (CYP19) genes in order to analyze genetic structure of Holstein and Limousine cattle populations. The reproductive cycle of all mammalian females is regulated mainly by oestrogen hormones. They are synthesized through the aromatization of androgens, which are mainly of ovarian origin. A key role in this complex has the P450 aromatase, encoded by CYP19 gene. In total 119 genomic DNA samples of Holstein and 82 samples of Limousin cattle were used to genotyping of ER1 and CYP19 genes by means of PCR-RFLP method. After digestion with restriction enzymes (ER1/SnaBI and CYP19/PwuII) were detected in Holstein and Limousine cattle the allele frequencies: 0.97/0.95 and 0.03/0.05 for G and A ER1 variants, and 0.91/0.70 and 0.09/0.30 for A and B CYP19 variants, respectively. In both analyzed populations the high frequency for homozygous genotypes $ER1^{GG}$ and $CYP19^{AA}$ were found. The $ER1^{AA}$ homozygotes were not identified. The average observed heterozygosity (0.11) was relative low, therefore shows loci small or median level of polymorphic information content (ER1 0.06/010 and CYP19 0.16/0.33). Within population variance estimate were observed negative values of fixation index ($F_{IS}$) -0.03/-0.06 and -0.11 for ER1 and CYP19 genes, respectively. For CYP19 gene in population of Limousin cattle was observed positive value of $F_{IS}$, which indicates deficit of heterozygote animals. The ER1 and CYP19 genes were significant associated with reproductive performance mainly in dairy cattle, therefore the next step of our work will be confirm this relationship through associations analyses. The animals breeding selection assisted with genetic markers as ER1 or CYP19 can also increase the production traits or optimize reproduction performance in dairy cattle.

---



**Use of high-performance computing in animal breeding**

*J. Vandenplas[1,2] and N. Gengler[2]*
*[1]National Fund for Scientific Research, Rue d'Egmont, 5, 1000 Brussels, Belgium, [2]University of Liege – Gembloux Agro-Bio Tech, Passage des Déportés, 2, 5030 Gembloux, Belgium; jvandenplas@ulg.ac.be*

High-perfomance computing facilities proposing shared-memory and distributed-memory multiprocessors are becoming available. With those clusters, parallel computing could lead to increased performances and problem sizes. However, to our knowledge and especially for variance components estimations, most software available in animal breeding, based on sparse matrices computations, do not allow parallel computing and are limited by memory accessible by the central processing unit, or allow parallel computing only for options with dense matrices computations, which limits anyway problem sizes due to storage of dense matrices. The aim was to propose simple and effective modifications for the BLUPF90 family of programs to reduce computing time with consideration of required memory. Modifications were based on academic free packages proposing solver and sparse inversion for sparse symmetric indefinite linear systems. First, modifications concerned the sparse inversion subroutine implemented in the package FSPAK. Rearrangements of 'do' loops to allow optimizations of computer operations by some compilers and addition of OpenMP directives were performed. The ordering operation was modified to more easily compare a multiple minimum degree algorithm (MMD; implemented in FSPAK) and a multilevel nested dissection algorithm (implemented in METIS 4.0.3). Second, the package PARDISO Version 5.0.0 was used instead of FSPAK. This package proposes in particular a parallel solver and sparse inversion on shared-memory multiprocessors. Modified FSPAK and PARDISO were compared to original FSPAK using MMD through REMLF90. Different models, such as univariate or bivariate (random regressions) test-day animal and single-step genomic models, were tested. All jobs were run 5 times. With an appropriate ordering algorithm, speedup for each REMLF90 iteration were up to 7.5 for modified FSPAK and up to 22.8 for PARDISO with 2 threads. With 4 threads, speedup increased to 8.3 and 32.5, respectively.

**Evaluation of genetic diversity in Iberian chicken breeds based on microsatellite markers**

*M.G. Gil, S.G. Dávila, P. Resino and J.L. Campo*
*Instituto Nacional Investigación y Tecnología Agraria y Alimentaria, Departamento Mejora Genética Animal, Carretera Coruna Km 7,5, 28040 Madrid, Spain; ggil@inia.es*

Spain and Portugal have a rich genetic diversity of native chickens, nowadays under threat of extinction because the utilization of industrial hybrids to produce eggs and meat. A total of 13 Spanish breeds are maintained in a conservation program at the Experimental Station of El Encín (Madrid, Spain), whereas three different Portuguese breeds have been reported. The objective of this study was to evaluate the genetic variability of these breeds using 24 microsatellites. These markers were chosen based on their location and their degree of polymorphism. A total of 384 hens were used. Number of alleles, polymorphism information content, observed and expected heterozygosity, F-statistics ($F_{IS}$, $F_{ST}$ and $F_{IT}$) were estimated using POPGENE, FSTAT, and CERVUS packages. A total of 167 alleles for all loci were found across the breeds. Most of the microsatellite markers showed a high degree of polymorphism, the average number of alleles per locus being higher than FAO recommendation for at least 4 alleles per locus to reduce errors. None marker was monomorphic, the mean number of alleles by locus being 6.958±0.088. The mean polymorphic information content was 0.605±0.022. The expected heterozygosity was higher than 0.5 in most of the microsatellite markers. The observed heterozygosity was lower than the expected heterozygosity for all loci, the mean values being 0.468±0.022 and 0.648±0.021, respectively. Mean value of $F_{IS}$ was 0.082±0.024, and mean $F_{ST}$ was 0.225±0.015, the mean $F_{IT}$ being 0.279±0.023. Five of the Spanish populations and the three Portuguese breeds had a total of 16 private alleles at 12 microsatellites. A total of seven private alleles had frequencies higher than 10% (six in Spanish and one in Portuguese breeds). In conclusion the panel of microsatellite markers was of good usefulness in studying the genetic diversity of Spanish and Portuguese breeds, they showing a high number of private alleles.

---



**Genetic variability of Girgentana goat breed using molecular markers**

*M.T. Sardina, S. Mastrangelo, L. Tortorici, R. Di Gerlando and B. Portolano*
*Università degli Studi di Palermo, Scienze Agrarie e Forestali, Viale delle Scienze, 90128 Palermo, Italy; baldassare.portolano@unipa.it*

Assessing genetic diversity of local breeds through information provided by neutral molecular markers allows determination of their extinction risk and designing of strategies for their management and conservation. The aim of this work was to quantify the levels of genetic variability in Girgentana goat breed using microsatellite markers and genetic polymorphisms at the casein genes. A total of 264 individuals were genotyped for 20 microsatellites, selected as suggested by ISAG and FAO. Moreover, the same individuals were characterized for the casein loci (CSN1S1, CSN2, CSN1S2, and CSN3) using PCR and sequencing protocols. Several genetic diversity indexes were estimated. A total of 130 alleles were observed considering the 20 analyzed microsatellites. The observed and expected heterozygosity mean across loci were 0.586±0.204 and 0.604±0.203, respectively. Furthermore, the PIC was equal to 0.563±0.202 and heterozygosity deficit, as measured by Wright's Fis, was 0.035. The results showed moderate levels of variability for microsatellite markers in Girgentana goat breed, and may be explained considering the reduced effective population size. This breed is, indeed, reared in a restricted area of Sicily and nowadays about 650 animals are enrolled in the herd book. Based on the genetic structure of casein genes, the Girgentana goat breed showed high polymorphisms, as most dairy goat breeds from the Mediterranean area, and was characterized by the predominance of strong alleles, which play an essential role in quantity, composition, and technological properties of milk. Moreover, for CSN3 locus, two new genetic variants were identified, and named D' and N. In order to maintain genetic diversity, breeding strategies that increase effective population size minimizing genetic drift effect should be implemented. This work was supported by PSR Sicilia 2007-2013, Misura 1.2.4, CUPG66D11000039999.

**Inbreeding, coancestry and effective population size in Sicilian cattle breeds using SNPs markers**

*S. Mastrangelo[1], M. Saura[2], M.T. Sardina[1], M. Serrano[2] and B. Portolano[1]*
*[1]Università degli Studi di Palermo, Scienze Agrarie e Forestali, Viale delle Scienze, 90128 Palermo, Italy, [2]INIA, Mejora Genetica Animal, Carretera la Coruña Km 7.5, 28040 Madrid, Spain; salvatore.mastrangelo@unipa.it*

Maintaining the highest levels of genetic diversity and limiting the increase of inbreeding is the premise of most conservation programs. The aim of this work was to estimate the inbreeding (F), coancestry (f), and effective population size ($N_e$) in two Sicilian cattle breeds, Cinisara (CIN) and Modicana (MOD). Rate of molecular inbreeding and coancestry were used to estimate the $N_e$. A total of 144 animals were genotyped using the Illumina Bovine SNP50K v2 BeadChip. The average molecular F and f coefficients were 0.68±0.02 and 0.67±0.03 in CIN, and 0.69±0.02 and 0.70±0.03 in MOD cattle breeds, respectively. The results were not unexpected considering the reduced number of reared animals and the farming system, where natural mating is the common practice and the exchange of bulls among flocks is quite unusual; therefore, mating with close relatives can be quite frequent. Rates of molecular inbreeding (ΔF) and coancestry (Δf) per year were 0.004 and 0.022 in CIN, and 0.007 and 0.010 in MOD cattle breeds, respectively. $N_e$ values estimated from the ΔF were about 19 animals in CIN and 12 animals in MOD, and those calculated from the Δf, were about 4 and 8 animals in CIN and MOD cattle breeds, respectively. In animal breeding, the recommendation is to maintain a $N_e$ of at least 50 to 100. High values of F and f in local breeds, with low population size, as CIN and MOD, can compromise the viability of the populations. The control of F and f would restrict inbreeding depression, the probability of losing beneficial rare alleles, and therefore, the risk of extinction. The information generated from this study could have important implications for the development of conservation programs. This research was financed by PON01_02249, CUP: B11C11000430005 funded by MIUR.

---



**Genotype imputation in Gir (*Bos indicus*): comparing different commercially available SNP chips**

*S.A. Boison[1], D. Santos[2], A. Utsunomiya[2], F. Garcia[3], R. Verneque[4], M.V.B. Silva[4] and J. Sölkner[1]*
*[1]BOKU University, Gregor-Mendel 33, 1180, Austria, [2]UNESP, Jaboticabal, 14884-900, Brazil, [3]UNESP, Aracatuba, 16015-050, Brazil, [4]Embrapa Dairy Cattle, Juiz de Fora, 36038-330, Brazil; solomon.boison@students.boku.ac.at*

We evaluated genotype imputation of HD genotypes from three commercially available SNP chips (Illumina 50 k and two GeneSeek *Bos indicus* chips: 20 ki and 90 ki) using 452 Gir bulls. Imputation from Ilumina 3 k and 7 k to 50 k was also evaluated for 1,609 cows. The effects of genomic relationships between imputed and reference sets on imputation accuracy were studied. Genotype data after quality control consisted of 490,009 SNPs for HD, 63,044 SNPs for 90 ki, 19,657 SNPs for 50 k, 15,265 SNPs for 20 ki, 3,952 SNPs for 7 k and 1,717 SNPs for 3 k. The true HD genotypes of 302 young bulls were masked and imputed thus 150 HD genotyped sires were used as reference. Genotypes of cows were imputed to 50 k using all available sire (n=452) genotypes. Population based imputation algorithms implemented in FImpute v2 and Beagle v4 were used. Imputation accuracy (Rcor) was assessed as the correlation between the imputed allele dosage and the true dosage. The average of the top 5 genomic relationships of an imputed animal and individuals of the reference population was calculated (rel5) as an indicator of connectedness. Rcor were similar for Beagle and FImpute in all scenarios. For the sire imputation scenarios, Rcor was >0.97 with 90 ki having the highest accuracy (>0.99). Rcor were similar between 20 ki (0.972) and 50 k (0.979). For the cow scenarios, Rcor was 0.90 (0.91) and 0.96 (0.95) imputing from 3 k and 7 k to 50 k with FImpute (Beagle). There was a strong relationship between rel5 and Rcor. Most animals with rel5 <0.05 had Rcor <0.85 for almost all scenarios. Beagle v4 outperformed the older version Beagle v3.2 when tested. Geneseek 20 ki can be used as an alternative to Illumina 50 k to help reduce genotyping cost of genetic improvement programs. Average rel5 can be used as a criterion for discarding animals with potentially lower imputation accuracies.

**Expression study of fourteen genes influencing lipid and energy metabolism in two pig breeds**

*R. Davoli, S. Braglia, M. Bigi, M. Zappaterra and P. Zambonelli*
*University of Bologna, Department of Agricultural and Food Sciences, Via Fanin 46, 40127 Bologna, Italy; roberta.davoli@unibo.it*

Meat and carcass quality traits are influenced by many parameters, mainly by fat content, fatty acid composition and lean cut weight. Up to now, there is poor knowledge on the expression level of genes involved in and regulating lipid and energy metabolism in porcine skeletal muscle and fat tissues. To this aim we studied the expression level of fourteen genes (ACACA, ACLY, ACVRC1, CES3, ENO3, FASN, INSIG2, LMNA, MTTP, NAMPT, PLIN1, PLIN2, PLTP and SORT1) mapped on different chromosomes (1, 4, 6, 7, 8, 9, 12, 15 and 17) and chosen for their involvement in lipid and/or energy metabolism in porcine muscle and backfat tissues. Tissue samples were collected at slaughterhouse from Italian Large White (ILW) and Italian Duroc (IDU) pigs with divergent breeding values for backfat thickness and frozen in liquid nitrogen. After extraction, mRNA was quantified by qRT-PCR and transcription levels of the analysed genes were compared between breeds within each tissue. In muscle samples differential gene expression between breeds was evident for ACACA, ACLY, PLIN1 and SORT1 genes (significant P values ranged from 0.001 to 0.045). In particular, IDU pigs showed higher expression levels than ILW. In backfat tissue differences between breeds were found for ACACA, ACLY, CES3 and LMNA genes (significant P values ranged from 0.007 to 0.048). In this tissue, the highest gene expression levels were found in ILW pigs. The results on different transcription profiles of these genes involved in fat and/or energy metabolism could be related to dissimilarities between the analysed breeds considering their divergent attitude for fat deposition in meat and carcass as reported in literature. The results obtained in this study may contribute to better understand the molecular basis of fat deposition in pig breeds. The Authors wish to thank ANAS for providing samples and data. This research was supported by project AGER-HEPIGET (grant N. 2011-0279).

---



**Genome wide association study for intramuscular fat in Italian Large White pigs using SNP60K chip**

*P. Zambonelli[1], D. Luise[1], S. Braglia[1], A. Serra[2], V. Russo[1] and R. Davoli[1]*
*[1]University of Bologna, Department of Agricultural and Food Sciences, Via Fanin 46, 40127 Bologna, Italy, [2]University of Pisa, Dipartimento di Scienze Agrarie, Alimentari e Agro-ambientali, Via del Borghetto 80, 56124 Pisa, Italy; roberta.davoli@unibo.it*

Intramuscular fat (IMF) is one of the most important traits influencing meat quality as its level affects juiciness, tastiness and, in general, consumer acceptance. IMF content can change with diet and is also under genetic control. Several studies reported in PigQTLdb showed the presence of significant QTL associated to intramuscular fat content but relevant causative mutations haven't been so far detected. With the aim to detect association between regions of porcine genome and fat content we performed a genome scan with the Illumina PorcineSNP60 BeadChip using 912 Italian Large White pigs included in the Sib Test genetic evaluation program of the Italian Association of Pig Breeders (ANAS). These pigs were reared on the same herd, feed on the same diet, and slaughtered on the same abattoir during year 2012. Intramuscular fat content was measured on Semimembranosus muscle using Ankom method based on Soxhlet technique. The average fat content was 1.98±1.09% and minimum and maximum values were 0.51% and 8.64%, respectively. The genotypes for 61,565 SNPs were obtained and these data were filtered using PLINK toolset to remove animals without genetic evaluation and MIND>0.1. Additional criteria were used to filter the SNP data according to MAF<0.01, GENO>0.1, HWE <0.001. The final dataset was composed by 890 pigs and 49,677 SNPs and population stratification was examined using IBS approach. Preliminary results showed that some of the most significant putative markers for IMF content mapped on chromosome 1. Further analyses are in progress in order to complete and validate the obtained results. The Authors wish to thank ANAS for providing samples and data. This research was supported by project AGER-HEPIGET (grant N. 2011-0279).

**Genetic ressources breeding and production systems ofcattle in algeria: present and perspectives**

*N. Adili[1], M. Melizi[1] and H. Belabbas[2]*
*[1]University of El-HadjLakhdar, Department of veterinary medicine, Institute of veterinary sciences and agricultural sciences, Batna, 05000, Algeria, [2]University of Mohamed BOUDIAF, Department of Microbiology and Biochimistry, M'sila, 28000, Algeria; anavet.belabbas@yahoo.fr*

This work represents a synthesis of numerous reports and preliminary studies whose aim is to make an identification of bovine breeds that are found in Algeria and to know the different management and rearing mechanisms. The study focuces mainly on the recognition of the genetic diversity and the breeding techniques. It describes the different production systems of cattle in Algeria. The analysis of the obtained results was performed using the Microsoft Office Excel 2010. This analysis allowed the classification of the Algerian cattle into three categories: local cattle (78% of the total bovine population), imported breeds and their breeding descendents (nearly 22%). Local cattle breeds belong to one group named the small brown cow of the Atlas. These bovine breed populations can be distinguished from each other sharply by their phenotypes. Therefore, we can observe the Guelmoise, the Cheurfa, the Setifian, the Chelifian, the Djerba, the Kabyle and the Chaouia. These cattle live in regions where the extensive systems are widely used. They are bred mainly for their meat production. ThePrim'Holstein, the Montbeliard and the brown of the Alps are the bovine populations regularly imported for milk production under intensive breeding conditions. The local cattle called improved breeds (local ximported) are used for a mixed production of milk and meat. The Algerian bovine herds process an important genetic diversity that is not exploited properly. Therefore, appropriate measures should be taken in order to preserve these genetic resources that would insure food supply security and genetic preservation of the universal genetic heritage.

---

Session 41 Poster 31

**Overview about the situation of local cattle genetic resources in Algeria**

*B. Abdeltif[1], F. Ghalmi[1], M. Laouadi[1,2], N. Kafidi[1,3] and S. Tennah[1]*
*[1]School Veterinary National Superior of Algiers, Algeria, Hacène Badi El Harrach, Algiers, B.P. 161, Algeria, [2]University of Amar Teliji, Faculty of science, Department of Agonomy, Laghouat, B.P. 37G, Algeria, [3]Canadian Food Inspection Agency, Ottawa, Canada, ON K2C, Canada; tensaf2004@yahoo.fr*

In Algeria the number of bovine local is the largest kind (840,000 head let 48% of the bovine total number), it is represented by the Brown breed of the Atlas and its varieties. This type of bovine has a low milk production approximately about 3 to 4 liters per cow. This breed is still exploited for its meat. Although the potentialities of production of this breed are not of a very high level, it represents nevertheless an interesting animal material in more than a title. First, it is about a rustic breed adapted in to specific living conditions in Algeria. Second, it is also about an important genetic heritage that is extremely necessary to preserve the variability. Several varieties derive from the breed Brown of the Atlas, by succession, presumably, to the various genetic mutations and the adaptations to the various environments. Currently these varieties are characterized by phenotypic differences (of size and coat color) to which we gave local names. Nearly 2/3 of the size are in the East of the country. This distribution and the adaptation to the respective environments offer a diversity of bovine populations in Algeria. The 'Guelmoise' and 'Cheurfa' make up the majority of the number of the varieties, then the 'Setifienne' and 'Chelifienne' with less important numbers. Four other varieties with smaller numbers are also listed: the 'Chaouiya', the 'Kabyle', the 'Tlemcenienne' and 'Jerba'. Despite this genetic diversity, few research works exist for the zootechnic and phenotypic characterization. Of more than one analysis of genetic polymorphism would be necessary to determine their degree of similarity and allow racial distribution of these populations. This study will present the zootechnic and phenotypic characteristics of the local bovine populations in Algeria.

Session 41 Poster 32

## Genome wide scan for total solids in Gyr dairy cattle

*A.T.H. Utsunomiya[1], D.J.A. Santos[1], S.A. Boison[2], M.A. Machado[3], R.S. Verneque[3], J. Sölkner[2], R. Fonseca[1] and M.V.G.B. Silva[3]*
*[1]UNESP, Jaboticabal-SP, 14884-900, Brazil, [2]BOKU University, Vienna, 1180, Austria, [3]Embrapa Dairy Cattle, Juiz de Fora-MG, 36038-330, Brazil; adamtaiti@gmail.com*

The aim of this study was to identify candidate genomic regions explaining differences in total solids yield of milk in Gyr dairy cattle (*Bos indicus*) via genome-wide scan. Deregressed breeding values of total solids were considered as phenotypes to be analyzed. A total of 457 bulls were genotyped using Illumina® BovineHD BeadChip (HD) and 1,684 cows were genotyped using Illumina® BovineSNP50 BeadChip (50K). We imputed the cow data from 50K to HD using genotypes from sires. FImpute v2 imputation software was used with pedigree information. Markers and samples were filtered according to the following criteria: Markers – Call rate<0.95; GenCall score<0.50; minor allele frequency <0.02 and HWE with P-value $<1e^{-06}$ were removed; Samples – Call rate <0.90 were excluded. After filtering and imputation, 1185 individuals with phenotypes remained along with 490,009 SNP markers. The effect of each SNP was estimated using Bayes C as implemented in GS3 software. The $\pi$ was fixed at 0.99 and the Monte Carlo Markov Chain was run for 160,000 iterations, with a burn-in of 20,000 iterations and thin interval of 20. The posterior inclusion probability was used as criteria of detection of candidate genomic regions. Information about genes surrounding the SNP (±250 kb) were investigated using the Ensembl Biomart database. Three markers, on chromosomes 2, 13 and 25 with highest allele substitution effect were analyzed. The gene closest to the marker on BTA2 encodes an uncharacterized protein while the marker on BTA25 was in an intronic region of a gene with an unknown function. There was no gene surrounding the marker on BTA25. Allele substitution effect were very small, thus signifying that total solids in milk is influenced by many genes of small effect.

---

Session 41 Poster 33

## Genome-wide association on milk production at peak and end of lactation in Guzerá (*Bos indicus*)

*D. Santos[1], A. Utsunomiya[1], S.A. Boison[2], M.G. Peixoto[3], J. Sölkner[2], H. Tonhati[1] and M.V. Silva[3]*
*[1]UNESP-FCAV, Zootecnia, Via de Acesso Prof. Paulo Donato Castellane s/n, 14884-900 Jaboticabal-SP, Brazil, [2]BOKU University, Gregor-Mendel 33, 1180 Vienna, Austria, [3]Embrapa Dairy Cattle, Rua Eugênio do Nascimento, 610 – Dom Bosco, 36038-330 Juiz de Fora-MG, Brazil; daniel_jordan2008@hotmail.com*

GWAS studies are of great importance in animal breeding due to the possibility of including detected QTLs in genetic evaluation. Thus, the aim of the study was to identify loci related to milk production at peak (records at second month – MY2) and at the end (tenth month – MY10) of lactation. Deregressed breeding values and 50K Illumina genotypes of 45 bulls and 856 Guzera cows were used. Samples and SNP quality control were done according to the following criteria; samples with call rate >0.90 and with heterozygosity levels ±3 SD of the mean were removed. SNPs with call rate <0.98, MAF <2% and HWE $<10^{-6}$ were discarded. After editing, 25,024 SNPs were left. Statistical analyses were performed using the GenABEL package. Pseudo-phenotypes were corrected for population substructure by means of principal components obtained from genomic similarity matrix. The additive genetic model was used. None of the SNPs exceeded the Bonferroni significance threshold of $2\times10^{-6}$. However, the most associated SNPs were investigated in detail. For the MY2, 1 Mb regions around the positions 31.0 Mb on BTA3, 26.3 Mb on BTA20 and 33.5 Mb on BTA27 were studied. For MY10 the 1 Mb regions around 42.3 Mb on BTA3, 2.37 Mb on BTA15 and 24.1 Mb on BTA20 were also studied. The genes associated with the most significant SNPs were CTTNBP2NL and NTNG1 to MY2 and MY10, repectively. These two genes are related to expression of messenger RNA for different periods of lactation in humans. The CTTNBP2NL gene is also related to decreased fertility in dairy cows, this possibly explains part the relation between lactation peak and return to estrus. The results suggest that different genes have different effects on different periods of bovine lactation.

## Session 41 Poster 34

**Association of OLR1 and SCD1 genes with milk production traits in Iranian Holstein dairy cattle**

*M. Hosseinpour Mashhadi*
*Department of Animal Science, College of Agriculture, Mashhad Branch, Islamic Azad University, Mashhad, Iran; mojtaba_h_m@yahoo.com*

The present study was carried out to investigate the association of A/C SNP in the 3' untranslated region of OLR1 gene and C/T SNP in exon 5 of SCD1 gene with milk production traits in Iranian Holstein dairy Cattle. The OLR1 gene encodes a vascular endothelial cell surface receptor that binds and degrades the oxidized forms of low-density lipoproteins (oxLDL). The SCD1 gene expressed in adiposities mammary glands and also in other tissues, encodes a key enzyme which is responsible for converting palmitic and stearic acids into their monounsaturated forms. The blood samples of 136(for OLR1) and 274(for SCD1) dairy cattle from three different farms were used for genotyping. A 146 bp fragment of 3'-untranslated region of OLR1 gene and a 400 bp fragment of exon 5 of SCD1 gene was amplified by standard PCR. Single nucleotide polymorphism of OLR1 and SCD1 gene was determined by PCR-RFLP technique. The association between genotypes of OLR1 and SCD1 genes with milk production traits was studied by GLM procedure of SAS package and Duncan multiple range test was used for compare means of traits. On OLR1 gene, the means for milk yield (CC: 8,273 kg, AC: 8,344 kg, AA 7178 kg), fat yield (CC: 276.3 kg, AC: 277.6 kg, AA 239.7 kg) and protein yield (CC: 286.7 kg, AC: 290.5 kg, AA 253 kg) for CC and AC were more than AA genotype ($P<0.05$). On SCD1, the mean of fat percentage between genotypes of VV (3.43%) and AA (3.33) were significant ($P<0.05$).

## Session 41 Poster 35

**Genome structure in Sicilian local cattle breeds**

*S. Mastrangelo, M.T. Sardina, M. Tolone, R. Di Gerlando and B. Portolano*
*Università degli Studi di Palermo, Scienze Agrarie e Forestali, Viale delle Scienze, 90128 Palermo, Italy; salvatore.mastrangelo@unipa.it*

Genomic technologies provide background information concerning genome structure in domestic animals. The aim of this work was to investigate the genetic structure and the patterns of linkage disequilibrium (LD), using Bovine SNP50K v2 BeadChip, in two Sicilian local cattle breeds, Cinisara (CIN) and Modicana (MOD). Genotypes from animals of Italian Holstein (HOL) breed were also used to investigate the relationship among breeds. Structure software was used to analyze the genetic structure and assign the individuals to each cluster. The genetic relationship between individuals was estimated by Principal Components Analysis (PCA) of genetic distance. A standard descriptive LD parameter ($r^2$) was estimated between adjacent SNPs and for all pairwise combinations of SNPs on each chromosome. PCA and Bayesian clustering algorithm showed that animals from the three breeds formed non-overlapping clusters and are clearly separated populations. Between the Sicilian cattle breeds, the MOD was the most differentiated breed, whereas the CIN animals showed a lowest value of assignment, the presence of substructure and genetic links occurred between both breeds. Considering the levels of LD between adjacent SNPs, the average $r^2$ in MOD ($0.200\pm0.247$) was comparable to those reported for others cattle breeds, whereas CIN showed a lower value ($0.162\pm0.222$). The results may be explained considering the influence of selection on LD; in fact the CIN breed is not subject to breeding programs whereas the MOD breed is characterized by low selection pressure. Average $r^2$ ranging from 0.192 and 0.234 for SNPs located up to 50 kb apart, to 0.021 and 0.027 for SNPs separated by more than 5000 kb, for CIN and MOD breeds, respectively. The information generated from this study could have important implications for the design and applications of selection breeding programs in these two local cattle breeds. This research was financed by PON01_02249, CUP: B11C11000430005 funded by MIUR.

## Session 41 Poster 36

**Genetic diversity of Pannon White rabbits as revealed by pedigree analysis**

*H. Nagyné-Kiszlinger, J. Farkas, Z.S. Szendrő and I. Nagy*
*University Kaposvár, Animal genetics and biotechnology, Guba S. u. 40, 7400 Kaposvár, Hungary; kiszlinger.henrietta@ke.hu*

The objective of this study was to analyze the population structure of Pannon White rabbit population of the Kaposvár University. The analysis was performed by the pedigree analysis of breeding rabbits born between 1992 and 2012. The following population characteristics were computed using the software ENDOG 4.8: pedigree completeness, generation intervals, inbreeding coefficient, average relatedness and effective population size, effective number of founders and effective number of ancestors. The pedigree contained 5,819 different individuals. These parameters were calculated for different reference populations (rabbits born in the same year belonged to the same reference population). The number of complete generation equivalents was 16.00 for rabbits born in 2012. For these rabbits the pedigree was more than 90% complete until the $12^{th}$ generation. The mean generation interval for buck pathways was 1.5 while that for doe pathways was 1.2. The mean inbreeding calculating for each year was slightly increasing with a regression coefficient of 0.0054. It's value for the year 2012 was 8.65%. The average relatedness reached 7.27% by the last year of the observed period. The number of breeding rabbits during the period 2008-2012 was unbalanced and at the end of the observed time interval it decreased to 160. The realized effective population size for the reference population 2012 was 83.21, while the family variance effective population size could only be calculated for the penultimate year 2011 and was 105.96. Both the number of founders and that of ancestors remained stable during the last five years. These were 52 and 25 respectively for the reference population 2012. Number of ancestors explaining the 50% of genetic variability reduced to 9. Based on our results it could be concluded that the genetic diversity of this rabbit breed for the last five years corresponded with that of our previous studies showing continuous slight increase of inbreeding coefficients.

---

## Session 41 Poster 37

**Linkage disequilibrium and genetic diversity estimation in three Sicilian dairy sheep breeds**

*R. Di Gerlando, S. Mastrangelo, M.T. Sardina and B. Portolano*
*Università degli Studi di Palermo, Scienze Agrarie e Forestali, Viale delle Scienze, 90128 Palermo, Italy; rosalia.digerlando@unipa.it*

Understanding genetic structure is essential for achieving genetic improvement through genome-wide association studies, genomic selection and the dissection of quantitative traits. In the present study, we used the OvineSNP50K BeadChip to characterize linkage disequilibrium (LD), identify haplotype blocks and to analyze genetic diversity in the Valle del Belice (VDB), Comisana (COM), and Pinzirita (PIN) dairy sheep breeds. LD between adjacent SNPs and for all pairwise combinations of SNPs on each chromosome was measured using $r^2$. Haplotype blocks were estimated using D' based method. Genetic diversity and Principal Component Analysis (PCA) were calculated using PLINK. Small differences in average LD value for adjacent SNPs on each chromosome were observed between VDB and COM (0.155±0.204 and 0.156±0.208, respectively), whereas a low value was found in PIN breed (0.128±0.188). The same differences were observed across the genome; for SNPs separated by 50-100 kb, the average $r^2$ was 0.117, 0.113 and 0.084, respectively. In all breeds, the chromosome showing the highest number of blocks, the highest number of SNPs involved in the blocks, and the highest value of $r^2$ was OAR2. This result suggesting that there is similarity of conserved haplotypes among Sicilian breeds. PCA indicated that while VDB and PIN are clearly separated populations, the individuals of COM breed showing the presence of substructure that could be due to introgression of genes from other breeds. The low level of inbreeding and high genetic diversity (expected heterozigosity) were found in PIN breed, and this reflect the short extent of LD. Our results for average LD within chromosome and breed are in agreement with the block structure across the genome; moreover, the different levels of LD among chromosomes and breeds, indicate different selection pressure. This research was financed by PON01_02249, CUP: B11C11000430005 funded by MIUR.

## Session 41 Poster 38

**The A allele of CCKAR g.420 C**

*K. Rikimaru[1], D. Takahashi[1], M. Komatsu[1], T. Ohkubo[2] and H. Takahashi[3]*
*[1]Akita Prefectural Livestock Experiment Station, Jingu-ji, Daisen, Akita, 019-1701, Japan, [2]College of Agriculture, Ibaraki University, Ami, Ibaraki, 300-0393, Japan, [3]National Institute of Livestock and Grassland Science, Tsukuba, Ibaraki, 305-0901, Japan; Rikimaru-Kazuhiro@pref.akita.lg.jp*

We previously reported the association between a single nucleotide polymorphism (SNP; g.420 C<A) in the cholecystokinin type A receptor gene (CCKAR) and growth traits in the Hinai-dori which is a breed of chicken native to Japan, and we showed that the A allele had a superior effect on growth traits compared to the C allele. In the present study, we demonstrated that this SNP improves the growth rate using the commercial Hinai-jidori chicken (a cross between Hinai-dori sires and Rhode Island Red dams) which is a slow-growing Japanese meat type chicken. Individuals of three the genotypes (A/A, A/C, and C/C) were raised free-range until 23 weeks of age. The body weight of each individual was measured at 4, 10, 14, 18, and 23 weeks of age, and the mean daily gain was calculated from the body weights. Feed conversion ratio was calculated from the mean daily gain and feed intake. The data showed that body weight at 23 weeks of age and the average daily gain between 4 and 23 weeks of age of A/A individuals were significantly greater than those of C/C individuals ($P<0.05$). There were no significant differences in feed intake among the three genotypes. The feed conversion ratio between 4 and 23 weeks of age in A/A individuals showed lowest value among of the three groups. We conclude that the A allele of the g.420 SNP in CCKAR improves the growth rate of the commercial meat type chicken.

---

## Session 42 Theatre 1

**Livestock production: vulnerability to population growth and climate change**

*O.F. Godber and R. Wall*
*University of Bristol, Veterinary Parasitology and Ecology, School of Biological Sciences, Woodland Road, Bristol, BS8 1UG, United Kingdom; olivia.godber@bristol.ac.uk*

Livestock production is an important contributor to sustainable food security for many nations, particularly in low income areas and marginal habitats that are unsuitable for crop production. Vulnerability analysis has been widely used in global change science to predict impacts on food security and famine. It is a particularly useful tool for predicting the sensitivity of coupled human-environment systems and can be used to inform policy decision-making and direct the targeting of interventions. Using a range of indicators derived from FAOSTAT and World Bank statistics, the relative vulnerability of nations are modelled at the global scale to predicted climate and population changes which are likely to impact on their use of grazing livestock for food. The model shows that sub-Saharan Africa, particularly in the Sahel region, and some Asian nations are the most vulnerable. Livestock-based food security is already compromised in many areas on these continents and suffers constraints from current climate in addition to the lack of economic and technical support allowing mitigation of predicted climate change impacts. Governance is shown to be a highly influential factor and, paradoxically, current self-sufficiency is shown to increase future potential vulnerability because trade networks are poorly developed. The latter may be mitigated through freer trade of food products, which is also associated with improved governance. The study suggests that policy decisions, support and interventions will need to be targeted at the most vulnerable nations but, given the strong influence of governance, effective implementation will require considerable care in the management of underlying structural reform.

**Designing a multicriteria index at farm-scale to assess dairy farm abandonment risk in mountain area**

*A.L. Jacquot[1,2] and C. Laurent[1]*
*[1]VetAgro Sup, UMR 1213 Herbivores, BP 10443, 63000 Clermont-Ferrand, France, [2]Agrocampus Ouest, UMR1348 INRA-Agrocampus PEGASE, 35590 Saint-Gilles, France; anne-lise.jacquot@agrocampus-ouest.fr*

Grass-based dairy farming systems located in mountains provide multiple services to these territories by preserving landscapes, biodiversity... However, the number of dairy farms decreases since the 1980's. This abandonment is mainly due to a lack of intergenerational transmission (IT) explained by farm features (structure constraints, economic assets) and on-farm working conditions (FWC) which are considered as major critical issues of sustainability. So, assessing those aspects can lead to detect possible leeway to improve sustainability of dairy farms and to maintain milk production in mountains. Yet, among the many existing sustainability indexes at farm-scale, few integrate social and economic aspects including FWC and IT. Those indexes do not allow getting a precise overview of working conditions and transmission possibilities by being too restrictive or too subjective. So the aim of this study is to design a multicriteria index focused on those aspects. To design a relevant and meaningful tool to the target users – the farmers – a bottom-up approach has been chosen. First, a focus-group, including farmers, advisors and researchers, defined criteria covering the different aspects of FWC, IT. Then, they selected or created appropriate indicators out of a list of possible existing indicators. Afterwards, the index has been built by weighting indicators and criteria. This process results in an index made up with: (1) 5 criteria for working condition described by 40 indicators including original criteria such as mental dimension of work; (2) 6 criteria for farm transmission described by 21 indicators covering structure and production features, economic assets and transmission perspectives. The index will be further tested on dairy farms.

---



**Environmental and economic consequences of feeding increased amounts of solid feed to veal calves**

*H. Mollenhorst[1,2], P.B.M. Berentsen[1], H. Berends[3], W.J.J. Gerrits[3] and I.J.M. De Boer[2]*
*[1]Business Economics group, Wageningen University, P.O. Box 8130, 6700 EW Wageningen, the Netherlands, [2]Animal Production Systems group, Wageningen University, P.O. Box 338, 6700 AH Wageningen, the Netherlands, [3]Animal Nutrition group, Wageningen University, P.O. Box 338, 6700 AH Wageningen, the Netherlands; erwin.mollenhorst@wur.nl*

Veal calves, traditionally, receive the majority of their nutrients from milk replacer (MR). Nowadays, however, increasing amounts of solid feed (i.e. concentrates and roughages) are fed, partly because it is enforced by legislation, partly because of increasing prices of MR. To determine the potential and the effects of substitution of MR by solid feed, an experiment was set up with 160 calves, aiming at equal carcass gain on eight different diets. Two different solid feed mixtures were fed at four predefined levels, ranging from 20 to 250 kg dry matter. Mixture-1 (M1) contained 80% concentrates, 10% maize silage and 10% straw on dry matter basis, whereas mixture-2 (M2) contained 50% concentrates, 25% maize silage and 25% straw. Additionally, MR was fed to achieve equal rates of carcass gain on all diets. The reduction in MR provision required to achieve equal rates of carcass gain at increasing levels of solid feed was 0.70 kg MR/kg solid feed for M1 and 0.62 kg/kg for M2. Costs of the solid feed needed to substitute one kg MR are 53-55% lower for both mixtures, whereas global warming potential was 40% higher for M1 and 34% higher for M2. The higher global warming potential mainly resulted from an increase in methane emission from rumen fermentation. Lower energy use, due to less water to be heated, and lower emissions from feed production (for M2) could not compensate for this. Other environmental impacts, for example acidification and land use, however, still need to be considered. From this study we conclude that replacing part of the MR in diets of veal calves with solid feed is economically sound, when only feeding costs are considered, but results in increased global warming potential, mainly due to increased methane emissions.

### A dominance analysis of greenhouse gas emissions, beef output and land use of German dairy farms

*M. Zehetmeier[1], H. Hoffmann[1], J. Sauer[1] and D. O'Brien[2]*
*[1]Technische Universität München, Alte Akademie 14, 85350 Freising, Germany, [2] Teagasc, Moorepark, Fermoy, Ireland; monika.zehetmeier@tum.de*

Three main approaches can be distinguished when searching for greenhouse gas (GHG) mitigation options at farm level: technical mitigation options, change of production systems and improved management within farming systems. The last one was addressed in this study. To identify GHG mitigation options within farming systems those parameters have to be identified that explain variation of GHG emissions between farms. We used dominance analysis to identify most important parameters. Besides GHG emissions, potential beef output and land use/kg of FPCM was investigated. 27 dairy farms from south Germany with dual purpose Fleckvieh cows (South-FV) and 26 dairy farms from west Germany with Holstein-Friesian cows (West-HF) both feeding total mixed rations were assessed. The results showed that South-FV dairy farms emitted greater GHG emissions/kg of FPCM ($P<0.01$) than higher yielding West-HF dairy farms. A wide range in GHG emissions within region was found from 0.90-1.25 kg $CO_2$ eq/kg of FPCM for South-FV German farms and 0.79-1.20 kg $CO_2$ eq/kg of FPCM for West-HF German farms. Average beef output/kg of FPCM of West-HF dairy farms was statistically significant lower compared to South-FV dairy farms. Outcomes of variable importance analysis showed that milk yield and replacement rate had a high impact on variation of GHG emissions and beef output of both dairy farm groups. A trade off between GHG emissions/kg of FPCM and beef output/kg FPCM was shown in the case of increasing milk yield and reducing replacement rate. No difference between the groups was found in case of land use/kg of FPCM. The analysis is a first approach identifying the parameters of dairy farms that are key contributors to GHG emissions/kg of FPCM and are also highly variable between farms. It was also shown that it is important to identify those parameters that have a negative impact on beef output to avoid shifting of GHG emissions between production systems.

---



### Effects of dietary nitrate and lipid on methane emissions from beef cattle are basal diet dependant

*J.A. Rooke[1], R.J. Wallace[2], C.-A. Duthie[1], S. Troy[1], J.J. Hyslop[1], D.W. Ross[1], T. Waterhouse[1] and R. Roehe[1]*
*[1]SRUC, West Mains Road, Edinburgh EH9 3JG, United Kingdom, [2]Rowett Institute of Nutrition and Health, Bucksburn, Aberdeen AB21 9SB, United Kingdom; john.rooke@sruc.ac.uk*

The potential for nitrate and lipid to reduce methane ($CH_4$) emissions was tested using two contrasting basal diets. The experiment had a 2×2×3 factorial design where the factors were steer genotype (crossbred Charolais or purebred Luing), basal diet (concentrate, CON (920 g concentrate dry matter(DM)/kg DM) or forage, FOR (500 g concentrate/kg DM)) and $CH_4$ mitigation treatment (control, C (barley and rapeseed meal), lipid, L (155 g cold-pressed rape cake/kg DM) and nitrate, N (Calcinit, 25 g/kg DM) giving 6 nutritional treatments: CON-C, CON-L, CON-N, FOR-C, FOR-L, FOR-N. $CH_4$ emissions were measured using 6 open-circuit respiration chambers. Steers (12 per nutritional treatment; mean live-weight, 698 kg) were assigned to chambers using a randomised block design in 12 weekly periods. Data were analysed using general linear models with fixed effects of breed, basal diet and treatment and random effects of chamber and block. DM intake did not differ between treatments. Genotype had no effect on $CH_4$ emissions. Daily $CH_4$ output (g/d, CON-C 153; CON-L 147; CON-N 136; FOR-C 245; FOR-L 240; FOR-N 212; sed, 15.8) was lower on CON than FOR ($P<0.001$) and decreased by $CH_4$ mitigation treatment ($P<0.05$). When expressed as g $CH_4$/kg DM intake (CON-C 14.7; CON-L 15.3; CON-N 15.3; FOR-C 25.3; FOR-L 23.4; FOR-N 20.6, sed 1.51), $CH_4$ output was lower on CON than FOR ($P<0.001$). $CH_4$ mitigation treatment had no effect on CON but on FOR, nitrate and lipid reduced $CH_4$ emissions by 19% ($P<0.05$) and 8% ($P>0.05$) respectively. In conclusion, $CH_4$ emissions were substantially lower when the high concentrate diet was fed. However, whereas nitrate and to a lesser extent lipid reduced $CH_4$ emissions by steers fed a mixed forage: concentrate diet, these treatments were ineffective when the high concentrate diet was fed. Study supported by EBLEX and Scottish Government.

**Innovative lactation stage specific prediction of $CH_4$ from milk MIR spectra**

*A. Vanlierde[1], M.-L. Vanrobays[2], F. Dehareng[1], E. Froidmont[1], N. Gengler[2], H. Soyeurt[2], S. McParland[3], E. Lewis[3], M.H. Deighton[3] and P. Dardenne[1]*
*[1]Walloon Agricultural Research Centre, 5030, Gembloux, Belgium, [2]University of Liège, Gembloux Agro-Bio Tech, 5030, Gembloux, Belgium, [3]Animal and Grassland Research & Innovation Centre, Moorepark, Cork, Ireland; a.vanlierde@cra.wallonie.be*

Previous research has shown that $CH_4$ emissions of dairy cows are linked to milk composition and particularly to fatty acids (FA). We showed that mid-infrared (MIR) prediction equations can be used to obtain individual enteric $CH_4$ emissions from the milk MIR spectra. However body tissue mobilisation alters milk FA and potentially links between $CH_4$ and MIR spectra. Therefore to reflect the expected metabolic status during lactation, a method was developed to consider days in milk (DIM) in the MIR based prediction equation. A total of 446 $CH_4$ reference data were obtained using the $SF_6$ method on 146 Jersey, Holstein and Holstein-Jersey cows. Linear (P1) and quadratic (P2) Legendre polynomials were computed from DIM of $CH_4$ measurements. A first derivative was applied to the MIR spectra. The calibration model was developed using as independent variables first derivative, first derivative × P1, first derivative × P2 and a modified PLS regression. The $CH_4$ emission prediction (g $CH_4$/day) showed a calibration coefficient of determination ($R^2c$) of 0.75, a cross-validation coefficient of determination ($R^2cv$) of 0.67 and the standard error of calibration (SEC) was 63 g/day. In order to check if this new equation showed an expected and biological meaningful behavior, it was applied to the milk MIR spectra database of the Walloon Region of Belgium (1,804,476 records). The resulting trend across lactation was similar to what was expected, with increasing averaged $CH_4$ up to DIM 83 and a slight decrease after. This pattern was a clear improvement when compared to predictions from previous equations. Results indicate that this innovative approach with integration of DIM information could be a good strategy to improve the equation by taking better account of the metabolism of the cows.

---

Session 42 Theatre 7

**Life cycle assessment of heavy pig production in a sample of Italian farms**

*G. Pirlo[1], S. Carè[2], G. Della Casa[2], R. Marchetti[2], G. Ponzoni[2], V. Faedi[2], V. Fantin[3], P. Masoni[3], P. Buttol[3] and F. Falconi[4]*
*[1]Consiglio per la ricerca e sperimentazione in agricoltura, CRA-FLC, via Porcellasco 7, 26100 Cremona, Italy, [2]Consiglio per la ricerca e sperimentazione in agricoltura, CRA-SUI, via Beccastecca, 41018 San Cesario, Italy, [3]ENEA, via Martiri di Monte Sole, 40129 Bologna, Italy, [4]LCA-Lab SRL, via Martiri di Monte Sole, 40129 Bologna, Italy; giacomo.pirlo@entecra.it*

The purpose of this study was to estimate the environmental impact of breeding and fattening of heavy pig used for Italian cured ham production. For this purpose, a life cycle assessment (LCA) was used in two samples of 4 breeding farms and of 8 fattening farms. The functional unit was 1 kg of body weight of fattened pig. The system was subdivided into breeding and fattening phases. Impact categories were: global warming (GW), abiotic resource depletion (AD), photochemical ozone formation (PO), acidification (AC), and eutrophication (EU). Reference units were kg of $CO_2$ eq for GW, kg of Sbeq for AD, kg of $C_2H_4$ eq for PO, kg of $SO_2$eq for AC, and kg of PO43 eq for EU. Farm activities were: (1) on-farm energy consumption (EC); (2) manure management (MM); (3) manure application (MA); (4) on -farm feed production (ONFP); (5) off-farm feed production (OFFP); (6) enteric fermentation (ENF); (7) transports (TR); (8) recyclable products (RP). Average final body weight was 167±5.2. LCA was performed with the support of SimaPro 7.3.3 software. The average environmental impacts associated with 1 kg of fattened pig body weight were: GW 3.43 kg $CO_2$ eq, AD 3.13 E-3 kg Sbeq, PO 1.78 E-3 kg $C_2H_4$ eq, AC 5.37 E-2 kg $SO_2$ eq, EU 3.20 E-2 kg PO43 eq. Contribution of breeding and fattening phases on GW, AC, PO, AC, and EU were: 30 and 70; 23 and 77; 31 and 69; 37and 63;30 and 70 respectively. Normalization analysis showed that the major contributions to environmental impact of 1 kg of fattened pig body weight come from GW, AC and EU whereas AD and PO are negligible.

**Sustainability of the chicken supply chain in Lebanon: an evaluation system**

*R. El Balaa[1] and C. Tannoury[2]*
*[1]Balamand University, Koura, P.O. Box:100 Tripoli, Lebanon, [2]Lebanese University, Dekwaneh, Beirut, Lebanon; rodrigue.elbalaa@balamand.edu.lb*

This study proposes a supply chain sustainability evaluation system to identify major weaknesses of the poultry sector and eventually propose concrete solutions. The system includes a series of fourteen indicators covering the sustainability of producers, processors and distributors. At the environmental level, indicators are energy consumption, greenhouse gas emissions, nitrogenous and organic effluents, water use and packaging material. The social indicators are gender equity, salaries, employees' turn-over rate, training and injuries percentage. Finally, the economic indicators are productivity, profit growth, internal investment and added value. The system was validated through a testing over representatives of the producers (2), processors (2) and distributors (5), then a unified scoring system was established with a score varying between 0 and 10, 0 being the weakest performance and 10 the best one, and consequently 5 being the threshold of minimum acceptance. At the environmental level, one of the major problems were energy production and greenhouse gases production caused by a lack of a general environmental management. At the social level, there is a low rate of female employees with a need to introduce measures favouring equal gender opportunities. At the economic level, imported feed was responsible for 65 to 70% of the production cost added to the cash sales strategies, burdening small and medium sized producers. The evaluation system succeeded in identifying major weaknesses and threats to the local poultry sector and allowed the proposition of important suggestions to improve the supply chain's sustainability. The system will be further tested on more components of the Lebanese poultry supply chain to improve its efficiency and performance.



**Cost per kg of live pig obtained from a new information system of livestock activities in Mexico**

*E.E. Pérez-Martínez, R. Trueta and C. López*
*FMVZ Universidad Nacional Autónoma de México, Av. Universidad no 3000 Del. Coyoacán, 04510, Mexico; trueta@unam.mx*

Availability of information on the cost production per kg of live pig sold (c/kg) in pig production is essential for farmers as well as in public policy decision-making. The objective of this study is to provide this information obtained from the data of a new information system. The information used for this research was obtained from the Cost Efficiency and Competitiveness Information System of livestock activities in Mexico (SICEC in Spanish). 208 farms were selected of the largest cattle registry in the country and interviewed in the 2012 SICEC poll. The selection was randomly made in each of the following stratums: less than 49; 50 to 100 sows, 101 to 200, 201 to 500 and 501 or more sows. After removing farms selling piglets and introducing information filters, 82 farms were analyzed. The c/kg was calculated considering the costs of food, health, labor, equipment, facilities, services, reproduction and replacement. Comparison of means of the five stratums was performed using ANOVA with SPSS19® software. The total mean for c/kg was \$21.48 (±0.8) and the means in each stratum from lowest to highest were 23.5, 22.2, 20.6, 21.2, 21.1 pesos. When comparing the means between layers no significant difference was found (P=0.39). Breaking down the c/kg, the feeding costs represents the highest percentage with 81.26%. The c/kg obtained is similar to that reported by technified companies in Latin America in 2012 (\$20.9), and at reported by FAO for Mexico in 2011 (\$20.3). The c/kg average of \$21.48 is \$1 greater that average selling price of live pig kg in 2012. This reflects the great challenges that had to face the Mexican pork industry in 2012. The failure to find significant difference in costs of production between stratums denotes the absence of scale economies. In conclusion the SICEC proves to be a system that complements the few sources of economic and technical information existing in México.

## Session 42 Poster 10

**Are camel wrestling events able to save extinction of camel population in Turkey?**

*O. Yilmaz*[1] *and M. Ertugrul*[2]

[1]*Canakkale Onsekiz Mart University, Faculty of Agriculture, Department of Animal Science, Terzioglu Yerleskesi, Canakkale, 17020, Turkey,* [2]*Ankara University, Faculty of Agriculture, Department of Animal Science, Diskapi, Ankara, 06110, Turkey; zileliorhan@gmail.com*

During thousands of years of Turkish History, camels were always important in their life. In the past camels were used as transport, pack, ride, war, food, and sport animal by Turks. After industrialization and modernization since 20th century, camel lost their importance and nowadays they are only a sport and tourism material in Turkey. Hence the camel population was about 118,211 in 1937, but it decreased to 808 in 2003. Later than the number of camel population increased to 1.315 in 2012 because of popularity of camel wrestling events. The camel population is mostly used for camel wrestling events in Aegean Region and partially in Marmara and Mediterranean Regions. A small number of camels are used by nomadic Yoruk Turks in provinces of Antalya, Mersin and Mugla in Mediterranean Region. The camel wrestling events are organized about in 60-70 places annually during winter season when camels are in heat season. Wrestling events are on Sundays and follewed by not only men spectatorsi but also women and children. Because of this side, camel wrestling events is a family sport. Although camel wrestling equipments, accessories, ornaments, wages of takecarers, transport for wrestling from city to city, accommadation, catering are quite expensive, camel owners are not so rich people, but low or middle income people. Hence, those organizations and camel owners should be supported by the state more in order to survive this traditional event.

---

## Session 42 Poster 11

**What do young French people think about livestock productions in 2014?**

*A.L. Jacquot, Y. Le Cozler and C. Disenhaus*

*UMR1348 Agrocampus-Ouest INRA PEGASE, 65 rue de St-Brieuc, 35042 Rennes, France; anne-lise.jacquot@agrocampus-ouest.fr*

In France, as in many other countries, the gap between farmer's reality and citizen's expectations is increasing. As citizens and consumers are more and more concerned with livestock farm evolutions, decreasing this gap is of importance, and this should start at school level. However, information on feelings and knowledge of young people about livestock productions are scarce. Surveys and analysis were then performed by 32 master's level students from Agrocampus-Ouest. They conducted their work from 13th to 24th January 2014, to collect such informations. A closed-questionnaire was full-filled by French secondary school pupils (SSP) who will get their 'baccalauréat' (GCE A-levels) in June 2014. SSP were selected since they are soon becoming adult consumers and citizens (around 18 years of age). A total of 1,083 SSP from 28 different general (77%), farming (21%) and technical (7%) high-schools, located almost everywhere in France, were inquired. 46% of them lived in rural area, and 78% of them visited at least once a livestock farm (dairy or beef cattle, mainly). Knowledge on livestock originated from media (25%), family or neighbourhood (23%) or school (18%). Their opinion about livestock farming was good (25%) or neutral (45%). Two third of SSP estimated that faming is a good job. Regarding animal well-fare, they considered that broilers conditions are the worst (35%), whereas 38% of SSP thought that horses had the best living conditions. According to them, livestock has a negative effect on water (72%) and air (57%) quality, but a positive impact on tourism (38%) and landscape (37%). For them, eating meat is important and has to be regular (94%). The main concerns for them regarding livestock productions are those about animal welfare (80%), environment protection (54%) and food production (48%). Finally, 96% of them estimated that keeping livestock productions is of importance in France.

**Variation and repeatability of methane emission from dairy cows measured in two subsequent years**

*M.N. Haque, C. Cornou and J. Madsen*
*University of Copenhagen, Faculty of Health and Medical Sciences, Department of Large Animal Sciences, Groennegaardsvej 2, 1870 Frederiksberg C, Denmark; naha@sund.ku.dk*

The objective of this study was to investigate the individual variation and repeatability of $CH_4$ production in dairy cows measured in two different years. A total of 23 dairy cows were used, with an average body weight of 621±0.50 and 640±0.50 kg (mean ± se) and an energy corrected milk (ECM) production of 31.8±0.33 and 34.1±0.31 kg/day (mean ± se) during the 1$^{st}$ and 2$^{nd}$ year, respectively. Cows were offered a total mixed ration *ad libitum* and concentrate in the milking robot based on their milk production in both years. Dry matter intake (DMI) kg/day and ECM (kg/day) were significantly different ($P<0.001$) in the two years. $CH_4$ in l/day, l/kg DMI and l/kg ECM were also different in the two years ($P<0.001$). Mixed model results showed that most of the variation in $CH_4$ (l/day) is due to within cow variation (76-77%), whereas the variation between cows was significantly lower (23-24%) in both years ($P<0.001$). The diurnal pattern of $CH_4$ emissions (l/day) was very similar during the two years, and no difference was found for the emission among the days. When the amount of $CH_4$ was expressed in terms of ECM or DMI, the within cow variation was still greater than the between cows variation. $CH_4$ (l/kg ECM) was negatively correlated with the daily ECM (kg) production, whereas $CH_4$ (l/kg DMI) showed a positive correlation with DMI (kg/day). The repeatability of $CH_4$ (l/day) emission was higher (52%) during 2$^{nd}$ year as compared to the 1$^{st}$ year (47%). According to DMI (kg/day), ECM (kg/day) and $CH_4$ (l/day) in both years, no systematic ranking were observed for the cows. Only 13% of the cows were identified as low methane emitter. In conclusion, most of the variation in $CH_4$ (l/day) emission is associated to within cow variation and is mostly influenced by DMI and ECM production. The ranking of the cows indicates that methane measurements should be taken from a larger number of animals in order to select low emitter cows.

---



**Environmental impact of high standards of animal health and welfare in Dutch meat production chains**

*D. De Rooij[1], M. Smits[1], J. Scholten[2], A. Moerkerken[3] and J.W.G.M. Swinkels[1]*
*[1]HAS University of Applied Science, Onderwijsboulevard 221, 5223 DE 's Hertogenbosch, the Netherlands, [2]Blonk Consultants, Gravin Beatrixstraat 34, 2805 PJ Gouda, the Netherlands, [3]Netherlands Enterprise Agency, Croeselaan 15, 3521 BJ Utrecht, the Netherlands; d.derooij@has.nl*

Concerns about animal health and welfare in meat production chains are rising in Dutch society. In 2010, Dutch retail signed the agreement 'All fresh meat sustainable in 2020'.Together with the animal production chain they are working towards a more sustainable production of meat for the Dutch market. Moreover, in 2008 the Dutch government has signed an agreement with the Dutch agrosector to reduce GHG emissions and energy use with respectively 20% and annually 2% in 2020 versus 1990. The objective of this scenario study was to gain insight in the effects of high standards of animal health and welfare on GHG and energy use. For the production chains veal, pork and broilers three possible scenarios of improved animal health and welfare segments in retail where developed for the year 2020. Costs of implementing these scenarios were not part of this study. The baseline (normal production) of each meat production chain covers the situation in 2020 with no changes in way of production compared to 2010 except foreseen legal measures. Using Life Cycle Assessment methodology the environmental impact was calculated for the scenarios and the baseline. All phases of the production chain were in scope (cradle-to-gate), but the focus was on farm level. Compared to the baseline it was found that meat with high standards of animal health and welfare can be produced without a negative impact on GHG and energy use on farm level. High standards of animal health and welfare in general improved the gain to feed ratio. The improved gain to feed ratio is responsible for a reduced environmental impact when expressed per kg meat. However, additional environmental measures are needed to attain the reduction goals of GHG and energy use as agreed upon between the Dutch government and the agrosector.

**Partial life cycle assessment of greenhouse gases emissions in dairy cattle farms of Southern Italy**

*A.S. Atzori[1], M.G. Serra[1], G. Todde[1], P. Giola[1], L. Murgia[1], F. Giunta[1], G. Zanirato[2] and A. Cannas[1]*
*[1]University of Sassari, Dipartimento di Agraria, viale Italia 39, 07100 Sassari, Italy, [2]Filiera AQ srl, via Cadriano 27, 40127 Bologna, Italy; gabriserra@uniss.it*

Accurate estimates of carbon footprint (CF) at local level are needed to assess inventories and to plan mitigation strategies. In particular, little information is available on the CF of dairy cattle farms in Mediterranean production and climatic conditions. Thus, this work assessed the CF of dairy farms operating in Southern Italy (n=285; Sardinia, 83; Puglia and Basilicata, 88; Calabria, 44; Siciliy, 70). Farm data were collected with a direct in deep survey and included: farm characteristics, consistency and diets of each animal category, crop data and management, facilities and equipment, and fuel and energy consumption. The CF was estimated using a partial life cycle assessment approach (from cradle to farm gate) and applying a modified Tier 2 approach based on IPCC (2006) guidelines. Surveyed farms included 20,298 mature cows and 141.46 kiloton of fat and protein corrected milk (FCPM) sold. On average, farms had 71.2±4.6 mature cows, 44.4±2.9 ha, and produced 16.9±1.5 ton of milk per ha. Average milk yield was 6,100±0.12 (range: 1,170-11,100) kg/year per present cow. Feed efficiency was 1.16±0.01 kg of milk sold per kg of DM consumed by raised cattle. 95% of GHG were attributed to milk (energetic allocation). Average farm emissions were 1.66±0.4 kg of $CO_2$eq/kg of FPCM. Contribution of emissions sources was estimated at 0.71±0.014, 0.33±0.007, 0.18±0.023, 0.14±0.005 and 0.30±0.005 kg of $CO_2$eq/kg of FPCM for enteric production, manure management, energy, soil management, and purchased feeds, respectively. Farms with low CF also produced high amount of milk than others, thus the weighted mean of CF per kg of milk collected by the cooperatives was equal to 1.35 kg of $CO_2$eq/kg of FPCM. Herd production level was the variable that explained the most part of the variability in the dataset. This project was supported by FILIERA A.Q. srl.



**The effect of different dietary energy levels during rearing and mid-gestation on gilt performance**

*S.L. Thingnes[1], E. Hallenstvedt[2], E. Sandberg[3] and T. Framstad[3]*
*[1]Norsvin, P.O. Box 504, 2304 Hamar, Norway, [2]Felleskjøpet Agri SA, Flyporten, 2060 Gardermoen, Norway, [3]Norwegian University of Life Sciences, P.O. Box 5003, 1432 Ås, Norway; signe-lovise.thingnes@norsvin.no*

In an effective pig production, proper feeding management during rearing and first gestation is important to ensure a good lifetime performance in terms of a high number of weaned pigs per sow lifetime. Dietary energy level is one factor that can effect both development and production performance in gilts. The aim of the current study was to investigate the effect of different dietary energy levels offered during rearing and mid-gestation on development and performance of gilts, in a commercial herd. Five hundred Norwegian Landrace × Yorkshire gilts were followed from 25 kg live weight and until rebreeding or culling after weaning of their first litter. The study was designed as a 2×2 factorial, with two dietary energy levels offered during rearing; 13.2-29.0 MJ NE/d (High) or 10.6-22.9 MJ NE/d (Norm), and two dietary energy levels during mid-gestation (day 42-94); 27.3 MJ NE/d (High) or 22.3 MJ NE/d (Norm). This gave four gilt development strategies, High/High (HH), High/Norm (HN), Norm/High (NH), and Norm/Norm (NN). Data collection included individual gilt weight, age, backfat, productive and reproductive performance. Gilts offered the higher energy diet during rearing were younger ($P<0.001$) and had more backfat ($P<0.0001$) at selection for mating compared to norm reared gilts. Three weeks before expected parturition NN gilts were older ($P<0.001$) and leaner ($P<0.05$) compared to the HH gilts, with the HN and NH gilts at intermediate levels. Productive and reproductive performance was not affected by gilt development strategy, but the probability of death/removal was lower among the HH gilts compared to the other gilt development strategies ($P<0.05$). In conclusion, although maternal performance was not affected, gilt development strategy affected both development and survival probability.

**The effect of dietary energy levels during rearing and first gestation on sow lifetime performance**

*S.L. Thingnes[1], A.H. Gaustad[1], N.P. Kjos[2], E. Sandberg[2] and T. Framstad[2]*
*[1]Norsvin, P.O. Box 504, 2304 Hamar, Norway, [2]Norwegian University of Life Sciences, P.O. Box 5003, 1432 Ås, Norway; signe-lovise.thingnes@norsvin.no*

There is an unfortunate conflict between the selection for leaner animals, and the requirement of having breeding females in a condition that promotes and maintains reproductive performance through successive parities. The aim of the current study was to investigate how first parity manipulations of dietary energy level would affect sow lifetime performance and longevity. Five hundred Norwegian Landrace × Yorkshire sows were followed from 25 kg live weight and until rebreeding or culling. The study was designed as a 2×2 factorial, with two dietary energy levels offered during rearing; 13.2-29.0 MJ NE/d (High) or 10.6-22.9 MJ NE/d (Norm), and two dietary energy levels during mid-gestation in parity one (day 42-94); 27.3 MJ NE/d (High) or 22.3 MJ NE/d (Norm). This gave four gilt development strategies, High/High (HH), High/Norm (HN), Norm/High (NH), and Norm/Norm (NN). Data collection included measures of individual sow weight, age, backfat, productive and reproductive performance throughout successive parities. The HH sows reached all defined reproductive stages at lower ages compared to the NN sows ($P<0.05$), with the HN and NH sows at intermediate levels. Throughout successive parities the HN sows remained the overall lightest and leanest group of sows. The HH sows produced more born alive piglets per sow lifetime compared to the HN sows ($P<0.05$), with the NH and NN sows at intermediate levels. Lifetime performance, in terms of litter weights and weaning-to-service interval, was not majorly influenced by dietary treatment. The overall survival probability among sows did not differ between dietary treatments ($P>0.05$), but less HH sows were culled in lower parities compared to the other treatment groups. In conclusion, the HH development strategy ensured the highest number of born alive piglets per sow lifetime, and these sows also had a higher survival probability in lower parities.



**Nutrient balances of energy, lysine and nitrogen in late gestating and early lactating sows**

*T. Feyera and P.K. Theil*
*Aarhus University, Animal Science, Blichers allé 20, 8830 Tjele, Denmark; peter.theil@agrsci.dk*

Nutrition during the transition period is important for reproductive success. In this study, requirements (req) for dietary energy (E), nitrogen (N) and lysine for foetal growth, mammary growth, colostrum production, uterine tissues (including placenta, fluids and membranes), milk production and sow maintenance were calculated factorially for transition sows based on available data and new data on colostrum and milk yields. One week before parturition, E for maintenance accounted for 79% of the total req, whereas E required for colostrum and foetal growth amounted to 9% and 6%, respectively, and mammary growth and uterine tissues to 4% and 1%. Colostrum production (32%), mammary growth (27%), foetal growth (22%) and sow maintenance (16%) had the highest impact on the total lysine req, whereas uterine tissues (3%) were of minor importance. Urinary loss of N (40%) had a major impact on N req, while colostrum production (23%), foetal growth (22%) and mammary growth (12%) contributed less to the overall N req and uterine tissues only required 3%. From d 101 to 115 of gestation, E, N and lysine req increased by 3%, 16% and 27%, respectively. During the first week of lactation, E-, N and lysine requirements increased daily due to increasing milk production. At day 7 of lactation, 69% of the total E req was associated with milk production (54% secreted + 15% heat), while 30% was required for sow maintenance. Furthermore, 70% of the total N-req was associated with milk production and 29% was lost in urine. In contrast, 94% of the lysine was required for milk production and only 5% for sow maintenance. Finally, the regressing uterus contributed with approx. 13% and 9% of the daily N and lysine req, respectively, during the first week after parturition. From parturition to d 7 of lactation, E-, N- and lysine req increased as much as 3, 6 and 20-fold, respectively. In conclusion, dramatic changes in nutrient req occur during transition.

## Session 43 Theatre 4

### The ratio between energy and lysine is essential to improve milk yield in sows

*C. Flummer[1], T. Feyera[1], U. Krogh[1], T. Hinrichsen[2], T.S. Bruun[3] and P.K. Theil[1]*
*[1]Aarhus University, Blichers Allé 20, 8830 Tjele, Denmark, [2]DLG, Vesterbrogade 4 A, 1620 København V, Denmark, [3]Danish Pig Research Center, Axeltorv 3, 1609 Copenhagen, Denmark; christine.flummer@agrsci.dk*

Nutrients supplied through the feed and body mobilization are important for milk production in sows. However, it is unknown whether mobilization of body fat or body protein is most beneficial for achieving a high milk yield. The present study was conducted to study how mobilization of body protein and fat affects sow milk yield. Twelve sows were fed 70% of calculated energy requirements and supplied with either 70% (LL) or 86% (HL) of calculated lysine requirements. Both groups were designed to mobilize from body tissues; the HL mainly from fat depots and the LL from both fat and muscle. Sows were fed a basal diet and a lactation supplement, targeting the nutrient requirements for sow maintenance and milk production, respectively. On average, the daily feed intake was 6.13 and 5.49 kg/d, and supplied 798 and 807 g/d of SID crude protein, and 47 and 52 g/d of SID lysine, for the LL and HL group, respectively. The HL group lost numerically more backfat (3.7 mm) than LL sows (2.0 mm), although it was not significantly different (P=0.26). The LL sows weaned 13.2 pigs per litter, had higher milk production throughout lactation, and an average daily milk production of 13.0 kg, compared to 11.6 kg milk/d (P=0.01) in the HL group (12.2 weaned pigs per litter). Sow body weight loss from day 2 to 28 of lactation was similar (32 kg in both groups). All LL and HL sows returned to estrus within 5 days after weaning. The current results demonstrated that body mobilization is important for the milk yield. Furthermore, data indicate that sows are not able to compensate for insufficient energy supply. Milk yield seems to depend on an optimal ratio between fat and protein mobilization, which can be achieved by selecting the right lysine to energy ratio.

---

## Session 43 Theatre 5

### The effect of dietary valine-to-lysine ratio of lactating sows on milk composition and piglet gain

*A.V. Hansen[1], T.S. Bruun[2] and C.F. Hansen[1]*
*[1]University of Copenhagen, Grønnegaardsvej 2, 1870 Frederiksberg, Denmark, Denmark, [2]Danish Pig Research Centre, Danish Agriculture & Food Council, Axeltorv 3, 1609 Copenhagen, Denmark; avha@sund.ku.dk*

The aim of this study was to determine the effect of dietary valine-to-lysine ratios (Val:Lys) during lactation on milk composition and litter growth rate of high producing lactating sows. Sixty $2^{end}$ parity sows were allotted to one of six dietary treatments varying in standard ileal digestible Val:Lys (76, 79, 82, 85, 91, and 97%) from day 2 postpartum. The sows were fed semi *ad libitum* 2 times per day until day 10 and from day 10 three times per day. Litters were equalized at day 2 to 14 piglets and weaned at day 28. A milk sample was obtained on day 17-18. The litters were weighed at equalization and at weaning and the average daily gain (ADG, kg/day) was calculated. The litter was removed from the sow 45 min before milk sampling and after 45 min an intramuscular injection with 2 ml oxytocin was given. Milk was sampled from 3-5 glands. The milk was analyzed for dry matter (DM), protein, fat, lactose and urea (MilkoScan FT2). Data was analyzed in R using a linear model testing the effect of dietary treatment and block. Pearson's correlations were calculated to test for correlations between response variables. Preliminary results on milk composition were available for 14 sows. There was no effect of block (P>0.05). There was no effect of dietary Val:Lys on milk composition (P>0.05). Dry matter content was correlated with lactose (r=0.79, P<0.05) and fat (r=0.77, P<0.05) percentage. Sows fed the diet with lowest Val:Lys (76%) had the highest fat (9.2%) and DM (26.2%) content of the milk and sows fed the diet with the highest Val:Lys (97%) had the lowest fat (7.4%) and DM (22.0%) content in milk. The highest ADG of piglets (3.49 kg/d) was seen in the lowest Val:Lys (76%) and the lowest ADG was in the second highest Val:Lys group (91%). In conclusion, the preliminary results of the study shows no effects of dietary Val:Lys ratio on milk composition and ADG of the piglets.

**Preweaning mortality in SWAP farrowing pens**

*J. Hales[1], V.A. Moustsen[2], M.B.F. Nielsen[2] and C.F. Hansen[1]*
*[1]University of Copenhagen, Department of Large Animal Sciences, Groennegaardsvej 2, 1870 Frederiksberg, Denmark, [2]Pig Research Centre, Axeltorv 3, 1609 Copenhagen, Denmark; cfh@sund.ku.dk*

Confining sows in farrowing crates increases stress and affects welfare negatively. However, piglets may be at greater risk of dying in loose systems, especially in early lactation. The objective of this study was to investigate if short term confinement around farrowing would decrease piglet mortality and increase number of weaned piglets in a farrowing pen where sows could be confined temporarily (SWAP=Sow Welfare And Piglet protection). The study was conducted in a 1,200 sow piggery where sows were randomly allocated to one of three experimental treatments: loose-loose (LL, n=376), loose-confined (LC, n=369) or confined-confined (CC, n=354). In LL sows were loose from placement in the farrowing unit to weaning, LC sows were loose to birth of the last piglet and thereafter confined to day 4 post farrowing, and CC sows were confined from day 114 of gestation to day 4. All sows were loose housed from day 4 to weaning. Live born and stillborn piglets, equalised litter size and dead piglets were recorded. All data were analysed using linear models in SAS. As expected, the number of total born piglets per litter (17.4±0.10; P=0.55) did not differ between treatments. However, the number of dead piglets per litter before equalisation was greater in LL (2.0 piglets/litter (1.8-2.1)) and LC (1.8 piglets/litter (1.7-1.9)) than in CC (1.3 piglets/litter (1.2-1.4); P<0.001). Piglet mortality from litter equalisation to day 4 was higher in LL (7.6% (7.1-8.7)) than in CC (6.0% (5.3-6.8); P<0.001) but did not differ between LC (7.0% (6.2-7.8)) and the other two treatments. From day 4 to weaning the piglet mortality rate was 5.8% in all treatments (P=0.98). Consequently there were more weaned piglets in CC (13.7±0.19) than in LL (12.9±0.19; P=0.005) but there was no difference between LC (13.1±0.19) and LL or CC (P>0.05). In conclusion, confinement of sows around farrowing reduced piglet mortality compared to loose housed sows.

---



**Saliva cortisol levels of nurse sows and ordinary sows through lactation**

*C. Amdi[1], V.A. Moustsen[2], G. Sørensen[2], L.C. Oxholm[2] and C.F. Hansen[1]*
*[1]University of Copenhagen, Grønnegårdsvej 2, Frb C, Denmark, [2]Pig Research Centre, Axeltorv 3, Kbh V, Denmark; ca@sund.ku.dk*

Nurse sows are used in piggeries with hyper-prolific sows to manage large litters. It is, however, not known if nurse sows experience prolonged stress by having to stay in farrowing crates beyond the normal weaning time. The aim of this study was to quantify the long-term saliva cortisol response as a measurement of the nurse sows stress level compared to ordinary sows (OSOW) weaning their piglets at d 25. In Denmark cascade fostering using two lactating sows are normally performed. The first nurse sow (NURSE1) has her own piglets removed after a week and receives surplus newborn piglets that she fosters until weaning. The second nurse sow (NURSE2) weans her own litter after 21 days and receives the litter from NURSE1 which she rears until weaning. In total 60 sows (n=20) were randomly allocated to become an OSOW, NURSE1 or NURSE2. Saliva cortisol was collected on d 6, 13, 20 and 24 postpartum at 10 h, 13 h and 16 h for all sows and pooled on a daily basis for analyses. Additional samples were taken on d 31 for NURSE1 and NURSE2 and d 38 for NURSE2. Saliva samples were assayed for cortisol levels using a Salivary Cortisol EIA kit (Salimetrics, UK). Cortisol data were log transformed and analysed using proc mixed in SAS. Results showed that there was no effect of group but there was an effect of day (P<0.001) with saliva cortisol declining throughout lactation with values of 19.9, 17.1, 12.8 and 10.4 nmol/l (back transformed values) at d 6, 13, 20 at d 24, respectively. NURSE1 tended to have lower values (8.3 nmol/l) on d 31 than on d 24 (11.5 nmol/l; P=0.08). NURSE2 had lower cortisol values on d 38 (7.4 nmol/l) and on d 31 (7.5 nmol/l) than on d 24 (11.1 nmol/l; P<0.05). Results indicate that saliva cortisol levels decline throughout lactation and that there is no difference in saliva cortisol levels between OSOWS and nurse sows. In conclusion, saliva cortisol levels indicate no additional long-term stress of being selected as a nurse sow.

**Effect of radiant heating at the births place of newborn piglet**

*H.M.-L. Andersen and L.J. Pedersen*
*Aarhus University, Dept. of Animal Science, Blichers Allé 20, Postboks 50, 8830 Tjele, Denmark; heidimai.andersen@agrsci.dk*

It has been documented that heat in the floor of the farrowing area in loose housed sows improves survival of piglets significantly. However, today the majority of farrowing pens are designed with crating of sows and slatted floor. We therefore investigated if radiant heating behind the sow reduced hypothermia in new born piglets and increased milk intake measured as growth during the first wk. Second parity LY sows (n=36) were randomly divided into two groups 'Control' and 'Heat' (radiant heat panels mounted above the slatted floor behind the sow). The heater was turned on at birth of first piglet and turned off 12 h after. Birth time, time to reach heated area, time to first contact with udder and time to first suckling was registered. The piglet's rectal temperature (RT) was measured 15, 30, 60, 120, 180, 240 min after birth and 12 and 14 h after birth of first piglet. Piglets were weighed 1, 24 and 48 h and 7 d after birth. Data was analysed in a mixed model in SAS. The preliminary results showed that RT was significantly higher in Heat compared to control piglets from 30 min to 4 h after birth ($P<0.05$). Heat piglets also stayed longer in the heated zone behind the sow than control piglets (H: 12.7±0.02 min. vs C: 7.8±0.02 min., $P=0.02$), whereas time to reach the udder (H: 19.2±0.02 min. vs C: 16.8±0.02 min., $P=0.39$) and to suckle (H: 43.7±0.02 min. vs C: 47.8±0.02 min., $P=0.42$) did not differ. Heat piglets compared to control piglets were 147 g smaller at birth ($P=0.05$) while they were heavier at 24 h (C: 1272±30 g vs H: 1325±31 g, $P=0.05$) and 48 h after birth (C: 1305±53 g vs H: 1441±54 g, $P=0.02$) while no difference was found at 7 d. ($P=0.31$). Low birth weight strongly reduced RT, increased time to first suckle and reduced weight at all time measured ($P<0.0001$). The results showed that radiant heating behind the sows reduced hypothermia in new born piglets and improved initial milk intake.



**The protein level fed to mink dams before implantation affects fetal survival rate**

*C.F. Matthiesen and A.-H. Tauson*
*University of Copenhagen, Department of Veterinary Clinical and Animal Sciences. The faculty of Health and Medical Sciences, Grønnegårdsvej 3, 1st floor, 1870 Frederiksberg C, Denmark; cmt@sund.ku.dk*

The mink is a seasonal breeder with one annual breeding season. It has induced ovulation and delayed implantation. Once implantation has taken place the gestation is completed by the true gestation within 30 days. The mink is a strict carnivore and has high requirements for dietary protein which varies with life stage. The protein and amino acid requirements are however, still not completely known in all parts of the mink production cycle. According to Danish legislation there is a demand to reduce the overall nitrogen emission and it is therefore important to investigate whether the levels of protein commonly used today can be reduced. Our objective was to determine the protein requirement before implantation needed to support a good reproductive performance. Six levels of protein (20, 25, 30, 35, 40 and 45% of metabolisable energy – ME- from protein) were fed to 96 female mink from the 24 February until the 10 April. Three females from each treatment were euthanized the 16 April to investigate the implantation rate, fetal survival, fetus length and weight. The remaining dams were used to measure the reproductive performance, kit survival rate and pre-weaning kit growth. The number of implanted fetuses was not affected by the protein provision whereas the survival rate of implanted fetuses was significantly ($P=0.02$) lower in the group fed the 20% of ME from protein compared to the other groups. The protein requirement before implantation, estimated by a broken line linear regression approach from the fetus survival rate, indicate that the requirement was 30.5% of ME from protein. The reproductive performance was not inferior in the group fed 25% compared to the groups fed 30% and 35% of ME from protein. Inexplicably was however, an instance of a significantly higher number of barren dams and a tendency towards increased kit loss in the group fed 30% of ME from protein compared to the other groups.

**Behaviour and performance of sows housed with continuous or temporary confinement during lactation**

*C. Lambertz[1], M. Petig[1], A. Elkmann[2] and M. Gauly[1,3]*
*[1]University of Göttingen, Department of Animal Sciences, Albrecht-Thaer-Weg 3, 37075 Göttingen, Germany, [2]Big Dutchman Pig Equipment GmbH, Auf der Lage 2, 49377 Calveslage, Germany, [3]Free University of Bolzano, Universitätsplatz 5I, 39100 Bozen, Italy; clamber2@gwdg.de*

Alternatives to farrowing crates with continuous confinement of sows are needed because the animal welfare is negatively impacted. The aim of this study was to investigate the effect of fixation for different periods post partum (pp) on the sow behaviour and piglet performance. Gilts from a conventional herd were observed from 5 to 26 days pp and housed in farrowing crates (1.80×2.50 m) that could be altered between conventional farrowing crates and free-farrowing pens. Animals were divided into 3 groups, that were either crated continuously (group A; n=55), until 14 days pp (group B; n=54) or 7 days pp (group C; n=59). During lactation piglets were weighed weekly and mortality rates and causes as well as weight of gilts recorded. Additionally, the behaviour (lying in lateral, abdominal or semi-abdominal position, standing and sitting) of 6 gilts per group was video recorded and analysed with the time-sampling technique. Weight gains of piglets ranged during the observation period between 220 to 225 g/d and gilts lost on average 31.8 kg during lactation without differences between groups (P>0.05). Mortality rates were 11.4, 12.9 and 13.3% for groups A, B and C (P>0.05). Piglet losses were mainly caused by crushing the piglet to death (85%) and occurred in the first week of life, in which sows of all groups were crated. Furthermore there were no differences between the groups in sow behaviour (P>0.05). Sows spent 72 to 76% of the time lying in lateral, 14 to 17% lying in abdominal or semi-abdominal position, 9 to 10% standing and 1 to 3% sitting. Only sows in group B were lying longer in week 3 and 4 of lactation compared to A- and C-sows (P<0.05). In conclusion, loose-housing of gilts after a pp crating period did not affect piglet's performance and mortality and the behaviour of the gilts.

---



**Effects of arginine and carnitine supplementation on growth performance in low birth weight piglets**

*J.G. Madsen[1,2], S. Müller[1,2], M. Kreuzer[1] and G. Bee[2]*
*[1]ETH Zurich, Department of Environmental Systems Science, Universitätstrasse 2, 8092 Zurich, Switzerland, [2]Agroscope, Milk and Meat Production Research Department, P.O. Box 64, Route de la Tioleyre 4., 1725 Posieux, Switzerland; johannes.madsen@agroscope.admin.ch*

Selection for prolificacy in sows has increased litter size at birth. This and decreased average birth weight (BtW) has resulted in low survival rates within the first week of the piglets' life. The objective of this study was to compare 2 dietary supplements to an unsupplemented control offered in the early postnatal period on growth performance, organ and semitendinosus muscle (STM) development of low BtW piglets. A total of 30 female and castrated male piglets were weaned at d 7 of age and assigned for 21 d to 3 milk replacer diets containing either no supplements (CTRL), carnitine (CAR, 0.4 g/d) or arginine (ARG, 0.1 g/kg body weight/d). Piglets were weighed daily and supplements were adjusted accordingly. Intake of feed, which was offered 6 times per day, was determined daily. At d 28 of age piglets were euthanized and complete STM and organs were weighed. Data were analysed using the lmer and lm procedures in R. CAR piglets showed the greatest growth rate in the first and second week (128 vs 95, 84 g/d, P<0.05 and 158 vs 129, 112 g/d, P=0.10) and numerically displayed greatest final body weight (4.85 vs 4.71, 4.27 kg.), which is in line with the greatest STM weight (12.9 vs 12.7, 11.8 g), greatest growth rate (144 vs 141, 122 g/d) and greatest feed efficiency (2.10 vs 2.29, 2.36) compared to ARG and CTRL, respectively. Treatments had no effect on final body weight (P=0.32), overall growth rate (P=0.30), feed intake (P=0.39), feed efficiency (P=0.55) and weight of the STM (P=0.73), organs (P>0.17) and brain (P=0.57). However, regardless of the dietary treatment, brain:STM ratio was greater (P<0.05) in male than female piglets. Although not conclusive, these results suggest that carnitine may be a promising candidate to improve postnatal development of low BtW piglets.

## Session 43 Poster 12

**_Scutellaria baïcalensis_ extract improves milk production in hyperprolific sows**

*L. Boudal and L. Roger*
*CCPA Group, Swine Department, ZA du bois de Teillay, 35150 Janze, France; lboudal@ccpa.fr*

In hyperprolificity context, the amount of milk available per piglet decreases, compromising the survival, health and performances of the litter, with a special negative impact on lighter piglets. In dairy cows, inflammation increases after parturition and some experiments show that this inflammation process can be negatively linked to milk production. A trial was conducted to evaluate the relation between inflammation and milk production of hyperprolific sows. The potential effects of a plant extract (Sb) on lactation performances were studied. 43 LW × L hyperprolific sows were assigned into 2 different groups depending on backfat thickness and parity: 21 sows in the trial group (Sb) and 22 in the control group (CO). From the entry in the farrowing facility, the lactation feed of Sb group was supplemented with a *Scutellaria baïcalensis* extract, selected for its anti-inflammatory properties. Blood samples collected from each sow 6 days after farrowing were used to determine concentrations of serum amyloid A (SAA). Sow rectal temperature was measured 1, 2 and 3 days after farrowing. Individual body weight of piglets was measured at birth, 48 h and weaning. The average daily weight (ADG) was calculated from birth to 48 h (b) and from 48 h to weaning (w). Data were analyzed by ANOVA using the MIXED MODEL procedure. In multiparous sows, high SAA concentration was correlated with a decrease in ADGw of piglets ($P<0.05$). Sow rectal temperature decreased faster in Sb group than in CO group ($P<0.05$). Piglets of the Sb group showed higher ADGb ($P<0.001$) and ADGw ($P<0.001$) (126.04 g/d; 264.04 g/d) compared with CO group (107.96 g/d; 235.34 g/d). Body weight birth classes analysis showed an increase of ADGw for each category in Sb group. These results indicate that inflammation occurring the days following farrowing has a negative impact on milk production and the use of a specific *S. baïcalensis* extract improves growth of piglets through an enhanced milk production of sows.

---

## Session 43 Poster 13

**Review of scientific knowledge on the nutrition of the highly prolific modern sow**

*A. Craig[1,2] and E. Magowan[1,2]*
*[1]Queens University, Belfast, BT7, 1NN, Northern Ireland, United Kingdom, [2]Agri-food and Bioscience Institute, Hillsborough, BT26 6DR, Northern Ireland, United Kingdom; aleslie02@qub.ac.uk*

Genetic selection for prolificy in sows has increased litter size, now commonly exceeding 12 pigs weaned per litter. A systematic literature search was conducted to identify nutrient requirements for sows rearing 12 or more pigs. The search terms 'sow' AND 'lactation' AND 'diet' OR 'nutrition' were entered into Web of Science, returning 725 records. The terms 'sow' AND 'lactation' AND 'birth' OR 'wean×weight' were also searched, returning 699 results. The combined searches gave a total of 1,052 records. The data set was first refined to records dated from 2012, as it was hypothesised these papers would have the best probability of reporting nutritional effects on high litter sizes. From these 131 papers only 3 references reported nutritonal effects on piglet performance based on litter sizes of 11 or more pigs weaned per sow. In terms of what was being investigated, additional fat in lactating sow diets has been a popular area of research from 2010. However, research on protein and amino acid requirements has been lacking in recent years. The majority of papers studying protein were published from 2000-2009 when litter sizes averaged 9 weaned per sow. As litter sizes have increased by 30%, more research on the protein needs of sows is required. This is critical, as the limited research in this area suggests that amino acids other than lysine may become a limiting factor for mammary tissue synthesis and milk production. It is known that a high weaning weight will support optimum lifetime performance and that piglets have potential to grow at extraordinary rates pre weaning provided milk is available. Lactation diets to maximise mammary structure and milk yield for sows rearing large litters (13 or more piglets) requires more investigation. It is suggested that protein requirements, in particular the ratio of amino acids, should be focal.

## Session 43 Poster 14

### Piglet colostrum intake and birth weight are equally important for weaning weight

*U. Krogh[1], T.S. Bruun[2], C. Amdi[3], J. Poulsen[2,4], C. Flummer[1] and P.K. Theil[1]*
*[1]Aarhus University, Blichers Allé 20, 8830, Denmark, [2]Danish Pig Research Center, Axeltorv 3, 1609, Denmark, [3]University of Copenhagen, Grønnegårdsvej 2, 1870, Denmark, [4]DLG, Vesterbrogade 4A, 1620, Denmark; uffek.larsen@agrsci.dk*

The colostrum period is vital for short term piglet survival, but may also be important for performance on a longer term. Data from 166 piglets was used to study characteristics of neonatal piglets and their possible effect on piglet weaning weight, as a part of a nutrition experiment on transition and lactating sows. Colostrum intake (CI) of individual piglets was estimated based on birth weight ($BW_{Birth}$) and piglet weight gain during the 24 h period after onset of parturition. Blood was sampled by jugular vein venipuncture at birth and immediately analyzed for glucose, lactate, oxygen, carbon dioxide and pH on a blood analyzer. Piglets were weighed at weaning (d 28). Relations between variables were evaluated by Pearson coefficient of correlation using the Proc Corr procedure of SAS. The response variable, piglet weaning weight (BWd28), was analyzed using the Mixed Procedure of SAS including block and sow as random effects. Piglet $BW_{Birth}$ ($r=0.51$, $P<0.001$), CI ($r=0.49$, $P<0.001$) and lactate concentration ($r=-0.16$, $P<0.05$) were correlated with $BW_{d28}$. Piglet $BW_{Birth}$ ($P<0.001$), CI ($P=0.001$) and lactate ($P=0.107$) were included in the full model (SE in parenthesis). No interactions were significant. $BW_{d28}$ (g) = 2,575 (807) + 2.71 (0.71) × $BW_{Birth}$ (g) + 5.05 (1.50) × CI (g) – 77.72 (48) × lactate (mM) Addition of glucose, oxygen, carbon dioxide and pH to the model did not contribute with additionally information to $BW_{d28}$. The trend of lactate concentration on $BW_{d28}$ may be linked to birth related problems. Though a correlation between $BW_{Birth}$ and CI was observed ($r=0.67$, $P<0.001$), data suggest distinct effects of $BW_{Birth}$ and CI on $BW_{d28}$. In conclusion $BW_{Birth}$ and CI were equally important for piglet weaning weight.

## Session 43 Poster 15

### Evaluation of an on-farm method to assess colostrum quality in sows

*A. Balzani[1], H.J. Cordell[2] and S.A. Edwards[1]*
*[1]School of Agriculture, Food and Rural Development, Newcastle University, NE1 7RU, United Kingdom, [2]Institute of Genetic Medicine, Newcastle University, NE1 3BZ, Newcastle upon Tyne, United Kingdom; a.balzani@ncl.ac.uk*

The increase in sow prolificacy over the last decade, as a result of genetic selection strategy in specialised damlines, now poses challenges for piglet survival. Improved transfer of passive immunity from colostrum can improve piglet survival and subsequent health. The aim of this experiment was to compare the radial immunodiffusion (RID) method for determination of colostrum Ig content with an inexpensive and quick on-farm method, the Brix refractometer, used previously in cattle. Colostrum samples (10 ml) were collected across all teats at ±8 h before and after birth of the first piglet on 52 Landrace × Large White sows of mixed parities. After the samples were drawn, they were divided into three portions; one was immediately analysed, a second was refrigerated for 24 hours and then analysed, and the remaining sample was frozen and stored at -20 °C until analysis. Fresh and refrigerated colostrum samples were analysed at the farm with a Brix refractometer. Frozen samples were shipped to the laboratory and IgG was analysed by using a commercially available RID kit. The Brix scores ranged from 18.3% to 33.2%, showing the variability of colostrum composition. The fresh, refrigerated and frozen samples had a Brix percentage (mean ± SD) of 24.1±2.9, 23.2±3.11, 24.2±3.39, respectively ($P>0.05$). The correlation between readings for fresh samples and those stored at different temperature was very strong ($r=0.93$, $P<0.001$). IgG concentration measured by RID (22.4±4.08) ranged from 13.27 to 35.08 mg/ml. The Brix percentage was strongly correlated ($P<0.001$) with RID results (regression equation: RID = 1.32 (±0.2) Brix + 9.50 (±4.7); $R^2=0.52$). The results of this study indicate that the Brix refractometer provides a simple, rapid, inexpensive measure of colostrum quality in sows. The ease and low cost of this technique make it ideal as a tool that could be used in a commercial setting as well as a research environment.

### The volume and economic efficiency of production in reconstructed piggery using hyperprolific sows

*M. Sviben*
*Freelance consultant, Siget 22B, 10020 Zagreb, Croatia; marijan.sviben@zg.t-com.hr*

In 2008 the farm in Gradec, Croatia, was reconstructed in order to improve the production volume and economic efficiency. The process included increasing the number of farrowing pens and usage of more prolific sows. After the reconstruction the number of farrowing pens was increased from 500 to 630 and as the breeding material the hybrid hyperprolific sows were used. According to the data, published by Croatian Agricultural Agency in their yearbooks, the production volume of 53,219 weaners was achieved on an average and the mean of 11.661 weaners was attained in years 2010-2012. In years 1986-1991 the swine reproduction method I was performed. The achieved production volume was 52,055 weaners a year. Within those six years the registered mean number of weaned piglets per litter was 8.390. In years 1992-1999 the swine reproduction method changed. Attained mean annual production volume was 52,801 weaners. The average of 8.205 weaners per litter was achieved. Obviously, 26% higher number of the farrowing pens and 39-42% higher number of piglets weaned per litter in years 2010-2012 did not result with 26-79% higher production volume. The aim of this research was to establish the indices of the production economic efficiency and to explain why the higher number of pens and more prolific females did not give more productive results. Indices of the economic efficiency ($E_P$: number of weaned piglets/pen/year) were as follows: 84.87 in reconstructed farm, 104.1 in years 1986-1991 and 105.6 in years 1992-1999. The reason for diminished production economic efficiency in percentage between 18.47 and 19.63% was the increase of the occupation period duration of farrowing pens per shift. Smaller number of shifts per pen a year (41.62-43.71%) with more weaners per litter (38.99-42.12%) gave only a slightly higher production volume (0.79-2.24%) than the one before the farm reconstruction and the one required for the investment in the reconstruction to be paid out (103 weaned piglets/pen/year).

---



### Fatty acids contents in NEFA, in triglycerides and in the red blood cells in horses offered linseed oil

*S. Patoux, Y. Mehdi and L. Istasse*
*Nutrition Unit, Animal Porduction, B 43, Veterinary Faculty, University of Liège, 4000 Liège, Belgium; spatoux@ulg.ac.be*

Diets high in concentrates are usually offered to sport horses during the training and the racing seasons. In order to increase the energy content and to provide essential fatty acids, fats are incorporated with the cereals. By doing so, it is expected that the starch content and the associated disturbances are reduced. In this study 8 adult horses were used during 4 months. The diet was made of 50% grass hay and 50% concentrate. The control concentrate was composed of 48% of whole spelt, 48% of rolled barley, 3% of molasses and 1% of mineral mixture. In the treated group with oil, 8% barley was substituted by 8% first pressure linseed oil. Individual fatty acids were determined in the NEFA, in the triglycerides and in the red blood cells by HPLC. There were no effects of the oil supplementation on the fatty acids profiles in the NEFA fraction except for a significant increase in the C18:3 n-3 (0.22 vs 0.06 mg/dl, $P<0.001$). By contrast linseed oil increased or reduced significantly the content of most of the fatty acids both in the triglycerides and in the red blod cells fractions ($P<0.05$, $P<0.01$ or $P<0.001$). Interestingly in the triglycerides, the C18:2 n-6 increased from 80.6 to 94.8 mg/dl and the C18:3 n-3 from 2.62 to 7.47 mg/dl. In the red blood cells fraction the most significant effects were reductions in the saturated fatty acids contents except for C18:0. There were also reductions in the n-6 unsaturated fatty acids fractions; in C 20:3 (0.31 vs 0.22 mg/g) and in C20:4 (1.78 vs 1.09 mg/g). In the n-3 unsaturated fatty acids of the red blood cells, linseed oil induced a significant increase in C18:3 (1.67 vs 2.65 mg/g) and a significant reduction in C22: 6 (0.10 vs 0.07 mg/g). It can be concluded that the feeding of sport horses with linseed oil influenced the fatty acids profiles both in the plasma lipid components and in the red blod cells membranes.

**Effect of dietary omega-3 fatty acids on insulin and ovarian dynamics in transition dairy cows**

*A. Towhidi, S. Cas-Aghaee, M. Ganjkhanlou, H. Kohram and H. Javaheri Barforoushi*
*University of Tehran, Department of Animal Scienec, College of Agriculture and Natural resources, Karaj, 3158777871, Iran; atowhidi@ut.ac.ir*

The purpose of this study was to evaluate the ovarian and insulin dynamic in transition dairy cows fed omega-3 fatty acid source. 10 multiparous Holstein cows were selected and after adaptation period the cows were randomly given diets containing palm oil as a source of saturated fatty acids (PO, n=5) and fish oil as a source of omega-3 fatty acids (FO, n=5). The diets were given from day -40 to 60 after calving. For the determination of the glucose-induced insulin responses of the dairy cows at the different time points relative to calving, intravenous glucose tolerance tests were performed on days 14 and 42 after calving. Plasma insulin and glucose responses to the glucose load was estimated by the peak concentration, the area under the curve (AUC), and the clearance rates of glucose. The cows were synchronized using two injections of prostaglandin on day 20 and 34 after calving and consequently, ovarian dynamics were assessed using an ultrasonography instrument. Blood samples were obtained from coccygeal vein on days -35, -21, 0, 21, 42 and 63 days after calvinge. Experimental diets did induce any significant effect on the plasma concentration of cholesterol, triglycerides, HDL cholesterol. However, FO significantly decreased the levels of LDL cholesterol ($P<0.05$). FO induced decrease in glucose tolerance as reflected by lower glucose-induced insulin AUC, peak concentration of glucose and increase of glucose inclearance. Both reatments did not affect milk production, although milk yield was non-significantly higher in the fish oil group. Milk fat percentage was significantly decreased in the FO group compared to PO group ($P<0.05$). There were no significant differences between two groups in the number of small (less than 6 mm), medium (between 7 and 9 mm) and large (greater than 10 mm) follicles. According to the results, we suggested that feeding fish oil could improve insulin resistance in transition dairy cows.

---



**Effects of a mycotoxins-binder on plasma biochemistry in early lactating dairy cattle**

*F. Abeni, F. Petrera, A. Dal Prà, A. Gubbiotti, G. Brusa and M. Capelletti*
*Consiglio per la Ricerca e la Sperimentazione in Agricoltura, Centro di Ricerca per le Produzioni Foraggere e Lattiero-casearie, Via Porcellasco 7, 26100 Cremona, Italy; fabiopalmiro.abeni@entecra.it*

The contamination of feeds with mycotoxins from Fusarium spp is a problem of growing interest for dairy cattle. On-farm, a widely adopted solution to reduce the toxic effects of mycotoxin in cows is the addition of a mycotoxin-binder to the contaminated feed. Production and health problems from Fusarium mycotoxins can be managed by the introduction of polymeric glucomannan-based adsorbents (PGA) in the feed preparation. Many adsorbents show the ability to bind one or more mycotoxins, but these feed additives may also have some adverse effects, for example, interfering in the availability of essential nutrients (generally minerals) to the animal. A preliminary evaluation of possible secondary effects from the introduction of a PGA in the diet of early-lactating dairy cows was performed on two groups of 16 cows each, during the first 6 wk of lactation, homogeneous for age at calving and parity. Cows were fed a corn silage-based diet supplemented (top dressing method) with 500 g of a barley flour-flaked corn mixture with or without 20 g/d per cow of a commercial PGA. All the concentrate feeds were bought in the respect of the EU limits for the contamination with undesirable substances in animal feed. Blood samples were drawn from the jugular vein at 7 d intervals starting the first week after calving. Plasma was analysed for mineral profile and enzyme activities. No significant differences were evidenced for the investigated blood parameters between groups. The addition of the tested binder had no significant effect on plasma Ca, inorganic P, Na, K, Cl, nor iron concentrations, suggesting the PGA did not interfere with mineral bioavailability in early lactating cow. This research was supported by CANADAIR project (MiPAAF).

## Session 44 Theatre 4

**Comparative multi-tissue gene expression profiles related to milk performance in cattle**

*N. Melzer[1], J. Friedrich[2], G. Nürnberg[1], D. Repsilber[3], M. Schwerin[1,2] and B. Brand[1]*
*[1]Leibniz Institute for Farm Animal Biology, Wilhelm-Stahl-Allee 2, 18196 Dummerstorf, Germany, [2]University of Rostock, Justus-von-Liebig-Weg 6, 18059 Rostock, Germany, [3]Örebro University, Fakultetsgatan 1, 701 82 Örebro, Sweden; melzer@fbn-dummerstorf.de*

Gene expression profiling in farm animal science becomes more and more important to gain deeper insights into molecular mechanisms of complex traits. In the recent literature the analysis of gene expression data focused frequently on tissues important for energy metabolism and health. Due to their relevance in udder health, milk production, energy metabolism and meat quality in cattle, primarily the gene expression of mammary gland, liver and adipose tissues is analyzed. In this study, gene expression profiles (Affymetrix GeneChip®Bovine Genome Array) of three different tissues (mammary gland, liver, adrenal cortex) crucial for energy metabolism and milk production were analyzed applying different integrative bioinformatics approaches to identify molecular networks important for the regulation of milk production and their relevance in each tissue and across tissues. First results indicated tissue specific gene expression. 4,987 genes were expressed across all tissues, whereas 213, 397 and 1,001 genes were exclusively expressed in mammary gland, liver and adrenal cortex tissue, respectively. Ingenuity pathway analysis of tissue specific genes highlighted the functional active networks of each tissue. Genes involved in glucocorticoid biosynthesis were particularly expressed in the adrenal cortex, whereas genes involved in FXR/ RXR activation were predominantly expressed in liver tissue. Further investigations integrating milk performance data, using partial least squares regression and correlation analysis, might help to understand the complex regulation of lactation within and across the different tissues.

---

## Session 44 Theatre 5

**Absorption and intermediary metabolism of purines and pyrimidines in lactating dairy cows**

*C. Stentoft, S.K. Jensen, M. Vestergaard and M. Larsen*
*Aarhus University, Foulum, 8830 Tjele, Denmark; charlottes.nielsen@agrsci.dk*

More than 20% of total ruminal microbial nitrogen (N) derive from purine and pyrimidines (PP), the main constituents of DNA/RNA. An improved understanding of the absorption and intermediary metabolism of PP nucleosides (NS), bases (BS) and degradation products (DP), could yield insight to improve the N efficiency in dairy cows. The objective was to examine the postprandial pattern of portal-drained visceral (PDV) and liver metabolism of the PP. Blood was collected from four multicatheterized (artery, hepatic portal, hepatic, gastrosplenic vein) and ruminally cannulated Holstein cows. Cows were fed a basal TMR (CP: 126 g/kg DM) and continuously infused with urea (ventral rumen) to achieve a total protein content of 150 g/kg DM. Sampling was at d 14 with eight hourly sampling times beginning ½ h pre-feeding. Data was analysed considering sampling time as fixed effect and repeated measure and cow as random effect. Net fluxes were determined across the PDV, hepatic tissue and total splanchnic tissue (TSP). The net PDV release of the purine NS and BS (guanosine, inosine, deoxyguanosine, deoxyinosine, adenine, guanine, hypoxanthine, xanthine) were greater than zero ($P<0.05$) ranging from 2 to 1204 μmol/h. The liver removal was almost equivalent to the net PDV, resulting in no net TSP release. In comparison, net PDV release of the purine DP, uric acid and allantoin, were greater (6,730 and 7,765 μmol/h). In the case of allantoin, an endogenous contribution resulted in a net negative TSP (-6057 μmol/h). In the case of uric acid, a net positive TSP (7916 μmol/h) arose from metabolism of NS/BS. The net PDV release of all pyrimidine NS/ BS (cytidine, uridine, thymidine, deoxyuridine, cytosine, thymine) and DP (β-alanine, β-ureidopropionic acid, β-aminoisobutyric acid) were greater than zero ($P<0.05$) ranging from 47 to 2788 μmol/h, but the liver removal resulted in net TSP releases of -2,193 to 648 μmol/h. No prominent effect of feeding time was observed for any of the metabolites ($P>0.05$).

**Straw particle size in calf starters: effects on rumen fermentation and rumen development**

*F.X. Suarez-Mena[1], A.J. Heinrichs[1], C.M. Jones[1], T.M. Hill[2] and J.D. Quigley[2]*
*[1]The Pennsylvania State University, 324 Henning Building, 16802 University Park, PA, USA, [2]Provimi North America, 10 Collective Way, 45309 Brookville, OH, USA; fxs18@psu.edu*

This experiment studied effects of straw particle size in calf starter on rumen fermentation and rumen development. Male Holstein calves (n=17; 44.1±3.3 kg BW at birth) were housed in individual pens; bedding was covered with landscape fabric. Water was offered free choice, and milk replacer was fed at 12% of birth BW. Calves were randomly assigned to four treatments differing in geometric mean particle length (Xgm) of straw comprising 5% of starter dry matter. Straw was provided within the pellet (0.82 mm Xgm; 18% CP, 15% NDF) or mixed with the pellet (18.7% CP, 12.7% NDF) at Xgm of 3.04 mm, 7.10 mm or 12.7 mm. Rumen cannulas were fitted in 12 calves (3/treatment) by wk 2 of life. A fixed amount of starter was offered; orts were fed through the cannula in cannulated calves, except in wk 6. Rumen contents were sampled weekly through the cannula at -8, -4, 0, 2, 4, 8, and 12 h after starter feeding for pH and VFA determination. Calves were euthanized 6 wk after starter was offered; organs were harvested, emptied, rinsed, and weighed. Data were analyzed as completely randomized design. Starter intake was not different between treatments by design (P>0.05); weekly intakes were 393, 956, 1,764, 2,852, 5,583, 9,803±364 g for wk 1 to 6 of starter feeding. Rumen digesta pH linearly decreased with age while VFA concentration linearly increased (P>0.05). Molar proportion of acetate linearly decreased with age while propionate proportion increased. Reticulorumen, omasum, and abomasum weight, and rumen papillae length did not differ (P>0.05) between diets. Omasum % of BW at harvest linearly decreased (P<0.05) as straw particle size increased. Under the conditions of this study straw particle length in starter grain did not affect rumen fermentation parameters and rumen development. Rumen pH and fermentation changes with age were likely effects of greater starter intake.

---



**Rearing protocol during postnatal life affects the histology of pancreatic β-cells in male calves**

*L. Prokop[1], R. Lucius[2], H.-J. Kunz[3], M. Kaske[4] and S. Wiedemann[1]*
*[1]CAU, Institute of Animal Breeding and Husbandry, Olshausenstr. 40, 24098 Kiel, Germany, [2]CAU, Institute of Anatomy, Olshausenstr. 40, 24098 Kiel, Germany, [3]Chamber of Agriculture of Schleswig-Holstein, LVZ Futterkamp, Gutshof, 24327 Blekendorf, Germany, [4]University of Zurich, Department for Farm Animals, Vetsuisse Faculty, Winterthurer Str. 260, 8057 Zurich, Switzerland; lprokop@tierzucht.uni-kiel.de*

Primarily in humans and rodents it has been reported that early postnatal development has permanent effects on growth and function of tissues and organs. The objective of this study was to compare the impact of two different rearing strategies during the first 3 weeks of life on morphology of insulin producing pancreatic beta cells in male Holstein calves. Calves were fed either *ad libitum* (INT; n=21) or restrictively (10% of body weight per day; CG; n=21) during the first 3 weeks. INT was kept in single hutches throughout the first 3 weeks of life; CG was also kept in single hutches during the 1st wk of life and thereafter in group pens. After the 3rd wk of life all calves were housed and fed similarly. INT calves had higher weight gains during the first 10 weeks of life and increased insulin concentrations in blood during the 2nd and 3rd week of life. Calves were slaughtered at an age of 8 months and pancreatic tissue from the medium body was removed and fixed in formaldehyde. Number of islets of Langerhans and beta-cell area (n=9 of each group) were examined histologically. A generalized linear model was applied (SAS). The mean number of islets of Langerhans was higher in INT calves compared to the CG calves (9.1±0.3 vs 7.8±0.3; P=0.002). An intensified rearing also resulted in a trend towards a higher area of pancreatic beta cells compared to rearing according to a standard protocol (102,799±8,193 vs 85,699±8,193 μm²; P=0.14). In conclusion, differences in weight gain as well as in endocrine and metabolic traits indicate that calves can be programmed metabolically by an altered postnatal feeding intensity.

**Monochromatic red light stimulation in broiler breeders**

*A. Rosenstrauch[1] and I. Rozenboim[2]*
*[1]Achva Academiv College, Life Sciences, MP Shikmim, 7980400, Israel, [2]Hebrew University of Jerusalem, Faculty of Agriculture, Anima Sciences, Rehovot, 7654901, Israel; aviro@macam.ac.il*

Many studies had been conducted on the role of monochromatic light in birds. Many of the results are being implemented both in Israel and abroad. In general green and blue monochromatic light stimulate growth, while red light stimulate reproductive activities. Preliminary results showed that rearing broiler breeders under red monochromatic during the last 20 weeks of production improves productivity by 5%, compare to non-treated control. In the present study seven houses with commercial breeders were used. In five houses birds were reared under conventional white florescent lamps (40 W, Warm White) according to the primary breeders recommendations, and in two houses birds were reared under filterized monochromatic red light. Light intensity was 90 and 17 Lux for white and red light respectively. The experiment was conducted during the first season and daily reproductive activities, weekly body weight, and monthly fertility and hatchability were recorded. Results for reproductive activities: Eggs/housed hen: 131.5 (Red light); 126.5 (White Control). Fertility: 96.0 (Red light); 96.0 (White Control). Hatchability: 88.6 (Red light); 88.5 (White Control). Mortality: 12.8 (Red light); 13.5 (White Control). Feed intake/hen: 36.5 (Red light); 36.5 (White Control). Egg production was delayed by one week in the red treated groups, however egg production was significantly higher (+3%) in the red treated compare to the white control treated birds. This experiment presents a possible use of low intensity red light stimulation for broiler breeders that increase productivity.

---



**Effects of *Aloe vera* and wild marjoram oregano on broilers blood factors and carcass quailty**

*B. Shokrollahi[1], N. Faraji[1], M. Amiri Andi[1] and F. Kheirollahi[2]*
*[1]Sanandaj branch, Islamic Azad University, Department of Animal Science, Kurdistan, Sanandaj-6616935391, Iran, [2]Razi University, Kermanshah, Daneshgahe razi, 73441-81746 Kermanshahira, Iran; Borhansh@yahoo.com*

This study was made to evaluate the effect of *Aloe vera* and wild marjoram oregano supplementation on some blood biochemical parameters and carcass characteristics of broiler chickens. Two hundred and sixteen day-old broiler chicks (Ross 308) were randomly assigned to the six treatment groups, each with three replicates. Birds were housed in replicate pens each containing 12 birds (6 male and 6 female). The six treatments were as follows: (1) Basal diet (Control); (2) Basal diet + 1% Aloe vera powder; (3) Basal diet + 1% wild marjoram oregano powder; (4) 0.25% *A. vera* powder and 0.75% wild marjoram oregano powder; (5) 0.5% *A. vera* powder and 0.5% wild marjoram oregano powder; (6) 0.75% *A. vera* powder and 0.25% wild marjoram oregano powder. At 21 and 42 days of age, blood samples were collected for biochemical and hematological analysis. Two birds per replicate were slaughtered for the determination of carcass and organ weights at 42 days. The relative weight of breast, thigh and abdominal fat was positively affected by herbal supplemented treatments in comparison with control treatment. Experimental treatments notably reduced blood glucose, total cholesterol and LDL and significantly increased HDL ($P<0.01$) but did not affect total protein concentrations. It is concluded that *A. vera* and wild marjoram oregano have beneficial effects on health of broiler chicks.

## Session 44 — Poster 10

**Relation between hormone secretion and partial moulting in domestic goose**

*R. Mueller[1], A. Weissmann[2], H. Pingel[1], A. Einspanier[2], J. Gottschalk[2] and M. Waehner[1]*
*[1]Anhalt University of Applied Sciences, Agriculture, Strenzfelder Allee 28, 06406 Bernburg, Germany, [2]University of Leipzig, Faculty of Veterinary Medicine, Institute of Physiological Chemistry, An den Tierkliniken 1, 04103 Leipzig, Germany; m.waehner@loel.hs-anhalt.de*

Downs and feathers of geese are a valuable renewable natural product. They are utilizable as natural insulating and filling material for bedspreads as well as for high-quality winter-clothing. With regard to harvesting of feathers and downs from living geese more knowledge is necessary on cycling processes of the partial moult and on their endocrine control. Therefore, the aim of the present study was to assess the secretion of moult-associated hormones thyroxine (T4), progesterone (P), estradiol-17β (E2) and testosterone (T) as well as their potential role as a trigger of partial moult with special regard on feather growth. The study was performed with each six male and female goslings. Blood samples were collected weekly during three different time periods of moult: first partial moult ($7^{th}$ to $10^{th}$ week of life, WL), second ($15^{th}$ to $17^{th}$ WL) and third ($23^{rd}$ to $25^{th}$ WL). Plasma hormone concentrations were analyzed by enzyme and radioimmunoassay, respectively. A strong correlation between feather growth and hormone concentration could be only derived for T4, but not for P, E2 and T. During periods of maximum T4 values, the growth of feathers and downs has been most widely finished and is directly followed by the subsequent period of new feather growth. In conclusion, this study was helpful to obtain more details about endocrine situation during the partial moult in both sexes of growing geese.

## Session 44 — Poster 11

**Comparison of growth, carcass traits and blood composition between two populations of rabbits**

*N. Benali, H. Ainbaziz, Y. Dahmani, D. Saidj and S. Temim*
*High National School Veterinary, Algiers, BP 161, Hacene Badi, El Hrrach, 16000, Algeria; na.benali@yahoo.fr*

The aim was to compare growth performances, carcass traits and blood composition between two rabbit populations: the White population coming from an imported line (W) and the Kabyle population coming from an Algerian rabbit line (K). The average mature weight of the animals in the two lines ranges between 2 and 2.2 kg, which classify them as light breeds. Live weight, daily weight gain, feed intake and feed conversion ratio were measured in 70 females rabbits (35 W and 35 K) between 42 and 91 days of age. The carcass yield was determined on 47 animals (24 W, 23 K) and the blood composition on 14 rabbit rabbits (7 W and 7 K). The rabbits were fed with the same standard granulated food. No significant difference between the two populations were seen in daily growth rate, average daily feed intake and feed conversion ratio. However, weight gain between 42 and 63 days of age was significantly higher in W compared to K rabbits (+8%, $P<0.05$). Between 70 and 91 days of age, the average difference in weight gain between the two populations tended to persist in favor of the W population (+6%, $P>0.05$). The carcass component yields were not significantly different between W and K rabbits, except for the weight and proportion of liver, which was significantly higher (+71%, $P<0.05$ and +3.4%, $P<0.05$, respectively) in the W population in comparison with the K population. The plasma concentration of glucose, total protein, cholesterol, triglycerides and urea were not significantly different. However, serum creatinine was significantly higher (+23%; $P<0.05$%) in the W population compared to the K population.

**Effect of colostrum heating on immunoglobulins' absorption and performance and health of calves**

*F. Moslemipur, F. Vakili Saleh, Y. Mostafaloo and R. Ghorbani*
*University of Gonbad Kavoos, Animal Production, Shahid Fallahi St., Gonbad Kavoos, Golestan Province, Iran, 4971799151, Iran; farid.moslemipur@gmail.com*

The immunity and health of newborn calves are related to immunoglobulins' uptake from colostrum. In this study, the effect of heat-treatment of colostrum and(or) antibiotic addition on immunoglobulins' absorption and on performance and health parameters of calves was investigated. The colostrum pool was prepared from multiparious dams and frozen after the treatments. Colostrums were fed immediately after birth three times a day for 48 h. Sixteen newborn Holstein calves were divided into four treatment groups: (1) raw colostrum (control); (2) heat-treated colostrum; (3) colostrum with oxytetracyclin; and (4) heat-treated colostrum with oxytetracyclin. Heat-treated colostrum increased immunoglobulins' uptake and their apparent efficiency of absorption ($P<0.01$). No significant differences between treatment groups were observed in weight gain, weaning weight and the time elapsed between the start of the experiment and the time of beginning starter intake. Feed conversion ratio was not affected by the treatments. There were no significant differences in blood cells counts between groups. Fecal consitency score was not affected by the treatments but clinical health score was better in calves receiving heat-treated colostrum where the scour incidence was reduced. In conclusion, our results demonstrate that controlled heat-treated colostrum is an effective and practical method to improve immunoglobulins' absorption and reduce scour incidence in calves.

---



**Evolution of some biochemical parameters of the energy status during the peri-partum period in dairy cows**

*M. Laouadi[1], T. Madani[2], S. Sadi[3] and S. Tennah[4]*
*[1]University Amar Telidji of Laghouat, Agronomy, BP 37G, Route de Ghardaïa. Laghouat – Algeria, 03000, Algeria, [2]University Ferhat Abbes, Agronomy, setif, 19000, Algeria, [3]Technical Institute The Breedings, Baba ALI, 16200, Algeria, [4]Superior National Veterinary School of Algiers, BP 161 El-Harrach, 16200, Algeria; laouadi.mourad@yahoo.fr*

Our work aimed at describing the evolution of some biochemical parameters characterizing the energy status (glycaemia and cholesterolemia) in dairy cows. A total of 27 Holstein and Montbeliarde cows were followed during a period of one year in order to trace biochemical profiles from the time of drying off to the third month post-partum. Glycaemia decreased gradually from the time of drying off until reaching its minimum level one month after calving (0.50±0.11 g/l) and then increased slightly during the second and third month post-partum. For cholesterolemia, a minimum level was observed at calving (0.74±0.13 g/l), then, it increased gradually to finally stabilize by the second and the third month of lactation (1.60 g/l). The calving season showed a significant effect ($P<0.05$) on the evolution of the two biochemical parameters during the peri-partum period with lower values in the summer season. This study has broadened our knowledge about the profiles of glyceamia and cholesterol levels during the peri-partum under Algerian conditions.

## Session 44 — Poster 14

**Blood parameters of dairy cows supplemented with canola oil**

*K.C. Welter[1], B.R.T. Reis[1], G.F. Gambagorte[1], M. Maturana Filho[2], T. Santin[2] and A. Saran Netto[1]*
*[1]Faculdade de Zootecnia e Engenharia de Alimentos/USP, Departamento de Zootecnia, Rua Duque de Caxias Norte, 225, 13365-900, Brazil, [2]Faculdade de Medicina Veterinária e Zootecnia/USP, Departamento de Reprodução Animal, Rua Duque de Caxias Norte, 225, 13635-900, Brazil; saranetto@usp.br*

The objective of this study was to evaluate the effect of canola oil inclusion in the diet of dairy cows on some blood parameters. Eighteen lactating Holstein cows in the middle stage of lactation were distributed to 6 contemporary 3×3 Latin square designs, 3 periods and 3 treatments:T1=control diet, T2=inclusion of 3% of canola oil and T3 = inclusion of 6% of canola oil in the diet (based on dry matter). The blood samples were collected via coccygeal vein puncture before the offering of the morning diet. The samples were centrifuged at 2,000 g for 15 minutes at 4 °C and blood parameters were determined using a commercial kit (Randox®), where the reading was performed on a blood biochemistry automatic analyzer (SBA System for Automated Biochemistry -200-CEL®). The results were analyzed using the MIXED procedure of SAS, and use of a significance level at $P \leq 0.05$. The inclusion of canola oil (0, 3% and 6%) did not change the blood concentrations of glucose (P=0.647), triglycerides (P=0.404), non-esterified fatty acids (P=0.068) and VLDL (P=0.404). However, the inclusion of canola oil linearly increased the concentration of (1) total cholesterol (P<0.0001); (2) HDL (<0.0001); and (3) LDL (0.005), but reduced linearly the blood concentration of (4) β-hydroxybutyrate (P=0.012), according to the following equations: (1) Y(mg/dl) = 142.28(13.09) + 5.8823(1.1913)X; (2) Y(mg/dl) = 94.06(4.62) + 4.0219(0.7639)X; (3) Y(mg/dl) = 43.64(9.92) + 2.0415(0.6542)X; (4) Y(mmol/l) = 0.42(0.03) – 0.01884(0.007263)X. In conclusion, the inclusion of canola oil in the diet of dairy cows increases the blood concentration of total cholesterol, but does not alter glucose and triglycerides without affecting the health of the cows.

---

## Session 44 — Poster 15

**Effect of essential oils inclusion in the horse's diet on the digestibility of nutrients**

*A.A.O. Gobesso, M.A. Palagi, R. Françoso, I.V.F. Gonzaga and T.N. Centini*
*School of Veterinary Medicine and Animal Science, São Paulo University, Animal Production and Nutrition, Avenida Duque de Caxias Norte, 225, 13630.900 Pirassununga, SP, Brazil; gobesso.fmvz@usp.br*

The aim of this study was to evaluate the effect of different inclusion levels of essential oil in the diet of horses on the digestibility of nutrients, such as dry matter, organic matter, crude protein, ether extract, ash, neutral detergent fiber and acid detergent fiber. The study composed eight geldings ponies, a mini-horse breed, aged 42±6 months, with an average initial weight of 135±15 kg. The ponies were fed a diet with high proportion of concentrate, which was consisted of 60% commercial concentrate and 40% hay. The variation between treatments was the level of inclusion of a compound of essential oils at levels of 0 (control), 100, 200 and 300 mg/kg. The essentia oil was composed of 7% carvacrol. At each meal, the fraction corresponding to treatment was added to the concentrate at the individual feeder. Water and mineral supplement were provided *ad libitum*. The experimental design used was a 4×4 Latin square with repeated measures. The experiment was composed of four periods of 23 days each, with 15 days of adaptation to the diet, five days to collect samples and 3-day interval between periods. The apparent digestibility of nutrients (dry matter, organic matter, crude protein, ether extract, ash, neutral detergent fiber and acid detergent fiber) was estimated by the total collection method. No differences were found between the studied variables between the different treatments. It may conclude that under the present experimental conditions, dietary supplementation of horses with essential oil carvacrol did not influence the apparent digestibility of their diet.

## Session 44 Poster 16

**Effects of perinatal dietary lipid supplementation on fatty acids status of dairy goats and kids**

*M. Ferroni, D. Cattaneo, A. Agazzi, G. Invernizzi, F. Bellagamba, F. Caprino, V.M. Moretti and G. Savoini*
*University of Milan, Department of Health, Animal Science and Food Safety, Via Celoria, 10, 20134 Milano, Italy; giovanni.savoini@unimi.it*

Aims of the study were to determine the effects of perinatal dietary supplementation of fish oil or stearic acid on plasma,colostrum and milk fatty acids (FA) profiles of dairy goats and plasma FA status of their kids. From the last week of gestation until 3 weeks after kidding 24 multiparous Alpine goats were divided into 3 groups: control diet (C) no lipid supplement; fish oil (FO) rich in EPA (10.4%) and DHA (7.8%); calcium stearate (ST) rich in saturated FA C16:0 (26%) and C18:0 (69.4%). FO and ST diets supplied 30 or 50 g/head/d of FA respectively before and after kidding. Goat blood samples were collected weekly from 130 d of gestation until 21 d of lactation. Colostrum was sampled within the first 24 h postpartum and milk samples were collected at 7 and 21 d of lactation. Kids blood samples were collected at 0, 7 and 21 d. All samples were analysed by GC to determine FA composition. Statistical analysis on colostrum and milk FA profile was performed by GLM or MIXED repeated procedures of SAS v9.2. Plasma FA profile were evaluated by SPSS v17 using ANOVA and Student-Newman-Kelus post hoc test. Maternal mean plasma content of EPA (2.06%) and DHA (1.61%) was significantly higher in FO than C and ST. EPA and DHA were increased in colostrum ($P \leq 0.05$) and in milk ($P \leq 0.01$) of FO vs ST and C groups. Average milk secretion of EPA and DHA during the first 3 weeks of lactation was 0.40 g/d and 0.31 g/d respectively, with an apparent transfer efficiency of 7.69% (EPA) and 8% (DHA). Preliminary data on kids FA profile show an higher content of EPA and DHA at 0 d in FO kids. This tendency seems to persist at 7 and 21 d. In conclusion perinatal lipid supplementation produced valuable variations of colostrum, milk and plasma FA profile both in dams and kids, especially for EPA and DHA contents in FO group.

---

## Session 44 Poster 17

**Prolactin and progesterone concentrations around farrowing influence sow colostrum yield**

*F. Loisel[1,2], C. Farmer[3], H. Van Hees[2] and H. Quesnel[1]*
*[1]INRA, UMR1348 PEGASE, 35590 Saint-Gilles, France, [2]Nutreco R & D, 5832 AE, Boxmeer, the Netherlands, [3]Agriculture and Agri-Food Canada, Dairy and Swine R & D Centre, Sherbrooke, QC, J1M 0C8, Canada; helene.quesnel@rennes.inra.fr*

In swine, colostrum production is induced by the drop of progesterone concentrations which leads to the prepartum peak of prolactin. Prolactin regulates mammary cell turnover and stimulates lacteal nutrient synthesis. Progesterone inhibits prolactin secretion and down-regulates the prolactin receptor in the mammary gland. The aim of the study was to determine if the relative peripartal concentrations of progesterone and prolactin (PRL/P4 ratio) influence sow colostrum production. Twenty-nine Landrace × Large White primiparous sows were used. Colostrum yield was estimated during 24 h starting at the onset of parturition (T0) using litter weight gains. Colostrum was collected at T0 and 24 h later (T24). Repeated jugular blood samples were collected during the peripartum period (i.e. from -72 to +24 h related to farrowing) and were assayed for progesterone and prolactin. Sows were retrospectively categorized according to their PRL/P4 ratio 24 h before farrowing: $<2$ (LowPRL/P4, n=16) or $>3$ (HighPRL/P4, n=13). Data were analyzed by ANOVA using the MIXED procedure (SAS Inst.), except for piglet mortality (GENMOD procedure). During the peripartum period, the circulating concentrations of progesterone were lower ($P<0.05$) while those of prolactin tended to be greater ($P<0.10$) in HighPRL/P4 compared with LowPRL/P4 sows. Colostrum yield was greater in HighPRL/P4 compared with LowPRL/P4 sows (4.1 versus 3.5 kg [RMSE=0.7], $P<0.05$). Colostrum gross composition and IgG and IgA concentrations did not differ between the two groups of sows ($P>0.10$). Piglet mortality between birth and T24 averaged 10.0% in LowPRL/P4 litters and 7.0% in HighPRL/P4 litters ($P=0.29$). In conclusion, a higher PRL/P4 ratio 24 h prepartum, characterized by lower progesterone concentrations and a trend for higher prolactin concentrations peripartum, led to a greater colostrum yield.

**Nitrogen utilisation and carcass composition in finisher pigs fed different content of crude protein**

*J.V. Nørgaard, M.J. Hansen, E.A. Soumeh, A.P.S. Adamsen and H.D. Poulsen*
*Aarhus University, Blichers Alle 20, Foulum, 8830 Tjele, Denmark; jan.noergaard@agrsci.dk*

The effect of diets with different crude protein (CP) concentrations on pig performance and physiology was studied using 256 female pigs weighing 53.9 kg and allotted to 4 dietary treatments of 136, 148, 159 and 168 g CP/kg diet, all fulfilling the amino acid recommendations. The latter diet was the 159 g CP/kg diet supplemented with dispensable amino acids corresponding to 9 g CP extra per kg. The experiment had a duration of 6 weeks, and gain and feed intake were recorded at the $3^{rd}$ and $6^{th}$ week. During week 4, a digestibility and balance study using 6 pigs from each treatment was carried out in metabolic cages for 5 days. At the end of the experiment, blood samples were collected from the jugular vein from half of the pigs, and all pigs were slaughtered and carcass characteristics were studied. Data was analysed using mixed linear models. The average daily feed intake was not affected by the dietary treatments. There was a quadratic effect (P=0.02) of CP levels on average daily gain during the first 3 weeks with the 148 g CP/kg diet resulting in the best (P<0.05) performance. However, no effect on gain was observed during the last 3 weeks and the overall period. Feed utilisation was not different among the dietary treatments. Both plasma urea concentration (P<0.001) and urinary N excretion (P<0.01) decreased linearly by lowering the CP concentration. The weight of the carcass increased linearly (P=0.01) by reducing the CP concentration, but neither nitrogen retention nor carcass meat percentage and back fat depth differed among treatments. The results showed that dietary CP concentration could be decreased to 136 g/kg without impairing animal performance and carcass characteristics significantly, when the amino acid profile of the diet was adjusted by adding crystalline amino acids according to the requirement. Leanness was not positively affected by excess CP by the supplementation of dispensable amino acids.



**Lysine restriction in finishing phase affects growth performance and quality traits of heavy pigs**

*J. Suárez-Belloch, J.A. Guada, M. Fondevila and M.A. Latorre*
*IUCA. University of Zaragoza, Miguel Servet 177, 50013, Spain; jsuarezbelloch@gmail.com*

A total of 160 Duroc × (Landrace × Large White) pigs, 50% barrows and 50% gilts, with 28.3±4.52 kg body weight (BW) and 73±3 days of age, were used to study the influence of dietary lysine (Lys) content during finishing phase on growth performance and carcass and meat (Logissimus thoracis muscle) quality. During the growing period (30-90 kg BW), all pigs were fed a common commercial diet. During the finishing period (90-130 kg BW), diets were based on cereals and soybean meal and contained 3,100 kcal ME/kg and 0.77, 0.69, 0.53 and 0.42% total lysine (Lys) (19.2, 16.6, 14.3 and 11.9% crude protein, respectively). There were 8 treatments with 2 sexes and 4 levels of Lys which were replicated 5 times in a randomized block design (the experimental unit was the pen with 4 pigs). During the finishing phase, the barrows had higher daily gain (P=0.0004) and also feed intake (P<0.0001) but tended to be less efficient (P=0.055) than gilts. Sex did not affect carcass yield but carcass length (P=0.001) and ham length (P=0.09) were longer in gilts than in barrows. Also, the carcasses from barrows had wider backfat depth (P<0.01) and had less weight of main cuts (hams + shoulders + loins) (P=0.04) than those of gilts. Meat of barrows tended to have higher intramuscular fat (P=0.08) and lower moisture (P=0.09) contents than that of gilts. The Lys restriction linearly reduced BW gain and feed consumption and increased feed conversion ratio (P<0.0001). In addition, the dietary Lys reduction decreased quadratically the main cuts (P=0.04) and affected meat quality; it increased intramuscular fat (P=0.003) and reduced protein (P<0.0001) contents. In summary, the restriction of dietary Lys level from 90 to 130 kg BW in pigs impaired productive performance but improved some meat traits especially desirable for dry-cured products processing. It could be concluded that 0.69% is the optimum total Lys level to include in the finishing diet for heavy barrows and gilts.

## Session 45 Theatre 3

**Growth response of pigs to the supply of phenylalanine and tyrosine offered low-protein diets**

*M. Gloaguen[1], N. Le Floc'h[1], Y. Primot[2], E. Corrent[2] and J. Van Milgen[1]*
*[1]INRA-Agrocampus Ouest, UMR1348 PEGASE, Domaine de la Prise, 35590 Saint-Gilles, France, [2]Ajinomoto Eurolysine S.A.S., 153, rue de Courcelles, 75817 Paris, France; jaap.vanmilgen@rennes.inra.fr*

The adoption of low-CP diets for pigs depends on the knowledge of the minimum levels of indispensable amino acids that maximize growth. Phe and Tyr are indispensable amino acids, although Tyr can be provided by biosynthesis from Phe. Both amino acids are constituents of body protein while Tyr is also a precursor for catecholamines, neurotransmitters and thyroid hormones. Knowledge about the requirements for Phe and Tyr is very limited. The objectives of this study were to estimate these requirements in 10-20 kg pig and to determine the extent to which Phe can be used to cover the Tyr requirement. In three dose-response studies, pigs were assigned to diets with 14.5% CP with a sub-limiting level of standardized ileal digestible (SID) Lys of 1.00%. In experiment 1, the SID Phe:Lys requirement estimate was assessed by supplementing a Phe-deficient diet with different levels of L-Phe to attain 33, 39, 46, 52, 58, and 65% SID Phe:Lys, using diets with a sufficient Tyr supply. A similar approach was used in experiment 2 with six levels of L-Tyr supplementation to attain 21, 27, 33, 39, 45 and 52% SID Tyr:Lys, while Phe was supplied at a level sufficient to sustain maximum growth (estimated in experiment 1). The SID Phe:Lys and Tyr:Lys requirements for maximizing daily gain were 54 and 40% using a curvilinear-plateau model, while a 10% deficiency reduced daily gain by 3.0 and 0.7% for Phe and Tyr, respectively. In experiment 3, the effect of an equimolar substitution of dietary SID Tyr by Phe to obtain 50, 57, and 64% SID Phe:(Phe+Tyr) was assessed. From 57 to 64% SID Phe:(Phe+Tyr), performance was slightly reduced indicating that Phe cannot be used on an equimolar basis to provide Tyr. It is therefore recommended not to use a Phe+Tyr requirement in the ideal amino acid profile but to use a SID Phe:Lys of 54% and a SID Tyr:Lys of 40% to support maximal growth.

## Session 45 Theatre 4

**Isoleucine, valine, and leucine requirements in 7 to 19 kg pigs**

*E.L.H. Assadi Soumeh[1], J.N. Værum Nørgaard[1], H.N. Damgaard Poulsen[1], N.L.M. Sloth[2], J.P. Van Milgen[3] and E.T.N. Corrent[4]*
*[1]Århus univrsity, Animal Science, Blichers Alle 20, 8830 Tjele, Denmark, [2]Pig Research Centre, Agro Food Park 15, 8200 Aarhus, Denmark, [3]INRA, UMR1348 PEGASE, 35590 Rennes, France, [4]Ajinomoto Eurolysine s.a.s., 153 rue de Courcelles, 75817 Paris cedex 17, France; elhama.soumeh@agrsci.dk*

The objective of this study was to determine requirements of standardized ileal digestible (SID) Ile, Val, and Leu relative to Lys that support maximum animal performance. This was studied in 3 dose-response experiments with individually penned piglets starting 1 week after weaning. Experimental diets were sub-limiting in Lys and included 0.42, 0.46, 0.50, 0.54, 0.58, and 0.62 of SID Ile:Lys, 0.58, 0.62, 0.66, 0.70, 0.74, and 0.78 of SID Val:Lys, and 0.70, 0.80, 0.90, 1.00, 1.10, and 1.20 of SID Leu:Lys. Each of the 6 diets were fed to 16 pigs for 2 weeks, and average daily feed intake (ADFI), average daily gain (ADG), and gain to feed (G:F) were used as response criteria and were recorded at the end of each week. Blood and urine samples were taken at days 8 and 15 of the experiment and analyzed for metabolites. In Ile study, the highest ADFI and ADG were obtained at 0.50 SID Ile:Lys but there was a decline in both ADFI and ADG from 0.50 to 0.62 SID Ile:Lys. Experimental diets did not affect G:F. Increasing dietary SID Ile, Val, and Leu linearly increased ($P<0.05$) concentration of free plasma Ile, Val, and Leu, respectively. Urea content in urine and plasma was not affected by experimental diets. Response criteria were fitted to different models to estimate the requirements. A quadratic regression model was used for Ile and estimated the requirement at 0.53 SID Ile:Lys. The data from both the Val and Leu studies were fitted to a curve-linear plateau model and requirements were estimated at 0.71 SID Val:Lys and 1.00 SID Leu:Lys. An Ile supply 10% below or above the requirement caused a 10% decline in animal response. No response was observed for excess Val and Leu, but 10% below Val requirement and 20% below Leu requirement resulted in 10 to 20% decrease in animal response.

**The effect of dietary valine-to-lysine ratio on sow performance and piglet growth during lactation**

*A.V. Hansen[1], T.S. Bruun[2] and C.F. Hansen[1]*
*[1]University of Copenhagen, Grønnegaardsvej 2, 1870 Frederiksberg, Denmark, [2]Danish Pig Research Centre, Danish Agriculture & Food Council, Axeltorv 3, 1609 Copenhagen, Denmark; avha@sund.ku.dk*

The aim of the study was to determine the effect of dietary valine-to-lysine ratio (Val:Lys) on BW, backfat thickness (BF) and litter gain of high producing lactating sows. Sixty second parity sows were allotted to one of six dietary treatments varying in standardized ileal digestible Val:Lys (76, 79, 82, 85, 91, and 97%) from day 2 postpartum. The sows were fed semi *ad libitum* 2 times per day until day 10 and from day 10 three times per day. Litters were equalized at day 2 to14 piglets and weaned at day 28. Sow BW and BF were recorded at day -7, 2, 18 and 28 of lactation. Litter weight and size were recorded at day 2, 10, 18 and at weaning. Differences in sow BW and BF were calculated as total change (day 2-28, n=26), change from day 2- 18 (n=38) and from day 18-27 (n=26). The ADG (kg/day) of the litter were calculated as total ADG (day 2-28, n=26) and for early (day 2-10, n=48), mid (day 10-18, n=38) and late (day 18-28, n=26) lactation. Data was analyzed in R using a linear model testing the effect of dietary treatment and block. The BW, BF and litter weight at standardization was used as covariate. There was no effect of treatment, block and litter weight at day 2 on litter size at weaning and ADG ($P>0.05$). The mean ADG was 3.08 kg/day (SD=0.52) and the mean litter size at weaning was 13.08 piglet (SD=1.02). The change in BF from day 2-18 ($P<0.01$) and day 18-28 ($P<0.05$) was affected by BF at day 2. Higher BF at standardization resulted in higher BF loss. There was a tendency for a dietary effect on BF change from day 2-18 ($P=0.08$). There was no dietary effect on change in BW ($P>0.05$). Sows fed the highest Val:Lys had the highest ADG (3.3 kg/day), BF loss (-6 mm) and BW loss (-40 kg). And sows fed a Val:Lys of 79% had the lowest ADG (2.9 kg/day), BF loss (-2.2 mm) and BW loss (-18.8 kg). In conclusion, the prelimenary results shows no effect of Val:Lys on piglet ADG, litter size and sow body condition.

---

Session 45 Theatre 6

**Use of esterified palm acid oils with different acylglycerol structure in fattening pig diets**

*E. Vilarrasa[1], A.C. Barroeta[1] and E. Esteve-Garcia[2]*
*[1]Universitat Autònoma de Barcelona, Animal Nutrition and Welfare Service, Edifici V, 08193 Bellaterra, Spain, [2]IRTA, Monogastric Nutrition, Ctra. Reus-El Morell Km 4.5, 43120 Constantí, Spain; ester.vilarrasa@uab.cat*

Esterified acid oils are obtained by reacting acid oils with glycerol. Because these technical fats have an increased proportion of saturated fatty acids located at the acylglycerol sn-2 position, and a higher amount of mono- (MAG) and diacylglycerol (DAG) molecules than native oils, it was hypothesized that esterified acid oils could have higher nutritive value than their corresponding acid oils and even than their corresponding native oils for pigs. The aim of the present study was to compare the effects of esterified palm acid oils with different acylglycerol structure, with their corresponding acid (negative control) and native (positive control) oils in fattening pig diets. For this purpose, 72 pigs (36 boars and 36 gilts of 24.7±0.30 kg) were ranked by 9 blocks of initial weight, housed in adjacent individual boxes, and fed one of the 4 dietary treatments, which were the result of a basal diet supplemented with 4% of native palm oil (P-N), acid palm oil (P-A), esterified palm acid oil low in MAG and DAG (P-EL) or esterified palm acid oil high in MAG and DAG (P-EH). In addition, a balance study was conducted from d 6 to 9 using an inert marker. At the end of the trial (100 d), 6 pigs per treatment were slaughtered, and backfat was analyzed for fatty acid composition. Esterified palm acid oils achieved higher total fatty acid apparent absorption than their respective native and acid oils, mainly due to the increased saturated fatty acid apparent absorption. This resulted to an improved growth performance and an increased saturated fatty acid deposition in pigs fed with P-EH when compared to those fed P-A.

**Feeding replacement gilts as finishers or less?**

*M. Neil[1], L. Eliasson-Selling[2] and K. Sigfridson[3]*
*[1]SLU, Animal Nutrition and Management, P.O. Box 7024, 75007 Uppsala, Sweden, [2]Svenska Djurhälsovården, Kungsängens gård hus 6B, 75323 Uppsala, Sweden, [3]Lantmännen Lantbruk, Boplatsgatan 8B, 20503 Malmö, Sweden; maria.neil@slu.se*

About 50% of the breeding sows have to be replaced each year, so obviously sow longevity needs to improve. The feeding regime during rearing might be one tool to affect longevity. Current goals are to produce gilts for mating at 2$^{nd}$-3$^{rd}$ oestrus, 230 days of age, 140 kg live weight (LW) and 12-13 mm ultrasonic backfat (BF). The present study was undertaken to investigate effects of energy allowance and feed lysine level on growth, BF accretion, and early reproduction. The study comprised 4 batches of 20 female L×Y pigs, born in June (batch 1-2) and December-January (batch 3-4). Litter sisters were allocated to 4 feeding treatments in a 2×2 factorial design. The factors were high (HL) or low (LL) lysine level (0.83 or 0.57 g sid lys per MJ NE) and high (he) or low (le) daily energy allowance (increasing from 14.5 MJ NE at 30 kg LW to 25.9 MJ NE at 60 kg which equals the standard for finishers, or a slower increase to 23.3 MJ NE at 60 kg which is about 10% less). At the age of 9 weeks and LW approximately 30 kg, gilts were placed in groups of 10 in pens with feeding stalls and were fed individually after training. Gilts were weighed and BF was measured repeatedly. Oestrus detection started at 150 days of age. At 151 days of age mean LW was 102, 104, 108 and 111 kg, and BF 8.3, 8.6, 7.7 and 7.8 mm, in LLle, LLhe, HLle and HLhe gilts respectively. HL gilts were heavier ($P<0.001$) and tended to have lower BF ($P=0.08$) than LL gilts. No difference in LW or BF at 151 days was detected between le and he gilts. Lysine level and feed allowance had no effect on age at first oestrus, whereas batch had a strong effect ($P<0.001$) with mean age of 186 days in batch 1-2, and 165 days only in batch 3-4. Thus, the applied difference in feed lysine level affected LW and BF in gilts whereas the applied difference in energy allowance did not. To affect LW and BF in gilts the energy allowance may need to differ more.



**Distillers dried grains with solubles (DDGS) as protein feed to lactating dairy cows**

*J. Sehested[1], M.T. Sørensen[1], A. Basar[1], M. Vestergaard[1], H. Martinussen[2] and M.R. Weisbjerg[1]*
*[1]Aarhus University, Dep. Animal Science, AU-Foulum, P.O. Box 50, 8830 Tjele, Denmark, [2]Knowledge Centre for Agriculture, Dairy & Cattle Farming, Agro Food Park 15, 8200 Aarhus N, Denmark; jakob.sehested@agrsci.dk*

Distillers dried grains with soluble (DDGS) is a byproduct of the bio-ethanol industry. This byproduct is high in protein (~30% of DM) and can be used in diets for livestock as an alternative or supplement to for example soybean meal and rapeseed cake. The effect of feeding increasing dietary levels of DDGS on feed intake and milk production was studied in the present experiment. Fifty lactating Danish Holstein dairy cows were used in a 3×3 Latin squares design where cows were fed DDGS at 4, 13 and 22% of diet dry matter (DM). The DDGS substituted on DM basis matching mixtures of soybean meal, rapeseed cake and dried beet pulp. The diets also contained (DM basis) whole-crop maize silage (45%), grass silage (15%), NaOH treated wheat (5%), barley (4%) and a concentrate mixture allocated in the milking robot (8%). The DDGS contained 33.9% crude protein of DM and 7.9 g lysine per kg DM and all rations were balanced at 16.7% crude protein. Daily DM intake were 22.7, 22.8 and 23.3 kg at DDGS levels of 4, 13 and 22% of DM ($P>0.05$). Daily energy corrected milk (ECM) yield were 37.7, 37.8 and 36.8 kg ($P>0.05$) and milking frequencies were 2.83, 2.77 and 2.66 ($P=0.04$) at DDGS levels of 4, 13 and 22% of DM. There were slight decreases in milking frequency, daily milk, daily milk protein and daily milk lactose yield at 22% DDGS. We conclude that DDGS can substitute soybean meal and rapeseed cake in diets for lactating dairy cows and be included at up to 22% of dry matter without significant effects on feed intake and ECM yield. However, yield of milk, milk protein and milk lactose are slightly decreased at the highest dietary DDGS level.

**Reducing soybean meal import for dairy cattle by using DDGS or rumen protected soybean meal**

*L. Vandaele, J. De Boever, D. De Brabander and S. De Campeneere*
*Institute for Agricultural and Fischeries Research (ILVO), Animal Sciences, Scheldeweg 68, 9090, Belgium; leen.vandaele@ilvo.vlaanderen.be*

Despite many public concerns, soybean meal (SBM) remains the predominant protein source in European dairy concentrates. We evaluated DDGS (dried distiller grains and solubles) and rumen-protected SBM as alternative protein sources in three (3×3) Latin square experiments (exp) with 18 to 21 lactating Holstein cows per exp. The control diets (CTRL) of the 3 experiments contained on average 2 kg DM of SBM and were compared with the following treatment diets: Exp1: wheat/maize DDGS at two doses ($DDGS_L$: 2.2 kg and $DDGS_H$: 3.5 kg DM of DDGS); Exp2: 2.5 kg DM pure wheat DDGS ($DDGS_W$) and 2.0 kg DM pure maize DDGS ($DDGS_M$) both in a diet without SBM; Exp3 1.1 kg DM heat+xylose treated SBM and 1.1 kg DM formaldehyde treated SBM. All diets were formulated according to the energy and protein requirements of the individual cows with ad lib roughage intake. For the treatment diets a maximal replacement of soybean meal was pursued during formulation by using different additional concentrates. On average, the use of DDGS or rumen-protected SBM allowed a 1.5 kg DM/cow/day lower use of SBM. Data of each trial were subjected to an analysis of variance followed by post-hoc Scheffé test ($P<0.05$). Milk production significantly increased in Exp 1 for $DDGS_H$ (CTRL: 28.4; $DDGS_L$: 28.5 and $DDGS_H$: 29.2 kg/day) and in Exp 2 for $DDGS_W$ and $DDGS_M$ (CTRL: 31.5, $DDGS_W$: 33.2 and $DDGS_M$: 32.8 kg/d). Daily protein production was clearly stimulated when including DDGS in the ration (Exp 1: CTRL: 993, $DDGS_L$: 1,001 and $DDGS_H$: 1,015 g/d and Exp2: CTRL: 1,030, $DDGS_W$: 1,105; $DDGS_M$: 1,083 g/d). In Exp3 daily milk and protein yield was not significantly different between treatments. Daily milk fat production was never significantly different. In addition, with DDGS the amount of concentrates could be reduced substantially. In conclusion, DDGS or treated SBM can successfully be used for high producing dairy cattle and have a strong potential to decrease the import of SBM.

---



**The influence of mechanical processing of pea seeds on nutrients digestibility in sheep**

*M. Šimko, D. Bíro, M. Juráček, B. Gálik, M. Rolinec and O. Pastierik*
*Slovak University of Agriculture in Nitra, Faculty of Agrobiology and Food Resources, Tr. A. Hlinku 2, 949 76 Nitra, Slovak Republic; milan.simko@uniag.sk*

We investigated the effects of mechanical processing of pea seeds (Pisum sativum L.) on apparent total-tract nutrients digestibility and nitrogen balance in sheep. The basis of the diets was meadow hay (0.36 kg DM/d) supplemented with pea seed (0.30 kg DM/d), barley grain (0.17 kg DM/d) and oat grain (0.18 kg DM/d). The first diet contained whole pea, ground barley and ground oat (diet a). In the second diet the pea seed was grounded (diet b). The experiment was performed on six castrated male cigaja sheep with mean live weight of 43.2 kg. They had been castrated at about one month of age. Animals were fed twice a day – at 6.30 a.m. and 6.30 p.m. The 11-day adaptation period was followed by the 5-day experimental period. Diet samples were taken and analyzed for nutrient contents according to the Regulation of the Ministry of Agriculture of the Slovak Republic no. 2145/2004-100. During the experimental period, feces and urine were collected. From the amounts collected per 24 hour, average samples were taken for chemical analysis. Significance of differences between the diets was evaluated by Fischer`s LSD test using the program Statgraphics, ver. 5.0. The higher apparent total-tract digestibility of crude protein (78%) crude fat (70%) and organic matter (80%) was detected for the diet b. Compared to the diet a the differences were not significant. When we fed whole pea seeds (diet a), the digestibility of crude protein was 76%, crude fat 67% and organic matter 79%. There were no differences between diet a and diet b in the digestibility of crude fibre (63%) and nitrogen free extract (85%). Retention of nitrogen did not differ significantly between the diets a and b. Mechanical processing of pea seed did not significant affect apparent total-tract nutrients digestibility and nitrogen balance.

## Session 45 Poster 11

**Use of esterified palm acid oils with different acylglycerol structure in weaning piglet diets**

*E. Vilarrasa[1], A.C. Barroeta[1] and E. Esteve-Garcia[2]*
*[1]Universitat Autònoma de Barcelona, Animal Nutrition and Welfare Service, Edifici V, 08193 Bellaterra, Spain, [2]IRTA, Monogastric Nutrition, Ctra. Reus-El Morell Km 4.5, 43120 Constantí, Spain; ester.vilarrasa@uab.cat*

Esterified acid oils are obtained by reacting acid oils with glycerol. Because these technical fats have an increased proportion of saturated fatty acids located at the acylglycerol sn-2 position, and a higher amount of mono- (MAG) and diacylglycerol (DAG) molecules than native oils, it was hypothesized that esterified acid oils could have higher nutritive value than their corresponding acid oils and even than their corresponding native oils for piglets. The aim of the present study was to compare the effects of esterified palm acid oils with different acylglycerol structure, with their corresponding acid (negative control) and native (positive control) oils, and also with an unsaturated fat source in weaning piglet diets. For this purpose, 120 weaning piglets (8.52±0.163 kg) were randomly distributed into 30 pens. The experiment was designed as a randomized complete block design with 3 blocks of initial live weight and 5 dietary treatments, which were the result of a basal diet supplemented with 10% of native soybean oil (S-N), native palm oil (P-N), acid palm oil (P-A), esterified palm acid oil low in MAG and DAG (P-EL) or esterified palm acid oil high in MAG and DAG (P-EH). In addition, a balance study was conducted from d 7 to 10 using an inert marker, and blood samples were drawn by jugular venopuncture at d 23. The P-A treatment showed the lowest total fatty acid apparent absorption due to their high free saturated fatty acid content, while the S-N treatment showed the highest total fatty acid absorption due to their high unsaturated fatty acid content. Although no differences were found among P treatments, P-EH achieved the highest total fatty acid digestibility value, and not different from S-N. There were no significant differences in performance and serum lipid profile, although the P-EH treatment showed the greatest ADG and the S-N treatment showed the lowest FCR.

---

## Session 45 Poster 12

**Effect of diet, gender and body weight on bone traits in Large White pigs**

*I. Ruiz[1,2], P. Stoll[1], P. Schlegel[1], M. Kreuzer[2] and G. Bee[1]*
*[1]Agroscope, Tioleyre 4, 1725 Posieux, Switzerland, [2]ETH Zürich, Universitätstrasse 2, 8092 Zürich, Switzerland; isabel.ruiz@agroscope.admin.ch*

Breeding continuously alters growth rate and body composition of slaughter pigs. This may have affected their nutrient requirements and oppose the use of low-protein diets which are favoured for environmental reasons. The current study aims at updating nutritional recommendations for Swiss Large White pigs. An experiment was performed with 30 entire males, 30 barrows and 30 females. Half of each gender group were allocated to a control grower-finisher diet, designed according to the Swiss feeding recommendations, or a low protein grower-finisher diet (80% of control). Pigs had *ad libitum* access to the experimental diets. From weaning to 20 kg body weight (BW) pigs were offered the same starter diet, then the experimental diets. Using the serial slaughter technique, whole body composition was determined at 10, 20, 40, 60, 80, 100, 120 and 140 kg BW on 2 pigs per each of the six subgroups. One day post-mortem, the third and fourth metacarpal bones were collected. Dry matter (DM) and ash content, density, breaking resistance (BR) and mineral composition (Ca, P, Mg, K, Na, Cu, Fe, Mn and Zn) of the bones were assessed. Data were analysed using the GLM procedure of Systat 13 considering BW, gender and diet as effects. Diet had no ($P>0.05$) effect on any of the bone traits. Bone DM and ash content were influenced by BW and gender. Bone DM was higher at 40, 60, 100 and 120 kg BW than with the other BW classes. Compared to entire males, bones from females and barrows had higher ($P<0.05$) ash content, which was also high at 60, 100 and 140 kg BW than at the other BW classes. An 88% increase in BR was found from 20 to 140 kg BW. The BR was 11% higher ($P<0.05$) in females than entire males. Barrows were intermediate. In conclusion, low-protein diets do not seem to impair bone quality. It was of special interest that BR not only increased with BW but also that it was highest in female pigs. This is especially important for animals designated as breeding sows.

## Effect of forage fiber intake on rumen motility in heifers fed highly digestible grass/clover silage

*A.K.S. Schulze[1], M.R. Weisbjerg[2] and P. Nørgaard[1]*
*[1]University of Copenhagen, Dept. of Veterinary Clinical and Animal Sciences, 1870 Frb., Denmark, [2]Aarhus University, Foulum, Dept. of Animal Science, 8830 Tjele, Denmark; pen@sund.ku.dk*

The objective was to assess effects of forage NDF content and time after feeding on primary rumen contraction cycle (PCC) frequency. Four grass/clover silages were harvested in 2009 at different regrowth stages, resulting in NDF contents of 312, 360, 371, and 446 g/kg DM. All silages were chopped to 19 mm and ensiled in round bales without use of ensiling additives. Four rumen-fistulated Jersey heifers (343±32 kg BW) were fed silage at 90% of *ad libitum* level in a 4×4 Latin square design, replicated with further restricted feeding levels (50%, 60%, 70%, or 80% of *ad libitum*) in a balanced 4×4×4 Greco-Latin square design. Silage rations were fed twice daily, half at 0800 h and at 1530 h. Rumen contractions were recorded from 0700 h to 1600 h by a pressure transducer (Honeywell, 24PCAFA6D, NJ, USA) placed in the reticulum through the rumen fistula. The frequency of PCC was assessed in eight 30-min time intervals around each half-hour, the first interval from 07.15 h to 07.45 h, and the last interval from 14.15 h to 14.45 h. The PCC frequency was analyzed with fixed effects of silage treatment, period, feeding level, and sampling time as autoregressive repeated measure, and random effect of heifer×period. The mean daily intake of NDF at 90% of *ad libitum* level was between 2.3 and 2.8 kg, and ranged from 0.4% of BW at 50% feeding level to 0.8% of BW at 90% feeding level. The frequency of PCC was affected by sample time ($P<0.001$); 1.52 PCC/min before morning feeding, and 1.62 PCC/min during the morning meal decreasing to 1.44 PCC/min 7 hours after feeding. The mean frequency of PCC (1.5±0.05 (SE) PCC/min) was unaffected by NDF content of silage ($P>0.05$), while greater feeding level tended to increase mean frequency of PCC ($P=0.06$). This study indicated that stimulation of rumen motility is stimulated by increased rumen fill, but unaffected by NDF concentration of grass/clover silages.

---



## Biohydrogenation of C18:2n-2 and C18:3n-3 is reduced when esterified to complex lipid frac-tions

*M.B. Petersen[1] and S.K. Jensen[2]*
*[1]AgroTech, Agro Food Park 15, 8200 Aarhus N, Denmark, [2]Aarhus University, Department of Animal Science, Blichers Allé 20, P.O. Box 50, 8830 Tjele, Denmark; mbp@agrotech.dk*

Manipulation of rumen biohydrogenation (BH) is of great interest, as decreased BH of n-3 and n-6 fatty acids (FA) is linked to increased milk content of these beneficial polyunsaturated FAs (PUFA). We hypothesized that PUFAs embedded in the complex lipid fractions in plants are less prone to BH of PUFAs compared to PUFAs embedded in the simple lipid fractions. *In vitro* rumen BH of C18:3n-3 and C18:2n-6 was investigated as free fatty acids (FFA) and as esterified to triglycerides (TG), cholesterol esters (CE) and phospholipids (PL). Eleven ml of buffer, 11 ml of strained rumen fluid, and either 5 mg FFA, 10 mg of TG and CE or 15 mg of PL were transferred to test tubes and incubated for 0, 0.5, 1, 2, 4, 6, and 11 hours and C18 FAs were quantified on GC. *In vitro* BH was quantified according to Michaelis-Menten kinetics. Modeling intermediates of BH should be done with care, as they can be formed during multiple steps through the complex BH pathways. Modeling of the apparent appearance of the intermediate C18:1t11 was done to elucidate possible differences in the BH pathways between fractions. Regardless of FA the order of BH was; FFA, TG, PL and CE, with the fastest *in vitro* BH rate occurring when FAs were added as FFAs and slowest when they were esterified to CE ($P<0.005$). There was no significant effect in appearance of the end product of BH, C18:0, between the different lipid fractions. Apparent appearance of C18:1t11 was highest in free C18:2n-6 and TG with C18:2n-6 ($P<0.005$). Interestingly apparent appearance of C18:1t11 was not different between PL, FFAs and TG with C18:2n-6. These preliminary results show that different lipid fractions could impact BH of n-3 and n-6 acids.

**The effect of feed restriction on growth, meat quality and carcass traits of Czech White rabbits**

*Z. Volek[1], L. Volková[1], E. Tůmová[2], D. Chodová[2] and L. Uhlířová[1,2]*
*[1]Institute of Animal Science, Přátelství 815, 104 00 Prague, Czech Republic, [2]Czech University of Life Sciences Prague, Kamýcká 129, 165 21 Prague, Czech Republic; volek.zdenek@vuzv.cz*

The aim of the present work was to evaluate the effect of a feed restriction on a feed cost, growth performance and meat quality and carcass traits of Czech White rabbits. This breed is one of 7 Czech rabbit breeds included in the National Program of Rabbit Genetic Resources. A total of 72 weaned rabbits (42 days old) were allocated to two groups (ADL or R rabbits) (36 rabbits per group) and fed pelleted feed (CP 16.9%, NDF 32.8%, EE 3.4%, starch 13.4%; 3-mm pellets 5-10 mm long). ADL rabbits were fed *ad libitum* over the whole fattening period, whereas R rabbits had the feed restriction (50% of ADL group) between 56 and 63 and 84 and 87 day of age. After a restriction period R rabbits were fed *ad libitum*. At the end of experiment (91 days of age), animals were slaughtered and used for the evaluation of carcass and meat traits. Over the whole fattening period, the feed restriction reduced significantly the feed conversion ratio (3.10 and 3.53 for R and ADL rabbits, respectively; P=0.009), and thereby a margin on the feed cost was improved by 10.3%. Carcass characteristics and meat quality were not greatly affected by the feed restriction, except for a profile of fatty acids. Hind leg meat of R rabbits contained more myristic acid (P=0.005), palmitic acid (P=0.007), oleic acid (P=0.020), eicosapentaenoic acid (P=0.053) and docosahexaenoic acid (P=0.079) than the ADL rabbits. However, no significant effect of the feed restriction on the saturated fatty acids/ polyunsaturated fatty acids ratio was observed. This study was supported by the Ministry of Agriculture of the Czech Republic (Project No. QI101A164).

---



**Beneficial effects of dietary Camelina cakes on cholesterol concentration in pig muscle and liver**

*M. Ropota, N. Lefter, E. Ghita and M. Habeanu*
*National Research Development Institute for Animal Biology and Nutrition, Chemistry and Nutrition Physiology, Calea Bucuresti nr.1, Balotesti, Ilfov, 077015, Romania; elena.ghita@ibna.ro*

The purpose of this study was to investigate during a 30-day trial the effects of three different protein-oleaginous sources (camelina cakes, sunflower meal and rapeseed meal) on the cholesterol level in two types of muscles (longissimus dorsi, LD and Semitendinosus, ST) and in liver. A total of 36 finishing hybrid Topigs [(Large White × Pietrain)×(Talent)] were assigned randomly to three homogenous groups (n=12 pigs/group): C (animals were fed with sunflower meal-based diets); E1 (animals were fed with rapeseed meal-based diets) and E2 (animals were fed with camelina cakes-based diets). The three types of meals were included in the same amount (12%) in the diets for the three groups. The cholesterol in muscle samples was determined by gas chromatography method using Gas chromatograph device Perkin Elmer Clarus -500 equipped and Elite 5 capillary column (30 m, 0.32 mm ID, 0.1 µm. thick film). Irrespective the type of tissue, the camelina cakes decreased highly significantly (P<0.0001) the cholesterol level (by 38% in LD muscle, 40% in ST muscle and 24% in the liver, compared to the sunflower meal). The rapeseed meal also had a beneficial effect, but less powerful (17% and 18% lower cholesterol level in ST and LD muscles, respectively, and 12% in the liver, compared to the sunflower meal, P=0.003). Liver cholesterol concentration was higher than the muscular cholesterol level (ranging between 45.39 mg/100 g DM for the camelina cakes and 59.76 mg/100 g DM for the sunflower meal, compared to 8 and 14 mg/100 g DM in LD and ST muscles, respectively). Camelina cakes, probably to its higher content of n-3 fatty acids, decreases highly significantly the muscular and hepatic tissue cholesterol.

**Bio-availability of two types of rumen bypass fat**

*J.L. De Boever[1], L. Segers[2], L. Vandaele[1] and S. De Campeneere[1]*
*[1]ILVO (Institute for Agriculture and Fisheries Research), Animal Sciences Unit, Scheldeweg 68, 9090 Melle, Belgium, [2]Orffa International, Vierlingstraat 51, 4251 LC Werkendam, the Netherlands; johan.deboever@ilvo.vlaanderen.be*

In early lactation dairy cows are not able to ingest enough energy to meet requirements. One may fill this energy gap by supplying rumen-protected fats. Nowadays, there are several bypass products on the market, but it is difficult for the farmer to judge their efficacy. We examined the bio-availability of 2 calcium-soaps of palm fats, Orffa Fat (OF) and Megalac (ML) and 2 palm fat prills, Bergafat F-100 (B100) and Bergafat T-300 (B300) containing fatty acids and fat, respectively. Rumen bypass and gut availability was determined by means of the *in situ* nylon bag technique. Per product eight bags (L×W: 7×3 cm, pore size: 37 μm) were filled with 1 g sample. The bags were put in a perforated cylinder and incubated in the rumen of a lactating cow for 8 h. After incubation, bags were machine-washed, freeze-dried and weighed. A few days later, the bags with bypass product were first soaked in a solution of pepsin in 0.1 N HCl during 1 h and then immediately inserted in the duodenum of the same cow. After passage through the intestines, which lasted for 9 to 11 h, bags were recuperated in the feces, intensively washed, freeze-dried and individually weighed again. On DM-basis, rumen bypass amounted to 95% for OF and to 100% for the other products, meaning that protection against rumen degradation was very effective. Gut availability on the other hand was very variable among products amounting to 83.5±4.0, 43.7±3.4, 12.1±4.2 and 13.9±2.0% for OF, ML, B100 and B300, respectively. Based on the fat content in the product, it can be calculated that OF, ML, B100 and B300 provide 17.2, 8.9, 2.9 and 2.8 MJ/kg net energy for lactation, respectively. These results indicate a clearly better efficacy of calcium-soaps than of hardened fat prills. It was hypothesized that the higher availability of OF as compared to ML is related with its finer particle size, amounting on average to 1.15 and 2.53 mm, respectively.

---



**Chemical composition and nutritive value of whole crop maize silage: effect of water shortage**

*F. Masoero, S. Bruschi, P. Fortunati, C. Cerioli, M. Moschini, G. Giuberti and A. Gallo*
*Università Cattolica del Sacro Cuore, Istituto di Scienze degli Alimenti e della Nutrizione, Via Emilia Parmense 84, 29122, Italy; francesco.masoero@unicatt.it*

Maize silage represents the main forage used in dairy cow diets. The growing of maize requires high amount of inputs for maximizing forage yield, such as irrigation water. Global climate changes could unavoidably intensify problems of water scarcity. Thus, the improvement of knowledge on the effect of irrigation water shortage on yield and nutritive value of maize silage is essential. Two commercial maize hybrids (FAO class 700) were grown on 2012/2013 in a location of Po Valley under two irrigation regimes: fully irrigated (FI) and water restricted (WR) conditions. The total amount of water was 440 mm for FI and 332 mm for WR, of which rainfall accounted for 240 mm. A split-plot factorial arrangement in a randomized complete block design with two main plots (i.e. FI and WS) and two sub plots (i.e. the two maize hybrids) with 6 replications/treatment was carried out. Individual plots consisted of 16 rows of plants seeded at 70.0 and 21.0 cm inter and within rows distances, respectively. Studied parameters were dry matter (DM) yield, chemical composition, rumen *in situ* DM and neutral detergent fiber (NDF) disappearances (DMD and NDFD, respectively). Whole-plant yield differed ($P<0.05$) between water irrigation regimes, being 24.3 t DM/ha in FI vs 17.8 t DM/ha in WR. The DM at harvest was 29% higher ($P<0.05$) in WR than in FI. Crops grown under WR showed a higher NDF (50.3 vs 48.2 g/kg DM; $P<0.05$) and a lower starch (22.6 vs 27.3 g/kg DM; $P<0.05$) contents. The DMD and NDFD were lower (54.5 vs 57.0 g/kg as fed and 60.2 vs 63.4 g/kg DM, respectively; $P<0.05$) in WR than in FI. Tested parameters marginally differed between hybrids. Concluding, the lower yield per hectare and the minor availability of nutrients suggest silages obtained from maize grown under WR regime could be less suited for high milk yield, representing an economical loss for farmers.

**Influence of the inclusion of standard tapioca meal in balanced diets on pig performance**

*E. Royer[1] and R. Granier[2]*
*[1]IFIP-Institut du porc, 34 bd de la Gare, 31500 Toulouse, France, [2]IFIP-Institut du porc, Les Cabrières, 12200 Villefranche-de-Rouergue, France; eric.royer@ifip.asso.fr*

Due to their high starch content, tapioca (cassava) roots (*Manihot esculenta* Crantz) are an excellent source of energy for pigs. However, lower inclusion rates in pig fattening diets are often proposed for standard tapioca pellets containing 62.5% starch than for high quality tapioca meal or chips containing 70% starch, as a result from higher ash and fiber contents and lower energy digestibility. An experiment was undertaken to examine the effect of the tapioca pellet proportion in feed on pig performance. A grower then finisher control diet based on sorghum, barley, soybean- and rapeseed- meals was compared with dietary proportions of tapioca meal of 10, 20 and 25% in the growing phase, then 15, 25 and 35% in the finishing phase. A standard tapioca meal with low cyanide content (<10 mg/kg of hydrocyanic acid), good bacteriological quality and containing 63.7% starch, 5.5% crude fiber, 5.5% ash and 3.1% HCL insoluble ash was used. All diets were balanced for net energy concentration (9.50 MJ/kg) and amino acids (0.90 and 0.80 g dig. lys. per MJ NE, for growing then finishing period, respectively). A total of 160 castrate male and female pigs (LW×Ld)×P76 were blocked, with 8 single-sex pens per treatment and 5 pigs per pen. Feed was distributed as meal mixed with water in the trough (1:1 water:feed), and pigs were fed according to plan. The pig herd had a poor health status with PMWS disease, but the number of clinical signs was not influenced by the diets. Feed intake (mean: 2.31 kg/d), daily gain (816 g/d), feed conversion rate (2.83 kg:kg) and carcass parameters were not affected by the tapioca meal inclusion rate. With these results, it can be concluded that tapioca meal can be used at a 30-35% incorporation rate in amino-acid balanced diets for growing–finishing pigs. Further investigations could be done to study the effects of higher tapioca levels on performance of high productivity herds.

---



**Relationship between iNDF and chemical and digestibility measures in concentrates**

*M. Johansen and M.R. Weisbjerg*
*Aarhus University, AU Foulum, Department of Animal Science, P.O. Box 50, 8830 Tjele, Denmark; marianne.johansen@agrsci.dk*

Indigestible NDF (iNDF) is an important feed characteristic in up to date feed evaluation systems, e.g. the Nordic NorFor. Today iNDF is determined using the *in situ* method with 288 h incubation in the rumen, which requires access to rumen fistulated animals. The aim of this study was to study the relationship between iNDF in concentrate feeds and chemical and digestibility measures, in the search for laboratory methods for iNDF prediction. 52 different concentrate feeds were analysed for iNDF content, chemical composition (NDF, ADF and ADL) and *in vitro* enzyme digestibility (EDOM). The simple linear relationship between iNDF and NDF, ADF, ADL and EDOM, respectively, were weak except for EDOM ($R^2$=0.32, 0.28, 0.52, 0.84, respectively). By separating the concentrates into types (cereals, oil, protein and fibre rich residues), the $R^2$ increased for all combined regressions ($R^2$=0.75, 0.80, 0.63 and 0.93, respectively). These results showed that the correlation between iNDF content and chemical composition was higher within uniform feedstuffs, whereas the correlation across different concentrate types was not satisfactory for a general prediction. By combining the chemical measures in multiple prediction equations the $R^2$ could be increased heavily, but including complexity and with a risk of over parameterisation of the model. The average iNDF:ADL ratio for all concentrates was 1.85 ranging from 0.32-6.63. For the different types of concentrates the average iNDF:ADL ratio was 2.47 (0.72-6.63), 1.58 (0.73-3.03), 0.98 (0.32-1.76), and 1.94 (1.37-2.36) for cereals, oil, protein and fibre rich residues, respectively. Therefore, the fixed factor approach as used in the CNCPS system (iNDF = 2.4×ADL) seems problematic.

**Effects of feeding of glycerine on performance, carcass traits, meat quality, and rumen metabolites**

*L. Barton, D. Bures, P. Homolka, F. Jancik and M. Marounek*
*Institute of Animal Science, Nutrition and feeding of farm animals, Pratelství 815, 104 00 Prague 10, Czech Republic; homolka.petr@vuzv.cz*

The objective of this study was to examine performance, carcass traits, muscle chemical composition, and rumen metabolites of bulls fed diets with different levels of crude glycerine. A total of 48 Fleckvieh bulls (initial age and live weight 222±16 d and 232±29 kg, respectively) were divided into four dietary treatment groups with different levels of glycerine supplementation: C (without glycerine), G5 (4.7% of glycerine), G10 (9.3% of glycerine) for the entire experimental period (266±38 d), and CG10 (0% for 118 d and then 9.3% of glycerine until slaughter). The diets were similar in their energy and protein contents with glycerine proportionally substituting barley meal. Feed intakes were recorded daily and the bulls were weighed every two weeks until slaughter (592±29 kg of live weight). After slaughter, rumen fluid was collected, carcass characteristics were recorded, and m. longissimus lumborum composition was determined. No significant effect of glycerine inclusion was observed in any of the growth performance, carcass and meat quality traits studied. Also, no apparent effects on rumen metabolites were detected. We conclude that crude glycerine can be used as a long-term substitution for barley meal up to the level of approximately 10% of dry matter in the diets of finishing bulls.

---

Session 45 Poster 22

**Levels of residual oil and extrusion on *in situ* rumen degradability of corn grain and corn germ**

*J.M.B. Ezequiel, S.L. Meirelles, R.L. Galati, R.N. Ferreira and E.H.C.B. Van Cleef*
*UNESP, Dept. of Animal Science, Jaboticabal, São Paulo, 14884-900, Brazil; janembe_fcav@yahoo.com.br*

The objective of this study was to evaluate the effect of levels of residual oil and subsequent extrusion on the corn and corn germ dry matter and protein *in situ* rumen degradability. The treatments were: (1) Corn; (2) Extruded corn; (3) Corn germ 7% EE; (4) Extruded corn germ 7% EE; (5) Corn germ 10% EE; (6) Extruded corn germ 10% EE. A ruminally cannulated Holstein steer (750 kg BW) was fed twice daily for 69 d a diet composed of 70% coast cross grass hay, 22.2% corn germ with 7% EE, 7.2% soybean meal, and 0.6% minerals. Each experimental period was composed of 21 d adaptation and 2 d data collection. The tested ingredients were ground to 2 mm, placed in 45-μm nylon bags (n=4/ingredient/time/period), and incubated in the rumen for 1, 3, 6, 12, 18, 24, 36, and 48 h. The potential degradability was calculated with the model: $p = a + b (1 - e^{kt})$, and the effective degradability with the model: $P = a + b \times [k / (k + K_p)]$. The soluble fraction was determined by washing bags in water and insoluble fraction considered the residue after 48 h of incubation. Data were analyzed as a completely randomized design with a 3×2 factorial arrangement of treatments, with the PROC GLM of SAS. The extrusion of corn germ 10% EE caused severe reduction in the solubility, but increased 48.6% the fraction 'b'. The DM effective degradability was increased with the extrusion for corn. The fraction 'a' of corn CP was increased, but it was decreased in corn germ 7% EE and corn germ 10% EE. The fraction 'b' of CP was improved with the extrusion in both treatments. The effective degradability of CP did not vary among extruded treatments, but it was decreased for non-extruded corn. It was concluded that the content of residual oil do not affect ruminal degradability of CP of corn germ, but it improves corn degradability, when fed to animals in maintenance ration. Furthermore, the extrusion process is efficient for corn, but it is not for the corn germs studied.

**Effect of linseed or algae on lamb meat quality, fatty acid profile, and FASD1 and FADS2 expression**

*O. Urrutia, A. Arana, K. Insausti, J.A. Mendizabal, S. Soret and A. Purroy*
*ETSIA, Universidad Pública de Navarra, Departamento de Producción Agraria, Campus de Arrosadía, 31006 Pamplona, Spain; olaia.urrutia@unavarra.es*

Omega-3 polyunsaturated fatty acids (PUFA) have beneficial effects on human health; nevertheless, an increase of PUFA in meat could enhance the development of organoleptic problems and cause a higher susceptibility to oxidation. For this reason this work aimed to study the effect of linseed, rich in α-linolenic acid (ALA), or linseed and microalgae, rich in eicosapentaenoic acid (EPA) and docosahexaenoic acid (DHA), on meat quality traits, fatty acid (FA) composition and the expression of fatty acid desaturase 1 (FADS1) and 2 (FADS2) in longissimus dorsi (LD) muscle. Thirty three Navarra breed male lambs were allocated to one of three diets: control, C (barley and soya-based concentrate); L (concentrate plus 10% linseed) and LA (concentrate plus 5% linseed and 3.7% microalgae). All lambs had *ad libitum* access to feeds from 16 to 26-27 kg LW. Samples of LD were taken for meat quality traits, FA composition and gene expression analysis. The pH of LD did not differ among different treatments (P>0.05). Feeding linseed did not affect LD oxidative stability and the organoleptic quality of meat. However, feeding linseed and algae increased TBARS concentration and reduced the ratings for odour, flavour and overall acceptance of meat (P<0.05). Results confirmed that feeding linseed or linseed and algae increased the content of ALA and trans-11 C18:1 and decreased cis-9 C18:1 and n6/n3 ratio in meat (P<0.001). Both L and LA diets increased contents of EPA and docosapentaenoic acid, but only the LA diet increased DHA content. Dietary PUFA caused a transcriptional regulation of genes involved in LCPUFA synthesis, with reduced expression of FADS1 and FADS2 genes (P<0.01). To conclude, algae supplementation at levels studied in this work, in contrast to linseed addition, would not be the most suitable tool to enrich lamb meat in PUFA considering the negative effect on sensory properties and oxidation susceptibility of meat.

---



**Meat quality of lambs fed four types of legumes silage**

*M. Girard[1,2], F. Dohme-Meier[1], M. Kreuzer[2] and G. Bee[1]*
*[1]Agroscope, Tioleyre 4, 1725 Posieux, Switzerland, [2]ETHZ, Universitätstrasse 2, 8092 Zurich, Switzerland; marion.girard@agroscope.admin.ch*

Meat of lambs finished on pasture may have a pastoral off-flavour. One reason could be the rapid ruminal degradation of protein to compounds like skatole (S) and indole (I). Plants prevalent in secondary metabolites such as condensed tannins (CT) or enzymes like polyphenol oxidase (PPO) are promising to slow down ruminal protein degradation and ultimately modify meat quality. A feeding experiment was conducted with 48 ram lambs (56-d old). They were assigned to 4 dietary groups with similar average initial body weight. Each group was offered either lucerne (LU; control), red clover (RC; containing PPO), or 2 CT-containing plants like sainfoin (SF) or birdsfoot trefoil (BT). Individual body weight was recorded weekly. The rams were slaughtered at an average age of 182 d. During slaughter, organs and carcasses were weighed. Perirenal fat was removed to determine S and I content. The longissimus dorsi was used to assess the fatty acidy profile of the intramuscular fat and to perform the sensory analysis by a trained panel. Compared to LU and RC, SF and BT lambs grew slower (P<0.01) and reached a lower slaughter and carcass weight as well as carcass yield (P<0.01). In SF lambs, relative liver and perirenal fat (expressed per kg carcass weight) was lower compared to BT lambs. By contrast, relative heart weight was greater (P<0.05) in SF and BT compared to RC lambs, with intermediate values for LU lambs. The intramuscular fat of lambs offered CT-rich plants contained more polyenoic fatty acids (FA), especially C18:3, C20:5 and C22:5, and less saturated FA (P<0.01). The S, but not I, content was 2-fold lower (P<0.05) in the SF than the LU group. The panelists found lower intensity for the descriptors 'livery' and 'sheepy' and a stronger intensity for the descriptor 'grassy' in the meat of the SF lambs compared to BT and RC groups. In conclusion, this study showed the potential of CT-rich plants to improve nutritional value and to reduce pastoral off-flavour of lamb meat.

**Fatty acid of nutritional interest in young Holstein bulls fed linseed and CLA enriched diets**

*I. Gomez[1], J.A. Mendizabal[1], M.V. Sarries[1], K. Insausti[1], P. Alberti[2], C. Realini[3], A. Purroy[1] and M.J. Beriain[1]*

*[1]ETISA, Campus Arrosadia s/n, Edif. Los Olivos, Universidad Publica de Navarra, 31006, Pamplona, Spain, [2]CITA de Aragón, Avda. Montañana, 930, 50059 Zaragoza, Spain, [3]Centro IRTA, Finca Camps i Arnet, 17121 Monells, Spain; inma.gomez@unavarra.es*

The aim of this work was to study the content of polyunsaturated fatty acid (PUFA) of nutritional interest in beef from cattle fed with concentrates enriched with whole linseed and protected conjugated linoleic acid (CLA). Forty-eight Holstein bulls were randomly assigned to one of four dietary treatments: Control, fed on corn, barley and soybean meal concentrate (C, n=12, 0% linseed and 0% CLA); Linseed (L, n=12, 10% linseed and 0% CLA); CLA (CLA, n=12, 0% linseed and 2% CLA); Linseed plus CLA (L+CLA, n=12, 10% linseed and 2% CLA). All the diets were isoenergetic (3.34 Mcal EM/kg) and isoproteic (16.9% CP). Animals were fattened from 239.8±6.61 to 458.6±9.79 kg body weight (322±6.0 d old at slaugther). Lipid profile was analyzed by GC and fatty acids were quantified using tricosanoic acid methyl ester (C23:0) as an internal standard. The contents of some PUFA of nutritional interest in bulls fed L and L+CLA diets were higher than those fed Control and CLA diets: C18:3n3 (ALA, 12.28 and 13.50 vs 1.84 and 2.08 mg /100 g muscle), C20:5n3 (EPA, 4.91 and 5.91 vs 2.39 and 2.54 mg /100 g muscle) and C22:5n3 (DPA, 11.78 and 13.26 vs 6.62 and 6.91 mg /100 g muscle). Addition of 10% whole linseed or 2% CLA improved the CLA content of beef compared with the beef from bulls fed Control diet (7.30 and 7.43 vs 5.23 mg /100 g muscle). Furthermore, linseed plus CLA showed a synergistic effect achieving the highest content of CLA (9.86 mg /100 g muscle). In conclusion, dietary supplementation with 10% linseed and 2% CLA would be recommended to increase the contents of some of the PUFA of nutritional interest (omega 3 and CLA) in beef from young Holstein bulls.



**Meat and fat quality of pigs given diets with increasing levels of dry pitted olive cake**

*M. Joven, M.E. Pintos, M.A. Latorre, J. Suárez-Belloch and M. Fondevila*

*IUCA. University of Zaragoza, Animal Production and Food Science, Miguel Servet 177, 50013 Zaragoza, Spain; jsuarezbelloch@gmail.com*

Olive cake is a by-product from olive oil industry characterised by a high fibrous content and an important proportion of fat, mostly oleic acid. A total of 60 Duroc × (Landrace × Large White) gilts, with 69.5±5.02 kg of body weight (BW) and 126±3 days of age were used. Increasing dietary proportions of depitted olive cake (OC; 5, 10 and 15%) were contrasted with a control (0%) diet, based on barley and soybean meal as major ingredients. Control diet rendered (per kg of feed) 3.01 Mcal ME/kg, 169 g crude protein and 7.9 g total Lys. Trial lasted 35 days and pigs were slaughtered with 96.7±7.45 kg BW. From each carcass, a sample of longissimus thoracis muscle (200±25 g) and a sample of subcutaneous fat at the coxal region were taken to be analysed. There were five replicates per treatment, being the pen with three pigs considered as the replicate. Chemical composition, water holding capacity and hardness were not affected by the experimental diet. However, lightness decreased linearly (P=0.02) and yellowness also tended to decrease linearly (P=0.06) with the dietary OC content. The increase in OC level promoted a linear reduction of total saturated fatty acid (SFA) proportion (P=0.01) due to a quadratic reduction of C17:0 (P=0.02) and a linear decrease of C18:0 (P=0.01). In addition, it was observed a linear increase in the total monounsaturated (MUFA) percentage (P=0.02) because of linear increases (P=0.01) in C18:1 and C20:1. The total polyunsaturated fatty acid content was not affected by dietary treatment. We can conclude that the effect of increasing level of OC in diet for growing pigs had scarce effect on meat characteristics but improved the fat composition by decreasing the SFA and increasing the MUFA, notably the proportion in oleic acid, which is desirable from a quality and healthy point of view for consumers. More variables should be considered to recommend the optimum level.

## Session 45 Poster 27

**Apparent digestibility of nutrients in growing pigs with diets high in cull chickpeas**

*J.M. Uriarte, H.R. Guemez, J.A. Romo, R. Barajas, J.M. Romo, J.F. Nuñez and N.A. López*
*Universidad Autónoma de Sinaloa, Facultad de Medicina Veterinaria y Zootecnia, Gral. Angel Flores s/n Poniente Col. Centro, 80000, Culiacán, Sinaloa, Mexico; jumanul@uas.edu.mx*

To determine the effect of the substitution of soybean meal and sorghum for cull chickpeas on apparent digestibility of nutrients in growing diets for pigs; six pigs (BW=36.15±1.5 kg; Large White × Landrace × Large White × Pietrain) were used in a replicated Latin Square Design. Pigs were assigned to consume one of three diets: (1) Diet with 14.8% NDF, 18.2% CP and 3.28 Mcal ME/kg, containing sorghum 68.6%, soybean meal 27.4%, and premix 4.0% (CONT); (2) Diet with 17.53% NDF, 18.19% CP and 3.28 Mcal ME/kg with sorghum 48.5%, cull chickpeas 35%, soybean meal 10.0%, vegetable oil 2.5%, and premix 4.0% (CHP35); and (3) Diet with 19.22% NDF, 18.24% CP and 3.26 Mcal ME/kg with sorghum 17.0%, cull chickpeas 70%, soybean meal 5.0%, vegetable oil 4.0%, and premix 4.0% (CHP70). Pigs were individually placed in metabolic crates (0.6×1.2 m). The adaptation period was 6 days and sample collection period was 4 days. From each diet and period, one kg of diet was taken as a sample and the total fecal production was collected. Feed Intake (1.88, 1.86 and 1.68 kg/day) was not affected by treatments (P=0.10) for CONT, CHP35 and CHP70, respectively. Apparent digestibility of DM (82.99, 83.76 and 82.92%) was similar (P=0.81) across treatments. Apparent digestibility of crude protein was not altered (P=0.35) by CHP inclusion (78.46, 77.57 and 76.04%). These results suggest that cull chickpeas can be used up to 70% in growing pig diets without affecting nutrient digestibility.

## Session 45 Poster 28

**Effect of essential oils, monensin and live yeast supplements on rumen fermentation in Jersey cows**

*L.J. Erasmus[1], A.P. Meiring[1], R. Meeske[2], R. Venter[3] and W.A. Van Niekerk[1]*
*[1]University of Pretoria, Animal and Wildlife Sciences, Private Bag X20, 0028 Hatfield, South Africa, [2]Outeniqua Research Farm, Western Cape, Department of Agriculture, P.O. Box 249, 6530 George, South Africa, [3]Vitam International, P.O. Box 9362, 0046 Centurion, South Africa; willem.vanniekerk@up.ac.za*

The prohibition of the use of growth promoting antibiotics in animal feeds within the EU and some other countries has led to increased interest in alternative means of manipulating rumen fermentation. The objectives of our study were to determine the effects of alternative natural feed additives, essential oils (EO) and rumen specific live yeast (LY), to monensin (MO) and their effects on rumen fermentation and in sacco NDF, starch and N disappearance. Four rumen cannulated lactating Jersey cows were used in a 4×4 Latin square design experiment. The four experimental treatments were: (1) Control (C), a Lucerne hay/maize based TMR (10.7 MJ ME/kg DM, 17% CP, 32.1% NDF); (2) C+MO; (3) C+EO; (4) C+EO+LY. The experimental periods were 25 days with the last 4 days for rumen sampling and the in sacco study. Mean ruminal pH ranged between 5.80 and 6.02 and tended (P<0.10) to be lower in the control cows when compared to the other treatments. Total VFA production, acetate:propionate ratio and rumen ammonia N concentration did not differ (P>0.05) but lactic acid concentration tended to be lower for the C+EO treatment when compared to MO supplemented cows. In sacco TMR starch disappearance at 24 h were higher for the EO treatment when compared to the C and C+EO+LY treatment groups (P<0.5). Nitrogen disappearance at 24 h was lower for EO+LY supplemented cows when compared to the MO and EO supplemented cows (P<0.05). Milk production and DMI did not differ (P>0.05) but fat % was higher for EO supplemented cows (4.52%) when compared to the C cows (4.32%; P<0.05). Results suggest a negative interaction between the essential oil and live yeast products used in this study. More research is needed on potential complimentary effects and interactions between feed additives.

### *Scutellaria baïcalensis* extract improves milk production in dairy cows

*F. Robert[1], L. Leboeuf[2] and E. Dupuis[2]*
*[1]CCPA Group, R&D department, ZA du bois de Teillay, 35150 Janze, France, [2]CCPA Group, Ruminants department, ZA du bois de Teillay, 35150 Janze, France; frobert@ccpa.fr*

In dairy cows, inflammation increases after parturition. Pro-inflammatory cytokines (TNFα, IL1β, IL6) may have a direct impact on feed intake, nutrients parturition and lead to decreased milk production. Anti-inflammatory drugs like salicylate used the days after parturition have been shown to have long-term effects increasing milk production. Nevertheless, the use of anti-inflammatory drugs to stimulate milk production is an off-label use not accurate for routine farming procedures. Flavonoids extracts have been shown *in vitro* and *in vivo* to have anti-inflammatory effects. *Scutellaria baïcalensis* extracts for instance reduces in a dosis dependent manner the level of TNFα, IL1β, IL6 on immune-stimulated animals. In order to evaluate the effects of this natural extract, we performed a trial in a prim'holstein dairy cows herd. 24 dairy cows were distributed in 2 groups according to expected calving date, lactation rank, milk production, fat and protein level at 305 days at previous lactation for multiparous cows and genetic indexes for primiparous cows. All the cows received a diet based on corn silage, wrapped grass and concentrate. The cows of the *Scutellaria* group (SCU) receive the plant extract in the concentrate daily from day 1 to day 60 post-partum. Milk production was recorded for each cow daily during 60 days post-partum. There was no significant difference in milk production in the 2 groups during the first month post-partum (41.7 and 42.8 kg/cow/day respectively for the control and SCU group). In the second month post-partum milk production was significantly increased in the SCU group (42.9 and 45.9 kg/cow/day respectively for the control and SCU group; P=0.015). The delayed effect observed in our trial has been described with salicylate. We observed the same effect of *S. baïcalensis* extract on lactation in sows. In conclusion, *S. baïcalensis* extract distributed to dairy cows after calving improves milk production.

---



### Joint Nordic genetic evaluation of sport horses increase accuracies of breeding values for stallions

*S. Furre[1], Å. Viklund[2], J. Philipsson[2], B. Heringstad[1] and O. Vangen[1]*
*[1]Norwegian University of Life Sciences, Department of Animal and Aquacultural Sciences, P.O. Box 5003, 1432, Ås, Norway, [2]Swedish University of Agricultural Sciences, Animal Breeding and Genetics, P.O. Box 7023, 75007 Uppsala, Sweden, Sweden; siri.furre@nmbu.no*

Young Horse Test (YHT) results from 41,216 horses (123,108 horses in pedigree) in the Nordic countries, Denmark (DEN), Finland (FIN), Norway (NOR) and Sweden (SWE) were included in a joint genetic analyses and breeding value estimation (EBV). FIN and NOR have too few tested horses yearly (20-30) to conduct a national genetic evaluation of stallions, while DEN and SWE publish national EBVs yearly. Performance traits included in the analyses were walk, trot, canter, jumping-performance, and two temperament traits (rideability during gaits test and attitude during jumping test). Fixed effects were sex, age, country and place×year and genetic analyses were done in DMU. Due to the limited number of tested horses in FIN and NOR, and the similarity between the YHT in these countries and SWE, test-data from the three countries were combined into one dataset (FNS). Genetic correlations between identical traits in FNS and DEN were moderate to high and in the range of 0.5 to 0.9. The lowest correlation was for rideability gaits (0.51) and the highest for the attitude during the jumping test (0.95). There were 343 common stallions had ≥15 offspring with YHT-results in total in two or more countries. Accuracies (rTI) of EBV for these stallions were computed with: (1) univariate analyses within country; (2) joint tri-variate analyses between similar traits in FNS and DEN; and (3) joint univariate analyses for each trait. Both (2) and (3) gave an average increase in rTI compared to (1), with (2) giving the largest increase (19-60%). All Nordic studbooks benefits from a joint genetic analyses as DEN and SWE can provide EBVs to their breeders for more than twice as many stallions as in the current national genetic evaluation, and FIN and NOR can start to publish EBVs for common stallions.

**Genetic parameters for eventing in Irish Sport Horses**

*K.M. Quinn-Brady, D. Harty and A. Corbally*
*Horse Sport Ireland, Beech House, Millennium Park, Osberstown, Naas, Co. Kildare, Ireland; kbrady@horsesportireland.ie*

The breeding objective of the Irish Sport Horse studbook is to produce a performance horse that is sound, athletic with good paces and suitable temperament and capable of winning at the highest international level in FEI disciplines. Eventing and showjumping are the main disciplines prioritised by breeders. In 2013, the Irish Sport Horse won the WBFSH studbook rankings for the fourteenth time. Although, genetic evaluations of showjumping ability have been carried out in Ireland since 1995, they have not been available for eventing ability. A study was carried out to estimate genetic parameters for eventing. Data used in the study included pedigree data on Irish Sport Horses (n=57,136) and a database of national and international performances constructed from FEI data, manually retrieved international performances and national performance data (n=102,950). The Lifetime Performance Rating (LPR) of 2,406 horses was assessed as the 'highest level' successfully achieved. Success was defined as the highest level of competition at which a horse has achieved a Minimum Eligibility Requirement as set down by the Federation Equestre Internationale (FEI). There are thirteen possible levels of competition ranging from national training (e.g. EI 80 or BE 80) competitions to CCI4* events. Multivariate analyses of competition results using a normal score based on the penalty points from each phase of the eventing competiton – dressage, showjumping and cross country and the overall competition score were also carried out. Genetic parameters were estimated in uni- and multivariate animal models with Asreml software. The heritability of LRP for eventing was estimated at 0.28. Genetic parameters between levels of competition were moderate to high showing that results from younger horses could predict a horse's performance at advanced level.



**Estimation of breeding values and genetic changes of Posavje horse in Slovenia**

*B. Bračič and K. Potočnik*
*University of Ljubljana, Biotechnical Faculty, Department of Animal Science, Groblje 3, 1230 Domžale, Slovenia; klemen.potocnik@bf.uni-lj.si*

Posavje horse is an indigenous horse breed in Slovenia as well as in Croatia, traditionally reared in the surroundings of the river Sava. Slovenian population of Posavje horse was estimated on 1,550 horses in the year 2013. Breeding program has defined type traits scoring which are not linear and uses these scores as the main selection criteria. Breeding values (BV) for Posavje horse were estimated for the first time in 2013. The aim of the study was to consider the usefulness of estimated BV in selection. To make the BV estimation for the 23 body traits (7 measured and 16 scored traits) 764 Posavje horses were used, 50 stallions and 714 mares born between 1999 and 2010. Pedigree records counted 1220 animals. The mixed model were used for the genetic evaluation of type traits considering gender, age and birth year as fixed effects as well as animal as random effect. Variance components were estimated with using VCE-5 package. Heritabilities were low (0.04-0.10) for forelegs, hind legs, gaits correctness and gaits efficacious. Moderate heritabilities (0.11-0.25) were found for neck, front and middle part of the body as well as overall score. High heritabilities (0.26-0.61) were estimated for wither height, chest girth, width and chest depth, croup height and width, body length, cannon bone circumference, breed type, head, rear part of the body, and four indices like cannon bone circumference, chest with, chest depth, croup width divided by wither height. Genetic correlations among all included traits were positive, except between forelegs-hind legs and forelegs-gaits correctness. Most of genetic trends of type traits were low. This could be explained that selection on phenotype values was not efficient. For the near future, introduction of joint prediction of breeding values for Slovenian and Croatian Posavje horse populations would be performed.

**Genetics of lifetime reproductive performance in Italian Heavy Draught Horse mares**

*R. Mantovani[1], F. Folla[1] and G. Pigozzi[2]*
*[1]Dept. of Agronomy Food Natural resources Animals and Environment, Viale dell'Universita', 16, 35020 Legnaro (PD), Italy, [2]Italian Heavy Draught Horse Breeders Association, Via Verona, 90, 37068 Vigasio (VR), Italy; roberto.mantovani@unipd.it*

The aims of this study were to find out a phenotypic value to express mares' lifetime reproductive performance in Italian Heavy Draught Horse breed and to analyze the genetic components of this trait. A first step of the study involved 1,487 mares born after 1990 for which all reproductive events were extracted from the studbook. Mares had at least 6 subsequent registered reproductive seasons (RS), and belonged to environmental units with at least 2 observations (group of studs in the same geographical area and common rearing system by year of birth). This training dataset was used to implement a set of predictive equations to estimate foal production (FP) at the 6$^{th}$ RS depending on previous FP (i.e. after 3, 4 or 5 RS), and accounting or not for the age at first foaling (3 or 4 years when considered). The validation of predictive equations was carried out by analyzing the mean biases (i.e. differences between actual and estimated FP). A further dataset accounting for 2,904 mares following the same editing restriction as above, but with at least 3 registered RS, was used to obtain the foaling rate at the 6$^{th}$ RS (i.e. foals after 6 RS on potential foals). This dataset contained actual (n=1,487) and estimated (n=1,417) foaling rates and was used to estimate the heritability of the trait accounting for 6,593 animals in the pedigree file. The predictive equations resulted well validated when separate linear regression equations accounting for the age at first foaling were used to estimate FP at 6$^{th}$ RS, with minimum percentage square biases (1.2%), mean absolute deviation (0.45) and residual standard deviation (0.54). The estimated foaling rate showed a moderate but significant genetic variation, and the heritability of the trait resulted 0.244.



**Systematization of recording and use of equine health data and its potential for horse breeding**

*K.F. Stock[1], S. Sarnowski[1], E. Kalm[2] and R. Reents[1]*
*[1]Vereinigte Informationssysteme Tierhaltung w.V., Heideweg 1, 27283 Verden, Germany, [2]Christian-Albrechts-University of Kiel, Institute for Animal Breeding and Husbandry, Olshausenstrasse 40, 24098 Kiel, Germany; friederike.katharina.stock@vit.de*

Appropriate consideration of equine health and welfare in breeding programs has started to become a factor of competitiveness for the studbooks. However, the low level of systematization of health data recording and major concerns regarding assess and use of health-related information have interfered with developing strategies for efficient improvement of equine health traits by breeding. In Germany, initiatives of improving the quality of routinely collected phenotype data of riding horses have been intensified in 2013 with special focus on health, so first results of the work of a consortium of stakeholders in Warmblood breeding, including veterinarians and studbook representatives, can be presented. Because harmonization of health phenotypes is base requirement for population-wide analyses, a comprehensive recording standard for equine health data was developed with distinct sections for diagnoses, radiographic findings and clinical findings. Utmost flexibility of use was ensured by hierarchical structure which allows simple as well as highly detailed recording with consistent coding, such that design of integrative systems for equine health data is facilitated. Veterinary support is considered a key factor of success of health-oriented measures in horse breeding, and the new recording standard can serve as reference in user-friendly software applications. Logistics around a central equine health data base involve data transfer from veterinary software and information exchange with breeding data bases and must meet high demands on data security. Interdisciplinary collaboration has enabled recent development towards systematic recording and use of equine health data, which will benefit research and practical breeding by increase of epidemiologic knowledge and approaches to targeted improvement of equine health and welfare.

## Session 46 Theatre 6

**SNPs associated with osteochondrosis in horses on different stage of training**

*D. Lewczuk[1], M. Hecold[2], A. Ruść[3], M. Frąszczak[4], A. Bereznowski[2], A. Korwin-Kossakowska[1], S. Kamiński[3] and J. Szyda[4]*

*[1]IGAB PAS Jastrzębiec, ul. Postępu 36A, 05-552 Magdalenka, Poland, [2]Warsaw Uniwersity of Live Sciences-SGGW, ul. Nowoursynowska 166, 02-787 Warszawa, Poland, [3]University of Warmia and Mazury Olsztyn, ul. M. Oczapowskiego 2, 11-041 Olsztyn, Poland, [4]Wrocław University of Environmental and Live Sciences, ul. C.K. Norwida 25, 50-375 Wrocław, Poland; d.lewczuk@ighz.pl*

The aim of the study was to identify single nucleoid polimorphysm (SNP) associated with osteochondrosis dissecants (OCD) independent on the stage of horse training. The group of 201 horses was x-rayed twice (before and after 100 days performance test) using digital equipment. Three kind of joints (4 fetlock, 2 stifle, 2 knee) were tested for the overall presence of OCD and horses were grouped as affected with OCD and health. DNA isolated from blood was used to genotype of each horse genome by the Illumina EquineSNP50 genotyping BeadChip (consisting of 54,602 SNPs). Statistical analysis including Cochran-Armitage test and logistic regression assuming an additive model of inheritance – except SNP effect, also training centre, sex, age, breeder and pedigree information were used. The SNP effects were estimated twice: before and after training and differed between investigations. Only four SNPs out of 20 observed as important for each term were involved in OCD reveal at each stage of training. This research were founded by NCN grant 2011/01/B/NZ2/00893.

## Session 46 Theatre 7

**Morphological characterization in Ayvacik Pony: a nearly extinct horse breed in Turkey**

*O. Yilmaz[1] and M. Ertugrul[2]*

*[1]Canakkale Onsekiz Mart University, Faculty of Agriculture, Department of Animal Science, Terzioglu Yerleskesi, Canakkale, 17020, Turkey, [2]Ankara University, Faculty of Agriculture, Department of Animal Science, Diskapi, Ankara, 06110, Turkey; zileliorhan@gmail.com*

The target of this study was to analyze the body coat colour and some morphological traits of Turkish Ayvacik horses used by smallholders in Canakkale regions. A total of 198 Ayvacik horses (128 males and 70 females) was analyzed divided in three age groups (4-6, 7-9, and 10-26 years old). The descriptive statistics gave the following means for the analyzed traits: 126.1±0.37 cm for height at withers, 125.4±0.37 cm for height at rump, 133.4±0.52 cm for body length, 151.4±0.57 cm for heart girth circumference, 56.6±0.21 cm for chest depth, 16.9±0.07 cm for cannon circumference, 52.7±0.19 cm for head length, and 13.3±0.09 cm for ear length. In the sampled horses the frequencies of body coat colour were: 63.1% for bay, 17.2% for black 10.6% for grey, 6.6% for chestnut, 1.1% for Roan, and 1.0% for buckskin. It can be concluded that Turkish Ayvacik Horses reach mature body size at 4 years old being the smallest size horse breed of Turkey. This breed is increasing the census because of the continuous crossbreeding with other horse breeds.

**Tools to better manage the genetic variability of French breeds**

*S. Danvy*[1], *M. Sabbagh*[1], *C. Blouin*[2] *and A. Ricard*[2]
[1]*IFCE, Recherche et Innovation, 61310 Exmes, France,* [2]*INRA, UMR1313, 78352 Jouy-en-Josas, France; sophie.danvy@ifce.fr*

The French territories diversity is also represented by its equine breeds (28 French breeds) that constitute a real landscape management value and are greatly appreciated by the general public. The current economic climate and the withdrawal of the public stallions have led to a significant decline in births -22% in 5 years without Trotters. Analysis of their genetic variability is facilitated by the existence of an unique national database (SIRE). However, significant differences in pedigree depth between studbooks are observed: depending on the breed, between 3 to 9 generations are filled up more than 50%. Those differences have a big impact on the relevance of genetic variability indicators. Through the analysis of these genealogies, inbreeding coefficients, major ancestors and racial composition are regularly updated. The average inbreeding coefficient calculated for all the horses born in 2011 is equal to 1.3% for the Selle Français, 2.2% for the Trotteur Français and 6.9% for the Poitevin Mulassier. In 2013, thanks to funding provided by CASDAR (French Minister of Agriculture, project VARUME), a simulation tool which calculates the inbreeding of an unborn foal has been developed and is available on the internet. Some studbooks are using these new and genetic variability indicators and integrate them into their breeding program. For instance, these tools are implemented during the stallions' agreement process or by encouraging financially the breeders to reason their breeding. Others are looking for solutions to imbalance the genetic bottlenecks created by using too much a narrow number of sires. In all breeds, there is an awareness of the importance of managing genetic variability. However, the implementation of genetic management to overcome this problem is still complicated. There is still a large amateurism of the French breeders (over 64% of breeders only put 1 mare for breeding), and their individualism leads to difficulties in applying a collective breeding program.

---



**France commits in the preservation of its horses and donkeys genetic resources**

*M. Magistrini*[1], *P. Milon*[1], *M. Caillaud*[2], *M. Sabbagh*[3], *F. Mea*[4] *and S. Danvy*[3]
[1]*INRA, UMR85 Physiologie de la reproduction et des comportements, 37380 Nouzilly, France,* [2]*IFCE, Formation, Jumenterie du Pin, 61310 Exmes, France,* [3]*IFCE, Recherche et Innovation, Jumenterie du Pin, 61310 Exmes, France,* [4]*IFCE, DAFAT, Haras national de Blois, BP14309, 41043 Blois cedex, France; michele.magistrini@tours.inra.fr*

The French genetic resources are very diverse and are a real wealth that must be maintained. However, given the significant changes in breeding farms and the use of horses in recent decades, the potential risk of disappearance of these genetic resources has become real. To help the preservation of this diversity, a major research project (CRB-anim) was built in 2010, for 11 species, aiming to develop, consolidate and enrich the already existing biological resources centres for domestic animals in France. This project is funded by the National Research Agency ('Investissements d'Avenir', ANR-11-INBS-0003); it has started in 2013 and will finish in 2020. A close collaboration between INRA and IFCE has started and two main axes will be conducted: (1) A research axis to remove technical locks to the conservation of biological samples: (a) to improve the fertility using donkey frozen semen; the new freezing extenders recently finalized for stallion sperm will be tested, (b) to develop a reliable technique for equine embryo freezing; the first approach will be to evaluate the effect of a deposit of prostaglandin E2 to the uterus-tubal junction on the arrival of the embryo in the uterus. The objective is to get small embryos (morula stage), more readily freezable by slow freezing technique. (2) A second axis encompasses a further collections of biological material in our National Cryobank; so the most endangered breeds in terms of number or genetic variability are properly represented. A genetic analysis of French horse breeds is conducted to determine which populations were given priority for preservation. Sperm collections for donkey breeds will be processed as soon as the research will have removed the lock on semen freezing in this species.

**Morphometric evaluation of Arabian horses in Algeria: proposal of equation for body weight estimate**

*S. Tennah[1,2], N. Antoine-Moussiaux[2,3], N. Kafidi[1], D.D. Luc[4], N. Moula[2,3], C. Michaux[2], P. Leroy[2,3] and F. Farnir[2]*

*[1]School Veterinary National Superior of Algiers, Haecène Badi, Algiers, B.P. 161, Algeria, [2]Faculty of Veterinary Medicine, University of Liège, FARAH (Fundamental and Applied Research for Animal Health), 4000 Liège, Belgium, [3]Tropical Veterinary Institute, Faculty of Veterinary Medicine, University of Liege, 4000 Liege, Belgium, [4]Hanoi Agricultural University, Hanoi, UAH, Viet Nam; f.farnir@ulg.ac.be*

This study consists in morphobiometric characterization of 54 stallions and 44 mares over three years old, distributed on three racetracks of Algeria (Zemmouri, Tiaret and Caroubier) and belonging to 77 owners-breeders. Nineteen body measurements were recorded and the body weight was also measured. The average weight is 456.2±43.0 kg, varying from 335 kg to 545 kg. The selection of variables to be included in the equation for body weight estimation was performed using the stepwise procedure of the SAS software. Four measurements were selected among nineteen for the proposed estimation of body weight (BW) horses equations: the thoracic perimeter (TP), the height at the croup (HC), the length of the neck (LN), and the girth of the neck (GN). Equations estimated for stallions and mares are, respectively: $BW = 7.024 \times TP - 787.119$ ($R^2$=0.99), $BW = 6.207 \times TP + 0.633 \times HC + 0.668 \times GN - 0.878 \times LN - 746.370$ ($R^2$=0.96). The results of this study should allow owners-breeders and the trainers to easily track the weight of their horses. This monitoring is necessary to adapt the activity and feeding of horses and favor their race performance.



**Relationship between linear conformation traits and dressage performance scores in Pura Raza Español**

*M.J. Sánchez[1], A. Molina[2], M.D. Gómez[3], I. Cervantes[1] and M. Valera[3]*

*[1]Departamento de Producción Animal, Universidad Complutense de Madrid, 28040, Spain, [2]Departamento de Genética, Universidad de Córdoba, 14071, Spain, [3]Departamento de Ciencias Agroforestales, Universidad de Sevilla, 41013, Spain; icervantes@vet.ucm.es*

In horse breeding, functional animals competing at the highest competition level are demanded. Therefore, it is continuously necessary to develop methodologies for the morphofunctional improvement of animals and to incorporate the obtained advances into the breeding programs. Conformation assesses the structure of an animal in relation to its function and its breed pattern. Therefore, relationship between dressage performance and conformation has to be analyzed. There were 10,127 conformation records collected with a linear evaluation methodology; and 11,442 Dressage records obtained in the young horse performance tests from 1,518 horses, 90% of Dressage participant parents had an offspring with linear morphological records. A total of 567 horses had both records. The pedigree matrix contained 37,229 individuals. A BLUP univariate animal model was applied for the final score in the test and the scores of walk, trot and canter as performance traits in Dressage. The model included age (3 levels), sex (2), stud of birth (358) and event (249) as fixed effects; and rider, match (rider-horse), permanent environmental effect, animal and residual as random effects. For the 35 linear score traits collected the model included: age (8 levels), sex (2), stud of birth zone (49) and combination event-appraiser (461) as fixed effects; and animal and residual as random effects. Correlations between breeding values of Dressage and conformation were calculated. The 93.57% of them were significant, ranged between 0.01 and 0.35; and the 16.43% had a low-medium magnitude significant correlation, ranged between 0.20 and 0.35. It could be highlighted correlations between Dressage and hoof-limb direction, schium-stifle distance, length and angle of croup with the breeding values obtained for walk, trot and canter.

**The use of Thurstonian models in genetic evaluations of ranking in endurance races**

*I. Cervantes[1], M.J. Sánchez[1], M. Valera[2], J.P. Gutiérrez[1] and L. Varona[3]*
*[1]Departamento de Producción Animal, Universidad Complutense de Madrid, 28040, Spain, [2]Departamento de Ciencias Agroforestales, Universidad de Sevilla, 41013, Spain, [3]Área de Genética Cuantitativa y Mejora Animal, Universidad de Zaragoza, 50013, Spain; icervantes@vet.ucm.es*

The discontinuous nature of the ranks in competitions has promoted the use of different transformations in the genetic models. Nowadays, there are statistical tools that allow using the full rank without any transformation e.g. threshold models, where there is a tricky question to define the number of thresholds. Moreover, given that the rank of the horse in the competition is independent from the level of such competition, the model of choice should be able to adjust the results for the competitive level of each race. The endurance competitions are long distance races that vary from 40 to 160 km per day. The aim of this study was to use a full Bayesian approach to estimate variance parameters for ranks using a Thurstonian model. Results were compared with a threshold model with 8 categories (1$^{st}$ for eliminated, 2$^{nd}$ for ranks higher than 7, and the others the inverted rank). A total of 3,365 records from 812 animals (68% Arab horses) from competitions between 2000 and 2012 were used. Pedigree contained 6,643 animals and 10 genetic groups. Sex, age, and competition were fitted as systematic effects, and rider-horse interaction, permanent environmental effect and additive genetic random effects were fitted besides residual. Heritability and repeatability were 0.075 (0.036) and 0.259 (0.046) for Thurstonian model, and 0.053 (0.025) and 0.083 (0.017) for threshold model. The ratio of rider-horse interaction (variance of rider-horse effect/phenotypic variance) was lower in the Thurstonian (0.057 (0.036)) than in the threshold model (0.133 (0.033)). The correlation and the rank correlation between breeding values predicted with both models were high (0.98 and 0.97). Genetic parameters from Thrustonian model resulted more accurate and allowed comparison of animals not competing together.

---

Session 46 Poster 13

**Genetic diversity of Slovak Sport Pony based on genealogical information**

*O. Kadlečík and R. Kasarda*
*Slovak University of Agriculture in Nitra, Department of Animal Genetics and Breeding Biology, Tr. A. hlinku 2, 949 76 Nitra, Slovak Republic; radovan.kasarda@uniag.sk*

The aim of this study was the assessment of genetic diversity parameters in Slovak Sport Pony using genealogical information. The origin of Slovak Sport Pony was related with the ambition to create a multi-purpose horse preferentially used for preparatory riding of children, tourism and recreation riding. This breed has been established in 1980, based on the combination of crossing warm blood horses bred in Slovakia and Wells Pony stallions. The breeding strategy under development includes management of genetic diversity to control inbreeding using a specific mating program based on optimum contribution tools. The analysis included 179 individuals (63 stallions, 116 mares) in the pedigree (PP) and 45 (4 stallions, 41 mares) in the reference population (RP). In parent and grand-parent generations 100% of individuals were known and it decreases to 64.03% in 5$^{th}$ generation. The pedigree completeness in PP was lower, with an average of 67.70% with a substantial decrease to only 31% of known individuals in 4$^{th}$ generation. The average of maximum number of generation was 6.93 and the number of fully traced generations was 4.0. The average of equivalent complete generations in RP was 4.97±0.56, average value of inbreeding was 3.33%, average relationship coefficient was 9.44%, $\Delta F_i$ was 0.78% and realised effective population size had a value of 23.27±0.94. The evaluated population was derived from 57 founders. The effective number of founders was 35, and the effective numbers of ancestors was 7. There are 3 ancestors explaining 50% of genetic diversity and 16 ancestors contributing to RP. Branco, Shalom and Karneol-70 were the most important ancestors. High average values of the inbreeding and its relatively high increase per generation near 1%, the high relatedness of individuals as well as the low effective population size support the necessity of the monitoring and mating management of Slovak Sport Pony nucleus population with the aim to increase the effective population size.

**Pedigree analysis of Old Kladruber horse**

*H. Vostra-Vydrova[1,2], L. Vostry[1,2], B. Hofmanova[1] and I. Majzlik[1]*
*[1]Czech University of Life Sciences Prague, Faculty of Agrobiology, Food and Natural Resources, Kamycka 129, 165 21 Prague, Czech Republic, [2]Institute of Animal Science, Pratelstvi 815, 10401 Prague, Czech Republic; vydrova@pef.czu.cz*

The Old Kladruber horse is the oldest original Czech horse breed included among genetic resources, with pedigree records spanning three centuries. Because the population is closed, there is a concern about the loss of genetic variation. In order to estimate effective population size, generation interval and the trend of inbreeding coefficients ($F_x$) in two colour variants (grey and black) Old Kladruber individuals born in the Czech Republic in the period 1990-2013 were analysed. The average values of generation interval (in years) between parents and their offspring were: 12.00 sir-son, 12.28 sir-daughter, 10.43 dam-son and 10.47 dam-daughter. Average values of effective population size were estimated to be: 61.64 in grey variant and 44.76 in black variant. The average values of inbreeding coefficient were 11.09% in grey stallions and 11.16% in grey mares and 14.89% in black stallions and 14.87% in black mares. Overall averages of $F_x$ were: 11.13% for the grey and 14.88% for the black variant. These results suggest the possible reduction in genetic variability in the population. Reduction in genetic variability could be controlled increasing the number of stallions used for breeding.



**Population study of three breeds of cold blooded horses in the Czech Republic**

*B. Hofmanová[1], M. Přibáňová[2], I. Majzlík[1], L. Vostrý[1] and H. Vostrá[1]*
*[1]Czech University of Life Sciences Prague, Faculty of Agrobiology, Food and Natural Resources, Kamycka 129, 165 21 Prague, Czech Republic, [2]Czech-Moravian Breeders' Corporation, Hradištko p.M. 123, 252 09, Czech Republic; hofmanova@af.czu.cz*

The genetic diversity of cold-blooded breeds Silesian Noriker (SN), Noriker (N) and Czech-Moravian Belgian Horse (CMB) was studied using a microsatellite set. CMB has been formed in the CR territory on the basis of Belgian and Walloon stallions. SN has originated from imports of original Noriker and Bavarian cold-blooded stallions. N was originally used for the formation of SN breed, these two breeds were further separated geographically and in 1991 were recognized as separate breeds. The population size for each breed is: SN (♂ 43, ♀464) with an effective population size ($N_e$) of 43.1, N (♂ 68, ♀ 967) with $N_e$ of 79.1, and CMB (♂ 61, ♀ 1012) with $N_e$ of 52.2. Breeds SN and CMB belong to critically endangered horse gene resources of the CR. The study includes 149 SN heads, 246 N heads and 328 CMB heads. There were 13 microsatellite loci (VHL20, HTG4, AHT4, HMS7, HTG6, AHT5,HMS6, ASB2, HTG10, HTG7, HMS3, HMS2, HMS1) analyzed using TFPGA software (Miller, 1997). Results showed an average allelic number of 8.1 with minimum of 4 (HTG7) and a maximum of 12 (HMS1). The average heterozygosity is similar over all loci: 0.68 for SN, 0.71 for N, 0.69 for CMB. F statistics values were: $F_{IS=}$-0.007, $F_{IT}$=0.04 and $F_{ST}$=0.047. Genetic distances were evaluated according to Nei as 0.03 between SN – N, 0.17 between SN – CMB and 0.14 between N – CMB. For Nei's original distance the UPMGA Cluster was constructed.

**Can suckling increase cheese yield and welfare in the Swedish dairy goat?**

*M. Högberg, A. Andrén, E. Hartmann, E. Sandberg and K. Dahlborn*
*Swedish University of Agricultural Sciences, Anatomy, Physiology and Biochemistry, Box 7011, 750 07 Uppsala, Sweden; kristina.dahlborn@slu.se*

Swedish dairy goats produce milk with low contents of fat and casein which affects the cheese making properties negatively. Goats store large volumes of milk in the gland cisterns that can be obtained without oxytocin (OT) induced milk ejections. The cisternal milk has in contrast to the alveolar milk a poor fat concentration. The aim was therefore to investigate if milk fat and thereby cheese yield could be improved by suckling. Swedish dairy goats were kept together with one kid each in MIX-systems (milking combined with suckling). In study I (n=8), milk samples were collected from one teat before, during and after suckling (S, day one) and milking only (M, day 2). Blood samples were collected on both days (before, during and after S/M), plasma levels of OT was analyzed by ELISA. In study II, milk composition and individual cheese yield was measured when 12 goats participated in a cross-over design with 4 treatments. In treatment 1 and 2, does and kids were separated for 8 h a day and were either suckled before milking (S-8) or milked only (M-8). The same procedure was applied for treatment 3 and 4, except that doe and kid were separated for 16 h a day (S-16) and (M-16). Milk composition was analyzed by mid-infrared spectroscopy, the casein content and cheese yield were measured by a rennet-coagulation method. Proc Mixed (SAS®) was used for statistical analyses. In study I, plasma levels of OT increased during S, but not during M. The milk fat concentration was higher during S ($P<0.05$) than during M. In study II, fat concentration and individual cheese yield were higher during S-8 and S-16 than during M-8 and M-16. Suckling in connection with milking improved udder emptying and increased both milk fat and cheese yield.

---



**Prevalence and risk factors for limb and foot lesions in piglets**

*L. Boyle[1], A. Quinn[1], A. Kilbride[2] and L. Green[2]*
*[1]Teagasc, Pig Development Department, Moorepark, Fermoy, Co. Cork, Ireland, [2]Life Sciences, University of Warwick, Coventry, CV4 7AL, United Kingdom; laura.boyle@teagasc.ie*

A survey of 68 farms was conducted to determine the prevalence and risk factors for foot and limb lesions in 2,948 piglets from 272 litters. One litter was selected per 4 age (from 3 to 28 d) categories per farm and all piglets in each litter were examined for abrasions, joint swellings, sole bruising and erosion, claw swellings and coronary band injuries which were scored from 0-3 based on size. 58 environmental parameters (e.g. crate, pen dimensions and characteristics [material etc.]) were recorded for each litter. A questionnaire was completed on management, health and performance factors for each farm. Lesion prevalence was calculated and multilevel mixed effect logistic regression for 145 factors was conducted using MlwiN 2.27. Prevalence: sole bruising=61.5%, sole erosion=34.1%, coronary band injuries=11.3%, abrasions=55.7%, swollen joints=2.4% and swollen claws=4.4%. The only risk factors with a significant impact on these lesions were age and the nature of the openings in the floors in the piglet area of the farrowing pen. Age was a risk factor for sole bruising (OR 0.043; CI 0.41, 0.48) and coronary band damage (OR 0.66; CI 0.59, 0.74) (decreased with age), and skin abrasions (OR 1.16; CI 1.08, 1.24) which increased with age. There was a reduced risk of sole bruising (OR 0.24; CI 0.18, 0.32) and limb abrasions (OR 0.48; CI 0.35, 0.65) in floors with oval shaped openings compared to floors with rectangular shaped openings. Oval shaped openings were predominately associated with plastic coated woven wire floors while rectangular openings were typical of slatted steel floors which are highly abrasive. The low number of risk factors identified was related to the uniformity of the farrowing systems used. However, the high prevalence of lesions indicates that the physical features of farrowing pens currently in use pose a serious risk to piglet welfare and that this risk is exacerbated by slatted steel flooring.

## The influence of flooring type before and after 10 weeks of age on osteochondrosis in gilts

*D.B. De Koning[1], E.M. Van Grevenhof[2], B.F.A. Laurenssen[1], P.R. Van Weeren[3], W. Hazeleger[1] and B. Kemp[1]*
*[1]Wageningen University and Research centre, Adaptation Physiology Group, De Elst 1, 6700 AH Wageningen, the Netherlands, [2]Wageningen University and Research centre, Animal Breeding and Genomics centre, De Elst 1, 6700 AH Wageningen, the Netherlands, [3]Faculty of Veterinary Medicine, Utrecht University, Department of Equine Sciences, Yalelaan 112, 3508 TD Utrecht, the Netherlands; danny.dekoning@wur.nl*

Osteochondrosis (OC) involves formation of abnormal cartilage in joints in growing gilts and is associated with lameness and premature culling at older age. Flooring type in fattening pigs seems to affect OC prevalence. However, it is unknown if flooring type affects OC prevalence in breeding gilts which have to last several parities. This study investigated age dependent effects of flooring type, conventional concrete partially slatted versus wood shavings, on OC prevalence in breeding gilts. A total of 212 'Topigs 20' gilts were divided over 4 treatments: from weaning (4 weeks of age) till slaughter (24 weeks of age) on conventional flooring (CC); from weaning until slaughter on wood shavings (WW); from weaning until 10 weeks of age on conventional flooring and thereafter on wood shavings until slaughter (CW); from weaning until 10 weeks of age on wood shavings and thereafter on conventional flooring until slaughter (WC). After slaughter the elbow, hock, and knee joints were macroscopically scored on a 5 point grading scale for OC. Total OC prevalence did not differ significantly between treatments. Gilts kept on wood shavings before or after 10 weeks of age, however, had higher odds ($P \leq 0.05$) to have higher OC scores than gilts kept continuously on partially slatted flooring (OR=2.3 for CW; OR=2.6 for WC; OR=3.7 for WW). A clear age dependent effect of flooring type on OC prevalence was not present. The results suggest that breeding gilts should not be kept on wood shavings during rearing to avoid development of severe OC lesions, thus increasing leg quality and reducing lameness at older age during productive life of breeding sows.



## Transmission and control of the parasite *Ascaris suum* (large round worm) in Danish organic farms

*H. Mejer, K.K. Katakam, S. Gautam, A. Dalsgaard and S.M. Thamsborg*
*University of Copenhagen, Department of Veterinary Disease Biology, Dyrlaegevej 100, 1870 Frederiksberg C, Denmark; hem@sund.ku.dk*

*Ascaris suum* is the most common intestinal worm in pigs irrespective of production system. However, organic farms which promote animal welfare (e.g. by providing pastures and bedding material) may further promote survival of the eggs through which the parasite is dispersed to new hosts. The parasite has a negative effect on production results and possibly impairs vaccines. A one year survey of 5 Danish organic farms therefore aimed at mapping the occurrence of parasite eggs in the farm environment, supplemented with laboratory and field studies, including serial necropsies of different age groups of pigs from 2 farms to monitor infection levels. The combined results showed that eggs were present on farrowing and weaning pastures, ensuring continuous exposure to the parasite outdoors. The farms did not have enough land for effective long-term rotation schemes to allow eggs to die and thus disappear naturally, as they can live for at least 9 years in the soil. Consequently, young infected pigs brought the parasite with them into the stables. Here, large numbers of eggs accumulated in the bedding material, but the majority of eggs died as the result of high temperatures and ammonia levels due to bacterial degradation of fecal matter. Nevertheless, infective eggs were present in the bedding material and pigs continued to be exposed and become infected in weaner/starter and fattening pig pens up to slaughter (22 weeks old). As the eggs may need several weeks or months to develop in the environment depending on season, some of the infective eggs in the pens may have been deposited by previous groups of pigs, as not all pens were cleaned between batches of pigs. In conclusion, exposure to *A. suum* is currently difficult to prevent outdoors in Denmark. The best approach to reduce overall exposure is thus to improve cleaning and disinfection procedures indoors and to compost and store manure/slurry sufficiently to inactivate eggs.

**Detection of a quantitative trait locus associated with resistance to whipworm infections in pigs**

*P. Skallerup[1,2], S.M. Thamsborg[2], C.B. Jørgensen[1], H.H. Göring[3], M. Fredholm[1] and P. Nejsum[1,2]*
*[1]University of Copenhagen, Department of Veterinary Clinical and Animal Sciences, Grønnegårdsvej 3, 1870 Frederiksberg C, Denmark, [2]University of Copenhagen, Department of Veterinary Disease Biology, Dyrlægevej 100, 1870 Frederiksberg C, Denmark, [3]Texas Biomedical Research Institute, 7620 N. W. Loop 410, San Antonio, TX, USA; pesk@sund.ku.dk*

Whipworms (*Trichuris* spp.) infect a variety of hosts, including production animals and humans. *Trichuris suis* has a global distribution with highest prevalence in organic production systems. High infection levels may cause growth retardation, anaemia and haemorrhagic diarrhoea. Up to 73% of the variation in *T. suis* faecal egg count (FEC) has been attributed to the host's genetic make-up. The aim of the present study was to identify genetic marker(s) associated with resistance to *T. suis* in pigs. We used single nucleotide polymorphism (SNP) markers to perform a whole-genome scan of an $F_1$ resource population (n=195) trickle-infected with *T. suis*. A measured genotype analysis revealed a putative quantitative trait locus (QTL) for *T. suis* FEC on chromosome 13 covering ~3.5 Mbp, although none of the SNPs reached genome-wide significance. We next validated our results by genotyping three SNPs within our QTL in unrelated pigs exposed to either experimental or natural *T. suis* infections and from which we had FEC (n=113) or worm counts (n=178). In these validation studies, two of the SNPs (IL, ST) were associated with FEC (P<0.01), thus confirming our initial findings. However, we did not find any of the SNPs to be associated with *T. suis* worm burden in these populations. In conclusion, our study suggests that genetic markers for resistance to *T. suis* as determined by low FEC or worm counts can be identified in pigs.



***In vivo* efficacy of chicory silage against parasitic nematodes in cattle**

*M. Peña[1], O. Desrues[2], A. Williams[2], S. Thamsborg[2] and H. Enemark[1]*
*[1]Technical University of Denmark, National Veterinary Institute, Bülowsvej 27, 1870 Frederiksberg C, Denmark, [2]University of Copenhagen, Department of Veterinary Disease Biology, Dyrlægevej 100, 1870 Frederiksberg C, Denmark; miap@vet.dtu.dk*

Infections with parasitic nematodes are a persistent risk for grazing cattle, which is met with widespread use of synthetic anthelmintics. However, heavy reliance on these drugs may lead to selection of drug-resistant nematodes. Integrated control strategies, including anti-parasitic crops, are needed. Chicory (*Cichorium intybus*) can have an impact against sheep nematodes, but this has not been investigated in cattle. Our study aimed to assess the effect of chicory on infections with abomasal (*Ostertagia ostertagi*) and intestinal (*Cooperia oncophora*) nematodes in stabled calves. Chicory (cv. Spadona) was ensiled and fermented for 5-12 weeks. 15 Jersey male calves (2-4 months) were stratified by live weight and randomly allocated to 2 groups: chicory (CHI, n=9) and control (CON, n=6). CHI and CON calves were fed with chicory silage and hay *ad libitum*, resp., for 8 weeks. Daily intake was monitored and protein/energy intake in both groups was equalized during the study. After 2 weeks on the diet all calves were infected with approx. 10,000 *O. ostertagi* and 65,000 *C. oncophora* third-stage larvae. Fecal excretion of nematode eggs was calculated as number of eggs per g of dried faeces (FECDM). Individual live weight was measured once a week. 6 weeks after infection calves were slaughtered for worm recovery. FECDM and live weights were analysed by ANOVA using repeated measurements. Log-transformed worm counts were analysed by ANOVA. Mean FECDM were not different between groups (P=0.14) but live weight gains during the study were significantly higher in CHI calves (+ 35%; P<0.05). Mean worm counts for *O. ostertagi* adults were 1599 and 3752 in CHI and CON, respectively (P<0.01). Worm counting of *C. oncophora* is undergoing. Based on these preliminary results we conclude that chicory may have an anti-parasitic effect against *O. ostertagi*.

**Evaluation of *in vitro* inhibitory effect of Enoxacin on *Babesia* and *Theileria* parasites**

*M.A. Mohammed*
*SouthValley University, Parasitology, Qena, 83523, Egypt; dr_mosab2081@yahoo.com*

Enoxacin is a broad-spectrum 6-fluoronaphthyridinone antibacterial agent (fluoroquinolones) structurally related to nalidixic acid used mainly in the treatment of urinary tract infections and gonorrhea. Also it has been shown recently that it may have cancer inhibiting effect. The primary antibabesial effect of Enoxacin is due to inhibition of DNA gyrase subunit A, and DNA topoisomerase. In the present study, enoxacin was tested as a potent inhibitor against the *in vitro* growth of bovine and equine Piroplasms. The *in vitro* growth of five *Babesia* species that were tested was significantly inhibited ($P<0.05$) by micro molar concentrations of enoxacin (IC50 values = 13.5, 7.2, 7.5 and 24.2 μM for *Babesia bovis*, *Babesia bigemina*, *Babesia caballi*, and *Theileria equi*, respectively). Enoxacin IC50 values for *Babesia* and *Theileria* parasites were satisfactory as the drug is potent antibacterial drug with minimum side effects.



***Salmonella* risk assessment in home-mixed feeds for pigs**

*E. Royer*
*IFIP-Institut du porc, 34 bd de la Gare, 31500 Toulouse, France; eric.royer@ifip.asso.fr*

Epidemiological studies have shown that *Salmonella* infection risk in pig herds was lower for farms manufacturing their own feeds. However, it was highlighted by EFSA experts that too little attention has been paid about the risk of introduction of *Salmonella* from home-mixed contaminated feed ingredients. A study investigated procedures and feedstuffs used in France. The supply structure of home-mixing farms was analyzed from previous surveys. Whereas cereals represented 75% of total feedstuffs, almost 75% of the farms used two or three sources of oil seed meals. Purchased raw materials were about 66% of the total mixed volume. A total of 50 on-farm milling facilities were visited and 154 feed materials and 84 compound feeds were collected. Analyses for *Salmonella* contamination in 100-g samples showed the contamination of two oilseed meals (soybean- and rapeseed-meals), one bread by-product, and one gestation diet. The serotypes (Cerro, Mbandaka, Veneziana and Arizonae, respectively) were not major serotypes in human salmonellosis. This confirms that oilseed meals appear as the major risk feed material. Particle sizes analyzed in 52 compound feeds were slightly higher than usual recommendations, but are probably not coarse enough to decrease the *Salmonella* development in the digestive tract. Thus, the favorable effect of on-farm mixed feed on *Salmonella* occurrence could mainly be explained by the distribution as meal and by the frequent use of liquid feeding. Biosecurity of facilities were assessed using 419 questionnaire forms from 3 pig farm surveys. Implementation of biosecurity practices was satisfactory regarding control and cleaning operations as well as storage and processing of feeds. However, in 48% of the farms, raw ingredient storage units were not out of reach of birds. It is concluded that the risk of *Salmonella* transmission from home-mixed feeds to pigs seems generally restrained. The compliance with the good hygiene and manufacturing recommendations should ensure a low risk of *Salmonella* contamination.

### Aflatoxin control in feed by *Trametes versicolor*

*C. Bello[1], M. Reverberi[2], C. Fanelli[2], A.A. Fabbri[2], M. Scarpari[2], C. Dell Asta[3], F. Righi[3], S. Zjalic[4], A. Angelucci[1] and L. Bertocchi[1]*
*[1]Istituto Zooprofilattico Sperimentale della Lombardia e dell'Emilia Romagna 'B. Ubertini', via Bianchi 9, 25124 Brescia, Italy, [2]Sapienza Università di Roma, Dipartimento di Biologia Ambientale, Largo Cristina di Svezia 24, 00165 Roma, Italy, [3]Università di Parma, Dipartimento di Scienze degli Alimenti, Parco Area delle Scienze, 59/A., 43124 Parma, Italy, [4]University of Zadar, M. Pavlinovica bb, 23000 Zadar, Croatia; szjalic@unizd.hr*

*Aspergillus flavus* are well known ubiquitous fungi able to contaminate both in pre- and postharvest period different feed and food commodities. During their growth these fungi can synthesise aflatoxins, secondary metabolites highly hazardous for animal and human health. These mycotoxins can enter the human food chain by the direct ingestion of contaminated food and by the consumption of animal products coming from livestock fed with contaminated silages. In this work the effect of the basidiomycete *Trametes versicolor* on the aflatoxin production by *A. flavus* in maize, was investigated. *T. versicolor* was grown on pure sugar beet pulp for 14 days at 25 °C in dark conditions. The obtained compost was dehydrated at 40 °C o/n and powdered. The addition of this fine powder at 1% w/v in maize kernels (moistened up to aw 0.85) significantly (Mann-Whitney test, $P<0.001$) inhibited up to 85% the production of aflatoxin B1 in maize by *A. flavus*. Furthermore, treatment of contaminated maize with culture filtrates of *T. versicolor* containing ligninolytic enzymes, namely laccases at 1 U/mg protein, showed a significant reduction of the content of aflatoxin B1 (-70%; $P<0.01$). Finally, treated and control maize samples were also compared under *in vitro* ruminal digestive condition to simulate the possible releasing of aflatoxins upon cow's digestion. *T. versicolor* could represent an eco-friendly tool for a significant control of aflatoxin production, in order to obtain feedstuffs and feeds with a higher quality standards.



### Climatic conditions and mycotoxins in grains and feed: the Croatian case

*S. Zjalic[1], V. Brlek[2], L. Coso[1,3], I. Sabljak[2], M. Grubelic[2], J. Gavran[2] and M. Matek Saric[1]*
*[1]University of Zadar, M. Pavlinovica bb, 23000 Zadar, Croatia, [2]Croatiakontrola doo, Karlovacka c. 4L, 10000 Zagreb, Croatia, [3]Croatia Control Ltd, P.O. Box 45, 10045 Zagreb-Airport, Croatia; szjalic@unizd.hr*

Mycotoxins present a serious hazard in animal production due to their effects on animal health and growth as well as on the quality and commercial value of animal products. The influence of environmental conditions on biosynthesis of different mycotoxins is known and well reported. Contamination of crops by some mycotoxigenic fungi (i.e. *Fusarium*) occurs mainly in field while for the others (i.e. *Aspergillus* sp.) the contamination can occur both in field and during the storage and processing of crops, feed and food. In Europe, due to climatic conditions unfavourable for *Aspergillus flavus* pathogenesis, aflatoxins (AF) were considered more a problem of imported feed and food stuff or bad storage conditions. In last decades few outbreaks of aflatoxin contamination in European corn, feed and milk were reported. The data indicate that AF contamination of corn occurred in field, most probably due to unusually extreme climatic conditions registered in some of past summers. In Croatian milk AF M1 levels higher than allowed by EU legislation were detected in 2013. The economical losses for dairy farmers and industry were huge. In this paper the correlation of contamination of grains and feed stuff produced in Croatia region of Slavonia and whether conditions is evaluated. During the last 5 years AF and deoxynivalenol (DON) content in more than 100 samples of each group (maize seeds, maize derived feed, wheat and other cereals) each year were analysed and compared to the temperature and humidity registered in the fields from June to October. The positive correlation between drought and high temperatures and AF contamination of grains has been confirmed. Lower temperatures and high humidity in early summer period seemed to have better correlation with DON contamination.

## Session 47 Poster 11

**Establishment of gastro-intestinal helminth infections in free-range chickens**

*K. Wongrak, G. Das, E. Moors, B. Sohnrey and M. Gauly*
*University of Göttingen, Department of Animal Sciences, Albrecht-Thaer-Weg 3, 37075, Germany; gdas@gwdg.de*

This study monitored establishment and development of gastro-intestinal helminth infections in chickens over two production years (PY) on a free-range farm in Lower Saxony, Germany. Laying hens were kept in two mobile stalls that moved to new positions at regular intervals. In both PY1 and PY2, 20 individual faecal samples per stall were randomly collected at monthly intervals in order to calculate the number of internal parasite eggs per gram of faeces (EPG). At the end of the laying periods, approximately 10% (n=42) or more than 50% (n=265) of hens were subjected to post-mortem parasitological examinations in PY1 and PY2, respectively. No parasite eggs were found in the faecal samples during PY1, whereas almost all of the hens (97.6%) were infected with *Heterakis gallinarum* at the end of the period. In PY2, nematode eggs in faeces were found from the third month onwards at a low level, increasing considerably towards the final three months. Mortality rate ranged from 18.3 to 27.4% but did not differ significantly between genotypes or production years. Average worm burden was 207 worms/hen in PY2. The most prevalent species were *H. gallinarum* (98.5%) followed by *Ascaridia galli* (96.2%) and *Capillaria* spp. (86.1%). Furthermore, three *Capillaria* species, *C. obsignata*, *C. bursata* and *C. caudinflata* were differentiated. In conclusion chickens kept on free-range farms are exposed to high risks of nematode infections and have high mortality rates with no obvious link to parasite infections. Once the farm environment is contaminated with the nematode eggs, establishment and further spread of nematodes to the hens is a matter of time. This will latest be the case in the second production period, even if the hens are kept in a rotation system and in small herd sizes. This underlines the importance of nematode infections in all free-range systems.

---

## Session 47 Poster 12

***In vitro* anti-parasitic effects of sesquiterpene lactones from chicory against cattle nematodes**

*M. Peña[1], A. Williams[2], U. Boas[1], S. Thamsborg[2] and H. Enemark[1]*
*[1]Technical University of Denmark, National Veterinary Institute, Bülowsvej 27, 1870 Frederiksberg C, Denmark, [2]University of Copenhagen, Department of Veterinary Disease Biology, Dyrlægevej 100, 1870 Frederiksberg C, Denmark; miap@vet.dtu.dk*

Chicory (*Cichorium intybus*) has the potential as an anti-parasitic crop for ruminants. However, the mechanisms behind observed *in vivo* effects are poorly understood but it is likely that plant secondary metabolites like sesquiterpene lactones (SL) play a role. In this study we tested the effect of SL-rich extracts from 2 chicory cultivars on the viability of first-stage larvae (L1) of *Ostertagia ostertagi*, a pathogenic cattle nematode. Chicory Spadona and Puna II were grown at the same farm and leaves were sampled the same day. 1 g of freeze-dried leaves was extracted in methanol/water. Resulting extracts were incubated with cellulase enzymes, recovered in ethyl acetate and purified by normal solid-phase extraction. Obtained extracts were dissolved in 100% dimethyl sulfoxide (DMSO). A calf infected with *O. ostertagi* served as donor of nematode eggs. Eggs were hatched and L1 obtained were incubated in 8 extract concentrations for each cultivar (in duplicates) ranging from 2000 μg to 16 μg dry matter (DM) extract/ml (final concentration 1% DMSO in phosphate buffered saline-PBS). Ivermectin (1 mg/ml) and 1% DMSO in PBS were used as positive and negative controls, respectively. Viability of L1 was evaluated morphologically after 12 h of incubation (25 °C) and was expressed as the number of live L1 to the total number of L1. Spadona-SL dramatically decreased the survival of *O. ostertagi* L1, with a mortality of 99% at concentrations ≥500 μg/ml and $EC_{50}$ of 132.8 μg/ml (CI=117.5-150.2 μg/ml). Conversely, Puna-SL induced a larval mortality of only 37% at the highest concentration tested (2,000 μg/ml), thus estimation of $EC_{50}$ was not possible. Results showed a marked difference in the anti-parasitic activity of SL-rich extracts from 2 different chicory cultivars. Further biochemical analyses of the extracts may reveal the responsible compounds.

**A cross-sectional study on risk factors of gestating sow leg disorders in group-housing systems**

*C. Cador[1], F. Pol[2], M. Hamoniaux[3], V. Dorenlor[1], E. Eveno[1], C. Guyomarc'h[3], N. Rose[1] and C. Fablet[1]*
*[1]Anses, Unité Epidémiologie et Bien-Etre du Porc, B.P. 53, 22440 Ploufragan, France, [2]Anses, Unité Epidémiologie et Bien-Etre en Aviculture et Cuniculture, B.P. 53, 22440 Ploufragan, France, [3]Cooperl Arc Atlantique, 7 Rue de la Jeannaie, 22400 Lamballe, France; christelle.fablet@anses.fr*

Group-housing, in place of individual-housing systems, is mandatory for gestating sows in the European Union (2008/120/EEC). However, leg problems occur more frequently in group-housing than in individual-housing systems and are a welfare and health concern. A cross-sectional epidemiological study involving 108 farrow-to-finish farms in western France has been carried out to see whether the type of the 4 main group-housing systems (i.e. large groups with electronic feeder station in stable or in dynamic groups, small groups in walk-in lock-in stalls or partial feeding stalls), the type of floor, and other husbandry practices, were associated with leg disorders. On each farm, the sows were examined visually for claw lesions, scored for lameness and their breeding characteristics were recorded. Our results showed that lameness was positively correlated with heel lesions and dewclaw lesions. Concrete slatted floor was a major risk factor (unadjusted relative risk (RR)=9.9; 95% confidence interval (CI): 4.4-34.5) compared to straw bedding. Walk-in lock-in stalls was found to be the most protective system. Using a multivariable logistic regression model, housing in large groups (RR=1.5; CI [1.1; 2.4]), dirty floors (RR=1.6; CI [1.0; 2.9]), high level of ammonia (RR=1.5; CI [1.1; 2.1]), strong feeding restriction in particular at the last stage of pregnancy (RR=1.5; CI [1.0; 2.1]) and a high number of sows per stockman (RR=1.5; CI [1.0; 2.4]) were identified as factors significantly associated with leg troubles.

---



**Evaluation of the pathogenicity of *Eimeria magna* in the rabbit of local population**

*M.S. Bachene[1], S. Maziz-Betahar[1], S. Temim[2], M. Aissi[2] and H. Ainbaziz[2]*
*[1]University Saad Dahleb, Institute of veterinary sciences, Blida, 25000, Algeria, [2]High Veterinary School, Laboratory Research of Animal health and production, Algiers, 16000, Algeria; ainbaziz@yahoo.fr*

The aim of this study was to evaluate the effect of Eimeria magna infection induced through three different inoculum sizes ($1\times10^4$; $2.5\times10^4$ and $5\times10^4$ oocysts) on growth performance of local rabbits. A total of 16 male rabbits were distributed in 4 groups of four rabbits each at an age of 4 weeks. The rabbits were previously treated with an anticoccidial between 21 and 24 days of age. Oocyst excretion and body weight gain were determined for 8 weeks post-infection (p.i.). Oocyst excretion data revealed a power multiplication equal on average to $1.10^6$ oocysts. Concerning the pathogenicity of E. magna, the results showed a prepatent period of 6 days p.i. and a peak excretion of oocysts at 9 days p.i. At 12 weeks of age, the rabbits inoculated with the higher dose ($5\times10^4$oocysts) showed significantly lower daily gains (-35%; $P<0.05$) compared to uninfected controls. The weight loss was less than 13% in rabbits inoculated with the lower inoculums ($1\times10^4$ and $2.5\times10^4$ oocysts; $P>0.05$). In conclusion, growth of local rabbit was significantly imparied by Eimeria magna.

**Presence of *Toxoplasma gondii* in common moles (*Talpa europaea*)**

*I.M. Krijger[1], J.B.W.J. Cornelissen[2], H.J. Wisselink[2] and B.G. Meerburg[1]*
*[1]Wageningen UR, Livestock research, Edelhertweg 15, 8219 PH Lelystad, the Netherlands, [2]Central Veterinary Institute of Wageningen UR, P.O. Box 65, 8200 AB Lelystad, the Netherlands; inge.krijger@wur.nl*

The presence of *Toxoplasma gondii* in common moles, *Talpa europaea*, was investigated in order to establish if moles could serve as an indicator species for *T. gondii* infections within livestock. In total, 86 moles were caught from 25 different sites in the Netherlands. Five different habitat types were distinguished: pasture, garden, forest, roadside, or natural recreation area. No positive samples (brain cysts) were found during microscopic detection (n=70). Using the Latex Agglutination Test (LAT), sera of 70 moles were examined, whereby no sample reacted with *T. gondii* antigen. Real Time-PCR test on brain tissue samples indicated 2 moles as positive (2.3%). Because of the low number of positive samples in our study, no relationship was demonstrated between the *T. gondii* prevalence in common moles, their gender, habitat type or trapping location. Consequently, the use of the common mole as an indicator species for livestock is currently not to be recommended.



**Prevalence of antibodies against *Chlamydiae* in Austrian pigs**

*F. Schmoll[1], T. Sattler[2], L. Fischer[3], M. Schindler[3] and M. Müller[1]*
*[1]AGES, Institute for Veterinary Disease Control, Robert-Koch-Gasse 17, 2340 Mödling, Austria, [2]University Leipzig, Large Animal Clinic for Internal Medicine, An den Tierkliniken 11, 04103 Leipzig, Germany, [3]LaboVet GmbH, Campus Vienna Biocenter 3, 1030 Vienna, Austria; friedrich.schmoll@ages.at*

*Chlamydiae* are widely distributed and associated with a multitude of diseases especially respiratory and reproductive failures. Experimental studies document that the agent is detectable in multiple tissues and organs of the affected animal in a high percentage of sows and boars. Since *Chlamydiae* are ubiquitous, it is not clear if a positive result means infection or contamination. An infection should cause the development of antibodies. Aim of the study was to show the seroprevalence of antibodies against *Chlamydiae* in pigs in several regions of Austria. The study included a total of 2,935 sera from adult pigs from Austria. All sera were tested in two different labs for antibodies against *Chlamydiae* with complement binding reaction (KBR). Laboratory 1 (Labovet GmbH, Vienna) tested mostly samples of sows (n=1,219, sample year 2013), laboratory 2 (AGES, Mödling) tested mostly samples of young breeding boars (n=1,737, sample years 2009-2013). Titers were considered positive if equal or higher than 1:16 (lab 1) or 1:10 (lab 2), respectively. Antibodies against *Chlamydiae* were found in 3% of the samples tested in lab 1(sows) and in 4% of the samples tested in lab 2 (boars). Additional 3% of the sera were anticomplementary and the results of these samples therefore not interpretable. Titers of the positive sera in lab 1 ranged between 1:16 and 1:128, in lab 2 between 1:10 and 1:40, whereas the lower titers were more often found. The results suggest that an infection with *Chlamydiae* that leads to development of antibodies is rare.

**Herd level factors associated with H1N1 or H1N2 influenza virus infections in finishing pigs**

*C. Fablet, G. Simon, V. Dorenlor, E. Eveno, S. Gorin, S. Queguiner, F. Madec and N. Rose*
*Anses, B.P. 53, 22440 Ploufragan, France; christelle.fablet@anses.fr*

Herd-level factors associated with European H1N1 or H1N2 swine influenza virus (SIV) infections were assessed in 125 French herds by means of a cross-sectional study carried out from 2006 to 2008. In each herd serum samples from 15 finishing pigs were tested by haemagglutination inhibition. Data related to herd characteristics, biosecurity, management and housing conditions were collected by questionnaire during a farm visit. Climatic conditions in the nursery and finishing rooms, where the sampled pigs were housed, were measured over 20 hours. Factors associated with H1N1 or H1N2 sero-positive status of the herd were identified by logistic regressions. For both subtypes, the odds for a herd to be SIV sero-positive increased if there were more than two pig herds in the vicinity (OR=3.2 and OR=3.5, P<0.01 for H1N1 and H1N2 respectively). Other factors were specifically associated with either H1N1 or H1N2 SIV infections. The odds for a herd to be H1N1 sero-positive were significantly increased by having more than 28 pigs per nursery pen (OR=3.2, P<0.02), temperature setpoints below 25 °C (OR=2.6, P=0.03) and below 24 °C (OR=2.6, P=0.03) for the heating device in the farrowing room and the ventilation controller in nursery, respectively. Moving the pigs to the finishing facility via a room housing older pigs also significantly increased the odds of being H1N1 sero-positive (OR=3.3, P=0.03). A H1N2 sero-positive status was associated with a brief down period in the farrowing room (OR=2.6, P=0.03), small floor area per nursery pig (<0.35 $m^2$/pig, OR=2.9, P=0.02), large-sized finishing room (>110 pigs, OR=2.5, P=0.03), lack of all-in all-out management in the finishing room (OR=2.4, P=0.04) and a temperature range of less than 5 °C controlling the ventilation in the finishing facilities (OR=3.2, P<0.01). Factors related to external and internal biosecurity and to the control of inside cimatic conditions should be considered together when implementing programmes to better control SIV infections.

---



**Emergence of canine distemper virus mutated strains, identified from vaccinated carnivores in China**

*J.J. Zhao, C.C. Zhu, Y.G. Sun and X.J. Yan*
*Institute of Special Animal and Plant Sciences, Chinese Academy of Agricultural Sciences, Division of Infectious Dieases, No. 4899 Juye Street, 130112, China, P.R.; zhaojianjun@caas.cn*

Canine distemper (CD) is a highly contagious disease for several carnivores caused by Canine distemper virus (CDV). The CDV haemagglutinin (H) protein is the main target for the host immune system and plays a key role in virus entry. Changes in N- linked glycosylation of the H protein of distinct CDV strains have been associated with changes in CDV virulence and antigenicity. Between 2011 and 2013, a canine distemper-like disease was observed in vaccinated breeding minks, foxes and raccoon dogs from four provinces in the Northeast of China. RNA was extracted directly from sixteen CD tissue samples of minks, foxes and raccoon dogs from four provinces. RT-PCR for full-length H genes of sixteen wild-type CDV was carried. The sequences and phylogenetic tree were determined and aligned with other CDV H protein sequences retrieved from GenBank using the MEGA5.0 software package. The potential N-linked glycosylation sites of the H protein were determined using the software NetNGlyc1.0. Upon phylogenetic analysis, all the sixteen CDV strains segregated into the Asia-1 genotype, along with other Chinese wild-type CDV strains. Amino acid (aa) sequence identity in the full-length H protein among the sixteen strains ranged from 98.4 to 100%. Two amino acids located polymorphisms, 542 Ile to Asn and 549 Tyr to His were identified in the receptor (SLAM) binding region of the H protein of the ten CDV strains. Interestingly, the change at residue 542 generated a novel potential N-glycosylation site. The mutation 542 Ile to Phe was also observed in two simian CDV isolates, suggesting a possible role of this residue in the adaptation of CDV to new hosts. The novel N- linked glycosylation site 542-544 identified in the H protein of the CDV strains might be one of the mechanisms/explanations for the observed CDV vaccine breakthroughs.

**Effects of acupuncture treatment on stress markers in Japanese black calves after transportation**

*M. Noguchi[1], I. Hatazoe[2], T. Iriki[2], N. Miura[1], S. Hobo[1], A. Tanimoto[3] and H. Kawaguchi[1]*
*[1]Kagoshima University, Joint Faculty of Veterinary Medicine, Korimoto 1-21-24, 890-0065, Kagoshima, Japan, [2]Kagoshima Prefectural Economic Federation of Agricultural Co-operatives, Kamoike-shinmachi 15, 890-8515, Kagoshima, Japan, [3]Kagoshima University, Graduate School of Medical and Dental Sciences, Sakuragaoka 8-35-1, 890-8544, Kagoshima, Japan; k0426835@kadai.jp*

The aim of this study is to investigate effects of acupuncture treatment on stress marker levels in Japanese Black calves after short-term transportation. Eighty-three calves (age: 12.7±0.5 days [Ave ± SEM], weight: 38.4±1.2 kg) judged healthy from the results of body temperature, and respiratory and heart rate measurements were randomly assigned to a control (n=41) or treatment (n=42) group after collection of a blood sample from the jugular vein. Animals in the treatment group were treated with acupuncture needles (diameter × length = 0.22×1.5 mm, PYONEX, Seirin Co., Shizuoka, Japan) at meridians (called 'Jisen') on the peripheries of each ears at 83.4±2.4 min. before transportation. The animals were transferred to a truck for 61.1±0.8 min. transportation, immediately after which blood samples were collected. Plasma concentrations of cortisol and catecholamine (adrenaline, noradrenaline and dopamine) were determined as stress markers by enzyme-linked immunoassay and high performance liquid chromatography, respectively. No interaction was found between treatment and time (before or after transportation). Plasma noradrenaline and dopamine concentrations in the treatment group were lower ($P<0.05$) than those in the control group. An increase ($P<0.05$) in cortisol levels was observed after transportation in all calves. These results suggest that acupuncture treatment suppresses the central catecholaminergic system but does not affect the suppression of the hypothalamic-pituitary-adrenal function in short-term transportation.

---



**Cause of death of suckling piglets in SWAP farrowing pens**

*J. Hales[1], V.A. Moustsen[2], M.B.F. Nielsen[2] and C.F. Hansen[1]*
*[1]University of Copenhagen, Groennegaardsvej 2, 1870 Frederiksberg, Denmark, [2]Pig Research Centre, Axeltorv 3, 1609 Copenhagen, Denmark; vam@lf.dk*

In loose housed systems sow welfare is improved, however, piglet mortality is higher than in crates. Identifying the cause of death can aid in establishing beneficial interventions to reduce piglet mortality. The objective of this study was to investigate cause of piglet mortality in farrowing pens where sows could be confined for a few days after farrowing (SWAP=Sow Welfare And Piglet protection). The study was conducted in a 1,200 sow piggery where sows were randomly allocated to loose-loose (LL, n=376), loose-confined (LC, n=369) or confined-confined (CC, n=354). In LL sows were loose from placement in the farrowing unit to weaning, LC-sows were loose to birth of the last piglet and then confined to day 4 post farrowing, and CC-sows were confined from day 114 of gestation to day 4. All sows were loose housed from day 4 to weaning. Total born piglets was recorded and dead piglets were noted and collected for post mortem examination. Dead piglets were assigned a cause of death (stillborn, euthanized, crushed, weak or other), given a score for intrauterine growth restriction (IUGR) and a score for stomach content (empty, <half full, >half full or full) if they were live born. All data were analysed using chi square tests. Preliminary results showed that 22%, 20% and 23% were stillborn in LL, LC and CC, respectively ($P<0.04$). Crushing was the main cause of live born deaths (LL: 53%, LC: 49% and CC: 45%, $P<0.01$) and a large proportion of live born dead piglets died with no or little stomach contents in all treatments (LL: 75%, LC: 80% and LL: 76%, $P<0.01$). The proportion of piglets with signs of IUGR was similar in all groups (LL: 20%, LC: 21% and CC: 21%, $P=0.75$). In conclusion sows that were loose housed had more crushed piglets than sows that were confined in SWAP pens.

### Observations on well-managed weaner producers in terms of antibiotic use, health and productivity

*M. Fertner[1], N. Dupont[2], A. Boklund[1], H. Stege[2] and N. Toft[1]*
*[1]National Veterinary Institute, Section for Epidemiology, Bülowsvej 27, build. 1, 1870 Frederiksberg C, Denmark, [2]University of Copenhagen, Department of Large Animal Sciences, Grønnegårdsvej 2, 1870 Frederiksberg C, Denmark; memun@vet.dtu.dk*

Use of antibiotics in the swine industry is a major issue of discussion in Denmark and has caused several political initiatives serving a reduced antibiotic consumption. However, the discontinuation of growth promoters was observed to have a negative impact on health, especially for weaners. Hence, the latest Danish political initiative has prompted debates on whether the strict political approach is at the expense of animal health. This study focuses on well-managed Danish weaner producers; farms with a low use of antibiotics and concurrently high health and productivity. The objective was to identify management-related factors characterizing these well-managed farms. Eleven veterinarians working in swine practice recommended their best conventional weaner producers. Participating farms had to be among the best 50% regarding use of antibiotics (<8.2 $ADD_{15}$/100 weaner/day), mortality (<2.8%) and daily growth (>443 g/day). Eleven farms fulfilled the criteria and agreed to participate. Management practices influencing the level of administered antibiotics were addressed in a semi-open ended interview, adressing: employees, stable, management, hygiene, feed, biosecurity, vaccination and medication procedures. This qualitative study presents the variation in management practices of swine farmers. Practices that influence the antibiotic consumption at the farm level, but which cannot be incoorporated in studies purely based on register data. A few management practices were rediscovered in most of the farms (e.g. sorting, sectioning). However, the farms divided into various groups having each of their focus of interest, e.g. feeding, method of medication, alertness in the stable. These results indicate that various strategies in the management can induce successful results, highly influenced by the personal approach of the farmer.

---



### Effect of feed ration on milk production and quality

*M.K. Larsen, S.K. Jensen, T. Kristensen and M.R. Weisbjerg*
*Aarhus University, AU Foulum, 8830 Tjele, Denmark; mette.larsen@agrsci.dk*

The relation between feeding, composition and quality of dairy products has achieved increased attention in order both to produce high quality bulk milk and to produce special dairy products with a unique composition. Feed rations are most often optimised to balance maximum milk production against least cost. Concentrate feed items increase the energy density and ensure proteins supply. Concentrates include grains, oilseeds, and by products from the production of vegetable oils, sugar and bio fuels. Forages contribute to the physical structure of the ration. Typical forages are maize silage and grass/legume crops which may be fed preserved (hay, silage) or by grazing. Increased ration energy concentration increase milk protein concentration. Vitamin and lipid composition of milk is to a large extent influenced by concentration in the ration fed. Vitamin A, E and carotenoids depend on the supply from feed often with green forages as main source. For human nutrition the amount of saturated fat should be reduced and unsaturated fatty acids, in particular polyunsaturated omega 3 fatty acids, should be increased. This is mainly obtained from linolenic acid rich green forages. Transfer of flavour compounds from feed to milk affects the sensory properties and may result in distinct flavours of dairy products of specific origin, or in feed related off-flavours. Also oxidation of polyunsaturated fatty acids causes off-flavours. Feeding influences technological properties of dairy products, where more unsaturated fat gives a softer texture of butter and cheeses, and the carotenoid content determines the yellow colour of the lipid fraction. Awareness of environmental problems in relation to milk production has attained increasing interest, and labelling according to production method (organic) or after carbon foot print is increasing. Development within this area is foreseen, but require sound methods for estimation of the claimed effect of specific feed, feeding strategies e.g. on sustainability in the total chain for feed to food.

Session 48 Theatre 2

**Affecting milk composition of Jersey cows in grass silage of different cuts with or without rapeseed**

*S. Vogdanou[1], M.R. Weisbjerg[2], J. Dijkstra[3] and M.K. Larsen[1]*
*[1]Aarhus University, AU Foulum, Department of Food Science, Blichers Alle 20, P.O. Box 50, 8830 Tjele, Denmark, [2]Aarhus University, AU Foulum, Department of Animal Science, Blichers Alle 20, P.O. Box 50, 8830 Tjele, Denmark, [3]Wageningen University, Animal Nutrition Group, Department of Animal Sciences, P.O. Box 338, 6700 AH, the Netherlands; stefi-seth@hotmail.com*

The objective of this study was to investigate how to alter milk composition of fatty acids, carotenoids and tocopherols by feeding four different treatments to cows in order to improve milk for human health. Thirty-six Jersey cows were used in a 4×4 Latin square design, for 4 periods of 3 weeks and assigned to 4 treatments as a 2×2 factorial design: spring grass silage from spring growth (harvest in May) or autumn grass silage which was a mixture of 3$^{rd}$ regrowth (harvest in September) and 4$^{th}$ regrowth (harvest in October), with or without rapeseed supplementation. Milk FA, carotenoids and tocopherols were affected by treatments. Compared with feeding cows spring grass silage, feeding autumn grass silage moderately increased the concentration of C18:1cis9, C18:2n6 and β-carotene and, to a greater extent, C18:3n3 concentration in milk whereas the content of C16:0 slightly decreased. Inclusion of rapeseed in feed greatly increased all C18 FAs in milk, except C18:2n6 and C18:3n3, and α-tocopherol content whereas the content of C11:0 to C16:0 FAs and β-carotene were considerably reduced. Apparent recovery of C18:3n3 was slightly increased with autumn grass silage and rapeseed inclusion in the feed whereas the synthesis of C16:0 was reduced with the same treatment. It was concluded that autumn grass silage and rapeseed supplementation represented useful strategies for enhancing MUFA and PUFA concentrations in milk, accompanied by an increase in antioxidants.

Session 48 Theatre 3

**Can milk composition analysis predict grass content in rations?**

*A. Elgersma*
*P.O. Box 323, 6700 AH Wageningen, the Netherlands; anjo.elgersma@hotmail.com*

In the Netherlands, in 2012 the label 'Weidemelk' ('pasture milk') was introduced for milk and dairy products produced by cows that are kept at pasture at least four months per year and 6 hours per day, and dairy cooperatives started paying their farmers a premium for such milk (€ 0.50/100 kg milk). It is well recognized that the milk from cows fed green forages contains higher contents of unsaturated fatty acids than with conserved feeds and concentrates, but feeding e.g. linseed also has this effect. A marker for authenitifcation of pasture-based milk is thus desirable for fraud detection. Phytanic acid (PHY) is a branched-chain fatty acid, produced by bacteria from enzymatic degradation of chlorophyl in the rumen. PHY cannot be synthesised de novo by mammals, and thus its presence in milk and animal-derived products depends exclusively on feed, especially on green materials, i.e. grass and forage which contain chlorophyll. Since the proportion of grass in animal rations often is higher in organically raised cattle, PHY has been proposed by German researchers as a marker for for pasture-fed milk and dairy products, as well as for organic dairy products. However, Danish findings indicated that total PHY content is not necessarily directly correlated with the intake of green feed, being linked in a more complex way to the whole diet. Recently, a Dutch study showed that PHY content was not a suitable indicator of milk produced by pasture grazing or organic/biodynamic farming systems. The aim of this review is to compare various studies, further explore the effect of fresh grass in cows' diet and of organic/biodynamic farming systems on PHY and its isomers (SRR and RRR) in milk, and discuss implications for the verification of farm management system.

**Low protein degradability and precision protein feeding improve nitrogen efficiency of dairy cows**

*E. Cutullic[1,2], L. Delaby[1,2], N. Edouard[1,2] and P. Faverdin[1,2]*
*[1]Agrocampus Ouest, UMR1348 PEGASE, Rennes, 35000, France, [2]INRA, UMR1348 PEGASE, St-Gilles, 35590, France; philippe.faverdin@rennes.inra.fr*

The low efficiency of nitrogen (N) intake in dairy cows can increase feed costs and ammonia emissions. At the rumen level, decreasing degradable N to improve the urea reuse in microbial protein synthesis, can increase N efficiency (Neff=Nmilk/Nintake). Moreover, at the animal level, the metabolisable protein (MP) efficiency varies greatly between dairy cows. Applying precision feeding, with adjustment of the diet according to individual observations, could increase this MPeff. The objective of this study was to improve both N and MP efficiencies by combining a reduction in rumen degradable protein (RDP, factor N) with an individual adjustment of MP content in the diet (factor E) in a 2×2 factorial design with 4 groups of 11 cows for 12 weeks fed maize-based diet. Factor N was either at the reference level of INRA for RDP (N0) or at a reduced level (N-10) of -1.4 g of degradable protein per MJ of net energy for lactation (NEL). Factor E was either constant with 14.8 g of MP per MJ NEL (EC), or variable (EV), from 12.0 to 15.5 g MP/MJ NEL and calculated for each cow according to its MPeff on a 4-week reference period. The lower the efficiency was, the lower the applied MP/NEL ratio was. No interaction occurred between the 2 factors. The decrease in RDP had no significant effect on intake and production. The milk production of EV was 1.7 kg/day lower than EC, but the energy efficiency (kg of milk/kg of DMI) was unchanged. The RDP reduction improved Neff (35.7 vs 32.2%) and individual protein feeding mainly improved MPeff (69.1 vs 59.3%). Between extreme treatments (N0_EC vs N-10_EV), crude protein content decreased from 14.9 to 12.5%, Neff varied from 32.1% to 36.4%, and estimated urine urea-N excretion was reduced by 56%. Reducing degradable N content of the diet and applying precision feeding to dairy cows fed indoors is thus a way to better use proteins and to reduce ammonia emissions.

---



**Animal performance and carcass characteristics of cattle fed different levels of yerba mate**

*A. Berndt[1], M.P. Vidal[1], R.T. Nassu[1], R.R. Tullio[1], M.E. Picharillo[1] and D.R. Cardoso[2]*
*[1]EMBRAPA Southeast Livestock, Rod. Washington Luiz, 234, P.O. Box 339, 13560970, Brazil, [2]Chemistry Institute of São Carlos – IQSC/USP, Av. Trab. São Carlense, 400, P.O. Box 780, 13560970, Brazil; alexandre.berndt@embrapa.br*

Animal performance and carcass characteristics of cattle fed diets containing different levels of yerba mate (*Ilex paraguariensis*, St. Hilaire) were evaluated in order to obtain a healthier and higher oxidative stability meat. Forty-eight Nellore steers with an average age of 21 months and an initial weight of 419 kg were individually fed during 94 days with the same base diet, differing only by yerba mate levels (0, 0.5, 1.0 and 1.5% w/w). Diets were composed of maize silage and concentrate (60:40 w/w) with 11% crude protein and 72% of total digestible nutrients. Diets were balanced using 0, 0.5, 1.0 and 1.5% w/w of Kaolinite (kaolin), an inert ingredient. The experimental design was completely randomized with 4 treatments and 12 replications, with diet as a fixed factor. Dry matter intake (DMI) was calculated by weighting offer and leftover daily. Animal live weights were obtained every 28 days. Hot carcass weights were collected immediately after slaughter and the rib eye area and backfat thickness were measured in longissimus muscle between the 12$^{th}$ and 13$^{th}$ ribs. Statistical data treatment was performed using XLSTAT statistical program with a significance level, P-value, at 0.05. There were no significant differences (P-value>0.05) between diets in DMI, average daily gain, feed conversion, fasted slaughter weight, hot carcass weight, carcass yield, backfat thickness and rib eye area. The addition of yerba mate extract, at vaying levels from 0.5 to 1.5% w/w, in the diet of feedlot cattle did not affect performance and carcass characteristics of the animals, showing expected values for feedlot-finished Nellore cattle. These results were positive as the use of the yerba mate is recent and there is a lack of information about the use of yerba mate as an antioxidant in diets for beef cattle.

**Cow health and welfare in bedded pack dairy barns**

*W. Ouweltjes and P. Galama*
*Wageningen UR, Livestock Research, P.O. Box 65, 8200 AB Lelystad, the Netherlands; wijbrand.ouweltjes@wur.nl*

Cubicle barns are the most common type of housing for dairy cattle in the Netherlands. Some less favourable aspects of cubicle barns have become apparent, particularly regarding cow health and welfare. Walking floors are often wet, hard and slippery, this is linked to suboptimal claw health. Cubicles inherently restrict the animals in their lying behaviour, and it is difficult to optimise the cubicle dimensions for animals of different size. It is quite common to house vulnerable cows (ill, transition) temporarily in straw yards instead of cubicle barns to improve their recovery. Despite improvements of cubicle housing that are still ongoing, about 40 Dutch farmers have built a new barn with a bedded pack lying area. An important motivation of these farmers for their decision was to improve cow comfort and increase cow longevity, despite that there are some other issues about these barns that are still to be solved (particularly ammonia emission and sporeforming bacteria). It takes time to increase the average herd life of the cows and most bedded pack barns are only in use for a limited time by now. Benefits with regard to health and welfare appear much quicker however, and we have investigated some aspects of animal health and welfare that can be related to housing in ten bedded pack barns. Mastitis incidence was below the Dutch average, while the bulk milk cell counts did not change after transition. Some farmers reported substantial improvements in claw health, and beneficial effects were seen with regard to welfare. The farmers realised a low usage of antibiotics. We did not find any evidence for increased risk of heat stress for cows kept in a barn with a bedding of composting woodchips (with a temperature of over 50 °C in the bedding).



**Bedding and housing in relation to cow comfort, milk quality and emissions**

*P.J. Galama[1], F. Driehuis[2], H.J. Van Dooren[1], H. De Boer[1] and W. Ouweltjes[1]*
*[1]Wageningen UR, Livestock Research, Edelhertweg 15, 8200 AB Lelystad, the Netherlands, [2]NIZO food research B.V., Food Research, Kernhemseweg 2, 6718 ZB Ede, the Netherlands; paul.galama@wur.nl*

Freestalls are the most frequently used housing system and sawdust is mostly used as bedding material in the Netherlands. However dairy farmers are looking for alternatives, because of increasing costs of sawdust and trends in dairy housing related to sustainability. Alternative bedding materials in freestalls, such as green waste compost or cattle manure solids, may introduce new sources of microbial contaminants of milk. An alternative housing system called bedded pack barns gives the cows more space and a soft floor for walking and lying. This increases the cow comfort due to less injuries, less problems with legs and claws, less time to lie down and more natural behaviour. In these barns, farmers use beddings with different kinds of compost from municipal biowaste or waste from gardens. Other farmers compost woodchips, mostly with forced mechanical ventilation. Bedding and milk sample analysis of 22 freestalls and 10 bedded pack barns show that compost and composted woodchips are an important source of bacterial spores, in particular of spores of thermophilic aerobic sporeformers. These sporeformers can cause non-sterility problems in sterilized dairy products. Results of 175 measurements with a mobile gas emission measurement chamber on 10 farms show that the ammonia emission per m2 in a bedded pack barn is 10 to 60% compared to emission from a slatted floor in a free stall. However, the ammonia emission per cow is usually much higher (up to 250% higher), due to more m2 per cow. However, when woodchips are used as bedding material, in combination with a low emission floor along the feed alley, the emission per cow has potential to be lower than the freestall. The composted materials from a bedded-pack barn have a negligible N emission after application to farmland and has a high content of stable organic matter which makes this compost very suitable for soil improvement.

## Generating test-day methane emissions as a basis for genetic studies with random regression models

*T. Yin[1], T. Pinent[1], K. Brügemann[1], H. Simianer[2] and S. König[1]*
*[1]Department of Animal Breeding, University of Kassel, Nordbahnhofstraße 1a, 37213 Witzenhausen, Germany, [2]Animal Breeding and Genetics Group, Georg-August-Universität Göttingen, Albrecht-Thaer-Weg 3, 37075 Göttingen, Germany; tong.yin@uni-kassel.de*

The objective of this study was to infer genetic (co)variance components for methane emissions (ME) by alterations of days in milk. Data for test-day production (milk yield (MY), fat%, protein%, milk urea nitrogen), conformation (wither height, hip width, body condition score), fertility (days open (DO), calving interval (CI), and stillbirth (SB)) and health (clinic mastitis (CM)) traits were available from 961 first lactation Brown Swiss cows kept in 41 low input farms in Switzerland. Test-day ME were predicted based on the the real data and on two deterministic equaitons, resulting in ME1 and ME2. Stochastic simulations were used to simulate concentrates and energy content of roughage as required when calculating ME2. Bivariate RRM were applied for all combinations of ME with production traits. For inferring genetic corrlations between ME with fertility and with health traits, RRM were used for ME, but animal models (AM) were applied to fertility and health traits. Threshold AM were applied to binary traits SB and CM. Predicted ME had moderate heritabilities in the range from 0.15 and 0.37. Antagonistic genetic correlations of 0.70 to 0.92 were found between ME2 and MY. Genetic correlations between ME with other production traits varied across lactation. Genetic correlations between ME with DO and with CI increased from 0.10 at the beginning to 0.90 at the end of lactation. Varied genetic relationships (-0.24 to 0.37) were found between ME2 and SB. Genetic relationships between ME2 and CM ranged from 0.02 to 0.49. In the current study, ME were predicted combining real data with deterministic equations and stochastic simulations, and the genetic background of ME revealed a broad range of genetic correlations with other traits of the overall dairy cattle breeding goal.

---



## Effect of reduced dietary protein and concentrate:forage ratio when cows enters deposition phase

*L. Alstrup, L. Hymøller and M.R. Weisbjerg*
*Aarhus University, AU, Foulum, Denmark; lene.alstrup@agrsci.dk*

The objective was to increase production efficiency and minimise use of protein supplements and thereby environmental impact by reducing concentrate:forage ratio (ratio) and protein concentration in the ration for dairy cows during deposition phase of lactation. Jersey and Holstein cows were assigned to 4 mixed rations (MR) in 2×2 factorial design with 2 ratios (40:60 and 30:70) and 2 protein levels (160 and 140 g/kg DM), high ratio-normal protein (HC-NP), high ratio-low protein (HC-LP), low ratio-normal protein (LC-NP), and low ratio-low protein (LC-LP). Cows were milked in an automatic milking unit, where they were also fed concentrate and weighed. After calving, all cows were fed HC-NP until they entered deposition defined as 11 kg (Jersey) or 15 kg (Holstein) weight gain from their minimum weight after calving. Then cows either remained on HC-NP or changed to one of the other MR. Independent of weight gain, no cows changed ration before 56 days in milk. Data were analysed in the MIXED procedure of SAS, using recordings during week 1-8 after calving as covariate. The model was applied to data obtained during week 9-30 after calving in Holstein (n=107) and Jersey (n=61) cows, respectively. Results showed higher dry matter intake on HC than LC and on NP than LP. The NEL intake was affected by ratio and protein with a higher NEL intake on HC and NP than on LC and LP. No interactions were found between ratio and protein for any feed intake parameters. Energy corrected milk yields were higher on NP than LP, whereas there were no effects of ratio. Milk composition did not differ among strategies, but the protein:fat ratio of the milk was higher on HC than LC and tended to be lower on NP than LP. Energy efficiency was higher on LC than HC and N efficiency was higher on LP than NP. It was concluded that in rations for dairy production it is possible to substitute concentrate in the MR with high quality forage during deposition phase without negative effects on milk yield, when a sufficient protein level is secured.

**Integrated modeling of strategies to reduce greenhouse gas (GHG) emissions from dairy farming**

*C.E. Van Middelaar, P.B.M. Berentsen, J. Dijkstra and I.J.M. De Boer*
*Wageningen University, De Elst 1, 6708 WD, the Netherlands; corina.vanmiddelaar@wur.nl*

Feeding and breeding strategies offer potential to reduce GHG emissions from dairy farming. We evaluated the impact of several strategies on GHG emissions at chain level (i.e. from cradle to farm gate) and on labor income at farm level. A whole-farm optimization model was combined with a mechanistic model for enteric methane and a life cycle approach (LCA). Feeding strategies included supplementation of diets with an extruded linseed product (56% linseed; 1 kg/cow per day in summer and 2 kg/cow per day in winter), supplementation of diets with nitrate (75% nitrate; 1% of DM intake), and reducing maturity stage of grass and grass silage (grazing at 1,400 instead of 1,700 kg DM/ha and harvesting at 3,000 instead of 3,500 kg DM/ha). In case of breeding, we assessed the impact of one genetic standard deviation improvement in milk yield and longevity. Strategies were evaluated for an average Dutch dairy farm under current (feeding strategies) or future (breeding strategies) production circumstances. A whole-farm model based on maximizing income was used to define the reference situations. Subsequently, strategies were implemented and the model was optimized again to determine the new economically optimal farm situation. Emissions were calculated based on mechanistic modeling and LCA. Total GHG emissions in the reference situation of the current farm were 840 kg $CO_2$e/t FPCM. Supplementation of linseed reduced emissions by 9 kg $CO_2$e/t FPCM, supplementation of nitrate by 32 kg $CO_2$e/t FPCM, and reducing grass maturity by 11 kg $CO_2$e/t FPCM. Of all three strategies, reducing grass maturity was most cost-effective (i.e. GHG reduction costs were €57/t $CO_2$e compared to €241/t $CO_2$e for nitrate and €2,594/t $CO_2$e for linseed). Total GHG emissions in the reference situation of the future farm were 796 kg $CO_2$e/t FPCM. Increasing milk yield reduced emissions by 27 kg $CO_2$e/t FPCM, increasing longevity by 23 kg $CO_2$e/t FPCM. Results can be used by breeding organizations that want to include GHG values in their breeding goal.

---

Session 48 Theatre 11

**Effects of dietary forage proportion on energy utilisation of lactating dairy cows**

*L.F. Dong[1], T. Yan[1], C.P. Ferris[1] and D.A. McDowell[2]*
*[1]Agri-Food and Biosciences Institute, AFBI Hillsborough, Large Park, Co. Down, BT26 6DP, United Kingdom, [2]University of Ulster, University of Ulster, Jordanstown, Co. Antrim, BT37 0QB, United Kingdom; lifeng.dong@afbini.gov.uk*

A total of 924 data of lactating dairy cows were collated from 32 calorimeter studies involving Holstein-Friesian (HF, n=814), Norwegian (n=48), Jersey × HF (n=46), and Norwegian × HF (n=16) cows at this institute from 1992 to 2010. The overall data were divided into three groups according to dietary forage proportion (FP, kg/kg DM), i.e. FP<0.30 (n=55),=0.30-0.99 (n=804), and = 1.00 (n=65). The REML model was used to examine if there was any significant difference among the three groups in terms of ME requirement for maintenance ($ME_m$) and the efficiency of ME use for lactation ($k_l$), with the effects of days in milk and experiment removed. The linear regression of ME intake ($MJ/kg^{0.75}$) against milk energy output adjusted to zero energy balance ($MJ/kg^{0.75}$) indicated that increasing FP from <0.30, 0.30-0.99 to 1.00 had no significant effects on $k_l$, while increased $ME_m$ which was calculated from the constant divided by the coefficient (0.645 vs 0.681 vs 0.738 $MJ/kg^{0.75}$). In addition, the effects of FP ratios on maintenance requirement was undertaken using linear regression of ME intake against estimated $ME_m$ for individual cows, with the latter calculated from heat production minus energy loses from the inefficiencies of ME use for lactation, body weight change and pregnancy. The results found that $ME_m$ ($MJ/kg^{0.75}$) was not a constant, as adopted in the majority of current energy feeding systems across the world, but increased with increasing ME intake, i.e. $ME_m$ increased by 0.059 $MJ/kg^{0.75}$ with one unit increase in ME intake ($MJ/kg^{0.75}$) with a constant of 0.543, 0.567 or 0.617 $MJ/kg^{0.75}$ for group of FP<0.30, 0.30-0.99 or 1.00. These results indicate that dairy cows managed under the low input (high FP ratio) systems may require more energy per unit of body weight to meet their basal metabolic rate.

**Effect of different rotational stocking management of *Brachiaria brizantha* on weight gain of heifers**

*M.L.P. Lima, F.F. Simili, E.G. Ribeiro, A. Giacomini, A.E. Vercesi Filho, J.M.C. Silveira and L.C. Roma Junior*
*Department of Agriculture, São Paulo State Government, Rod Carlos Tonani, km 94, Sertãozinho SP, 14160-970, Brazil; lucia.plima@hotmail.com*

This study evaluates how different management used on rotational system of *Brachiaria brizantha* (Hochst. ex A. Rich.) swards affects weight gain (WG) of crossbred heifers, all year long, in Central Brazil. Three stocking systems were tested, in which cattle were allowed to graze: (1) cycles of 28 days ($T_{28d}$); (2) when the sward reached a pre-grazing height of 30 cm ($T_{0.3m}$); and (3) irrigated when the sward reached a pre-grazing height of 30 cm ($Tirrig_{0.3m}$). Eight grazing cycles with sward recovery period followed by 2 days of grazing were evaluated using a completely randomized block design and split plot design with 2 repetitions per stocking system and 4 paddocks per system. The variables were pre-grazing forage mass (FM), leaf proportion (L%) and heifer's WG. The heifers in $T_{28d}$ and $T_{0.3m}$ were supplemented during the dry season. The stocking systems did not affect (P=0.4559) FM mass and L%. On the other hand, forage mass (FM) production was higher (P<0.0001) during summer while L% (P<0.0001) was lower L% in the winter. The FM production results were 9,966, 8,212, 8,814 and 11,856 kg of DM/ha and for L%, 29.1, 21.8, 30.24 and 31.5% in fall, winter, spring and summer, respectively. There was effect on heifer WG of stocking system (P=0.0055), season (P<0.001) and stocking systems and season interaction (P<0.001). The lowest WG was observed for heifers in T-$irrig_{0.3m}$ during winter. The WG values for heifers in $T_{28d}$, $T_{0.3m}$ and T-$irrig_{0.3m}$ were 0.42, 0.44 and 0.39 kg/day during fall; 0.45, 0.47, 0.26 kg/day, in winter; 0.46, 0.52, 054 kg/day, in spring; and 0.49, 0.55, 0.55 kg/day, in summer, respectively. During winter, lower FM and L% decreased the WG gain of heifers fed exclusively with irrigated pasture.

---



**Follicular characteristics of lactating Nellore cows submitted to the 5-d EB+CIDR program administer**

*M.V.C. Ferraz Jr.[1], A.V. Pires[2], M.V. Biehl[1], R. Sartori Filho[2], M.H. Santos[1], L.H. Cruppe[3] and M.L. Day[3]*
*[1]University of São Paulo-FMVZ, Av. Duque de Caxias, 225, 13635-900 Pirassununga, SP, Brazil, [2]University of São Paulo-ESALQ, Av. Padua Dias, 11, 13418-900 Piracicaba, SP, Brazil, [3]The Ohio State University, 2027 Coffey Rd, 43210 Columbus, OH, USA; pires.1@usp.br*

The objective was to evaluate dose effect of estradiol benzoate (EB) at the onset of the 5-d EB+CIDR program on follicle development and ovulation rate in 85 lactating Nellore cows. Animals were randomly assigned to receive either 1 mg EB (EB-1; n=42) or 2 mg EB (EB-2; n=43) at CIDR® insertion on d0 of the experiment. On d5, CIDR was removed and 25 mg PGF2α (Lutalyse®), 0.6 mg estradiol cypionate (EPC®) and 300 IU equine chorionic gonadotropin (eCG; Novormon®) were administered to all cows. Estrotect® heat detector was applied on d5 to aid estrus detection performed twice daily for 96 h after CIDR removal. Ovarian ultrasonography (US) was performed on d5 and d7. Cows with follicles ≥9 mm on d5 (EB-1, n=13/43; EB-2, n=11/42) were removed from the experiment to be considered not synchronized. Follicle growth was calculated between d5 and d7. Ovulation was confirmed by US 11 d after CIDR removal. Largest follicle diameter on d5 (EB-1, 7.4±1.2; EB-2, 7.3±1.0 mm), and on d 7 (EB-1, 10.0±2.2; EB-2, 10.3±1.4 mm), follicle growth (EB-1, 1.3±0.7; EB-2, 1.4±0.7 mm/d), estrus response (EB-1, 72.0%; EB-2, 78.5%), ovulation rate (EB-1, 88.3%; EB-2, 90.4%) and incidence of double ovulation (EB-1, 4.6%; EB-2, 7.1%) did not differ between treatments. In conclusion, 1 mg of EB at CIDR insertion is as effective as 2 mg to induce emergence of a new follicular wave in lactating Nellore cows submitted to the 5-d EB+CIDR program.

## Influence of dietary protein source on milk clotting time and rennet coagulation ability

*M.R. Yossifov*
*Institute of Animal Science, Animal nutrition and feed technologies, station Pochivka 1, 2232, Bulgaria; m_vet@abv.bg*

Rennet curd cheese is sensitive process based on milk coagulation and formation of rennet-induced gel. This study compared the effect of dietary protein sources with vegetable origin (sunflower meal (SFM) vs dried distillers' grains with solubles (DDGS)) in ewe's diets on milk clotting time (MCT, s), rennet coagulation ability (score 1-3) at 6 (RCA6) and 12 h (RCA12), and index of milk clotting time (IMCT, %). Simultaneously, it's examined the relationship between suggested parameters and effect of lactating day (27, 35, 42, 49, 56, 63 and 70-d). It's studied earlier lactation period (4-10 weeks) in 18 ewes (n=9), 126 individual milk samples of regional breed Synthetic Bulgarian Dairy Population (SBDP). Animals were allocated into two dietary treatments (n=9 per diet) formuled to contain equal amounts of fiber, energy, protein digestible in small intestines (PDI) and calcium:phosphorus ratio (Ca:P). The rates of rennet induced ewe milk coagulation (s) were measured as time from rennet addition to onset of rennet-induced gel. Rennet coagulation ability was measured subjectively based on scale of coagulum (shape, homogeneity, firmness, porosity) and serum (limpidity, turbidity) characteristics (score 1-3). Obtained results sugests significant effect ($P<0.05$) of feeding supplement (DDGS vs SFM) on RCA12 and tendency at RCA6 ($P=0.04$). The influence on MCT and IMCT also showed such tendency ($P=0.57$). Deduced Pearson's correlation coefficients between evaluated parameters values and diet supplements at different lactation stage (d). In conclusion, the obtained results revealed that investigated parameters (RCA6, RCA12, MCT, IMCT) were influenced by dietary protein source (SFM vs DDGSw). Firmer coagulum (RCA6, RCA12), but longer MCT were observed at milk, collected from ewes fed SFM- vs DDGSw- based diet.

---



## ACE-inhibitory activity of two hard cheeses at different ripening stages

*L. Basiricò, P. Morera, E. Catalani, U. Bernabucci and A. Nardone*
*University of Tuscia, Department of Agriculture, Forests, Nature and Energy, via S. C. De Lellis, s.n.c., 01100 Viterbo, Italy; nardone@unitus.it*

Parmigiano Reggiano (PR) and Grana Padano (GP) are two Italian hard cheeses. During the cheese ripening caseins are degraded to peptides, most of them show biological activities. VPP and IPP are known as angiotensin-converting enzyme (ACE)-inhibiting peptides founded in several cheese varieties. The aim of this work was to compare the ACE-inhibitory activity and the amount of VPP and IPP in PR and GP cheeses at 12 (GP12 and PR12) and 32 (GP32 and PR32) months of ripening. Twelve samples (6 for each ripening time) for each cheese (PR and GP) were analysed. Two samples of water-soluble extract (WSE) were extracted from cheeses and the fraction <3,000 Da was obtained by ultrafiltration. ACE-inhibitory activity was tested for all different samples and was evaluated as IC50 (50% inhibitory concentration expressed as µg of total peptides/ml) using a commercial kit. The IC50 for each sample was estimated in triplicate preparing an inhibition curve using the sample concentration (x axis) and ACE-inhibitory activity (inhibition rate %; y axis). The concentration of VPP and IPP was determined in triplicate by LC/ESI-HRMS analysis (Thermo Scientific, San Jose, CA) and were expressed as mg/kg of cheese. A two-ways ANOVA statistical model was used for the analysis of data. PR and GP WSE showed a consistent ACE-inhibitory activity and IC50 values of PR12 and GP12 were lower ($P<0.05$) than that of PR32 and GP32 (7.9 and 7.6 vs 9.6 and 8.4 µg/ml for PR12 and GP12 vs PR32 and GP32, respectively). VPP and IPP concentration was lower ($P<0.01$) in PR32 (1.4 and 0.7 mg/kg, VPP and IPP respectively) and GP32 (1.7 and 0.6) compared with PR12 (9.2 and 4.4) and GP12 (6.2 and 3.4). ACE-inhibitory activity was consistent with VPP and IPP concentration in both cheeses. As reported for other cheeses the increase in ripening time reduces the concentration of VPP and IPP ACE-inhibiting tripeptides.

# Session 49 Theatre 1

**The role of small ruminants in meeting the global challenges for sustainable intensification**

*J. Conington, S. Mucha, N. Lambe, C. Morgan-Davies and L. Bunger*
*SRUC, W. Mains Rd, Ediburgh, EH9 3JG Scotland, United Kingdom; jo.conington@sruc.ac.uk*

There are an estimated 1.16 bn and 996 m sheep and goats in the world respectively, with the majority of these located in developing countries for smallholder farming systems. In Europe, meat, milk and wool come from around 98 m sheep and 13 m goats which are managed in a diverse range of production systems. The main challenge to intensify production is similar for both parts of the world and all systems, which is to reduce inefficiencies without harming the ability to produce in the future, to meet projected increases in demand for livestock products. This paper addresses key inefficiencies in sheep meat and goat milk production to quantify the opportunity costs of low production levels. Examples will be given from research and practice on new opportunities to integrate technologies for more precision-led livestock farming for small ruminants. These include genetic and genomic selection for improved health and production, electronic identification and its role in large-scale precision farming, targeted selective treatment (of anthelmintics) for the sustainable control of parasites, novel methods for grazing control in extensive environments and abattoir feedback mechanisms on product quality for more efficient breeding and selection programmes. The lessons will be highlighted from the use of benchmarking performance and monitor farm networks in practice, that facilitate the uptake of new techniques to improve the efficiency of the livestock sector.

# Session 49 Theatre 2

**Adaptation goat production systems to climate change**

*N. Koluman Darcan[1] and I. Daskiran[2]*
*[1]Cukurova University, Department of Animal Science, 01330 Adana, Turkey, [2]Ministry of Agriculture and Animal Production, Animal Production Research Center, Ankara, Turkey; nazankoluman@gmail.com*

Extreme atmospheric events and climate change have been become a current global issue and parallel to signing Kyoto Protocol by Turkish government, comprehensive studies and international cooperation has been started on the necessary strategies which must be urgently taken and possible projections for the future. Animal production is contribute to global warming with causing to 9% of human welded $CO_2$ emission, 35-40% of $CH_4$ emission, 65% of $N_2O$ emission and 64% of $NH_3$ emission. On the other hand, the effects of global warming (high temperature and drought) has adversely affects on animal production. In the corresponding studies which are trying to bring up livestock-based scenarios according to assumptions for the future and climate modeling, it is very important that the projection of which feed and which goat are more appropriate for different regions. Economic importance of goat production rise up during last decades in Turkey. Goats have numerous advantages for maintain its production in extreme climate conditions. Especially, goats are effectively convert some feed sources when they compare to other farm animals. Also methane emission is less than the others. Therefore, some recommendations developed which are based on methane emissions and feed stocks and the adaptation of goat breeding against to climate change discussed in this article.

**Sustainable and low input dairy goat production in Greece**

*G. Arsenos, A. Gelasakis, R. Giannakou, S. Termazidou, M. Karatzia, K. Soufleri, P. Kazana and A. Aggelidis*
*Aristotle University of Thessaloniki, Veterinary Faculty, Box 393, 54124, Greece; arsenosg@vet.auth.gr*

The aim was to assess phenotypic differences between goat breeds in Greece which are perceived to be adapted to low input systems in terms of productivity, animal health and welfare and milk quality. Greece has the highest number of goats per head of population in Europe. The Greek national flock is ranked as the largest national dairy goat flock in the EU, counting more than 3.5 million female dairy goats (about 45% of the goats in EU); from them, around 3.1 millions goats (about 90%) belong to the local type and they are reared either under the semi-intensive system (about 82%), using common indigenous (upland) pastures but also providing indoors feeding and housing, or under the extensive pastoral system (about 8%), using mainly summer grazing on mountain pastures. The remaining population (10%) comprises of high producing goats of various indigenous (Skopelos) and foreign dairy breeds (Damascus, Alpine and Saanen) and their crosses. Two different dairy goat breeds (Greek autochthonous and Skopelos) were compared with a conventional breed (Damascus) within a dry, low input environment in Greec. Approximately 900 milking goats (n=300 per breed) in 7 herds, were monitored for two years, with milk yield, milk quality, incidence of mastitis (based on milk somatic cell counts and microbiological analyses), body tissue changes, and fertility status, recorded for each individual monthly. Results created a database comprised 8,600 individual milk yield records and 8,300 records of milk quality (fat, protein, lactose, SNF, cells and TVC). Also, it includes data of about 1,350 milk samples from individual goats that had been cultured for pathogens such as CNS, *Staphylococcus aureus*, *Streptococcus* spp, *Listeria* and Coliforms. Moreover, 2,000 parasitological examinations (from individual goats) have been performed including coprocultures for the identification of nematode genera. Funding was received by EC grant n° FP7-266367 (SOLID).



**Emission of rumen methane from sheep in silvopastoral system**

*F.O. Alari[1], N. Andrade[1], N.C. Meister[1], E.B. Malheiros[2] and A.C. Ruggieri[1]*
*[1]Faculdade de Ciências Agrarias e Veterinárias, Unesp, Animal Science, Via de Acesso Prof Paulo Donato Castellane, sn, 14884-900, Jaboticabal, São Paulo, Brazil, [2]Faculdade de Ciências Agrarias e Veterinárias, Unesp, Exact Science, Via de Acesso Prof Paulo Donato Castellane, sn, 14884-900, Jaboticabal, São Paulo, Brazil; fernandoalari7@gmail.com*

The emission of greenhouse gases by agricultural activity has been the focus of study in the world, so the knowledge of such emission, mainly of methane produced by ruminants, is of vital importance. The objective of this research was to determine how enteric methane emissions from sheep kept on pastures of massai-grass behave, with or without the presence of trees in rotational stocking (stocking flashing). The methane evaluation was made at Universidade Estadual Paulista, UNESP Jaboticabal, São Paulo, Brazil, in March 2013. Treatments were of two arrangements of eucalyptus spacing (6.0×1.5 m and 12.0×1.5 m) and one of untreated trees. The entry of animals to grazing was at 95% light interception and the animals remained to a height of 15 cm. The determination of ruminal methane production was performed by the methodology of sulfur hexafluoride gas ($SF_6$) tracer, with modifications for the determination in small ruminants. Samplings of gases were performed for six consecutive days in five sheep with an average weight of 29 kg, in each treatment, from where emissions per animal and per kilogram of body weight were calculated. The emission of ruminal methane per animal and per Kilogram of body weight was not altered by the three arboreal arrangements studied (P>0.05). Average ruminal methane emissions per animal per kilogram of body weight were of 5.8, 7.1 and 6.2 g/day, and 0,21, 0,25 and 0,22 g/day/kg, respectively in arboreal arrangements without eucalyptus and with spacing eucalyptus of 6 and 12 m. The silvopastoral system did not alter ruminal methane emission, and allied to the uncountable benefits that this system provides, it demonstrates its potential to replace the traditional pastoral system in monoculture.

## Session 49 Theatre 5

**Influence of dietary lipid supplements on methanogenesis, digestion and milk production in dairy cow**

*A.R. Bayat, H. Leskinen, I. Tapio, J. Vilkki and K.J. Shingfield*
*MTT Agriffod Research Finland, Jokioinen, 31600, Finland; alireza.bayat@mtt.fi*

Five Finnish Ayrshire cows in early lactation fitted with rumen cannulae were used in a 5×5 Latin Square with 28 d experimental periods. Treatments comprised a control diet, or the same diet supplemented with 50 g/kg diet DM of myristic acid (MA), rapeseed (RO), safflower (SO) or linseed (LO) oils. Diets were based on grass silage (0.60 diet DM), barley and rapeseed meal. Samples of rumen fluid (n=8) were collected during d 23 of each period. Enteric methane ($CH_4$) production was determined using the sulphur hexafluoride technique from d 16 to 21. Data were analyzed using with a model that included the fixed effects of period and treatment and random effect of cow. Compared with the control, all lipid supplements lowered DM intake ($P<0.01$), with MA causing the largest decrease (21.8, 15.0, 19.1, 20.4 and 20.1 kg/d for control, MA, RO, SO and LO, respectively). Yields of milk (28.7, 22.1, 29.0, 29.6 and 28.6 kg/d for control, MA, RO, SO and LO, respectively), energy corrected milk, protein and lactose were lower ($P<0.01$) for MA compared with other treatments. Lipid supplements had no influence on rumen fermentation, other than an increase ($P<0.01$) in ammonia-N concentration on MA compared with other treatments. Daily enteric $CH_4$ production was lowest ($P<0.01$) for MA and highest on the control (614, 403, 475, 488 and 484 g/d for control, MA, RO, SO and LO, respectively). All lipid supplements decreased ($P<0.01$) g $CH_4$ production/kg energy corrected milk (22.7, 18.8, 17.5, 17.5 and 17.5 for control, MA, RO, SO and LO, respectively), but did not affect ($P>0.05$) $CH_4$ production per unit of total organic matter digested. Treatments had no influence ($P>0.05$) on organic matter or fibre digestion. Dietary supplements of plant oils can be used to lower enteric $CH_4$ production of dairy cows, without compromising milk production or diet digestibility. Even though MA lowered $CH_4$ production, the decrease was accompanied by adverse effects on intake and animal performance.

## Session 49 Theatre 6

**Genetic characterization and polymorphism detection of casein genes in Egyptian sheep breeds**

*O. Othman[1], S. El-Fiky[1], N. Hassan[2], E. Mahfouz[1] and E. Balabel[1]*
*[1]National Research Center, Cell Biology Department, El Buhouth St., Dokki, 12311, Egypt, [2]Faculty of Science, Ain Shams University, Zoology Department, Ain Shams, Cairo, 12622, Egypt; othmanmah@yahoo.com*

This study aimed to detect the genetic polymorphism of casein genes; alpha 1-, alpha 2-, beta- and kappa-casein in three Egyptian sheep breeds; Rahmani, Barki and Ossimi. PCR-SSCP and PCR-RFLP were used to detect the genetic polymorphism of these casein genes. SSCP of 223-bp amplified fragment of alpha 1-casein gene recorded three different patterns; TT, TC and CC. The sequence analysis of two homologous patterns showed a SNP (T/C) at position 170. The frequencies of these patterns were 43.33%, 50.00%, and 6.67% in Rahmani; 83.33%, 13.33%, and 3.33% in Ossimi and 74.07%, 22.22%, and 3.70% in Barki, respectively. The restriction digestion of alpha 2-casein PCR product (1,300-bp) by Tru1I endonuclease revealed three different genotypes; AA, AG and GG with frequencies of 66.67%, 30.00%, and 3.33% in Rahmani; 96.67%, 3.33%, and 0.00% in Ossimi and 96.15%, 3.85%, and 0.00% in Barki, respectively. The sequence analysis revealed the presence of a SNP (A/G) in intron 6 of alpha 2-casein gene. SSCP of beta-casein gene revealed two different patterns in tested animals. The sequence analysis of the PCR product (299-bp) of these two different patterns showed two single nucleotide substitutions; A/C and C/T. The frequencies of these two different patterns were 96.67% and 3.33% in Rahmani; 65.52% and 34.48% in Ossimi and 88.46% and 11.54% in Barki, respectively. The genetic polymorphism of kappa-casein gene was examined using PCR-SSCP technique. PCR amplified a fragment with 406-bp in this gene. SSCP results showed that all tested sheep animals are monomorphic. The nucleotide sequences of alpha 1- T and C alleles, alpha 2-, beta- and kappa-casein in Egyptian tested sheep animals were submitted to GenBank with the accession numbers KF018339, KF018340, JX080380, JX080379 and JX050176, respectively.

**Effect of forage type and protein supplementation on chewing and faecal particle size in sheep**

*P. Nørgaard[1], J.I. Gerdinum[1], C. Helander[2] and E. Nadeau[2]*
*[1]University of Copenhagen, Dept. of Veterinary Clinical and Animal Sciences, 1870 Frb., Denmark, [2]SLU, Dept. of Animal Environment and Health, 532 23 Skara, Sweden; elisabet.nadeau@slu.se*

The objective was to assess effects of forage type and protein supplementation on chewing activity and faecal particle size in sheep. A duplicated 5×5 Latin square design with 10 ram lambs (73±7 kg) fed five different forages with or without daily supplementation of 0.15 kg rapeseed meal was used. The sheep were kept in individual pens during the first three weeks of each period; 2 weeks of adaptation and one week of *ad libitum* intake registrations. The sheep were fed at 80% of *ad libitum* intake in metabolic cages during the last week of each period. The forages were first cut grass silage, third cut red clover (75%)-grass (25%) silage without or with a bacterial inoculant (*Lactobacillus plantarum* and *Lactobacillus buchneri*) and whole-crop maize silage at dough and dent stage of maturity with NDF contents (% of DM) of 49, 45, 46, 39 and 40, respectively, and a theoretical chopping length of 20 mm. Measurements of chewing activity were based on continuously recorded jaw movement oscillations for 96 h from day 25. Faeces were collected for 4 days, washed, freeze dried and particle DM (PDM) was sieved (1, 0.5, 0.212, 0.106 mm and bottom bowl). Data were statistically analyzed using a mixed model with fixed effects of forage type, protein supplementation and period, and a random effect of sheep nested within protein supplementation. The 80% of *ad libitum* intake of forage DM per BW ($P<0.01$), daily ruminating time ($P<0.001$), ruminating time per kg forage NDF intake (NDFI) ($P<0.0001$), number of JM during rumination per kg forage NDFI ($P<0.0001$), faeces PDM per kg DM ($P<0.0001$) and mean particle size ($P<0.0001$) were affected by forage type. Protein supplementation did not significantly affect eating activity, rumination activity or faeces characteristics.



**Effect of aS1-casein genotypes on yield and chemical composition of milk from Skopelos goats**

*G. Arsenos, A. Gelasakis and M.S. Kalamaki*
*Aristotle University of Thessaloniki, Veterinary Faculty, Box 393, 54124, Greece; arsenosg@vet.auth.gr*

The aim was to assess the effect of as1-casein genotypes on milk yield (MY) and milk composition (MC) of 235 Skopelos dairy goats, randomly selected from two semi-extensively raised herds. Individual milk samples were collected monthly for two successive milking periods and analysed for protein (PRO), fat (FAT), lactose (LAC), solid non-fat (SNF) concentration and total viable count (TVC). Genotyping at the CSN1S1 locus was performed using different allele-specific protocols. For identifying the CSN1S1 A* (including A, G, I, H), B* (including B1, B2, B3, B4 and C), F and N alleles, a well established PCR-RFLP method was used. A general linear model was used to estimate the fixed effect of genotype on MY and MC traits. Genotype NN was used as reference. Farm, stage of milking period and year were the fixed effects; MY and log transformed values of TVC were used as covariates. A total of 9 genotypes were identified in the studied population. Genotype A*B* was the most common (23.8%) followed by genotypes B*B* and A*A* (21.7% and 21.3%, respectively). The less frequent genotype was FB* (1.3%). From the model, significantly higher PRO (0.25%, $P<0.001$), FAT (0.36%, $P<0.01$) and SNF (0.34%, $P<0.05$) were observed for A*A* compared to NN genotype. PRO was, also higher for B*B* genotype (0.20%, $P<0.01$), whereas FAT was higher for B*B* (0.27%, $P<0.05$), FB* (0.59%, $P<0.05$) and FF (0.80%, $P<0.01$) genotypes. MY was, also, favoured by FA*, FB* and NA* genotypes (171.1 gr, $P<0.001$, 131.6 gr, $P<0.05$ and 85.2 gr, $P<0.01$, respectively). High frequency of genotypes A*A*, A*B* and B*B* suggests a different genetic backround for Skopelos goats comparing to improved European dairy goat breeds. Knowledge of the genotypic effects on MY and MC traits for goats reared under semi-extensive systems, may facilitate targeted genetic selection for sustainable production when available feed resources are limited. Acknowledgement: Funding was received by EC grant n° FP7-266367 (SOLID).

**Liver gene expression in periparturient dairy goats fed diets enriched with stearate or PUFA**

*J.M. Caputo[1], G. Invernizzi[1], M. Ferroni[1], A. Agazzi[1], J.J. Loor[2], V. Dell'Orto[1] and G. Savoini[1]*
*[1]University of Milan, Department of Health, Animal Science and Food Safety, 10, via Celoria, 20133 Milan, Italy, [2]University of Illinois, Department of Animal Sciences & Division of Nutritional Sciences, 1207, W Gregory dr, 61801 Urbana, IL, USA; giovanni.savoini@unimi.it*

To study the hepatic expression of genes involved in lipid metabolism in periparturient dairy goats fed diets enriched with saturated or unsaturated fatty acids was the main aim of the trial. Twenty three second parity alpine dairy goats were assigned to three treatments: (C; n=8) fed a non fat-supplemented basal diet, (ST; n=7) fed a basal diet supplemented with stearic acid and (FO; n=8) fed a basal diet supplemented with fish oil. The supplementation started from the last week of gestation and lasted 21 days after kidding and supplied 30 g/head/d extra fatty acids during the dry period and 50 g/head/d during lactation. Liver biopsied samples were harvested at day -7, 7 and 21 relative to kidding date and immediately snap-frozen. Blood samples were collected on days -7, -2, 0, 2, 7, 14 and 21. Quantitative real-time RT-PCR of ACAA1, ACOX1, SCD, SOD, CAT, SREBF2 and PPARA was performed. Data obtained were analyzed using a MIXED procedure of SAS. No differences were observed on milk production, milk composition, body weight and body condition score. SCD expression was significantly (1.8-fold; $P<0.01$) up-regulated in ST at days 7 and 21 while the opposite trend was observed in FO. ACOX1 pattern was similar for both fat-supplemented diets peaking at day 7 (ST 1.6-fold; $P<0.05$) and returning close to neutrality at 21 day. NEFA serum content was high in FO and ST at day 2 and remained higher in FO at days 7 and 14 ($P<0.05$). Cholesterol peaked in C at day 7 and in FO at days 14 and 21 ($P<0.05$). SCD expression pattern could explain higher levels of free cholesterol, since has been proposed its role in cholesterol esterification to MUFA. Results support a modulation of hepatic lipid metabolism by saturated and unsaturated dietary fat supplements.

---



**Quantifying multifunctionality of pasture-based livestock systems in Mediterranean mountains**

*A. Bernués[1], T. Rodríguez-Ortega[2], R. Ripoll-Bosch[2,3] and F. Alfnes[1]*
*[1]Norwegian University of Life Sciences, P.O. Box 5003, 1432 Ås, Norway, [2]CITA Aragón, Avda. Montañana 930, 50059 Zaragoza, Spain, [3]Wageningen University, P.O. Box 338, 6700 AH Wageningen, the Netherlands; alberto.bernues@nmbu.no*

Mountain agroecosystems, mostly pasture-based livestock systems (PBLS), are highly multifunctional. Many of these functions constitute public (non-market) goods, and farmers have little incentive to provide them. The application of the Ecosystem Services (ES) framework to PBLS enables the simultaneous assessment of all goods and services, both provisioning (food products) and non-provisioning (regulating, supporting and cultural), at the same priority level, allowing for a further integration of agricultural and environmental policies. However, the incorporation of non-provisioning ES to policy design is challenging, being their quantification one of the main problems. Based on a previous socio-cultural valuation of ES delivered by PBLS in Spanish Mediterranean mountains, we used a survey-based stated-preference method (choice modelling) to obtain a ranking of ES and their economic value according to the willingness to pay (WTP) of the local (residents of the study area) and general (region where the study area is located) populations. For the general population, the prevention of forest fires (53% of total WTP) was valued as a key ES, followed by the production of specific quality products linked to the territory (20%), the conservation of biodiversity (18%) and cultural landscapes (8%). For the local population, the prevention of forest fires was also valued in the first place (40%), followed by the production of quality products (26%). However, the value attached to biodiversity and cultural landscapes reversed (9% and 25%, respectively). The link between ES and different components of the Total Economic Value taxonomy allowed giving an economic value to mountain agroecosystems, which was 120 €/person/year for the general population (three times the current level of support of CAP agro-environmental policies) and 197 € for the local population.

**Effects of levels of chicory on feedlot performance of Fars native lambs**

*M.R. Hashemi, B. Eilami and A.H. Karimi*
*Fars Research Center for Agriculture and Natural Resources, Animal Science, Shiraz, 71555-617, Iran; eilamibahman@mail.com*

In this experiment, with five treatments in a completely randomized design, effects of different levels of chicory plant hay on fattening performance and carcass characteristics of Fars native lambs were studied. Thirty lambs were randomly divided into five groups. Treatments were: T1: control; T2: 25%; T3: 50%; T4: 75%; and T5: 100% chicory replacement with alfalfa. Each ration contained 60% roughage and 40% concentrate. One and three mounts were considered for duration of adaptation and main periods of experiments respectively. Means of final live body weight, feed intake and daily gains were significant different between treatments. Feed conversion ratios were not significantly different between treatments. Final live body weights for groups 1 to 5 were 48.6, 44.0, 40.1, 36.9, 36.5 kg, daily feed intake were 1.45, 1.03, 0.94, 0.75, 0.75 kg and feed conversion ratios were 7.38, 6.67, 7.27, 8.28, 9.37 kg feed to kg gain respectively. Carcass characteristics were not influenced by the treatments. Due to results of this study, use of 25 and 50 percentage of chicory could be recommended.



**Milk fat content and fatty acid profile of heat-stressed dairy goats supplemented with soybean oil**

*S. Hamzaoui, A.A.K. Salama, G. Caja, E. Albanell and X. Such*
*Universitat Autònoma de Barcelona, Animal and Food Sciences, Group of Ruminant Research (G2R), Campus universitari, Edifici V, 08193 Bellaterra, Spain; gerardo.caja@uab.es*

The objective was to test whether supplementation with soybean oil is useful to enhance milk fat depression in heat stressed dairy goats. Multiparous Murciano-Granadina dairy goats (n=8; 42.8±1.3 kg BW; 99±1 DIM; 2.0±0.1 l/d) kept in metabolic cages were used in a replicated 4×4 Latin square with 4 periods (14 d adaptation, 5 d measurements, 2 d transition between periods). Goats were allocated to one of 4 treatments in a 2×2 factorial arrangement. Factors were no oil (C) or 4% soybean oil (SB), and thermal neutral (TN; 15 to 20 °C) or heat stress (HS; 12 h/d at 37 °C and 12 h/d at 30 °C) conditions, and treatments were: TN-C, TN-SB, HS-C, and HS-SB. Relative humidity was constant (40±5%). Feed intake, milk yield and digestibility were measured, and milk composition and milk fatty acids profile analyzed. Compared to TN, the HS goats had lower ($P<0.05$) feed intake (-38%), milk yield (-9%) and milk protein (-15%). SB varied ($P<0.01$) milk fat content (+30%), de novo <C16 (-26%), C16+C16:1 (-31%), preformed >C16 (+71%) and rumenic fatty acid (+390%), regardless the ambient temperature conditions. Goats under HS had greater digestibility coefficients of DM, CP, NDF and ADF (5 to 9 points; $P<0.05$) than TN goats, being unafected by SB supplementation. In conclusion, feeding soybean oil to heat-stressed dairy goats was a useful way to increase digestibility, milk fat, preformed fatty acids and conjugated linoleic acid, without altering intake, milk yield or milk protein content, and it is recommended for feeding goats under heat-stress conditions in practice.

## Session 49 Poster 13

**Lambing performances and maternal behavior of Najdi ewes under intensive conditions**

*M. Alshaikh[1], R. Aljumaah[1], M. Ayadi[1], M. Alfriji[1], A. Alhaidary[1] and G. Caja[2]*
*[1]King Saud University, Animal Production, College of Food and Agriculture Sciences, King Saud University, P.O. Box 2460, Riyadh 11451, Saudi Arabia, [2]Universitat Autònoma de Barcelona, Department of Animal and Food Sciences, Group of Ruminant Research (G2R), 08193 Bellaterra de Barcelona, Spain; alshaikh@ksu.edu.sa*

The study aimed to evaluate the maternal behavior of Najdi ewes lambing during winter and spring on a commercial farm under intensive conditions. A total of 242 lambings were monitored and daily time, litter size (LS), dystocia, parturition length (PL), maternal behavior measured as lamb acceptance and licking (LD, licking delay; LL, licking length), and concurrent parturitions in the same pen were recorded. Najdi ewes showed greater lambing during spring (59.8%) than during winter (40.2%) and moderate LS (1.31 lambs/ewe), which increased according to parity and age of the ewe. Despite being known as a poor motherhood sheep breed, low dystocia (4.5%), fast and easy parturition (28.5 min) were observed in Najdi ewes in our study. Moreover, the ewes take good care of their lambs, licking them early (LD, 1.2 min) and for enough time (LL, 28.5 min) after parturition. On the contrary, the studied flock showed a high incidence of lamb rejections (20.7%) which increased in primiparous ewes. No effects of LS, season or dystocia were detected on lamb acceptance, but concurrent parturition was identified as the main cause of lamb rejections. Use of isolating facilities shortly after parturition, especially in the case of primiparous ewes, was recommended.

## Session 49 Poster 14

**Karyotypes studies of three different types of *Ovis* species hybrids**

*M.A. Zhilinsky, V.A. Bagirov, P.M. Klenovitsky, B.S. Iolchiev, V.P. Kononov, V.V. Shpak and N.A. Zinovieva*
*All-Russian Research Institute of Animal Husbandry, Dubrovitsy settlement, Podolsk District, Moscow Region, 142132, Russian Federation; naitkin888@mail.ru*

Wild species are the resource of genetic material for new forms of animals. The frozen epididymal sperm of *Ovis nivicola borealis* (Severtzov,1873) was used to produce the hybrids with *Ovis aries* (L.,1758) of Romanov breed and *Ovis ammon pholli* (Blyth,1841). The sperm of *O.a. pholli* was used to produce the hybrids with Romanov ewes. The insemination was performed into the oviduct of the synchronized ewes using surgical method. The karyotypes of the parent species and their hybrids were investigated. Taking in a point the different number of bi-armed autosomes in the parent species we used the ideogram based on goat karyotype supplemented by metacentric chromosomes to perform the numbering of the chromosomes. The number of chromosomes in *O.n. borealis* × *O. aries* hybrids was 53. Metacentric chromosomes inherited from *O. aries* were presented by three pairs of homologues chromosomes corresponding to 1/3, 2/8 and 11/5 of basic karyotype and by unpaired metacentric chromosome 19/9 inherited from *O.n. borealis*. The *O.n. borealis* × *O.a. pholli* hybrids had 54 chromosomes. Metacentric chromosomes were presented by two pairs of homologues chromosomes corresponding to 1/3 and 2/8 of the real karyotype inherited from both parent species and two unpaired metacentric chromosomes 5/11 and 9/19 obtained from *O.n. borealis*. The *O.a. pholli* × *O. aries* hybrids carried out 5 bi-armed chromosomes. Metacentric homologues corresponding to 1/3, 2/8 chromosomes of basic karyotype were inherited from both parent species and unpaired chromosomes 11/15 were obtained from the domestic sheep. Results of studies have shown that first-generation of hybrids had a good viability, the normal growing capacity and fertile. Hybridization of domestic and wild species of the genus *Ovis* allows to enrich the gene pool of culture sheep breeds by valuable genetic qualities of a wild animals.

## Session 49 Poster 15

**Effect of chicory in diet on carcass characteristics of Fars native lambs**

*B. Eilami, M.R. Hashemi and A.H. Karimi*
*Fars Research Center for Agriculture and Natural Resource, Animal Science, Shiraz, 71555-617, Iran; eilamibahman@mail.com*

In this experiment, with five treatments in a completely randomized design, effects of different levels of chicory plant hay on carcass characteristics of Fars native lambs were studied. Treatments were T1: control; T2: 25%; T3: 50%; T4: 75%; and T5: 100% (chicory replacement with alfalfa). Each ration contained 60% roughage and 40% concentrate. Thirty Torky ghashghaii male lambs were divided into five groups. One and three mounts were considered for duration of adaptation and main periods of experiments respectively. Final live body weights for groups 1 to 5 were 48.4, 42.7, 41.2, 35.6, 38.2 kg respectively. Cold carcass percentages were 56.8, 53.5, 50.4, 49.6, 50.8, Carcass lean percentages were 44.0, 45.4, 49.8, 55.4, 51.2, fat depots percentages were 13.7, 12.9, 11.7, 7.73, 9.40, fattail percentages were 25/1, 21/2, 17/5, 13/7, 17/5 and bone percentages were 15.5, 17.8, 18.7, 20.7, 19.5 for groups 1 to 5, respectively ($P<0.05$). With increase of chicory in diets, leg weight percentages significantly increased. Lean percentages of hand, loin, breast and flap increased with increasing of chiory in diets and fat percentages of leg, hand, loin, breast and flap decreased. Back fat thickness were significantly different between treatments.

## Session 50 Theatre 1

**Genetic and genomic analysis of longitudinal egg-production data in laying hens**

*A. Wolc[1,2], J. Arango[1], P. Settar[1], J.E. Fulton[1], N.P. O'Sullivan[1], D.J. Garrick[2], R. Fernando[2] and J.C.M. Dekkers[2]*
*[1]Hy-Line International, 2583 240th St, Dallas Center IA 50063, USA, [2]Iowa State University, Department of Animal Science, 239 Kildee Hall, Ames IA 50011, USA; awolc@jay.up.poznan.pl*

Egg production is expressed throughout the productive life of a laying hen, during which both genetic and environmental factors affecting productivity may vary. Thus, random regression models (RRM) are expected to provide a more accurate evaluation of the genetic potential of a bird than a simple analysis of cumulative egg number. Further improvements in accuracy are expected by including genomic information. The objectives of this study were to estimate genetic parameters for egg production over the age trajectory and to evaluate the potential of genomic RRM to improve accuracy of selection. Pedigree and phenotypic data were available for 6 generations of a brown egg layer line and 3,908 animals were genotyped using a 42K Illumina SNP chip. Daily egg production records were pooled into 2-wk periods. Data were analyzed using ASReml with the average production curve within hatch as fixed effect and random animal genetic and permanent environmental effects modeled with linear or quadratic Legendre polynomials. A genomic relationship matrix was used in a reduced animal model to incorporate phenotypes of non-genotyped offspring of genotyped parents. Estimates of heritability by 2-wk period increased with age. Predictions of egg number in validation across the 46-week production period and for most 2-wk periods were more accurate with genomic than pedigree relationships. In conclusion, random regression reduced animal models can be utilized in breeding programs using genomic information, resulting in substantial improvements in accuracy of selection for longitudinal traits. Breeding values for slope from a model with a linear polynomial can be used to directly select for persistency, but such a model may not be optimal for all populations, especially when using genomic information.

**Random regression analyses using Legendre polynomial to model body weight growth in mink**

*M. Shirali[1], V.H. Nielsen[1], S.H. Møller[2] and J. Jensen[1]*
*[1]Center for Quantitative Genetics, Department of Molecular Biology and Genetics, Aarhus University, Blichers Allé 20, Postboks 50, 8830 Tjele, Denmark, [2]Department of Animal Science, Epidemiology and management, Blichers Allé 20, Postboks 50, 8830 Tjele, Denmark; mahmoud.shirali@agrsci.dk*

The aim of this study was to determine genetic background of longitudinal body weight growth in farmed mink using random regression methods considering heterogeneous residual variances. In this study, 2139 male mink and the same number of females were used with eight body weight measurements being recorded every three weeks for each mink from 9 to 30 weeks of age. A univariate random regression containing animal genetic effect and permanent environmental effect of litter along with heterogeneous residual variances using Legendre polynomial was performed for body weight of male and female mink, separately. The results showed that the heritability of body weight growth increases with age and has moderate to high range for male (0.32±0.03 to 0.85±0.02) and female (0.22±0.04 to 0.86±0.02). The genetic correlations for body weights within early and late stages of growth were high (0.50 to 1.00) for male and female mink, however, the genetic correlation for body weights in early with late growth were low and not significantly different from zero. The result of current study showed that body weight in mink is highly heritable, suggesting potential for high selection response for this trait, especially at later stages of growth. The genetic correlations suggested that body weights early or late in the growth period are not genetically correlated; therefore, body weight recording and selection should be carried out throughout growth to ensure optimum genetic improvement. In addition, random regression models are shown to be promising for genetic dissection of body weight growth in mink.



**Genetic evaluation of lamb survival in the conditions of the Czech Republic**

*M. Milerski, J. Schmidová and L. Vostrý*
*Institute of Animal Science, Přátelství 815, 104 00 Prague, Czech Republic; milerski.michal@vuzv.cz*

Genetic and non-genetic effects affecting survival at birth up to 24 h after birth (SB) and survival from 2 to 14 days of the age (S2-14) were evaluated in Charollais, Romney, Merinolandschaf, Romanov, Sumava, Suffolk and Texel sheep breeds. The model for the estimation of genetic parameters included fixed effects: dam's age, sex, litter size. The random effects are: direct genetic effect, maternal genetic effect, effect of the contemporary group, effect of litter and residual. Direct heritability were estimated for both traits (SB: from 0.024 to 0.043; S2-14: from 0.033 to 0.053). Maternal heritability reached similar values like the coefficient of direct heritability (SB: from 0.024 to 0.043; S2-14: from 0.034 to 0.056). Correlations between direct and maternal effect for the studied traits were in this range: SB: from -0.203 to -0.166; S2-14: from -0.268 to -0.188). Results suggest that low additive and maternal genetic variances of survival and negative correlations between them considerably limit the possibility of breeding in sheep populations according to this trait. Nevertheless the economic value of lamb survival is very high, according to a bio-economic model for extensive sheep production systems. Standardized economic weights were calculated as the product of the economic weight and the genetic standard deviation of each trait or trait component. The total economic importance was defined as the sum of the absolute standardized economic weights over all traits and trait components, and relative economic weights were then computed as the standardized economic weights expressed as percentages of total economic importance. The direct and mathernal components of survival rate at birth and till weaning had the highest relative economic weights despite the low genetic variances and together they represent above 30% of total economic importance of all considered traits.

**EBV trend validation for survival traits from a linear six-trait model in German dairy cattle**

*J. Wiebelitz[1,2], F. Reinhardt[1], H. Täubert[1], Z. Liu[1], M. Erbe[2] and H. Simianer[2]*
*[1]Vereinigte Informationssysteme Tierhaltung w.V. (vit), Heideweg 1, 27283 Verden, Germany, [2]Georg-August-University, Department of Animal Sciences, Albrecht-Thaer-Weg 3, 37075 Goettingen, Germany; johannes.wiebelitz@vit.de*

Genetic evaluation of longevity has to deal with censored data which may cause biased estimated breeding values (EBVs) of young sires in routine genetic evaluations. An accepted method to estimate unbiasedness of EBVs is the Interbull trend validation method III which was applied to EBVs from a linear six-trait sire model for survival traits. Traits were defined as the survival in the first three lactations which were split into two periods: before or after 150 days from calving (L1.1, L1.2, …, L3.2). A period was considered to be survived if the cow appeared in the subsequent period. Fixed effect was herd×quota-year of calving. Genetic parameters from a previous variance component estimation were used ($h^2$: 0.026 (L1.2) - 0.046 (L3.2); genetic correlations: 0.59 ($r_{L1.1,L3.2}$) - 0.96 ($r_{L2.1,L3.1}$)). For the trend validation, two data sets were created. Full data (FD) consisted of lactation records from 3,615,938 cows with at least one calving between 1998 and 2012. Other than right-censored data were removed. The full data set was truncated at the end of 2008 in order to create the second data set (TD). EBVs of sires were estimated from both data sets separately ($EBV_{TD}$ and $EBV_{FD}$). A measure for additional information (t) between TD and FD was computed for each sire. A regression of $EBV_{FD}$ on $EBV_{TD}$ and t with birth year of the sire as fixed effect was conducted. The expectation for the coefficient of t is zero and Interbull validation is passed if it is smaller than 2% of a genetic standard deviation. In this study, it ranged between 0.66% (L1.1) and 2.78% (L3.2) for the different traits. Three traits (L1.1, L1.2 and L2.1) were below 2%. These preliminary results are promising for further optimization of the current national routine genetic evaluation.

---



**Statistical tools to dissect the genetic architecture of longitudinal data**

*M.J. Sillanpää*
*University of Oulu, Departments of Mathematical Sciences and Biology, Biocenter Oulu, P.O. Box 3000, 90014 Oulu, Finland; mjs@rolf.helsinki.fi*

The dynamic quantitative traits such as production or animal growth traits, are often repeatedly measured over time. In quantitative trait loci (QTL) mapping, it is often beneficial to simultaneously relate the repeated trait measurements over time to large panels of SNP data in order to improve the power to detect QTLs. Such a simultaneous functional mapping aims at modeling the time-varying genetic effects of each SNP as a smooth function (or curve). In this presentation, few different approaches are described and some data examples are shown. The fuctional models considered include multilevel model, varying coefficient model utilizing spline smoothing and linear mixed effect model. In all of these models, Bayesian variable selection is utilized to detect a subset of important functional trait loci.

**Double hierarchical generalized linear models for micro-sensitivity in daily feed intake**

*A.B. Strathe[1,2], T. Mark[2], B. Nielsen[1], D.N. Do[2], H.N. Kadarmideen[2] and J. Jensen[3]*
*[1]Pig Research Centre, Department of breeding and genetics, Axeltorv 3, 1609 Copenhagen V, Denmark, [2]University of Copenhagen, Department of Clinical Veterinary and Animal Sciences, Groennegaardsvej 3, 1870 Frederiksberg C, Denmark, [3]Aarhus University, Center for Quantitative Genetics and Genomics, Blichers Alle 20, 8830 Tjele, Denmark; strathe@sund.ku.dk*

Micro-environmental sensitivity can be studied by estimating the genetic variance in the residual dispersion. Therefore, the objective of this study was to develop double hierarchical generalized linear models for daily feed intake (DFI) in Danish Duroc pigs. This enabled genetic effects for initial, time-trend and residual dispersion in feed intake to be estimated, including their correlations. The resulting model was a natural extension of random regression models, allowing the genetic effect on heterogeneity of residual variance to be estimated. The phenotypic feed intake curve, based on Legendre polynomials, was decomposed into a fixed curve, being specific to the barn-year-season effect and curves associated with the random permanent animal and additive genetic effects. A total of 570,901 DFI records were available on 8,804 Danish Duroc boars between 70 to 160 days of age. Out-of-sample predictions suggested that Legendre polynomials of first order were enough to model the data. The heritability of DFI ranges from 0.08 to 0.12, while the repeatability ranged from 0.30 to 0.35 along the age trajectory. The genetic standard deviation for residual variance was 0.15 for DFI, which indicated a low to moderate genetic variance for residual variance and implied that one unit change in the genetic standard deviation for the dispersion would alter the residual variance by 15%. The genetic correlations between the breeding value for residual variance and breeding values for initial and time trend in feed intake were 0.51 and 0.92, respectively, suggesting effects of scale. Therefore, issues related to scale effects and Box-Cox transformations will be discussed.

---

Session 50 Theatre 7

**Use of carbon dioxide as a breath marker**

*J. Madsen*
*University of Copenhagen, Large Animal Sciences, Groennegaardsvej 2, 1870 Frederiksberg C., Denmark; jom@sund.ku.dk*

For a century experiments with cattle in respiration chambers, feed experiments and feed evaluation experiments have produced knowledge about the feed requirements, feed intake and utilization of different feeds for different production. These results can also be used to calculate the carbon dioxide production of the cattle and the quantitative carbon dioxide excretion of the cattle can be used as a marker for the amount of gasses released by the cattle. By doing this a simple marker that can be used in many different situations has been established. It can be used as a simple, fast and cheap method to estimate the methane ($CH_4$) production from animals, but it can also be used to quantify the excretion of other gasses as for example acetone to potentially give an indication of a ketosis situation of dairy cows. The amount of carbon in the excreted $CO_2$ can basically be calculated from the intake of carbon minus the undigested carbon and the carbon in weight gain, the produced milk and in the methane. In practice the $CO_2$ excretion can be calculated from the intake of metabolizable energy minus the energy in the weight gain or milk produced, as there is close relation between heat production and $CO_2$ excretion. Cows carbon dioxide production can be calculated which make it recommendable to use carbon dioxide as a breath marker, – like the $SF_6$ -, when excretion of different gasses are going to be quantified. Different feeds and supplements can be tested in experiments. Moreover, the method can be used large scale to establish the differences between cows. In experiments where the objective is selection for example for reduced methane excretion then the differences between cows most likely have to be corrected for differences in the factors that influence both the methane production and the carbon dioxide production. In short term experiments this is most likely feed intake, if it is known, or milk production.

## Session 50 Poster 8

### Effects of season, parity and herd size on reproductive performance in Japanese Black cattle

*Y. Sasaki, M. Uematsu and M. Sueyoshi*
*University of Miyazaki, 1-1 Gakeunkibanadai-nishi, Miyazaki 889-2192, Japan; ysksk@cc.miyazaki-u.ac.jp*

The Japanese Black is the most common breed of beef cattle in Japan. However, there is only limited data on the effect of season, parity and herd size on reproductive performance in Japanese Black cattle. Therefore, the objective of the present study was to quantify the reproductive performance such as calving to first service interval and first service conception rate in Japanese Black cattle, and to determine the effect of season, parity and herd size on the reproductive performance. Data were collected from 34,763 calvings in 13,186 animals from 826 farms. The mean number of adult cows per farm was 18 (range 1-454). All of the animals were bred by artificial insemination (AI). Period of calving and AI was divided according to season; winter (December to February), spring (March to May), summer (June to August) and autumn (September to November). Herd size was classified into three groups: small (≤10 cows), intermediate (11-50 cows) and large herds (≥51 cows). The mean ± SD of the parity, calving to first service interval and the first service conception rate were 4.9±2.88, 80.0±46.21 days and 53.5%±49.88%, respectively. Cows calved in the spring showed the longest calving to first service interval ($P<0.05$). Calving to first service interval in first parity cows was shorter than those in cows at parity 7 or higher ($P<0.05$). Cows in large herds had approximately 10 days shorter calving to first service interval than those in small herds ($P<0.05$). Cows inseminated in winter or spring had approximately 5% lower first service conception rate than those in summer or autumn ($P<0.05$). As the parity increased from 1 to 9, first service conception rate decreased from 60.0% to 43.1% ($P<0.05$). Cows in small herds had lower first service conception rate than those in intermediate and large herds ($P<0.05$). In summary, low reproductive performance in Japanese Black cattle was associated with calving in spring, AI in winter and spring, high parity and small herd size.

## Session 50 Poster 9

### Estimation of genetic covariance functions using random regression test-day models

*D.C.B. Scalez[1], S. König[2], H.H.R. Neves[1], E. Gernand[3], D.F. Cardoso[1] and S.A. Queiroz[1]*
*[1]State University of São Paulo – Unesp, Via de Acesso Prof. Paulo Donato Castellane, 14884-900 Jaboticabal, Brazil, [2]University of Kassel, Nordbahnhofstr. 1a, 37213 Witzenhausen, Germany, [3]Thuringian State Institute of Agriculture, August-Bebel-Str. 2, 36433 Bad Salzungen, Germany; daiane-becker@hotmail.com*

The aim of this study was to estimate genetic covariance functions to first lactation test-day milk yields of German Holstein cows using random regressions on orthogonal Legendre polynomials for days in milk. We used 139,102 test-day records for milk yield from 17,244 Holstein cows in first parity, belonging to 12 large-scale contract herds located in the federal state of Thuringia in Germany. Test-day yields from 5 to 305 days in milk were grouped in 10 monthly classes. Contemporary groups (CG) were defined as herd-year-test month. Fixed effects included the CG, linear and quadratic covariate effects of cow age and the average lactation curve of the population modeled with Legendre polynomials of order 4. Random regressions on Legendre polynomials for days in milk for additive genetic as well as for permanent environmental effects were considered with orders 3 to 6. Two different scenarios for modeling residual variances (VE) were compared: the scenario HOM assuming homogeneous VE and the scenario HET assuming heterogeneous VE with 10 classes. Variance components were estimated by REML using the 'Wombat' package. Results from Akaike's (AIC) and Schwarz's Bayesian (BIC) information criteria indicated that HET provided better fit to the data than scenario HOM. AIC and BIC favored the HET model using polynomials of order 6 for both additive genetic and permanent environmental effects (HET_6_6, 52 parameters). Daily heritabilities estimated with this model ranged from 0.23 to 0.40, with highest values in the middle lactation. The worse fit was obtained with the less parameterized model. Although the model HET_6_6 demanded more computing time, it was parsimonious and fitted German Holstein cows test-day milk records adequately.

## Session 50 Poster 10

**Relationship between conformation traits and longevity in Slovak Simmental cows**

*E. Strapáková, J. Candrák and A. Trakovická*
*Slovak University of Agriculture in Nitra, Tr. A. Hlinku 2, 949 76 Nitra, Slovak Republic; eva.strapakova@uniag.sk*

Longevity is a highly desirable trait affecting overall profitability in dairy production. Direct measurement of longevity takes very long time. Conformation traits are recorded relatively early in life, most often in the first lactation and they are more heritable than longevity. The relationships between 27 conformation traits and functional productive life were analyzed in 11,285 Slovak Simmental cows first calved from 1992 to 2010. Cows were scored during the first lactation. Functional productive life was defined as the number of days from first calving to culling or censoring. Analyses were performed for each conformation trait one at a time using Weibull proportional hazard model, hence 27 analyses were carried out. Rear udder attachment, teats placement, pastern and teats arrangement had a highly statistically significant association with functional longevity of cows. Each of these traits had more than 40% relative contribution to the likelihood (-2logL). Rear udder attachment show a linear relationship with functional longevity. Cows with optimum score 6-8 points had 0.2-0.7 times lower risk of culling than cows in reference class (5 points). Teats placement had non-linear relationship with functional productive life. Optimum score 7 for teats placement was associated with a lower risk of culling than were low scores (cows with far centred teats). Optimum for teats arrangement is 6-8 points. Cows with score 7 had 1.21 times lower risk of culling than cows in reference class (score 4). Opposite this, cows with score 9 had the highest risk of culling, 2.5 times higher compared to reference class. Formation pastern is important trait of feet and legs. The highest risk of culling reached cows with score 1, 1.6 times higher in comparison to optimum of trait (score 7). On the basis our results, cows with high attachment rear udder, moderate inwards placed teats, equally arranged teats and upright pastern achieved longer productive life.

## Session 50 Poster 11

**Genetic analysis of litter size in different sheep breeds using linear and threshold model**

*J. Schmidova[1,2], M. Milerski[2], A. Svitakova[1,2] and L. Vostry[1,2]*
*[1]Czech University of Life Sciences in Prague, Faculty of Agrobiology, Food and Natural Resources, Kamýcká 129, 165 21, Prague – 6 Suchdol, Czech Republic, [2]Institute of Animal Science, Přátelství 815, 104 01, Prague-Uhříněves, Czech Republic; schmidova.jitka@vuzv.cz*

The objective of this investigation was to estimate genetic parameters for litter size in Charollais, Romney, Merinolandschaf, Romanov, Suffolk, Šumava and Texel breeds of sheep and to compare the results obtained from the linear and threshold models. A total of 143,896 lambing records from 1990 to 2012 were analyzed. Variance components and genetic parameters for litter size were estimated separately for each breed using the linear model (LM) and the threshold model (TM) (the burn data were 20,000 iterations). The model equation for both models contained ewe age as a fixed effect and random effects of contemporary group, permanent environment and direct additive genetic effect of the animal. Heritability for LM ranged from 0.062 to 0.109, while the ratio between $\sigma_a^2$ and $\sigma_P^2$ estimated by TM was 1.09-3.68 times higher ($h^2$=0.119-0.228). Estimates of phenotype variance by LM increased across breeds ($\sigma_P^2$=0.236-0.779) with increasing average of litter size (not in TM). Variance component estimates for permanent environment of the ewe were low for both models for all breeds except Šumava by TM ($\sigma_{pe}^2$=0.027). Variance component estimations by LM indicate that the contemporary group effect had a somewhat higher influence on the variability of lambs born per ewe ($CG^2$=0.075-0.131) than additive genetic effects in all breeds except Merinolandschaf and Texel. While the estimation by TM indicate that the contemporary group had lower influence on the variability ($CG^2$=0.007-0.143), except Romney. We conclude that genetic parameters did differ among the studied breeds, which should be taken into account in breeding value estimation. The linear model seems more suitable for practical application in breeding value prediction.

## Session 50 Poster 12

**Evaluation of growth records of bulls from two different station tests using linear splines**

*A. Svitáková[1,2], Z. Veselá[2], J. Schmidová[1,2] and L. Vostrý[1,2]*
*[1]Czech University of Life Sciences Prague, Faculty of Agrobiology, Food and Natural Resources, Kamycka 129, 165 21 Prague 6 – Suchdol, Czech Republic, [2]Institute of Animal Science, Pratelstvi 815, 10400 Prague – Uhrineves, Czech Republic; vostry@af.czu.cz*

The aim of this study was to join the data from two station tests into one routine genetic evaluation of meat production. Growth records of Czech Fleckvieh bulls from progeny and performance test stations were evaluated by a common growth curve. Bulls in performance test stations were weighed every one month from 120 days of age to 430 days of age. Bulls in progeny test stations were weighed at 150 days of age and 530 days of age. The evaluated trait was live weight during a test. To join the data from two stations, a test day model with random regression was used. Regression coefficients were estimated using linear splines. Heritability gradually increased and peaked at the age of 430 days (0.26) and then decreased to 0.17 by day 530. The decreased heritability in the last recorded period was due to an increase in the permanent environmental variance. The results confirmed the possibility of using random regression in routine genetic evaluation combining data of two station tests that differ in age of weight measurements.

## Session 50 Poster 13

**Cross-sectional and longitudinal study on udder conformation measures in sows**

*A. Balzani[1], H.J. Cordell[2] and S.A. Edwards[1]*
*[1]School of Agriculture, Food and Rural Development, Newcastle University, NE1 7RU, United Kingdom, [2]Institute of Genetic Medicine, Newcastle University, NE1 3BZ, Newcastle upon Tyne, United Kingdom; a.balzani@ncl.ac.uk*

The aim of this study was to develop a method to define udder morphology and study the changes of udder traits during the sow life. Teat length, diameter, orientation, distance to the abdominal mid line (AML), and intra-teat distance within the same row (SR) were measured in 124 sows. Udder traits were recorded shortly prior to farrowing when the sow was in a lying down posture. 24 sows were re-sampled at a subsequent farrowing to assess changes with parity. Teat length was greater in the anterior four teats than in the posterior ones (18.3 v 15.0 se 0.13 mm). Sow parity had a significant ($P<0.001$) effect on teat length, which was 22% higher in multiparous animals, whereas SR did not differ between parities. Teat position had a significant ($P<0.001$) effect on SR, (mean 109.8 mm, s.e. 0.94), with less distance between teat pairs in the fourth and fifth position. Teat diameter (mean 10.6 mm, s.e. 0.07) was greater ($P<0.001$) in the middle teats; parity explained 21% of circumference variation, with first parity sows having a smaller diameter ($P<0.001$) than multiparous sows. AML (mean 75.5 mm, s.e. 0.66) was significantly shorter ($P<0.001$) in the rear teats compared with the front and middle teats. 70% of teat were orientated perpendicular to the udder (n=619); there was a significant association between the teat orientation and teat position ($x^2=16.96$, $P<0.05$). The proportion of teats not perpendicular was higher in the anterior part (19% v 11%). Young sows had higher proportion of teats perpendicular to the udder compared with multiparous sows ($x^2=42.49$, $P<0.001$). Thus, teat characteristics were related to the position on the udder according to three sectors: pairs 1-2, 3-5 and 6-8. Furthermore teat length and diameter increased after the first parity, but only minor changes occurred after the third parity. These traits are now being related to ease of colostrum extraction and piglet suckling.

Session 50 Poster 14

**Analyses of calving interval in SA Holstein cows using random regression models**

*M.D. Fair, F.W.C. Neser and J.B. Van Wyk*
*University of the Free State, Animal, Wildlife and Grassland Sciences, P.O. Box 339, 9300 Bloemfontein, South Africa; fairmd@ufs.ac.za*

Random regression models to analyse longitudinal or repeated measures data have become common practice amongst animal breeders. The aim of the study was to analyse the intercalving period (ICP) of the SA Holstein cows using random regression models to estimate variance components and heritabilities. Data of pedigree and performance records collected from 1947 to 2012 for 3,010,971 animals were edited resulting in n=330,636 (ICP 1), 232,400 (ICP 2) and 153,136 (ICP 3) records of cows with four parities respectively allowing three intercalving periods per individual cow to be analysed. Contemporary groups with less than 5 records were removed from the data. A random regression model using ASREML was used to model the three ICP"s by fitting the fixed effect of 28,314 contemporary groups (a concatenation of breeder, keeper and year) and the direct genetic additive random effect. The direct genetic effect was fitted as a random linear regression with intercept, for parity two, three and four of the cow. The means and standard deviations of the three intercalving periods for the cows were 403.14±75.55; 400.01±71.32 and 400.45±71.27 respectively, ranging from 300-750 days. Resultant heritability estimates for the three ICP"s were 0.05, 0.06 and 0.11 respectively. It can be speculated that the increase in heritability over parities could be the result of that later parities being from a selected productive group. These heritability values correspond to those found in literature and although low indicate that genetic improvement through selection would still be possible and that random regression models can be used to estimate heritabilities for the repeated measures trait of ICP successfully.

---

Session 52 Theatre 1

**Estimation of nutritive value of seaweed species and extracted seaweed in animal diets**

*P. Bikker[1], M.M. Van Krimpen[1], W.A. Brandenburg[2], A.M. López-Contreras[3], P. Van Wikselaar[1], R.A. Dekker[1] and G. Van Duinkerken[1]*
*[1]Wageningen UR Livestock Research, P.O. Box 338, 6700 AH Wageningen, the Netherlands, [2]Wageningen UR Plant Research International, P.O. Box 616, 6700 AP Wageningen, the Netherlands, [3]Wageningen UR Food & Biobased Research, P.O. Box 17, 6700 AA Wageningen, the Netherlands; paul.bikker@wur.nl*

The development of the world population increases the requirement of biomass for human and animal consumption. At present, seaweed is not used as feed material to any significant extent and information on the feeding value is scarce. Nonetheless, seaweed cultivation is of interest because cultivation does not compete for land-use. Optimal use may include biorefinery and application of the residue in animal feed. The aim of this study was to evaluate the nutritive value of different seaweed species from various European locations and the influence of a biorefinery process on the value of the residue. The observations included (1) contents of protein and amino acids, fatty acids, carbohydrates, fibre (NDF, ADF, ADL), and minerals; (2) *in vitro* ileal and total tract digestibility of protein and organic matter using a modified Boisen method; and (3) fermentation in a gas production system (GPT) to simulate rumen degradation of organic matter. Soybean meal (SBM) and grass silage were used as reference substrates. Preliminary results indicate large differences in nutritive value between seaweed species (e.g. lysine 2-11 g, fat 3-23 g, CF 27-80 g, P 0.6-3.4 g per kg) and relevant similarities within species from different locations. *In vitro* ileal (18-69%) and total tract (36-68%) dry matter digestibility varied drastically, with some species exceeding the values obtained for SBM (45% ileal and 73% total tract) in this system. Cumulative gas production of the seaweed species in the GPT varied from 10-100% of that of grass silage. Results emphasize the importance of adequate selection of species and suggest a promising nutritive value of specific species, to be confirmed in *in vivo* digestibility and performance studies.

**Comparison of fractionation methods for nitrogen and starch in maize and grass silages**

*M. Ali[1,2], L.H. De Jonge[2], J.W. Cone[2], G. Van Duinkerken[3] and W.H. Hendriks[2]*
*[1]Institute of Animal Nutrition and Feed Technology, University of Agriculture, Faisalabad, 38040, Pakistan,*
*[2]Animal Nutrition Group, Wageningen University, P.O. Box 338, 6700 AH Wageningen, the Netherlands,*
*[3]Wageningen UR Livestock Research, 8200 AB, Lelystad, the Netherlands; gert.vanduinkerken@wur.nl*

The nylon bag technique used in many feed evaluation systems, prescribes the use of a washing machine method (WMM) to determine the washout (W) fraction and to wash the rumen incubated nylon bags. As this WMM has some disadvantages, an alternate modified method (MM) was recently introduced. The aim of this study was to determine and compare the W and non-washout (D+U) fractions of nitrogen (N) and starch of maize and grass silages, as determined by the WMM and MM. In addition, regression equations were derived to estimate the different fractions of N and starch from the chemical composition of the silages using both methods. Ninety-nine maize silage and 99 grass silage samples, with a broad range in chemical composition, were selected. The values of the N fractions of maize and grass silages were significantly ($P<0.05$) different between the two methods. The regression analysis showed that the D+U values for N of maize and grass silages determined with the WMM and the MM were not aligned with each other. Large differences ($P<0.001$) were found in the D+U fraction of starch of the maize silages, which might be due to the different methodological approaches, such as different rinsing procedures (washing vs shaking), duration of rinsing (60 vs 40 min) and different solvents (water vs buffer solution). The large differences in the insoluble washout (W-S) and D+U fractions of starch can lead to different estimates for the effective rumen starch degradation. Regression analysis showed relationships, with an $R^2$ ranging from 0.31 to 0.86, between the N and starch fractions and the chemical composition of the maize and grass silages.

---

Session 52 Theatre 3

***In vitro* incubation of dairy cow diets – 1. Degradability, total gas and methane yield**

*L. Maccarana[1], M. Cattani[1], F. Tagliapietra[1], S. Schiavon[1], H.H. Hansen[2] and L. Bailoni[1]*
*[1]University of Padova, BCA and DAFNAE depts., Viale dell'Università 16, 35020 Legnaro, Italy,*
*[2]University of Copenhagen, Large Animal Sciences dept., Grønnegårdsvej 2, 1870 Frederiksberg, Denmark; laura.maccarana@studenti.unipd.it*

This study investigated effects of 8 diets, commonly used in the dairy farms of northern Italy, on *in vitro* gas (GP) and methane ($CH_4$) production, using an automated GP system. A reference diet (RD) was used (15.2% CP; 37.0% NDF; 3.3% lipids). The other diets had a smaller or a greater proportion of CP (11.7 or 18.7%) or NDF (33.1 or 44.8%) or lipids (2.2 or 5.5%) compared to RD. These dietary constituents were not correlated. Each diet (1 g) was incubated for 24 or 48 h into bottles with 150 ml of buffered rumen fluid, gas venting occurred at 6.8 kPa, and GP was recorded. The experimental design was: 2 runs×2 incubation times×8 diets×5 replicates, plus 20 blanks (bottles without feed sample), for a total of 180 bottles. Methane production was assessed from GP, and $CH_4$ was analyzed by gas chromatography (GC) from gas samples taken from bottle headspace. Degradability of NDF was measured, and true dry matter degradability (TDMd) was computed. Data were analyzed by multiple regression considering CP, NDF and lipid contents as independent variables. The $CH_4$ production (ml/g DM or ml/g TDMd) was reduced with decreasing dietary CP ($P<0.01$). The $CH_4$/g DM was increased ($P<0.01$), but $CH_4$/g TDMd was reduced ($P<0.01$) with decreasing dietary NDF. Both $CH_4$/g DM and $CH_4$/g TDMd were increased ($P<0.01$) with decreasing dietary lipid. Results suggest that dietary chemical composition had a small but significant influence on *in vitro* $CH_4$ production as each percentage point of dietary CP, NDF, and lipid reduction induced a change of -0.55 (-1.00%), -0.21 (-0.38%) and +0.55 (+1.00%) ml $CH_4$/g TDMd, respectively. A reduction of dietary CP was found to reduce $CH_4$ production. Although this effect was small, it suggests that such a strategy, adopted to decrease N excretion, might also mitigate $CH_4$ emission.

**_In vitro_ incubation of dairy cow diets – 2. Relations between measures and estimates of methane yield**

*M. Cattani[1], L. Maccarana[1], F. Tagliapietra[1], S. Schiavon[1], H.H. Hansen[2] and L. Bailoni[1]*
*[1]University of Padova, BCA and DAFNAE depts., Viale dell'Università 16, 35020 Legnaro, Italy, [2]University of Copenhagen, Large Animal Sciences dept., Grønnegårdsvej 2, 1870 Frederiksberg, Denmark; mirko.cattani@unipd.it*

This study investigated the relationships between measures and stoichiometrical estimates of *in vitro* gas (GP) and methane ($CH_4$) production of 8 diets for dairy cows. A basal diet (BD) was used as reference (15.2% CP; 37.0% NDF; 3.3% lipids). The other diets contained a smaller or a greater proportion of CP (11.7 or 18.7%) or NDF (33.1 or 44.8%) or lipids (2.2 or 5.5%) compared to BD. Dietary contents of these constituents were not correlated. Each diet (1 g) was incubated for 24 or 48 h into bottles with 150 ml of buffered rumen fluid. Bottles were vented at 6.8 kPa and GP was recorded. Experimental design was as follows: 2 runs×2 incubation times×8 diets×5 replicates, plus 20 blanks (bottles without feed sample), for a total of 180 bottles. Methane production was assessed from GP, and $CH_4$ was anayzed by gas-chromatography (GC) from gas samples taken from bottle headspace. Fluids of fermentation were analyzed for volatile fatty acids (VFA) content by GC. Values of GP and $CH_4$ were also estimated from VFA yields by stoichiometry. Data were analyzed considering diet and incubation time as fixed independent factors and run as a random effect. The GP measures (ml/g DM) were always greater than the GP estimates ($P<0.01$), and less variable as the SEM were 2.5 or 6.8 ml/g DM for measures or estimates, respectively. The $CH_4$ measures (ml/g DM) were on average 10% greater than the estimates ($P<0.01$), and less variable as the SEM were 0.45 or 1.28 ml $CH_4$/g DM for measures or estimates, respectively. Incubation time had no effect on the measures/estimates ratio for $CH_4$, that averaged 0.90 both at 24 and 48 h of incubation. Direct measures of GP and $CH_4$ production are more reliable than predictions based on VFA analysis.



**Evaluation of five models predicting feed intake by dairy cows fed total mixed rations**

*L.M. Jensen[1], P. Nørgaard[1], N.I. Nielsen[2], B. Markussen[3] and E. Nadeau[4]*
*[1]University of Copenhagen, Dept. of Vet. Clin. and Anim. Sci., 1870 Frb, Denmark, [2]Knowledge Centre for Agriculture – cattle, Agro Food Park 15, 8200 Aarhus, Denmark, [3]University of Copenhagen, Lab of Applied Stat, 2100 Copenhagen, Denmark, [4]SLU, Dept. of Animal Environment and Health, 532 23 Skara, Sweden; lauramie.jensen@gmail.com*

The objective was to evaluate the accuracy and robustness of five different feed intake models predicting DMI in dairy cows fed total mixed rations. The five evaluated models were: US National Research Council (NRC), the Nordic feed evaluation system, the total dry matter index (TDMI), the DMI model (Zom), and the DLG model. The NRC uses only animal characteristics; the NorFor, TDMI, DLG, and the Zom models include interactions between animal and dietary characteristics, however the Zom model excludes milk yield and BW. Accuracy and robustness was evaluated by MSPE and RMSPE in percentage of observed DMI, respectively. RMSPE<10% of mean DMI indicated satisfactory prediction. The evaluation was performed on data from 14 Scandinavian experiments with a total of 1055 dairy cows in 96 treatment means. The NorFor model was evaluated on only 10 of the experiments, 4 experiments had been used in the development. The five models predicted DMI with an RMSPE ranging from 1.2 to 3.2 kg DM per day. The error due to central tendency and the error due to regression ranged between 0% and 64% and between 4% and 36% of MSPE, respectively. Error due to disturbance (ED) ranged between 32% and 92% of MSPE. These models were expected to give satisfactory predictions on the Scandinavian data due to similar animal and dietary characteristics, however models cannot be considered universally valid. The NorFor model predicted DMI most robustly as 10 out of 10 experiments were predicted satisfactorily. The DLG and NRC models performed almost as robustly. The DLG model predicted DMI most accurately with negligible mean and line biases resulting in a 92% ED.

**Prediction of manure nitrogen output from dry cows fed freshly cut grass**

*S. Stergiadis and T. Yan*
*Agri-Food and Biosciences Institute, Large Park, Hillsborough, County Down, BT26 6DR, United Kingdom; sokratis.stergiadis@afbini.gov.uk*

Decreasing manure (faeces and urine) nitrogen (N) output by improving N utilisation efficiency can advance economic performance and reduce environmental footprint (water pollution and nitrous oxide emissions) of pasture-based dairy systems. Although prediction of N output from silage-fed animals has been extensively reported, equations developed by feeding fresh grass are not commonly available. The objective of this study was therefore to develop prediction equations for manure N output using N intake data of 32 non-lactating dairy cows, fed solely freshly cut grass at maintenance level. N utilisation data were measured daily for grass harvested from 9 perennial ryegrass swards from early to late growth over 6 weeks. Three day averaged data (n=464) were used to generate prediction equations, using simple linear regression analysis, with effects of variation of grass swards removed. Under maintenance feeding, the majority of N intake was excreted in urine rather than in faeces (0.634 vs 0.267 g/g). Prediction of manure or urine N output (g/d) using N intake (g/d) as a sole predictor showed slightly stronger relationships than prediction of faeces N output ($R^2$=0.60 or 0.57 vs 0.50), indicating a weaker influence of N intake on the latter under maintenance feeding. The $R^2$ values from the current relationships between N intake and output are relatively low compared to published relationships generated under adlib silage intake, maybe due to differences in feeding management which was at the maintenance level in this study. Results also show that mitigation strategies which aim to decrease N intake in dry cows in pasture-based systems may have higher impact on urine than faeces output. This is desirable provided the higher nitrous oxide emission potential from urine than faeces N.

---



**Relationship between intake of metabolizable energy and chewing index of diets fed to pregnant ewes**

*M.V. Nielsen[1], E. Nadeau[2], B. Markussen[3], P. Nørgaard[1], C. Helander[2], Å.T. Randby[4] and M. Eknæs[4]*
*[1]University of Copenhagen, Dept. of Veterinary Clinical and Animal Science, 1870 Frederiksberg, Denmark, [2]Swedish University of Agricultural Sciences, Dept. of Animal Environment and Health, 532 23 Skara, Sweden, [3]University of Copenhagen, Lab. of Applied Statistics, 2100 Copenhagen, Denmark, [4]Norwegian University of Life Sciences, Dept. of Animal and Aquacultural Sciences, 1432 Ås, Norway; mettevn@life.ku.dk*

The objective was to study whether a linear relationship exists between metabolizable energy intake (MEI) of pregnant ewes and the chewing index (CI) of their rations, and proportionality between the slopes and squared intercepts. The relationship was studied using 5 feeding trials with intake data from 107 pregnant ewes, 4 to 1 week before parturition. All ewes were fed grass silage *ad libitum*, supplemented with concentrates either separately or in a total mixed ration. The ewes were from different breeds between two and seven years old, had a mean body weight of 99.6 kg (SD=10.90), and were bearing 1 to 3 lambs. The MEI was estimated based on *in vitro* digestible OM and ranged between 11.2 and 49.0 MJ/d. The NorFor CI (min/MJ ME) of the feeds was estimated based on iNDF (g/kgNDF), NDF (g/kgDM) and theoretical particle length (mm) of the forage. Chewing index was adjusted to BW of the ewes and deviation from 0.7% forage NDF intake per kg BW. Relationship between MEI and CI was analyzed using linear mixed effects modeling. Effect of random variation on slope and intercept was tested by model reduction, and likelihood ratio test statistics was used for comparison. Direct proportionality between slopes and squared intercepts were analyzed by Pearson's product-moment correlation and Wald test. The MEI declined with increasing dietary CI ($P<0.001$) and different intercepts in relation to week prior to parturition was found ($P<0.001$). The Wald test showed that there was a direct proportionality between slopes and squared intercepts ($P=0.055$), indicating the squared intercepts to be proportional with the slope value.

**Fibrous agroindustrial by-products in pig diets: effects on nutrient balance and gas emission**

*A. Cerisuelo[1,2], P. Ferrer[3], A. Beccaccia[4], S. Calvet[3], M. Cambra-López[3], F. Estellés[3], W. Antezana[3], P. García-Rebollar[4] and C. De Blas[4]*
*[1]Universidad CEU, Cardenal Herrera, 46115 Alfara del Patriarca, Spain, [2]Centro de Investigación y Tecnología Animal (CITA-IVIA), Pol. La Esperanza, 100, 12400 Segorbe, Spain, [3]Universitat Politècnica de València, Camino de Vera s/n, 46022 Valencia, Spain, [4]Universidad Politécnica de Madrid, Avenida Complutense 3, 28040 Madrid, Spain; cerisuelo_alb@gva.es*

The effect of including orange (OP) and carob (CP) pulps in feeds on nutrient digestibility, slurry composition and ammonia ($NH_3$) and methane ($CH_4$) emissions was studied. Thirty finishing pigs (85.4±12.3 kg) were fed five diets: a commercial wheat/barley diet (C), two diets with OP (7.5%, OP7.5 and 15%, OP15) and two diets with CP (7.5%, CP7.5 and 15%, CP15). After a 14-day adaptation period, faeces and urine were collected individually for 7 days in metabolic pens. Feed and faeces were chemically analysed for dry matter (DM), ash, energy (E), ether extract (EE), nitrogen (N) and fibre fractions. The DM, ash, N, N-NH4+, pH, volatile fatty acids content and potential $NH_3$ and $CH_4$ emissions from slurry (faeces+urine) were determined. Animals fed with CP diets excreted a higher amount of organic matter (OM) (g/pig and day; P<0.05), but no differences were detected in slurry production. Energy, DM, OM and N digestibility were lower (P<0.05) in animals fed CP compared with OP and C diets. Neutral detergent fibre digestibility was the highest in OP15, lowest in CP15 and intermediate in C groups. Fibre content in faeces was greater in CP15 group and EE and N content in faeces was higher in OP15. Slurry characteristics were similar among treatments. Animals fed OP15 tended (P<0.10) to show lower $NH_3$ emissions (mg/l of slurry and day). Biochemical $CH_4$ potential and $CH_4$ yield per animal and day did not differ among treatments. Thus, the fibre sources and levels tested affected nutrient digestion and faecal composition, showing potential changes in $NH_3$ emissions but not on $CH_4$ yield. Project funded by the Spanish Ministry of Science and Innovation (AGL2011-30023-C03) and the Valencian Government (ACOMP/2013/118).

---



**Amino acid digestibility and energy concentration in soybean and rapeseed products fed to pigs**

*D.M.D.L. Navarro[1], Y. Liu[1], T.S. Bruun[2] and H.H. Stein[1]*
*[1]University of Illinois at Urbana-Champaign, 1207 W. Gregory Dr., Urbana, IL, 61801, USA, [2]Danish Pig Research Centre, Danish Agriculture & Food Council, Axeltorv 3, 1609, Copenhagen, Denmark; navarro3@illinois.edu*

Two experiments were conducted to determine standardized ileal digestibility (SID) of AA and concentration of DE and ME in 2 sources of enzyme-treated soybean meal (ESBM-1 and ESBM-2), extruded soybean meal (SBM-EX), soy protein concentrate (SPC), conventional soybean meal (SBM-CV), conventional 00-rapeseed expellers (RSE), and a fermented co-product mixture (FCM). In Exp. 1, 27 barrows (initial BW: 9.29±0.58 kg) equipped with a T-cannula in the distal ileum were randomly allotted to three 9×5 Youden squares with 9 pigs and 5 periods in each square. In each square, pigs were fed one of 7 diets containing each of the 7 ingredients and 2 pigs were fed a N-free diet. The SID of Lys was greater (P<0.05) in SBM-CV than in ESBM-2, and SID of Met was greater (P<0.05) in ESBM-1 than in SPC. The SID of Thr and Trp was not different among the soybean products. The SID of most AA in RSE and SID of all AA in FCM were less (P<0.05) than in all soybean products, but SID of all AA in RSE was greater (P<0.05) than in FCM. In Exp. 2, 64 barrows (initial BW: 19.81±0.90 kg) were placed in metabolism cages and allotted to a randomized complete block design with 8 diets and 8 pigs per diet. A corn-based diet and 7 diets containing corn and each of the 7 ingredients were prepared. Results indicated that ME in ESBM-1, SBM-EX, SPC, and SBM-CV (4,158, 4,240, 4,226, and 4,044 kcal/kg DM) were not different, but greater (P<0.05) than in ESBM-2 (3,782 kcal/kg DM). There was no difference in ME between RSE and FCM (3,522 and 3,364 kcal/kg DM), but these values were less (P<0.05) than in all soybean products. It is concluded that there are differences among processed soybean products with some having greater SID of AA, DE, and ME than others. Furthermore, SID of AA, DE, and ME in all soybean products were greater than in rapeseed expellers and the fermented co-product mixture.

**Is it possible to overcome the post weaning growth check**

*E. Magowan*
*AFBI, Hillsborough, BT26 6DR, United Kingdom; elizabeth.magowan@afbini.gov.uk*

This trial aimed to investigate the separate effects of the stressors imposed at weaning. Ten treatments were designed. Sows were weaned from piglets when they were 28 days of age (±2 days). In five treatments (Ts) pigs were moved to stage 1 accommodation on the day of weaning and pigs on the other five Ts remained in the farrowing crate for 4 days after which they were moved to stage 1 accommodation. Within each set of 5 Ts one litter did not receive creep pre weaning and was not mixed with other pigs post weaning, they were offered feed from a small circular hopper (SCH) after weaning. Creep was offered pre weaning in the other 4 Ts and pigs were mixed post weaning in these Ts. In two of these Ts creep was offered from a SCH pre weaning and in the other two Ts creep was offered on the floor of the creep area. Post weaning one of these Ts offered feed from a SCH only and in the other feed was offered from both the SCH and a larger dry multi space hopper. Wean weight averaged 9.4 kg and feed intake pre weaning averaged 49.7 g/pig/day and 7.9 g/pig/day when offered on the floor or with the SCH respectively. The post weaning intake of all pigs was similar across the Ts (Av. of 0.11 kg/pen (10 pigs) in day 1 after weaning) except when pigs remained in the farrowing crate and were not mixed with others (0.40 kg/pen (10 pigs), $P<0.05$). The overall intake (4 to 10 weeks of age) of the pigs which remained in the farrowing accommodation post weaning and were not mixed with others was not significantly different to that of pigs on some of the other Ts but these pigs had a significantly ($P<0.001$) greater 10 week weight (32.3 kg) than pigs on any other Ts (Av. 29.2 kg). This study demonstrates that feed intake post weaning can be significantly increased post weaning by the elimination of the stressors resulting from moving to different accommodation and mixing. Also, the introduction of creep at the point of weaning perhaps stimulated 'intake'. This study also demonstrates the importance of achieving a high feed intake immediately after weaning in the aim of optimising growth performance.

---



**Effect of dietary crude fibre level on productivity of indigenous Venda chickens aged one to 91 days**

*M.M. Ginindza, J.W. Ng'Ambi and D. Norris*
*university of Limpopo, Agricultural Economics and Animal Production, School of Agric. and Environmental Sciences, 0727, Sovenga, South Africa; jones.ngambi@ul.ac.za*

A study was conducted to determine dietary crude fibre levels for optimal intake, digestibility and productivity of indigenous Venda chickens aged 1 to 91 days. In the first phase of the experiment (1-49 days), 120 unsexed day-old chicks were assigned to five dietary treatments having crude fibre (CF) levels of 2 (VF2), 3 (VF3), 5 (VF5), 6 (VF6) and 8 (VF8) %. Each treatment was replicated four times in a completely randomised design. In the second phase (49-91 days), 80 male Venda chickens were assigned to the above five dietary treatments. Each treatment was replicated four times in a completely randomised design. Diets used in both trials were isocaloric and isonitrogenous (12 MJ ME/kg and 18% CP). A quadratic equation was used to determine dietary CF levels for optimal feed intake, growth rate, live weight, feed conversion ratio and nitrogen retention. Linear regressions were also used to determine relationships between dietary CF levels and production parameters. Increases in dietary CF levels resulted in lower ($P<0.05$) feed intake and nutrient digestibility in unsexed Venda chickens. Growth rate, feed conversion ratio and live weight of unsexed Venda chickens were optimized at dietary CF levels of 3.4, 4.0 and 3.0%, respectively. Increases in CF level resulted in poor ($P<0.05$) feed conversion ratio and digestibility in male chickens. Feed intake, growth rate, live weight, and nitrogen retention of male Venda chickens were optimized at different dietary crude fibre levels of 6.5, 4.3, 3.2 and 4.4%, respectively. Thus, different levels of dietary crude fibre optimized different production parameters. This has implications on diet formulation for these chickens.

**Digestive efficiency of the growing rabbit according to restriction strategy and dietary energy conc**

*T. Gidenne, L. Fortun-Lamothe and S. Combes*
*INRA, UMR 1388 GenPhySE, CS 52627, 31326 Castanet-Tolosan, France; thierry.gidenne@toulouse.inra.fr*

Post-weaning restriction strategies are used in French rabbit breeding systems to improve the health status, the digestion and the feed efficiency. However, this effect could be affected by the dietary chemical composition. Thus, we aimed to study the digestive response of the growing rabbit according to a 2×2 experimental design: the intake level (free=100 vs limited to 80% of the free intake) and the digestible energy 'DE' concentration of the diet (control 'C' or high 'H'). Four groups of 12 rabbits were housed individually in metabolism cages and allotted at weaning (35 d) to four treatments: C100, C80, H100, H80. The C and H diets were formulated to differ by 16% for their DE content (9.69 vs 11.24 MJ DE/kg as fed, resp.), by substituting starch to lipid (141 vs 91 g/kg and 1.9 vs 7.2 g/kg), while their NDF level was similar (337 vs 345 g/kg). Animals were fed the experimental diets from 35 to 70 d old. For T80 and E80 groups the intake was limited during four weeks after weaning. Digestive efficiency was assessed after two weeks of adaptation to the treatments (49-53 d). For rabbits fed freely, the energy digestion was 2.2 pts higher for H diet ($P<0.05$), leading to a DE level of 10.90 MJ/kg (vs 9.71 for C100 group). The protein digestion was improved by 1.7 pts for H diet, because more protein was supplied by soya bean meal. Digestive efficiency of restricted rabbits was higher for all nutrients and for the two diets, but especially for H diet. The energy digestion was 2.7 pts higher for C80 (64.4 vs 61.7% for C100) but 5.3 pts higher for E80 (69.2 vs 63.9% for E100), leading to a significant interaction ($P=0.012$ for diet × restriction). The digestibility of crude protein was meanly 6.1 pts higher for restricted rabbits (78.7 vs 72.6%, $P<0.001$) without interaction with the diet. The restriction also improved the NDF digestion by 6 pts. In conclusion, the digestion of the growing rabbit was improved when restricted fed, but the extent of this effect depends of the dietary composition.



**A mobile device based ration preparation software**

*M. Boga[1], K.K. Çevik[1], H.R. Kutlu[2] and H. Önder[3]*
*[1]University of Nigde, Bor Vocational School, University of Nigde, Bor Vocational School, 51700 Nigde, Turkey, [2]University of Cukurova, Department of Animal Science, University of Cukurova, Faculty of Agriculture, 01130 Adana, Turkey, [3]University of Ondokuz Mayis, Department of Animal Science, University of Ondokuz Mayis, Faculty of Agriculture, 55139 Samsun, Turkey; mboga@nigde.edu.tr*

Rapid growth of world's human population increases in proteins of animal origin demand every year. To satisfy that needs number of animal enterprises and animals per enterprise are increase day by day. To increase the profitability and production efficiency of enterprises, it is essential to exactly meet the nutriment requirements of animals. Produced feeds should be as nutritious and cheap as possible with items on hand. To address the mentioned requirements an Internet and GSM based ration preparation software was developed. The mentioned software can solve the optimum ration both in nutrition and price. Properties (price etc.) of feed stuff can be updated and new feed stuff can be added by user. Rations of concentrate, roughage and TMR; mixed feed of concentrate and roughage can be solved. To achieve most appropriate solution, genetic algorithm as a member of artificial intelligence were used. Thanks to this algorithm, it is optimized best fit with minimum cost. The software was developed with eclipse Java editor. That software can also be used with mobile devices which are independent of time and space. Common usage of cell phones is a main advantage of developed software, taking in to account that everybody has at least one mobile device with Android operating systems.

## Session 52 Poster 14

**The effect of polyethylene glycol on gas production and microbial protein synthesis**

*F.N. Fon[1], M.J. Foley[2] and I.V. Nsahlai[3]*
*[1]University of Zululand, Agriculture, KwaDlangezwa Campus, 3886, South Africa, [2]University of KwaZulu Natal, Animal and Poultry Science, Pietermaritzburg Campus, 3201, South Africa, [3]University of KwaZulu Natal, Animal and Poultry Science, Pietermaritzburg Campus, 3201, South Africa; fonf@unizulu.ac.za*

During the long dry or winter seasons, ruminants often graze pastures or cop residues that are poor in nitrogen especially in the rural poor communities. However, legumes supplementation has often been used to improve protein availability to these animals. *Sericea lespedeza* is regularly used as supplements but has always been limited by its relatively high tannin concentrations. This study investigated the effect of polyethylene glycol 4000 (PEG) on gas production, microbial protein synthesis, and on tannins present in *S. lespedeza*. The samples were evaluated using *in vitro* gas production technique. Five levels of PEG (0, 15, 30, 60 and 90 mg) each containing 1 g *S. lespedeza* in 77 ml salivary buffer was inoculated with 33 ml of rumen fluid and incubated for 72 h at 38 °C. The total gas production and rate of gas production, apparent digestibility (APD), true digestibility (TD), microbial yield (MY) and volatile fatty acid (VFA) percentages were determined. Total gas produced and kinetic parameters, TD, Total VFA and MY were subjected to analysis of variance (ANOVA) using the general linear model of SAS. There was no significant difference with the total gas produced and total VFA. However, the rate of gas production from *S. lespedeza* fermentation increased ($P<0.05$) at 90 mg PEG (0.031 $h^{-1}$) compared to the control (0.027 $h^{-1}$). No significant differences were observed for APD, but TD (607 g/kg) and MY (255 g/kg) tended to increase ($P=0.052$) at 90 mg compared to the control (TD=536 g/kg and MY=177 g/kg). Although, based on statistical analysis, there was no conclusive evidence suggesting that PEG increased gas production, TD and MY, the results are very promising for future research on PEG in reducing tannin effect in tanniferous feeds.

---

## Session 52 Poster 15

**Digestibility evaluation of micronized full fat soybean in weaned piglets**

*J.C. Dadalt[1], G.V. Polycarpo[1], C. Gallardo[1], D. Malagoni[2], F.E.L. Budiño[2] and M.A. Trindade Neto[1]*
*[1]University of São Paulo, Department of Animal Nutrition e Production, Duque de Caxias Norte, 225, 13635-900 Pirassununga, SP, Brazil, [2]Institute of Animal Science, Department of Animal Science, Heitor Penteado, 56, 13460-000 Nova Odessa, SP, Brazil; julio@zootecnista.com.br*

The objective of the current study was to evaluate the digestibility of dry matter (DM), crude protein (CP), mineral matter (MM), digestible energy (DE) and, metabolizable energy (ME) and nitrogen retention (NR) of micronized full fat soybean, with or without exogenous enzymes inclusion, in weaned piglets. Twenty-five cross-bred castrated piglets with initial weight 8.53 kg±1.48 kg, were allotted in a completely randomized design under five treatments and five replicates. The experimental unit was represented by one pig within its respective metabolic cage. The experimental period was 10 d, in which 5 for cage adaptation and 5 for feces and urine collection. The treatments were: Control diet (CD); CD + 30% of micronized full fat soybean (MFFS); CD + 30% (MFFS) + 200 mg/kg of carbohydrase (MFFS+Carb); CD + 30% MFFS + 50 mg/kg of phytase (MFFS+Phy); CD + 30% MFFS+ Carb + Phy (MFFS+Carb+Phy). The control diet based on corn (61.33%), milk skim dried (10%) and whey dried (15%). Supplemented enzymes were: commercial phytase (10,000 FTU/g); carbohydrase providing 10% of Galactomananase, 10% of Xilanase, 10% of β-glucanase, 60% of malted barley and 10% of α-galactosidase. Isolated or combined carbohydrase, phytase did not influence ($P>0.05$) the digestibility of MFFS nutrients. The DE (4,682 kcal/kg) and ME (4,501 kcal/kg) were lower to Brazilian tables for poultry and pigs (2011). The apparent digestibility was: 85.57% DM; 85.44% CP; 50.73% MM and 63.35% NR. These results were similar some results from literature with growing pigs. The current study suggests that used exogenous enzymes have not an effective improve on nutritional value of MFFS for weaned piglets.

**Digestibility evaluation of rice grain polished and broken in weaned piglets**

*J.C. Dadalt[1], D. Malagoni[2], G.V. Polycarpo[1], C. Gallardo[1], F.E.L. Budiño[2] and M.A. Trindade Neto[1]*
*[1]University of São Paulo, Department of Animal Nutrition e Production, Duque de Caxias Norte, 225, 13635-900 Pirassununga SP, Brazil, [2]Institute of Animal Science, Animal Science, Heitor Penteado, 56, 13460-000 Nova Odessa SP, Brazil; julio@zootecnista.com.br*

The objective of the current study was to evaluate the digestibility of dry matter (DM), crude protein (CP), digestible energy (DE) and metabolizable energy (ME) of rice grain polished and broken, with or without exogenous enzyme, in weaned piglets. Twenty-five crossbred castrated piglets with initial weight 8.81 kg±1.52 kg, were allotted in a completely randomized design under five treatments and five replicates and experimental unit represented by one pig within its respective metabolic cage. The experimental period was 10 d, in which 5 d of adaptation and 5 d for feces and urine collection. The treatments were: Control diet (CD); CD + 30% rice grain polished and broken (RGPB); CD + 30% RGPB + 200 mg/kg of carbohydrase (RGPB + Carb); CD + 30% RGPB + 50 mg/kg of phytase (RGPB+Phy); CD + 30% RGPB + Carb + Phy (RGPB+Carb+Phy). The control diet was based on corn (61.33%), milk skim dried (10%) e whey dried (15%). The addictive enzymes were: commercial phytase (10,000 FTU/g); carbohydrase providing 10% of galactomananase, 10% of xilanase, 10% of β-glucanase, 60% of malted barley and 10% of α-galactosidase. Carbohydrase and phytase combination improved the digestibility coefficients (P=0.006) of CP (84.55%), however, without statistical difference compared to RGPB+Carb (80.64%). Lower digestibility of CP occurred with RGPB (76.83%) without enzymes inclusion followed by RGPB+Phy (77.52%), which was similar to RGPB+Carb. In this situation enzymes combination with RGPB did not show statistical difference for DM digestibility (93.72) and for DE (3,742 kcal/kg) and ME (3,720 kcal/kg). The exogenous enzyme propitiated an effective improve on digestibility of CP in RGPB for young pigs after weaning.

---



***In situ* degradability and rumen parameters of cows fed two sugarcane varieties and soybean hulls**

*F.F. Simili, M.L.P. Lima, M.I.M. Medeiros, G. Balieiro Neto, E.G. Ribeiro and C.C.P. Paz*
*Instituto de Zootecnia SAA/APTA, (Department of Agriculture, São Paulo State Government), C. Postal 63, Sertãozinho SP, Brazil, CEP 14160-970, Brazil; flaviasimili@gmail.com*

Two sugarcane varieties (IAC2480 and IAC2195) combined or not with soybean hulls (SH), partially replacing corn in concentrate, were included in the diet of dairy cows to evaluate the effects on *in situ* degradability and rumen parameters. The experiment was arranged in a 4×4 latin square design with 4 rumen-fistulated cows and at 3, 6, 12, 24, 48, 72 and 96 hours of incubation for *in situ* degradability. The parameters soluble fraction (1), potentially degradable insoluble fraction (2), insoluble fraction (3) and degradation rate (KD) were evaluated. The effective degradation (ED) was estimated to passage rate of 5%/hour. The parameters studied were ruminal pH, ammoniacal nitrogen (N-$NH_3$), acetic acid, propionic acid, butyric acid and total volatile fatty acids (VFA). Sugarcane variety and concentrate composition had an interaction effect (P<0.05) on (1), (2), (3), KD and ED. The results for (1), (2), (3), KD and ED were 54.8%; 19.3%; 25.8%; 3.8%/h and 63.2% for IAC2480 with SH; 57.1%; 18.4%; 24.5%; 3.6%/h and 64.8% for IAC2480 without SH; 50.1%; 23.0%; 26.8%; 4.9%/h and 61.5% for IAC2195 with SH and 54.4%; 13.4%; 32.2%; 2.6%/h and 59.1% for IAC2195 without SH, respectively. The diet with IAC862480 or IAC912195 affected (P<0.001) ruminal pH (6.47 and 6.61, respectively) and ammoniacal nitrogen (12.9 and 16.6 mg/dl, respectively). Sugarcane variety and concentrate composition had an interaction effect (P<0.05) on VFA levels and acetic acid/propionic acid ratio. The results were 93.62 mM and 2.54 for IAC862480 with SH; 106.70 mM and 2.41 for IAC862480 without SH; 115.70 mM and 3.30 for IAC912195 with SH; 93.21 mM and 1.81for IAC912195 without SH, respectively. Sugarcane variety can change ruminal pH and ammoniacal nitrogen The SH inclusion improved the degradability of the IAC2195 sugarcane and influenced ruminal VFA levels, thus improving its nutritional quality.

**Comparative digestibility of nutrients from bio-ethanol byproducts by sheep and pigs**

*J.L. De Boever, S. Millet and S. De Campeneere*
*ILVO (Institue for Agriculture and Fisheries Research), Animal Sciences Unit, Scheldeweg 68, 9090 Melle, Belgium; johan.deboever@ilvo.vlaanderen.be*

Ruminants are known to better digest fibrous feeds than pigs. Byproducts from bio-ethanol production based on grains are not only high in protein and fat, but consist for about one third of fibrous substances. Particularly dried distillers grains and solubles (DDGS) are rich in cell walls and are fed to both ruminants and pigs. One may question which animal is most efficient in digesting these feedstuffs. In total 18 byproducts were studied: 12 batches of DDGS (3 pure wheat, 3 pure maize and 6 mixed, originating from 7 European plants), 5 batches of condensed distillers solubles (CDS) (from 5 plants) and one mixture of CDS and wheat bran. Byproducts were fed in combination with hay or maize silage to sheep (40/60 on DM-basis) or combined with a concentrate for pigs (30/70 ratio) and apparent digestibility was calculated by difference. Determination was with 5 mature sheep by total fecal collection and with 6 growing pigs after adding chromium oxide as indicator to the feed and by spot sampling of the feces. Digestion coefficients of crude protein (CP), crude fat (CFat) and organic matter (OM) were determined with both species, whereas fibre digestibility was based on NDF for sheep and on non starch polysaccharides (NSP) for pigs. The byproducts contained on average (per kg dry matter): 315 g CP (range: 205-495), 90 g CFat (60-148), 256 g NDF (43-366) and 326 g NSP (194-406). Mean digestibility of CP amounted to 75.2 vs 76.2%, of CFat to 87.8 vs 83.1%, of NDF/NSP to 68.5 vs 52.5% and of OM to 81.4 vs 76.2% in sheep vs pigs respectively. The difference in digestibility as analysed by a paired t-test was not significant ($P>0.05$) for CP, but highly significant ($P<0.001$) for the other nutrients. These results confirm the better fibre digestibility by ruminants as compared with pigs, but the higher fat digestibility by sheep is rather unexpected. Moreover, the difference in fat digestibility increased linearly ($P=0.014$) with higher NSP content.

---



**Evaluation of raw and processed soybean on blood parameters in Japanese quails (*Coturnix japonica*)**

*N. Vali and R. Jafari Dehkordi*
*Islamic Azad University, Shahrekord Branch, Animal Sciences Department, Faculty of Agriculture, 8813733395, Iran; nasrollah.vali@gmail.com*

The aim of this study was to investigate the effects of different types of processed (toasted, autoclaved and extruded soybean) soybean meal and raw soya seed on feed efficiency, carcass traits and blood parameters in Japanese quails (*Coturnix japonica*). A total of 189, twelve days old male and female Japanese quails were randomly assigned to 9 treatments containing three replicates in each and 7 birds per replicate. The first group (control group) was fed with a standard diet of quails with soybean meal, and other treatments (T2-T9) included 10 or 20% of raw (T2 and T3), autoclaved (T4 and T5), and toasted (T6 and T7) and extruded soybean (T8 and T9). Feed conversion ratio (FCR), live weight, carcass characteristics and blood parameters (concentration of blood cholesterol, triglyceride and iron) were recorded. Data were analyzed based on completely randomized design using GLM procedure of software SAS (statistical analysis system). Results indicated that Live weight, weight without feather, carcass weight, gizzard weight and concentration of iron (Fe) and triglyceride were significantly different between male and female birds (363.49±32.91 (mg/cc×100 Fe, 376.71±61.09 mg/dl triglyceride for female and 199.27±32.91 mg/cc×100 Fe, 154.32±54.97 mg/dl triglyceride for male), but in other traits no significant different ($P>0.05$) was observed. And also results indicated that the highest serum Fe concentration belonged to T5 group (514.41±80.62 mg/dl) and the lowest amount belonged to T9 (184.45±55.93 mg/dl) and T6 (185.09±55.93 mg/dl) groups ($P<0.05$). In conclusion, the results of present study showed that increasing processed soybean in the diet up to 20% especially extruded soya improved birds feed efficiency. The consumption of raw soybean caused negative effects on feed efficiency.

**Rapeseed meal as an alternative to soybean meal proteinaceous feed for growing-finishing pigs**

*I. Giannenas[1], E. Bonos[1], E. Christaki[1], P. Florou-Paneri[1], I. Skoufos[2], A. Tsinas[2], A. Tzora[2] and J. Peng[3]*
*[1]Aristotle University of Thessaloniki, Laboratory of Nutrition, Faculty of Veterinary Medicine, Laboratory of Nutrition, Faculty of Veterinary Medicine, Aristotle University of Thessaloniki, 54124 Thessaloniki, Greece, [2]Technological Institute of Epirus, Department of Animal Production, Department of Animal Production, Technological Institute of Epirus, 47100 Arta, Greece, [3]Huazhong Agricultural University, Laboratory of Animal Nutrition and Feed Science, Animal Nutrition and Feed Science Department, Shizishan Street 1, 430070, Wuhan, China, P.R.; ppaneri@vet.auth.gr*

A total of 120 pigs, 3 months old, were allocated to two equal groups (4 replications per group), to investigate the effect of the dietary inclusion of rapeseed meal (R group) of Greek origin instead of imported soybean meal (S group) on growth performance and certain carcass characteristics, such as chemical composition, lipid oxidation and fatty acid profile, during growing and fattening periods. The whole experimentation lasted 90 days. The substitution of soybean meal by rapeseed meal did not change (P>0.05) body weight gain in growing [S:13.9 kg vs R:12.9 kg] or in fattening period [S:46.1 kg vs R:43.1 kg]. Lipid oxidative stability was similar (P>0.05) among the experimental groups after 4 days of storage for pork loin [S:55.3 ng MDA/g vs R:45.6 ng MDA/g] and rib steak [S:45.8 ng MDA/g vs R:56.3 ng MDA/g]. However, significant differences (P<0.05) were noticed on the carcass fat content in pork loin [S:1.3% vs R:4.2%] and rib steak [S:2.1% vs R:1.4%] and on the fatty acid profile in rib steak for SFA [S:38.0% vs R:35.1%], MUFA [S:44.7% vs R:50.2%], and PUFA [S:15.2% vs R:12.4%]. Consequently, rapeseed meal could be suggested as a potential alternative proteinaceous feed for fattening pigs without adverse effects on performance or carcass quality. This work is part of the project 12CHN91, of the Bilateral R&D Cooperation between Greece and China 2012-2014, funded by the Hellenic GSRT.



**Effects of rice shape difference on productive performance in broiler chickens**

*K. Tatsuda and S. Ishikawa*
*Hyogo Prefectural Institute of Agriculture, Forestry and Fisheries, Animal Science Department, 1533 Befucho Kasai, Hyogo, 679-0198, Japan; ken_tatsuda@pref.hyogo.lg.jp*

This experiment was conducted to determine the effects of rice shape difference on productive performance in broiler chickens. Two hundreds Chunky birds were divided into four experimental groups and a control group. The experimental groups were bred under different rice shape using whole paddy rice (WPR), crushed paddy rice (CPR), whole dehulled rice (WDR) and crushed dehulled rice (CDR). Thirty percent of the corn in the formula diet for the experimental groups was replaced by each shape of rice and fed during the later term of fattening. The formula diet for the control group contained 65% of corn with no rice. The following results were obtained. All rice shapes suited the birds' taste and the condition of the birds' health was good. The average body weight was not different among the five groups. The feed conversion rate tended to be better in dehulled rice groups. The meat yield and the abdominal fat ratio were not different among the five groups. The ratio of gizzard weight to live body weight was significantly higher in the WPR group than in the other groups (WPR:1.45%, CPR:1.19%, WDR:1.12%, CDR:1.06%, control:1.09%). The fatty acid composition in the thigh meat was not different among the five groups. The total amino acid yield was significantly higher in the crushed groups than in the control group (WPR:54.1 μmol/g, CPR:55.4 μmol/g, WDR:54.0 μmol/g, CDR:55.0 μmol/g, control:51.0 μmol/g). Cost economies in all the experimental groups were higher than in the control group. We suggest that rice shape does not significantly influence the productive performance and increase the economic benefit in broiler chickens.

**Effect of differences in feed intake of Polish Landrace pigs on some fattening parameters**

*M. Tyra, R. Eckert, G. Żak and A. Bereta*
*National Research Institute of Animal Production, Department of Animal Genetics and Breeding, ul. Sarego 2, 31-047 Krakow, Poland; miroslaw.tyra@izoo.krakow.pl*

Feed costs are the largest cost driver in pig production. They are estimated to account for 70-75% of all costs incurred in pork production. At the same time, pigs vary considerably in daily feed intake as a result of several factors affecting the living environment. The experiment used 122 Polish Landrace gilts, which were tested in performance stations. Animals were kept individually and were tested according to test procedures on reaching 30 and 100 kg body weight. Animals were fed *ad libitum* standard pelleted feed. Based on the results of feed intake control, the animals were divided into groups according to daily feed intake (up to 2.38 kg, from 2.39 to 2.58 kg, and over 2.58 kg). The size of these groups was respectively 39, 48 and 35 gilts. The feed intake of the studied animals was found to be correlated to fattening parameters. Animals with the highest feed intake (over 2.58 kg per day) were characterized by the highest rate of growth and were quicker to reach slaughter weight. These animals reached a 100-kg slaughter weight around 8 days earlier than the group of animals with the lowest feed intake (less than 2.4 kg of feed per day) but when considering the entire fattening period, they consumed the greatest amount of feed (almost 20 kg more than others) until reaching the slaughter weight of 100 kg, and thus had the poorest feed conversion efficiency (kg feed/kg gain). Higher feed intake in animals results in a lower efficiency of its use, especially in the final fattening period. Probably due to the capacity limitations associated with protein deposition in the body.



**Fermentation and ruminal microbiota under the influence of different strains of *Bacillus thuringiens***

*F.C. Campos[1], P.S. Corrêa[2], C. Nazato[1], R. Monnerat[3], C.M. McManus[4], A.L. Abdalla[1] and H. Louvandini[1]*
*[1]CENA/USP, Laboratory of Animal Nutrition, Avenida Centenário 303, 13400-970, Piracicaba, São Paulo, Brazil, [2]CENA/USP, Cellular and Molecular Biology Laboratory and GRASP, Avenida Centenário 303, 13400-970, Piracicaba, São Paulo, Brazil, [3]EMBRAPA, CENARGEN, Av. W5 Norte (final), 70770-917, Brasilia, DF, Brazil, [4]UnB, Asa Norte, 70910-900, Brasilia, DF, Brazil; louvandini@cena.usp.br*

*Bacillus thuringiensis* (Bt) produces crystalline protein inclusions that have been reported to be toxic to larvae and adult gastrointestinal nematodes of ruminants. However, little is known about the interference of this bacillus in ruminal fermentation. Therefore, we used *in vitro* gas production to evaluate the effect of strains 907, 1192, 2036, 2493, 2496 and S1185 of Bt on degradability in the diet and possible toxicity of the microbial population from the rumen fermentation process. 0.5 g of substrate were added in glass bottles (160 ml) along with 25 ml of inoculum, 50 ml of buffered solution and $5.7\times10^6$ spores of different Bt strains. The control group received no strain of Bt. Subsequently the bottles were sealed with rubber stoppers, agitated and brought to 39 °C incubator for 24 hours. Fermentation was stopped and the bags were retrieved for the dry matter degradability (DMD) evaluation. The pH of the liquid from the bottles was measured and sampled for analyzes of ammonia nitrogen, short-chain fatty acids, protozoa and by quantifying the relative abundance of Fibrobacter succinogenes, Ruminococcus flavefaciens, anaerobic fungi and methanogenic archaea assessed by qPCR. The use of Bt 907 strain resulted in a decrease of DMD compared to the control without interfering with the population of F. succinogenes, while other strains (Bt 1192, 2036, 2493, 2496 and S1185) reduced this bacterium population. The other variables investigated showed no differences among treatments. Except for strain Bt 907, other strains of Bt (1192, 2036, 2493, 2496 and S1185) did not interfere in the degradability of the diets and ruminal fermentation.

## Session 52 — Poster 24

**The effect of different silage additives on mycotoxin concentrations in high moisture corn silages**

*D. Bíro, B. Gálik, M. Juráček, M. Šimko, M. Rolinec and M. Majlát*
*Slovak University of Agriculture in Nitra, Faculty of Agrobiology and Food Resources, Tr. A. Hlinku 2, 949 76 Nitra, Slovak Republic; daniel.biro@uniag.sk*

The target of the experiment was to analyse the effect of biological and biochemical silage additives on the mycotoxins concentrations in ensiled high moisture crimped corn. Experiment was realized in cooperation with University Experimental Farm of the Slovak University of Agriculture in Nitra. Corn was harvested in dry matter content form 608.9 till 613.3 g/kg, immediately after the harvest was corn crimped. After additives addition were variants storage in laboratory conditions in silage bags with capacity 50 $dm^3$. Total four different variants were used in experiment. In untreated variant (control) was crimped corn conserved without additives, in variant A1 and A2 were used additives with active units of lactic acid bacteria, and in variant B combined additive (lactic acid bacteria, benzoate sodium, cellulases). The lowest zearalenone concentration in silage of high moisture crimped corn was found in variant B (29.83 µg/kg), in variants A1 and A2 was a tendency (P>0.05) of higher zearalenone concentrations in comparison to untreated control variant. In deoxynivalenol concentration was the highest value (0.133 µg/kg) found in untreated control variant. In total fumonisins concentration values ranges from 42.93 µg/kg (variant A1) to 70.2 µg/kg (control variant). In T-2 toxin was found significantly (P<0.05) differences between untreated control variant (232.17 µg/kg) and variants A1 (192.7 µg/kg) and A2 (280.28 µg/kg) respectively. In total aflatoxins concentrations ranged from 2.13 µg/kg (B variant) and 2.97 µg/kg (variant A1). Differences between A1 and B variants were significantly different (P<0.05). Concentration of ochratoxins in high moisture crimped corn silages ranged from 0.533 µg/kg (variant A2) till 2.333 µg/kg (untreated control variant). Differences were significantly different (P<0.05).

---

## Session 52 — Poster 25

**The effect of dietary plant extract on nutrients digestibility of warmblood sport horses**

*B. Gálik, D. Bíro, M. Rolinec, M. Šimko, M. Juráček, M. Halo, E. Mlyneková, Z. Schubertová and R. Herkeľ*
*Slovak University of Agriculture in Nitra, Faculty of Agrobiology and Food Resources, Tr. A. Hlinku 2, 949 76 Nitra, Slovak Republic; branislav.galik@uniag.sk*

The aim of the study was to determine the effect of dietary phytogenic additive on organic and inorganic nutrients digestibility in sport horses. In the trial, 10 adult healthy geldings were used (Slovak warmblood bred, average body weight 550 kg). Two groups (5 geldings per group) were analysed. In control group of sport geldings, feed ration from crimped barley, meadow hay and mineral feed mixture was used. In experimental group of sport geldings, feed ration was the same; however plant additive (a blend of essential oils from Origanum vulgare L. and Pimpinella anisum) was supplemented in dosage 1 g per 1 kg of concentred feed. The experiment lasted 14 days. After preparatory period (7 days) was during experimental period (7 days) individual total faeces collection method used. In experimental group of geldings, significantly (P<0.05) higher dry matter digestibility was found in comparison with control group (63.29 vs 57.68%), as well as in organic matter digestibility (71.33% in experimental group and 67.23% in control group of geldings). A tendency (P>0.05) of higher digestibility of crude protein (76.28 vs 70.63%) was found. In experimental group of geldings was found the significant effect (P<0.05) in crude fibre faecal digestibility in comparison with control group of geldings (51.61 vs 47.29%). In NDF faecal digestibility was detected significantly higher (P<0.05) coefficient in experimental group (45.37%) in comparison with control group of geldings (41.42%). In experimental group of geldings were analysed significantly (P<0.05) higher faecal digestibility of Ca (70.11%) and P (54.72%). A tendency of higher Mg, Na and K faecal digestibility was detected after plant additive supplementation.

## Session 52 — Poster 26

**A Markov model for optimal feeding and marketing decisions in production units of growing pigs**

*R. Pourmoayed[1], L.R. Nielsen[1] and A.R. Kristensen[2]*

*[1]Aarhus University, Department of Economics and Business, Fuglesangs Allé 4, 8210 Aarhus, Denmark, [2]University of Copenhagen, Department of Large Animal Sciences, Grønnegårdsvej 2, 1870 Frederiksberg C, Denmark; rpourmoayed@econ.au.dk*

In the pig production system, feeding is the most important factor that has a direct influence on the quality of the meat and the costs of the production unit. Another important element is the timing of slaughter (the marketing decision) that refers to a sequence of culling decisions until the production unit is emptied. In the production system, the economic optimizations of feeding and marketing decisions are interrelated and there is a need for a simultaneous analysis. In this study a stochastic optimization model is developed at pen level. In the model the state of the system is updated using a Bayesian approach based on on-line data obtained from a set of sensors in the pen. More precisely, two statistical models are used to update estimates of live weight and growth rate on a weekly basis. The proposed models are tested based on the data from a Danish farm to show the model may be used as decision support in finding optimal feeding and marketing decisions.

## Session 52 — Poster 27

**Evaluation of chemical, microbiological and sensory properties of the papaya pulp whey beverages**

*A.G. Mohamed[1], A.F. Zayan[2] and N.S. Abdrabw[1]*

*[1]National Research Center, Elbohoos st. Dokki Gizaa, 11268, Egypt, [2]ARC, cairo uni. st. Giza, 11268, Egypt; ashrafare@yahoo.com*

Six blends of different ratios from sweet whey and papaya pulp were prepared to produce new beverage blends. Prepared samples named blend1 (100% sweet whey), blend2 (10%papaya + 90% sweet whey), blend3 (20% papaya + 80% sweet whey), blend4 (30% papaya + 70% sweet whey) blend5 (40% papaya + 60% sweet whey) and blend 6 (50% papaya + 50% sweet whey). Chemical constituents, physical properties, microbial counts and sensory characteristics were determined for prepared beverages directly after preparing and after storage for a month at refrigerator. The acidity of beverages samples increased gradually during storage periods. Meanwhile, the loss percentage of ascorbic acid contents increased with increasing storage time. Lactose and calcium contents of the beverages samples slightly decreased. Microbial analysis of the prepared beverages revealed that, bacterial counts, yeast and molds counts gradually decreased with increasing storage period. On the other hand, sensory attributes of the whey beverages samples directly after preparation and after one month of storage indicated that, the beverage blend 5 had the highest organoleptic scores followed by blend 6, blend 4 blend 3, blend 2 and Blend 1 respectively.

**Physiochemical and sensory evaluation of yoghurt fortified with dietary fiber and phenolic compounds**

*A.G. Mohamed[1], A.F. Zayan[2] and H. Abbas[1]*
*[1]National Research Center, Dairy Science, Elbohoos st. Dokki Gizaa, 11268, Egypt, [2]ARC, Cairo Uni. st. Giza, 11268, Egypt; ashrafare@yahoo.com*

The effect of adding different level of dried grape pomace as a source of antioxidants and fiber at level 1.0 to 5.0% on physiochemical and sensory properties of yoghurt. The physiochemical properties of yoghurt were determined during 0, 7, 14 and 21 days of cold storage at 4C. Increasing the ratios of dried grape pomace had no effect on yoghurt pH values and acidity. During storage periods the pH values were decreased gradually and increasing gradually in acidity in all treatments throughout storage period of 21 d. Dried grape pomace addition up to 2.0% did not affect the synersis value, but the synersis value decreased, when 4.0 to 5.0% dried grape pomace was added to the samples. Yogurt prepared with 1 and 2% dried grape pomace had similar textural properties as control yogurts. Fortifying yogurt with 3, 4 and 5% dried grape pomace had significant effect on the textural properties. Hardness, gumminess, and springiness increased and adhesiveness and cohesiveness decreased significantly. The dietary fiber content of yoghurt increased with increasing of dried grape pomace in the yoghurt samples. Total phenolic content of dried grape pomace fortified products increased along with increasing dried grape pomace concentration in the product, 75, 102,132,168 and 190 mg GAE/100 g yoghurt with 1, 2, 3, 4 and 5% dried grape pomace, respectively. Yoghurt prepared with 5% dried grape pomace (w/w yoghurt) sample received the highest RSA of 84.72 mg AAE/100 g yoghurt initially, followed by yoghurt with 4, 3, 2 and 1% of dried grape pomace respectively. Yoghurt fortified with up to 3% dried grape pomace DGP had similar appearance, flavor, texture, consistency, and overall acceptance ratings as control yoghurt. Increasing dried grape pomace fortification level to 4 and 5% decreased appearance, flavor, texture, consistency and overall acceptance ratings significantly compared with control yoghurt.

---



**The effect of dietary *Taraxum officinale* on boars reproductive performance**

*M. Rolinec, D. Bíro, B. Gálik, M. Šimko, M. Juráček, Z. Schubertová, L. Guláčiová and O. Hanušovský*
*Slovak University of Agriculture, Faculty of Agrobiology and Food Resources, Tr. A. Hlinku 2, 94976 Nitra, Slovak Republic; michal.rolinec@uniag.sk*

The aim of this study was to analyze the effect of *Taraxum officinale* in feed ration of boars with incorrect nutrition. In the past as well as nowadays at many pig farms, the boars have been fed by a diet similar to that of the pregnant sows, even though their requirements are quite different. Total six 19.96±11.8 months-old boars of breed White improved from University Experimental Farm of Slovak University of Agriculture in Nitra were used in the experiment, which lasted 120 days. The boars were fed during all their production life till the beginning of experiment with feed mixture intended for pregnant sows. Feed mixture for pregnant sows contains less crude protein 130 vs 160 g/kg, lysine 6.5 vs 7.5 g/kg, methionine + cysteine 3.8 vs 4.5 g/kg, threonine 4 vs 5 g/kg, natrium 1.5 vs 2 g/kg, manganese 25 vs 30 mg/kg as well as copper 6 vs 8 mg/kg than feed mixture used for boars. Despite this minus in nutrition, the reproductive performance of boars was: 12.11±1.98 total born, 10.93±2.11 live born and 10.30±2.41 weaned pigs. During experiment (April 2013 to July 2013) the boars were fed with the same feed mixture intended for pregnant sows but with addition of 3 g of dried aboveground of *T. officinale*. The reproductive performance of boars during the experiment was: 13.00±4.96 total born, 10.00±2.1 live born and 9.25±1.7 weaned pigs. From these parameters only the number of total born pigs is affected by boars. This reproductive parameter was higher during experiment than it was before experiment ($P>0.05$). Supplementation of *T. officinale* affected positively the total born piglets. The results confirmed, that feeding boars with feed mixture for pregnant sows decreased their potential reproductive performance. For good reproductive performance of boars it is very important to feed them by a diet which always meets their nutrient needs.

## Session 52 Poster 30

**Prediction of dry matter intake by beef steers fed grass silage only diets**

*D. Hynes, S. Stergiadis and T. Yan*
*Agri-Food and Biosciences Institute, Large Park, Hillsborough, Co Down, BT26 6DR, United Kingdom; deborah.hynes@afbini.gov.uk*

Grass silage quality varies greatly, resulting in contrasting DM intake (DMI) by beef steers and challenging issues regarding prediction of DMI. Accurate estimate of feed intake is essential to improve production efficiency and reduce environmental impact of beef production. The main focus of this study was therefore to develop prediction models for DMI using data from a large scale study. This study evaluated effects of silage quality on *ad libitum* intake of beef steers (no concentrates offered) using 136 grass silages collected from commercial farms across Northern Ireland. Measurements lasted for 2 years with 17 three-week periods (12 periods in year 1 and 5 in year 2). Eight grass silages were evaluated in each period using 10 beef steers/silage. The animals were on average 415 kg liveweight (LW) and 14 months old. Stepwise regression analysis was performed to relate DMI against animal LW, chemical composition and fermentation parameters of silage. Liveweight was used in all equations. $R^2$ increased from 0.35 (when LW was used as a sole predictor) to 0.55 with the addition of chemical composition parameters. Alcohol corrected toluene DM, water soluble carbohydrates contents and non-protein N of total N were positively related to DMI while silage ash and lignin contents showed a negative relationship. Ammonia N and propionate were the only of the silage fermentation parameters that were highly related to DMI and their inclusion into the equation only marginally increased $R^2$ (0.57). These findings allow the manipulation of diets to achieve high DMI to maximise performance and to minimise environmental impact.

## Session 53 Theatre 1

**Genomic selection: changes and challenges in cattle breeding**

*S. Borchersen*
*VikingGenetics, Ebeltoftvej 16, 8960 Randers, Denmark; sobor@vikinggenetics.com*

Genomic Selection (GS) allows us to select young breeding candidates more accurate. Several factors influencing the efficiency of GS have been stated: size of reference population, quality of performance recording and efficiency of the methodology including chip technology. Efficiency in GS also depends on genetic diversity and customized chips: breed and trait specific including genetic defects etc. In future females is expected to be added to reference to get higher reliabilities on genomic enhanced breeding values (GEBVs) and for maintaining genetic diversity. GS has set new standards for optimal breeding schemes. Genotyping effort, number of sires tested and use of A.I. in practice, use of best females more intensively and handling inbreeding in new ways including marker information are all key factors for optimization. In addition GS initiate new possibilities for customized breeding schemes at herd level. A crucial factor influencing the reliability of genomic predictions has shown to be the size of reference population. A joined Holstein sire reference population is established in the Eurogenomics consortium that now exceeds 25,000 reliable proven bulls. The future aim is continuously exchange of genotypes and supporting developments of methods by strong research collaboration. In addition VikingRed and VikingJersey have respectively established agreements with Norway and North America for bull references. Less tested young A.I. bulls have a huge impact on practical issues like reduction of capacity for housing bulls at A.I.-Stations. Genomic test of heifers in combination with use of sexed semen on high pedigree animals and insemination of low pedigree animals with beef bulls expect to enhance economic benefit in dairy cattle production. Development of cost effective customized low density chips for large scale genotyping of females. Future steps will be made to use all available genomic marker information from full genome sequence information. Further benefit from GS is the ability to provide genetic evaluations for low heritable traits and in future also including new traits as feed efficiency.

**In vivo characterization of ham and loin morphology in 120 kg live pigs of four different sexes**

*M. Gispert, A. Carabus, A. Brun, J. Soler, C. Pedernera and M. Font-i-Furnols*
*IRTA, Product Quality, Finca Camps i Armet, IRTA, Spain; marina.gispert@irta.es*

Pig sex has an important effect on carcass and cuts characteristics. The aim of this trial was to characterize several traits in the loin and in the ham of pigs from different sexes by means of measures obtained by computed tomography (CT) images taken at 120 kg live pigs. For this purpose 12 females (FE), 12 castrated (CM), 12 immunocastrated (IM) and 11 entire (EM) males pigs of Pietrain × (Duroc × Large White) genotype were enrolled. Pigs were born in the same week and fed *ad libitum* the same diet. Immunocastration vaccine Improvac® was injected at 12 and 18 weeks of age. When pigs reached the target weight (120.6+2.57 kg) were CT scanned with GE HiSpeed Zx/i device (140 kV, 145 mA, axial 1 s and 10 mm thick). Two tomograms were obtained, one in the ham area (junction between the femur and pubis bones) and another in the loin area (between the 3rd and 4th last ribs). After that pigs were slaughtered and carcass weight and killing out percentage were calculated. Meat quality measurements (pH at 45 min. and ultimate pH and electrical conductivity) were also taken. Ham's and loin's images were analyzed and measurements of fat thickness, lengths and areas were obtained. Mixed analysis was performed to determine differences between sexes in the different measurements obtained. Results show no differences neither in live and carcass weight nor in killing out percentage among sexes. Area and maximum length of the longissimus muscle was significantly ($P<0.05$) higher in FE than IM, while the other sexes were in between and not significantly different. Loin subcutaneous fat was higher in CM than FE and EM, IM being in between. No differences have been found among sexes in ham's width, maximum high and area values. However, subcutaneous fat was higher in CM and IM than EM, FE being in between. Differences are probably related to different fat deposition growth among sexes and anatomical region. No differences in meat quality were observed.



**Assessment of quality of pig carcasses in Ukraine according to national and European classification**

*O. Kravchenko[1], A. Loza[2] and A. Getya[3]*
*[1]Poltava state agrarian academy, Production and Processing Technologies of Animal Products, str. 1/3 Skovorody, 36003 Poltava, Ukraine, [2]Association of Pig Producers of Ukraine, str. M. Kotsiubynsky 1, 01030 Kiev, Ukraine, [3]Ministry of agrarian policy and food of Ukraine, Department of animal production, str. Khreschatyk 24, 01002 Kiev, Ukraine; oksanakravchenko@ukr.net*

The existing Ukrainian system of classification of pork carcasses was established in the Soviet Union. At slaughterhouses conducted a visual assessment of the degree of muscles development, measure of carcass weight and back fat thickness (6-7 thoracic vertebrae), but sex of an animal is not considered as well as lean meat percentage. At the slaughterhouse 'Yubileinyi', Dnipropetrovsk region of Ukraine, 160 hybrid pigs were slaughtered (Landrace × Yorkshire × Peitrain, Yorkshire × Landrace × Duroc) and evaluated according to both Ukrainian and European classification. For EUROP assessment the French device CGM was used which involves measuring between last 3 and 4 thoracic vertebrae. According to Ukrainian classification all 160 carcasses were rated as the second category. The second category includes carcasses with weight from 47 to 102 kg and the thickness of fat from10 to 30 mm at the level of 6-7 thoracic vertebrae. Test assessment of carcasses according to European system has the following classification: 39 heads – class S, 88 heads – class E, 28 heads – class U and 5 heads – class R. It was determined that the biggest part of slaughtered pigs were castrated; 116 heads, or 72.5%. Slaughtered female have the highest MLD thickness (63.2±0.33 mm) and lean meat percentage (60.52±0.38%) at the same time have the thinner back fat (14,9±0,13 mm). In whole animal group lean meat percentage was 58.14±0.26%. Nowadays in Ukraine conducted work from the adaptation and implementation of the assessment system and payment of carcasses at slaughterhouses based on lean meat percentage of the carcasses.

**Intramuscular and subcutaneous fat quality in the indigenous Krškopolje pig**

*M. Žemva[1], T. Manu Ngapo[2], Š. Malovrh[1], A. Levart[1], M. Kovač[1] and T. Flisar[1]*
*[1]University of Ljubljana, Biotechnical Faculty, Animal Science, Groblje 3, 1230 Domžale, Slovenia, [2]FRDC, Agriculture and Agri-Food Canada, Saint-Hyacinthe, Quebec, J2S 8E3, Canada; tina.flisar@bf.uni-lj.si*

The search for fat composition and content favourable fatty acids in intramuscular fat IMF for fresh meat quality and at the same time in subcutaneous adipose tissue SCF for processed meat products, within the same breed, is reflected in the growing interest in native pig breeds. The aim of this study was to determine the fatty acid profile of the IMF of M. longissimus dorsi (LD) and adjacent SCF from the Slovenian indigenous Krškopolje pig. The type of fat from the 42 Krškopolje pig significantly affected the proportions of 29 of the 35 fatty acids measured here. The SCF contained higher proportions of SFA and PUFA and correspondingly less MUFA than the IMF. Taking the extremes, the IMF at 6-10% of the LD had higher proportions of SFA and MUFA and less PUFA than the IMF at 2-3%. The thicker SCF (57-67 mm) was higher in SFA and lower in PUFA than thinner (30-40 mm). Sex and age also effected fatty acid composition; the fat from barrows was higher in MUFA and lower in PUFA than that from gilts, and with increasing age, MUFA content increased.



**Nurse sows' reproductive performance in the subsequent litter: a farm survey**

*T.S. Bruun[1], J. Vinther[1], A.B. Strathe[2], C. Amdi[3] and C.F. Hansen[3]*
*[1]Danish Pig Research Centre, Danish Agriculture & Food Council, Axeltorv 3, 1609 Copenhagen, Denmark, [2]University of Copenhagen, Department of Clinical Veterinary and Animal Sciences, Groennegaardsvej 7, 1870 Frederiksberg, Denmark, [3]University of Copenhagen, Department of Large Animal Science, Groennegaardsvej 2, 1870 Frederiksberg, Denmark; tch@lf.dk*

The use of nurse sows in Danish herds is a common practice due to the large litter sizes; however, the effect of being selected as a nurse sow on subsequent reproductive performance is unknown. Therefore, the aim of this farm survey was to quantify a nurse sow's reproductive performance in the subsequent litter. Nurse sows (NSOWS) were defined as sows weaning their own litter at least 21 days post-partum and thereafter nursing another litter (nurse litter). Data (2012-2013) from 20 herds with more than 14.5 live born piglets per litter, stable herd size with a stable distribution of sows among parities over time were selected. Results from 83,928 litters were analyzed using the mixed procedure of SAS. Herd and the interaction herd×year×month were included as random effects, while parity and season were included as fixed effects. The average lengths of lactation were 39.8 and 27.6 days ($P<0.001$) for NSOWS and ordinary sows (OSOWS), respectively. Nurse sows weaned on average 12.3 piglets and subsequently 11.5 nurse piglets, whereas OSOWS weaned fewer piglets ($P<0.001$) in their single weaning (11.5). There was no difference in rate of sows returning to heat in the subsequent reproductive cycle between NSOWS and OSOWS ($P=0.74$). Subsequent litter size in the next reproductive cycle was higher ($P<0.001$) for NSOWS (18.4 total born piglets) than for OSOWS (17.8 total born piglets). Results indicate that nurse sows were selected among sows nursing large litters, and therefore may indicate that these sows were representing the best percentile of sows in a given herd. In conclusion, this survey indicated no negative effects on the subsequent reproductive results of being selected as a nurse sow.

**Nurse sows for supernumerous piglets**

*F. Thorup*
*Danish Pig Research Centre, Feeding & Reproduction, Axeltorv 3, 1609 Copenhagen, Denmark; ft@lf.dk*

When a group of sows farrow more piglets than the number of milk glands in these sows, the supernumerous piglets need to be handled. Weaning piglets before 21 days is prohibited by EU legislation. Thus technical solutions for artificial nursing are irrelevant. Using a nurse sow, the piglets can be nursed until weaning after 21 days. The optimal nurse sow was investigated. In a number of comparative studies, supernumerous piglets were randomly allotted to paired nurse sows that were randomly allotted to two 'treatments'. The following questions were investigated: Should the nurse sow be young or old? Is a two-step nurse sow superior to a one-step nurse? Does an injection of oxytocin facilitate the nurse situation? Is growth and survival affected in nurse piglets? How is the fertility of the nurse sow in the subsequent parity? The results revealed significantly higher survival rates among nurse piglets allotted to 1$^{st}$ parity sows than piglets allotted to nurse sows older than 1$^{st}$ parity. Nurse piglets allotted to a two-step nurse sow had significantly higher survival and growth rates than if allotted to a one-step nurse sow. Oxytocin injection to the nurse sow gave a non-significant decrease in weaning weight. Nurse piglets allotted to a two-step nurse of 1$^{st}$ or 2$^{nd}$ parity had the same survival and weaning weight as piglets staying with the dam. A two-step nurse sow receiving the nurse piglets 4 to 7 days after farrowing did not differ from other sows in respect to fertility. One-step nurse sows receiving nurse piglets >20 days after farrowing had a significantly lower farrowing rate, but farrowed significantly more piglets in next litter (17.1 piglets born versus 15.1 piglets in control sows). It is concluded that, if the nurse sows are young two-step nurse sows, supernumerous piglets will achieve the same growth and survival if they are transferred to nurse sows. Treatment of the sows with oxytocin does not improve the results. Being a nurse sow did improve litter size in the subsequent parity in one-step nurses, but had a negative effect on farrowing rate.

---



**Effects of number of feeder spaces on wean-finish performance of two genotypes**

*D. Carrion, B. Melody, A. Manabat and S. Jungst*
*Pig Improvement Company, Avda Ragull 80 2°, 08173, Sant Cugat del Vales, Spain;*
*domingo.carrion@genusplc.com*

A trial was conducted to investigate the effects of number of feeder spaces for two genotypes on wean-finish performance. Progeny of Camborough (Landrace × Large White) sows sired to PIC327L (Hampshire) and PIC337RG (Synthetic) boars were used. After weaning, pigs were allotted to pens by sire line and gender (gilts and castrates). Eighty pens in total were stocked with 25 pigs yielding a floor space of 0.68 m$^2$/pig. Pigs had access to two or four feeder holes per pen resulting in two feeder space treatments of 12.5 or 6.25 pigs/feeder space. The feeders were 142 cm in length with 35.5 cm of head room between feeder holes and an available linear length of feeder space of 5.68 and 2.84 cm/pig for the pens with four and two open feeder holes, respectively. Pan coverage was targeted for approximately 50%. Each pen had two cup drinkers. Pens of pigs were weighed each time diet changes occurred and at 168 days of age. Three pelleted nursery diets and seven meal grow–finish diets were fed. There was no statistical difference between the four and two feeder holes treatments for average daily gain, daily feed intake, feed conversion, full-value pigs marketed percentage, or mortality percentage (P>0.10). Carcasses from pigs assigned to the four feeder hole treatment grew 19 g/d faster(517 vs 508) and had 0.6 mm more loin depth (60,4 vs 59.8) than carcasses from pigs assigned to two feeder holes treatment (P<0.05). Genotype × number of feeder hole interactions were detected for carcass backfat thickness (P=0.04) resulting in an increases of 0.2 mm for PIC327L whereas a decrease of 0.4 mm for PIC337RG when 4 compared to the 2 feeders holes were available Based on these results, there is little evidence these two sire lines required different feeding spaces. Twelve pigs/feeder hole is adequate under described trial conditions.

**Effect of xylanase and phytase on the *in vitro* analysis of enzyme digestible organic matter at ileum**

*D.K. Rasmussen*
*Danish Agriculture & Food Council, Pig Research Centre, Nutrition & Reproduction, Axelborg, Axeltorv 3, 1609 Kbh V, Denmark; dkr@lf.dk*

Research with pigs in Denmark has demonstrated a positive effect of xylanase on daily gain and FCR of approx. 3% and a 0.6-0.7 unit increase in the *in vitro* analysis of enzyme digestible organic matter at ileum (EDOMi) when feed does not contain phytase. When both xylanase and phytase are added to pig feed, the effect drops to around 1% and the effect on EDOMi drops as well. Digestibility trials with pigs have also demonstrated that phytase as well as xylanase may increase the digestibility of protein and amino acids, but that the effect of both products combined is not significantly greater than the effect of the products individually. It may therefore be that when a diet also contains phytase, the effect of xylanase is reduced. The aim is to demonstrate, *in vitro*, whether the addition of xylanase or phytase to pig feed increases EDOMi and whether this effect varies among the different xylanases and phytases available in Denmark. It will also be clarified if the addition of phytase has an impact on the effect of xylanase in EDOMi in pig feed. This is a two-factor trial with xylanase and phytase as the two factors. Products from two Danish manufacturers are being investigated and results are analysed separately for each product. Laboratory analyses are currently underway, and results will be presented at the conference.

---

Session 54 Theatre 1

**Identifying and monitoring pain in pigs and ruminants**

*A. Prunier[1], M. Kluivers-Poodt[2] and L. Mounier[3]*
*[1]INRA, UMR 1348, PEGASE, 35590 Saint-Gilles, France, [2]Wageningen UR Livestock Research, P.O. Box 65, 8200 AB Lelystad, the Netherlands, [3]VetAgro Sup, Campus Vétérinaire de Lyon, 1 avenue Bourgelat, 69280 Marcy l'étoile, France; armelle.prunier@rennes.inra.fr*

One important requirement for animal welfare is to maintain animals free from pain, injury or disease. Therefore, detecting and evaluating the intensity of animal pain is crucial. Since animals cannot directly communicate their feelings sensitive and specific indicators easy to use have to be identified. Behavioural changes associated with pain are one of the techniques most frequently used on farms to assess the degree of pain experienced by pigs or ruminants. Pain, as a stressor, is associated to variations in the hypothalamic-pituitary-adrenal axis as well as in the sympathetic and immune systems that can be used to identify the presence of pain rapidly after it started. However, most of these measures need sophisticated equipment for their assessment and they are mainly adapted to experimental situations. Injuries and other indicators of lesions give information on the sources of pain and are convenient to use in all types of situations. Histo-pathological analyses can identify sources of pain and are especially useful to detect long term effects of mutilation practises that are performed in farms. When pronounced and/or long-lasting, the pain-induced behavioural and physiological changes can decrease production performance. Some indicators are very specific and sensitive to pain whereas others are more generally related to stressful situations. The latter can be used to indicate that animals are suffering from something which may be pain and can constitute a warning indicator. Even if in some cases, a single indicator, usually a behavioural indicator, can be sufficient to detect pain, combining various types of indicators increases sensitivity and specificity of pain assessment.

**Tail docking in pigs: is there any possibility of renunciation?**

*C. Veit, I. Traulsen and J. Krieter*
*Institute of Animal Breeding and Husbandry, Christian-Albrechts-University, Olshausenstr. 40, 24098 Kiel, Germany; cveit@tierzucht.uni-kiel.de*

Tail biting in pigs is a welfare concern in intensive pig husbandry and leads to economic losses. The aim of this study was to reveal the effects of provision of rootable material on tail biting outbreaks in long-tailed pigs. The experimental setup took place on two commercial farms (1+2) and one research farm (3) from 09/2013 to 01/2014. Different substrates (straw (n=350), hay (n=350), peat (n=50), dried corn silage (n=350), alfalfa hay (n=250)) were provided to the pigs once per day during rearing. Two control groups with docked (n=650) and long (n=350) tails were kept without provision of raw material. Each tail was scored with regard to bite occurrence and tail losses once per week with a four-point-score. To investigate the behaviour of the animals in regard to a tail biting outbreak and to analyse the duration of occupation with raw material, 40% of the pigs of one farm were observed by video. Tail biting occurred in 75-100% of the groups, except for the docked control groups (0.5%). The severity of an outbreak and its time after weaning differed between the experimental groups. Notably, in the long-tailed-control groups tail biting started mostly in the second week after weaning. The provision of raw material delayed an outbreak about one till two weeks and decreased the severity (e.g. on farm 3 partial and complete losses of tails were reduced by half), but could not prevent it totally. On farm 1 85% of the piglets of the raw material groups had the original length of their tail at the end of rearing. On farm 2 only 12% of the piglets of the raw material groups reached the end of rearing with original length of tail, which can be explained by health problems in the livestock. In comparison with farm 3 56% of the pigs out of the occupied groups reached the end of rearing with the original length of tail. However, rearing of long tailed pigs requires an intensive animal observation, provision of raw material is useful, and a proper health status is important.

---



**The MC4R gene affects puberty attainment in gilts but not in boars**

*A. Van Den Broeke[1], M. Aluwé[1], F. Tuyttens[1], S. Janssens[2], A. Coussé[2], N. Buys[2] and S. Millet[1]*
*[1]The Institute for Agricultural and Fisheries Research (ILVO), Animal Sciences Unit, Scheldeweg 68, 9090 Melle, Belgium, [2]KU Leuven, Livestock Genetics, Department of Biosystems, Kasteelpark Arenberg 30, 3001 Leuven, Belgium; alice.vandenbroeke@ilvo.vlaanderen.be*

Boars are routinely castrated to prevent boar taint. In the past, genetic selection against boar taint influenced the testicular steroid production and delayed the onset of puberty. In this study, we assessed with a 2×2 design (genotype × gender) if the MC4R gene, a gene linked to daily gain and a possible marker for boar taint, has negative effects on reproduction, behaviour and skin lesions in gilts and boars. During the entire rearing period, boars showed more aggression (P<0.001), mounting behaviour (P<0.001) and anogenital sniffing (P<0.001) compared to gilts. As a consequence, boars had more skin lesions at their anterior part (P<0.001) and tail (P=0.016). Genotype affected boar behaviour when they were 22 weeks of age: boars of the GG genotype showed more aggressive behavior compared to AA boars (P=0.002). In addition, at 20 weeks of age, GG boars, but not gilts, had more skin lesions on their back, anterior and caudal part than AA boars. Around that age, we observed a peak in testosterone levels in both genotypes. Gender and genotype had no effect on lameness. The attainment of puberty of the boars was estimated by the presence of spermatozoa in a preputial smear sample collected weekly, from 10 weeks of age to slaughter. Genotype did not affect age (149±12 days) or weight (94±16 kg) of the boars at the onset of puberty, nor testestosterone level. 31% of the GG gilts, however, had cycling ovaries at slaughter compared to 8% of the AA gilts. In conclusion, the polymorphism in the MC4R did not affect reproduction characteristics in boars but may influence puberty onset in gilts. There seems to be more aggression at the age of puberty of the GG boars compared to the AA boars, but the general effect of the marker on aggression remains unclear.

**Reliability of the Welfare Quality® animal welfare assessment protocol for growing pigs**

*I. Czycholl[1], C. Kniese[2], L. Schrader[2] and J. Krieter[1]*
*[1]Christian-Albrechts-University, Institute of Animal Breeding and Husbandry, Olshausenstr. 40, 24098 Kiel, Germany, [2]Friedrich Loeffler Institute, Institute of Animal Welfare and Animal Husbandry, Doernbergstr. 25/27, 29223 Celle, Germany; iczycholl@tierzucht.uni-kiel.de*

The Welfare Quality® Animal Welfare Assessment Protocol (AWAP) is an animal based measurement tool for welfare on farm, including a Qualitative Behaviour Assessment (QBA), direct behaviour observations (BO) by instantaneous scan sampling, a Human Animal Relationship Test (HAR) and checks for different individual parameters e.g. presence of tail biting, wounds and bursitis. An Interobserver Reliability study was performed by comparing the results of in total 29 assessments carried out by 3 trained observers, whereby 19 combined assessments were fulfilled by observers 1 and 2 and 10 by observers 1 and 3. The observations took place at the same time and on the same animals, but completely independent from each other. For comparison, Spearman Rankcorrelation Coefficient (RS), Intraclass Correlation Coefficient (ICC) and Smallest Detectable Change (SDC) were calculated. The study showed no agreement in the QBA (RS 0.17, ICC 0.0, SDC 0.3) and good to moderate agreement in the BO (e.g. social: RS 0.45, ICC 0.5, SDC 0.05, use of enrichment material: RS: 0.75, ICC 0.75, SDC 0.04). The rather low agreement of the HAR (RS 0.47, ICC 0.5, SDC 0.14) can be explained by the fact that the persons entered the pens one after the other to minimise mutual interference which influenced the reaction towards the second intruder. Overall, observers agreed well in the individual parameters, e.g. tail biting (RS 0.52, ICC 0.47; SDC 0.02), wounds (RS 0.43, ICC 0.56, SDC 0.03). However, the parameter bursitis showed great differences (RS 0.45; ICC 0.27, SDC 0.14), which can be explained by difficulties in the assessment when the animals are moving fast or the legs are dirty. In conclusion, the Interobserver Reliability was good in the BO and most individual parameters, but not for the parameter bursitis and the QBA.

---

Session 54 Theatre 5

**Social hierarchy formation when piglets are mixed in different group compositions after weaning**

*M. Fels[1], N. Kemper[1] and S. Hoy[2]*
*[1]University of Veterinary Medicine Hannover, Foundation, Institute for Animal Hygiene, Animal Welfare and Farm Animal Behaviour, Bischofsholer Damm 15, 30173 Hannover, Germany, [2]Justus Liebig University Giessen, Department of Animal Breeding and Genetics, Leihgesterner Weg 52, 35392 Giessen, Germany; michaela.fels@tiho-hannover.de*

The aim of this study was to investigate whether mixing of piglets in different group compositions after weaning can affect aggressive behaviour and the establishment of social hierarchy. We studied homogeneous and heterogeneous weight groups, groups of piglets originating from two or six origin litters and single sex groups (12 piglets per group). Furthermore, different group sizes (six and 12 piglets per group) were analysed. In 38 groups, the behaviour of piglets was video recorded for 72 h after mixing and the outcome of all fights was analysed. Sociometric parameters (Landaus's index, Kendall's coefficient and Directional consistency index) were calculated. ANOVA analyses followed by SNK test were carried out to find significant differences. Regardless of group composition, in groups of 12 piglets, quasi-linear hierarchies were established with means of Landau's index h'=0.67 and Kendall's index K=0.62. In groups of six, individuals fought significantly less than in groups of 12 (52 vs 64 fights per piglet) and the social hierarchy was almost totally linear indicated by h'= 0.9 and K=0.88. Concerning the number of fights, there was no further difference between group compositions. For all groups there was a majority of one-way-relationships (63%, $P<0.05$), i.e. dyads with wins for only one individual, followed by two-way relationships with wins for both individuals (21%), unknown (15%) and tied relationships (3%). Groups of six piglets showed the highest percentage of one-way-relationships (74%) and the lowest percentage of tied relationships (1%). We conclude that piglets are motivated to form a linear hierarchy after mixing regardless of group composition; however, when increasing group size the degree of linearity decreases.

**Social network analysis: investigation of agonistic behaviour in pig production**

*K. Büttner, K. Scheffler and J. Krieter*
*Institute of Animal Breeding and Husbandry, CAU, Olshausenstr. 40, 24098, Germany; kbuettner@tierzucht.uni-kiel.de*

In pig production, a reduction of agonistic behaviour positively affects animal health, welfare aspects as well as production parameters. Social network analysis provides an insight in animal societies and can help to detect structures of agonistic behaviour. Therefore, video observation data of 148 animals of three different age levels (weaned pigs, growing pigs and gilts) were investigated for 2 days directly after mixing. The aim of the study was to identify the development of specific centrality parameters regarding direct agonistic interactions in different age levels. The in-degree describes the number of received agonistic interactions and the out-degree measures the number of initiated agonistic interactions. Social network analysis was performed for both existence (unweighted) and frequency (weighted) of agonistic interactions between different animals illustrated as weighted and unweighted in-and out-degree. The results showed that weaned pigs had significant higher values for the weighted and unweighted in- and out-degree than growing pigs and gilts (Kruskal-Wallis-Test, $P<0.05$). The Spearman-Rank-Correlation between the unweighted and weighted in- and out-degree showed high values (0.93-0.96) for growing pigs and gilts, whereas weaned pigs had smaller values (0.74-0.81). In contrast, the correlation between the unweighted (weighted) in- and out-degree was higher for weaned pigs with 0.71 (0.64) than for growing pigs with 0.46 (0.47) and gilts with 0.60 (0.54). An explanation is the more stable rank order of growing pigs and gilts due to the increased familiarity. Here, the choice of antagonists was more specific (animals with a similar position in rank order) than in weaned pigs. By quantifying the important aspects of group structure in different production stages, information based on social network analysis could help identifying and understanding the formation of behavioural patterns, e.g. agonistic interactions, tail biting and access to resources.

---



**Analysis of static and dynamic position determination accuracy of Location System Ubisense**

*T. Rose[1,2], I. Traulsen[2], U. Hellmuth[1], H. Georg[3] and J. Krieter[2]*
*[1]University of Applied Sciences Kiel, Department of Agriculture, Grüner Kamp 11, 24783 Osterrönfeld, Germany, [2]Institute of Animal Breeding and Husbandry, Christian-Albrechts-University, Olshausenstr. 40, 24098 Kiel, Germany, [3]Thuenen-Institute of Organic Farming, Trenthorst 32, 23847 Westerau, Germany; trose@tierzucht.uni-kiel.de*

Aim of this laboratory investigation was to determine the main influences on the accuracy of indoor position determination using the location system 'Ubisense Series 7000' to increase the achievable precision of automatic behavior observations of dairy cows. The Ubisense system consists of a sensor network and allows a three dimensional position determination of tags via an ultra-wideband signal (6-8GHz) and the methods 'time difference of arrival' and 'angle of arrival'. To evaluate the best sensor configuration, 2-8 sensors were used in 16 different combinations and 2 heights. In a 16×8 m environment 8 squares of equal size were defined and one tag was placed in each square (static position). A model railroad (speed 0.4, 0.5 and 0.6 m/s) and motion capturing were used to determine the position of three tags, each placed on one of the three carriages of the railroad (dynamic position). The euclidean distance (ed) between the measurement of the location system and the respective reference point was analyzed using a mixed model. The best dynamic accuracy was reached using 2 sensors in a diagonal configuration (ed=0.16 m, se=0.064, $P<0.0001$, static: ed=0.09 m, se=0.063, $P<0.0001$). With the same number of sensors but a one sided configuration the dynamic accuracy fell off to 0.37 m (se=0.064, $P<0.0001$). A higher speed resulted in a declining accuracy (0.4 m/s: ed=0.19 m, se=0.028, $P<0.0001$, 0.6 m/s: ed=0.24 m, se=0.028, $P<0.0001$). This effect decreased with an increasing number of sensors. Under laboratory conditions, the Ubisense location system is qualified for high accuracy indoor positioning. The results can be used to optimize localization accuracy of cows.

## Session 54 Theatre 8

**Timing of transfer after mating influence dam cortisol and maternal care in farm mink**

*J. Malmkvist[1] and R. Palme[2]*

*[1]Aarhus University, Animal Science, Blichers alle 20, 8830 Tjele, Denmark, [2]University of Veterinary Medicine Vienna, Biomedical Sciences, Veterinärplatz 1, 1210 Vienna, Austria; jens.malmkvist@agrsci.dk*

Farmed mammals are typically transferred to another housing environment prior to the delivery. We investigated whether timing of transfer before delivery – early (day -36), intermediate (day -18) or late (day -3) relative to the expected day of birth (day 0) – affects maternal stress, maternal care and the early kit vitality in farm mink (*Neovison vison*) We hypothesized that early transfer is beneficial for mink mothers and their offspring, in comparison to intermediate or late transfer closer to delivery, which is the current practise. We used 180 double mated female yearlings in three equal-sized groups (n=60): (1) 'Early', transfer to maternity unit immediately after the end of the mating period, March 23; (2) 'Intermediate', transfer in the middle of the period, April 10; (3) 'Late', transfer late in the pregnancy period, April 25. Data collection included weekly determination of Faecal Cortisol Metabolites (FCM) and evaluation of maternal care: nest building, in-nest temperature, as well as kit-retrieval behaviour, kit mortality and growth day 0 to 7 postpartum (pp). We found that mated mink females build and maintain a nest at least 1 month prior to delivery, when given free access to nest building material. During the weeks before delivery, Intermediate females had 50% higher FCM concentration than the other two groups (P=0.002), indicative of stress. Late moved females had in average 2.7 °C colder nest compared to early moved females (P=0.002). Additionally, the mortality in group Late tended to be higher in affected litters (P=0.085). Kits from early moved females displayed less vocalisation (17% vs 40-41%, P=0.015), when tested away from the nest day 5 pp. This indicates enhanced vitality from offspring of early moved females. In conclusion, transfer immediately after mating, rather than later during the sensitive pregnancy period, reduce stress and increase maternal care in farmed mink.

---

## Session 54 Theatre 9

**Dividing large litters reduce number of bites among mink kits (*Neovison vison*)**

*P.F. Larsen and T.N. Clausen*

*Kopenhagen Fur, Agro Food Park 15, 8200 Aarhus N, Denmark; pfl@kopenhagenfur.com*

Recent studies have demonstrated that postponed weaning age from 6 weeks to 8 weeks in mink increases the number of bites among mink kits especially within the large litters. Therefore a series of studies have been conducted in order to quantify effects of weaning age on mortality, number of bites and growth rates in mink kits weaned at respectively 42, 49 and 56 days after birth from 2011 to 2013. Results showed that the occurrences of bites among mink kits were reduced with 40% in the large litters (6-9 kits per litter) by dividing the litter on day 42 after birth. Moreover the study showed that weaning day 42 besides reducing bites, also increased the growth rate for male mink kits in the large litters. These results are highly important for ensuring optimal welfare in mink production and dividing of large litters could be considered in the Danish legislation to improve mink welfare.

**Early weaning and separation to group-housing limit aggression resulting in bite marks in juvenile**

*S.H. Møller and S.W. Hansen*
*Aarhus University, Department of Animal Science, Blichers Allé 20, P.O. Box 50, 8830 Tjele, Denmark; steenh.moller@agrsci.dk*

Housing more than one juvenile mink of each sex in the same cage (group housing) has been shown to increase aggression, and bite- marks has been shown to be an ideal indicator of this aggression. This investigation set out to test two hypothesises for bite-marks in the leather side of mink pelts: (1) It is possible to reduce the bite-mark score by group selection, compared to an unselected control. (2) Early weaning and separation of juveniles into group-housing decreases the bite-mark score compared to late separation. We scored the number of bite-marks in a group selection line against bite-marks and an unselected control line of brown, group-housed mink over four successive years. The fifth year we also investigated the effect of early and late weaning and separation of litters into group-housing, in combination with continued selection. Early and late weaning and separation into group-housing was also tested on a private farm. The bite-mark score increased in the unselected control line in the first three generations, while it dropped significantly in generation four. The difference in bite-mark score between selection and control increased with each generation and the bite-mark score is now the double of that in the selection line. We accept our first hypothesis, that it is possible to reduce the bite-mark score by group selection compared to the unselected control line. At the private farm the bite mark score was significantly lower in the group housed juveniles from early than from late weaning and separation, while this was not the case on the experimental farm. Therefore, we cannot accept our second hypothesis, that early weaning and separation of juveniles into group housing reduces the number of bite marks compared to late weaning and separation. In conclusion, the welfare of juvenile group housed mink can be improved during the growth season by group selection, while the welfare effects of early weaning and separation is inconclusive.

Session 54 Poster 11

**Assessment of fitness of dairy cows for short distance road transportation**

*K. Dahl-Pedersen[1], M.S. Herskin[1], H. Houe[2] and P.T. Thomsen[1]*
*[1]Aarhus University, Department of Animal Science, Blichers Allé 20, 8830 Tjele, Denmark, [2]University of Copenhagen, Department of Large Animal Sciences, Grønnegårdsvej 8, 1870 Frederiksberg C, Denmark; kirstin.dahlpedersen@agrsci.dk*

Approximately 200,000 dairy cows destined for slaughter are transported by road every year in Denmark. Assessing the fitness for transport of individual cows can be a difficult challenge for the farmer, the driver and even veterinary professionals. It is important to be able to assess fitness for transportation correctly in order to avoid transportation of un-fit cows, which could result in unnecessary stress, exhaustion, pain and potentially violate animal welfare legislation. The aim of the study is to develop a tool to assess the fitness for transportation of individual cows as well as gain knowledge of the impact of transportation on dairy cow welfare. The project (2014 through 2016) includes an observational study including approximately 700 Danish dairy cows destined for slaughter. Each cow will be given a meticulous examination by a trained veterinarian before and after transportation to the abattoir. Examinations will include both behavioural and clinical observations, e.g. grinding of teeth, lameness and body condition score. Also other information such as gestational age will be recorded. Factors relating to the vehicles used for transportation and the transportation itself e.g. handling and loading procedures, use of partitions between animals and driving quality will also be recorded. The study is expected to identify behavioural and clinical manifestations likely to have significant adverse effect on the welfare of dairy cows during and after short distance road transportation. As part of this work, a reliable and easy manageable tool in the form of a scoring system for assessing fitness for transportation of individual cows will be developed. In the future, such a scoring system may be used by farmers, drivers or local veterinarians to identify cows which are fit for transportation and cows which should not be transported.

## Session 54 Poster 12

**Anise smell enrichment impact to aggressive behavior in piglets**

*P. Juhás, O. Bučko, J. Petrák and O. Debrecéni*
*Slovak University of Agriculture, Faculty of Agrobiology and Food Resources, Department of Animal Husbandry, Tr. A. Hlinku 2, 94901 Nitra, Slovak Republic; peter.juhas@uniag.sk*

Aim of research was investigate the smell enrichment of pen for weaning to aggressive behavior in piglets. Piglets are under high excitation after weaning, separation from sow and transfer to new pen and show high level of aggression to each other in group. In experiment we have tested the smell of anise essence as enrichment to reduce aggressive behavior. Anise essence was painted to back of pregnant sow week before farrowing. Farrowing pen was treated with anise essence on 10$^{th}$ and 20$^{th}$ day after farrowing by painting walls with essence. The pen for weaning was treated by same essence at morning on day of weaning. We have expected the essence smell remind piglets 'home' environment in farrowing pen. We have used 3 experimental litters (EL) (EL1 n=10, EL2 n=9, EL3 n=13). Average weight of piglets in EL was 11.47, 10.68 and 6.29 kg resp. Three litters without enrichment were used as control group (CL) in same pen (CL1 n=10, CL2 n=6, CL3 n=11). Average weight of piglets in CL was 7.98, 7.94 and 9.20 kg resp. Size of pen was 2,380×1,840 mm. Behavior was recorded for 8 hours after transfer to new pen. We have evaluated total number of attacks per 1$^{st}$ hour per piglet (TNA1H), total number of attacks per 8 hours per piglet (TNA8H) and difference among particular hours (D1-8) in number of attacks. Average TNA1H after transfer to new pen was 7.903, SD=3.139 in EL and 7.263, SD=0.426 in CL. Average TNA8H was 16.83, SD=4.388 in EL and 12.38, SD=1.824. There was no difference between EL and CL groups in number of attacks. Difference was recorded only in TNA1H and number of attacks in other hours ($F_{7,15}$=3.781, P=0.0009). We have found correlation between TNA8H and number of piglets in group (R=0.812, P=0.5) and area per pig (R=-0.812, P=0.5). The research was supported by VEGA No.: 1/2717/12 and ECOVA and ECOVA Plus projects.

## Session 54 Poster 13

**Effect of the presence of males on fear and stress of domestic hens**

*S.G. Dávila, M.G. Gil and J.L. Campo*
*Instituto Nacional Investigación y Tecnología Agraria y Alimentaria, Mejora Genética Animal, Ctra. Coruña, Km. 7,5, 28040 Madrid, Spain; sgdavila@inia.es*

The males may have an important role on the female behaviour in chickens, the presence of males suppressing aggressive behaviour towards other subordinate hens. The purpose of this study was to analyse the effect of the presence of males on two measurements of fear and stress (tonic immobility duration and heterophil to lymphocyte ratio) in females of several breeds of layers at 9 months of age. A total of 200 females from five Spanish breeds, which were housed in litter pens with or without the presence of males, were used. Tonic immobility duration was logarithm transformed before analysis, being assigned a maximum score of 600 s for the latency duration. One hundred leukocytes were counted, and the heterophil to lymphocyte ratio was calculated. The ratios were square root transformed before analysis. A two-way analysis of variance was used with the model: $x_{ijk} = m + G_i + B_j + BG_{ij} + e_{ijk}$ where $x_{ijk}$ is the analyzed measurement, m is the overall mean, $G_i$ is the effect of the group (presence vs absence of males), $B_j$ is the effect of breed (j=1-5), $GB_{ij}$ is the interaction, and $e_{ijk}$ is the residual (k=1-20). The SAS statistical package, GLM procedure, was used. Although there was a trend (P=0.09; F=2.92), the effect of the presence of males on tonic immobility duration was not significant (130 vs 157±13 s), indicating that the presence of males is not associated with the fearfulness of females. The interaction between group and breed for the heterophil to lymphocyte ratio was significant (P<0.05; F=2.64), the values being lower in females housed with the presence of males in the Quail Castellana (0.42 vs 0.65±0.06), the opposite being true in the Red Villafranquina (0.50 vs 0.32±0.06). In conclusion, females at 9 months of age in flocks containing males had similar fear levels as those in flocks without males, although they were less or more stressed depending on the breed as indicated by the heterophil to lymphocyte ratio.

**Measurement of floor space covered by fattening pigs during the finishing period**

*H. Arndt, J. Hartung, B. Spindler, A. Briese, M. Fels and N. Kemper*
*TiHo Hannover, Institute for Animal Hygiene, Animal Welfare and Farm Animal Behaviour, Bischofsholer Damm 15, 30173 Hannover, Germany; michaela.fels@tiho-hannover.de*

An important question in intensive pig fattening is to define the optimal stocking density on the farm as well as during transport to the slaughterhouse. To determine the space requirement of the individual pig in the pen and on the animal transporter, it is important to know the actual body dimensions of the pig in different positions. The aim of this study was to measure the floor space covered by a pig's body in different standing and lying positions. Therefore, pigs (n=232, hybrids of DanBred sows and Pietrain boar db.77® line, Ø 109±11.45 kg, 108 ♂ castr./ 124 ♀) in the finishing period were weighed and photographed from top view in a specially prepared 'Planimetrie-Box'. Using the software 'KobaPlan', the images were analysed. Based on the number of pixels of the photo area occupied by the animal, the floor space covered by this animal was calculated automatically in relation to a reference surface. It was shown that finishing pigs with an average live weight of 115±0.56 kg in standing position with the head down occupied a floor area of 0.3±0.018 $m^2$ on average. Comparing these results to the legal requirements in the European Union (1 $m^2$/pig with more than 110 kg), it can be concluded that the remaining area for locomotion is about 0.7 $m^2$. Comparing the results with the European requirements on pig transport (235 kg/$m^2$, i.e. 0.47-0.51 $m^2$/pig from 110 to120 kg) this means a remaining area of 0.17-0.21 $m^2$. For pigs in standing position with the head raised or moved to the side, or in a lying position, values for covered floor space were higher, exemplary going up to 0.5 $m^2$. Correspondingly lower is the remaining free floor area. In order to evaluate whether the remaining areas are sufficient to perform natural behaviours, further studies are needed.



**Fitness for transport of Danish broiler chickens**

*S.N. Andreasen and H.H. Kristensen*
*University of Copenhagen, Department of Large Animal Sciences, Section for Animal Welfare and Disease Control, Groennegaardsvej 8, 1870 Frederiksberg C, Denmark; hek@sund.ku.dk*

Approximately 100 million broilers are mechanically harvested, transported and slaughtered in Denmark each year. The Council Regulation (EC) No 1/2005 on the protection of animals during transport and related operations, states that 'No animal shall be transported unless it is fit for the intended journey, and all animals shall be transported in conditions guaranteed not to cause them injury or unnecessary suffering'. To be able to secure the overall welfare of broilers submitted to transport an evaluation of their fitness for transport should be conducted. No such validated practical evaluation currently exists. The current project is in its initial stages and aims at generating knowledge, which can be used in the development of a tool to assess the transport suitability of Danish broiler flocks. Overall, the project will be carried out as observational studies under commercial conditions to maximize the applicability of the results to actual transport conditions. The project will comprise several stages: (1) Collection of information on risk factors relevant for the welfare of broilers transported for slaughter under Danish conditions. This is data concerning the flock and management (bird density, litter conditions, etc), the harvesting (duration, speed, etc), the transport conditions (e.g. temperature and humidity), the lairage (duration, temperature, etc) and processing results (e.g. dead-on-arrival and meat inspection); (2) generation of simple clinical criteria for assessing transport suitability in the flock, including tests of observer reliabilities; (3) clinical examination of broilers before harvesting, immediately after harvesting, at unloading and just before slaughter; (4) application of the clinical data to confirm the relevance of risk factors during the different stages of the transport process. The project is financed by the Ministry of Food, Agriculture and Fisheries, the Danish Veterinary and Food Administration and will be finalised ultimo 2016.

## Session 55 Theatre 1

**The Iberian pig production systems and their products**

*R. Lizardo and E. Esteve-García*
*IRTA, Monogastric Nutrition, Ctra Reus – El Morell, km 3.8, 43120 Constantí, Spain; rosil.lizardo@irta.es*

The Iberian pig is a slow-growing breed native from the Iberian Peninsula which is perfectly adapted to the Mediterranean natural ecosystem and known for the high quality and consumer acceptability of its dry-cured products. From a population of around 2 million heads about 20% were raised on a pastoral setting where the land is particularly rich in natural resources, in this case acorns from oaks. The finishing phase or 'montanera' corresponds to Autumn-Winter seasons where 92-115 kg pigs are raised for 60 d and gaining 46 kg bodyweight, at least. Pigs are then slaughtered at a heavy weight with carcasses weighing 108-115 kg minimum. After dissection, the main cuts, like ham, shoulder and loin are used to produce the typical dry-cured products. The Iberian pig exhibits a good appetite and a great capacity to accumulate intramuscular fat or marbling. This together with acorns based feeding, is what makes their products of high sensorial quality.

## Session 55 Theatre 2

**Improving meat quality in endangered pig breeds using *in vivo* indicator traits**

*A.D.M. Biermann, T. Yin, K. Rübesam, B. Kuhn and S. König*
*University of Kassel, Department of Animal Breeding, Nordbahnhofstraße 1a, 37213 Witzenhausen, Germany; a.biermann@agrar.uni-kassel.de*

The object of this study was to evaluate breeding strategies for improving meat quality traits within the endangered German pig breed 'Bunte Bentheimer' using backfat thickness measured *in vivo* by ultrasound ($BF_{iv}$), the same trait obtained from carcass (BF), meat quality traits measured on meat samples (electric conductivity (EC), brightness (Opto), drip loss (DL) and intramuscular fat content (IMF)), and the MHS status. 713 ultrasound measurements, 613 carcass and meat quality records, and 700 genotyped animals were available (1) to assess the impact of gene variants at the MHS locus on meat quality using a selective genotyping approach, (2) to estimate genetic (co)variance components, and (3) to evaluate breeding strategies on meat quality by combining results from quantitative-genetic and molecular-genetic approaches. The presence of the unfavourable P allele was associated with increased EC, paler meat, and higher DL, reflecting typical characteristics of pale, soft and exudative (PSE) meat. The allele substitution effect on IMF was unfavourable, indicated lower IMF when the P allele is present. Heritabilities for $BF_{iv}$ and BF were moderate ($h^2$=0.27 and $h^2$=0.48). Estimates for quality traits ranged from $h^2$=0.15 (Opto) to $h^2$=0.78 (IMF). Moderate and significant phenotypic correlations were found between IMF and $BF_{iv}$ ($r_p$=0.25) and BF ($r_p$=0.24). The genetic correlation between IMF and $BF_{iv}$ was higher than estimates between IMF and BF ($r_g$=0.39 vs $r_g$=0.25), supporting a selection strategy based on ultrasound measurements to improve IMF. The breeding strategy based on own performance ($BF_{iv}$) combined with a genetic marker (MHS) achieved higher accuracy and selection response (0.24 and 0.11%) compared to the selection strategy only based on $BF_{iv}$ (0.20 and 0.09%). Implementations of breeding strategies based on $BF_{iv}$ in combination with MHS status is recommended to improve meat quality in the BB population.

### Muscle biomarkers to differentiate pork quality categories based on industry and consumer demands

*B. Lebret, A. Vincent and J. Faure*
*INRA, UMR1348 PEGASE, 35590 Saint-Gilles, France; benedicte.lebret@rennes.inra.fr*

Pork quality is a complex phenotype assessed by a combination of physical, biochemical, and occasionally sensory indicators measured one or even few days after slaughter using generally costly and/or invasive analyses. Early post-mortem (p.m.) biomarkers of pork quality have been identified by functional genomics however they usually refer to a single indicator and not to the overall quality level of a pork sample. This study aimed at determining pork quality categories combining both sensory and technological dimensions, and at identifying biomarkers discriminating these categories to further develop tools to monitor and predict pork quality in slaughterhouses. Sensory, technological and gene expression data were collected on the longissimus muscle of 100 pigs of two different breeds reared in various production systems and thereby exhibiting a wide and gradual variability in meat quality. Scientific expertise and statistical (multiple correspondence analysis) approaches were combined to select indicators and their thresholds specifying quality categories: low (impaired) quality (L; pH30<6.10 or pH 24 h<5.50, n=37); and among non-defective pork, acceptable (A; drip≥1% or intramuscular fat (IMF)<2.5%, n=26) and extra (E; drip<1% and IMF≥2.5%; n=35). The L category had the highest drip loss and lightness (P<0.001) and the lowest tenderness (P<0.001) and juiciness (P<0.05) scores, whereas E had the highest values of pH30, pH24, red colour and tenderness score (P<0.001). Levels of previously identified biomarkers of single quality traits, determined in longissimus sampled 30 min p.m., highly differed according to quality categories. Two genes were up-regulated (P–0.003) specifically in L, 5 genes were up-regulated (P<0.007) and 4 down-regulated (P≤0.009) specifically in E, and 2 genes had gradual expression level between L, A, and E categories (P<0.0001). Biomarkers differentiating pork quality categories will be validated on a commercial pig population.

---



### Forage preferences of free-ranging UK pigs in deciduous woodland and pasture

*A.L. Wealleans, I. Mueller-Harvey, D.I. Givens and J.C. Litten-Brown*
*University of Reading, School of Agriculture, Policy and Development, Reading, Berkshire, RG6 6AR, United Kingdom; a.l.wealleans@pgr.reading.ac.uk*

Iberian pigs, which exist in a regimented and planted ecosystem, are reported to consume acorns and grass, almost exclusively. However as pannage pigs in the UK are often turned out into mixed woodland, where there are more diverse feedstuffs available, the present study aimed to quantify the forage preferences of a group of pannage UK woodland pigs. Three adult sows (with piglets at foot) and two juvenile males were penned in mixed woodland and pasture, without access to concentrate feeds. Pigs were observed for 25 days between September and December 2011; piglets at foot were not observed. Focal observation of individual pigs recorded total number of bites taken per hour and bite composition, using the method set out by Rodriguez-Estevez *et al.* Means separation was conducted using the GLM procedure of Minitab 16, with age/gender as a fixed effect. There were clear differences in feed preference between the two age groups, with older animals taking significantly more bites of grass (P<0.01) and less of mast (P<0.05) than the juvenile males. Acorns and grass contributed significantly to the total number of bites recorded: grass comprised 63.2 and 35.0% of total bites for sows and juveniles respectively, with acorns comprising 17.9 and 32.5%. As expected, however, acorns were not the sole mast species consumed by pigs in UK woodland, with beech mast contributing 14% of the bite intake of sows and 25% of juvenile bites. This may affect the meat quality of pork produced in mixed woodland systems, reducing the 'acorn aroma' assigned to Iberian pigs.

## Session 55 — Theatre 5

**Effect of genotype and feed allowance on behavior, performance and meat quality of free-range pigs**

*A.G. Kongsted and M. Therkildsen*
*Aarhus University, Agroecology, Food Science, Blichers Allé 20, 8830 Tjele, Denmark; anneg.kongsted@agrsci.dk*

Use of traditional breeds and free-range systems based on natural foraging may be one way to improve the quality of organic pork in a broad sense and thereby justify the larger price. The effect of genotype on pig behaviour, performance and meat quality was investigated in 48 growing pigs fed either restrictedly (RES) or according to recommendations (NORM). Genotypes were a modern crossbreed of Duroc, Yorkshire and Landrace (DYL) and a traditional crossbreed of Tamworth, Yorkshire and Landrace (TYL). Each pig had access to 406 $m^2$ of grass-clover and root chicory. The average daily consumption of feed was 2.7 and 1.8 kg for NORM and RES, respectively. There were no significant interactions between genotype and feed allowance on behavior or performance. Compared to the NORM pigs, the RES pigs spent significant more time foraging, had significant lower daily weight gain (666 vs 863 g) and used significant less feed per kg live weight gain (2.8 vs 3.3 kg). There were no significant differences between genotypes in foraging behavior but TYL pigs had significant lower daily weight gain (645 vs 801 g), poorer feed conversion ratios (3.3 vs 2.7 kg) and lower meat percentage (54.5 vs 63.6) compared to the DYL pigs. The loin (LD) and a ham muscle (BF) were used for sensory evaluation of the pork following 4 days of ageing. The loin from DYL had more intense acidic and metal taste, and less sweet taste than the loin from TYL pigs. RES feeding reduced the tenderness in the loin significantly. DYL ham was significant more tender than TYL ham. In conclusion, restricted feeding encouraged the pigs to forage in the range area and this improved feed efficiency substantially but the texture of the meat need to be improved e.g. through post mortem handling. It is questionable whether the observed differences in eating quality between genotypes are large enough to justify a premium market price that could counterbalance the lower performance of the traditional crossbreed used in this study.

---

## Session 55 — Theatre 6

**Stakeholder participation in practical pig welfare research**

*K.H. De Greef and P.L. De Wolf*
*Wageningen UR, P.O. Box 65, 8200 AB Lelystad, the Netherlands; karel.degreef@wur.nl*

As a response to the societal unease regarding farm animal welfare, the Pigs in Comfort Class initiative was organised by the major Dutch farmers organisation and the major Dutch animal protection NGO. A welfare view was developed and a housing system was tested which fully meets the needs of the animals. The studies confirmed that meeting the (theory based) needs of pigs is of considerable distance to conventional standards, but also indicated practical welfare improvement opportunities on stocking density, environmental enrichment and tail management. The initiative was based on active stakeholder involvement in both programming the research and in the dissemination of results. This closely fits the new requirements for the EU Programme Horizon 2020. For this reason, the Comfort Class casus was analysed on strategies to enhance collaboration among stakeholder groups using four theoretical frameworks. Theories and guidelines from (1) strategic niche management, (2) boundary work, (3) stakeholder management and (4) strong and weak ties were projected on the Comfort Class initiative. Key conclusion from the analysis is that the specific combination of stakeholders (having 'weak ties'), the joint product (their joint 'boundary object') and the creation of clear cut small initiatives (practical farmers testing in 'niches') provided a context where research results did not need a translation into practice, but were accepted and used as a part of practical initiatives and farmers experimentation. The implication of the analysis is that combining 'classical' animal science with stakeholder involvement in a well planned innovation program is effective. The tension between cost price orientation and animal welfare ambitions was bridged by shared ownership of the research studies. The project reframed animal welfare from a legal obligation towards a (costly!) ambition and opened up the practical acceptance of increased space allowance in a NGO-labelled market product, currently comprising about 1 million pigs/year.

### Effect of feed presentation on growth performance of entire male pig and boar taint risk

*N. Quiniou[1], A.S. Valable[1], F. Montagnon[2] and T. Mener[2]*
*[1]IFIP-Institut du Porc, BP 35, 35650 Le Rheu, France, [2]Cooperl Arc Atlantique, 1 rue de la Gare, 22640 Plestan, France; nathalie.quiniou@ifip.asso.fr*

The aim of the trial was to compare the performance and boar taint risk when entire male pigs (40 per treatment) were fed with diets presented as mash (M) or ground pellets (G) over the 22-109 BW range. A liquid feeding system that allows for a simultaneous feeding was used to control the daily feed allowance per pen (5 pigs/pen) that corresponded to 4.5% of the initial BW per pen on the first day, thereafter +25 g/d/pig up to 2.60 kg/d. Dietary nutrient contents were calculated from chemical characteristics of ingredients and corresponding nutritional values assessed from www.evapig.com. The growing (<65 kg BW) and the finishing diets were formulated to contain 159 and 152 g/kg of crude protein, 9.2 and 8.2 g/kg of digestible lysine, 157 and 145 g/kg of total dietary fiber. The calculated net energy content was fixed at 9.64 MJ/kg. The daily feed intake tended to be lower with pellets (G: 1.95, M: 1.98 kg/d, P=0.06). As the average daily gain averaged 882 g/d (P=0.99), the feed conversion ratio was lower with pellets (G: 2.20, M: 2.26, P=0.05). No differences were noticed in carcass characteristics. Proportion of pigs with a low androstenone level (<1 µg/g pure fat) averaged 96% for both groups. Less pigs presented a skatole level above the minimum detectable concentration (MDC, 30 ng/g) with pellets (G: 78, M: 97%, P=0.01) and their skatole concentration tended to be reduced (G: 64, M: 94 ng/g, P=0.07). When both skatole (S) and indol (I) levels were above MDC, the ratio S/I+S was significantly lower with pellets (G: 51, M: 61%, P=0.01), but the coefficient of correlation between S and S/I+S was similar for both treatments (r=0.78). The improved feed conversion ratio associated with a stable carcass composition at slaughter would result from an increased digestibility of nutrients induced by the technologies used in the pelleting process, which would contribute to reduce boar taint risk in entire male pigs.

---



### Carcase weight is not a reliable tool to minimize the consumer acceptance risk of boar taint

*D.N. D'Souza[1], F.R. Dunshea[2], R.J.E. Hewitt[3], B.G. Luxford[4], R.J. Smits[4], R.J. Van Barneveld[3] and H.A. Channon[1]*
*[1]Australian Pork Limited, Brisbane Ave, Barton, ACT 2600, Australia, [2]The University of Melbourne, Grattan St, Parkville, Victoria 3010, Australia, [3]CHM Alliance Pty Ltd, Riverland Drv, Loganholme QLD 4129, Australia, [4]Rivalea Australia, Redlands Rd, Corowa, NSW 2646, Australia; darryl.dsouza@australianpork.com.au*

This study tested the hypothesis that carcase weight was not a reliable tool to minimise the consumer acceptance risk of boar taint and was based on an Australian survey (2011) reporting a poor correlation between boar taint compounds in fat of entire male pigs of differing carcase weights. A total of 10 pigs per treatment were used in this 2×2 factorial study. The main treatments were (1) Carcase weight: Porker (P), 62±7 kg and Baconer (B), 80±10 kg and (2) Boar taint levels: Low: 0.1 µg/g androstenone; <0.2 µg/g skatole and High >1 µg/g androstenone, >0.2 µg/g skatole). The pork samples were sourced from a larger survey (n=400 pigs. A total of 40 M. longissimus steaks per treatment (4 steaks per pig) were evaluated, with 320 evaluations made across all four treatments. Steaks were cooked to an internal temperature of 70 °C. Overall liking was assessed on a continuous line scale (0=Dislike extremely; 100=Like extremely). Consumers rated each steak for quality grade (QG) (1=Unsatisfactory; 5=Excellent) and re-purchase intention (RPI) (1=Definitely would not buy; 5=Would definitely buy). Fail rates for QG and RPI were determined as the percentage of steaks/treatment with scores <3. Overall liking scores were not influenced (P>0.05) by either carcase weight or boar taint level. Boar taint level influenced fail rates of both QG and RPI (27 (high) vs 19% (low) for QG and 31 (high) and 23% (low) for RPI, respectively) to a greater extent than carcase weight (24 (B) vs 22% (P) for QG and 28 (B) and 26% (P) for RPI, respectively). The results support the hypothesis that carcase weight is not a reliable tool to minimize the consumer acceptance risk of boar taint.

# Session 55 Theatre 9

**Detection of boar taint at slaughter**

*J. Trautmann, J. Gertheiss, M. Wicke and D. Mörlein*
*Georg-August-University of Goettingen, Department of Animal Sciences, Albrecht-Thear-Weg 3, 37073 Goettingen, Germany; johanna.trautmann@agr.uni-goettingen.de*

Due to animal welfare concerns the production of entire male pigs is one viable alternative to surgical castration. The lack of technical methods requires controlling boar taint using sensory evaluation. While the need for control measures with respect to boar taint has been clearly stated EU legislation, no specific requirements for selection of assessors have yet been documented. This study proposes tests for psychophysical evaluation of olfactory acuity to androstenone and skatole for sensory quality control. Odor detection thresholds for key volatiles contributing to boar taint are assessed as well as the subject's ability to identify odorants at various levels through the use of paper smell strips. Subsequently, fat samples are rated by the assessors, and the accuracy of boar taint evaluation is studied. Considerable variation of olfactory performance is observed demonstrating the need for objective criteria to select assessors. The sensory ratings of fat samples are significantly affected by the subjects' androstenone detection threshold ($P<0.01$). A lower androstenone threshold, i.e. a higher dilution sufficient for detection, is significantly associated with an increased probability of scoring a given fat sample as being 'deviant' (i.e. boar tainted). Through the application of detection threshold testing and identification tasks in which various dilutions of key volatile components are used, it is possible to objectively characterize the performance of the subjects participating. Considerable inter-individual as well as intra-individual variability is revealed for androstenone as well as for skatole. Olfactory acuity to key odorants, i.e. androstenone detection threshold affects the probability of rating a fat sample as boar tainted. The lack of agreement between chemical and sensory characterization of boar taint needs further investigation. Future studies are needed to establish normative criteria for selecting assessors.

# Session 55 Theatre 10

**Comparison of meat quality parameters in surgically castrated versus vaccinated Iberian pigs**

*M. Martinez-Macipe[1], P. Rodriguez[1], M. Izquierdo[2], M. Gispert[1], X. Manteca[3], F.I. Hernandez[2] and A. Dalmau[1]*
*[1]IRTA, Veïnat de Sies s/n, 17121 Monells, Spain, [2]la Orden-Valdesequera, A-5 KM 372, 06187, Guadajira, Spain, [3]UAB, Veterinary School, 08193 Cerdanyola Del Valles, Spain; sundarimmm@gmail.com*

When Iberian pigs are reared free-range, females on heat can be an attraction to wild boars, so farmers prefer to spay them. Males are also castrated to avoid boar taint. The vaccination by means of an anti-GnRF can be considered an animal friendly alternative to the surgical castration. The aim of our study was to compare several meat and carcass quality parameters in 37 immunocastrated Iberian pigs (16 females, ICF, and 21 males, ICM) with 38 surgically castrated pigs (19 females, CF, and 19 males, CM). We also added 8 entire females (EF). Vaccination with Improvac® was applied in all animals at 11, 12 and 14 months old. Pigs were slaughtered at 16 months old and quality measures were evaluated at the slaughterhouse. To perform the statistical analysis we used Proc Mixed of SAS (9.2) adjusted with Tukey. In females we didn't find any significant differences among groups except for the proportion of the loin (5.84, 6.39 and 5.96%) and shoulder (13.82, 14.55 and 14.07%) for CF, ICF and EF, respectively, being in both cases higher ($P<0.05$) in ICF than CF. For males, we found ($P<0.05$) significant differences in the proportion of ham (19.98 vs 21.08%) and in the proportion of loin (6.03 vs 6.83%)) in CM and ICM respectively. Also, ICM obtained higher results of shear force than CM (5.08 vs 4.28 kg; $P<0.05$), and lower ($P<0.05$) intramuscular fat content (7.15 vs 9.05%), backfat thickness at the Gluteus medius (41.71 vs 47.16 mm), and in the shoulder (65.14 vs 78.32 mm), pH at 24 h pm in the Semimembranosus (5.65 vs 5.81) and in the longissimus thoracis (5.64 vs 5.79). Our results suggest that immunocastration can be used as an alternative to surgical castration in females keeping similar quality traits. However, as vaccination may change some parameters in males, this has to be considered when deciding the type of castration.

## Session 55 Theatre 11

**Incidence of lameness in sows housed in dynamic or static groups at commercial farms**

*E.-J. Bos[1,2], M.M.J. Van Riet[1,2], S. Millet[2], B. Ampe[2], G.P.J. Janssens[1], D. Maes[1] and F.A.M. Tuyttens[1,2]*
*[1]Ghent University, Veterinary Medicine, Salisburylaan 133-D4, 9820 Merelbeke, Belgium, [2]Institute for Agricultural and Fisheries Research, Scheldeweg 68, 9090 Melle, Belgium; emiliejulie.bos@ilvo.vlaanderen.be*

Housing system, feeding system and method of grouping may influence the level of aggression in a group housing system. This may affect lameness and thereby sow welfare and performance. There is more agonistic behaviour in dynamic groups, where individuals are regularly replaced, compared to static groups where group composition remains stable during pregnancy. The aim of this study is to compare the incidence of lameness in static versus dynamic group housed sows at different stages of the reproduction cycle. In addition, we investigated number of skin lesions, stocking density, method of feeding and availability of straw bedding. On 10 farms (5 static and 5 dynamic), a total of 250 group housed sows were monitored during two reproductive cycles. Sows were visually assessed for lameness and skin lesions as indicator for aggression 3 times per cycle: at insemination (A), three days after grouping (B) and at the end of group housing (C). Using a linear model, with feeding (*ad libitum* or restricted), availability of straw, space per sow ($m^2$/sow), and method of grouping (static vs dynamic) as fixed effects, the mean incidence of lameness per phase in the reproductive cycle was estimated. The risk of becoming lame per sow during grouping period (A to B) tended to be lower for sows housed in static groups (23%) compared to dynamic groups (40%; P=0.08). No significant differences were seen for skin lesions. The incidence of lameness in period A-B decreases for dynamic groups, in pens with less available space per sow (P<0.01). This could be due to the design of the pen e.g. open area vs available barriers to use as hiding place. Preliminary results suggest that gestating sows kept in static groups have a reduced risk to become lame at grouping. Dynamic groups with less than average (2.28 $m^2$/sow) space showed a lower incidence of lameness.

## Session 55 Theatre 12

**Grass silage in diets for organic growing-finishing pigs**

*P. Bikker[1], G.P. Binnendijk[1], C.M.C. Van Der Peet-Schwering[1], H.M. Vermeer[1], E. Koivunen[2] and H. Siljander-Rasi[2]*
*[1]Wageningen UR Livestock Research, Animal Nutrition, P.O. Box 338, 6700 AH Wageningen, the Netherlands, [2]MTT Agrifood Research, Tervamäentie 179, 05840, Finland; paul.bikker@wur.nl*

Organic pig farms need to include roughage in the daily ration of the animals. Inclusion of early harvested grass silage in the diet of growing finishing pigs may reduce the need for import of protein rich feed ingredients and contribute to the closure of regional nutrient cycles. However, information on the nutritive value of grass silage, the potential inclusion level in the diet and consequences for growth performance is scarce. The present study was conducted to determine the effect of grass silage in a completely mixed ration, on performance and nutrient utilisation of organically raised growing finishing pigs. Control pigs received compound feed only, whereas silage fed pigs received an increasing proportion of grass silage up to 10 and 20% dry matter (DM) in the daily ration in the grower and finisher period, respectively. The experiment comprised 256 (Pietrain × (Dutch Landrace × GY)) pigs from 26 to 119 kg body weight in 2 treatments with 8 pens of 16 mixed sex pigs each. Performance and carcass data were analysed with ANOVA for a randomised block design with pen as experimental unit. The pigs receiving a mixture of grass silage and compound feed ingested on average 0.3 kg DM/d (13% of their daily ration) as grass silage and realised a similar daily net energy intake as pigs fed compound feed only. However, the silage fed pigs realised a 37 g/d lower daily gain (P=0.002) and a lower calculated net energy utilisation for gain (1.6 MJ/kg body gain, P=0.02) and a 1.1% lower dressing percentage (P=0.03) of the carcass. This may indicate that the nutritive value of grass silage was overestimated. Ileal and faecal nutrient digestibility in grass silage will be available to verify this aspect. The optimal feeding system for the supply of grass silage to growing pigs requires further investigation.

**Carcass and meat quality of Majorcan Black Pig at two slaughter weights and a crossbred with Duroc**

*J. Gonzalez[1], J. Jaume[2], M. Gispert[1], J. Tibau[1] and A. Oliver[1]*
*[1]IRTA, Finca Camps i Armet, 17121 Monells, Spain, [2]SEMILLA, Esperanto, 8, 07198 Son Ferriol, Palma, Spain; marina.gispert@irta.cat*

The Majorcan Black Pig (MBP) is a native breed from Mallorca, Balearic Islands, reared in extensive conditions. Animals are reared during almost one year and a large resource investment is needed to achieve the required slaughter weight. Local farmers and meat industry seeks more efficient alternatives for producing MBP meat and meat products to match market demands. Crossbreeding MBP with commercial breeds could produce animals with high meat quality obtaining better productive efficiency. Two classifications were made according to MBP carcass weight: light (L-MBP; 103.9±8.2 kg) and heavy (H-MBP; 127.5±11.1 kg). The aim of the study was to characterise L-MBP (n=32) and H-MBP (n=34) and a MBP crossbred with Duroc (75%) (n=20), to evaluate their potentiality for producing adequate carcass and meat quality. Crossbreds were reared in intensive conditions, presenting a mean carcass weight of 89.5±11.3 kg. All animals were slaughtered at the same abattoir following the commercial procedures. Subcutaneous fat at Gluteus medius was 23.3±8.2 mm in crossbred, 65.0±7.7 mm in L-MBP and 74.6±11.0 mm in H-MBP. Meat quality was focused on instrumental colour and intramuscular fat content, being the most valued parameters by local meat chain. Mean values for L* were 50.1±4.1 in crossbreds, 44.2±3.8 in L-MBP and 43.7±4.4 in H-MBP; a* value was 6.6±0.8, 10.7±1.5 and 10.9±1.3 for the same categories; and intramuscular fat content was 3.0±0.7%, 7.8±3.2% and 9.7±3.7%. Meat from both MBP categories was similar between them but crossbreds showed meat with notable differences. In conclusion, although the crossbreds could not be used to elaborate the same products as MBP purebreds, they could match the industry requirements fulfilling niche markets demanding for high quality meat products.

---



**Comparison of slaughter performance and meat quality in F1 crosses of PLW×Duroc and PL×Duroc pigs**

*G. Żak[1], M. Tyra[1] and T. Blicharski[2]*
*[1]National Research Institute of Animal Production, Department of Animal Genetics and Breeding, ul. Sarego 2, 31-047 Krakow, Poland, [2]Institute of Genetics and Animal Breeding of the Polish Academy of Sciences, Jastrzebiec, ul. Postepu 36A, 05-552 Magdalenka, Poland; grzegorz.zak@izoo.krakow.pl*

The study was performed with F1 crosses of PLW × Duroc (100 pigs) and PL × Duroc (100 pigs) according to the procedures of Polish pig testing stations. Evaluation was made of slaughter traits (backfat thickness, loin eye area, carcass lean content) and meat quality traits (pH1, pH2, L*a*b* meat colour, water holding capacity, intramuscular fat content). Evaluation of meat quality included longissimus dorsi muscle. Carcass lean content exceeded 57% for PLW × Duroc and 58.2% for PL × Duroc. Differences were found in the thickness of external fat, with considerably thicker backfat in PLW × Duroc (2.12 cm) compared to PL × Duroc (1.73 cm). As regards the parameters of meat quality, pH1 of loin was higher in PLW × Duroc compared to PL × Duroc animals (6.63 vs 6.44). Both values are considered normal for pork meat. Meat pH measured 24 h postmortem levelled off in both types of crossbreds. Also meat colour and water holding capacity were similar in both crosses. Differences were found in the level of intramuscular fat (IMF), which was 1.74% in PLW × Duroc and 1.93% in PL × Duroc F1 pigs. It is therefore concluded that PL × Duroc F1 pigs are more desirable in terms of meat quality (taste properties). However, IMF levels in meat from both crosses differ from the optimal value of approx. 2.5%. The study was carried out as part of the project 'BIOFOOD – innovative, functional products of animal origin' no. POIG.01.01.02-14-090/09-00

**Correlations of technological properties with meat quality and slaughter traits in Pietrain pigs**

*G. Żak[1], M. Pieszka[2], M. Tyra[1] and W. Migdał[3]*
*[1]National Research Institute of Animal Production, Department of Animal Genetics and Breeding, ul. Sarego 2, 31-047 Krakow, Poland, [2]National Research Institute of Animal Production, Department of Animal Nutrition and Feed Science, ul. Sarego 2, 31-047 Krakow, Poland, [3]University of Agriculture in Krakow, Department of Animal Products Technology, Balicka 122, 31-149 Krakow, Poland; grzegorz.zak@izoo.krakow.pl*

The experiment used Pietrain pigs from nucleus herds in Poland. Analysis was made on the relationships between technological properties (shear force, toughness, springiness, cohesiveness, chewiness, resilience) and intramuscular fat content (IMF), drip loss, backfat thickness, loin eye area, carcass lean content, and daily gain. Correlation analysis showed that shear force was negatively ($P \leq 0.01$) correlated to daily gain ($r=-0.39$), and positively ($P \leq 0.05$) to IMF ($r=0.31$) and loin eye area ($r=0.29$). In addition, daily gain was negatively ($P \leq 0.01$) correlated to springiness ($r=-0.42$), and resilience ($r=-0.37$) and ($P \leq 0.05$) to cohesiveness ($r=-0.31$), and chewiness ($r=-0.27$), which evidence the importance of daily gain on technological properties of meat. Resilience was correlated ($P \leq 0.01$) to backfat thickness ($r=-0.39$), daily gain ($r=-0.37$), carcass lean percentage ($r=0.37$) and ($P \leq 0.05$) to loin eye area ($r=0.31$). Cohesiveness was significantly ($P \leq 0.01$) correlated to loin eye area ($r=0.33$) and ($P \leq 0.05$) to carcass lean content ($r=0.29$). It should be noted that many of the estimated correlations are unfavourable for improvement of many meat quality parameters when using breeding programs aimed exclusively at improving slaughter and fattening performance. Also no or low correlations between fattening and slaughter traits prevent the improvement of some meat quality parameters under such breeding programmes.

---



**A new recombinant immunocastration vaccine for male pigs**

*L. Sáenz, D. Siel, J. Álvarez, S. Vidal and M. Maino*
*Faculty of Veterinary Medicine, University of Chile, Department of Animal Production, Santa Rosa 11735, La Pintana, Santiago, 8820808, Chile; mmaino@uchile.cl*

A new recombinant vaccine for immunocastration created in the Faculty of Veterinary Medicine, University of Chile has achieved these same effects that surgical castration, improving organoleptic characteristics of meat by decreasing levels of testosterone, androstenone and skatole by blocking the hypothalamic-pituitary axis and subsequent secretion of GnRH hormone. It is also consistent with animal welfare avoiding the stress of managing the surgically castrated animals. The objective of the study was to determine the levels of antibodies against GnrH and testosterone as well as the associated testicular atrophy and sensory characteristics of meat from immunocastrated or surgically castrate pigs. At the age of 130 days and an average weight of 65 kg, a total of 30 male pigs of PIC origin, were included in the experiment. In group A 20 castrated pigs and in group B 10 non-castrated pigs vaccinated subcutaneously with 2 ml of the recombinant vaccine were included. Indirect ELISA was performed to evaluate antibody response and competition ELISA to evaluate testosterone levels. At the end of the study the histological analysis of the testes was performed and a sensory analysis was carried out. Statistical analysis was made by analysis of variance and a P-value $\leq 0.05$ was considered indicative of a statistically significant difference. The ELISA results revealed an increase ($P \leq 0.05$) of specific antibodies and a decrease ($P \leq 0.05$) in serum levels of testosterone associated with testicular atrophy in immunocastrated animals. Sensory analysis showed no ($P \geq 0.05$) differences in organoleptic characteristics of the meat from immunocastrated and surgically castrated pigs. These results demonstrated the immunological effectiveness of the recombinant vaccine, inducing production of specific antibodies against GnRH, inducing testes atrophy associated with a decrease of testosterone levels and elimination of boar taint in the meat.

# Session 55 Poster 17

**Chemical composition of meat and fat in fattening boars, barrows and gilts**

*O. Bucko, I. Bahelka, A. Lehotayova and M. Margetin*
*Slovak University of Agriculture in Nitra, Department of Animal Husbandry, Tr. A. Hlinku 2, 94976 Nitra, Slovak Republic; ondrej.bucko@uniag.sk*

In recent years, piglet castration has become a controversial issue. In Slovakia, for many years pigs have been castrated to prevent the so-called boar taint, which is typical for meat of boars after they reach sexual maturity. Negative effects of castration are a lower growth rate and lower feed conversion compared to boars fed intensively. In our experiment, we determined the differences in the chemical composition and nutritional value of meat and fat produced by boars, barrows and gilts. The experiment included 21 fattening hybrid pigs (7 boars,7 barrows and 7 gilts). During the fattening period, pigs were housed in identical conditions and fed by the same diet. Twenty-four hours after slaughter at 105 kg BW, longissimus thoracis muscle (LT) and belly samples were collected for the determination of the chemical composition. The samples were analysed by FTIR method using the Nicolet 6700. The results showed that pork from boars contained a lower ($P<0,05$) portion of intramuscular fat and a greater water content ($P<0,05$) compared with barrows. The cholesterol content in LT was also lower in the boar compared with the barrow group, but the differences were not statistically significant. With respect to human nutrient requirements, the greater content of ω3 fatty acids and the lower content of saturated fatty acids in the LT and bellies were more optimal in the boar compared with barrow group. This work was supported by projects VEGA 1/0493/12, VEGA 1/2717/12, ECACB -ITMS 26220120015 and ECACB Plus – ITMS: 26220120032.

---

# Session 55 Poster 18

**Feeding entire male pig with amino acid supplies adapted to requirements of entire or castrated male**

*N. Quiniou and P. Chevillon*
*IFIP, BP 35, 35650 Le Rheu, France; nathalie.quiniou@ifip.asso.fr*

This study was performed to evaluate the growth, economic and environmental performances of entire male pigs (EM) when fed *ad libitum* with biphase diets formulated to meet their amino acid (AA) requirements (group E) or those of castrated males (CM, group C). In a previous study, the requirement in standardized ileal digestible lysine per unit of net energy (NE) was estimated at the beginning of the growing and the finishing periods to 0.94 and 0.71 g/MJ for EM, and to 0.84 and 0.71 g/MJ for CM, respectively. These levels were used in the trial, other AA being supplied proportionally to lysine based on the ideal protein concept. During both periods, the crude protein (165 and 145 g/kg, respectively), and total dietary fiber (150 and 141 g/kg) levels were similar for E and C diets. The NE content was fixed to 9.7 MJ/kg for all diets. Pigs were from a (Large White × Landrace) × (Large White × Pietrain) crossbreed, group-housed (5/pen) and studied over the 24-112 kg BW range. Daily feed intake (C: 2.13, E: 2.06 kg/d; $P=0.18$), BW gain (C: 886, E: 898 g/d; $P=0.67$) and backfat thickness at slaughter (C: 12.7, E: 11.6 mm; $P=0.11$) were not significantly influenced by AA level, in contrast with feed conversion ratio (C: 2.39, E: 2.28; $P=0.02$) and muscle thickness (C: 55, E: 58 mm, $P=0.02$). No differences were noticed on boar taint risk level assessed from average concentration of androstenone (C: 810, E: 732 ng/g, $P=0.54$) or skatole (C: 160, E: 160 ng/g, $P=0.98$) in pure liquid fat, or from partition of pigs in different categories of risk. From average performances, estimated N output increased by 5% with diets C (C: 2838, E: 2699 g/pig). With price of ingredients available in January 2014, diets E were 7 € more expensive than diets C, but it was counterbalanced by the improved feed efficiency, and finally resulted in a lower feed cost (-1 €). Maximizing performances of EM by adequate AA supplies is interesting both with regard to the margin and environment, providing that EM are fed separately from other genders.

**Consumer's acceptance regarding alternatives to piglet castration without anesthesia – a meta-analysis**

*T. Sattler[1], K. Sinemus[2] and L. Niggemann[3]*
*[1]University Leipzig, Large Animal Clinic for Internal Medicine, An den Tierkliniken 11, 04103 Leipzig, Germany, [2]Quadriga University of Applied Sciences, Werderscher Markt 13, 10117 Berlin, Germany, [3]Reputation Advise GmbH, Schiessstraße 44a, 40549 Düsseldorf, Germany; tasat@vetmed.uni-leipzig.de*

The in most European countries used practice of piglet castration without anesthesia to prevent boar taint will be forbidden with the year 2018. Several alternative methods such as vaccination against boar taint, surgical castration with anesthesia and fattening of entire boars are discussed. Which method will be used depends also on the acceptance of consumers. A lot of consumer's studies were conducted with different results depending on the country, the way of questioning and the focus group. The aim of the present study was to evaluate the consumer's opinion on the basis of recently conducted studies in German-speaking states. Consumer surveys regarding alternatives to surgical piglet castration without anesthesia conducted during the last six years in Germany, Austria and Switzerland were analyzed and results compared. A previously unpublished web based survey of consumer perception and discussion was included in the analysis. This survey was designed by the Quadriga University of Applied Sciences Berlin and based on data from 1,796 posts in German language. In sum, vaccination to avoid boar taint is strongly preferred to fattening of entire boars. Alternatives without castration were more often accepted than surgical castration. Especially the acceptance of vaccination depends on the information level of the consumers. The more information is given the better the acceptance. Consumers that buy pork from organic farms accept vaccination and are willing to pay more for the meat. The on-line survey shows that in the web based discussion of alternatives to surgical castration vaccination is equally valued to fattening of entire boars. If the term Improvac® is mentioned at all (in 18% of the scientific discussions), then mostly positively.

---



**Effect of pig body weight variability at the end of the growing period on marketing body weigh**

*S. López-Vergé, D. Solà-Oriol and J. Gasa*
*Animal Nutrition and Welfare Service, Universitat Autònoma de Barcelona, 08193, Bellatera, Spain; sergilv1@gmail.com*

Batch homogeneity of pigs' live weight at slaughtering is a huge issue in order to properly manage growing-finishing facilities. Hence, it is essential to ensure the homogeneity of pigs' BW at the end of the finishing period. The objective was to study the effect of BW variability at the end of the growing period on marketing weight and the time to reach the slaughterhouse. A total of 262 commercial crossbreed pigs ([Duroc × Landrace] × Pietrain), 120 days old at the end of the growing period, were distributed in three groups (H: high, M: medium, L: low) and weekly weighted during the finishing period. Pigs were fed *ad libitum* the same finishing diet. Individual BW data was adjusted using a linear model BW = a + bx being a the intercept, b the slope and x, the time (d); The BW at day 162 was analyzed with ANOVA by using the GLM procedure of SAS. Higher BW was observed for animals corresponding to the group H (102.2 kg; P<0.001) and M (99.7 kg; P=0.002) in contrast to L (94.3 kg). No differences were observed between groups H and M (P=0.151). Similar results were observed for the time to reach the marketing weight (100 kg) where pigs in groups H and M, spent less time (average: 40.2 d and 43.6 d; P<0.05, respectively) than those in the L category (51.0 d). Again, no differences were observed between groups H and M (P=0.171). However, the statistical model only explained a 15.2% of the total variability, indicating that marketing weight is affected by other factors than those related with the finishing period. Thus, a strong relationship was observed between BW at 120 d and at 162 d (r=0.83, P<0.001, Pearson Correlation Coefficient (r), CORR procedure of SAS) showing that other events occurred during lactation, nursery and growing periods which play a key role on BW development and age at slaughter. Results suggest that pigs with low BW at the end of the growing period are the main responsible for both carcass commercial depreciation at the slaughterhouse and the variability on farm occupation.

## Session 55 Poster 21

**Relationships between slaughter traits measured on live gilts and their reproductive performance**

*J. Ptak[1], A. Mucha[2] and M. Różycki[2]*
*[1]Polish Pig Breeders and Producers Association POLSUS, ul. Ryżowa 90, 02-495 Warszawa, Poland, [2]National Research Institute of Animal Production, Department of Animal Genetics and Breeding, ul. Sarego 2, 31-047 Kraków, Poland; aurelia.mucha@izoo.krakow.pl*

The objective of the study was to determine current relationships between slaughter traits measured on live gilts and their subsequent reproductive performance. Subjects were 5468 Polish Landrace gilts from nucleus farms. Meat percentage was estimated based on measurements of backfat thickness and longissimus dorsi muscle performed with a PIGLOG 105 device in the age range of 150-210 days. Reproductive traits of the gilts were determined from the number of piglets born alive per litter and the number of piglets reared until 21 days of age in parity 1, parity 1 to 2, parity 1 to 3, and parity 1 to 4. The animals were divided twice: into 3 groups according to backfat thickness, and into 3 groups depending on carcass leaneness. The first group was made up by herds in which animals showed lowest performance (-½δ). The second group included herds with results between >-½δ and <+½δ. The third group contained herds with highest results that exceeded the average by more than ½δ. The present study showed that sows from herds in which live measured animals achieve best results for slaughter traits, are characterized by better reproductive performance. The presence of differences in sow reproductive traits between herds differing in the means for leaneness and backfat thickness suggests that herds managed using intense selection for leaneness also have better environmental conditions, as evidenced by the higher values of reproductive traits.

## Session 55 Poster 22

**The influence of Arimil on the reproductive performance of the breeding boars**

*L. Liepa, V. Antane and M. Mangale*
*LUA, Faculty of Veterinary Medicine, Helmana 8, 3004 Jelgava, Latvia; laima.liepa@llu.lv*

The aim of the study was to investigate the influence of a new veterinary medication Arimil on the quality of ejaculate and the concentration of testosterone in serum (T) of clinically healthy adult boars with good and bad breeding performance in the seasonal T decrease period (March-May). The new medication Arimil consists of meldonium (improves metabolism, promotes synthesis of gamma-beta-betaine, stimulates functions of testis) and anastrozolum (inhibitor of aromatase, stabilizer of T). The first experiment (EI) was realized with ten clinically healthy adult breeding boars at the Artificial Insemination Company. The experimental group: five boars received daily 2 g meldonium and 4 mg anastrozolum with the feed for 14 days. The control group: five boars got only feed. Before the experiment (D1) and on D10, D21 and D51 boars were examined for T and quality of ejaculate assessed by traits like weight (W), concentration (C), motility and viability (V) during 4 days of storage. The second experiment (EII) was carried out at the same place with three boars which had low T and low quality of ejaculate: all boars on D1 received daily 3 g meldonium and 4 mg anastrozolum with the feed for 14 days. The results on D1 of these boars were compared to results on D7, D14, D51. Data of EI were statistically analysed with SPSS 11.5. Under the influence of Arimil, the W and C and T were not (P>0.05) altered in EI. There was a constant numerical (P>0.05) increase in V (D10, D21) on the second and third day of ejaculate storing. In EII, feeding the Arimil to the three boars with low quality of ejaculate, the T numerically (P>0.05) elevated from 0.5±0.19 ng/ml (D1) till 2.0±0.75 ng/ml (D7), 1.00±0.47 ng/ml (D14) and 1.4±0.51 ng/ml (D51). After 14 days of treatment with Arimil the V was better on day 3 and 4. In conclusion, Arimil could be used for treatment of testosterone deficiency syndrome of boars and for improvement of viability of ejaculate, but more experiments are necessary with different dozes of active components of Arimil.

# Session 55 Poster 23

**Pig video tracking methods: a review and new trends**

*G.J. Tu and E. Jørgensen*
*Department of Animal Science, Aarhus University, Blichers Alle 20, 8830 Tjele, Denmark; gangjun.tu@agrsci.dk*

Light changes (e.g. switch on/off in the pig house), dynamic background (e.g. straw in the pig pen) and motionless foreground (e.g. pigs don't move) are three main factors that are always encountered in complex pig pens. These factors seriously affect the process of pig video tracking. The objective of this paper is to review existing pig video tracking methods with the aim of revealing the strengths and weaknesses of the methodologies based on the above three problems, and introduce the new trends of methodological development. In order to develop effective solutions to these important issues in pig video tracking, we give an overview of the processing framework of pig tracking in general: image pre-processing, object segmentation or object modelling, object tracking. Then, we categorize the existing methods into two groups (i.e. based on object segmentation or object modelling), provide detailed descriptions of representative methods in each group. The potential in resolving the three major obstacles above discussed has been examined for each method, and this analysis suggests that methods based on wavelet transform exhibit a good potential to overcome the light changes, dynamic background and motionless foreground in the complex pen, since local texture difference measures between a current image and a reference image can be calculated by using the wavelet gradient that are less sensitive to illumination changes, and the dynamic background can be regularized to a relatively static background by using wavelet decomposition. Our research demonstrates that wavelet transform has great potential in pig tracking as it provides a rich representation of a pig. In conclusion, the comparative analyses of existing video tracking methods presented in this review will help the researchers to fulfil their monitoring pig behaviour in complex pig pens and help to further optimize wavelet transform-based algorithms for their application in video tracking of animal behaviour.

# Session 55 Poster 24

**Segregation of small pigs as strategy to reduce the occupation time of growing-fattening facilities**

*D. Solà-Oriol, S. López-Vergé and J. Gasa*
*Animal Nutrition and Welfare Service, Universitat Autònoma de Barcelona, 08193, Bellatera, Spain; sergilv1@gmail.com*

One important factor affecting production costs in the swine industry is the occupation time of the growing-fattening facilities. This is mainly driven by the desired market body weight (MBW) but also by its variability, which affects the time of emptying the barn facility. Moreover, this fact also affects the quality classification and quotation of carcasses. The aim of this study was to know whether segregation of pigs with lower body weight at the start of the finishing period (BWS) could help to reduce farm occupation time without severely affecting carcasses classification. A total of 262 commercial crossbreed pigs ([Duroc × Landrace] × Pietrain) fed *ad libitum* the same finishing diet were used. Actually, pigs were slaughtered and marketed in three times, every 14 days, in a total period of 28 days between the first and the last batch. Variation of MBW was studied according to 0, 10, 20, 30 and 40% segregation of the animals with lower BW at the start of the finishing period. The study shows that, assuming only two slaughtering batches (reducing in 14 days the occupation of facilities); with no segregation 20.7% of the pigs were bellow the MBW (100 kg), but 4.2% were bellow 90 kg (10 pigs), causing a huge carcass depreciation. In contrast, by segregating a 10 or 20% of the pigs, values were reduced to 13.3 and 9.3% at MBW; 1.9% (4 pigs) and 1.1% (2 pigs) were below 90 kg, respectively. The results suggest that, in big swine operations of some thousands of sows, segregating between 10 to 20% of the smaller pigs at the beginning of the finishing period may results in a significant reduction in farm occupation time without severely affecting carcass depreciation at slaughterhouse of 80-90% of the pigs. At the same time the 10-20% of the remaining smallest pigs could be fed and managed separately in a better proper way. Performing segregation at the end of the nursery period would be even a more practical approach.

## Session 55 Poster 25

**Models for estimation of phosphorus excretion in heavy pigs**

*D. Biagini and C. Lazzaroni*
*University of Torino, Department of Agricultural, Forest and Food Sciences, Largo P. Braccini 2, 10095 Grugliasco, Italy; carla.lazzaroni@unito.it*

To estimate P potential excretion in heavy pigs' farms using mathematical models, 40 intensive farms (in Italian north-west plain) were studied for a fattening period (about 7 months) to evaluate the actual level of P excretion. Management data, feeding systems, feedstuff composition (including additives and integrators), feed consumption, and pigs' weight gain were recorded. Collected data were at first analysed altogether and P balance to animal level was calculated. Data were then analysed comparing diets with or without phytase or amino acids addition, number of feeding phases, and wet or dry rations. Correlations between farms variables and P excretion were also studied. No differences were found between P balance factors, except for P intake and retention in different phase feeding systems, but efficiency was not affected from such differences. Based on these results, and considering the different farming systems and the high number of farm variables reciprocally influenced and affecting P balance elements, three empirical models to estimate P excretion were calculated: (Eq. 1) daily excretion as $P_{excr\_day} = \sum (P_{feed} \times FCR_{head} \times ADG_{head}) - [0.009 + (8.830\ E^{-8})\ LW_{head} + (-\ 4.590\ E^{-5})\ LW^2_{head}]$; (Eq. 2) fattening period excretion as $P_{excr\_fat\_period} = P_{feed} \times FCR_{head} \times ADG_{head} \times D_{fattening} - (0.24 + 0.004\ LW_{final} - 0.007\ LW_{initial})$; and (Eq. 3) excretion (g/d) according to P intake (g/d) as $P_{excr} = -\ 0.467 + 0.905\ P_{intake}$; where: $P_{feed}$ = Phosphorus ration content (g/kg); FCR = Feed Conversion Rate (kg/kg); ADG = Average Daily Gain (kg); LW = Live Weight of head, at the beginning or at the end of the fattening period (kg); D = fattening days (n). Known the swine ADG and FCR, with Eq. 1 it is possible to estimate P excretion for each level of P ratio in feed, according to average live weight of animals. With the same data and the fattening period duration Eq. 2 estimates P excreted per head per fattening period. Finally Eq. 3, obtained by regression, estimates P excretion on P intake.

---

## Session 55 Poster 26

**The effect of wind shielding and pen position on growth rate and efficiency in grower/finisher pigs**

*D.B. Jensen[1], N. Toft[2] and C. Cornou[1]*
*[1]University of Copenhagen, Department for Large Animal Sciences, Grønnegårdsvej 2, 1870 Frederiksberg, Denmark, [2]the Technical University of Denmark, Veterinary Institute, Bülowsvej 27, 1870 Frederiksberg, Denmark; daj@sund.ku.dk*

Pigs are known to be particularly sensitive to temperature. Too cold and the pigs will grow less efficiently and be more susceptible to diseases such as pneumonia. Too hot and the pigs will tend to foul the pen, leading to additional risks of infection. In addition, previous research has demonstrated that pigs can be significantly affected by wind, even when not directly exposed to it. In response to this problem, some pig producers and research stations have implemented a shielding to prevent winds from blowing between separate sections. However, there seem to be no published studies which have investigated the efficacy of such shielding. To determine the impact of wind shielding, multivariate linear models were fitted to describe the average daily weight gain and feed conversion rate of 1,273 groups (14 individuals per group) of purebred Duroc, Yorkshire and Danish Landrace boars, as a function of shielding (yes/no), insert season (winter, spring, summer, autumn), start weight (large, small) and interaction effects between shielding and start weight and shielding and insert season. Shielding was not found to be a significant predictor of any outcome for any breed. To determine the effect of a group's placement relative to the central corridor, a similar model was fitted to the data for Duroc pigs, replacing shielding with distance from the corridor (1$^{st}$ 2$^{nd}$, 3$^{rd}$ or 4$^{th}$ pen). Significant differences could be seen in daily weight gain between groups of large pigs placed in the 1$^{st}$ and 4$^{th}$ pen (P=0.0001) and between groups in the 2$^{nd}$ and the 4$^{th}$ pen (P=0.028). A similar effect was not seen on smaller pigs. Pen placement appears to have no effect on feed conversion rate.

**Effect of high long-term temperature on the concentration of cortisol in serum of fattening pigs**

*J. Petrak, A. Lehotayova, O. Debreceni and K. Vavrisinova*
*Slovak University of Agriculture in Nitra, Department of Animal Husbandry, Tr. A. Hlinku 2, 949 76 Nitra, Slovak Republic; klara.vavrisinova@uniag.sk*

Regulated via the hypothalamic-pituitary-adrenal axis (HPA), cortisol is the primary hormone responsible for the different loads. Pigs are very sensitive to changes in environmental temperature. The effect of high constant long-term ambient temperature on the concentration of cortisol was studied in twenty-five fattening pigs of Large White breed. The pigs were divided into two groups and were housed in two environmental conditions: experimental group of pigs was housed in climatic chamber with constant long-term high temperature (30 °C) and control group of pigs was housed in standard conventional conditions in stall (20,25 °C in winter and 25 °C in summer). The experiment was realised from 30 kg and pigs were slaughtered at 100 kg body weight. Blood samples were collected from pigs of both experimental and control group in the end of experiment for determination of cortisol concentration. The optical absorbance was measured using Microplate Reader Model DV 990BV4 (UniEquip Deutschland). The results showed that pigs in conventional conditions had higher concentration of cortisol than the pigs housed in climatic chamber with constant long-term temperature (12,91 and 10,76 μg/dl). This suggest that high constant long-term ambient temperature in climatic chamber compared with varying temperatures in stall caused that the pigs can be adapted and the activity of HPA axis was lower. Therefore it is optimal to maintain temperature stability in the zone of thermal neutrality according to different categories of pigs which does not activate HPA axis.



**Cattle health and antibiotic use in Denmark with emphasis on udder health and milk quality**

*J. Katholm*
*Knowledge Center for Agriculture, Cattle, Team Animal Health, Welfare and Reproduction, Agro Food Park 15, 8200 Aarhus N, Denmark; jka@vfl.dk*

Dairy production in Denmark has changed dramatically since 2000. Herd size increased from 61 to 150 cows. Herds with Automatic milking systems (AMS) reached a maximum in 2011 with 920 AMS herds. Herds with AMS are now reducing. Milk producers in Denmark are ready for further expansion when the quota system ends in 2015. Use of antibiotics for production animals is reported annually in the DANMAP report. In 2009 total consumption of antibiotics for cattle was 13.1 ton of actives substance, which increased in 2010 to 14.0 ton. Since then it reduced to 12,1 ton. For cows simple penicillin accounts for 66% of total consumption. Mastitis plus dry cow therapy accounts for 68% of the antibiotics used in cows. Thus, udder health is highly influencing the total consumption. Simple penicillin accounts for 79% of the antibiotics used for mastitis therapy. The number of initiated mastitis treatments reported to the Danish cattle database was stable around 210,000-231,000 cases from 2001-2009. Since 2009 it has been reduced by 29%. The incidence in cases per 100 cows reduced by 31% from 41 to 28 cases from 2009-2013. From 2001-2009 milk quality measured by geometric somatic cell count varied between 221,000 to 244,000 cells/ml with a maximum in 2008. Since then it has decreased and in 2013 it was the lowest ever in Denmark with 212,100. The use of fluoroquinolon and 3+4 generations cephalosporin in production animals has great public concern. Use of fluoroquinolons is regulated by law and the amount used is practically zero. The amount of 3+4 generation cephalosporin used for cattle has been reduced by 50% since 2007, while the amount used for intramammary therapy in cows has been reduced by 78% to now only 6 kg active substance. Udder health, milk quality and antibiotics use for cattle is in a very positive development.

**Factors affecting antibiotics use in period 2005-2012 in dairy herds in The Netherlands**

*A. Kuipers[1] and H. Wemmenhove[2]*
*[1]Expertise Centre for Farm Management and Knowledge Transfer Wageningen UR, P.O. Box 35, 6700 AA Wageningen, the Netherlands, [2]Livestock Research Wageningen UR, P.O. Box 65, 8200 AB Lelystad, the Netherlands; abele.kuipers@wur.nl*

National goal was to reduce antibiotic use (AU) by 50% in 2013 compared to 2009. AU and attitudes of farmers were examined on 94 farms during 2005-2012. Number of Daily Dosages (NDD) indicates how many days/year an average cow in the herd is under treatment of antibiotics. Mean NDD was 5.86, SD 2.14. NDD rose in period 2005-2007, followed by a period of growing societal interest in AU and a significant reduction in AU in 2011-2012. 68% of NDD was applied to udder: 25% for clinical mastitis and 43% for dry-cow therapy. Other drugs than applied to udder health decreased the most. Farmers were reluctant to lower use of dry-cow tubes. Use of $3^{rd}/4^{th}$ gen drugs minimized from 18% of NDD in 2005-2010 to 1% in 2012. Drop in NDD varied between 3 groups of farmers. The guided study groups reduced AU from 2008 on, the environmental and incidental groups (not guided) had, respectively, a significant drop in 2011 and in 2012. Trend in AU was modelled applying the Rogers diffusion of innovation theory. A logistic function fitted to the adaptation process, illustrating the early adapters, i.e. study groups, and late majority group, i.e. incidental farmers. Farm factors affecting AU were studied with a step-wise regression procedure. Variation in total use and dry cow therapy were explained, respectively, for 39 and 46% by quota size, milk amount/cow, health status, cell count and calving interval. r cell count-NDD was -0.55. Farms reducing AU the most were compared to farms that showed an increase. Behavioural data about both groups were collected with questionnaire. The decreasing farms started at a high level and were more active in applying management practices to reduce use. The 'more successful and entrepreneurial ' farmers, with a good relation to the veterinarian, tended to use somewhat more antibiotics than the other colleagues. They were also able to adapt more easily to the new conditions.

---



**Evaluation of a tool for integration of economy in veterinary herd health programs for dairy herds**

*A.B. Kudahl[1], J.F. Ettema[2], S. Østergaard[1], I. Anneberg[1] and D.B. Jørgensen[3]*
*[1]Aarhus University, Animal Science, Blichers Alle 20, 8830 Tjele, Denmark, [2]SimHerd Inc, Niels Pedersens Alle 2, 8830 Tjele, Denmark, [3]Dyrlæger & Ko, Nattergalevej 20, 8860 Skanderborg, Denmark; anneb.kudahl@agrsci.dk*

Veterinarians often express frustrations when farmers don't follow their advices, and farmers sometimes shake their heads when they receive veterinary advices that are practically unfeasible. That is the background for the development of a focused 2-page economic report created in cooperation between veterinarians, farmers, advisors and researchers. Based on herd specific key-figures for management, the report presents the short and long-term economic effects of 23 management improvements. Simulations are performed by the dairy herd simulation model 'SimHerd'. Besides the economic effects, the report also presents changes in labor requirement, number of livestock of different age-groups and changes in methane emission from the herd as a result of each the management improvements. The aim is to assist the veterinarian in identifying the economically most favorable and feasible management improvements and thereby provide more relevant and prioritized advices to the farmer. The choices of the 23 scenarios were based on ideas and suggestions arising from two initial workshops – one with veterinarians and one with dairy farmers. They resulted in scenarios including reduction of a list of disease risks, improved reproduction and more complex scenarios like improved calf management and cow-longevity. Results from the workshops also described the demanded quality and content of the product like 'easy, quick and simple to use' (veterinarians' wish) and 'calculations of extra labor and space requirements in the stable' (farmers' wish). The report is tested by 12 veterinarians in 60 herds and subsequently it is evaluated how the tool affected the farmers' decisions. Whether this tool actually improves the communication between the farmer and the veterinarian will be presented at the conference.

## Low cost farming systems in the Mid-West of USA

*V.J. Haugen*
*University of Wisconsin-Extension, Department of Agriculture and Life Sciences, Cooperative Extension, 225 North Beaumont Rd, Suite 240, Prairie du Chien, WI 53821, USA; vance.haugen@ces.uwex.edu*

North American grass dairy systems have many facets that make them successful. The first is to have a great forage based system to produce milk and secondly having a relievable low cost milking system to harvest that milk. Given the size of most grass based dairies, less than 250 cows, milking systems cannot be overly expensive. This contribution will give an overview how new and existing dairies have implemented milking systems that are very cost effective without sacrificing safety, efficiency and heard health both with retrofitting and new construction.



## A non-invasive method for measuring mammary apoptosis in dairy animals

*G.E. Pollott, K. Wilson, R.C. Fowkes and C. Lawson*
*Royal Veterinary College, Royal College Street, NW1 0TU, London, United Kingdom; gpollott@rvc.ac.uk*

Milk production from dairy animals has been described in terms of three processes; the increase in secretory cell numbers in late pregnancy and early lactation, secretion rate of milk per cell and the decline in cell numbers as lactation progresses. This latter process is thought to be determined by the level of programmed cell death (apoptosis) found in the animal. Until now apoptosis has been measured by taking udder biopsies, using MRI scans or using animals post-mortem. This paper describes an alternative, non-invasive method for estimating apoptosis by measuring microparticles in milk samples. Microparticles are the product of a number of processes in dairy animals, including apoptosis. Milk samples from 12 Holstein cows, past their peak of lactation, were collected at 5 monthly intervals. The samples (n=57) were used to measure the number of microparticles and calculate microparticle density for four metrics; Annexin-V positive, MC540 positive, positive for both and total particles in both whole milk (WM) and spun milk (SM). Various measures of milk production were also recorded for the 12 cows, including daily milk yield (DMY), fat and protein % in the milk, somatic cell count (SCC) and the day of lactation (DIM) when the samples were taken. A high correlation was found between the 4 WM microparticle densities and DIM (0.46 to 0.64) and a moderate correlation between WM microparticle densities and DMY (-0.33 to -0.44). No relationships were found involving spun milk samples or SCC, fat and protein %. General linear model analyses revealed differences between cows for both level of microparticle density and its rate of change in late lactation ($P<0.05$). Persistency of lactation was also found to be correlated with the WM microparticle traits (-0.65 to -0.32). Since apoptosis is likely to be the major contributor to microparticle numbers in late lactation, this work has found a non-invasive method for estimating apoptosis that gave promising results.

## Session 56 Theatre 6

**How the farmer makes 'sense' from sensor data**

*A. Van Der Kamp, L. De Jong and A. Hempenius*
*Lely International, Farm Management Support, Cornelis van der Lelylaan 1, 3147 PB, Maassluis, the Netherlands; avanderkamp@lely.com*

Continuously new sensors are being developed and current available sensors are being improved. This results in more and more available data for dairy farmers. It is tried to make smart combinations of sensors in order to support the farmer with information instead of with data. But do all these sensors, data and information make sense to the famer? Sensors like milk yield measurement, milk conductivity and milk color measurement are used to help the farmer in detecting udder health problems in an early stage. They are farm management supporting sensors. Anomalies detected by these sensors are presented to the farmer as an action list, he is advised to have a closer look at the cows with these anomalies. A recent study shows that although from a research point of view cows with these anomalies need attention, only 3.5% of these cows are really checked by the farmer. So the famer makes a management decision to treat or not to treat the cow, which is not corresponding with the research point of view. Investigated in this research is the reason(s) why a farmer starts treating a cow, what is the attention of combination of attention(s) which trigger him. In this study, data of 29 farms is analyzed. This showed that loss of milk yield is an important trigger for a farmer to start treating a cow. Another reason for a farmer to start the treatment of a cow is the combinations of anomalies, a cow with an attention for an abnormal color of the milk and an attention for loss of milk production is two times more likely to be treated by a farmer than a cow with only a drop in milk production. Concluding, it is not enough to have good sensors available. Key to make a sensor successful, is to understand how a farmer interprets it and how he acts on it. Once this is known clear farm management tools can be developed to support sensors and farm management.

## Session 56 Theatre 7

**Knowing the normal variation in lameness parameters in order to build a warning system**

*A. Van Nuffel[1], G. Opsomer[2], S. Van Weyenberg[1], K.C. Mertens[1], J. Vangeyte[1], B. Sonck[1] and W. Saeys[3]*
*[1]ILVO, Burg. Van Gansberghelaan 115, 9820 Merelbeke, Belgium, [2]Ugent, Salisburylaan 133, 9820 Merelbeke, Belgium, [3]KULeuven, Kasteel Arenberg 30 – 2456, 3001 Heverlee, Belgium; annelies.vannuffel@ilvo.vlaanderen.be*

To tackle the lameness problem, techniques to measure and monitor cow gait on a daily basis are being developed to detect changes in cows' gait over time. Due to the large variation in the values of the gait variables observed between and within cows, an automatic lameness detection system based on fixed thresholds will either only detect severely lame cows or produce many false alarms. To develop a more intelligent detection algorithm, information on the effect of normal external and internal factors on cow gait changes is needed. In this study the effects of external factors (wet surfaces and dark environment), and internal factors (age, production level, lactation and gestation stage) on gait variables were investigated. Gait variable data were collected automatically with the GAITWISE system over a 5 month period at ILVO experimental farm (Melle, Belgium) for 30 non lame, healthy Holstein cows. Measurements in dark environments did not significantly influence any of the gait variables. On wet surfaces, however, cows did take smaller and more asymmetrical strides. In general, older cows had a more asymmetrical gait and they walked slower with more abduction. Production level only significantly influenced the force distribution between left and right legs. Age, lactation or gestation stage, all showed to be significantly associated with more asymmetrical and slower gait and less step overlap probably due to the heavy calf in the uterus. It can, thus be concluded, that this 'normal' variation in the gait parameters should be taken into consideration when analysing daily cow gait data for the detection of alterations caused by lameness. However, further research is needed to better understand these and many other possible influencing factors, such as trimming or conformation, on the cow gait variables.

**Changes in the mammary gland transcriptome due to diet-induced milk fat depression in dairy cows**

*S. Viitala[1], D. Fischer[1], L. Ventto[2], A.R. Bayat[2], H. Leskinen[2], K.J. Shingfield[2,3] and J. Vilkki[1]*
*[1]MTT Agrifood Research Finland, BEL, 31600 Jokioinen, Finland, [2]MTT Agrifood Research Finland, KEL, 31600 Jokioinen, Finland, [3]Aberystwyth University, IBERS, SY23 3EE, United Kingdom; sirja.viitala@mtt.fi*

The increase in milk yield in high genetic merit dairy cattle has coincided with an increase in metabolic and reproductive disorders. During late pregnancy and early lactation the cow is often unable to consume enough feed to meet energy requirements. This shortfall is met through the mobilization of body energy reserves, which increases the risk of metabolic and reproductive disorders. Decreases in milk energy secretion through nutritionally controlled reductions in milk fat content represents one strategy to manage negative energy balance during early lactation. Our objective was to elucidate effects of diet-induced milk fat depression on the mammary transcriptome. Four Finnish Ayrshire dairy cows were used in a 4×4 Latin square with 35 d experimental periods. Experimental diets consisted of total mixed rations based on grass silage containing either high (65:35) or low (35:65) forage to concentrate ratio, supplemented with either 0 g or 50 g/kg diet dry matter of sunflower oil. Feed intake, milk yield and milk composition were determined daily. Tissue biopsies of the mammary glands were taken with a Bard biopsy needle on days 1, 8, 15 and 26 of each experimental period. Treatments had no effect on milk yield, whereas the low forage diet containing sunflower oil decreased milk fat concentration and yield. For all cows and different diets the gene expression levels at distinguished time points of each experimental period were obtained using RNA-seq (50 M reads per sample, 100 bp paired end libraries, Illumina HiSeq). The Grape pipeline was applied for mapping and quantification of the reads. Based on this quantification, the mammary gland transcriptome was characterized according to diet. Results of the effects of milk fat depression on fatty acid composition and mammary gland transcriptome will be presented.

---



**Effect of climatic conditions on the conception rates of dairy cows in Central Europe**

*S. Ammer, F. Rüchel, C. Lambertz and M. Gauly*
*Georg-August-University, Livestock Production Systems, Department of Animal Science, Albrecht-Thaer Weg 3, 37075 Göttingen, Germany; sammer1@gwdg.de*

The objective of this study was to evaluate climatic effects on the conception rates of dairy cows under Central European conditions. Therefore, the reproductive performance of a Holstein-Friesian herd (75 cows), kept in a non-insulated loose-housing barn, was analyzed. The evaluation included the outcome of 2.650 artificial inseminations (AI) during the last two decades (1993 to 2013). Climatic data including hourly air temperature (°C) and relative humidity (%) were used to calculate the temperature-humidity-index (THI). Daily mean and maximum THI from 21 days before until 30 days after each service were correlated with conception rates. For the statistical analysis the daily mean and maximum THI values were divided into four classes (<40, ≥40<60, ≥60<70, ≥70). The effects of mean and maximum THI at the day of insemination on conception rate were not significant (P>0.05). When maximum THI-values on day 21 before insemination were >70 the first-service conception rates decreased by about 20% compared to class 2 (≥40<60) (P=0.05), while it increased by about 15% from THI class 3 (≥60<70) compared to THI class 2. Furthermore, the first-service conception rates increased when mean THI was at days 17 and 19 before insemination greater than 60 compared to classes lower than 60 (P<0.05). For the period after insemination the effects of mean and maximum THI were not significant. In the present study, climatic effects on the conception rates of dairy cows under Central European conditions were only found for a short period before insemination, while during the 30-day period after insemination the effects were not significant. However, it has to be kept in mind, that the results may reflect a specific farm situation as the data were only collected on one farm.

**Comparison of two gold standards for evaluation of a mastitis detection model in AMS**

*L.P. Sørensen[1], M. Bjerring[2] and P. Løvendahl[1]*
*[1]Aarhus University, Dept. Molecular Biology and Genetics, Research Centre Foulum, 8830 Tjele, Denmark, [2]Aarhus University, Dept. Animal Science, Research Center Foulum, 8830 Tjele, Denmark; larspeter.sorensen@agrsci.dk*

In automated milking systems (AMS) visual control of cow's milk is not possible; thus, automated sensor systems are used to detect abnormal milk for mastitis detection. Evaluation of mastitis detection systems is difficult because of the lack of a proper gold standard for mastitis. The aim of this study was to compare evaluations of a mastitis detection model using two possible gold standards for cows with mastitis: A) recorded mastitis treatments and B) screening results of PCR analysis of milk samples. The detection model was evaluated during 11 mo using a low and a high alert threshold. Records of mastitis treatments were matched with model alerts using time window analysis (48 h). This setup resulted in sensitivities (Se) of 91 and 78% for the low and high threshold, respectively. Also, two more practical validation parameters were calculated, success rate (SR) and false alert rate (FAR). For the low threshold, SR and FAR were 80% and 0.6 per 100 cows, and for the high threshold, 77% and 0.6. The second gold standard was based on PCR analysis of foremilk samples (n=1,084 from 6 herds). Again, time windows were used for matching PCR results positive for major pathogens with low and high model alerts. This approach gave much lower Se of 25 and 8% for the low and high thresholds. The SR was also lower with 71 and 68%. However, FAR was also lower with values of 0.4 and 0.2 for low and high thresholds. The effect of threshold level had similar impact on the validation parameters using the two gold standards. However, in the current setup, PCR was a poor gold standard for mastitis because of it's high sensitivity, which do not discriminate between live and non-viable pathogens and therefore define a larger number of sick cows. However, the objective nature of the PCR method makes it interesting for further studies.

---



**Milking behavior in dairy cows naturally infected with clinical mastitis**

*K.K. Fogsgaard, M.S. Herskin and T.W. Bennedsgaard*
*Aarhus University, Department of Animal Science, Blichers Allé 20, 8830 Tjele, Denmark; karine.kopfogsgaard@agrsci.dk*

Behavioral changes – the extent and duration – in response to naturally occurring mastitis are not well understood despite its importance for animal welfare. The aim of the present experiment was to describe the extent of behavioral changes during milking in a 10 d period after treatment for naturally occurring clinical mastitis. A free stall herd of Danish Holstein, milked in AMS, was followed during 6 months. The first 30 cows treated with antibiotic for naturally occurring mastitis – detected by presence of bacteria in the milk – were included in the dataset. Treatment day was defined as day 0. Each infected cow was paired with a control cow matched by lactation number and stage, yield and body condition. Daily clinical udder examination including hardness, redness, swelling, milk drip and soreness was turned into a score from 0-5, showing that the infected cows had a higher udder score than control cows during the whole 10 d period ($P<0.01$). Behavior during milking was recorded from d -2 to d10. Infected cows lifted their hoofs (2.3 vs 1.8 step/min, $P<0.01$) and kicked (0.3 vs 0.1 kick/min, $P<0.01$) more frequently than control cows, and the duration of each milking was shorter (7.0 vs 7.3 min, $P<0.01$). Kicking off teat cups was observed at a higher frequency for mastitic cows than for the control animals (0.04 vs 0.01 freq/min, $P<0.01$). Despite the lack of systemic reactions such as fever, local udder signs of inflammation were clear and persistent for more than 10 days after diagnosis. Infected cows showed a more restless behavior which could be a sign of discomfort and potentially even pain. Behavioral changes persisted for at least 7 days after the 3 d antibiotic treatment, showing that even in mild mastitis cases the infected animals are not symptom free after treatment. Knowledge about behavioral changes – their magnitude and duration – during naturally occurring clinical mastitis is important in order to optimize management and welfare of mastitic dairy cows.

**Post-partum Inflammation reduces milk production in dairy cows**

*F. Robert[1] and P. Faverdin[2]*
*[1]CCPA Group, R&D department, ZA du bois de Teillay, 35150 Janze, France, [2]INRA, UMR1348 PEGASE, Domaine de la Prise, 35590 St-Gilles, France; frobert@ccpa.fr*

More and more evidences show that inflammation induces a partition of nutrients towards immune processes, decreasing the production of farm animals. Furthermore, acute phases can decrease feed intake that, in turn, could affect the production and health. In dairy cows, inflammation increases after parturition and some experiments show that this inflammation process can be negatively linked to milk production. Three trials were conducted on dairy cows in order to evaluate the relationship between milk production and inflammation using the plasmatic level of haptoglobin. The first trial was conducted at INRA Mésujeaume, 23 Holstein dairy cows have been blood sampled during the dry period, 7, 28 and 56 days after calving. Haptoglobin, hydroperoxydes (dROM test) and biological antioxidant activity (BAP test) have been measured at each period. In order to validate the first observations, 2 following trials have been performed in 2 commercial farms; blood was taken from 40 cows in each farm at 7 and 21 days after calving. The trial 1 shows that haptoglobin increases after calving; the level at day 7 post-calving being significantly higher than the other periods. The oxidative stress index (OSI), evaluated by the ratio dROM/BAP, begins to increase the last month of gestation, remains high the week following calving and decrease thereafter. The haptoglobin level measured at 7 days after calving showed significant negative correlation with milk production during the first 4 weeks post calving in all the 3 farms. The Pearson correlations are moderate but homogenous (between 0.36 and 0.4). None of the other sampling dates showed any correlation with milk production. Inflammation occurring the days following calving has a negative impact on milk production at least during 4 weeks. Any action targeting the inflammation process the days following calving could have benefit on milk production and cows' health.

---



**Udder health during mid-lactation of Holstein cows raised under different housing systems in Greece**

*M.A. Karatzia[1], G. Arsenos[1], G.E. Valergakis[1], M. Kritsepi-Konstantinou[1], I.L. Oikonomidis[1] and T. Larsen[2]*
*[1]Aristotle University of Thessaloniki, Faculty of Veterinary Medicine, Box:393, 54124 Thessaloniki, Greece, [2]Aarhus University, Department of Animal Science, Blichers Allé 20, 8830 Tjele, Denmark; mkaratz@vet.auth.gr*

Housing conditions are predisposing factors for a variety of health disorders in modern dairy cow enterprises. The objective here was to assess the interaction between housing systems and udder health of cows by means of biomarker measurements. The study was carried out using 120 clinically healthy Holstein cows of 6 commercial farms located in Northern Greece. All cows were in mid-lactation. In 3 farms the housing shed had straw bedding (SB) whereas in the remaining farms it had cubicle housing (CH). In all farms the outdoor area was earthen yard. In each farm the cows were selected according to days in milk (DIM≥90) and were allocated in 2 groups following the results of a consecutive milk sampling at monthly intervals to assess somatic cell counts. This resulted in a low and a high group (L: n=54; H: n=66). Blood samples were obtained by tail venipuncture and individual milk samples were collected at evening milking at 3 monthly samplings. Blood serum concentrations of urea, uric acid and albumin were determined. NAGase, LDH and β-glucuronidase activities were measured, and levels of BOHB, isocitrate, Glu6P and free glucose were determined in milk. Repeated measures analysis was used for the analysis (SPSS© v.21). NAGase, LDH and β-glucuronidase activity in milk was significantly higher in H and CH cows (P<0.05). Udder health had a significant effect on BOHB, Glu6P and free glucose levels, with H animals displaying higher levels (P<0.05). SB cows exhibited significantly higher free glucose levels, whereas their urea concentrations were significantly lower (P<0.05). Albumin concentration was significantly lower in CH animals (P<0.05). Housing system appears to have an impact on udder health, however, the individual variability and adaptability of the cows merits further investigation.

# Session 56 Poster 14

**The effect of pulsation ratios on teat and udder health and productivity in dairy cows in AMS**

*S. Ferneborg and K.M. Svennersten Sjaunja*
*Swedish University of Agricultural Sciences, Animal Nutrition and Managemnt, Kungsängens Research Centre, 753 23 Uppsala, Sweden; kerstin.svennersten@slu.se*

The pulsation ratio of a milking machine affects not only milk flow and milking time, but has also been reported to cause oedema in the teats and increased somatic cell count. However, most of the studies comparing pulsation ratios have been performed on traditional cluster milking, and the effects are likely to be caused by over milking. A large part of the milking today is performed on quarter level using automatic milking systems (AMS), where the risk of over milking is reduced. The aim of this project was to investigate the effects of different pulsation ratios on teat and udder health, productivity and milking parameters in AMS, where each quarter is individually milked. 356 cows on 5 privately owned farms were included in the study, which was performed in a split-udder design where three pulsation ratios; 60:40, 70:30 and 75:25 were compared to the standard pulsation 65:35. Teat score and teat thickness were measured on three occasions during the trial. Milk samples were taken from strip milk for the analysis of milk SCC. Milk yield, milking time and milk flow were registered automatically by the AMS for each milking. All statistical analysis was performed on the differences between control and treated fore- and hind quarter respectively. Data was analyzed using procedures MIXED and GLIMMIX in SAS 9.3. The MIXED model included the fixed effects of treatment and week, as well as treatment by week interaction and the random effect of cow. We found that the pulsation ratios 70:30 and 75:25 increased peak and average flow, and that machine-on-time was shorter on 75:25, while both peak and average flow were lower and machine-on-time shorter on 60:40. We found no support for negative effects on teat or udder health from any of the pulsation ratios applied, and do therefore conclude that increased pulsation ratios can be used to increase productivity in automatic milking systems where quarter milking is applied.

---

# Session 56 Poster 15

**Productive and reproductive traits of cows in view of persistency value lactation**

*A.A. Sermyagin and V.I. Seltsov*
*All-Russian Research Institute of Animal Breeding, Laboratory for Breeding and Selection of Animals, Dubrovicy 60, Moscow region, 142132 Podolsk, Russian Federation; alex_sermyagin85@mail.ru*

The field data of 7,214 test-day milk yield (MY) records from 762 first lactation Russian Simmental cows, the daughters of 24 sires were used. The MY persistency ($P_{SD}$ and $P_{YV}$) and the persistency of chemical composition of milk were evaluated as proposed by J. Sölkner with the modification of N. Gengler, based on the variation of the test-day records by months or parts of lactations. The analysis of variance and variance-covariance components was performed using Stata/SE12 software. The effects of maximum MY (MaxY), MY for the first 100, 101-200 and 305 days of lactation, days open (DO) and calving interval (CI) were studied. The heritability of MY persistency for measures $P_{SD200}$, $P_{SD305}$ and $P_{YV305}$ ($h^2$=0.09-0.13) was higher comparing to DO and CI ($h^2$=0.04-0.05) but lower than for MY traits ($h^2$=0.16-0.24).The heritability of test-day MY by months of lactation was 0.09-0.28. The significant random effect of sire was shown for MY and persistency of daughters: F=1.7-2.9 (P<0.05-0.001). The genetic relationship between the MY and persistency was decreasing from 100 to 305 days of lactation. The $P_{SD200}$, $P_{SD305}$ and $P_{YV305}$ values were characterized by the low dependence from the MY of 305 days of lactation ($r_g$=-0.34...-0.14), whereas the moderate correlation with the MaxY values was found ($r_g$=-0.76...-0.58). The relationships between the reproductive traits and persistency were not identified: $r_g$=0.03-0.12 for DO and $r_g$=-0.01...0.12 for CI. Some relations between the MY persistency and persistency of the milk components were found which were r=0.24 for fat, r=0.29 for protein, r=0.38 for solids, r=0.09 for lactose, r=0.23 for SCS and r=0.16 for fat-to-protein ratio. The highest persistency value of milk components was for lactose, fat/protein and protein content. Thus, the evaluation of $P_{SD}$ or $P_{YV}$ in the breeding program of the Russian Simmental cattle will improve the milk production traits of cows, leveling impact of the reproductive traits.

## Session 56 Poster 16

**A report on the reproductive performance of Holstein heifers inseminated with sexed semen in Iran**

*A. Forouzandeh and S. Joezy-Shekalgorabi*
*Young Researchers Club, Shahr-e-Qods Branch, Islamic Azad University, Tehran, Iran; joezy5949@gmail.com*

The objective of the current study was to evaluate some reproductive performance of Holstein heifers inseminated with sex sorted semen in a commercial dairy herd in Iran. Data were collected from an industrial dairy farm in Zanjan province from 2008-2012. Data from 254 artificial inseminations were analyzed. Average rate of conception, stillbirth, abortion, and dystocia were 0.51, 0.10, 0.06, 0.368, respectively. Average gestation length was about 273±12.01 days and more than 98% of live births were monoparous. As expected, a large number (91.5%) of born calves were female. The result demonstrated a suitable reproductive performance by using sex sorted semen in the studied herd. Mean use of sex sorted semen for heifers had been increased during time which represent the improved popularity of this kind of semen in large herds.

---

## Session 56 Poster 17

**Evaluation of the *in situ* degradation of starch in the washout fraction of feedstuffs**

*L.H. De Jonge[1], H. Van Laar[2] and J. Dijkstra[1]*
*[1]Wageningen University, Animal Nutrition Group, De Elst 1, 6708 WD Wageningen, the Netherlands, [2]Nutreco, R&D, P.O. Box 220, 5830 AE Boxmeer, the Netherlands; leon.dejonge@wur.nl*

In the *in situ* method used to evaluate rumen degradation characteristics of feed components, small particles (i.e. washout fraction) are removed during rinsing after rumen incubation, and their fractional degradation rate ($k_d$) cannot be determined. This hampers evaluation of degradation in feed ingredients with a high fraction of small particles. Most feed evaluation systems assume that the $k_d$ of starch in the washout fraction is very high compared to that in the non-washout fraction. The aim of this study was to evaluate this assumption by comparing the $k_d$ of starch obtained with two rinsing techniques, either conventional rinsing with a washing machine or using a new rinsing method that reduces the washout fraction of starch by rinsing at 40 spm in a water bath. Six raw materials (barley, faba beans, maize, oats, peas, and wheat) were incubated in nylon bags during 0, 2, 4, 8, 12, 24, and 48 hours in four rumen fistulated dairy cows according to the all in all out principle. Bags were rinsed with the new or washing machine method. The $k_d$ values were estimated for each ingredient per animal using a first-order model. The effect of the rinsing method on the washout fraction and the $k_d$ of starch was analysed by ANOVA with rinsing method as fixed factor. For all products, the washout fraction of starch decreased ($P<0.05$) for all raw materials when using the new rinsing method and varied between 0.10 and 0.60 (washing machine method) and between 0.01 and 0.11 (new method). For faba beans, maize, and peas, the kd was not affected by method of washing, whereas for the other raw materials the $k_d$ of starch using the new method (mean 0.14 $h^{-1}$) was lower ($P<0.05$) than with the washing machine method (mean 0.32 $h^{-1}$). The assumption about the high $k_d$ of starch in the washout fraction relative to that in the non-washout fraction was not confirmed. The causes of this discrepancy need to be investigated.

**Study of environmental effects on Holstein milk performances under tunisian conditions**

*E. Meddeb[1], S. Bedhiaf[2], M. Khlifi[1], A. Hamrouni[1] and M. Djemali[1]*
*[1]Institut National Agronomique de Tunisie, Cité Mahrajène, 1082, Tunisia, [2]Laboratoire PAF. Institut National de la Recherche Agronomique de Tunisie, Rue Hédi Karray, 2049, Tunisia; bedhiaf.sonia@gmail.com*

This study was conducted on a dairy cattle under Tunisian subhumid climate and aimed to characterize the performance of dairy Holsteins and assess the adjustment coefficients for the duration of lactation, lactation number and month of calving. A total of 1,544 lactations recorded from 1998 to 2013 were analyzed. The variation in milk yield was studied by a linear model including the lactation number, month of calving, age at calving, year of calving and lactation length. The results showed that the average of the total milk production was 6,236±2,142 kg for an average length of lactation equal to 291±86 days, which corresponds to an average milk production at 305 days of 6,708±1,326 kg. The fat and protein content were 217 kg and 197 kg, respectively. The year of calving, month of calving, lactation number and duration of lactation significantly affected (P<0.01) milk production. The adjustment factors are calculated for the correct duration of lactation, lactation number and month of calving.



**Effect of workability traits on functional longevity in Polish Holstein-Friesian cows**

*M. Morek-Kopeć[1], A. Otwinowska-Mindur[1] and A. Zarnecki[2]*
*[1]University of Agriculture, Al. Mickiewicza 24/28, 30-059 Kraków, Poland, [2]National Research Institute of Animal Production, ul. Krakowska 1, 32-083 Balice k. Krakowa, Poland; rzmorek@cyf-kr.edu.pl*

Survival analysis was applied to evaluate the effect of workability traits on cow functional longevity. Data consisted of milking speed and temperament subjective scores recorded in the SYMLEK National Milk Recording System from 2006 to 2013 for 884,899 first-parity cows. Milking speed was classified as very slow (5.9%), slow (8.5%), medium (64.2%), fast (20.2%) and very fast (1.1%). Temperament was classified as calm (9.7%), average (84.8%) and nervous (5.4%). Based on SYMLEK production and disposal data, length of productive life (LPL) for cows was calculated as number of days from first calving to culling or censoring. The percentage of censored data was 60.4%. Mean LPL was 901.9 days for uncensored and 863.2 days for censored records. Functional longevity was defined as length of productive life corrected for production. Milking speed and temperament classes were included, one at a time, as time-independent fixed effect in the Weibull proportional hazard model together with time-dependent fixed effects of year-season, parity-stage of lactation, annual change in herd size, fat yield and protein yield, random herd-year-season and time-independent fixed effect of age at first calving. Likelihood ratio tests showed a highly significant but small overall effect of milking speed and an even smaller effect of temperament on functional longevity. For milking speed the relative risk of culling (RRC) associated with scores from very slow to very fast was 0.42, 0.99, 1.0, 0.98 and 0.98, respectively. RRC estimates for temperament were 0.60, 1.0 and 0.98 for calm, average, and nervous categories. The almost identical and close to 1 risk of culling for all classes except very slow and calm indicates the very small and limited effect of workability traits on longevity.

**mRNA expression of adiponectin system related genes in colostrum versus formula fed dairy calves**

*H. Sadri[1], H.M. Hammon[2] and H. Sauerwein[1]*
*[1]Institute of Animal Science, Physiology and Hygiene Group, University of Bonn, Katzenburgweg 7-9, 53115, Germany, [2]Leibniz Institute for Farm Animal Biology (FBN), Dummerstorf, 18196, Germany; hsadri@uni-bonn.de*

Adiponectin is an adipokine that regulates metabolism and insulin sensitivity through activation of its two receptors, AdipoR1 and AdipoR2. Adiponectin is present in colostrum and milk but is very low in formula and we thus hypothesized that the mRNA expression of adiponectin and its receptors will differ in metabolically relevant tissues of colostrum (COL) or formula (FOR) fed calves. Fourteen male German Holstein calves were randomly assigned to two experimental groups after birth. Calves were fed either pooled colostrum (COL) or formula (FOR; with similar nutrient composition as colostrum but no growth factors and hormones), until slaughter at d 4 of life. Expression of adiponectin mRNA in adipose tissue (AT) and of its receptors in AT, duodenum, ileum and liver was quantified by real-time RT-PCR. The mRNA abundance of adiponectin and AdipoR2 in AT did not differ between groups, whereas there was a trend ($P=0.13$) for greater AdipoR1 mRNA abundance in AT of COL versus FOR calves. Duodenal expression of AdipoR2 mRNA was greater in FOR versus COL calves as a trend ($P=0.09$), whereas COL and FOR animals did not differ in ileal AdipoR1 and AdipoR2 mRNA abundance. Hepatic AdipoR1 mRNA expression was greater ($P=0.04$) in FOR than in COL calves. The expression of adiponectin receptors in the small intestine of neonatal calves we report herein supports a role of adiponectin in gastrointestinal tract development. Our observation that liver and duodenum of FOR calves had more AdipoR1 (liver) or AdipoR2 (duodenum) mRNA than COL calves might indicate compensatory and tissue-specific mechanisms in response to FOR feeding. Early postnatal nutrition was shown to influence the mRNA expression of adiponectin receptors and may thus be able to modulate the functional maturation of the adiponectin system.

---

Session 56 Poster 21

**Hepatic mRNA abundance of genes related to vitamin E status in periparturient dairy cows**

*H. Sadri[1], S. Dänicke[2], J. Rehage[3], J. Frank[4] and H. Sauerwein[1]*
*[1]Institute of Animal Science, Physiology and Hygiene Group, University of Bonn, Katzenburgweg 7-9, 53115 Bonn, Germany, [2]Institute of Animal Nutrition, Friedrich-Loeffler-Institute (FLI), Braunschweig, 38116, Germany, [3]University for Veterinary Medicine, Foundation, 30173 Hannover, Germany, [4]University of Hohenheim, Institute of Biological Chemistry and Nutrition, 70599 Stuttgart, Germany; hsadri@uni-bonn.de*

α-tocopherol is absorbed in the small intestine and transported to the liver from where it is secreted into the systemic circulation. Hepatic vitamin E-binding proteins and metabolizing enzymes control vitamin E concentrations. We hypothesized that hepatic gene expression of vitamin E-binding proteins and the rate-limiting metabolic enzyme cytochrome P450 4F2 (CYP4F2) will change in dairy cows that are faced with high need for vitamin E with the onset of lactation. Liver biopsies and blood samples were taken from 10 German Holstein cows on days -21, 1, 21, 70, and 105 relative to calving. The mRNA abundance of α-tocopherol transfer protein (TPP), α-tocopherol associated protein (TAP), and CYP4F2 was quantified by real-time RT-PCR. Serum concentrations of α-tocopherol were quantified by HPLC. The mRNA abundance of TTP, which specifically binds α-tocopherol and facilitates its secretion via lipoproteins, increased from d -21 to d 105; d 105 values were higher than on the other days. The mRNA encoding TAP, a ligand-dependent transcriptional activator, followed a similar pattern to that observed for TTP. Expression of CYP4F2 mRNA, the enzyme that catalyzes the hydroxylation of the phytyl side chain of tocopherols, remained unchanged. Serum α-tocopherol increased post partum in line with the observed increase in TTP mRNA. Increasing mRNA expression of TPP with days in milk indicates improved hepatic release of vitamin E into circulation to support target tissues; this was indeed reflected by increasing serum α-tocopherol concentrations. The functional consequences of increasing TAP remain to be elucidated.

**In vivo effects of Sainfoin (Onobrychis viciifolia) on parasitic nematodes in calves**

*O. Desrues[1], M. Peña[2], T.V.A. Hansen[1], H.L. Enemark[2] and S.M. Thamsborg[1]*
*[1]University of Copenhagen, Veterinary Disease Biology, Dyrlægevej 100, 1870 Frederiksberg C, Denmark, [2]Technical University of Denmark, National Veterinary Institute, Bülowsvej 27, 1870 Frederiksberg C, Denmark; olivierd@sund.ku.dk*

Sainfoin (*Onobrychis viciifolia*) is a fodder legume containing condensed tannins known to improve protein self-sufficiency, animal health and environment. In addition, anthelmintic effects have been demonstrated *in vitro* against cattle nematodes, and *in vivo* against nematodes of small ruminants, but *in vivo* effects against gastro-intestinal parasites (GIN) of cattle still remains to be proven. Thus, the aim of the present investigation was to determine the *in vivo* effects of sainfoin against the most important trichostrongylids of cattle the brown stomach worm *Ostertagia ostertagi*, and the intestinal worm *Cooperia oncophora*. Jersey male calves (2-4 months) reared indoor and GIN naive, were stratified for live weight and randomly allocated into a test group (SF; n=9) which was fed 80% sainfoin (cv. Perly) pellets and hay, and a control group (CO; n=6) fed hay and compound feed. Daily intake was monitored and protein/energy intake was equalized in both groups throughout the 8 weeks study period. After 2 weeks adaptation, calves were inoculated with approximately 10,000 *O. ostertagi* and 65,000 *C. oncophora* infective third-stage larvae. Individual weight gain was registered weekly, whereas eggs excretion, calculated as number of eggs per g of faecal dry matter (FECDM) was recorded 3 times a week. The calves were slaughtered and adult worms recovered 6 weeks post infection. FECDM was analysed by ANOVA using repeated measurements, and worm counts by Kruskal Wallis test. The mean FECDM was not significantly lower in the SF group ($P>0.05$). Worm counting of *C. oncophora* is ongoing, whereas mean worm counts of *O. Ostertagi* revealed 1,300±837 worms in the SF group and 2,500±870 in the CO groups ($P<0.05$).Thus, the preliminary results demonstrate an anthelmintic effect of sainfoin against establishment of *O. Ostertagi* in cattle.

# Author index

**D**

**E**

**F**



**G**

**M**

**Q**



**R**

**S**

Printed in the United States
by Baker & Taylor Publisher Services